Applied Statistics and Probability for Engineers

Third Edition

Douglas C. Montgomery
Arizona State University

George C. Runger
Arizona State University

John Wiley & Sons, Inc.

ACQUISITIONS EDITOR *Wayne Anderson*
ASSISTANT EDITOR *Jenny Welter*
MARKETING MANAGER *Katherine Hepburn*
SENIOR PRODUCTION EDITOR *Norine M. Pigliucci*
COVER DESIGNER *Madelyn Lesure*
ILLUSTRATION EDITOR *Gene Aiello*
PRODUCTION MANAGEMENT SERVICES *TechBooks*
COVER IMAGE *Norm Christiansen*

This book was set in Times Roman by TechBooks and printed and bound by Donnelley/Willard. The cover was printed by Phoenix Color Corp.

This book is printed on acid-free paper.

Library of Congress Cataloging-in-Publication Data

Montgomery, Douglas C.
   Applied statistics and probability for engineers / Douglas C. Montgomery, George C. Runger.—3rd ed.
      p.   cm.
   Includes bibliographical references and index.
   ISBN 0-471-20454-4 (acid-free paper)
      1. Statistics. 2. Probabilities. I. Runger, George C. II. Title.

   QA276.12.M645 2002
   519.5—dc21

                                                                2002016765

Printed in the United States of America.

10 9 8 7 6 5 4 3

Summary of One-Sample Hypothesis-Testing Procedures

Case	Null Hypothesis	Test Statistic	Alternative Hypothesis	Criteria for Rejection	OC Curve Parameter	OC Curve Appendix Chart VI
1.	$H_0: \mu = \mu_0$ σ^2 known	$z_0 = \dfrac{\bar{x} - \mu_0}{\sigma/\sqrt{n}}$	$H_1: \mu \neq \mu_0$ $H_1: \mu > \mu_0$ $H_1: \mu < \mu_0$	$|z_0| > z_{\alpha/2}$ $z_0 > z_\alpha$ $z_0 < -z_\alpha$	$d = |\mu - \mu_0|/\sigma$ $d = (\mu - \mu_0)/\sigma$ $d = (\mu_0 - \mu)/\sigma$	a, b c, d c, d
2.	$H_0: \mu = \mu_0$ σ^2 unknown	$t_0 = \dfrac{\bar{x} - \mu_0}{s/\sqrt{n}}$	$H_1: \mu \neq \mu_0$ $H_1: \mu > \mu_0$ $H_1: \mu < \mu_0$	$|t_0| > t_{\alpha/2, n-1}$ $t_0 > t_{\alpha, n-1}$ $t_0 < -t_{\alpha, n-1}$	$d = |\mu - \mu_0|/\sigma$ $d = (\mu - \mu_0)/\sigma$ $d = (\mu_0 - \mu)/\sigma$	e, f g, h g, h
3.	$H_0: \sigma^2 = \sigma_0^2$	$\chi_0^2 = \dfrac{(n - 1)s^2}{\sigma_0^2}$	$H_1: \sigma^2 \neq \sigma_0^2$ $H_1: \sigma^2 > \sigma_0^2$ $H_1: \sigma^2 < \sigma_0^2$	$\chi_0^2 > \chi_{\alpha/2, n-1}^2$ or $\chi_0^2 < \chi_{1-\alpha/2, n-1}^2$ $\chi_0^2 > \chi_{\alpha, n-1}^2$ $\chi_0^2 < \chi_{1-\alpha, n-1}^2$	$\lambda = \sigma/\sigma_0$ $\lambda = \sigma/\sigma_0$ $\lambda = \sigma/\sigma_0$	i, j k, l m, n
4.	$H_0: p = p_0$	$z_0 = \dfrac{x - np_0}{\sqrt{np_0(1 - p_0)}}$	$H_1: p \neq p_0$ $H_1: p > p_0$ $H_1: p < p_0$	$|z_0| > z_{\alpha/2}$ $z_0 > z_\alpha$ $z_0 < -z_\alpha$	— — —	— — —

Summary of One-Sample Confidence Interval Procedures

Case	Problem Type	Point Estimate	Two-Sided $100(1 - \alpha)$ Percent Confidence Interval
1.	Mean μ, variance σ^2 known	$\bar{x}$	$\bar{x} - z_{\alpha/2}\sigma/\sqrt{n} \leq \mu \leq \bar{x} + z_{\alpha/2}\sigma/\sqrt{n}$
2.	Mean μ of a normal distribution, variance σ^2 unknown	$\bar{x}$	$\bar{x} - t_{\alpha/2, n-1}s/\sqrt{n} \leq \mu \leq \bar{x} + t_{\alpha/2, n-1}s/\sqrt{n}$
3.	Variance σ^2 of a normal distribution	s^2	$\dfrac{(n - 1)s^2}{\chi_{\alpha/2, n-1}^2} \leq \sigma^2 \leq \dfrac{(n - 1)s^2}{\chi_{1-\alpha/2, n-1}^2}$
4.	Proportion or parameter of a binomial distribution p	$\hat{p}$	$\hat{p} - z_{\alpha/2}\sqrt{\dfrac{\hat{p}(1 - \hat{p})}{n}} \leq p \leq \hat{p} + z_{\alpha/2}\sqrt{\dfrac{\hat{p}(1 - \hat{p})}{n}}$

To:

Meredith, Neil, Colin, and Cheryl

Rebecca, Elisa, George, and Taylor

Preface

This is an introductory textbook for a first course in applied statistics and probability for undergraduate students in engineering and the physical or chemical sciences. These individuals play a significant role in designing and developing new products and manufacturing systems and processes, and they also improve existing systems. Statistical methods are an important tool in these activities because they provide the engineer with both descriptive and analytical methods for dealing with the variability in observed data. Although many of the methods we present are fundamental to statistical analysis in other disciplines, such as business and management, the life sciences, and the social sciences, we have elected to focus on an engineering-oriented audience. We believe that this approach will best serve students in engineering and the chemical/physical sciences and will allow them to concentrate on the many applications of statistics in these disciplines. We have worked hard to ensure that our examples and exercises are engineering- and science-based, and in almost all cases we have used examples of real data—either taken from a published source or based on our consulting experiences.

We believe that engineers in all disciplines should take at least one course in statistics. Unfortunately, because of other requirements, most engineers will only take one statistics course. This book can be used for a single course, although we have provided enough material for two courses in the hope that more students will see the important applications of statistics in their everyday work and elect a second course. We believe that this book will also serve as a useful reference.

ORGANIZATION OF THE BOOK

We have retained the relatively modest mathematical level of the first two editions. We have found that engineering students who have completed one or two semesters of calculus should have no difficulty reading almost all of the text. It is our intent to give the reader an understanding of the methodology and how to apply it, not the mathematical theory. We have made many enhancements in this edition, including reorganizing and rewriting major portions of the book.

Perhaps the most common criticism of engineering statistics texts is that they are too long. Both instructors and students complain that it is impossible to cover all of the topics in the book in one or even two terms. For authors, this is a serious issue because there is great variety in both the content and level of these courses, and the decisions about what material to delete without limiting the value of the text are not easy. After struggling with these issues, we decided to divide the text into two components; a set of core topics, many of which are most likely to be covered in an engineering statistics course, and a set of supplementary topics, or topics that will be useful for some but not all courses. The core topics are in the printed book, and the complete text (both core and supplementary topics) is available on the CD that is included with the printed book. Decisions about topics to include in print and which to include only on the CD were made based on the results of a recent survey of instructors.

The *Interactive e-Text* consists of the complete text and a wealth of additional material and features. The text and links on the CD are navigated using Adobe Acrobat™. The links within the *Interactive e-Text* include the following: (1) from the Table of Contents to the selected *e-Text* sections, (2) from the Index to the selected topic within the *e-Text*, (3) from reference to a figure, table, or equation in one section to the actual figure, table, or equation in another section (all figures can be enlarged and printed), (4) from end-of-chapter Important Terms and Concepts to their definitions within the chapter, (5) from in-text **boldfaced terms** to their corresponding Glossary definitions and explanations, (6) from in-text references to the corresponding Appendix tables and charts, (7) from boxed-number end-of-chapter exercises (essentially most odd-numbered exercises) to their answers, (8) from some answers to the complete problem solution, and (9) from the opening splash screen to the textbook Web site.

Chapter 1 is an introduction to the field of statistics and how engineers use statistical methodology as part of the engineering problem-solving process. This chapter also introduces the reader to some engineering applications of statistics, including building empirical models, designing engineering experiments, and monitoring manufacturing processes. These topics are discussed in more depth in subsequent chapters.

Chapters 2, 3, 4, and 5 cover the basic concepts of probability, discrete and continuous random variables, probability distributions, expected values, joint probability distributions, and independence. We have given a reasonably complete treatment of these topics but have avoided many of the mathematical or more theoretical details.

Chapter 6 begins the treatment of statistical methods with random sampling; data summary and description techniques, including stem-and-leaf plots, histograms, box plots, and probability plotting; and several types of time series plots. Chapter 7 discusses point estimation of parameters. This chapter also introduces some of the important properties of estimators, the method of maximum likelihood, the method of moments, sampling distributions, and the central limit theorem.

Chapter 8 discusses interval estimation for a single sample. Topics included are confidence intervals for means, variances or standard deviations, and proportions and prediction and tolerance intervals. Chapter 9 discusses hypothesis tests for a single sample. Chapter 10 presents tests and confidence intervals for two samples. This material has been extensively rewritten and reorganized. There is detailed information and examples of methods for determining appropriate sample sizes. We want the student to become familiar with how these techniques are used to solve real-world engineering problems and to get some understanding of the concepts behind them. We give a logical, heuristic development of the procedures, rather than a formal mathematical one.

Chapters 11 and 12 present simple and multiple linear regression. We use matrix algebra throughout the multiple regression material (Chapter 12) because it is the only easy way to understand the concepts presented. Scalar arithmetic presentations of multiple regression are awkward at best, and we have found that undergraduate engineers are exposed to enough matrix algebra to understand the presentation of this material.

Chapters 13 and 14 deal with single- and multifactor experiments, respectively. The notions of randomization, blocking, factorial designs, interactions, graphical data analysis, and fractional factorials are emphasized. Chapter 15 gives a brief introduction to the methods and applications of nonparametric statistics, and Chapter 16 introduces statistical quality control, emphasizing the control chart and the fundamentals of statistical process control.

Each chapter has an extensive collection of exercises, including end-of-section exercises that emphasize the material in that section, supplemental exercises at the end of the chapter that cover the scope of chapter topics, and mind-expanding exercises that often require the student to extend the text material somewhat or to apply it in a novel situation. As noted above, answers are provided to most odd-numbered exercises and the *e-Text* contains complete solutions to selected exercises.

USING THE BOOK

This is a very flexible textbook because instructors' ideas about what should be in a first course on statistics for engineers vary widely, as do the abilities of different groups of students. Therefore, we hesitate to give too much advice but will explain how we use the book.

We believe that a first course in statistics for engineers should be primarily an applied statistics course, not a probability course. In our one-semester course we cover all of Chapter 1 (in one or two lectures); overview the material on probability, putting most of the emphasis on the normal distribution (six to eight lectures); discuss most of Chapters 6 though 10 on confidence intervals and tests (twelve to fourteen lectures); introduce regression models in Chapter 11 (four lectures); give an introduction to the design of experiments from Chapters 13 and 14 (six lectures); and present the basic concepts of statistical process control, including the Shewhart control chart from Chapter 16 (four lectures). This leaves about three to four periods for exams and review. Let us emphasize that the purpose of this course is to introduce engineers to how statistics can be used to solve real-world engineering problems, not to weed out the less mathematically gifted students. This course is not the "baby math-stat" course that is all too often given to engineers.

If a second semester is available, it is possible to cover the entire book, including much of the *e-Text* material, if appropriate for the audience. It would also be possible to assign and work many of the homework problems in class to reinforce the understanding of the concepts. Obviously, multiple regression and more design of experiments would be major topics in a second course.

USING THE COMPUTER

In practice, engineers use computers to apply statistical methods to solve problems. Therefore, we strongly recommend that the computer be integrated into the class. Throughout the book we have presented output from Minitab as typical examples of what can be done with modern statistical software. In teaching, we have used other software packages, including Statgraphics, JMP, and Statistica. We did not clutter up the book with examples from many different packages because how the instructor integrates the software into the class is ultimately more important than which package is used. All text data is available in electronic form on the *e-Text* CD. In some chapters, there are problems that we feel should be worked using computer software. We have marked these problems with a special icon in the margin.

In our own classrooms, we use the computer in almost every lecture and demonstrate how the technique is implemented in software as soon as it is discussed in the lecture. Student versions of many statistical software packages are available at low cost, and students can either purchase their own copy or use the products available on the PC local area networks. We have found that this greatly improves the pace of the course and student understanding of the material.

USING THE WEB

Additional resources for students and instructors can be found on the website for this book at www.wiley.com/college/montgomery/.

ACKNOWLEDGMENTS

We would like to express our grateful appreciation to the many organizations and individuals who have contributed to this book. Many instructors who used the first two editions provided excellent suggestions that we have tried to incorporate in this revision. We also thank Professors Manuel D. Rossetti (University of Arkansas), Bruce Schmeiser (Purdue University), Michael G. Akritas (Penn State University), and Arunkumar Pennathur (University of Texas at El Paso) for their insightful reviews of the manuscript of the third edition. We are also indebted to Dr. Smiley Cheng for permission to adapt many of the statistical tables from his excellent book (with Dr. James Fu), *Statistical Tables for Classroom and Exam Room.* John Wiley and Sons, Prentice Hall, the Institute of Mathematical Statistics, and the editors of Biometrics allowed us to use copyrighted material, for which we are grateful. Thanks are also due to Dr. Lora Zimmer, Dr. Connie Borror, and Dr. Alejandro Heredia-Langner for their outstanding work on the solutions to exercises.

Douglas C. Montgomery

George C. Runger

Contents

The Role of Statistics in Engineering

CHAPTER OUTLINE

LEARNING OBJECTIVES

After careful study of this chapter you should be able to do the following:

1. Identify the role that statistics can play in the engineering problem-solving process
2. Discuss how variability affects the data collected and used for making engineering decisions
3. Explain the difference between enumerative and analytical studies
4. Discuss the different methods that engineers use to collect data
5. Identify the advantages that designed experiments have in comparison to other methods of collecting engineering data
6. Explain the differences between mechanistic models and empirical models
7. Discuss how probability and probability models are used in engineering and science

CD MATERIAL

8. Explain the factorial experimental design.
9. Explain how factors can Interact.

Answers for most odd numbered exercises are at the end of the book. Answers to exercises whose numbers are surrounded by a box can be accessed in the e-Text by clicking on the box. Complete worked solutions to certain exercises are also available in the e-Text. These are indicated in the Answers to Selected Exercises section by a box around the exercise number. Exercises are also available for some of the text sections that appear on CD only. These exercises may be found within the e-Text immediately following the section they accompany.

1-1   THE ENGINEERING METHOD AND STATISTICAL THINKING

An engineer is someone who solves problems of interest to society by the efficient application of scientific principles. Engineers accomplish this by either refining an existing product or process or by designing a new product or process that meets customers' needs. The **engineering, or scientific, method** is the approach to formulating and solving these problems. The steps in the engineering method are as follows:

1.   Develop a clear and concise description of the problem.

2.   Identify, at least tentatively, the important factors that affect this problem or that may play a role in its solution.

3.   Propose a model for the problem, using scientific or engineering knowledge of the phenomenon being studied. State any limitations or assumptions of the model.

4.   Conduct appropriate experiments and collect data to test or validate the tentative model or conclusions made in steps 2 and 3.

5.   Refine the model on the basis of the observed data.

6.   Manipulate the model to assist in developing a solution to the problem.

7.   Conduct an appropriate experiment to confirm that the proposed solution to the problem is both effective and efficient.

8.   Draw conclusions or make recommendations based on the problem solution.

The steps in the engineering method are shown in Fig. 1-1. Notice that the engineering method features a strong interplay between the problem, the factors that may influence its solution, a model of the phenomenon, and experimentation to verify the adequacy of the model and the proposed solution to the problem. Steps 2–4 in Fig. 1-1 are enclosed in a box, indicating that several cycles or iterations of these steps may be required to obtain the final solution. Consequently, engineers must know how to efficiently plan experiments, collect data, analyze and interpret the data, and understand how the observed data are related to the model they have proposed for the problem under study.

The field of **statistics** deals with the collection, presentation, analysis, and use of data to make decisions, solve problems, and design products and processes. Because many aspects of engineering practice involve working with data, obviously some knowledge of statistics is important to any engineer. Specifically, statistical techniques can be a powerful aid in designing new products and systems, improving existing designs, and designing, developing, and improving production processes.

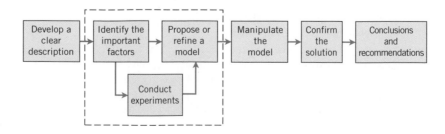

Figure 1-1   The engineering method.

Statistical methods are used to help us describe and understand **variability.** By variability, we mean that successive observations of a system or phenomenon do not produce exactly the same result. We all encounter variability in our everyday lives, and **statistical thinking** can give us a useful way to incorporate this variability into our decision-making processes. For example, consider the gasoline mileage performance of your car. Do you always get exactly the same mileage performance on every tank of fuel? Of course not—in fact, sometimes the mileage performance varies considerably. This observed variability in gasoline mileage depends on many factors, such as the type of driving that has occurred most recently (city versus highway), the changes in condition of the vehicle over time (which could include factors such as tire inflation, engine compression, or valve wear), the brand and/or octane number of the gasoline used, or possibly even the weather conditions that have been recently experienced. These factors represent potential **sources of variability** in the system. Statistics gives us a framework for describing this variability and for learning about which potential sources of variability are the most important or which have the greatest impact on the gasoline mileage performance.

We also encounter variability in dealing with engineering problems. For example, suppose that an engineer is designing a nylon connector to be used in an automotive engine application. The engineer is considering establishing the design specification on wall thickness at 3/32 inch but is somewhat uncertain about the effect of this decision on the connector pull-off force. If the pull-off force is too low, the connector may fail when it is installed in an engine. Eight prototype units are produced and their pull-off forces measured, resulting in the following data (in pounds): 12.6, 12.9, 13.4, 12.3, 13.6, 13.5, 12.6, 13.1. As we anticipated, not all of the prototypes have the same pull-off force. We say that there is variability in the pull-off force measurements. Because the pull-off force measurements exhibit variability, we consider the pull-off force to be a **random variable.** A convenient way to think of a random variable, say X, that represents a measurement, is by using the model

$$X = \mu + \epsilon \tag{1-1}$$

where μ is a constant and ϵ is a random disturbance. The constant remains the same with every measurement, but small changes in the environment, test equipment, differences in the individual parts themselves, and so forth change the value of ϵ. If there were no disturbances, ϵ would always equal zero and X would always be equal to the constant μ. However, this never happens in the real world, so the actual measurements X exhibit variability. We often need to describe, quantify and ultimately reduce variability.

Figure 1-2 presents a **dot diagram** of these data. The dot diagram is a very useful plot for displaying a small body of data—say, up to about 20 observations. This plot allows us to see easily two features of the data; the **location,** or the middle, and the **scatter** or **variability.** When the number of observations is small, it is usually difficult to identify any specific patterns in the variability, although the dot diagram is a convenient way to see any unusual data features.

The need for statistical thinking arises often in the solution of engineering problems. Consider the engineer designing the connector. From testing the prototypes, he knows that the average pull-off force is 13.0 pounds. However, he thinks that this may be too low for the

Figure 1-2   Dot diagram of the pull-off force data when wall thickness is 3/32 inch.

Figure 1-3   Dot diagram of pull-off force for two wall thicknesses.

intended application, so he decides to consider an alternative design with a greater wall thickness, 1/8 inch. Eight prototypes of this design are built, and the observed pull-off force measurements are 12.9, 13.7, 12.8, 13.9, 14.2, 13.2, 13.5, and 13.1. The average is 13.4. Results for both samples are plotted as dot diagrams in Fig. 1-3, page 3. This display gives the impression that increasing the wall thickness has led to an increase in pull-off force. However, there are some obvious questions to ask. For instance, how do we know that another sample of prototypes will not give different results? Is a sample of eight prototypes adequate to give reliable results? If we use the test results obtained so far to conclude that increasing the wall thickness increases the strength, what risks are associated with this decision? For example, is it possible that the apparent increase in pull-off force observed in the thicker prototypes is only due to the inherent variability in the system and that increasing the thickness of the part (and its cost) really has no effect on the pull-off force?

Often, physical laws (such as Ohm's law and the ideal gas law) are applied to help design products and processes. We are familiar with this reasoning from general laws to specific cases. But it is also important to reason from a specific set of measurements to more general cases to answer the previous questions. This reasoning is from a sample (such as the eight connectors) to a population (such as the connectors that will be sold to customers). The reasoning is referred to as **statistical inference.** See Fig. 1-4. Historically, measurements were obtained from a sample of people and generalized to a population, and the terminology has remained. Clearly, reasoning based on measurements from some objects to measurements on all objects can result in errors (called sampling errors). However, if the sample is selected properly, these risks can be quantified and an appropriate sample size can be determined.

In some cases, the sample is actually selected from a well-defined population. The sample is a subset of the population. For example, in a study of resistivity a sample of three wafers might be selected from a production lot of wafers in semiconductor manufacturing. Based on the resistivity data collected on the three wafers in the sample, we want to draw a conclusion about the resistivity of all of the wafers in the lot.

In other cases, the population is conceptual (such as with the connectors), but it might be thought of as future replicates of the objects in the sample. In this situation, the eight prototype connectors must be representative, in some sense, of the ones that will be manufactured in the future. Clearly, this analysis requires some notion of **stability** as an additional assumption. For example, it might be assumed that the sources of variability in the manufacture of the prototypes (such as temperature, pressure, and curing time) are the same as those for the connectors that will be manufactured in the future and ultimately sold to customers.

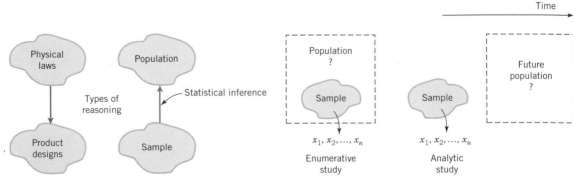

Figure 1-4   Statistical inference is one type of reasoning.

Figure 1-5   Enumerative versus analytic study.

The wafers-from-lots example is called an **enumerative** study. An enumerative sample is used to make an inference to the population from which the sample is selected. The connector example is called an **analytic** study. A sample is used to make an inference to a conceptual (future) population. The statistical analyses are usually the same in both cases, but an analytic study clearly requires an assumption of stability. See Fig. 1-5, on page 4.

1-2   COLLECTING ENGINEERING DATA

1-2.1   Basic Principles

In the previous section, we illustrated some simple methods for summarizing data. In the engineering environment, the data is almost always a sample that has been selected from some population. Three basic methods of collecting data are

- A **retrospective** study using historical data
- An **observational** study
- A **designed experiment**

An effective data collection procedure can greatly simplify the analysis and lead to improved understanding of the population or process that is being studied. We now consider some examples of these data collection methods.

1-2.2   Retrospective Study

Montgomery, Peck, and Vining (2001) describe an acetone-butyl alcohol distillation column for which concentration of acetone in the distillate or output product stream is an important variable. Factors that may affect the distillate are the reboil temperature, the condensate temperature, and the reflux rate. Production personnel obtain and archive the following records:

- The concentration of acetone in an hourly test sample of output product
- The reboil temperature log, which is a plot of the reboil temperature over time
- The condenser temperature controller log
- The nominal reflux rate each hour

The reflux rate should be held constant for this process. Consequently, production personnel change this very infrequently.

A retrospective study would use either all or a sample of the historical process data archived over some period of time. The study objective might be to discover the relationships among the two temperatures and the reflux rate on the acetone concentration in the output product stream. However, this type of study presents some problems:

1. We may not be able to see the relationship between the reflux rate and acetone concentration, because the reflux rate didn't change much over the historical period.

2. The archived data on the two temperatures (which are recorded almost continuously) do not correspond perfectly to the acetone concentration measurements (which are made hourly). It may not be obvious how to construct an approximate correspondence.

3. Production maintains the two temperatures as closely as possible to desired targets or set points. Because the temperatures change so little, it may be difficult to assess their real impact on acetone concentration.

4. Within the narrow ranges that they do vary, the condensate temperature tends to increase with the reboil temperature. Consequently, the effects of these two process variables on acetone concentration may be difficult to separate.

As you can see, a retrospective study may involve a lot of **data,** but that data may contain relatively little useful **information** about the problem. Furthermore, some of the relevant data may be missing, there may be transcription or recording errors resulting in **outliers** (or unusual values), or data on other important factors may not have been collected and archived. In the distillation column, for example, the specific concentrations of butyl alcohol and acetone in the input feed stream are a very important factor, but they are not archived because the concentrations are too hard to obtain on a routine basis. As a result of these types of issues, statistical analysis of historical data sometimes identify interesting phenomena, but solid and reliable explanations of these phenomena are often difficult to obtain.

1-2.3 Observational Study

In an observational study, the engineer observes the process or population, disturbing it as little as possible, and records the quantities of interest. Because these studies are usually conducted for a relatively short time period, sometimes variables that are not routinely measured can be included. In the distillation column, the engineer would design a form to record the two temperatures and the reflux rate when acetone concentration measurements are made. It may even be possible to measure the input feed stream concentrations so that the impact of this factor could be studied. Generally, an observational study tends to solve problems 1 and 2 above and goes a long way toward obtaining accurate and reliable data. However, observational studies may not help resolve problems 3 and 4.

1-2.4 Designed Experiments

In a designed experiment the engineer makes *deliberate* or *purposeful changes* in the controllable variables of the system or process, observes the resulting system output data, and then makes an inference or decision about which variables are responsible for the observed changes in output performance. The nylon connector example in Section 1-1 illustrates a designed experiment; that is, a deliberate change was made in the wall thickness of the connector with the objective of discovering whether or not a greater pull-off force could be obtained. Designed experiments play a very important role in engineering design and development and in the improvement of manufacturing processes. Generally, when products and processes are designed and developed with designed experiments, they enjoy better performance, higher reliability, and lower overall costs. Designed experiments also play a crucial role in reducing the lead time for engineering design and development activities.

For example, consider the problem involving the choice of wall thickness for the nylon connector. This is a simple illustration of a designed experiment. The engineer chose two wall thicknesses for the connector and performed a series of tests to obtain pull-off force measurements at each wall thickness. In this simple **comparative experiment,** the engineer is interested in determining if there is any difference between the 3/32- and 1/8-inch designs. An approach that could be used in analyzing the data from this experiment is to compare the mean pull-off force for the 3/32-inch design to the mean pull-off force for the 1/8-inch design using statistical **hypothesis testing,** which is discussed in detail in Chapters 9 and 10. Generally, a **hypothesis** is a statement about some aspect of the system in which we are interested. For example, the engineer might want to know if the mean pull-off force of a 3/32-inch design exceeds the typical maximum load expected to be encountered in this application, say 12.75 pounds. Thus, we would be interested in testing the hypothesis that the mean strength exceeds 12.75 pounds. This is called a **single-sample hypothesis testing problem.** It is also an example of an **analytic study.** Chapter 9 presents techniques for this type of problem. Alternatively, the engineer might be interested in testing the hypothesis that increasing the wall thickness from 3/32- to 1/8-inch results in an increase in mean pull-off force. Clearly, this is an **analytic study;** it is also an example of **a two-sample hypothesis testing problem.** Two-sample hypothesis testing problems are discussed in Chapter 10.

Designed experiments are a very powerful approach to studying complex systems, such as the distillation column. This process has three factors, the two temperatures and the reflux rate, and we want to investigate the effect of these three factors on output acetone concentration. A good experimental design for this problem must ensure that we can separate the effects of all three factors on the acetone concentration. The specified values of the three factors used in the experiment are called **factor levels.** Typically, we use a small number of levels for each factor, such as two or three. For the distillation column problem, suppose we use a "high," and "low," level (denoted $+1$ and -1, respectively) for each of the factors. We thus would use two levels for each of the three factors. A very reasonable experiment design strategy uses every possible combination of the factor levels to form a basic experiment with eight different settings for the process. This type of experiment is called a **factorial experiment.** Table 1-1 presents this experimental design.

Figure 1-6, on page 8, illustrates that this design forms a cube in terms of these high and low levels. With each setting of the process conditions, we allow the column to reach equilibrium, take a sample of the product stream, and determine the acetone concentration. We then can draw specific inferences about the effect of these factors. Such an approach allows us to proactively study a population or process. Designed experiments play a very important role in engineering and science. Chapters 13 and 14 discuss many of the important principles and techniques of experimental design.

Table 1-1   The Designed Experiment (Factorial Design) for the Distillation Column

Reboil Temp.	Condensate Temp.	Reflux Rate
-1	-1	-1
$+1$	-1	-1
-1	$+1$	-1
$+1$	$+1$	-1
-1	-1	$+1$
$+1$	-1	$+1$
-1	$+1$	$+1$
$+1$	$+1$	$+1$

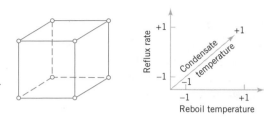

Figure 1-6   The factorial design for the distillation column.

1-2.5   A Factorial Experiment for the Connector Pull-off Force Problem (CD Only)

1-2.6   Observing Processes Over Time

Often data are collected over time. In this case, it is usually very helpful to plot the data versus time in a **time series** **plot.** Phenomena that might affect the system or process often become more visible in a time-oriented plot and the concept of stability can be better judged.

Figure 1-7 is a dot diagram of acetone concentration readings taken hourly from the distillation column described in Section 1-2.2. The large variation displayed on the dot diagram indicates a lot of variability in the concentration, but the chart does not help explain the reason for the variation. The time series plot is shown in Figure 1-8, on page 9. A shift in the process mean level is visible in the plot and an estimate of the time of the shift can be obtained.

W. Edwards **Deming,** a very influential industrial statistician, stressed that it is important to understand the nature of variability in processes and systems over time. He conducted an experiment in which he attempted to drop marbles as close as possible to a target on a table. He used a funnel mounted on a ring stand and the marbles were dropped into the funnel. See Fig. 1-9. The funnel was aligned as closely as possible with the center of the target. He then used two different strategies to operate the process. (1) He never moved the funnel. He just dropped one marble after another and recorded the distance from the target. (2) He dropped the first marble and recorded its location relative to the target. He then moved the funnel an equal and opposite distance in an attempt to compensate for the error. He continued to make this type of adjustment after each marble was dropped.

After both strategies were completed, he noticed that the variability of the distance from the target for strategy 2 was approximately 2 times larger than for strategy 1. The adjustments to the funnel increased the deviations from the target. The explanation is that the error (the deviation of the marble's position from the target) for one marble provides no information about the error that will occur for the next marble. Consequently, adjustments to the funnel do not decrease future errors. Instead, they tend to move the funnel farther from the target.

This interesting experiment points out that adjustments to a process based on random disturbances can actually *increase* the variation of the process. This is referred to as **overcontrol**

Figure 1-7   The dot diagram illustrates variation but does not identify the problem.

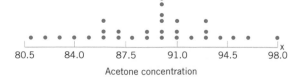

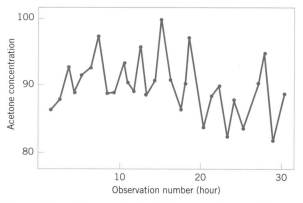

Figure 1-8   A time series plot of concentration provides more information than the dot diagram.

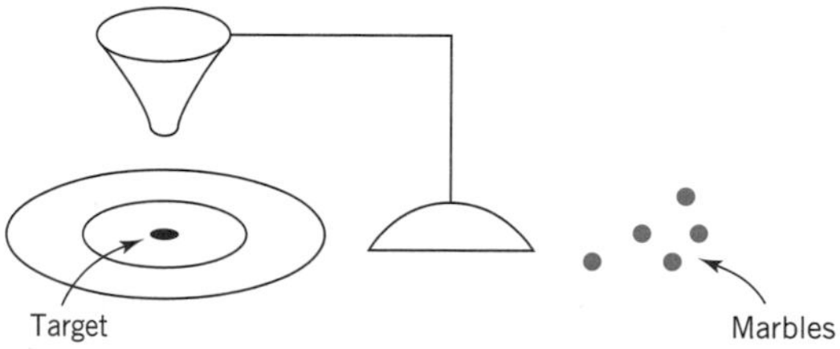

Figure 1-9   Deming's funnel experiment.

or **tampering.** Adjustments should be applied only to compensate for a nonrandom shift in the process—then they can help. A computer simulation can be used to demonstrate the lessons of the funnel experiment. Figure 1-10 displays a time plot of 100 measurements (denoted as y) from a process in which only random disturbances are present. The target value for the process is 10 units. The figure displays the data with and without adjustments that are applied to the process mean in an attempt to produce data closer to target. Each adjustment is equal and opposite to the deviation of the previous measurement from target. For example, when the measurement is 11 (one unit above target), the mean is reduced by one unit before the next measurement is generated. The overcontrol has increased the deviations from the target.

Figure 1-11 displays the data without adjustment from Fig. 1-10, except that the measurements after observation number 50 are increased by two units to simulate the effect of a shift in the mean of the process. When there is a true shift in the mean of a process, an adjustment can be useful. Figure 1-11 also displays the data obtained when one adjustment (a decrease of

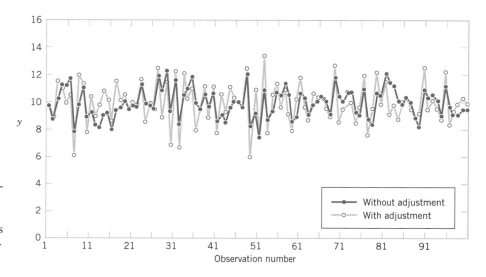

Figure 1-10   Adjustments applied to random disturbances overcontrol the process and increase the deviations from the target.

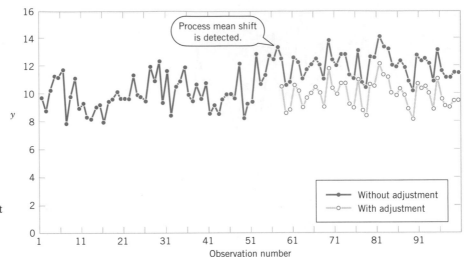

Figure 1-11   Process mean shift is detected at observation number 57, and one adjustment (a decrease of two units) reduces the deviations from target.

two units) is applied to the mean after the shift is detected (at observation number 57). Note that this adjustment decreases the deviations from target.

The question of when to apply adjustments (and by what amounts) begins with an understanding of the types of variation that affect a process. A **control chart** is an invaluable way to examine the variability in time-oriented data. Figure 1-12 presents a control chart for the concentration data from Fig. 1-8. The **center line** on the control chart is just the average of the concentration measurements for the first 20 samples ($\bar{x} = 91.5$ g/l) when the process is stable. The **upper control limit** and the **lower control limit** are a pair of statistically derived limits that reflect the inherent or natural variability in the process. These limits are located three standard deviations of the concentration values above and below the center line. If the process is operating as it should, without any external sources of variability present in the system, the concentration measurements should fluctuate randomly around the center line, and almost all of them should fall between the control limits.

In the control chart of Fig. 1-12, the visual frame of reference provided by the center line and the control limits indicates that some upset or disturbance has affected the process around sample 20 because all of the following observations are below the center line and two of them

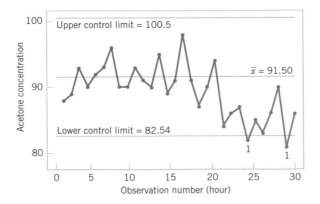

Figure 1-12   A control chart for the chemical process concentration data.

actually fall below the lower control limit. This is a very strong signal that corrective action is required in this process. If we can find and eliminate the underlying cause of this upset, we can improve process performance considerably.

Control charts are a very important application of statistics for monitoring, controlling, and improving a process. The branch of statistics that makes use of control charts is called **statistical process control,** or **SPC.** We will discuss SPC and control charts in Chapter 16.

1-3 MECHANISTIC AND EMPIRICAL MODELS

Models play an important role in the analysis of nearly all engineering problems. Much of the formal education of engineers involves learning about the models relevant to specific fields and the techniques for applying these models in problem formulation and solution. As a simple example, suppose we are measuring the flow of current in a thin copper wire. Our model for this phenomenon might be Ohm's law:

$$Current = voltage/resistance$$

or

$$I = E/R \tag{1-2}$$

We call this type of model a **mechanistic model** because it is built from our underlying knowledge of the basic physical mechanism that relates these variables. However, if we performed this measurement process more than once, perhaps at different times, or even on different days, the observed current could differ slightly because of small changes or variations in factors that are not completely controlled, such as changes in ambient temperature, fluctuations in performance of the gauge, small impurities present at different locations in the wire, and drifts in the voltage source. Consequently, a more realistic model of the observed current might be

$$I = E/R + \epsilon \tag{1-3}$$

where ϵ is a term added to the model to account for the fact that the observed values of current flow do not perfectly conform to the mechanistic model. We can think of ϵ as a term that includes the effects of all of the unmodeled sources of variability that affect this system.

Sometimes engineers work with problems for which there is no simple or well-understood mechanistic model that explains the phenomenon. For instance, suppose we are interested in the number average molecular weight (M_n) of a polymer. Now we know that M_n is related to the viscosity of the material (V), and it also depends on the amount of catalyst (C) and the temperature (T) in the polymerization reactor when the material is manufactured. The relationship between M_n and these variables is

$$M_n = f(V, C, T) \tag{1-4}$$

say, where the *form* of the function f is unknown. Perhaps a working model could be developed from a first-order Taylor series expansion, which would produce a model of the form

$$M_n = \beta_0 + \beta_1 V + \beta_2 C + \beta_3 T \tag{1-5}$$

where the β's are unknown parameters. Now just as in Ohm's law, this model will not exactly describe the phenomenon, so we should account for the other sources of variability that may affect the molecular weight by adding another term to the model; therefore

$$M_n = \beta_0 + \beta_1 V + \beta_2 C + \beta_3 T + \epsilon \qquad (1\text{-}6)$$

is the model that we will use to relate molecular weight to the other three variables. This type of model is called an **empirical model;** that is, it uses our engineering and scientific knowledge of the phenomenon, but it is not directly developed from our theoretical or first-principles understanding of the underlying mechanism.

To illustrate these ideas with a specific example, consider the data in Table 1-2. This table contains data on three variables that were collected in an observational study in a semiconductor manufacturing plant. In this plant, the finished semiconductor is wire bonded to a frame. The variables reported are pull strength (a measure of the amount of force required to break the bond), the wire length, and the height of the die. We would like to find a model relating pull strength to wire length and die height. Unfortunately, there is no physical mechanism that we can easily apply here, so it doesn't seem likely that a mechanistic modeling approach will be successful.

Table 1-2   Wire Bond Pull Strength Data

Observation Number	Pull Strength y	Wire Length x_1	Die Height x_2
1	9.95	2	50
2	24.45	8	110
3	31.75	11	120
4	35.00	10	550
5	25.02	8	295
6	16.86	4	200
7	14.38	2	375
8	9.60	2	52
9	24.35	9	100
10	27.50	8	300
11	17.08	4	412
12	37.00	11	400
13	41.95	12	500
14	11.66	2	360
15	21.65	4	205
16	17.89	4	400
17	69.00	20	600
18	10.30	1	585
19	34.93	10	540
20	46.59	15	250
21	44.88	15	290
22	54.12	16	510
23	56.63	17	590
24	22.13	6	100
25	21.15	5	400

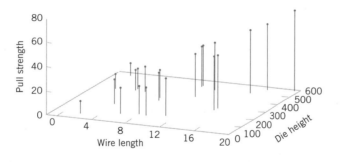

Figure 1-13 Three-dimensional plot of the wire and pull strength data.

Figure 1-13 presents a three-dimensional plot of all 25 observations on pull strength, wire length, and die height. From examination of this plot, we see that pull strength increases as both wire length and die height increase. Furthermore, it seems reasonable to think that a model such as

$$\text{Pull strength} = \beta_0 + \beta_1(\text{wire length}) + \beta_2(\text{die height}) + \epsilon$$

would be appropriate as an empirical model for this relationship. In general, this type of empirical model is called a **regression model.** In Chapters 11 and 12 we show how to build these models and test their adequacy as approximating functions. We will use a method for estimating the parameters in regression models, called the method of **least squares,** that traces its origins to work by Karl Gauss. Essentially, this method chooses the parameters in the empirical model (the β's) to minimize the sum of the squared distances between each data point and the plane represented by the model equation. Applying this technique to the data in Table 1-2 results in

$$\widehat{\text{Pull strength}} = 2.26 + 2.74(\text{wire length}) + 0.0125(\text{die height}) \qquad (1\text{-}7)$$

where the "hat," or circumflex, over pull strength indicates that this is an estimated or predicted quantity.

Figure 1-14 is a plot of the predicted values of pull strength versus wire length and die height obtained from Equation 1-7. Notice that the predicted values lie on a plane above the wire length–die height space. From the plot of the data in Fig. 1-13, this model does not appear unreasonable. The empirical model in Equation 1-7 could be used to predict values of pull strength for various combinations of wire length and die height that are of interest. Essentially, the empirical model could be used by an engineer in exactly the same way that a mechanistic model can be used.

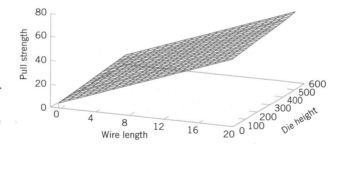

Figure 1-14 Plot of predicted values of pull strength from the empirical model.

1-4 PROBABILITY AND PROBABILITY MODELS

In Section 1-1, it was mentioned that decisions often need to be based on measurements from only a subset of objects selected in a sample. This process of reasoning from a sample of objects to conclusions for a population of objects was referred to as statistical inference. A sample of three wafers selected from a larger production lot of wafers in semiconductor manufacturing was an example mentioned. To make good decisions, an analysis of how well a sample represents a population is clearly necessary. If the lot contains defective wafers, how well will the sample detect this? How can we quantify the criterion to "detect well"? Basically, how can we quantify the risks of decisions based on samples? Furthermore, how should samples be selected to provide good decisions—ones with acceptable risks? **Probability** models help quantify the risks involved in statistical inference, that is, the risks involved in decisions made every day.

More details are useful to describe the role of probability models. Suppose a production lot contains 25 wafers. If all the wafers are defective or all are good, clearly any sample will generate all defective or all good wafers, respectively. However, suppose only one wafer in the lot is defective. Then a sample might or might not detect (include) the wafer. A probability model, along with a method to select the sample, can be used to quantify the risks that the defective wafer is or is not detected. Based on this analysis, the size of the sample might be increased (or decreased). The risk here can be interpreted as follows. Suppose a series of lots, each with exactly one defective wafer, are sampled. The details of the method used to select the sample are postponed until randomness is discussed in the next chapter. Nevertheless, assume that the same size sample (such as three wafers) is selected in the same manner from each lot. The proportion of the lots in which the defective wafer is included in the sample or, more specifically, the limit of this proportion as the number of lots in the series tends to infinity, is interpreted as the probability that the defective wafer is detected.

A probability model is used to calculate this proportion under reasonable assumptions for the manner in which the sample is selected. This is fortunate because we do not want to attempt to sample from an infinite series of lots. Problems of this type are worked in Chapters 2 and 3. More importantly, this probability provides valuable, quantitative information regarding any decision about lot quality based on the sample.

Recall from Section 1-1 that a population might be conceptual, as in an analytic study that applies statistical inference to future production based on the data from current production. When populations are extended in this manner, the role of statistical inference and the associated probability models becomes even more important.

In the previous example, each wafer in the sample was only classified as defective or not. Instead, a continuous measurement might be obtained from each wafer. In Section 1-2.6, concentration measurements were taken at periodic intervals from a production process. Figure 1-7 shows that variability is present in the measurements, and there might be concern that the process has moved from the target setting for concentration. Similar to the defective wafer, one might want to quantify our ability to detect a process change based on the sample data. Control limits were mentioned in Section 1-2.6 as decision rules for whether or not to adjust a process. The probability that a particular process change is detected can be calculated with a probability model for concentration measurements. Models for continous measurements are developed based on plausible assumptions for the data and a result known as the central limit theorem, and the associated normal distribution is a particularly valuable probability model for statistical inference. Of course, a check of assumptions is important. These types of probability models are discussed in Chapter 4. The objective is still to quantify the risks inherent in the inference made from the sample data.

Throughout Chapters 6 through 15, decisions are based statistical inference from sample data. Continuous probability models, specifically the normal distribution, are used extensively to quantify the risks in these decisions and to evaluate ways to collect the data and how large a sample should be selected.

IMPORTANT TERMS AND CONCEPTS

In the E-book, click on any term or concept below to go to that subject.

Analytic study
Designed experiment
Empirical model
Engineering method
Enumerative study

Mechanistic model
Observational study
Overcontrol
Population
Probability model
Problem-solving method
Retrospective study
Sample

Statistical inference
Statistical Process
    Control
Statistical thinking
Tampering
Variability

CD MATERIAL

Factorial Experiment
Fractional factorial
    experiment
Interaction

Probability

CHAPTER OUTLINE

LEARNING OBJECTIVES

After careful study of this chapter you should be able to do the following:

1. Understand and describe sample spaces and events for random experiments with graphs, tables, lists, or tree diagrams

2. Interpret probabilities and use probabilities of outcomes to calculate probabilities of events in discrete sample spaces

3. Calculate the probabilities of joint events such as unions and intersections from the probabilities of individual events

4. Interpret and calculate conditional probabilities of events

5. Determine the independence of events and use independence to calculate probabilities

6. Use Bayes' theorem to calculate conditional probabilities

7. Understand random variables

CD MATERIAL

8. Use permutation and combinations to count the number of outcomes in both an event and the sample space.

Answers for most odd numbered exercises are at the end of the book. Answers to exercises whose numbers are surrounded by a box can be accessed in the e-Text by clicking on the box. Complete worked solutions to certain exercises are also available in the e-Text. These are indicated in the Answers to Selected Exercises section by a box around the exercise number. Exercises are also available for some of the text sections that appear on CD only. These exercises may be found within the e-Text immediately following the section they accompany.

2-1   SAMPLE SPACES AND EVENTS

2-1.1   Random Experiments

If we measure the current in a thin copper wire, we are conducting an experiment. However, in day-to-day repetitions of the measurement the results can differ slightly because of small variations in variables that are not controlled in our experiment, including changes in ambient temperatures, slight variations in gauge and small impurities in the chemical composition of the wire if different locations are selected, and current source drifts. Consequently, this experiment (as well as many we conduct) is said to have a **random** component. In some cases, the random variations, are small enough, relative to our experimental goals, that they can be ignored. However, no matter how carefully our experiment is designed and conducted, the variation is almost always present, and its magnitude can be large enough that the important conclusions from our experiment are not obvious. In these cases, the methods presented in this book for modeling and analyzing experimental results are quite valuable.

Our goal is to understand, quantify, and model the type of variations that we often encounter. When we incorporate the variation into our thinking and analyses, we can make informed judgments from our results that are not invalidated by the variation.

Models and analyses that include variation are not different from models used in other areas of engineering and science. Figure 2-1 displays the important components. A mathematical model (or abstraction) of the physical system is developed. It need not be a perfect abstraction. For example, Newton's laws are not perfect descriptions of our physical universe. Still, they are useful models that can be studied and analyzed to approximately quantify the performance of a wide range of engineered products. Given a mathematical abstraction that is validated with measurements from our system, we can use the model to understand, describe, and quantify important aspects of the physical system and predict the response of the system to inputs.

Throughout this text, we discuss models that allow for variations in the outputs of a system, even though the variables that we control are not purposely changed during our study. Figure 2-2 graphically displays a model that incorporates uncontrollable inputs (noise) that combine with the controllable inputs to produce the output of our system. Because of the

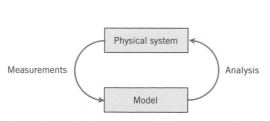

Figure 2-1   Continuous iteration between model and physical system.

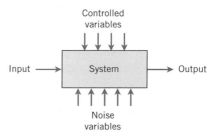

Figure 2-2   Noise variables affect the transformation of inputs to outputs.

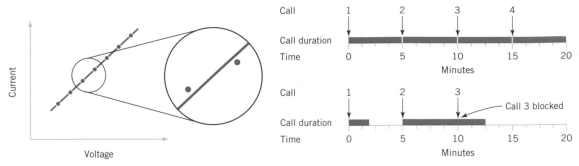

Figure 2-3   A closer examination of the system identifies deviations from the model.

Figure 2-4   Variation causes disruptions in the system.

uncontrollable inputs, the same settings for the controllable inputs do not result in identical outputs every time the system is measured.

Definition

> An experiment that can result in different outcomes, even though it is repeated in the same manner every time, is called a **random experiment.**

For the example of measuring current in a copper wire, our model for the system might simply be Ohm's law. Because of uncontrollable inputs, variations in measurements of current are expected. Ohm's law might be a suitable approximation. However, if the variations are large relative to the intended use of the device under study, we might need to extend our model to include the variation. See Fig. 2-3.

As another example, in the design of a communication system, such as a computer or voice communication network, the information capacity available to service individuals using the network is an important design consideration. For voice communication, sufficient external lines need to be purchased from the phone company to meet the requirements of a business. Assuming each line can carry only a single conversation, how many lines should be purchased? If too few lines are purchased, calls can be delayed or lost. The purchase of too many lines increases costs. Increasingly, design and product development is required to meet customer requirements *at a competitive cost.*

In the design of the voice communication system, a model is needed for the number of calls and the duration of calls. Even knowing that on average, calls occur every five minutes and that they last five minutes is not sufficient. If calls arrived precisely at five-minute intervals and lasted for precisely five minutes, one phone line would be sufficient. However, the slightest variation in call number or duration would result in some calls being blocked by others. See Fig. 2-4. A system designed without considering variation will be woefully inadequate for practical use. Our model for the number and duration of calls needs to include variation as an integral component. An analysis of models including variation is important for the design of the phone system.

2-1.2   Sample Spaces

To model and analyze a random experiment, we must understand the set of possible **out-comes** from the experiment. In this introduction to probability, we make use of the basic concepts of sets and operations on sets. It is assumed that the reader is familiar with these topics.

Definition

> The set of all possible outcomes of a random experiment is called the **sample space** of the experiment. The sample space is denoted as S.

A sample space is often defined based on the objectives of the analysis.

EXAMPLE 2-1   Consider an experiment in which you select a molded plastic part, such as a connector, and measure its thickness. The possible values for thickness depend on the resolution of the measuring instrument, and they also depend on upper and lower bounds for thickness. However, it might be convenient to define the sample space as simply the positive real line

$$S = R^+ = \{x \mid x > 0\}$$

because a negative value for thickness cannot occur.

If it is known that all connectors will be between 10 and 11 millimeters thick, the sample space could be

$$S = \{x \mid 10 < x < 11\}$$

If the objective of the analysis is to consider only whether a particular part is low, medium, or high for thickness, the sample space might be taken to be the set of three outcomes:

$$S = \{low, \ medium, \ high\}$$

If the objective of the analysis is to consider only whether or not a particular part conforms to the manufacturing specifications, the sample space might be simplified to the set of two outcomes

$$S = \{yes, \ no\}$$

that indicate whether or not the part conforms.

It is useful to distinguish between two types of sample spaces.

Definition

> A sample space is **discrete** if it consists of a finite or countable infinite set of outcomes. A sample space is **continuous** if it contains an interval (either finite or infinite) of real numbers.

In Example 2-1, the choice $S = R^+$ is an example of a continuous sample space, whereas $S = \{yes, \ no\}$ is a discrete sample space. As mentioned, the best choice of a sample space depends on the objectives of the study. As specific questions occur later in the book, appropriate sample spaces are discussed.

EXAMPLE 2-2

If two connectors are selected and measured, the extension of the positive real line R is to take the sample space to be the positive quadrant of the plane:

$$S = R^+ \times R^+$$

If the objective of the analysis is to consider only whether or not the parts conform to the manufacturing specifications, either part may or may not conform. We abbreviate *yes* and *no* as y and n. If the ordered pair yn indicates that the first connector conforms and the second does not, the sample space can be represented by the four outcomes:

$$S = \{yy, yn, ny, nn\}$$

If we are only interested in the number of conforming parts in the sample, we might summarize the sample space as

$$S = \{0, 1, 2\}$$

As another example, consider an experiment in which the thickness is measured until a connector fails to meet the specifications. The sample space can be represented as

$$S = \{n, yn, yyn, yyyn, yyyyn, \text{and so forth}\}$$

In random experiments in which items are selected from a batch, we will indicate whether or not a selected item is replaced before the next one is selected. For example, if the batch consists of three items $\{a, b, c\}$ and our experiment is to select two items **without replacement**, the sample space can be represented as

$$S_{\text{without}} = \{ab, ac, ba, bc, ca, cb\}$$

This description of the sample space maintains the order of the items selected so that the outcome ab and ba are separate elements in the sample space. A sample space with less detail only describes the two items selected $\{\{a, b\}, \{a, c\}, \{b, c\}\}$. This sample space is the possible subsets of two items. Sometimes the ordered outcomes are needed, but in other cases the simpler, unordered sample space is sufficient.

If items are replaced before the next one is selected, the sampling is referred to as **with replacement.** Then the possible ordered outcomes are

$$S_{\text{with}} = \{aa, ab, ac, ba, bb, bc, ca, cb, cc\}$$

The unordered description of the sample space is $\{\{a, a\}, \{a, b\}, \{a, c\}, \{b, b\}, \{b, c\}, \{c, c\}\}$. Sampling without replacement is more common for industrial applications.

Sometimes it is not necessary to specify the exact item selected, but only a property of the item. For example, suppose that there are 5 defective parts and 95 good parts in a batch. To study the quality of the batch, two are selected without replacement. Let g denote a good part and d denote a defective part. It might be sufficient to describe the sample space (ordered) in terms of quality of each part selected as

$$S = \{gg, gd, dg, dd\}$$

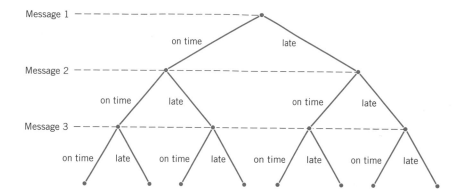

Figure 2-5 Tree diagram for three messages.

One must be cautious with this description of the sample space because there are many more pairs of items in which both are good than pairs in which both are defective. These differences must be accounted for when probabilities are computed later in this chapter. Still, this summary of the sample space will be convenient when conditional probabilities are used later in this chapter. Also, if there were only one defective part in the batch, there would be fewer possible outcomes

$$S = \{gg, gd, dg\}$$

because *dd* would be impossible. For sampling questions, sometimes the most important part of the solution is an appropriate description of the sample space.

Sample spaces can also be described graphically with **tree diagrams.** When a sample space can be constructed in several steps or stages, we can represent each of the n_1 ways of completing the first step as a branch of a tree. Each of the ways of completing the second step can be represented as n_2 branches starting from the ends of the original branches, and so forth.

EXAMPLE 2-3   Each message in a digital communication system is classified as to whether it is received within the time specified by the system design. If three messages are classified, use a tree diagram to represent the sample space of possible outcomes.

Each message can either be received on time or late. The possible results for three messages can be displayed by eight branches in the tree diagram shown in Fig. 2-5.

EXAMPLE 2-4   An automobile manufacturer provides vehicles equipped with selected options. Each vehicle is ordered

With or without an automatic transmission   With one of three choices of a stereo system

With or without air-conditioning   With one of four exterior colors

If the sample space consists of the set of all possible vehicle types, what is the number of outcomes in the sample space? The sample space contains 48 outcomes. The tree diagram for the different types of vehicles is displayed in Fig. 2-6.

EXAMPLE 2-5   Consider an extension of the automobile manufacturer illustration in the previous example in which another vehicle option is the interior color. There are four choices of interior color: red, black, blue, or brown. However,

With a red exterior, only a black or red interior can be chosen.

With a white exterior, any interior color can be chosen.

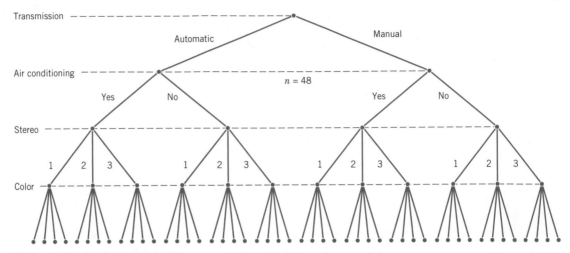

Figure 2-6   Tree diagram for different types of vehicles.

With a blue exterior, only a black, red, or blue interior can be chosen.

With a brown exterior, only a brown interior can be chosen.

In Fig. 2-6, there are 12 vehicle types with each exterior color, but the number of interior color choices depends on the exterior color. As shown in Fig. 2-7, the tree diagram can be extended to show that there are 120 different vehicle types in the sample space.

2-1.3   Events

Often we are interested in a collection of related outcomes from a random experiment.

Definition	An **event** is a subset of the sample space of a random experiment.

We can also be interested in describing new events from combinations of existing events. Because events are subsets, we can use basic set operations such as unions, intersections, and

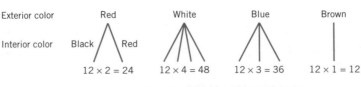

Figure 2-7   Tree diagram for different types of vehicles with interior colors.

complements to form other events of interest. Some of the basic set operations are summarized below in terms of events:

- The **union** of two events is the event that consists of all outcomes that are contained in either of the two events. We denote the union as $E_1 \cup E_2$.

- The **intersection** of two events is the event that consists of all outcomes that are contained in both of the two events. We denote the intersection as $E_1 \cap E_2$.

- The **complement** of an event in a sample space is the set of outcomes in the sample space that are not in the event. We denote the component of the event E as E'.

EXAMPLE 2-6

Consider the sample space $S = \{yy, yn, ny, nn\}$ in Example 2-2. Suppose that the set of all outcomes for which at least one part conforms is denoted as E_1. Then,

$$E_1 = \{yy, yn, ny\}$$

The event in which both parts do not conform, denoted as E_2, contains only the single outcome, $E_2 = \{nn\}$. Other examples of events are $E_3 = \varnothing$, the null set, and $E_4 = S$, the sample space. If $E_5 = \{yn, ny, nn\}$,

$$E_1 \cup E_5 = S \qquad E_1 \cap E_5 = \{yn, ny\} \qquad E_1' = \{nn\}$$

EXAMPLE 2-7

Measurements of the time needed to complete a chemical reaction might be modeled with the sample space $S = R^+$, the set of positive real numbers. Let

$$E_1 = \{x \mid 1 \le x < 10\} \qquad \text{and} \qquad E_2 = \{x \mid 3 < x < 118\}$$

Then,

$$E_1 \cup E_2 = \{x \mid 1 \le x < 118\} \qquad \text{and} \qquad E_1 \cap E_2 = \{x \mid 3 < x < 10\}$$

Also,

$$E_1' = \{x \mid x \ge 10\} \qquad \text{and} \qquad E_1' \cap E_2 = \{x \mid 10 \le x < 118\}$$

EXAMPLE 2-8

Samples of polycarbonate plastic are analyzed for scratch and shock resistance. The results from 50 samples are summarized as follows:

		shock resistance	
		high	low
scratch resistance	high	40	4
	low	1	5

Let A denote the event that a sample has high shock resistance, and let B denote the event that a sample has high scratch resistance. Determine the number of samples in $A \cap B$, A', and $A \cup B$.

The event $A \cap B$ consists of the 40 samples for which scratch and shock resistances are high. The event A' consists of the 9 samples in which the shock resistance is low. The event $A \cup B$ consists of the 45 samples in which the shock resistance, scratch resistance, or both are high.

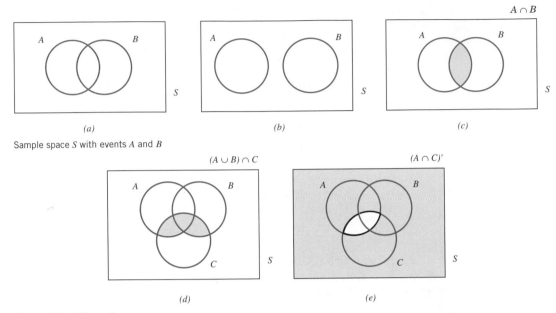

Figure 2-8   Venn diagrams.

Diagrams are often used to portray relationships between sets, and these diagrams are also used to describe relationships between events. We can use **Venn diagrams** to represent a sample space and events in a sample space. For example, in Fig. 2-8(a) the sample space of the random experiment is represented as the points in the rectangle S. The events A and B are the subsets of points in the indicated regions. Figure 2-8(b) illustrates two events with no common outcomes; Figs. 2-8(c) to 2-8(e) illustrate additional joint events.

Two events with no outcomes in common have an important relationship.

Definition

> Two events, denoted as E_1 and E_2, such that
>
> $$E_1 \cap E_2 = \varnothing$$
>
> are said to be **mutually exclusive.**

The two events in Fig. 2-8(b) are mutually exclusive, whereas the two events in Fig. 2-8(a) are not.

Additional results involving events are summarized below. The definition of the complement of an event implies that

$$(E')' = E$$

The distributive law for set operations implies that

$$(A \cup B) \cap C = (A \cap C) \cup (B \cap C), \quad \text{and} \quad (A \cap B) \cup C = (A \cup C) \cap (B \cup C)$$

DeMorgan's laws imply that

$$(A \cup B)' = A' \cap B' \quad \text{and} \quad (A \cap B)' = A' \cup B'$$

Also, remember that

$$A \cap B = B \cap A \quad \text{and} \quad A \cup B = B \cup A$$

2-1.4   Counting Techniques (CD Only)

As sample spaces become larger, complete enumeration is difficult. Instead, counts of the number of outcomes in the sample space and in various events are often used to analyze the random experiment. These methods are referred to as **counting techniques** and described on the CD.

EXERCISES FOR SECTION 2-1

Provide a reasonable description of the sample space for each of the random experiments in Exercises 2-1 to 2-18. There can be more than one acceptable interpretation of each experiment. Describe any assumptions you make.

2-1.   Each of three machined parts is classified as either above or below the target specification for the part.

2-2.   Each of four transmitted bits is classified as either in error or not in error.

2-3.   In the final inspection of electronic power supplies, three types of nonconformities might occur: functional, minor, or cosmetic. Power supplies that are defective are further classified as to type of nonconformity.

2-4.   In the manufacturing of digital recording tape, electronic testing is used to record the number of bits in error in a 350-foot reel.

2-5.   In the manufacturing of digital recording tape, each of 24 tracks is classified as containing or not containing one or more bits in error.

2-6.   An ammeter that displays three digits is used to measure current in milliamperes.

2-7.   A scale that displays two decimal places is used to measure material feeds in a chemical plant in tons.

2-8.   The following two questions appear on an employee survey questionnaire. Each answer is chosen from the five-point scale 1 (never), 2, 3, 4, 5 (always).

Is the corporation willing to listen to and fairly evaluate new ideas?

How often are my coworkers important in my overall job performance?

2-9.   The concentration of ozone to the nearest part per billion.

2-10.   The time until a tranaction service is requested of a computer to the nearest millisecond.

2-11.   The pH reading of a water sample to the nearest tenth of a unit.

2-12.   The voids in a ferrite slab are classified as small, medium, or large. The number of voids in each category is measured by an optical inspection of a sample.

2-13.   The time of a chemical reaction is recorded to the nearest millisecond.

2-14.   An order for an automobile can specify either an automatic or a standard transmission, either with or without air-conditioning, and any one of the four colors red, blue, black or white. Describe the set of possible orders for this experiment.

2-15.   A sampled injection-molded part could have been produced in either one of two presses and in any one of the eight cavities in each press.

2-16.   An order for a computer system can specify memory of 4, 8, or 12 gigabytes, and disk storage of 200, 300, or 400 gigabytes. Describe the set of possible orders.

2-17.   Calls are repeatedly placed to a busy phone line until a connect is achieved.

2-18.   In a magnetic storage device, three attempts are made to read data before an error recovery procedure that repositions the magnetic head is used. The error recovery procedure attempts three repositionings before an "abort" message is sent to the operator. Let

s denote the success of a read operation

f denote the failure of a read operation

F denote the failure of an error recovery procedure

S denote the success of an error recovery procedure

A denote an abort message sent to the operator.

Describe the sample space of this experiment with a tree diagram.

2-19. Three events are shown on the Venn diagram in the following figure:

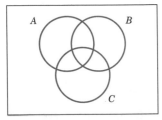

Reproduce the figure and shade the region that corresponds to each of the following events.
(a) A' (b) $A \cap B$
(c) $(A \cap B) \cup C$ (d) $(B \cup C)'$
(e) $(A \cap B)' \cup C$

2-20. Three events are shown on the Venn diagram in the following figure:

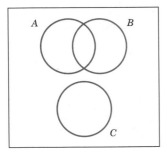

Reproduce the figure and shade the region that corresponds to each of the following events.
(a) A' (b) $(A \cap B) \cup (A \cap B')$
(c) $(A \cap B) \cup C$ (d) $(B \cup C)'$
(e) $(A \cap B)' \cup C$

2-21. A digital scale is used that provides weights to the nearest gram.
(a) What is the sample space for this experiment?
Let A denote the event that a weight exceeds 11 grams, let B denote the event that a weight is less than or equal to 15 grams, and let C denote the event that a weight is greater than or equal to 8 grams and less than 12 grams.
Describe the following events.
(b) $A \cup B$ (c) $A \cap B$
(d) A' (e) $A \cup B \cup C$

(f) $(A \cup C)'$ (g) $A \cap B \cap C$
(h) $B' \cap C$ (i) $A \cup (B \cap C)$

2-22. In an injection-molding operation, the length and width, denoted as X and Y, respectively, of each molded part are evaluated. Let

    A denote the event of $48 < X < 52$ centimeters

    B denote the event of $9 < Y < 11$ centimeters

    C denote the event that a critical length meets customer requirements.

Construct a Venn diagram that includes these events. Shade the areas that represent the following:
(a) A (b) $A \cap B$
(c) $A' \cup B$ (d) $A \cup B$
(e) If these events were mutually exclusive, how successful would this production operation be? Would the process produce parts with $X = 50$ centimeters and $Y = 10$ centimeters?

2-23. Four bits are transmitted over a digital communications channel. Each bit is either distorted or received without distortion. Let A_i denote the event that the ith bit is distorted, $i = 1, \ldots, 4$.
(a) Describe the sample space for this experiment.
(b) Are the A_i's mutually exclusive?
Describe the outcomes in each of the following events:
(c) A_1 (d) A_1'
(e) $A_1 \cap A_2 \cap A_3 \cap A_4$ (f) $(A_1 \cap A_2) \cup (A_3 \cap A_4)$

2-24. A sample of three calculators is selected from a manufacturing line, and each calculator is classified as either defective or acceptable. Let A, B, and C denote the events that the first, second, and third calculators respectively, are defective.
(a) Describe the sample space for this experiment with a tree diagram.
Use the tree diagram to describe each of the following events:
(b) A (c) B
(d) $A \cap B$ (e) $B \cup C$

2-25. A wireless garage door opener has a code determined by the up or down setting of 12 switches. How many outcomes are in the sample space of possible codes?

2-26. Disks of polycarbonate plastic from a supplier are analyzed for scratch and shock resistance. The results from 100 disks are summarized below:

		shock resistance	
		high	low
scratch	high	70	9
resistance	low	16	5

Let A denote the event that a disk has high shock resistance, and let B denote the event that a disk has high scratch resistance. Determine the number of disks in $A \cap B$, A', and $A \cup B$.

2-27. Samples of a cast aluminum part are classified on the basis of surface finish (in microinches) and edge finish. The results of 100 parts are summarized as follows:

		edge finish	
		excellent	good
surface	excellent	80	2
finish	good	10	8

(a) Let A denote the event that a sample has excellent surface finish, and let B denote the event that a sample has excellent edge finish. Determine the number of samples in $A' \cap B$, B', and $A \cup B$.

(b) Assume that each of two samples is to be classified on the basis of surface finish, either excellent or good, edge finish, either excellent or good. Use a tree diagram to represent the possible outcomes of this experiment.

2-28. Samples of emissions from three suppliers are classified for conformance to air-quality specifications. The results from 100 samples are summarized as follows:

		conforms	
		yes	no
	1	22	8
supplier	2	25	5
	3	30	10

Let A denote the event that a sample is from supplier 1, and let B denote the event that a sample conforms to specifications. Determine the number of samples in $A' \cap B$, B', and $A \cup B$.

2-29. The rise time of a reactor is measured in minutes (and fractions of minutes). Let the sample space be positive, real numbers. Define the events A and B as follows:

$$A = \{x \mid x < 72.5\}$$

and

$$B = \{x \mid x > 52.5\}$$

Describe each of the following events:
(a) A' (b) B'
(c) $A \cap B$ (d) $A \cup B$

2-30. A sample of two items is selected without replacement from a batch. Describe the (ordered) sample space for each of the following batches:
(a) The batch contains the items $\{a, b, c, d\}$.
(b) The batch contains the items $\{a, b, c, d, e, f, g\}$.
(c) The batch contains 4 defective items and 20 good items.
(d) The batch contains 1 defective item and 20 good items.

2-31. A sample of two printed circuit boards is selected without replacement from a batch. Describe the (ordered) sample space for each of the following batches:
(a) The batch contains 90 boards that are not defective, 8 boards with minor defects, and 2 boards with major defects.
(b) The batch contains 90 boards that are not defective, 8 boards with minor defects, and 1 board with major defects.

2-32. Counts of the Web pages provided by each of two computer servers in a selected hour of the day are recorded. Let A denote the event that at least 10 pages are provided by server 1 and let B denote the event that at least 20 pages are provided by server 2.
(a) Describe the sample space for the numbers of pages for two servers graphically.

Show each of the following events on the sample space graph:
(b) A (c) B
(d) $A \cap B$ (e) $A \cup B$

2-33. The rise time of a reactor is measured in minutes (and fractions of minutes). Let the sample space for the rise time of each batch be positive, real numbers. Consider the rise times of *two* batches. Let A denote the event that the rise time of batch 1 is less than 72.5 minutes, and let B denote the event that the rise time of batch 2 is greater than 52.5 minutes.

Describe the sample space for the rise time of two batches graphically and show each of the following events on a two-dimensional plot:
(a) A (b) B'
(c) $A \cap B$ (d) $A \cup B$

2-2   INTERPRETATIONS OF PROBABILITY

2-2.1   Introduction

In this chapter, we introduce probability for **discrete sample spaces**—those with only a finite (or countably infinite) set of outcomes. The restriction to these sample spaces enables us to simplify the concepts and the presentation without excessive mathematics.

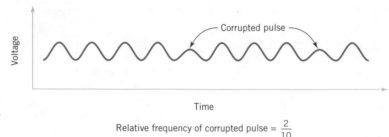

Figure 2-9   Relative frequency of corrupted pulses sent over a communication channel.

Probability is used to quantify the likelihood, or chance, that an outcome of a random experiment will occur. "The chance of rain today is 30%" is a statement that quantifies our feeling about the possibility of rain. The likelihood of an outcome is quantified by assigning a number from the interval [0, 1] to the outcome (or a percentage from 0 to 100%). Higher numbers indicate that the outcome is more likely than lower numbers. A 0 indicates an outcome will not occur. A probability of 1 indicates an outcome will occur with certainty.

The probability of an outcome can be interpreted as our subjective probability, or **degree of belief,** that the outcome will occur. Different individuals will no doubt assign different probabilities to the same outcomes. Another interpretation of probability is based on the conceptual model of repeated replications of the random experiment. The probability of an outcome is interpreted as the limiting value of the proportion of times the outcome occurs in n repetitions of the random experiment as n increases beyond all bounds. For example, if we assign probability 0.2 to the outcome that there is a corrupted pulse in a digital signal, we might interpret this assignment as implying that, if we analyze many pulses, approximately 20% of them will be corrupted. This example provides a **relative frequency** interpretation of probability. The proportion, or relative frequency, of replications of the experiment that result in the outcome is 0.2. Probabilities are chosen so that the sum of the probabilities of all outcomes in an experiment add up to 1. This convention facilitates the relative frequency interpretation of probability. Figure 2-9 illustrates the concept of relative frequency.

Probabilities for a random experiment are often assigned on the basis of a reasonable model of the system under study. One approach is to base probability assignments on the simple concept of equally likely outcomes.

For example, suppose that we will select one laser diode **randomly** from a batch of 100. The sample space is the set of 100 diodes. *Randomly* implies that it is reasonable to assume that each diode in the batch has an equal chance of being selected. Because the sum of the probabilities must equal 1, the probability model for this experiment assigns probability of 0.01 to each of the 100 outcomes. We can interpret the probability by imagining many replications of the experiment. Each time we start with all 100 diodes and select one at random. The probability 0.01 assigned to a particular diode represents the proportion of replicates in which a particular diode is selected.

When the model of **equally likely outcomes** is assumed, the probabilities are chosen to be equal.

> Whenever a sample space consists of N possible outcomes that are equally likely, the probability of each outcome is $1/N$.

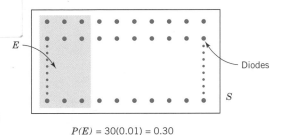

Figure 2-10
Probability of the
event E is the sum of
the probabilities of the
outcomes in E.

$P(E) = 30(0.01) = 0.30$

It is frequently necessary to assign probabilities to events that are composed of several outcomes from the sample space. This is straightforward for a discrete sample space.

EXAMPLE 2-9   Assume that 30% of the laser diodes in a batch of 100 meet the minimum power requirements of a specific customer. If a laser diode is selected randomly, that is, each laser diode is equally likely to be selected, our intuitive feeling is that the probability of meeting the customer's requirements is 0.30.

Let E denote the subset of 30 diodes that meet the customer's requirements. Because E contains 30 outcomes and each outcome has probability 0.01, we conclude that the probability of E is 0.3. The conclusion matches our intuition. Figure 2-10 illustrates this example.

For a discrete sample space, the probability of an event can be defined by the reasoning used in the example above.

Definition

> For a discrete sample space, the *probability of an event E,* denoted as $P(E)$, equals the sum of the probabilities of the outcomes in E.

EXAMPLE 2-10   A random experiment can result in one of the outcomes $\{a, b, c, d\}$ with probabilities 0.1, 0.3, 0.5, and 0.1, respectively. Let A denote the event $\{a, b\}$, B the event $\{b, c, d\}$, and C the event $\{d\}$. Then,

$$P(A) = 0.1 + 0.3 = 0.4$$
$$P(B) = 0.3 + 0.5 + 0.1 = 0.9$$
$$P(C) = 0.1$$

Also, $P(A') = 0.6$, $P(B') = 0.1$, and $P(C') = 0.9$. Furthermore, because $A \cap B = \{b\}$, $P(A \cap B) = 0.3$. Because $A \cup B = \{a, b, c, d\}$, $P(A \cup B) = 0.1 + 0.3 + 0.5 + 0.1 = 1$. Because $A \cap C$ is the null set, $P(A \cap C) = 0$.

EXAMPLE 2-11

A visual inspection of a location on wafers from a semiconductor manufacturing process resulted in the following table:

Number of Contamination Particles	Proportion of Wafers
0	0.40
1	0.20
2	0.15
3	0.10
4	0.05
5 or more	0.10

If one wafer is selected randomly from this process and the location is inspected, what is the probability that it contains no particles? If information were available for each wafer, we could define the sample space as the set of all wafers inspected and proceed as in the example with diodes. However, this level of detail is not needed in this case. We can consider the sample space to consist of the six categories that summarize the number of contamination particles on a wafer. Then, the event that there is no particle in the inspected location on the wafer, denoted as E, can be considered to be comprised of the single outcome, namely, $E = \{0\}$. Therefore,

$$P(E) = 0.4$$

What is the probability that a wafer contains three or more particles in the inspected location? Let E denote the event that a wafer contains three or more particles in the inspected location. Then, E consists of the three outcomes $\{3, 4, 5 \text{ or more}\}$. Therefore,

$$P(E) = 0.10 + 0.05 + 0.10 = 0.25$$

EXAMPLE 2-12

Suppose that a batch contains six parts with part numbers $\{a, b, c, d, e, f\}$. Suppose that two parts are selected without replacement. Let E denote the event that the part number of the first part selected is a. Then E can be written as $E = \{ab, ac, ad, ae, af\}$. The sample space can be enumerated. It has 30 outcomes. If each outcome is equally likely, $P(E) = 5/30 = 1/6$.

Also, if E_2 denotes the event that the second part selected is a, $E_2 = \{ba, ca, da, ea, fa\}$ and with equally likely outcomes, $P(E_2) = 5/30 = 1/6$.

2-2.2   Axioms of Probability

Now that the probability of an event has been defined, we can collect the assumptions that we have made concerning probabilities into a set of **axioms** that the probabilities in any random experiment must satisfy. The axioms ensure that the probabilities assigned in an experiment can be interpreted as relative frequencies and that the assignments are consistent with our intuitive understanding of relationships between relative frequencies. For example, if event A is contained in event B, we should have $P(A) \leq P(B)$. The **axioms do not determine probabilities;** the probabilities are assigned based on our knowledge of the system under study. However, the axioms enable us to easily calculate the probabilities of some events from knowledge of the probabilities of other events.

Axioms of Probability

Probability is a number that is assigned to each member of a collection of events from a random experiment that satisfies the following properties:

If S is the sample space and E is any event in a random experiment,

(1) $P(S) = 1$

(2) $0 \leq P(E) \leq 1$

(3) For two events E_1 and E_2 with $E_1 \cap E_2 = \varnothing$

$$P(E_1 \cup E_2) = P(E_1) + P(E_2)$$

The property that $0 \leq P(E) \leq 1$ is equivalent to the requirement that a relative frequency must be between 0 and 1. The property that $P(S) = 1$ is a consequence of the fact that an outcome from the sample space occurs on every trial of an experiment. Consequently, the relative frequency of S is 1. Property 3 implies that if the events E_1 and E_2 have no outcomes in common, the relative frequency of outcomes in $E_1 \cup E_2$ is the sum of the relative frequencies of the outcomes in E_1 and E_2.

These axioms imply the following results. The derivations are left as exercises at the end of this section. Now,

$$P(\varnothing) = 0$$

and for any event E,

$$P(E') = 1 - P(E)$$

For example, if the probability of the event E is 0.4, our interpretation of relative frequency implies that the probability of E' is 0.6. Furthermore, if the event E_1 is contained in the event E_2,

$$P(E_1) \leq P(E_2)$$

EXERCISES FOR SECTION 2-2

2-34. Each of the possible five outcomes of a random experiment is equally likely. The sample space is $\{a, b, c, d, e\}$. Let A denote the event $\{a, b\}$, and let B denote the event $\{c, d, e\}$. Determine the following:
(a) $P(A)$ (b) $P(B)$
(c) $P(A')$ (d) $P(A \cup B)$
(e) $P(A \cap B)$

2-35. The sample space of a random experiment is $\{a, b, c, d, e\}$ with probabilities 0.1, 0.1, 0.2, 0.4, and 0.2, respectively. Let A denote the event $\{a, b, c\}$, and let B denote the event $\{c, d, e\}$. Determine the following:
(a) $P(A)$ (b) $P(B)$
(c) $P(A')$ (d) $P(A \cup B)$
(e) $P(A \cap B)$

2-36. A part selected for testing is equally likely to have been produced on any one of six cutting tools.
(a) What is the sample space?
(b) What is the probability that the part is from tool 1?
(c) What is the probability that the part is from tool 3 or tool 5?
(d) What is the probability that the part is not from tool 4?

2-37. An injection-molded part is equally likely to be obtained from any one of the eight cavities on a mold.
(a) What is the sample space?
(b) What is the probability a part is from cavity 1 or 2?
(c) What is the probability that a part is neither from cavity 3 nor 4?

2-38. A sample space contains 20 equally likely outcomes. If the probability of event A is 0.3, how many outcomes are in event A?

2-39. Orders for a computer are summarized by the optional features that are requested as follows:

	proportion of orders
no optional features	0.3
one optional feature	0.5
more than one optional feature	0.2

(a) What is the probability that an order requests at least one optional feature?
(b) What is the probability that an order does not request more than one optional feature?

2-40. If the last digit of a weight measurement is equally likely to be any of the digits 0 through 9,

(a) What is the probability that the last digit is 0?
(b) What is the probability that the last digit is greater than or equal to 5?

2-41. A sample preparation for a chemical measurement is completed correctly by 25% of the lab technicians, completed with a minor error by 70%, and completed with a major error by 5%.

(a) If a technician is selected randomly to complete the preparation, what is the probability it is completed without error?
(b) What is the probability that it is completed with either a minor or a major error?

2-42. A credit card contains 16 digits between 0 and 9. However, only 100 million numbers are valid. If a number is entered randomly, what is the probability that it is a valid number?

2-43. Suppose your vehicle is licensed in a state that issues license plates that consist of three digits (between 0 and 9) followed by three letters (between A and Z). If a license number is selected randomly, what is the probability that yours is the one selected?

2-44. A message can follow different paths through servers on a network. The senders message can go to one of five servers for the first step, each of them can send to five servers at the second step, each of which can send to four servers at the third step, and then the message goes to the recipients server.

(a) How many paths are possible?
(b) If all paths are equally likely, what is the probability that a message passes through the first of four servers at the third step?

2-45. Disks of polycarbonate plastic from a supplier are analyzed for scratch and shock resistance. The results from 100 disks are summarized as follows:

		shock resistance	
		high	low
scratch	high	70	9
resistance	low	16	5

Let A denote the event that a disk has high shock resistance, and let B denote the event that a disk has high scratch resistance. If a disk is selected at random, determine the following probabilities:

(a) $P(A)$ (b) $P(B)$
(c) $P(A')$ (d) $P(A \cap B)$
(e) $P(A \cup B)$ (f) $P(A' \cup B)$

2-46. Samples of a cast aluminum part are classified on the basis of surface finish (in microinches) and edge finish. The results of 100 parts are summarized as follows:

		edge finish	
		excellent	good
surface	excellent	80	2
finish	good	10	8

Let A denote the event that a sample has excellent surface finish, and let B denote the event that a sample has excellent edge finish. If a part is selected at random, determine the following probabilities:

(a) $P(A)$ (b) $P(B)$
(c) $P(A')$ (d) $P(A \cap B)$
(e) $P(A \cup B)$ (f) $P(A' \cup B)$

2-47. Samples of emissions from three suppliers are classified for conformance to air-quality specifications. The results from 100 samples are summarized as follows:

		conforms	
		yes	no
	1	22	8
supplier	2	25	5
	3	30	10

Let A denote the event that a sample is from supplier 1, and let B denote the event that a sample conforms to specifications. If a sample is selected at random, determine the following probabilities:

(a) $P(A)$ (b) $P(B)$
(c) $P(A')$ (d) $P(A \cap B)$
(e) $P(A \cup B)$ (f) $P(A' \cup B)$

2-48. Use the axioms of probability to show the following:
(a) For any event E, $P(E') = 1 - P(E)$.
(b) $P(\varnothing) = 0$
(c) If A is contained in B, then $P(A) \le P(B)$

2-3 ADDITION RULES

Joint events are generated by applying basic set operations to individual events. Unions of events, such as $A \cup B$; intersections of events, such as $A \cap B$; and complements of events, such as A', are commonly of interest. The probability of a joint event can often be determined from the probabilities of the individual events that comprise it. Basic set operations are also sometimes helpful in determining the probability of a joint event. In this section the focus is on unions of events.

EXAMPLE 2-13

Table 2-1 lists the history of 940 wafers in a semiconductor manufacturing process. Suppose one wafer is selected at random. Let H denote the event that the wafer contains high levels of contamination. Then, $P(H) = 358/940$.

Let C denote the event that the wafer is in the center of a sputtering tool. Then, $P(C) = 626/940$. Also, $P(H \cap C)$ is the probability that the wafer is from the center of the sputtering tool and contains high levels of contamination. Therefore,

$$P(H \cap C) = 112/940$$

The event $H \cup C$ is the event that a wafer is from the center of the sputtering tool or contains high levels of contamination (or both). From the table, $P(H \cup C) = 872/940$. An alternative calculation of $P(H \cup C)$ can be obtained as follows. The 112 wafers that comprise the event $H \cap C$ are included once in the calculation of $P(H)$ and again in the calculation of $P(C)$. Therefore, $P(H \cup C)$ can be found to be

$$P(H \cup C) = P(H) + P(C) - P(H \cap C)$$
$$= 358/940 + 626/940 - 112/940 = 872/940$$

The preceding example illustrates that the probability of A or B is interpreted as $P(A \cup B)$ and that the following general **addition rule** applies.

$$P(A \cup B) = P(A) + P(B) - P(A \cap B) \qquad (2\text{-}1)$$

EXAMPLE 2-14

The wafers such as those described in Example 2-13 were further classified as either in the "center" or at the "edge" of the sputtering tool that was used in manufacturing, and by the degree of contamination. Table 2-2 shows the proportion of wafers in each category. What is

Table 2-1 Wafers in Semiconductor Manufacturing Classified by Contamination and Location

Contamination	Location in Sputtering Tool		Total
	Center	Edge	
Low	514	68	582
High	112	246	358
Total	626	314	

Table 2-2   Wafers Classified by Contamination and Location

Number of Contamination Particles	Center	Edge	Totals
0	0.30	0.10	0.40
1	0.15	0.05	0.20
2	0.10	0.05	0.15
3	0.06	0.04	0.10
4	0.04	0.01	0.05
5 or more	0.07	0.03	0.10
Totals	0.72	0.28	1.00

the probability that a wafer was either at the edge or that it contains four or more particles? Let E_1 denote the event that a wafer contains four or more particles, and let E_2 denote the event that a wafer is at the edge.

The requested probability is $P(E_1 \cup E_2)$. Now, $P(E_1) = 0.15$ and $P(E_2) = 0.28$. Also, from the table, $P(E_1 \cap E_2) = 0.04$. Therefore, using Equation 2-1, we find that

$$P(E_1 \cup E_2) = 0.15 + 0.28 - 0.04 = 0.39$$

What is the probability that a wafer contains less than two particles or that it is both at the edge and contains more than four particles? Let E_1 denote the event that a wafer contains less than two particles, and let E_2 denote the event that a wafer is both from the edge and contains more than four particles. The requested probability is $P(E_1 \cup E_2)$. Now, $P(E_1) = 0.60$ and $P(E_2) = 0.03$. Also, E_1 and E_2 are mutually exclusive. Consequently, there are no wafers in the intersection and $P(E_1 \cap E_2) = 0$. Therefore,

$$P(E_1 \cup E_2) = 0.60 + 0.03 = 0.63$$

Recall that two events A and B are said to be mutually exclusive if $A \cap B = \varnothing$. Then, $P(A \cap B) = 0$, and the general result for the probability of $A \cup B$ simplifies to the third axiom of probability.

If A and B are mutually exclusive events,

$$P(A \cup B) = P(A) + P(B) \tag{2-2}$$

Three or More Events
More complicated probabilities, such as $P(A \cup B \cup C)$, can be determined by repeated use of Equation 2-1 and by using some basic set operations. For example,

$$P(A \cup B \cup C) = P[(A \cup B) \cup C] = P(A \cup B) + P(C) - P[(A \cup B) \cap C]$$

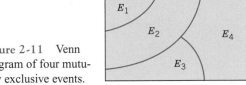

Figure 2-11   Venn diagram of four mutually exclusive events.

Upon expanding $P(A \cup B)$ by Equation 2-1 and using the distributed rule for set operations to simplify $P[(A \cup B) \cap C]$, we obtain

$$\begin{aligned}
P(A \cup B \cup C) &= P(A) + P(B) - P(A \cap B) + P(C) - P[(A \cap C) \cup (B \cap C)] \\
&= P(A) + P(B) - P(A \cap B) + P(C) \\
&\quad - [P(A \cap C) + P(B \cap C) - P(A \cap B \cap C)] \\
&= P(A) + P(B) + P(C) - P(A \cap B) - P(A \cap C) \\
&\quad - P(B \cap C) + P(A \cap B \cap C)
\end{aligned}$$

We have developed a formula for the probability of the union of three events. Formulas can be developed for the probability of the union of any number of events, although the formulas become very complex. As a summary, for the case of three events

$$\begin{aligned}
P(A \cup B \cup C) &= P(A) + P(B) + P(C) - P(A \cap B) \\
&\quad - P(A \cap C) - P(B \cap C) + P(A \cap B \cap C) \quad \text{(2-3)}
\end{aligned}$$

Results for three or more events simplify considerably if the events are mutually exclusive. In general, a collection of events, $E_1, E_2, \ldots, E_k$, is said to be mutually exclusive if there is no overlap among any of them.

The Venn diagram for several mutually exclusive events is shown in Fig. 2-11. By generalizing the reasoning for the union of two events, the following result can be obtained:

A collection of events, $E_1, E_2, \ldots, E_k$, is said to be **mutually exclusive** if for all pairs,

$$E_i \cap E_j = \varnothing$$

For a collection of mutually exclusive events,

$$P(E_1 \cup E_2 \cup \ldots \cup E_k) = P(E_1) + P(E_2) + \ldots P(E_k) \quad \text{(2-4)}$$

EXAMPLE 2-15   A simple example of mutually exclusive events will be used quite frequently. Let X denote the pH of a sample. Consider the event that X is greater than 6.5 but less than or equal to 7.8. This probability is the sum of any collection of mutually exclusive events with union equal to the same range for X. One example is

$$P(6.5 < X \le 7.8) = P(6.5 < X \le 7.0) + P(7.0 < X \le 7.5) + P(7.5 < X \le 7.8)$$

Another example is

$$P(6.5 < X \le 7.8) = P(6.5 < X \le 6.6) + P(6.6 < X \le 7.1) \\ + P(7.1 < X \le 7.4) + P(7.4 < X \le 7.8)$$

The best choice depends on the particular probabilities available.

EXERCISES FOR SECTION 2-3

2-49. If $P(A) = 0.3$, $P(B) = 0.2$, and $P(A \cap B) = 0.1$, determine the following probabilities:
(a) $P(A')$ (b) $P(A \cup B)$
(c) $P(A' \cap B)$ (d) $P(A \cap B')$
(e) $P[(A \cup B)']$ (f) $P(A' \cup B)$

2-50. If A, B, and C are mutually exclusive events with $P(A) = 0.2$, $P(B) = 0.3$, and $P(C) = 0.4$, determine the following probabilities:
(a) $P(A \cup B \cup C)$ (b) $P(A \cap B \cap C)$
(c) $P(A \cap B)$ (d) $P[(A \cup B) \cap C]$
(e) $P(A' \cap B' \cap C')$

2-51. If A, B, and C are mutually exclusive events, is it possible for $P(A) = 0.3$, $P(B) = 0.4$, and $P(C) = 0.5$? Why or why not?

2-52. Disks of polycarbonate plastic from a supplier are analyzed for scratch and shock resistance. The results from 100 disks are summarized as follows:

		shock resistance	
		high	low
scratch	high	70	9
resistance	low	16	5

(a) If a disk is selected at random, what is the probability that its scratch resistance is high and its shock resistance is high?
(b) If a disk is selected at random, what is the probability that its scratch resistance is high or its shock resistance is high?
(c) Consider the event that a disk has high scratch resistance and the event that a disk has high shock resistance. Are these two events mutually exclusive?

2-53. The analysis of shafts for a compressor is summarized by conformance to specifications.

		roundness conforms	
		yes	no
surface finish	yes	345	5
conforms	no	12	8

(a) If a shaft is selected at random, what is the probability that the shaft conforms to surface finish requirements?
(b) What is the probability that the selected shaft conforms to surface finish requirements or to roundness requirements?
(c) What is the probability that the selected shaft either conforms to surface finish requirements or does not conform to roundness requirements?
(d) What is the probability that the selected shaft conforms to both surface finish and roundness requirements?

2-54. Cooking oil is produced in two main varieties: mono- and polyunsaturated. Two common sources of cooking oil are corn and canola. The following table shows the number of bottles of these oils at a supermarket:

		type of oil	
		canola	corn
type of	mono	7	13
unsaturation	poly	93	77

(a) If a bottle of oil is selected at random, what is the probability that it belongs to the polyunsaturated category?
(b) What is the probability that the chosen bottle is monounsaturated canola oil?

2-55. A manufacturer of front lights for automobiles tests lamps under a high humidity, high temperature environment using intensity and useful life as the responses of interest. The following table shows the performance of 130 lamps:

		useful life	
		satisfactory	unsatisfactory
intensity	satisfactory	117	3
	unsatisfactory	8	2

(a) Find the probability that a randomly selected lamp will yield unsatisfactory results under any criteria.
(b) The customers for these lamps demand 95% satisfactory results. Can the lamp manufacturer meet this demand?

2-56. The shafts in Exercise 2-53 are further classified in terms of the machine tool that was used for manufacturing the shaft.

Tool 1

		roundness conforms	
		yes	no
surface finish	yes	200	1
conforms	no	4	2

Tool 2

		roundness conforms	
		yes	no
surface finish	yes	145	4
conforms	no	8	6

(a) If a shaft is selected at random, what is the probability that the shaft conforms to surface finish requirements or to roundness requirements or is from Tool 1?

(b) If a shaft is selected at random, what is the probability that the shaft conforms to surface finish requirements or does not conform to roundness requirements or is from Tool 2?

(c) If a shaft is selected at random, what is the probability that the shaft conforms to both surface finish and roundness requirements or the shaft is from Tool 2?

(d) If a shaft is selected at random, what is the probability that the shaft conforms to surface finish requirements or the shaft is from Tool 2?

2-4 CONDITIONAL PROBABILITY

A digital communication channel has an error rate of one bit per every thousand transmitted. Errors are rare, but when they occur, they tend to occur in bursts that affect many consecutive bits. If a single bit is transmitted, we might model the probability of an error as $1/1000$. However, if the previous bit was in error, because of the bursts, we might believe that the probability that the next bit is in error is greater than $1/1000$.

In a thin film manufacturing process, the proportion of parts that are not acceptable is 2%. However, the process is sensitive to contamination problems that can increase the rate of parts that are not acceptable. If we knew that during a particular shift there were problems with the filters used to control contamination, we would assess the probability of a part being unacceptable as higher than 2%.

In a manufacturing process, 10% of the parts contain visible surface flaws and 25% of the parts with surface flaws are (functionally) defective parts. However, only 5% of parts without surface flaws are defective parts. The probability of a defective part depends on our knowledge of the presence or absence of a surface flaw.

These examples illustrate that probabilities need to be reevaluated as additional information becomes available. The notation and details are further illustrated for this example.

Let D denote the event that a part is defective and let F denote the event that a part has a surface flaw. Then, we denote the probability of D given, or assuming, that a part has a surface flaw as $P(D|F)$. This notation is read as the **conditional probability** of D given F, and it is interpreted as the probability that a part is defective, given that the part has a surface flaw. Because 25% of the parts with surface flaws are defective, our conclusion can be stated as $P(D|F) = 0.25$. Furthermore, because F' denotes the event that a part does not have a surface flaw and because 5% of the parts without surface flaws are defective, we have that $P(D|F') = 0.05$. These results are shown graphically in Fig. 2-12.

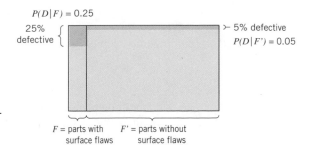

Figure 2-12 Conditional probabilities for parts with surface flaws.

Table 2-3   Parts Classified

		Surface Flaws		
		Yes (event F)	No	Total
Defective	Yes (event D)	10	18	38
	No	30	342	362
	Total	40	360	400

EXAMPLE 2-16   Table 2-3 provides an example of 400 parts classified by surface flaws and as (functionally) defective. For this table the conditional probabilities match those discussed previously in this section. For example, of the parts with surface flaws (40 parts) the number defective is 10. Therefore,

$$P(D|F) = 10/40 = 0.25$$

and of the parts without surface flaws (360 parts) the number defective is 18. Therefore,

$$P(D|F') = 18/360 = 0.05$$

In Example 2-16 conditional probabilities were calculated directly. These probabilities can also be determined from the formal definition of conditional probability.

Definition

The **conditional probability** of an event B given an event A, denoted as $P(B|A)$, is

$$P(B|A) = P(A \cap B)/P(A) \qquad (2\text{-}5)$$

for $P(A) > 0$.

This definition can be understood in a special case in which all outcomes of a random experiment are equally likely. If there are n total outcomes,

$$P(A) = (\text{number of outcomes in } A)/n$$

Also,

$$P(A \cap B) = (\text{number of outcomes in } A \cap B)/n$$

Consequently,

$$P(A \cap B)/P(A) = \frac{\text{number of outcomes in } A \cap B}{\text{number of outcomes in } A}$$

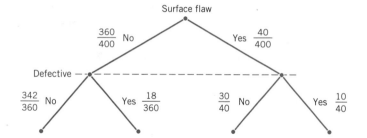

Figure 2-13 Tree diagram for parts classified

Therefore, $P(B|A)$ can be interpreted as the relative frequency of event B among the trials that produce an outcome in event A.

EXAMPLE 2-17

Again consider the 400 parts in Table 2-3. From this table

$$P(D|F) = P(D \cap F)/P(F) = \frac{10}{400} \Big/ \frac{40}{400} = \frac{10}{40}$$

Note that in this example all four of the following probabilities are different:

$$P(F) = 40/400 \qquad P(F|D) = 10/28$$
$$P(D) = 28/400 \qquad P(D|F) = 10/40$$

Here, $P(D)$ and $P(D|F)$ are probabilities of the same event, but they are computed under two different states of knowledge. Similarly, $P(F)$ and $P(F|D)$ are computed under two different states of knowledge.

The tree diagram in Fig. 2-13 can also be used to display conditional probabilities. The first branch is on surface flaw. Of the 40 parts with surface flaws, 10 are functionally defective and 30 are not. Therefore,

$$P(D|F) = 10/40 \qquad \text{and} \qquad P(D'|F) = 30/40$$

Of the 360 parts without surface flaws, 18 are functionally defective and 342 are not. Therefore,

$$P(D|F') = 342/360 \qquad \text{and} \qquad P(D'|F') = 18/360$$

Random Samples from a Batch
Recall that to select one item randomly from a batch implies that each item is equally likely. If more than one item is selected, *randomly* implies that each element of the sample space is equally likely. For example, when sample spaces were presented earlier in this chapter, sampling with and without replacement were defined and illustrated for the simple case of a batch with three items $\{a, b, c\}$. If two items are selected randomly from this batch without replacement, each of the six outcomes in the ordered sample space

$$S_{\text{without}} = \{ab, ac, ba, bc, ca, cb\}$$

has probability 1/6. If the unordered sample space is used, each of the three outcomes in $\{\{a, b\}, \{a, c\}, \{b, c\}\}$ has probability 1/3.

What is the conditional probability that b is selected second given that a is selected first? Because this question considers the results of each pick, the ordered sample space is used. The definition of conditional probability is applied as follows. Let E_1 denote the event that the first item selected is a and let E_2 denote the event that the second item selected is b. Then,

$$E_1 = \{ab, ac\} \quad \text{and} \quad E_2 = \{ab, cb\} \quad \text{and} \quad E_1 \cap E_2 = \{ab\}$$

and from the definition of conditional probability

$$P(E_2|E_1) = P(E_1 \cap E_2)/P(E_1) = \frac{1/6}{1/3} = 1/2$$

When the sample space is larger, an alternative calculation is usually more convenient. For example, suppose that a batch contains 10 parts from tool 1 and 40 parts from tool 2. If two parts are selected randomly, without replacement, what is the conditional probability that a part from tool 2 is selected second given that a part from tool 1 is selected first? There are 50 possible parts to select first and 49 to select second. Therefore, the (ordered) sample space has $50 \times 49 = 2450$ outcomes. Let E_1 denote the event that the first part is from tool 1 and E_2 denote the event that the second part is from tool 2. As above, a count of the number of outcomes in E_1 and the intersection is needed.

Although the answer can be determined from this start, this type of question can be answered more easily with the following result.

To select randomly implies that at each step of the sample, the items that remain in the batch are equally likely to be selected.

If a part from tool 1 were selected with the first pick, 49 items would remain, 9 from tool 1 and 40 from tool 2, and they would be equally likely to be picked. Therefore, the probability that a part from tool 2 would be selected with the second pick given this first pick is

$$P(E_2|E_1) = 40/49.$$

In this manner, other probabilities can also be simplified. For example, let the event E consist of the outcomes with the first selected part from tool 1 and the second part from tool 2. To determine the probability of E, consider each step. The probability that a part from tool 1 is selected with the first pick is $P(E_1) = 10/50$. The conditional probability that a part from tool 2 is selected with the second pick, given that a part from tool 1 is selected first is $P(E_2|E_1) = 40/49$. Therefore,

$$P(E) = P(E_2|E_1)P(E_1) = \frac{40}{49} \cdot \frac{10}{50} = 0.163$$

Sometimes a partition of the question into successive picks is an easier method to solve the problem.

EXAMPLE 2-18   A day's production of 850 manufactured parts contains 50 parts that do not meet customer requirements. Two parts are selected randomly without replacement from the batch. What is the probability that the second part is defective given that the first part is defective?

Let A denote the event that the first part selected is defective, and let B denote the event that the second part selected is defective. The probability needed can be expressed as $P(B|A)$. If the first part is defective, prior to selecting the second part, the batch contains 849 parts, of which 49 are defective, therefore

$$P(B|A) = 49/849$$

EXAMPLE 2-19   Continuing the previous example, if three parts are selected at random, what is the probability that the first two are defective and the third is not defective? This event can be described in shorthand notation as simply $P(ddn)$. We have

$$P(ddn) = \frac{50}{850} \cdot \frac{49}{849} \cdot \frac{800}{848} = 0.0032$$

The third term is obtained as follows. After the first two parts are selected, there are 848 remaining. Of the remaining parts, 800 are not defective. In this example, it is easy to obtain the solution with a conditional probability for each selection.

EXERCISES FOR SECTION 2-4

2-57. Disks of polycarbonate plastic from a supplier are analyzed for scratch and shock resistance. The results from 100 disks are summarized as follows:

		shock resistance	
		high	low
scratch	high	70	9
resistance	low	16	5

Let A denote the event that a disk has high shock resistance, and let B denote the event that a disk has high scratch resistance. Determine the following probabilities:
(a) $P(A)$ (b) $P(B)$
(c) $P(A|B)$ (d) $P(B|A)$

2-58. Samples of a cast aluminum part are classified on the basis of surface finish (in microinches) and length measurements. The results of 100 parts are summarized as follows:

		length	
		excellent	good
surface	excellent	80	2
finish	good	10	8

Let A denote the event that a sample has excellent surface finish, and let B denote the event that a sample has excellent length. Determine:
(a) $P(A)$ (b) $P(B)$
(c) $P(A|B)$ (d) $P(B|A)$
(e) If the selected part has excellent surface finish, what is the probability that the length is excellent?
(f) If the selected part has good length, what is the probability that the surface finish is excellent?

2-59. The analysis of shafts for a compressor is summarized by conformance to specifications:

		roundness conforms	
		yes	no
surface finish	yes	345	5
conforms	no	12	8

(a) If we know that a shaft conforms to roundness requirements, what is the probability that it conforms to surface finish requirements?
(b) If we know that a shaft does not conform to roundness requirements, what is the probability that it conforms to surface finish requirements?

2-60. The following table summarizes the analysis of samples of galvanized steel for coating weight and surface roughness:

		coating weight	
		high	low
surface	high	12	16
roughness	low	88	34

(a) If the coating weight of a sample is high, what is the probability that the surface roughness is high?
(b) If the surface roughness of a sample is high, what is the probability that the coating weight is high?
(c) If the surface roughness of a sample is low, what is the probability that the coating weight is low?

2-61. Consider the data on wafer contamination and location in the sputtering tool shown in Table 2-2. Assume that one wafer is selected at random from this set. Let A denote the event that a wafer contains four or more particles, and let B denote the event that a wafer is from the center of the sputtering tool. Determine:
(a) $P(A)$ (b) $P(A|B)$
(c) $P(B)$ (d) $P(B|A)$
(e) $P(A \cap B)$ (f) $P(A \cup B)$

2-62. A lot of 100 semiconductor chips contains 20 that are defective. Two are selected randomly, without replacement, from the lot.
(a) What is the probability that the first one selected is defective?
(b) What is the probability that the second one selected is defective given that the first one was defective?
(c) What is the probability that both are defective?
(d) How does the answer to part (b) change if chips selected were replaced prior to the next selection?

2-63. A lot contains 15 castings from a local supplier and 25 castings from a supplier in the next state. Two castings are selected randomly, without replacement, from the lot of 40. Let A be the event that the first casting selected is from the local supplier, and let B denote the event that the second casting is selected from the local supplier. Determine:
(a) $P(A)$ (b) $P(B|A)$
(c) $P(A \cap B)$ (d) $P(A \cup B)$

2-64. Continuation of Exercise 2-63. Suppose three castings are selected at random, without replacement, from the lot of 40. In addition to the definitions of events A and B, let C denote the event that the third casting selected is from the local supplier. Determine:
(a) $P(A \cap B \cap C)$
(b) $P(A \cap B \cap C')$

2-65. A batch of 500 containers for frozen orange juice contains 5 that are defective. Two are selected, at random, without replacement from the batch.
(a) What is the probability that the second one selected is defective given that the first one was defective?
(b) What is the probability that both are defective?
(c) What is the probability that both are acceptable?

2-66. Continuation of Exercise 2-65. Three containers are selected, at random, without replacement, from the batch.
(a) What is the probability that the third one selected is defective given that the first and second one selected were defective?
(b) What is the probability that the third one selected is defective given that the first one selected was defective and the second one selected was okay?
(c) What is the probability that all three are defective?

2-67. A maintenance firm has gathered the following information regarding the failure mechanisms for air conditioning systems:

		evidence of gas leaks	
		yes	no
evidence of	yes	55	17
electrical failure	no	32	3

The units without evidence of gas leaks or electrical failure showed other types of failure. If this is a representative sample of AC failure, find the probability
(a) That failure involves a gas leak
(b) That there is evidence of electrical failure given that there was a gas leak
(c) That there is evidence of a gas leak given that there is evidence of electrical failure

2-68. If $P(A|B) = 1$, must $A = B$? Draw a Venn diagram to explain your answer.

2-69. Suppose A and B are mutually exclusive events. Construct a Venn diagram that contains the three events A, B, and C such that $P(A|C) = 1$ and $P(B|C) = 0$?

2-5   MULTIPLICATION AND TOTAL PROBABILITY RULES

2-5.1   Multiplication Rule

The definition of conditional probability in Equation 2-5 can be rewritten to provide a general expression for the probability of the intersection of two events. This formula is referred to as a **multiplication rule** for probabilities.

Multiplication Rule

$$P(A \cap B) = P(B|A)P(A) = P(A|B)P(B) \qquad (2\text{-}6)$$

The last expression in Equation 2-6 is obtained by interchanging A and B.

EXAMPLE 2-20 The probability that an automobile battery subject to high engine compartment temperature suffers low charging current is 0.7. The probability that a battery is subject to high engine compartment temperature is 0.05.

Let C denote the event that a battery suffers low charging current, and let T denote the event that a battery is subject to high engine compartment temperature. The probability that a battery is subject to low charging current and high engine compartment temperature is

$$P(C \cap T) = P(C|T)P(T) = 0.7 \times 0.05 = 0.035$$

2-5.2 Total Probability Rule

The multiplication rule is useful for determining the probability of an event that depends on other events. For example, suppose that in semiconductor manufacturing the probability is 0.10 that a chip that is subjected to high levels of contamination during manufacturing causes a product failure. The probability is 0.005 that a chip that is not subjected to high contamination levels during manufacturing causes a product failure. In a particular production run, 20% of the chips are subject to high levels of contamination. What is the probability that a product using one of these chips fails?

Clearly, the requested probability depends on whether or not the chip was exposed to high levels of contamination. We can solve this problem by the following reasoning. For any event B, we can write B as the union of the part of B in A and the part of B in A'. That is,

$$B = (A \cap B) \cup (A' \cap B)$$

This result is shown in the Venn diagram in Fig. 2-14. Because A and A' are mutually exclusive, $A \cap B$ and $A' \cap B$ are mutually exclusive. Therefore, from the probability of the union of mutually exclusive events in Equation 2-2 and the Multiplication Rule in Equation 2-6, the following **total probability rule** is obtained.

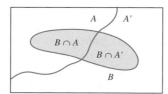

Figure 2-14 Partitioning an event into two mutually exclusive subsets.

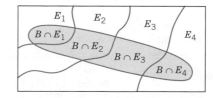

$B = (B \cap E_1) \cup (B \cap E_2) \cup (B \cap E_3) \cup (B \cap E_4)$

Figure 2-15 Partitioning an event into several mutually exclusive subsets.

Total Probability Rule (two events)

For any events A and B,

$$P(B) = P(B \cap A) + P(B \cap A') = P(B|A)P(A) + P(B|A')P(A') \qquad (2\text{-}7)$$

EXAMPLE 2-21

Consider the contamination discussion at the start of this section. Let F denote the event that the product fails, and let H denote the event that the chip is exposed to high levels of contamination. The requested probability is $P(F)$, and the information provided can be represented as

$$P(F|H) = 0.10 \qquad \text{and} \qquad P(F|H') = 0.005$$
$$P(H) = 0.20 \qquad \text{and} \qquad P(H') = 0.80$$

From Equation 2-7,

$$P(F) = 0.10(0.20) + 0.005(0.80) = 0.0235$$

which can be interpreted as just the weighted average of the two probabilities of failure.

The reasoning used to develop Equation 2-7 can be applied more generally. In the development of Equation 2-7, we only used the two mutually exclusive A and A'. However, the fact that $A \cup A' = S$, the entire sample space, was important. In general, a collection of sets $E_1, E_2, \ldots, E_k$ such that $E_1 \cup E_2 \cup \ldots \cup E_k = S$ is said to be **exhaustive.** A graphical display of partitioning an event B among a collection of mutually exclusive and exhaustive events is shown in Fig. 2-15 on page 43.

Total Probability Rule (multiple events)

Assume $E_1, E_2, \ldots, E_k$ are k mutually exclusive and exhaustive sets. Then

$$P(B) = P(B \cap E_1) + P(B \cap E_2) + \cdots + P(B \cap E_k)$$
$$= P(B|E_1)P(E_1) + P(B|E_2)P(E_2) + \cdots + P(B|E_k)P(E_k) \qquad (2\text{-}8)$$

EXAMPLE 2-22

Continuing with the semiconductor manufacturing example, assume the following probabilities for product failure subject to levels of contamination in manufacturing:

Probability of Failure	Level of Contamination
0.10	High
0.01	Medium
0.001	Low

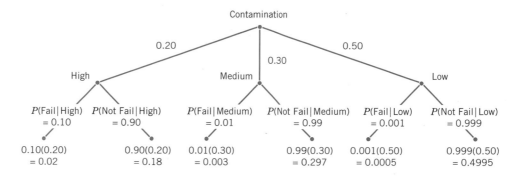

Figure 2-16   Tree diagram for Example 2-22.

$P(\text{Fail}) = 0.02 + 0.003 + 0.0005 = 0.0235$

In a particular production run, 20% of the chips are subjected to high levels of contamination, 30% to medium levels of contamination, and 50% to low levels of contamination. What is the probability that a product using one of these chips fails? Let

H denote the event that a chip is exposed to high levels of contamination

M denote the event that a chip is exposed to medium levels of contamination

L denote the event that a chip is exposed to low levels of contamination

Then,

$$P(F) = P(F\,|\,H)P(H) + P(F\,|\,M)P(M) + P(F\,|\,L)P(L)$$
$$= 0.10(0.20) + 0.01(0.30) + 0.001(0.50) = 0.0235$$

This problem is also conveniently solved using the tree diagram in Fig. 2-16.

EXERCISES FOR SECTION 2-5

2-70.   Suppose that $P(A\,|\,B) = 0.4$ and $P(B) = 0.5$. Determine the following:
(a) $P(A \cap B)$
(b) $P(A' \cap B)$

2-71.   Suppose that $P(A\,|\,B) = 0.2$, $P(A\,|\,B') = 0.3$, and $P(B) = 0.8$. What is $P(A)$?

2-72.   The probability is 1% that an electrical connector that is kept dry fails during the warranty period of a portable computer. If the connector is ever wet, the probability of a failure during the warranty period is 5%. If 90% of the connectors are kept dry and 10% are wet, what proportion of connectors fail during the warranty period?

2-73.   Suppose 2% of cotton fabric rolls and 3% of nylon fabric rolls contain flaws. Of the rolls used by a manufacturer, 70% are cotton and 30% are nylon. What is the probability that a randomly selected roll used by the manufacturer contains flaws?

2-74.   In the manufacturing of a chemical adhesive, 3% of all batches have raw materials from two different lots. This occurs when holding tanks are replenished and the remaining portion of a lot is insufficient to fill the tanks.

Only 5% of batches with material from a single lot require reprocessing. However, the viscosity of batches consisting of two or more lots of material is more difficult to control, and 40% of such batches require additional processing to achieve the required viscosity.

Let A denote the event that a batch is formed from two different lots, and let B denote the event that a lot requires additional processing. Determine the following probabilities:
(a) $P(A)$ (b) $P(A')$
(c) $P(B\,|\,A)$ (d) $P(B\,|\,A')$
(e) $P(A \cap B)$ (f) $P(A \cap B')$
(g) $P(B)$

2-75. The edge roughness of slit paper products increases as knife blades wear. Only 1% of products slit with new blades have rough edges, 3% of products slit with blades of average sharpness exhibit roughness, and 5% of products slit with worn blades exhibit roughness. If 25% of the blades in manufacturing are new, 60% are of average sharpness, and 15% are worn, what is the proportion of products that exhibit edge roughness?

2-76. Samples of laboratory glass are in small, light packaging or heavy, large packaging. Suppose that 2 and 1% of the sample shipped in small and large packages, respectively, break during transit. If 60% of the samples are shipped in large packages and 40% are shipped in small packages, what proportion of samples break during shipment?

2-77. Incoming calls to a customer service center are classified as complaints (75% of call) or requests for information (25% of calls). Of the complaints, 40% deal with computer equipment that does not respond and 57% deal with incomplete software installation; and in the remaining 3% of complaints the user has improperly followed the installation instructions. The requests for information are evenly divided on technical questions (50%) and requests to purchase more products (50%).

(a) What is the probability that an incoming call to the customer service center will be from a customer who has not followed installation instructions properly?

(b) Find the probability that an incoming call is a request for purchasing more products.

2-78. Computer keyboard failures are due to faulty electrical connects (12%) or mechanical defects (88%). Mechanical defects are related to loose keys (27%) or improper assembly (73%). Electrical connect defects are caused by defective wires (35%), improper connections (13%), or poorly welded wires (52%).

(a) Find the probability that a failure is due to loose keys.

(b) Find the probability that a failure is due to improperly connected or poorly welded wires.

2-79. A batch of 25 injection-molded parts contains 5 that have suffered excessive shrinkage.

(a) If two parts are selected at random, and without replacement, what is the probability that the second part selected is one with excessive shrinkage?

(b) If three parts are selected at random, and without replacement, what is the probability that the third part selected is one with excessive shrinkage?

2-80. A lot of 100 semiconductor chips contains 20 that are defective.

(a) Two are selected, at random, without replacement, from the lot. Determine the probability that the second chip selected is defective.

(b) Three are selected, at random, without replacement, from the lot. Determine the probability that all are defective.

2-6   INDEPENDENCE

In some cases, the conditional probability of $P(B|A)$ might equal $P(B)$. In this special case, knowledge that the outcome of the experiment is in event A does not affect the probability that the outcome is in event B.

EXAMPLE 2-23   Suppose a day's production of 850 manufactured parts contains 50 parts that do not meet customer requirements. Suppose two parts are selected from the batch, but the first part is replaced before the second part is selected. What is the probability that the second part is defective (denoted as B) given that the first part is defective (denoted as A)? The probability needed can be expressed as $P(B|A)$.

Because the first part is replaced prior to selecting the second part, the batch still contains 850 parts, of which 50 are defective. Therefore, the probability of B does not depend on whether or not the first part was defective. That is,

$$P(B|A) = 50/850$$

Also, the probability that both parts are defective is

$$P(A \cap B) = P(B|A)P(A) = \left(\frac{50}{850}\right) \cdot \left(\frac{50}{850}\right) = 0.0035$$

Table 2-4   Parts Classified

		Surface Flaws		
		Yes (event F)	No	Total
Defective	Yes (event D)	2	18	20
	No	38	342	380
	Total	40	360	400

EXAMPLE 2-24

The information in Table 2-3 related surface flaws to functionally defective parts. In that case, we determined that $P(D|F) = 10/40 = 0.25$ and $P(D) = 28/400 = 0.07$. Suppose that the situation is different and follows Table 2-4. Then,

$$P(D|F) = 2/40 = 0.05 \quad \text{and} \quad P(D) = 20/400 = 0.05$$

That is, the probability that the part is defective does not depend on whether it has surface flaws. Also,

$$P(F|D) = 2/20 = 0.10 \quad \text{and} \quad P(F) = 40/400 = 0.10$$

so the probability of a surface flaw does not depend on whether the part is defective. Furthermore, the definition of conditional probability implies that

$$P(F \cap D) = P(D|F)P(F)$$

but in the special case of this problem

$$P(F \cap D) = P(D)P(F) = \frac{2}{40} \cdot \frac{2}{20} = \frac{1}{200}$$

The preceding example illustrates the following conclusions. In the special case that $P(B|A) = P(B)$, we obtain

$$P(A \cap B) = P(B|A)P(A) = P(B)P(A)$$

and

$$P(A|B) = \frac{P(A \cap B)}{P(B)} = \frac{P(A)P(B)}{P(B)} = P(A)$$

These conclusions lead to an important definition.

Definition

Two events are **independent** if any one of the following equivalent statements is true:

(1) $P(A|B) = P(A)$

(2) $P(B|A) = P(B)$

(3) $P(A \cap B) = P(A)P(B)$

$\qquad\qquad$ (2-9)

It is left as a mind-expanding exercise to show that independence implies related results such as

$$P(A' \cap B') = P(A')P(B').$$

The concept of independence is an important relationship between events and is used throughout this text. A mutually exclusive relationship between two events is based only on the outcomes that comprise the events. However, an independence relationship depends on the probability model used for the random experiment. Often, independence is assumed to be part of the random experiment that describes the physical system under study.

EXAMPLE 2-25   A day's production of 850 manufactured parts contains 50 parts that do not meet customer requirements. Two parts are selected at random, without replacement, from the batch. Let A denote the event that the first part is defective, and let B denote the event that the second part is defective.

We suspect that these two events are not independent because knowledge that the first part is defective suggests that it is less likely that the second part selected is defective. Indeed, $P(B|A) = 49/849$. Now, what is $P(B)$? Finding the unconditional $P(B)$ is somewhat difficult because the possible values of the first selection need to be considered:

$$\begin{aligned}
P(B) &= P(B|A)P(A) + P(B|A')P(A') \\
&= (49/849)(50/850) + (50/849)(800/850) \\
&= 50/850
\end{aligned}$$

Interestingly, $P(B)$, the unconditional probability that the second part selected is defective, without any knowledge of the first part, is the same as the probability that the first part selected is defective. Yet, our goal is to assess independence. Because $P(B|A)$ does not equal $P(B)$, the two events are not independent, as we suspected.

When considering three or more events, we can extend the definition of independence with the following general result.

Definition

> The events $E_1, E_2, \ldots, E_n$ are independent if and only if for any subset of these events $E_{i_1}, E_{i_2}, \ldots, E_{i_k}$,
>
> $$P(E_{i_1} \cap E_{i_2} \cap \cdots \cap E_{i_k}) = P(E_{i_1}) \times P(E_{i_2}) \times \cdots \times P(E_{i_k}) \qquad \text{(2-10)}$$

This definition is typically used to calculate the probability that several events occur assuming that they are independent and the individual event probabilities are known. The knowledge that the events are independent usually comes from a fundamental understanding of the random experiment.

EXAMPLE 2-26   Assume that the probability that a wafer contains a large particle of contamination is 0.01 and that the wafers are independent; that is, the probability that a wafer contains a large particle is not dependent on the characteristics of any of the other wafers. If 15 wafers are analyzed, what is the probability that no large particles are found?

Let E_i denote the event that the ith wafer contains no large particles, $i = 1, 2, \ldots, 15$. Then, $P(E_i) = 0.99$. The probability requested can be represented as $P(E_1 \cap E_2 \cap \cdots \cap E_{15})$. From the independence assumption and Equation 2-10,

$$P(E_1 \cap E_2 \cap \cdots \cap E_{15}) = P(E_1) \times P(E_2) \times \cdots \times P(E_{15}) = 0.99^{15} = 0.86$$

EXAMPLE 2-27   The following circuit operates only if there is a path of functional devices from left to right. The probability that each device functions is shown on the graph. Assume that devices fail independently. What is the probability that the circuit operates?

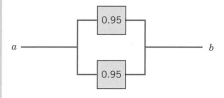

Let T and B denote the events that the top and bottom devices operate, respectively. There is a path if at least one device operates. The probability that the circuit operates is

$$P(T \text{ or } B) = 1 - P[(T \text{ or } B)'] = 1 - P(T' \text{ and } B')$$

a simple formula for the solution can be derived from the complements T' and B'. From the independence assumption,

$$P(T' \text{ and } B') = P(T')P(B') = (1 - 0.95)^2 = 0.05^2$$

so

$$P(T \text{ or } B) = 1 - 0.05^2 = 0.9975$$

$P(T) + P(B) - P(A \cap B)$
$= 0.95 + 0.95 - (0.95)^2$

EXAMPLE 2-28   The following circuit operates only if there is a path of functional devices from left to right. The probability that each device functions is shown on the graph. Assume that devices fail independently. What is the probability that the circuit operates?

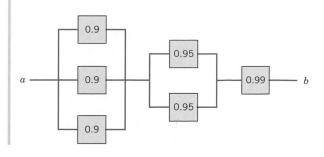

The solution can be obtained from a partition of the graph into three columns. The probability that there is a path of functional devices only through the three units on the left can be determined from the independence in a manner similar to the previous example. It is

$$1 - 0.1^3$$

Similarly, the probability that there is a path of functional devices only through the two units in the middle is

$$1 - 0.05^2$$

The probability that there is a path of functional devices only through the one unit on the right is simply the probability that the device functions, namely, 0.99. Therefore, with the independence assumption used again, the solution is

$$(1 - 0.1^3)(1 - 0.05^2)(0.99) = 0.987$$

EXERCISES FOR SECTION 2-6

2-81. If $P(A|B) = 0.4$, $P(B) = 0.8$, and $P(A) = 0.5$, are the events A and B independent?

2-82. If $P(A|B) = 0.3$, $P(B) = 0.8$, and $P(A) = 0.3$, are the events B and the complement of A independent?

2-83. Disks of polycarbonate plastic from a supplier are analyzed for scratch and shock resistance. The results from 100 disks are summarized as follows:

		shock resistance	
		high	low
scratch	high	70	9
resistance	low	16	5

Let A denote the event that a disk has high shock resistance, and let B denote the event that a disk has high scratch resistance. Are events A and B independent?

2-84. Samples of a cast aluminum part are classified on the basis of surface finish (in microinches) and length measurements. The results of 100 parts are summarized as follows:

		length	
		excellent	good
surface	excellent	80	2
finish	good	10	8

Let A denote the event that a sample has excellent surface finish, and let B denote the event that a sample has excellent length. Are events A and B independent?

2-85. Samples of emissions from three suppliers are classified for conformance to air-quality specifications. The results from 100 samples are summarized as follows:

		conforms	
		yes	no
	1	22	8
supplier	2	25	5
	3	30	10

Let A denote the event that a sample is from supplier 1, and let B denote the event that a sample conforms to specifications.
(a) Are events A and B independent?
(b) Determine $P(B|A)$.

2-86. If $P(A) = 0.2$, $P(B) = 0.2$, and A and B are mutually exclusive, are they independent?

2-87. The probability that a lab specimen contains high levels of contamination is 0.10. Five samples are checked, and the samples are independent.
(a) What is the probability that none contains high levels of contamination?
(b) What is the probability that exactly one contains high levels of contamination?
(c) What is the probability that at least one contains high levels of contamination?

2-88. In a test of a printed circuit board using a random test pattern, an array of 10 bits is equally likely to be 0 or 1. Assume the bits are independent.
(a) What is the probability that all bits are 1s?
(b) What is the probability that all bits are 0s?
(c) What is the probability that exactly five bits are 1s and five bits are 0s?

2-89. Eight cavities in an injection-molding tool produce plastic connectors that fall into a common stream. A sample is chosen every several minutes. Assume that the samples are independent.

(a) What is the probability that five successive samples were all produced in cavity one of the mold?

(b) What is the probability that five successive samples were all produced in the same cavity of the mold?

(c) What is the probability that four out of five successive samples were produced in cavity one of the mold?

2-90. The following circuit operates if and only if there is a path of functional devices from left to right. The probability that each device functions is as shown. Assume that the probability that a device is functional does not depend on whether or not other devices are functional. What is the probability that the circuit operates?

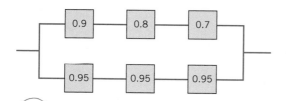

2-91. The following circuit operates if and only if there is a path of functional devices from left to right. The probability each device functions is as shown. Assume that the probability that a device functions does not depend on whether or not other devices are functional. What is the probability that the circuit operates?

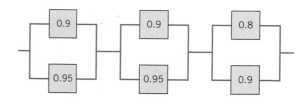

2-92. An optical storage device uses an error recovery procedure that requires an immediate satisfactory readback of any written data. If the readback is not successful after three writing operations, that sector of the disk is eliminated as unacceptable for data storage. On an acceptable portion of the disk, the probability of a satisfactory readback is 0.98. Assume the readbacks are independent. What is the probability that an acceptable portion of the disk is eliminated as unacceptable for data storage?

2-93. A batch of 500 containers for frozen orange juice contains 5 that are defective. Two are selected, at random, without replacement, from the batch. Let A and B denote the events that the first and second container selected is defective, respectively.

(a) Are A and B independent events?

(b) If the sampling were done with replacement, would A and B be independent?

2-7 BAYES' THEOREM

In some examples, we do not have a complete table of information such as the parts in Table 2-3. We might know one conditional probability but would like to calculate a different one. In the semiconductor contamination problem in Example 2-22, we might ask the following: If the semiconductor chip in the product fails, what is the probability that the chip was exposed to high levels of contamination?

From the definition of conditional probability,

$$P(A \cap B) = P(A|B)P(B) = P(B \cap A) = P(B|A)P(A)$$

Now considering the second and last terms in the expression above, we can write

$$P(A|B) = \frac{P(B|A)P(A)}{P(B)} \quad \text{for} \quad P(B) > 0 \qquad (2\text{-}11)$$

This is a useful result that enables us to solve for $P(A|B)$ in terms of $P(B|A)$.

EXAMPLE 2-29 We can answer the question posed at the start of this section as follows: The probability requested can be expressed as $P(H \mid F)$. Then,

$$P(H \mid F) = \frac{P(F \mid H)P(H)}{P(F)} = \frac{0.10(0.20)}{0.0235} = 0.85$$

The value of $P(F)$ in the denominator of our solution was found in Example 2-20.

In general, if $P(B)$ in the denominator of Equation 2-11 is written using the Total Probability Rule in Equation 2-8, we obtain the following general result, which is known as **Bayes' Theorem.**

Bayes' Theorem

If $E_1, E_2, \ldots, E_k$ are k mutually exclusive and exhaustive events and B is any event,

$$P(E_1 \mid B) = \frac{P(B \mid E_1)P(E_1)}{P(B \mid E_1)P(E_1) + P(B \mid E_2)P(E_2) + \cdots + P(B \mid E_k)P(E_k)} \qquad (2\text{-}12)$$

$$\text{for } P(B) > 0$$

EXAMPLE 2-30 Because a new medical procedure has been shown to be effective in the early detection of an illness, a medical screening of the population is proposed. The probability that the test correctly identifies someone with the illness as positive is 0.99, and the probability that the test correctly identifies someone without the illness as negative is 0.95. The incidence of the illness in the general population is 0.0001. You take the test, and the result is positive. What is the probability that you have the illness?

Let D denote the event that you have the illness, and let S denote the event that the test signals positive. The probability requested can be denoted as $P(D \mid S)$. The probability that the test correctly signals someone without the illness as negative is 0.95. Consequently, the probability of a positive test without the illness is

$$P(S \mid D') = 0.05$$

From Bayes' Theorem,

$$P(D \mid S) = P(S \mid D)P(D)/[P(S \mid D)P(D) + P(S \mid D')P(D')]$$
$$= 0.99(0.0001)/[0.99(0.0001) + 0.05(1 - 0.0001)]$$
$$= 1/506 = 0.002$$

Surprisingly, even though the test is effective, in the sense that $P(S \mid D)$ is high and $P(S \mid D')$ is low, because the incidence of the illness in the general population is low, the chances are quite small that you actually have the disease even if the test is positive.

EXERCISES FOR SECTION 2-7

2-94. Suppose that $P(A|B) = 0.7$, $P(A) = 0.5$, and $P(B) = 0.2$. Determine $P(B|A)$.

2-95. Software to detect fraud in consumer phone cards tracks the number of metropolitan areas where calls originate each day. It is found that 1% of the legitimate users originate calls from two or more metropolitan areas in a single day. However, 30% of fraudulent users originate calls from two or more metropolitan areas in a single day. The proportion of fraudulent users is 0.01%. If the same user originates calls from two or more metropolitan areas in a single day, what is the probability that the user is fraudulent?

2-96. Semiconductor lasers used in optical storage products require higher power levels for write operations than for read operations. High-power-level operations lower the useful life of the laser.

Lasers in products used for backup of higher speed magnetic disks primarily write, and the probability that the useful life exceeds five years is 0.95. Lasers that are in products that are used for main storage spend approximately an equal amount of time reading and writing, and the probability that the useful life exceeds five years is 0.995. Now, 25% of the products from a manufacturer are used for backup and 75% of the products are used for main storage.

Let A denote the event that a laser's useful life exceeds five years, and let B denote the event that a laser is in a product that is used for backup.

Use a tree diagram to determine the following:
(a) $P(B)$ (b) $P(A|B)$
(c) $P(A|B')$ (d) $P(A \cap B)$
(e) $P(A \cap B')$ (f) $P(A)$
(g) What is the probability that the useful life of a laser exceeds five years?
(h) What is the probability that a laser that failed before five years came from a product used for backup?

2-97. Customers are used to evaluate preliminary product designs. In the past, 95% of highly successful products received good reviews, 60% of moderately successful prod-

ucts received good reviews, and 10% of poor products received good reviews. In addition, 40% of products have been highly successful, 35% have been moderately successful, and 25% have been poor products.
(a) What is the probability that a product attains a good review?
(b) If a new design attains a good review, what is the probability that it will be a highly successful product?
(c) If a product does not attain a good review, what is the probability that it will be a highly successful product?

2-98. An inspector working for a manufacturing company has a 99% chance of correctly identifying defective items and a 0.5% chance of incorrectly classifying a good item as defective. The company has evidence that its line produces 0.9% of nonconforming items.
(a) What is the probability that an item selected for inspection is classified as defective?
(b) If an item selected at random is classified as nondefective, what is the probability that it is indeed good?

2-99. A new analytical method to detect pollutants in water is being tested. This new method of chemical analysis is important because, if adopted, it could be used to detect three different contaminants—organic pollutants, volatile solvents, and chlorinated compounds—instead of having to use a single test for each pollutant. The makers of the test claim that it can detect high levels of organic pollutants with 99.7% accuracy, volatile solvents with 99.95% accuracy, and chlorinated compounds with 89.7% accuracy. If a pollutant is not present, the test does not signal. Samples are prepared for the calibration of the test and 60% of them are contaminated with organic pollutants, 27% with volatile solvents, and 13% with traces of chlorinated compounds.

A test sample is selected randomly.
(a) What is the probability that the test will signal?
(b) If the test signals, what is the probability that chlorinated compounds are present?

2-8 RANDOM VARIABLES

We often summarize the outcome from a random experiment by a simple number. In many of the examples of random experiments that we have considered, the sample space has been a description of possible outcomes. In some cases, descriptions of outcomes are sufficient, but in other cases, it is useful to associate a number with each outcome in the sample space. Because the particular outcome of the experiment is not known in advance, the resulting value of our variable is not known in advance. For this reason, the variable that associates a number with the outcome of a random experiment is referred to as a **random variable.**

Definition

> A **random variable** is a function that assigns a real number to each outcome in the sample space of a random experiment.
>
> A random variable is denoted by an uppercase letter such as X. After an experiment is conducted, the measured value of the random variable is denoted by a lowercase letter such as $x = 70$ milliamperes.

Sometimes a measurement (such as current in a copper wire or length of a machined part) can assume any value in an interval of real numbers (at least theoretically). Then arbitrary precision in the measurement is possible. Of course, in practice, we might round off to the nearest tenth or hundredth of a unit. The random variable that represents this measurement is said to be a **continuous** random variable. The range of the random variable includes all values in an interval of real numbers; that is, the range can be thought of as a continuum.

In other experiments, we might record a count such as the number of transmitted bits that are received in error. Then the measurement is limited to integers. Or we might record that a proportion such as 0.0042 of the 10,000 transmitted bits were received in error. Then the measurement is fractional, but it is still limited to discrete points on the real line. Whenever the measurement is limited to discrete points on the real line, the random variable is said to be a **discrete** random variable.

Definition

> A **discrete** random variable is a random variable with a finite (or countably infinite) range.
>
> A **continuous** random variable is a random variable with an interval (either finite or infinite) of real numbers for its range.

In some cases, the random variable X is actually discrete but, because the range of possible values is so large, it might be more convenient to analyze X as a continuous random variable. For example, suppose that current measurements are read from a digital instrument that displays the current to the nearest one-hundredth of a milliampere. Because the possible measurements are limited, the random variable is discrete. However, it might be a more convenient, simple approximation to assume that the current measurements are values of a continuous random variable.

Examples of Random Variables

> Examples of **continuous** random variables:
>   electrical current, length, pressure, temperature, time, voltage, weight
>
> Examples of **discrete** random variables:
>   number of scratches on a surface, proportion of defective parts among 1000 tested, number of transmitted bits received in error.

EXERCISES FOR SECTION 2-8

2-100. Decide whether a discrete or continuous random variable is the best model for each of the following variables:

(a) The time until a projectile returns to earth.

(b) The number of times a transistor in a computer memory changes state in one operation.

(c) The volume of gasoline that is lost to evaporation during the filling of a gas tank.

(d) The outside diameter of a machined shaft.
(e) The number of cracks exceeding one-half inch in 10 miles of an interstate highway.
(f) The weight of an injection-molded plastic part.
(g) The number of molecules in a sample of gas.
(h) The concentration of output from a reactor.
(i) The current in an electronic circuit.

Supplemental Exercises

2-101. In circuit testing of printed circuit boards, each board either fails or does not fail the test. A board that fails the test is then checked further to determine which one of five defect types is the primary failure mode. Represent the sample space for this experiment.

2-102. The data from 200 machined parts are summarized as follows:

	depth of bore	
edge condition	above target	below target
coarse	15	10
moderate	25	20
smooth	50	80

(a) What is the probability that a part selected has a moderate edge condition and a below-target bore depth?
(b) What is the probability that a part selected has a moderate edge condition or a below-target bore depth?
(c) What is the probability that a part selected does not have a moderate edge condition or does not have a below-target bore depth?
(d) Construct a Venn diagram representation of the events in this sample space for part c.

2-103. Computers in a shipment of 100 units contain a portable hard drive, CD RW drive, or both according to the following table:

	portable hard drive	
CD RW	yes	no
yes	15	80
no	4	1

Let A denote the events that a computer has a portable hard drive and let B denote the event that a computer has a CD RW drive. If one computer is selected randomly, compute
(a) $P(A)$ (b) $P(A \cap B)$
(c) $P(A \cup B)$ (d) $P(A' \cap B)$
(e) $P(A|B)$

2-104. The probability that a customer's order is not shipped on time is 0.05. A particular customer places three orders, and the orders are placed far enough apart in time that they can be considered to be independent events.

(a) What is the probability that all are shipped on time?
(b) What is the probability that exactly one is not shipped on time?
(c) What is the probability that two or more orders are not shipped on time?

2-105. Let E_1, E_2, and E_3 denote the samples that conform to a percentage of solids specification, a molecular weight specification, and a color specification, respectively. A total of 240 samples are classified by the E_1, E_2, and E_3 specifications, where *yes* indicates that the sample conforms.

E_3 yes

		E_2		
		yes	no	Total
E_1	yes	200	1	201
	no	5	4	9
Total		205	5	210

E_3 no

		E_2		
		yes	no	
E_1	yes	20	4	24
	no	6	0	6
Total		26	4	30

(a) Are E_1, E_2, and E_3 mutually exclusive events?
(b) Are E'_1, E'_2, and E'_3 mutually exclusive events?
(c) What is $P(E'_1 \text{ or } E'_2 \text{ or } E'_3)$?
(d) What is the probability that a sample conforms to all three specifications?
(e) What is the probability that a sample conforms to the E_1 or E_3 specification?
(f) What is the probability that a sample conforms to the E_1 or E_2 or E_3 specification?

2-106. Transactions to a computer database are either new items or changes to previous items. The addition of an item can be completed less than 100 milliseconds 90% of the time, but only 20% of changes to a previous item can be completed in less than this time. If 30% of transactions are changes, what is the probability that a transaction can be completed in less than 100 milliseconds?

2-107. A steel plate contains 20 bolts. Assume that 5 bolts are not torqued to the proper limit. Four bolts are selected at random, without replacement, to be checked for torque.
(a) What is the probability that all four of the selected bolts are torqued to the proper limit?
(b) What is the probability that at least one of the selected bolts is not torqued to the proper limit?

2-108. The following circuit operates if and only if there is a path of functional devices from left to right. Assume devices fail independently and that the probability of *failure* of each device is as shown. What is the probability that the circuit operates?

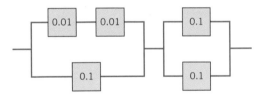

2-109. The probability of getting through by telephone to buy concert tickets is 0.92. For the same event, the probability of accessing the vendor's Web site is 0.95. Assume that these two ways to buy tickets are independent. What is the probability that someone who tries to buy tickets through the Internet and by phone will obtain tickets?

2-110. The British government has stepped up its information campaign regarding foot and mouth disease by mailing brochures to farmers around the country. It is estimated that 99% of Scottish farmers who receive the brochure possess enough information to deal with an outbreak of the disease, but only 90% of those without the brochure can deal with an outbreak. After the first three months of mailing, 95% of the farmers in Scotland received the informative brochure. Compute the probability that a randomly selected farmer will have enough information to deal effectively with an outbreak of the disease.

2-111. In an automated filling operation, the probability of an incorrect fill when the process is operated at a low speed is 0.001. When the process is operated at a high speed, the probability of an incorrect fill is 0.01. Assume that 30% of the containers are filled when the process is operated at a high speed and the remainder are filled when the process is operated at a low speed.
(a) What is the probability of an incorrectly filled container?
(b) If an incorrectly filled container is found, what is the probability that it was filled during the high-speed operation?

2-112. An encryption-decryption system consists of three elements: encode, transmit, and decode. A faulty encode occurs in 0.5% of the messages processed, transmission errors occur in 1% of the messages, and a decode error occurs in 0.1% of the messages. Assume the errors are independent.
(a) What is the probability of a completely defect-free message?
(b) What is the probability of a message that has either an encode or a decode error?

2-113. It is known that two defective copies of a commercial software program were erroneously sent to a shipping lot that has now a total of 75 copies of the program. A sample of copies will be selected from the lot without replacement.
(a) If three copies of the software are inspected, determine the probability that exactly one of the defective copies will be found.
(b) If three copies of the software are inspected, determine the probability that both defective copies will be found.

(c) If 73 copies are inspected, determine the probability that both copies will be found. Hint: Work with the copies that remain in the lot.

2-114. A robotic insertion tool contains 10 primary components. The probability that any component fails during the warranty period is 0.01. Assume that the components fail independently and that the tool fails if any component fails. What is the probability that the tool fails during the warranty period?

2-115. An e-mail message can travel through one of two server routes. The probability of transmission error in each of the servers and the proportion of messages that travel each route are shown in the following table. Assume that the servers are independent.

	percentage of messages	probability of error			
		server 1	server 2	server 3	server 4
route 1	30	0.01	0.015		
route 2	70			0.02	0.003

(a) What is the probability that a message will arrive without error?
(b) If a message arrives in error, what is the probability it was sent through route 1?

2-116. A machine tool is idle 15% of the time. You request immediate use of the tool on five different occasions during the year. Assume that your requests represent independent events.
(a) What is the probability that the tool is idle at the time of all of your requests?
(b) What is the probability that the machine is idle at the time of exactly four of your requests?
(c) What is the probability that the tool is idle at the time of at least three of your requests?

2-117. A lot of 50 spacing washers contains 30 washers that are thicker than the target dimension. Suppose that three washers are selected at random, without replacement, from the lot.
(a) What is the probability that all three washers are thicker than the target?
(b) What is the probability that the third washer selected is thicker than the target if the first two washers selected are thinner than the target?
(c) What is the probability that the third washer selected is thicker than the target?

2-118. Continuation of Exercise 2-117. Washers are selected from the lot at random, without replacement.
(a) What is the minimum number of washers that need to be selected so that the probability that all the washers are thinner than the target is less than 0.10?
(b) What is the minimum number of washers that need to be selected so that the probability that one or more washers are thicker than the target is at least 0.90?

2-119. The following table lists the history of 940 orders for features in an entry-level computer product.

		extra memory	
		no	yes
optional high-	no	514	68
speed processor	yes	112	246

Let A be the event that an order requests the optional high-speed processor, and let B be the event that an order requests extra memory. Determine the following probabilities:

(a) $P(A \cup B)$ (b) $P(A \cap B)$
(c) $P(A' \cup B)$ (d) $P(A' \cap B')$
(e) What is the probability that an order requests an optional high-speed processor given that the order requests extra memory?
(f) What is the probability that an order requests extra memory given that the order requests an optional high-speed processor?

2-120. The alignment between the magnetic tape and head in a magnetic tape storage system affects the performance of the system. Suppose that 10% of the read operations are degraded by skewed alignments, 5% of the read operations are degraded by off-center alignments, and the remaining read operations are properly aligned. The probability of a read error is 0.01 from a skewed alignment, 0.02 from an off-center alignment, and 0.001 from a proper alignment.

(a) What is the probability of a read error?
(b) If a read error occurs, what is the probability that it is due to a skewed alignment?

2-121. The following circuit operates if and only if there is a path of functional devices from left to right. Assume that devices fail independently and that the probability of *failure* of each device is as shown. What is the probability that the circuit does not operate?

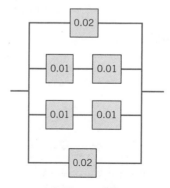

2-122. A company that tracks the use of its web site determined that the more pages a visitor views, the more likely the visitor is to provide contact information. Use the following tables to answer the questions:

Number of pages viewed:	1	2	3	4 or more
Percentage of visitors:	40	30	20	10
Percentage of visitors in each page-view catgory that provide contact information:	10	10	20	40

(a) What is the probability that a visitor to the web site provides contact information?
(b) If a visitor provides contact information, what is the probability that the visitor viewed four or more pages?

MIND-EXPANDING EXERCISES

2-123.   The alignment between the magnetic tape and head in a magnetic tape storage system affects the performance of the system. Suppose that 10% of the read operations are degraded by skewed alignments, 5% by off-center alignments, 1% by both skewness and off-center, and the remaining read operations are properly aligned. The probability of a read error is 0.01 from a skewed alignment, 0.02 from an off-center alignment, 0.06 from both conditions, and 0.001 from a proper alignment. What is the probability of a read error.

2-124.   Suppose that a lot of washers is large enough that it can be assumed that the sampling is done with replacement. Assume that 60% of the washers exceed the target thickness.

(a) What is the minimum number of washers that need to be selected so that the probability that all the washers are thinner than the target is less than 0.10?

(b) What is the minimum number of washers that need to be selected so that the probability that one or more washers are thicker than the target is at least 0.90?

2-125.   A biotechnology manufacturing firm can produce diagnostic test kits at a cost of $20. Each kit for which there is a demand in the week of production can be sold for $100. However, the half-life of components in the kit requires the kit to be scrapped if it is not sold in the week of production. The cost of scrapping the kit is $5. The weekly demand is summarized as follows:

weekly demand

Number of units	0	50	100	200
Probability of demand	0.05	0.4	0.3	0.25

How many kits should be produced each week to maximize the mean earnings of the firm?

2-126.   Assume the following characteristics of the inspection process in Exercise 2-107. If an operator checks a bolt, the probability that an incorrectly torqued bolt is identified is 0.95. If a checked bolt is correctly torqued, the operator's conclusion is always correct. What is the probability that at least one bolt in the sample of four is identified as being incorrectly torqued?

2-127.   If the events A and B are independent, show that A' and B' are independent.

2-128.   Suppose that a table of part counts is generalized as follows:

		conforms	
		yes	no
supplier	1	ka	kb
	2	a	b

where a, b, and k are positive integers. Let A denote the event that a part is from supplier 1 and let B denote the event that a part conforms to specifications. Show that A and B are independent events.

This exercise illustrates the result that whenever the rows of a table (with r rows and c columns) are proportional, an event defined by a row category and an event defined by a column category are independent.

IMPORTANT TERMS AND CONCEPTS

In the E-book, click on any term or concept below to go to that subject.

Addition rule
Axioms of probability
Bayes' theorem
Conditional probability
Equally likely outcomes

Event
Independence
Multiplication rule
Mutually exclusive events
Outcome
Random experiment

Random variables discrete and continuous
Sample spaces—discrete and continuous
Total probability rule
With or without replacement

CD MATERIAL

Permutation
Combination

Discrete Random Variables and Probability Distributions

CHAPTER OUTLINE

LEARNING OBJECTIVES

After careful study of this chapter you should be able to do the following:

1. Determine probabilities from probability mass functions and the reverse

2. Determine probabilities from cumulative distribution functions and cumulative distribution functions from probability mass functions, and the reverse

3. Calculate means and variances for discrete random variables

4. Understand the assumptions for each of the discrete probability distributions presented

5. Select an appropriate discrete probability distribution to calculate probabilities in specific applications

6. Calculate probabilities, determine means and variances for each of the discrete probability distributions presented

Answers for most odd numbered exercises are at the end of the book. Answers to exercises whose numbers are surrounded by a box can be accessed in the e-Text by clicking on the box. Complete worked solutions to certain exercises are also available in the e-Text. These are indicated in the Answers to Selected Exercises section by a box around the exercise number. Exercises are also available for some of the text sections that appear on CD only. These exercises may be found within the e-Text immediately following the section they accompany.

3-1 DISCRETE RANDOM VARIABLES

Many physical systems can be modeled by the same or similar random experiments and random variables. The distribution of the random variables involved in each of these common systems can be analyzed, and the results of that analysis can be used in different applications and examples. In this chapter, we present the analysis of several random experiments and **discrete random variables** that frequently arise in applications. We often omit a discussion of the underlying sample space of the random experiment and directly describe the distribution of a particular random variable.

EXAMPLE 3-1   A voice communication system for a business contains 48 external lines. At a particular time, the system is observed, and some of the lines are being used. Let the random variable X denote the number of lines in use. Then, X can assume any of the integer values 0 through 48. When the system is observed, if 10 lines are in use, $x = 10$.

EXAMPLE 3-2   In a semiconductor manufacturing process, two wafers from a lot are tested. Each wafer is classified as *pass* or *fail*. Assume that the probability that a wafer passes the test is 0.8 and that wafers are independent. The sample space for the experiment and associated probabilities are shown in Table 3-1. For example, because of the independence, the probability of the outcome that the first wafer tested passes and the second wafer tested fails, denoted as *pf*, is

$$P(pf) = 0.8(0.2) = 0.16$$

The random variable X is defined to be equal to the number of wafers that pass. The last column of the table shows the values of X that are assigned to each outcome in the experiment.

EXAMPLE 3-3   Define the random variable X to be the number of contamination particles on a wafer in semiconductor manufacturing. Although wafers possess a number of characteristics, the random variable X summarizes the wafer only in terms of the number of particles.

The possible values of X are integers from zero up to some large value that represents the maximum number of particles that can be found on one of the wafers. If this maximum number is very large, we might simply assume that the range of X is the set of integers from zero to infinity.

Note that more than one random variable can be defined on a sample space. In Example 3-3, we might define the random variable Y to be the number of chips from a wafer that fail the final test.

Table 3-1   Wafer Tests

Outcome		Probability	x
Wafer 1	Wafer 2		
Pass	Pass	0.64	2
Fail	Pass	0.16	1
Pass	Fail	0.16	1
Fail	Fail	0.04	0

EXERCISES FOR SECTION 3-1

For each of the following exercises, determine the range (possible values) of the random variable.

3-1. The random variable is the number of nonconforming solder connections on a printed circuit board with 1000 connections.

3-2. In a voice communication system with 50 lines, the random variable is the number of lines in use at a particular time.

3-3. An electronic scale that displays weights to the nearest pound is used to weigh packages. The display shows only five digits. Any weight greater than the display can indicate is shown as 99999. The random variable is the displayed weight.

3-4. A batch of 500 machined parts contains 10 that do not conform to customer requirements. The random variable is the number of parts in a sample of 5 parts that do not conform to customer requirements.

3-5. A batch of 500 machined parts contains 10 that do not conform to customer requirements. Parts are selected successively, without replacement, until a nonconforming part is obtained. The random variable is the number of parts selected.

3-6. The random variable is the moisture content of a lot of raw material, measured to the nearest percentage point.

3-7. The random variable is the number of surface flaws in a large coil of galvanized steel.

3-8. The random variable is the number of computer clock cycles required to complete a selected arithmetic calculation.

3-9. An order for an automobile can select the base model or add any number of 15 options. The random variable is the number of options selected in an order.

3-10. Wood paneling can be ordered in thicknesses of 1/8, 1/4, or 3/8 inch. The random variable is the total thickness of paneling in two orders.

3-11. A group of 10,000 people are tested for a gene called Ifi202 that has been found to increase the risk for lupus. The random variable is the number of people who carry the gene.

3-12. A software program has 5000 lines of code. The random variable is the number of lines with a fatal error.

3-2 PROBABILITY DISTRIBUTIONS AND PROBABILITY MASS FUNCTIONS

Random variables are so important in random experiments that sometimes we essentially ignore the original sample space of the experiment and focus on the probability distribution of the random variable. For example, in Example 3-1, our analysis might focus exclusively on the integers $\{0, 1, \ldots, 48\}$ in the range of X. In Example 3-2, we might summarize the random experiment in terms of the three possible values of X, namely $\{0, 1, 2\}$. In this manner, a random variable can simplify the description and analysis of a random experiment.

The **probability distribution** of a random variable X is a description of the probabilities associated with the possible values of X. For a discrete random variable, the distribution is often specified by just a list of the possible values along with the probability of each. In some cases, it is convenient to express the probability in terms of a formula.

EXAMPLE 3-4 There is a chance that a bit transmitted through a digital transmission channel is received in error. Let X equal the number of bits in error in the next four bits transmitted. The possible values for X are $\{0, 1, 2, 3, 4\}$. Based on a model for the errors that is presented in the following section, probabilities for these values will be determined. Suppose that the probabilities are

$$P(X = 0) = 0.6561 \qquad P(X = 1) = 0.2916 \qquad P(X = 2) = 0.0486$$

$$P(X = 3) = 0.0036 \qquad P(X = 4) = 0.0001$$

The probability distribution of X is specified by the possible values along with the probability of each. A graphical description of the probability distribution of X is shown in Fig. 3-1.

Suppose a loading on a long, thin beam places mass only at discrete points. See Fig. 3-2. The loading can be described by a function that specifies the mass at each of the discrete points. Similarly, for a discrete random variable X, its distribution can be described by a function that specifies the probability at each of the possible discrete values for X.

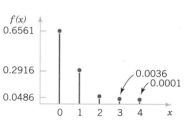

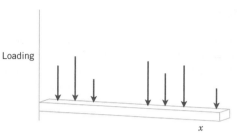

Figure 3-1   Probability distribution for bits in error.

Figure 3-2   Loadings at discrete points on a long, thin beam.

Definition

> For a discrete random variable X with possible values $x_1, x_2, \ldots, x_n$, a **probability mass function** is a function such that
>
> (1) $f(x_i) \geq 0$
>
> (2) $\displaystyle\sum_{i=1}^{n} f(x_i) = 1$
>
> (3) $f(x_i) = P(X = x_i)$ (3-1)

For example, in Example 3-4, $f(0) = 0.6561, f(1) = 0.2916, f(2) = 0.0486, f(3) = 0.0036$, and $f(4) = 0.0001$. Check that the sum of the probabilities in Example 3-4 is 1.

EXAMPLE 3-5

Let the random variable X denote the number of semiconductor wafers that need to be analyzed in order to detect a large particle of contamination. Assume that the probability that a wafer contains a large particle is 0.01 and that the wafers are independent. Determine the probability distribution of X.

Let p denote a wafer in which a large particle is present, and let a denote a wafer in which it is absent. The sample space of the experiment is infinite, and it can be represented as all possible sequences that start with a string of a's and end with p. That is,

$$s = \{p, ap, aap, aaap, aaaap, aaaaap, \text{ and so forth}\}$$

Consider a few special cases. We have $P(X = 1) = P(p) = 0.01$. Also, using the independence assumption

$$P(X = 2) = P(ap) = 0.99(0.01) = 0.0099$$

A general formula is

$$P(X = x) = \underbrace{P(aa \ldots ap)}_{(x-1)a\text{'s}} = 0.99^{x-1}(0.01), \qquad \text{for } x = 1, 2, 3, \ldots$$

Describing the probabilities associated with X in terms of this formula is the simplest method of describing the distribution of X in this example. Clearly $f(x) \geq 0$. The fact that the sum of the probabilities is 1 is left as an exercise. This is an example of a geometric random variable, and details are provided later in this chapter.

EXERCISES FOR SECTION 3-2

3-13. The sample space of a random experiment is $\{a, b, c, d, e, f\}$, and each outcome is equally likely. A random variable is defined as follows:

outcome	a	b	c	d	e	f
x	0	0	1.5	1.5	2	3

Determine the probability mass function of X.

3-14. Use the probability mass function in Exercise 3-13 to determine the following probabilities:

(a) $P(X = 1.5)$ (b) $P(0.5 < X < 2.7)$
(c) $P(X > 3)$ (d) $P(0 \leq X < 2)$
(e) $P(X = 0 \text{ or } X = 2)$

Verify that the following functions are probability mass functions, and determine the requested probabilities.

3-15.

x	-2	-1	0	1	2
$f(x)$	1/8	2/8	2/8	2/8	1/8

(a) $P(X \leq 2)$ (b) $P(X > -2)$
(c) $P(-1 \leq X \leq 1)$ (d) $P(X \leq -1 \text{ or } X = 2)$

3-16. $f(x) = (8/7)(1/2)^x$, $x = 1, 2, 3$

(a) $P(X \leq 1)$ (b) $P(X > 1)$
(c) $P(2 < X < 6)$ (d) $P(X \leq 1 \text{ or } X > 1)$

3-17. $f(x) = \dfrac{2x + 1}{25}$, $x = 0, 1, 2, 3, 4$

(a) $P(X = 4)$ (b) $P(X \leq 1)$
(c) $P(2 \leq X < 4)$ (d) $P(X > -10)$

3-18. $f(x) = (3/4)(1/4)^x$, $x = 0, 1, 2, \ldots$

(a) $P(X = 2)$ (b) $P(X \leq 2)$
(c) $P(X > 2)$ (d) $P(X \geq 1)$

3-19. Marketing estimates that a new instrument for the analysis of soil samples will be very successful, moderately successful, or unsuccessful, with probabilities 0.3, 0.6, and 0.1, respectively. The yearly revenue associated with a very successful, moderately successful, or unsuccessful product is $10 million, $5 million, and $1 million, respectively. Let the random variable X denote the yearly revenue of the product. Determine the probability mass function of X.

3-20. A disk drive manufacturer estimates that in five years a storage device with 1 terabyte of capacity will sell with probability 0.5, a storage device with 500 gigabytes capacity will sell with a probability 0.3, and a storage device with 100 gigabytes capacity will sell with probability 0.2. The revenue associated with the sales in that year are estimated to be $50 million, $25 million, and $10 million, respectively. Let X be the revenue of storage devices during that year. Determine the probability mass function of X.

3-21. An optical inspection system is to distinguish among different part types. The probability of a correct classification of any part is 0.98. Suppose that three parts are inspected and that the classifications are independent. Let the random variable X denote the number of parts that are correctly classified. Determine the probability mass function of X.

3-22. In a semiconductor manufacturing process, three wafers from a lot are tested. Each wafer is classified as *pass* or *fail*. Assume that the probability that a wafer passes the test is 0.8 and that wafers are independent. Determine the probability mass function of the number of wafers from a lot that pass the test.

3-23. The distributor of a machine for cytogenics has developed a new model. The company estimates that when it is introduced into the market, it will be very successful with a probability 0.6, moderately successful with a probability 0.3, and not successful with probability 0.1. The estimated yearly profit associated with the model being very successful is $15 million and being moderately successful is $5 million; not successful would result in a loss of $500,000. Let X be the yearly profit of the new model. Determine the probability mass function of X.

3-24. An assembly consists of two mechanical components. Suppose that the probabilities that the first and second components meet specifications are 0.95 and 0.98. Assume that the components are independent. Determine the probability mass function of the number of components in the assembly that meet specifications.

3-25. An assembly consists of three mechanical components. Suppose that the probabilities that the first, second, and third components meet specifications are 0.95, 0.98, and 0.99. Assume that the components are independent. Determine the probability mass function of the number of components in the assembly that meet specifications.

3-3 CUMULATIVE DISTRIBUTION FUNCTIONS

EXAMPLE 3-6

In Example 3-4, we might be interested in the probability of three or fewer bits being in error. This question can be expressed as $P(X \leq 3)$.

The event that $\{X \leq 3\}$ is the union of the events $\{X = 0\}$, $\{X = 1\}$, $\{X = 2\}$, and

$\{X = 3\}$. Clearly, these three events are mutually exclusive. Therefore,

$$P(X \leq 3) = P(X = 0) + P(X = 1) + P(X = 2) + P(X = 3)$$
$$= 0.6561 + 0.2916 + 0.0486 + 0.0036 = 0.9999$$

This approach can also be used to determine

$$P(X = 3) = P(X \leq 3) - P(X \leq 2) = 0.0036$$

Example 3-6 shows that it is sometimes useful to be able to provide **cumulative probabilities** such as $P(X \leq x)$ and that such probabilities can be used to find the probability mass function of a random variable. Therefore, using cumulative probabilities is an alternate method of describing the probability distribution of a random variable.

In general, for any discrete random variable with possible values $x_1, x_2, \ldots, x_n$, the events $\{X = x_1\}$, $\{X = x_2\}, \ldots, \{X = x_n\}$ are mutually exclusive. Therefore, $P(X \leq x) = \sum_{x_i \leq x} f(x_i)$.

Definition

> The **cumulative distribution function** of a discrete random variable X, denoted as $F(x)$, is
>
> $$F(x) = P(X \leq x) = \sum_{x_i \leq x} f(x_i)$$
>
> For a discrete random variable X, $F(x)$ satisfies the following properties.
>
> (1) $F(x) = P(X \leq x) = \sum_{x_i \leq x} f(x_i)$
> (2) $0 \leq F(x) \leq 1$
> (3) If $x \leq y$, then $F(x) \leq F(y)$ (3-2)

Like a probability mass function, a cumulative distribution function provides probabilities. Notice that even if the random variable X can only assume integer values, the cumulative distribution function can be defined at noninteger values. In Example 3-6, $F(1.5) = P(X \leq 1.5) = P\{X = 0\} + P(X = 1) = 0.6561 + 0.2916 = 0.9477$. Properties (1) and (2) of a cumulative distribution function follow from the definition. Property (3) follows from the fact that if $x \leq y$, the event that $\{X \leq x\}$ is contained in the event $\{X \leq y\}$.

The next example shows how the cumulative distribution function can be used to determine the probability mass function of a discrete random variable.

EXAMPLE 3-7

Determine the probability mass function of X from the following cumulative distribution function:

$$F(x) = \begin{cases} 0 & x < -2 \\ 0.2 & -2 \leq x < 0 \\ 0.7 & 0 \leq x < 2 \\ 1 & 2 \leq x \end{cases}$$

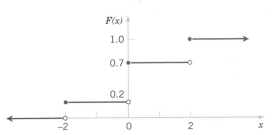

Figure 3-3   Cumulative distribution function for Example 3-7.

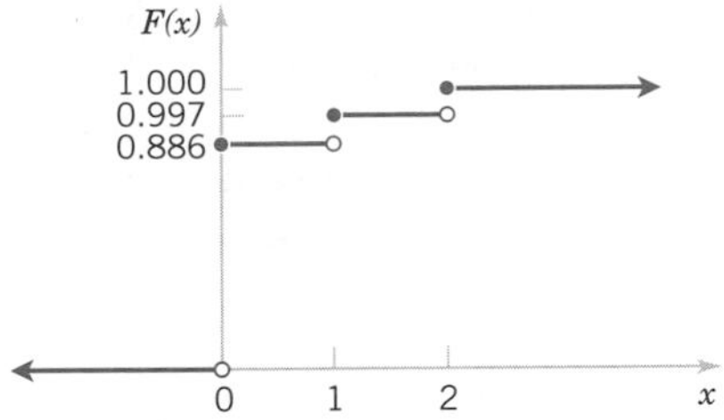

Figure 3-4   Cumulative distribution function for Example 3-8.

Figure 3-3 displays a plot of $F(x)$. From the plot, the only points that receive nonzero probability are -2, 0, and 2. The probability mass function at each point is the change in the cumulative distribution function at the point. Therefore,

$$f(-2) = 0.2 - 0 = 0.2 \quad f(0) = 0.7 - 0.2 = 0.5 \quad f(2) = 1.0 - 0.7 = 0.3$$

EXAMPLE 3-8   Suppose that a day's production of 850 manufactured parts contains 50 parts that do not conform to customer requirements. Two parts are selected at random, without replacement, from the batch. Let the random variable X equal the number of nonconforming parts in the sample. What is the cumulative distribution function of X?

The question can be answered by first finding the probability mass function of X.

$$P(X = 0) = \frac{800}{850} \cdot \frac{799}{849} = 0.886$$

$$P(X = 1) = 2 \cdot \frac{800}{850} \cdot \frac{50}{849} = 0.111$$

$$P(X = 2) = \frac{50}{850} \cdot \frac{49}{849} = 0.003$$

Therefore,

$$F(0) = P(X \le 0) = 0.886$$

$$F(1) = P(X \le 1) = 0.886 + 0.111 = 0.997$$

$$F(2) = P(X \le 2) = 1$$

The cumulative distribution function for this example is graphed in Fig. 3-4. Note that $F(x)$ is defined for all x from $-\infty < x < \infty$ and not only for 0, 1, and 2.

EXERCISES FOR SECTION 3-3

3-26.   Determine the cumulative distribution function of the random variable in Exercise 3-13.

3-27.   Determine the cumulative distribution function for the random variable in Exercise 3-15; also determine the following probabilities:
(a) $P(X \le 1.25)$ (b) $P(X \le 2.2)$

(c) $P(-1.1 < X \le 1)$ (d) $P(X > 0)$

3-28.   Determine the cumulative distribution function for the random variable in Exercise 3-17; also determine the following probabilities:
(a) $P(X < 1.5)$ (b) $P(X \le 3)$
(c) $P(X > 2)$ (d) $P(1 < X \le 2)$

3-29. Determine the cumulative distribution function for the random variable in Exercise 3-19.

3-30. Determine the cumulative distribution function for the random variable in Exercise 3-20.

3-31. Determine the cumulative distribution function for the random variable in Exercise 3-22.

3-32. Determine the cumulative distribution function for the variable in Exercise 3-23.

Verify that the following functions are cumulative distribution functions, and determine the probability mass function and the requested probabilities.

3-33.
$$F(x) = \begin{cases} 0 & x < 1 \\ 0.5 & 1 \le x < 3 \\ 1 & 3 \le x \end{cases}$$

(a) $P(X \le 3)$ (b) $P(X \le 2)$
(c) $P(1 \le X \le 2)$ (d) $P(X > 2)$

3-34. Errors in an experimental transmission channel are found when the transmission is checked by a certifier that detects missing pulses. The number of errors found in an eight-bit byte is a random variable with the following distribution:

$$F(x) = \begin{cases} 0 & x < 1 \\ 0.7 & 1 \le x < 4 \\ 0.9 & 4 \le x < 7 \\ 1 & 7 \le x \end{cases}$$

Determine each of the following probabilities:
(a) $P(X \le 4)$ (b) $P(X > 7)$
(c) $P(X \le 5)$ (d) $P(X > 4)$
(e) $P(X \le 2)$

3-35.
$$F(x) = \begin{cases} 0 & x < -10 \\ 0.25 & -10 \le x < 30 \\ 0.75 & 30 \le x < 50 \\ 1 & 50 \le x \end{cases}$$

(a) $P(X \le 50)$ (b) $P(X \le 40)$
(c) $P(40 \le X \le 60)$ (d) $P(X < 0)$
(e) $P(0 \le X < 10)$ (f) $P(-10 < X < 10)$

3-36. The thickness of wood paneling (in inches) that a customer orders is a random variable with the following cumulative distribution function:

$$F(x) = \begin{cases} 0 & x < 1/8 \\ 0.2 & 1/8 \le x < 1/4 \\ 0.9 & 1/4 \le x < 3/8 \\ 1 & 3/8 \le x \end{cases}$$

Determine the following probabilities:
(a) $P(X \le 1/18)$ (b) $P(X \le 1/4)$
(c) $P(X \le 5/16)$ (d) $P(X > 1/4)$
(e) $P(X \le 1/2)$

3-4 MEAN AND VARIANCE OF A DISCRETE RANDOM VARIABLE

Two numbers are often used to summarize a probability distribution for a random variable X. The mean is a measure of the center or middle of the probability distribution, and the variance is a measure of the dispersion, or variability in the distribution. These two measures do not uniquely identify a probability distribution. That is, two different distributions can have the same mean and variance. Still, these measures are simple, useful summaries of the probability distribution of X.

Definition

The **mean** or **expected value** of the discrete random variable X, denoted as μ or $E(X)$, is

$$\mu = E(X) = \sum_x xf(x) \tag{3-3}$$

The **variance** of X, denoted as σ^2 or $V(X)$, is

$$\sigma^2 = V(X) = E(X - \mu)^2 = \sum_x (x - \mu)^2 f(x) = \sum_x x^2 f(x) - \mu^2$$

The **standard deviation** of X is $\sigma = \sqrt{\sigma^2}$.

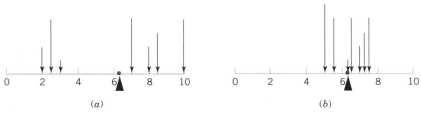

Figure 3-5   A probability distribution can be viewed as a loading with the mean equal to the balance point. Parts (a) and (b) illustrate equal means, but Part (a) illustrates a larger variance.

The mean of a discrete random variable X is a weighted average of the possible values of X, with weights equal to the probabilities. If $f(x)$ is the probability mass function of a loading on a long, thin beam, $E(X)$ is the point at which the beam balances. Consequently, $E(X)$ describes the "center" of the distribution of X in a manner similar to the balance point of a loading. See Fig. 3-5.

The variance of a random variable X is a measure of dispersion or scatter in the possible values for X. The variance of X uses weight $f(x)$ as the multiplier of each possible squared deviation $(x - \mu)^2$. Figure 3-5 illustrates probability distributions with equal means but different variances. Properties of summations and the definition of μ can be used to show the equality of the formulas for variance.

$$V(X) = \sum_x (x - \mu)^2 f(x) = \sum_x x^2 f(x) - 2\mu \sum_x x f(x) + \mu^2 \sum_x f(x)$$

$$= \sum_x x^2 f(x) - 2\mu^2 + \mu^2 = \boxed{\sum_x x^2 f(x) - \mu^2}$$

Either formula for $V(x)$ can be used. Figure 3-6 illustrates that two probability distributions can differ even though they have identical means and variances.

EXAMPLE 3-9   In Example 3-4, there is a chance that a bit transmitted through a digital transmission channel is received in error. Let X equal the number of bits in error in the next four bits transmitted. The possible values for X are $\{0, 1, 2, 3, 4\}$. Based on a model for the errors that is presented in the following section, probabilities for these values will be determined. Suppose that the probabilities are

$$P(X = 0) = 0.6561 \qquad P(X = 2) = 0.0486 \qquad P(X = 4) = 0.0001$$
$$P(X = 1) = 0.2916 \qquad P(X = 3) = 0.0036$$

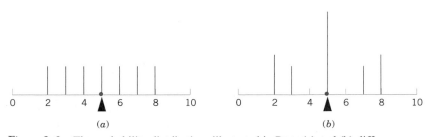

Figure 3-6   The probability distributions illustrated in Parts (a) and (b) differ even though they have equal means and equal variances.

Now

$$\mu = E(X) = 0f(0) + 1f(1) + 2f(2) + 3f(3) + 4f(4)$$
$$= 0(0.6561) + 1(0.2916) + 2(0.0486) + 3(0.0036) + 4(0.0001)$$
$$= 0.4$$

Although X never assumes the value 0.4, the weighted average of the possible values is 0.4. To calculate $V(X)$, a table is convenient.

x	$x - 0.4$	$(x - 0.4)^2$	$f(x)$	$f(x)(x - 0.4)^2$
0	−0.4	0.16	0.6561	0.104976
1	0.6	0.36	0.2916	0.104976
2	1.6	2.56	0.0486	0.124416
3	2.6	6.76	0.0036	0.024336
4	3.6	12.96	0.0001	0.001296

0.36

$$V(X) = \sigma^2 = \sum_{i=1}^{5} f(x_i)(x_i - 0.4)^2 = 0.36$$

The alternative formula for variance could also be used to obtain the same result.

EXAMPLE 3-10 Two new product designs are to be compared on the basis of revenue potential. Marketing feels that the revenue from design A can be predicted quite accurately to be \$3 million. The revenue potential of design B is more difficult to assess. Marketing concludes that there is a probability of 0.3 that the revenue from design B will be \$7 million, but there is a 0.7 probability that the revenue will be only \$2 million. Which design do you prefer?

Let X denote the revenue from design A. Because there is no uncertainty in the revenue from design A, we can model the distribution of the random variable X as \$3 million with probability 1. Therefore, $E(X) = \$3$ million.

Let Y denote the revenue from design B. The expected value of Y in millions of dollars is

$$E(Y) = \$7(0.3) + \$2(0.7) = \$3.5$$

Because $E(Y)$ exceeds $E(X)$, we might prefer design B. However, the variability of the result from design B is larger. That is,

$$\sigma^2 = (7 - 3.5)^2(0.3) + (2 - 3.5)^2(0.7)$$
$$= 5.25 \text{ millions of dollars squared}$$

Because the units of the variables in this example are millions of dollars, and because the variance of a random variable squares the deviations from the mean, the units of σ^2 are millions of dollars squared. These units make interpretation difficult.

Because the units of standard deviation are the same as the units of the random variable, the standard deviation σ is easier to interpret. In this example, we can summarize our results as "the average deviation of Y from its mean is \$2.29 million."

EXAMPLE 3-11   The number of messages sent per hour over a computer network has the following distribution:

x = number of messages	10	11	12	13	14	15
$f(x)$	0.08	0.15	0.30	0.20	0.20	0.07

Determine the mean and standard deviation of the number of messages sent per hour.

$$E(X) = 10(0.08) + 11(0.15) + \cdots + 15(0.07) = 12.5$$

$$V(X) = 10^2(0.08) + 11^2(0.15) + \cdots + 15^2(0.07) - 12.5^2 = 1.85$$

$$\sigma = \sqrt{V(X)} = \sqrt{1.85} = 1.36$$

The variance of a random variable X can be considered to be the expected value of a specific function of X, namely, $h(X) = (X - \mu)^2$. In general, the expected value of any function $h(X)$ of a discrete random variable is defined in a similar manner.

Expected Value of a Function of a Discrete Random Variable

If X is a discrete random variable with probability mass function $f(x)$,

$$E[h(X)] = \sum_x h(x)f(x) \tag{3-4}$$

EXAMPLE 3-12   In Example 3-9, X is the number of bits in error in the next four bits transmitted. What is the expected value of the square of the number of bits in error? Now, $h(X) = X^2$. Therefore,

$$E[h(X)] = 0^2 \times 0.6561 + 1^2 \times 0.2916 + 2^2 \times 0.0486$$
$$+ 3^2 \times 0.0036 + 4^2 \times 0.0001 = 0.52$$

In the previous example, the expected value of X^2 does not equal $E(X)$ squared. However, in the special case that $h(X) = aX + b$ for any constants a and b, $E[h(X)] = aE(X) + b$. This can be shown from the properties of sums in the definition in Equation 3-4.

EXERCISES FOR SECTION 3-4

3-37.   If the range of X is the set $\{0, 1, 2, 3, 4\}$ and $P(X = x) = 0.2$ determine the mean and variance of the random variable.

3-38.   Determine the mean and variance of the random variable in Exercise 3-13.

3-39.   Determine the mean and variance of the random variable in Exercise 3-15.

3-40.   Determine the mean and variance of the random variable in Exercise 3-17.

3-41.   Determine the mean and variance of the random variable in Exercise 3-19.

3-42.   Determine the mean and variance of the random variable in Exercise 3-20.

3-43.   Determine the mean and variance of the random variable in Exercise 3-22.

3-44.   Determine the mean and variance of the random variable in Exercise 3-23.

3-45.   The range of the random variable X is $[0, 1, 2, 3, x]$, where x is unknown. If each value is equally likely and the mean of X is 6, determine x.

3-5 DISCRETE UNIFORM DISTRIBUTION

The simplest discrete random variable is one that assumes only a finite number of possible values, each with equal probability. A random variable X that assumes each of the values $x_1, x_2, \ldots, x_n$, with equal probability $1/n$, is frequently of interest.

Definition

> A random variable X has a **discrete uniform distribution** if each of the n values in its range, say, $x_1, x_2, \ldots, x_n$, has equal probability. Then,
>
> $$f(x_i) = 1/n \tag{3-5}$$

EXAMPLE 3-13

The first digit of a part's serial number is equally likely to be any one of the digits 0 through 9. If one part is selected from a large batch and X is the first digit of the serial number, X has a discrete uniform distribution with probability 0.1 for each value in $R = \{0, 1, 2, \ldots, 9\}$. That is,

$$f(x) = 0.1$$

for each value in R. The probability mass function of X is shown in Fig. 3-7.

Suppose the range of the discrete random variable X is the consecutive integers a, $a + 1, a + 2, \ldots, b$, for $a \le b$. The range of X contains $b - a + 1$ values each with probability $1/(b - a + 1)$. Now,

$$\mu = \sum_{k=a}^{b} k\left(\frac{1}{b - a + 1}\right)$$

The algebraic identity $\displaystyle\sum_{k=a}^{b} k = \frac{b(b + 1) - (a - 1)a}{2}$ can be used to simplify the result to $\mu = (b + a)/2$. The derivation of the variance is left as an exercise.

> Suppose X is a discrete uniform random variable on the consecutive integers $a, a + 1, a + 2, \ldots, b$, for $a \le b$. The mean of X is
>
> $$\mu = E(X) = \frac{b + a}{2}$$
>
> The variance of X is
>
> $$\sigma^2 = \frac{(b - a + 1)^2 - 1}{12} \tag{3-6}$$

Figure 3-7   Probability mass function for a discrete uniform random variable.

EXAMPLE 3-14   As in Example 3-1, let the random variable X denote the number of the 48 voice lines that are in use at a particular time. Assume that X is a discrete uniform random variable with a range of 0 to 48. Then,

$$E(X) = (48 + 0)/2 = 24$$

and

$$\sigma = \{[(48 - 0 + 1)^2 - 1]/12\}^{1/2} = 14.14$$

Equation 3-6 is more useful than it might first appear. If all the values in the range of a random variable X are multiplied by a constant (without changing any probabilities), the mean and standard deviation of X are multiplied by the constant. You are asked to verify this result in an exercise. Because the variance of a random variable is the square of the standard deviation, the variance of X is multiplied by the constant squared. More general results of this type are discussed in Chapter 5.

EXAMPLE 3-15   Let the random variable Y denote the proportion of the 48 voice lines that are in use at a particular time, and X denotes the number of lines that are in use at a particular time. Then, $Y = X/48$. Therefore,

$$E(Y) = E(X)/48 = 0.5$$

and

$$V(Y) = V(X)/48^2 = 0.087$$

EXERCISES FOR SECTION 3-5

3-46.   Let the random variable X have a discrete uniform distribution on the integers $0 \leq x \leq 100$. Determine the mean and variance of X.

3-47.   Let the random variable X have a discrete uniform distribution on the integers $1 \leq x \leq 3$. Determine the mean and variance of X.

3-48.   Let the random variable X be equally likely to assume any of the values 1/8, 1/4, or 3/8. Determine the mean and variance of X.

3-49.   Thickness measurements of a coating process are made to the nearest hundredth of a millimeter. The thickness measurements are uniformly distributed with values 0.15, 0.16, 0.17, 0.18, and 0.19. Determine the mean and variance of the coating thickness for this process.

3-50.   Product codes of 2, 3, or 4 letters are equally likely. What is the mean and standard deviation of the number of letters in 100 codes?

3-51.   The lengths of plate glass parts are measured to the nearest tenth of a millimeter. The lengths are uniformly distributed, with values at every tenth of a millimeter starting at 590.0 and continuing through 590.9. Determine the mean and variance of lengths.

3-52.   Suppose that X has a discrete uniform distribution on the integers 0 through 9. Determine the mean, variance, and standard deviation of the random variable $Y = 5X$ and compare to the corresponding results for X.

3-53.   Show that for a discrete uniform random variable X, if each of the values in the range of X is multiplied by the constant c, the effect is to multiply the mean of X by c and the variance of X by c^2. That is, show that $E(cX) = cE(X)$ and $V(cX) = c^2V(X)$.

3-54.   The probability of an operator entering alphanumeric data incorrectly into a field in a database is equally likely. The random variable X is the number of fields on a data entry form with an error. The data entry form has 28 fields. Is X a discrete uniform random variable? Why or why not?

3-6   BINOMIAL DISTRIBUTION

Consider the following random experiments and random variables:

1. Flip a coin 10 times. Let X = number of heads obtained.

2. A worn machine tool produces 1% defective parts. Let X = number of defective parts in the next 25 parts produced.

3. Each sample of air has a 10% chance of containing a particular rare molecule. Let X = the number of air samples that contain the rare molecule in the next 18 samples analyzed.

4. Of all bits transmitted through a digital transmission channel, 10% are received in error. Let X = the number of bits in error in the next five bits transmitted.

5. A multiple choice test contains 10 questions, each with four choices, and you guess at each question. Let X = the number of questions answered correctly.

6. In the next 20 births at a hospital, let X = the number of female births.

7. Of all patients suffering a particular illness, 35% experience improvement from a particular medication. In the next 100 patients administered the medication, let X = the number of patients who experience improvement.

These examples illustrate that a general probability model that includes these experiments as particular cases would be very useful.

Each of these random experiments can be thought of as consisting of a series of repeated, random trials: 10 flips of the coin in experiment 1, the production of 25 parts in experiment 2, and so forth. The random variable in each case is a count of the number of trials that meet a specified criterion. The outcome from each trial either meets the criterion that X counts or it does not; consequently, each trial can be summarized as resulting in either a success or a failure. For example, in the multiple choice experiment, for each question, only the choice that is correct is considered a success. Choosing any one of the three incorrect choices results in the trial being summarized as a failure.

The terms *success* and *failure* are just labels. We can just as well use A and B or 0 or 1. Unfortunately, the usual labels can sometimes be misleading. In experiment 2, because X counts defective parts, the production of a defective part is called a success.

A trial with only two possible outcomes is used so frequently as a building block of a random experiment that it is called a **Bernoulli trial.** It is usually assumed that the trials that constitute the random experiment are **independent.** This implies that the outcome from one trial has no effect on the outcome to be obtained from any other trial. Furthermore, it is often reasonable to assume that the **probability of a success in each trial is constant.** In the multiple choice experiment, if the test taker has no knowledge of the material and just guesses at each question, we might assume that the probability of a correct answer is 1/4 *for each question.*

Factorial notation is used in this section. Recall that $n!$ denotes the product of the integers less than or equal to n:

$$n! = n(n-1)(n-2) \cdots (2)(1)$$

For example,

$$5! = (5)(4)(3)(2)(1) = 120 \qquad 1! = 1$$

and by definition $0! = 1$. We also use the combinatorial notation

$$\binom{n}{x} = \frac{n!}{x!\,(n-x)!}$$

For example,

$$\binom{5}{2} = \frac{5!}{2!\,3!} = \frac{120}{2 \cdot 6} = 10$$

See Section 2-1.4, CD material for Chapter 2, for further comments.

EXAMPLE 3-16

The chance that a bit transmitted through a digital transmission channel is received in error is 0.1. Also, assume that the transmission trials are independent. Let $X =$ the number of bits in error in the next four bits transmitted. Determine $P(X = 2)$.

Let the letter E denote a bit in error, and let the letter O denote that the bit is okay, that is, received without error. We can represent the outcomes of this experiment as a list of four letters that indicate the bits that are in error and those that are okay. For example, the outcome *OEOE* indicates that the second and fourth bits are in error and the other two bits are okay. The corresponding values for x are

Outcome	x	Outcome	x
OOOO	0	*EOOO*	1
OOOE	1	*EOOE*	2
OOEO	1	*EOEO*	2
OOEE	2	*EOEE*	3
OEOO	1	*EEOO*	2
OEOE	2	*EEOE*	3
OEEO	2	*EEEO*	3
OEEE	3	*EEEE*	4

The event that $X = 2$ consists of the six outcomes:

$$\{EEOO, EOEO, EOOE, OEEO, OEOE, OOEE\}$$

Using the assumption that the trials are independent, the probability of $\{EEOO\}$ is

$$P(EEOO) = P(E)P(E)P(O)P(O) = (0.1)^2(0.9)^2 = 0.0081$$

Also, any one of the six mutually exclusive outcomes for which $X = 2$ has the same probability of occurring. Therefore,

$$P(X = 2) = 6(0.0081) = 0.0486$$

In general,

$$P(X = x) = (\text{number of outcomes that result in } x \text{ errors}) \text{ times } (0.1)^x(0.9)^{4-x}$$

To complete a general probability formula, only an expression for the number of outcomes that contain x errors is needed. An outcome that contains x errors can be constructed by partitioning the four trials (letters) in the outcome into two groups. One group is of size x and contains the errors, and the other group is of size $n - x$ and consists of the trials that are okay. The number of ways of partitioning four objects into two groups, one of which is of size x, is

$\binom{4}{x} = \dfrac{4!}{x!(4 - x)!}$. Therefore, in this example

$$P(X = x) = \binom{4}{x}(0.1)^x(0.9)^{4-x}$$

Notice that $\binom{4}{2} = 4!/[2!\,2!] = 6$, as found above. The probability mass function of X was shown in Example 3-4 and Fig. 3-1.

The previous example motivates the following result.

Definition

> A random experiment consists of n Bernoulli trials such that
>
> (1)   The trials are independent
>
> (2)   Each trial results in only two possible outcomes, labeled as "success" and "failure"
>
> (3)   The probability of a success in each trial, denoted as p, remains constant
>
> The random variable X that equals the number of trials that result in a success has a **binomial random variable** with parameters $0 < p < 1$ and $n = 1, 2, \ldots$. The probability mass function of X is
>
> $$f(x) = \binom{n}{x}p^x(1 - p)^{n-x} \qquad x = 0, 1, \ldots, n \qquad (3\text{-}7)$$

As in Example 3-16, $\binom{n}{x}$ equals the total number of different sequences of trials that contain x successes and $n - x$ failures. The total number of different sequences that contain x successes and $n - x$ failures times the probability of each sequence equals $P(X = x)$.

The probability expression above is a very useful formula that can be applied in a number of examples. The name of the distribution is obtained from the *binomial expansion*. For constants a and b, the binomial expansion is

$$(a + b)^n = \sum_{k=0}^{n}\binom{n}{k}a^k b^{n-k}$$

Let p denote the probability of success on a single trial. Then, by using the binomial expansion with $a = p$ and $b = 1 - p$, we see that the sum of the probabilities for a binomial random variable is 1. Furthermore, because each trial in the experiment is classified into two outcomes, {success, failure}, the distribution is called a "bi"-nomial. A more

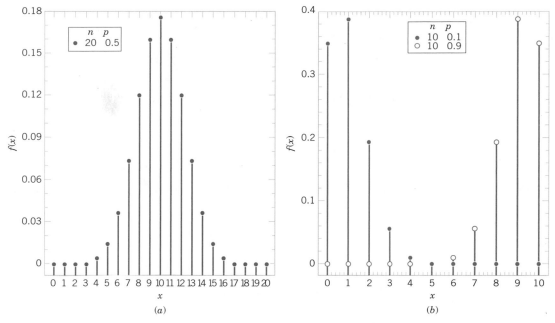

Figure 3-8   Binomial distributions for selected values of n and p.

general distribution, which includes the binomial as a special case, is the multinomial distribution.

Examples of binomial distributions are shown in Fig. 3-8. For a fixed n, the distribution becomes more symmetric as p increases from 0 to 0.5 or decreases from 1 to 0.5. For a fixed p, the distribution becomes more symmetric as n increases.

EXAMPLE 3-17

Several examples using the binomial coefficient $\binom{n}{x}$ follow.

$$\binom{10}{3} = 10!/[3!\ 7!] = (10 \cdot 9 \cdot 8)/(3 \cdot 2) = 120$$

$$\binom{15}{10} = 15!/[10!\ 5!] = (15 \cdot 14 \cdot 13 \cdot 12 \cdot 11)/(5 \cdot 4 \cdot 3 \cdot 2) = 3003$$

$$\binom{100}{4} = 100!/[4!\ 96!] = (100 \cdot 99 \cdot 98 \cdot 97)/(4 \cdot 3 \cdot 2) = 3,921,225$$

EXAMPLE 3-18

Each sample of water has a 10% chance of containing a particular organic pollutant. Assume that the samples are independent with regard to the presence of the pollutant. Find the probability that in the next 18 samples, exactly 2 contain the pollutant.

Let $X =$ the number of samples that contain the pollutant in the next 18 samples analyzed. Then X is a binomial random variable with $p = 0.1$ and $n = 18$. Therefore,

$$P(X = 2) = \binom{18}{2}(0.1)^2(0.9)^{16}$$

Now $\binom{18}{2} = 18!/[2!\ 16!] = 18(17)/2 = 153$. Therefore,

$$P(X = 2) = 153(0.1)^2(0.9)^{16} = 0.284$$

Determine the probability that at least four samples contain the pollutant. The requested probability is

$$P(X \geq 4) = \sum_{x=4}^{18} \binom{18}{x}(0.1)^x(0.9)^{18-x}$$

However, it is easier to use the complementary event,

$$P(X \geq 4) = 1 - P(X < 4) = 1 - \sum_{x=0}^{3} \binom{18}{x}(0.1)^x(0.9)^{18-x}$$

$$= 1 - [0.150 + 0.300 + 0.284 + 0.168] = 0.098$$

Determine the probability that $3 \leq X < 7$. Now

$$P(3 \leq X < 7) = \sum_{x=3}^{6} \binom{18}{x}(0.1)^x(0.9)^{18-x}$$

$$= 0.168 + 0.070 + 0.022 + 0.005$$

$$= 0.265$$

The mean and variance of a binomial random variable depend only on the parameters p and n. Formulas can be developed from moment generating functions, and details are provided in Section 5-8, part of the CD material for Chapter 5. The results are simply stated here.

Definition

> If X is a binomial random variable with parameters p and n,
>
> $$\mu = E(X) = np \quad \text{and} \quad \sigma^2 = V(X) = np(1 - p) \tag{3-8}$$

EXAMPLE 3-19   For the number of transmitted bits received in error in Example 3-16, $n = 4$ and $p = 0.1$, so

$$E(X) = 4(0.1) = 0.4 \quad \text{and} \quad V(X) = 4(0.1)(0.9) = 0.36$$

and these results match those obtained from a direct calculation in Example 3-9.

EXERCISES FOR SECTION 3-6

3-55. For each scenario described below, state whether or not the binomial distribution is a reasonable model for the random variable and why. State any assumptions you make.
(a) A production process produces thousands of temperature transducers. Let X denote the number of nonconforming

transducers in a sample of size 30 selected at random from the process.
(b) From a batch of 50 temperature transducers, a sample of size 30 is selected without replacement. Let X denote the number of nonconforming transducers in the sample.

(c) Four identical electronic components are wired to a controller that can switch from a failed component to one of the remaining spares. Let X denote the number of components that have failed after a specified period of operation.

(d) Let X denote the number of accidents that occur along the federal highways in Arizona during a one-month period.

(e) Let X denote the number of correct answers by a student taking a multiple choice exam in which a student can eliminate some of the choices as being incorrect in some questions and all of the incorrect choices in other questions.

(f) Defects occur randomly over the surface of a semiconductor chip. However, only 80% of defects can be found by testing. A sample of 40 chips with one defect each is tested. Let X denote the number of chips in which the test finds a defect.

(g) Reconsider the situation in part (f). Now, suppose the sample of 40 chips consists of chips with 1 and with 0 defects.

(h) A filling operation attempts to fill detergent packages to the advertised weight. Let X denote the number of detergent packages that are underfilled.

(i) Errors in a digital communication channel occur in bursts that affect several consecutive bits. Let X denote the number of bits in error in a transmission of 100,000 bits.

(j) Let X denote the number of surface flaws in a large coil of galvanized steel.

3-56. The random variable X has a binomial distribution with $n = 10$ and $p = 0.5$. Sketch the probability mass function of X.
(a) What value of X is most likely?
(b) What value(s) of X is(are) least likely?

3-57. The random variable X has a binomial distribution with $n = 10$ and $p = 0.5$. Determine the following probabilities:
(a) $P(X = 5)$ (b) $P(X \leq 2)$
(c) $P(X \geq 9)$ (d) $P(3 \leq X < 5)$

3-58. Sketch the probability mass function of a binomial distribution with $n = 10$ and $p = 0.01$ and comment on the shape of the distribution.
(a) What value of X is most likely?
(b) What value of X is least likely?

3-59. The random variable X has a binomial distribution with $n = 10$ and $p = 0.01$. Determine the following probabilities.
(a) $P(X = 5)$ (b) $P(X \leq 2)$
(c) $P(X \geq 9)$ (d) $P(3 \leq X < 5)$

3-60. Determine the cumulative distribution function of a binomial random variable with $n = 3$ and $p = 1/2$.

3-61. Determine the cumulative distribution function of a binomial random variable with $n = 3$ and $p = 1/4$.

3-62. An electronic product contains 40 integrated circuits. The probability that any integrated circuit is defective is 0.01, and the integrated circuits are independent. The product operates only if there are no defective integrated circuits. What is the probability that the product operates?

3-63. Let X denote the number of bits received in error in a digital communication channel, and assume that X is a bino-

mial random variable with $p = 0.001$. If 1000 bits are transmitted, determine the following:
(a) $P(X = 1)$ (b) $P(X \geq 1)$
(c) $P(X \leq 2)$ (d) mean and variance of X

3-64. The phone lines to an airline reservation system are occupied 40% of the time. Assume that the events that the lines are occupied on successive calls are independent. Assume that 10 calls are placed to the airline.
(a) What is the probability that for exactly three calls the lines are occupied?
(b) What is the probability that for at least one call the lines are not occupied?
(c) What is the expected number of calls in which the lines are all occupied?

3-65. Batches that consist of 50 coil springs from a production process are checked for conformance to customer requirements. The mean number of nonconforming coil springs in a batch is 5. Assume that the number of nonconforming springs in a batch, denoted as X, is a binomial random variable.
(a) What are n and p?
(b) What is $P(X \leq 2)$?
(c) What is $P(X \geq 49)$?

3-66. A statistical process control chart example. Samples of 20 parts from a metal punching process are selected every hour. Typically, 1% of the parts require rework. Let X denote the number of parts in the sample of 20 that require rework. A process problem is suspected if X exceeds its mean by more than three standard deviations.
(a) If the percentage of parts that require rework remains at 1%, what is the probability that X exceeds its mean by more than three standard deviations?
(b) If the rework percentage increases to 4%, what is the probability that X exceeds 1?
(c) If the rework percentage increases to 4%, what is the probability that X exceeds 1 in at least one of the next five hours of samples?

3-67. Because not all airline passengers show up for their reserved seat, an airline sells 125 tickets for a flight that holds only 120 passengers. The probability that a passenger does not show up is 0.10, and the passengers behave independently.
(a) What is the probability that every passenger who shows up can take the flight?
(b) What is the probability that the flight departs with empty seats?

3-68. This exercise illustrates that poor quality can affect schedules and costs. A manufacturing process has 100 customer orders to fill. Each order requires one component part that is purchased from a supplier. However, typically, 2% of the components are identified as defective, and the components can be assumed to be independent.
(a) If the manufacturer stocks 100 components, what is the probability that the 100 orders can be filled without reordering components?

(b) If the manufacturer stocks 102 components, what is the probability that the 100 orders can be filled without reordering components?

(c) If the manufacturer stocks 105 components, what is the probability that the 100 orders can be filled without reordering components?

3-69. A multiple choice test contains 25 questions, each with four answers. Assume a student just guesses on each question.

(a) What is the probability that the student answers more than 20 questions correctly?

(b) What is the probability the student answers less than 5 questions correctly?

3-70. A particularly long traffic light on your morning commute is green 20% of the time that you approach it. Assume that each morning represents an independent trial.

(a) Over five mornings, what is the probability that the light is green on exactly one day?

(b) Over 20 mornings, what is the probability that the light is green on exactly four days?

(c) Over 20 mornings, what is the probability that the light is green on more than four days?

3-7   GEOMETRIC AND NEGATIVE BINOMIAL DISTRIBUTIONS

3-7.1   Geometric Distribution

Consider a random experiment that is closely related to the one used in the definition of a binomial distribution. Again, assume a series of Bernoulli trials (independent trials with constant probability p of a success on each trial). However, instead of a fixed number of trials, trials are conducted until a success is obtained. Let the random variable X denote the number of trials until the first success. In Example 3-5, successive wafers are analyzed until a large particle is detected. Then, X is the number of wafers analyzed. In the transmission of bits, X might be the number of bits transmitted until an error occurs.

EXAMPLE 3-20   The probability that a bit transmitted through a digital transmission channel is received in error is 0.1. Assume the transmissions are independent events, and let the random variable X denote the number of bits transmitted *until* the first error.

Then, $P(X = 5)$ is the probability that the first four bits are transmitted correctly and the fifth bit is in error. This event can be denoted as $\{OOOOE\}$, where O denotes an okay bit. Because the trials are independent and the probability of a correct transmission is 0.9,

$$P(X = 5) = P(OOOOE) = 0.9^4 0.1 = 0.066$$

Note that there is some probability that X will equal any integer value. Also, if the first trial is a success, $X = 1$. Therefore, the range of X is $\{1, 2, 3, \dots \}$, that is, all positive integers.

Definition

> In a series of Bernoulli trials (independent trials with constant probability p of a success), let the random variable X denote the number of trials until the first success. Then X is a **geometric random variable** with parameter $0 < p < 1$ and
>
> $$f(x) = (1 - p)^{x-1}p \qquad x = 1, 2, \dots \tag{3-9}$$

Examples of the probability mass functions for geometric random variables are shown in Fig. 3-9. Note that the height of the line at x is $(1 - p)$ times the height of the line at $x - 1$. That is, the probabilities decrease in a geometric progression. The distribution acquires its name from this result.

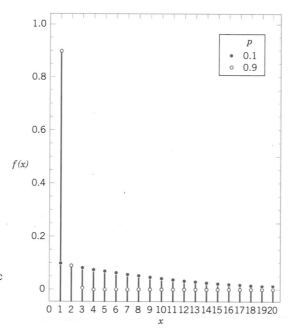

Figure 3-9 Geometric distributions for selected values of the parameter p.

EXAMPLE 3-21

The probability that a wafer contains a large particle of contamination is 0.01. If it is assumed that the wafers are independent, what is the probability that exactly 125 wafers need to be analyzed before a large particle is detected?

Let X denote the number of samples analyzed until a large particle is detected. Then X is a geometric random variable with $p = 0.01$. The requested probability is

$$P(X = 125) = (0.99)^{124}0.01 = 0.0029$$

The derivation of the mean and variance of a geometric random variable is left as an exercise. Note that $\sum_{k=1}^{\infty} k(1 - p)^{k-1}p$ can be shown to equal $1/p$. The results are as follows.

If X is a geometric random variable with parameter p,

$$\mu = E(X) = 1/p \quad \text{and} \quad \sigma^2 = V(X) = (1 - p)/p^2 \qquad (3\text{-}10)$$

EXAMPLE 3-22

Consider the transmission of bits in Example 3-20. Here, $p = 0.1$. The mean number of transmissions until the first error is $1/0.1 = 10$. The standard deviation of the number of transmissions before the first error is

$$\sigma = [(1 - 0.1)/0.1^2]^{1/2} = 9.49$$

Lack of Memory Property

A geometric random variable has been defined as the number of trials until the first success. However, because the trials are independent, the count of the number of trials until the next success can be started at any trial without changing the probability distribution of the random variable. For example, in the transmission of bits, if 100 bits are transmitted, the probability that the first error, after bit 100, occurs on bit 106 is the probability that the next six outcomes are *OOOOOE*. This probability is $(0.9)^5(0.1) = 0.059$, which is identical to the probability that the initial error occurs on bit 6.

The implication of using a geometric model is that the system presumably will not wear out. The probability of an error remains constant for all transmissions. In this sense, the geometric distribution is said to lack any memory. The **lack of memory property** will be discussed again in the context of an exponential random variable in Chapter 4.

EXAMPLE 3-23   In Example 3-20, the probability that a bit is transmitted in error is equal to 0.1. Suppose 50 bits have been transmitted. The mean number of bits until the next error is $1/0.1 = 10$—the same result as the mean number of bits until the first error.

3-7.2 Negative Binomial Distribution

A generalization of a geometric distribution in which the random variable is the number of Bernoulli trials required to obtain r successes results in the **negative binomial distribution.**

EXAMPLE 3-24   As in Example 3-20, suppose the probability that a bit transmitted through a digital transmission channel is received in error is 0.1. Assume the transmissions are independent events, and let the random variable X denote the number of bits transmitted until the *fourth* error.

Then, X has a negative binomial distribution with $r = 4$. Probabilities involving X can be found as follows. The $P(X = 10)$ is the probability that exactly three errors occur in the first nine trials and then trial 10 results in the fourth error. The probability that exactly three errors occur in the first nine trials is determined from the binomial distribution to be

$$\binom{9}{3}(0.1)^3(0.9)^6$$

Because the trials are independent, the probability that exactly three errors occur in the first 9 trials and trial 10 results in the fourth error is the product of the probabilities of these two events, namely,

$$\binom{9}{3}(0.1)^3(0.9)^6(0.1) = \binom{9}{3}(0.1)^4(0.9)^6$$

The previous result can be generalized as follows.

Definition

> In a series of Bernoulli trials (independent trials with constant probability p of a success), let the random variable X denote the number of trials until r successes occur. Then X is a **negative binomial random variable** with parameters $0 < p < 1$ and $r = 1, 2, 3, \ldots,$ and
>
> $$f(x) = \binom{x-1}{r-1}(1-p)^{x-r}p^r \qquad x = r, r+1, r+2, \ldots. \qquad (3\text{-}11)$$

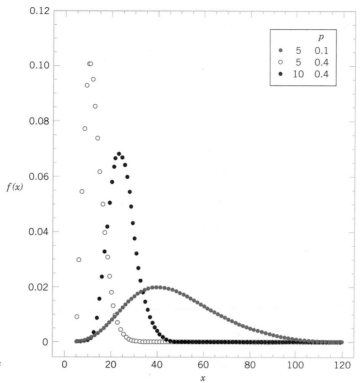

Figure 3-10 Negative binomial distributions for selected values of the parameters r and p.

Because at least r trials are required to obtain r successes, the range of X is from r to ∞. In the special case that $r = 1$, a negative binomial random variable is a geometric random variable. Selected negative binomial distributions are illustrated in Fig. 3-10.

The lack of memory property of a geometric random variable implies the following. Let X denote the total number of trials required to obtain r successes. Let X_1 denote the number of trials required to obtain the first success, let X_2 denote the number of extra trials required to obtain the second success, let X_3 denote the number of extra trials to obtain the third success, and so forth. Then, the total number of trials required to obtain r successes is $X = X_1 + X_2 + \cdots + X_r$. Because of the lack of memory property, each of the random variables $X_1, X_2, \ldots, X_r$ has a geometric distribution with the same value of p. Consequently, a negative binomial random variable can be interpreted as the sum of r geometric random variables. This concept is illustrated in Fig. 3-11.

Recall that a binomial random variable is a count of the number of successes in n Bernoulli trials. That is, the number of trials is predetermined, and the number of successes is random. A negative binomial random variable is a count of the number of trials required to

Figure 3-11 Negative binomial random variable represented as a sum of geometric random variables.

obtain r successes. That is, the number of successes is predetermined, and the number of trials is random. In this sense, a negative binomial random variable can be considered the opposite, or negative, of a binomial random variable.

The description of a negative binomial random variable as a sum of geometric random variables leads to the following results for the mean and variance. Sums of random variables are studied in Chapter 5.

If X is a negative binomial random variable with parameters p and r,

$$\mu = E(X) = r/p \quad \text{and} \quad \sigma^2 = V(X) = r(1 - p)/p^2 \qquad (3\text{-}12)$$

EXAMPLE 3-25   A Web site contains three identical computer servers. Only one is used to operate the site, and the other two are spares that can be activated in case the primary system fails. The probability of a failure in the primary computer (or any activated spare system) from a request for service is 0.0005. Assuming that each request represents an independent trial, what is the mean number of requests until failure of all three servers?

Let X denote the number of requests until all three servers fail, and let X_1, X_2, and X_3 denote the number of requests before a failure of the first, second, and third servers used, respectively. Now, $X = X_1 + X_2 + X_3$. Also, the requests are assumed to comprise independent trials with constant probability of failure $p = 0.0005$. Furthermore, a spare server is not affected by the number of requests before it is activated. Therefore, X has a negative binomial distribution with $p = 0.0005$ and $r = 3$. Consequently,

$$E(X) = 3/0.0005 = 6000 \text{ requests}$$

What is the probability that all three servers fail within five requests? The probability is $P(X \leq 5)$ and

$$P(X \leq 5) = P(X = 3) + P(X = 4) + P(X = 5)$$

$$= 0.0005^3 + \binom{3}{2}0.0005^3(0.9995) + \binom{4}{2}0.0005^3(0.9995)^2$$

$$= 1.25 \times 10^{-10} + 3.75 \times 10^{-10} + 7.49 \times 10^{-10}$$

$$= 1.249 \times 10^{-9}$$

EXERCISES FOR SECTION 3-7

3-71. Suppose the random variable X has a geometric distribution with $p = 0.5$. Determine the following probabilities:
(a) $P(X = 1)$ (b) $P(X = 4)$
(c) $P(X = 8)$ (d) $P(X \leq 2)$
(e) $P(X > 2)$

3-72. Suppose the random variable X has a geometric distribution with a mean of 2.5. Determine the following probabilities:
(a) $P(X = 1)$ (b) $P(X = 4)$
(c) $P(X = 5)$ (d) $P(X \leq 3)$
(e) $P(X > 3)$

3-73. The probability of a successful optical alignment in the assembly of an optical data storage product is 0.8. Assume the trials are independent.
(a) What is the probability that the first successful alignment requires exactly four trials?
(b) What is the probability that the first successful alignment requires at most four trials?
(c) What is the probability that the first successful alignment requires at least four trials?

3-74. In a clinical study, volunteers are tested for a gene that has been found to increase the risk for a disease. The probability that a person carries the gene is 0.1.
(a) What is the probability 4 or more people will have to be tested before 2 with the gene are detected?
(b) How many people are expected to be tested before 2 with the gene are detected?

3-75. Assume that each of your calls to a popular radio station has a probability of 0.02 of connecting, that is, of not obtaining a busy signal. Assume that your calls are independent.
(a) What is the probability that your first call that connects is your tenth call?
(b) What is the probability that it requires more than five calls for you to connect?
(c) What is the mean number of calls needed to connect?

3-76. In Exercise 3-70, recall that a particularly long traffic light on your morning commute is green 20% of the time that you approach it. Assume that each morning represents an independent trial.
(a) What is the probability that the first morning that the light is green is the fourth morning that you approach it?
(b) What is the probability that the light is not green for 10 consecutive mornings?

3-77. A trading company has eight computers that it uses to trade on the New York Stock Exchange (NYSE). The probability of a computer failing in a day is 0.005, and the computers fail independently. Computers are repaired in the evening and each day is an independent trial.
(a) What is the probability that all eight computers fail in a day?
(b) What is the mean number of days until a specific computer fails?
(c) What is the mean number of days until all eight computers fail in the same day?

3-78. In Exercise 3-66, recall that 20 parts are checked each hour and that X denotes the number of parts in the sample of 20 that require rework.
(a) If the percentage of parts that require rework remains at 1%, what is the probability that hour 10 is the first sample at which X exceeds 1?
(b) If the rework percentage increases to 4%, what is the probability that hour 10 is the first sample at which X exceeds 1?

(c) If the rework percentage increases to 4%, what is the expected number of hours until X exceeds 1?

3-79. Consider a sequence of independent Bernoulli trials with $p = 0.2$.
(a) What is the expected number of trials to obtain the first success?
(b) After the eighth success occurs, what is the expected number of trials to obtain the ninth success?

3-80. Show that the probability density function of a negative binomial random variable equals the probability density function of a geometric random variable when $r = 1$. Show that the formulas for the mean and variance of a negative binomial random variable equal the corresponding results for geometric random variable when $r = 1$.

3-81. Suppose that X is a negative binomial random variable with $p = 0.2$ and $r = 4$. Determine the following:
(a) $E(X)$ (b) $P(X = 20)$
(c) $P(X = 19)$ (d) $P(X = 21)$
(e) The most likely value for X

3-82. The probability is 0.6 that a calibration of a transducer in an electronic instrument conforms to specifications for the measurement system. Assume the calibration attempts are independent. What is the probability that at most three calibration attempts are required to meet the specifications for the measurement system?

3-83. An electronic scale in an automated filling operation stops the manufacturing line after three underweight packages are detected. Suppose that the probability of an underweight package is 0.001 and each fill is independent.
(a) What is the mean number of fills before the line is stopped?
(b) What is the standard deviation of the number of fills before the line is stopped?

3-84. A fault-tolerant system that processes transactions for a financial services firm uses three separate computers. If the operating computer fails, one of the two spares can be immediately switched online. After the second computer fails, the last computer can be immediately switched online. Assume that the probability of a failure during any transaction is 10^{-8} and that the transactions can be considered to be independent events.
(a) What is the mean number of transactions before all computers have failed?
(b) What is the variance of the number of transactions before all computers have failed?

3-85. Derive the expressions for the mean and variance of a geometric random variable with parameter p. (Formulas for infinite series are required.)

3-8 HYPERGEOMETRIC DISTRIBUTION

In Example 3-8, a day's production of 850 manufactured parts contains 50 parts that do not conform to customer requirements. Two parts are selected at random, without replacement from the day's production. That is, selected units are not replaced before the next selection is made. Let A and B denote the events that the first and second parts are nonconforming, respectively. In Chapter 2, we found $P(B|A) = 49/849$ and $P(A) = 50/850$. Consequently, knowledge that the first part is nonconforming suggests that it is less likely that the second part selected is nonconforming.

This experiment is fundamentally different from the examples based on the binomial distribution. In this experiment, the trials are not independent. Note that, in the unusual case that each unit selected is replaced before the next selection, the trials are independent and there is a constant probability of a nonconforming part on each trial. Then, the number of nonconforming parts in the sample is a binomial random variable.

Let X equal the number of nonconforming parts in the sample. Then

$$P(X = 0) = P(\text{both parts conform}) = (800/850)(799/849) = 0.886$$

$$
\begin{aligned}
P(X = 1) = P(&\text{first part selected conforms and the second part selected} \\
&\text{does not, or the first part selected does not and the second part} \\
&\text{selected conforms})
\end{aligned}
$$

$$= (800/850)(50/849) + (50/850)(800/849) = 0.111$$

$$P(X = 2) = P(\text{both parts do not conform}) = (50/850)(49/849) = 0.003$$

As in this example, samples are often selected without replacement. Although probabilities can be determined by the reasoning used in the example above, a general formula for computing probabilities when samples are selected without replacement is quite useful. The counting rules presented in Section 2-1.4, part of the CD material for Chapter 2, can be used to justify the formula given below.

Definition

A set of N objects contains

   K objects classified as successes

   $N - K$ objects classified as failures

A sample of size n objects is selected randomly (without replacement) from the N objects, where $K \leq N$ and $n \leq N$.

   Let the random variable X denote the number of successes in the sample. Then X is a **hypergeometric random variable** and

$$f(x) = \frac{\binom{K}{x}\binom{N-K}{n-x}}{\binom{N}{n}} \qquad x = \max\{0, n + K - N\} \text{ to } \min\{K, n\} \qquad (3\text{-}13)$$

The expression $\min\{K, n\}$ is used in the definition of the range of X because the maximum number of successes that can occur in the sample is the smaller of the sample size, n,

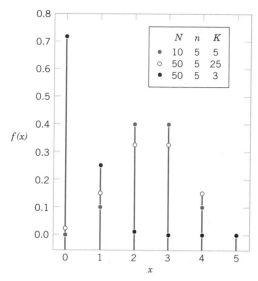

Figure 3-12   Hypergeometric distributions for selected values of parameters N, K, and n.

and the number of successes available, K. Also, if $n + K > N$, at least $n + K - N$ successes must occur in the sample. Selected hypergeometric distributions are illustrated in Fig. 3-12.

EXAMPLE 3-26

The example at the start of this section can be reanalyzed by using the general expression in the definition of a hypergeometric random variable. That is,

850 parts
50 Non Conf

$$P(X = 0) = \frac{\binom{50}{0}\binom{800}{2}}{\binom{850}{2}} = \frac{319600}{360825} = 0.886$$

$$P(X = 1) = \frac{\binom{50}{1}\binom{800}{1}}{\binom{850}{2}} = \frac{40000}{360825} = 0.111$$

$$P(X = 2) = \frac{\binom{50}{2}\binom{800}{0}}{\binom{850}{2}} = \frac{1225}{360825} = 0.003$$

EXAMPLE 3-27

A batch of parts contains 100 parts from a local supplier of tubing and 200 parts from a supplier of tubing in the next state. If four parts are selected randomly and without replacement, what is the probability they are all from the local supplier?

Let X equal the number of parts in the sample from the local supplier. Then, X has a hypergeometric distribution and the requested probability is $P(X = 4)$. Consequently,

$$P(X = 4) = \frac{\binom{100}{4}\binom{200}{0}}{\binom{300}{4}} = 0.0119$$

What is the probability that two or more parts in the sample are from the local supplier?

$$P(X \geq 2) = \frac{\binom{100}{2}\binom{200}{2}}{\binom{300}{4}} + \frac{\binom{100}{3}\binom{200}{1}}{\binom{300}{4}} + \frac{\binom{100}{4}\binom{200}{0}}{\binom{300}{4}}$$

$$= 0.298 + 0.098 + 0.0119 = 0.408$$

What is the probability that at least one part in the sample is from the local supplier?

$$P(X \geq 1) = 1 - P(X = 0) = 1 - \frac{\binom{100}{0}\binom{200}{4}}{\binom{300}{4}} = 0.804$$

The mean and variance of a hypergeometric random variable can be determined from the trials that comprise the experiment. However, the trials are not independent, and so the calculations are more difficult than for a binomial distribution. The results are stated as follows.

If X is a hypergeometric random variable with parameters N, K, and n, then

$$\mu = E(X) = np \quad \text{and} \quad \sigma^2 = V(X) = np(1 - p)\left(\frac{N - n}{N - 1}\right) \quad (3\text{-}14)$$

where $p = K/N$.

Here p is interpreted as the proportion of successes in the set of N objects.

EXAMPLE 3-28   In the previous example, the sample size is 4. The random variable X is the number of parts in the sample from the local supplier. Then, $p = 100/300 = 1/3$. Therefore,

$$E(X) = 4(100/300) = 1.33$$

and

$$V(X) = 4(1/3)(2/3)[(300 - 4)/299] = 0.88$$

For a hypergeometric random variable, $E(X)$ is similar to the mean of a binomial random variable. Also, $V(X)$ differs from the result for a binomial random variable only by the term shown below.

Finite Population Correction Factor

The term in the variance of a hypergeometric random variable

$$\frac{N - n}{N - 1}$$

is called the finite population correction factor.

Sampling with replacement is equivalent to sampling from an infinite set because the proportion of success remains constant for every trial in the experiment. As mentioned previously, if sampling were done with replacement, X would be a binomial random variable and its variance would be $np(1 - p)$. Consequently, the finite population correction represents the correction to the binomial variance that results because the sampling is without replacement from the finite set of size N.

If n is small relative to N, the correction is small and the hypergeometric distribution is similar to the binomial. In this case, a binomial distribution can effectively approximate the distribution of the number of units of a specified type in the sample. A case is illustrated in Fig. 3-13.

EXAMPLE 3-29   A listing of customer accounts at a large corporation contains 1000 customers. Of these, 700 have purchased at least one of the corporation's products in the last three months. To evaluate a new product design, 50 customers are sampled at random from the corporate listing. What is

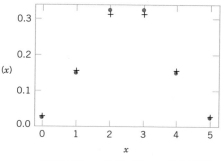

• Hypergeometric $N = 50$, $n = 5$, $K = 25$
+ Binomial $n = 5$, $p = 0.5$

Figure 3-13
Comparison of hypergeometric and binomial distributions.

	0	1	2	3	4	5
Hypergeometric probability	0.025	0.149	0.326	0.326	0.149	0.025
Binomial probability	0.031	0.156	0.321	0.312	0.156	0.031

the probability that more than 45 of the sampled customers have purchased from the corporation in the last three months?

The sampling is without replacement. However, because the sample size of 50 is small relative to the number of customer accounts, 1000, the probability of selecting a customer who has purchased from the corporation in the last three months remains approximately constant as the customers are chosen.

For example, let A denote the event that the first customer selected has purchased from the corporation in the last three months, and let B denote the event that the second customer selected has purchased from the corporation in the last three months. Then, $P(A) = 700/1000 = 0.7$ and $P(B|A) = 699/999 = 0.6997$. That is, the trials are approximately independent.

Let X denote the number of customers in the sample who have purchased from the corporation in the last three months. Then, X is a hypergeometric random variable with $N = 1000$, $n = 50$, and $K = 700$. Consequently, $p = K/N = 0.7$. The requested probability is $P(X > 45)$. Because the sample size is small relative to the batch size, the distribution of X can be approximated as binomial with $n = 50$ and $p = 0.7$. Using the binomial approximation to the distribution of X results in

$$P(X > 45) = \sum_{x=46}^{50} \binom{50}{x} 0.7^x (1 - 0.7)^{50-x} = 0.00017$$

The probability from the hypergeometric distribution is 0.000166, but this requires computer software. The result agrees well with the binomial approximation.

EXERCISES FOR SECTION 3-8

3-86.   Suppose X has a hypergeometric distribution with $N = 100$, $n = 4$, and $K = 20$. Determine the following:
(a) $P(X = 1)$ (b) $P(X = 6)$
(c) $P(X = 4)$ (d) Determine the mean and variance of X.

3-87.   Suppose X has a hypergeometric distribution with $N = 20$, $n = 4$, and $K = 4$. Determine the following:
(a) $P(X = 1)$ (b) $P(X = 4)$
(c) $P(X \le 2)$ (d) Determine the mean and variance of X.

3-88.   Suppose X has a hypergeometric distribution with $N = 10$, $n = 3$, and $K = 4$. Sketch the probability mass function of X.

3-89.   Determine the cumulative distribution function for X in Exercise 3-88.

3-90.   A lot of 75 washers contains 5 in which the variability in thickness around the circumference of the washer is unacceptable. A sample of 10 washers is selected at random, without replacement.
(a) What is the probability that none of the unacceptable washers is in the sample?
(b) What is the probability that at least one unacceptable washer is in the sample?
(c) What is the probability that exactly one unacceptable washer is in the sample?

(d) What is the mean number of unacceptable washers in the sample?

3-91.   A company employs 800 men under the age of 55. Suppose that 30% carry a marker on the male chromosome that indicates an increased risk for high blood pressure.
(a) If 10 men in the company are tested for the marker in this chromosome, what is the probability that exactly 1 man has the marker?
(b) If 10 men in the company are tested for the marker in this chromosome, what is the probability that more than 1 has the marker?

3-92.   Printed circuit cards are placed in a functional test after being populated with semiconductor chips. A lot contains 140 cards, and 20 are selected without replacement for functional testing.
(a) If 20 cards are defective, what is the probability that at least 1 defective card is in the sample?
(b) If 5 cards are defective, what is the probability that at least 1 defective card appears in the sample?

3-93.   Magnetic tape is slit into half-inch widths that are wound into cartridges. A slitter assembly contains 48 blades. Five blades are selected at random and evaluated each day for sharpness. If any dull blade is found, the assembly is replaced with a newly sharpened set of blades.

(a) If 10 of the blades in an assembly are dull, what is the probability that the assembly is replaced the first day it is evaluated?

(b) If 10 of the blades in an assembly are dull, what is the probability that the assembly is not replaced until the third day of evaluation? [*Hint:* Assume the daily decisions are independent, and use the geometric distribution.]

(c) Suppose on the first day of evaluation, two of the blades are dull, on the second day of evaluation six are dull, and on the third day of evaluation, ten are dull. What is the probability that the assembly is not replaced until the third day of evaluation? [*Hint:* Assume the daily decisions are independent. However, the probability of replacement changes every day.]

3-94. A state runs a lottery in which 6 numbers are randomly selected from 40, without replacement. A player chooses 6 numbers before the state's sample is selected.

(a) What is the probability that the 6 numbers chosen by a player match all 6 numbers in the state's sample?

(b) What is the probability that 5 of the 6 numbers chosen by a player appear in the state's sample?

(c) What is the probability that 4 of the 6 numbers chosen by a player appear in the state's sample?

(d) If a player enters one lottery each week, what is the expected number of weeks until a player matches all 6 numbers in the state's sample?

3-95. Continuation of Exercises 3-86 and 3-87.

(a) Calculate the finite population corrections for Exercises 3-86 and 3-87. For which exercise should the binomial approximation to the distribution of X be better?

(b) For Exercise 3-86, calculate $P(X = 1)$ and $P(X = 4)$ assuming that X has a binomial distribution and compare these results to results derived from the hypergeometric distribution.

(c) For Exercise 3-87, calculate $P(X = 1)$ and $P(X = 4)$ assuming that X has a binomial distribution and compare these results to the results derived from the hypergeometric distribution.

3-96. Use the binomial approximation to the hypergeometric distribution to approximate the probabilities in Exercise 3-92. What is the finite population correction in this exercise?

3-9 POISSON DISTRIBUTION

We introduce the Poisson distribution with an example.

EXAMPLE 3-30 Consider the transmission of n bits over a digital communication channel. Let the random variable X equal the number of bits in error. When the probability that a bit is in error is constant and the transmissions are independent, X has a binomial distribution. Let p denote the probability that a bit is in error. Let $\lambda = pn$. Then, $E(x) = pn = \lambda$ and

$$P(X = x) = \binom{n}{x} P^x(1 - p)^{n-x} = \binom{n}{x}\left(\frac{\lambda}{n}\right)^x\left(1 - \frac{\lambda}{n}\right)^{n-x}$$

Now, suppose that the number of bits transmitted increases and the probability of an error decreases exactly enough that pn remains equal to a constant. That is, n increases and p decreases accordingly, such that $E(X) = \lambda$ remains constant. Then, with some work, it can be shown that

$$\lim_{n\to\infty} P(X = x) = \frac{e^{-\lambda}\lambda^x}{x!}, \qquad x = 0, 1, 2, \dots$$

Also, because the number of bits transmitted tends to infinity, the number of errors can equal any nonnegative integer. Therefore, the range of X is the integers from zero to infinity.

The distribution obtained as the limit in the above example is more useful than the derivation above implies. The following example illustrates the broader applicability.

EXAMPLE 3-31

Flaws occur at random along the length of a thin copper wire. Let X denote the random variable that counts the number of flaws in a length of L millimeters of wire and suppose that the average number of flaws in L millimeters is λ.

The probability distribution of X can be found by reasoning in a manner similar to the previous example. Partition the length of wire into n subintervals of small length, say, 1 micrometer each. If the subinterval chosen is small enough, the probability that more than one flaw occurs in the subinterval is negligible. Furthermore, we can interpret the assumption that flaws occur at random to imply that every subinterval has the same probability of containing a flaw, say, p. Finally, if we assume that the probability that a subinterval contains a flaw is independent of other subintervals, we can model the distribution of X as approximately a binomial random variable. Because

$$E(X) = \lambda = np$$

we obtain

$$p = \lambda/n$$

That is, the probability that a subinterval contains a flaw is λ/n. With small enough subintervals, n is very large and p is very small. Therefore, the distribution of X is obtained as in the previous example.

Example 3-31 can be generalized to include a broad array of random experiments. The interval that was partitioned was a length of wire. However, the same reasoning can be applied to any interval, including an interval of time, an area, or a volume. For example, counts of (1) particles of contamination in semiconductor manufacturing, (2) flaws in rolls of textiles, (3) calls to a telephone exchange, (4) power outages, and (5) atomic particles emitted from a specimen have all been successfully modeled by the probability mass function in the following definition.

Definition

> Given an interval of real numbers, assume counts occur at random throughout the interval. If the interval can be partitioned into subintervals of small enough length such that
>
> (1) the probability of more than one count in a subinterval is zero,
>
> (2) the probability of one count in a subinterval is the same for all subintervals and proportional to the length of the subinterval, and
>
> (3) the count in each subinterval is independent of other subintervals, the random experiment is called a **Poisson process**.
>
> The random variable X that equals the number of counts in the interval is a **Poisson random variable** with parameter $0 < \lambda$, and the probability mass function of X is
>
> $$f(x) = \frac{e^{-\lambda}\lambda^x}{x!} \quad x = 0, 1, 2, \ldots \qquad (3\text{-}15)$$

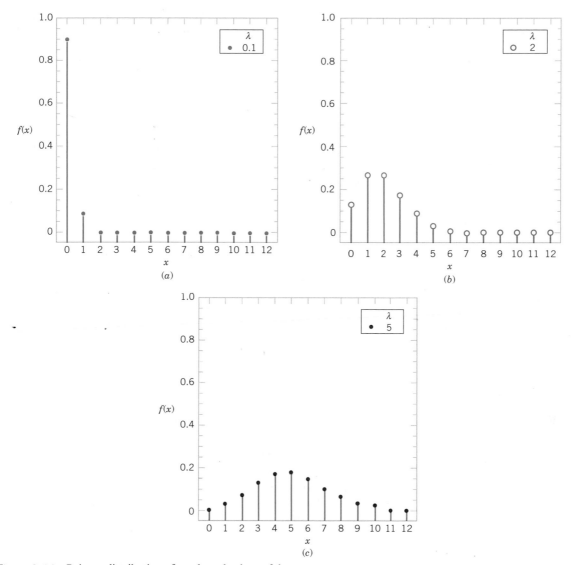

Figure 3-14   Poisson distributions for selected values of the parameters.

Historically, the term *process* has been used to suggest the observation of a system over time. In our example with the copper wire, we showed that the Poisson distribution could also apply to intervals such as lengths. Figure 3-14 provides graphs of selected Poisson distributions.

It is important to **use consistent units** in the calculation of probabilities, means, and variances involving Poisson random variables. The following example illustrates unit conversions. For example, if the

average number of flaws per millimeter of wire is 3.4, then the

average number of flaws in 10 millimeters of wire is 34, and the

average number of flaws in 100 millimeters of wire is 340.

If a Poisson random variable represents the number of counts in some interval, the mean of the random variable must equal the expected number of counts in the same length of interval.

EXAMPLE 3-32

For the case of the thin copper wire, suppose that the number of flaws follows a Poisson distribution with a mean of 2.3 flaws per millimeter. Determine the probability of exactly 2 flaws in 1 millimeter of wire.

Let X denote the number of flaws in 1 millimeter of wire. Then, $E(X) = 2.3$ flaws and

$$P(X = 2) = \frac{e^{-2.3}2.3^2}{2!} = 0.265$$

Determine the probability of 10 flaws in 5 millimeters of wire. Let X denote the number of flaws in 5 millimeters of wire. Then, X has a Poisson distribution with

$$E(X) = 5 \text{ mm} \times 2.3 \text{ flaws/mm} = 11.5 \text{ flaws}$$

Therefore,

$$P(X = 10) = e^{-11.5}\frac{11.5^{10}}{10!} = 0.113$$

Determine the probability of at least 1 flaw in 2 millimeters of wire. Let X denote the number of flaws in 2 millimeters of wire. Then, X has a Poisson distribution with

$$E(X) = 2 \text{ mm} \times 2.3 \text{ flaws/mm} = 4.6 \text{ flaws}$$

Therefore,

$$P(X \geq 1) = 1 - P(X = 0) = 1 - e^{-4.6} = 0.9899$$

EXAMPLE 3-33

Contamination is a problem in the manufacture of optical storage disks. The number of particles of contamination that occur on an optical disk has a Poisson distribution, and the average number of particles per centimeter squared of media surface is 0.1. The area of a disk under study is 100 squared centimeters. Find the probability that 12 particles occur in the area of a disk under study.

Let X denote the number of particles in the area of a disk under study. Because the mean number of particles is 0.1 particles per cm^2

$$E(X) = 100 \text{ cm}^2 \times 0.1 \text{ particles/cm}^2 = 10 \text{ particles}$$

Therefore,

$$P(X = 12) = \frac{e^{-10}10^{12}}{12!} = 0.095$$

The probability that zero particles occur in the area of the disk under study is

$$P(X = 0) = e^{-10} = 4.54 \times 10^{-5}$$

Determine the probability that 12 or fewer particles occur in the area of the disk under study. The probability is

$$P(X \leq 12) = P(X = 0) + P(X = 1) + \cdots + P(X = 12) = \sum_{i=0}^{12} \frac{e^{-10}10^i}{i!}$$

Because this sum is tedious to compute, many computer programs calculate cumulative Poisson probabilities. From one such program, $P(X \leq 12) = 0.791$.

The derivation of the mean and variance of a Poisson random variable is left as an exercise. The results are as follows.

If X is a Poisson random variable with parameter λ, then

$$\mu = E(X) = \lambda \quad \text{and} \quad \sigma^2 = V(X) = \lambda \tag{3-16}$$

The mean and variance of a Poisson random variable are equal. For example, if particle counts follow a Poisson distribution with a mean of 25 particles per square centimeter, the variance is also 25 and the standard deviation of the counts is 5 per square centimeter. Consequently, information on the variability is very easily obtained. Conversely, if the variance of count data is much greater than the mean of the same data, the Poisson distribution is not a good model for the distribution of the random variable.

EXERCISES FOR SECTION 3-9

3-97. Suppose X has a Poisson distribution with a mean of 4. Determine the following probabilities:
(a) $P(X = 0)$ (b) $P(X \leq 2)$
(c) $P(X = 4)$ (d) $P(X = 8)$

3-98. Suppose X has a Poisson distribution with a mean of 0.4. Determine the following probabilities:
(a) $P(X = 0)$ (b) $P(X \leq 2)$
(c) $P(X = 4)$ (d) $P(X = 8)$

3-99. Suppose that the number of customers that enter a bank in an hour is a Poisson random variable, and suppose that $P(X = 0) = 0.05$. Determine the mean and variance of X.

3-100. The number of telephone calls that arrive at a phone exchange is often modeled as a Poisson random variable. Assume that on the average there are 10 calls per hour.
(a) What is the probability that there are exactly 5 calls in one hour?
(b) What is the probability that there are 3 or less calls in one hour?
(c) What is the probability that there are exactly 15 calls in two hours?
(d) What is the probability that there are exactly 5 calls in 30 minutes?

3-101. The number of flaws in bolts of cloth in textile manufacturing is assumed to be Poisson distributed with a mean of 0.1 flaw per square meter.
(a) What is the probability that there are two flaws in 1 square meter of cloth?
(b) What is the probability that there is one flaw in 10 square meters of cloth?

(c) What is the probability that there are no flaws in 20 square meters of cloth?
(d) What is the probability that there are at least two flaws in 10 square meters of cloth?

3-102. When a computer disk manufacturer tests a disk, it writes to the disk and then tests it using a certifier. The certifier counts the number of missing pulses or errors. The number of errors on a test area on a disk has a Poisson distribution with $\lambda = 0.2$.
(a) What is the expected number of errors per test area?
(b) What percentage of test areas have two or fewer errors?

3-103. The number of cracks in a section of interstate highway that are significant enough to require repair is assumed to follow a Poisson distribution with a mean of two cracks per mile.
(a) What is the probability that there are no cracks that require repair in 5 miles of highway?
(b) What is the probability that at least one crack requires repair in 1/2 mile of highway?
(c) If the number of cracks is related to the vehicle load on the highway and some sections of the highway have a heavy load of vehicles whereas other sections carry a light load, how do you feel about the assumption of a Poisson distribution for the number of cracks that require repair?

3-104. The number of failures for a cytogenics machine from contamination in biological samples is a Poisson random variable with a mean of 0.01 per 100 samples.
(a) If the lab usually processes 500 samples per day, what is the expected number of failures per day?

(b) What is the probability that the machine will not fail during a study that includes 500 participants? (Assume one sample per participant.)

3-105. The number of surface flaws in plastic panels used in the interior of automobiles has a Poisson distribution with a mean of 0.05 flaws per square foot of plastic panel. Assume an automobile interior contains 10 square feet of plastic panel.
(a) What is the probability that there are no surface flaws in an auto's interior?
(b) If 10 cars are sold to a rental company, what is the probability that none of the 10 cars has any surface flaws?
(c) If 10 cars are sold to a rental company, what is the probability that at most one car has any surface flaws?

3-106. The number of failures of a testing instrument from contamination particles on the product is a Poisson random variable with a mean of 0.02 failure per hour.
(a) What is the probability that the instrument does not fail in an 8-hour shift?
(b) What is the probability of at least one failure in a 24-hour day?

Supplemental Exercises

3-107. A shipment of chemicals arrives in 15 totes. Three of the totes are selected at random, without replacement, for an inspection of purity. If two of the totes do not conform to purity requirements, what is the probability that at least one of the nonconforming totes is selected in the sample?

3-108. The probability that your call to a service line is answered in less than 30 seconds is 0.75. Assume that your calls are independent.
(a) If you call 10 times, what is the probability that exactly 9 of your calls are answered within 30 seconds?
(b) If you call 20 times, what is the probability that at least 16 calls are answered in less than 30 seconds?
(c) If you call 20 times, what is the mean number of calls that are answered in less than 30 seconds?

3-109. Continuation of Exercise 3-108.
(a) What is the probability that you must call four times to obtain the first answer in less than 30 seconds?
(b) What is the mean number of calls until you are answered in less than 30 seconds?

3-110. Continuation of Exercise 3-109.
(a) What is the probability that you must call six times in order for two of your calls to be answered in less than 30 seconds?
(b) What is the mean number of calls to obtain two answers in less than 30 seconds?

3-111. The number of messages sent to a computer bulletin board is a Poisson random variable with a mean of 5 messages per hour.
(a) What is the probability that 5 messages are received in 1 hour?

(b) What is the probability that 10 messages are received in 1.5 hours?
(c) What is the probability that less than two messages are received in one-half hour?

3-112. A Web site is operated by four identical computer servers. Only one is used to operate the site; the others are spares that can be activated in case the active server fails. The probability that a request to the Web site generates a failure in the active server is 0.0001. Assume that each request is an independent trial. What is the mean time until failure of all four computers?

3-113. The number of errors in a textbook follow a Poisson distribution with a mean of 0.01 error per page. What is the probability that there are three or less errors in 100 pages?

3-114. The probability that an individual recovers from an illness in a one-week time period without treatment is 0.1. Suppose that 20 independent individuals suffering from this illness are treated with a drug and 4 recover in a one-week time period. If the drug has no effect, what is the probability that 4 or more people recover in a one-week time period?

3-115. Patient response to a generic drug to control pain is scored on a 5-point scale, where a 5 indicates complete relief. Historically the distribution of scores is

1	2	3	4	5
0.05	0.1	0.2	0.25	0.4

Two patients, assumed to be independent, are each scored.
(a) What is the probability mass function of the total score?
(b) What is the probability mass function of the average score?

3-116. In a manufacturing process that laminates several ceramic layers, 1% of the assemblies are defective. Assume that the assemblies are independent.
(a) What is the mean number of assemblies that need to be checked to obtain five defective assemblies?
(b) What is the standard deviation of the number of assemblies that need to be checked to obtain five defective assemblies?

3-117. Continuation of Exercise 3-116. Determine the minimum number of assemblies that need to be checked so that the probability of at least one defective assembly exceeds 0.95.

3-118. Determine the constant c so that the following function is a probability mass function: $f(x) = cx$ for $x = 1, 2, 3, 4$.

3-119. A manufacturer of a consumer electronics product expects 2% of units to fail during the warranty period. A sample of 500 independent units is tracked for warranty performance.
(a) What is the probability that none fails during the warranty period?
(b) What is the expected number of failures during the warranty period?
(c) What is the probability that more than two units fail during the warranty period?

3-120. Messages that arrive at a service center for an information systems manufacturer have been classified on the basis of the number of keywords (used to help route messages) and the type of message, either email or voice. Also, 70% of the messages arrive via email and the rest are voice.

number of keywords	0	1	2	3	4
email	0.1	0.1	0.2	0.4	0.2
voice	0.3	0.4	0.2	0.1	0

Determine the probability mass function of the number of keywords in a message.

3-121. The random variable X has the following probability distribution:

x	2	3	5	8
probability	0.2	0.4	0.3	0.1

Determine the following:
(a) $P(X \leq 3)$ (b) $P(X > 2.5)$
(c) $P(2.7 < X < 5.1)$ (d) $E(X)$
(e) $V(X)$

3-122. Determine the probability mass function for the random variable with the following cumulative distribution function:

$$F(x) = \begin{cases} 0 & x < 2 \\ 0.2 & 2 \leq x < 5.7 \\ 0.5 & 5.7 \leq x < 6.5 \\ 0.8 & 6.5 \leq x < 8.5 \\ 1 & 8.5 \leq x \end{cases}$$

3-123. Each main bearing cap in an engine contains four bolts. The bolts are selected at random, without replacement, from a parts bin that contains 30 bolts from one supplier and 70 bolts from another.
(a) What is the probability that a main bearing cap contains all bolts from the same supplier?
(b) What is the probability that exactly three bolts are from the same supplier?

3-124. Assume the number of errors along a magnetic recording surface is a Poisson random variable with a mean of one error every 10^5 bits. A sector of data consists of 4096 eight-bit bytes.
(a) What is the probability of more than one error in a sector?
(b) What is the mean number of sectors until an error is found?

3-125. An installation technician for a specialized communication system is dispatched to a city only when three or more orders have been placed. Suppose orders follow a Poisson distribution with a mean of 0.25 per week for a city with a population of 100,000 and suppose your city contains a population of 800,000.
(a) What is the probability that a technician is required after a one-week period?
(b) If you are the first one in the city to place an order, what is the probability that you have to wait more than two weeks from the time you place your order until a technician is dispatched?

3-126. From 500 customers, a major appliance manufacturer will randomly select a sample without replacement. The company estimates that 25% of the customers will provide useful data. If this estimate is correct, what is the probability mass function of the number of customers that will provide useful data?
(a) Assume that the company samples 5 customers.
(b) Assume that the company samples 10 customers.

3-127. It is suspected that some of the totes containing chemicals purchased from a supplier exceed the moisture content target. Samples from 30 totes are to be tested for moisture content. Assume that the totes are independent. Determine the proportion of totes from the supplier that must exceed the moisture content target so that the probability is 0.90 that at least one tote in the sample of 30 fails the test.

3-128. Messages arrive to a computer server according to a Poisson distribution with a mean rate of 10 per hour. Determine the length of an interval of time such that the probability that no messages arrive during this interval is 0.90.

3-129. Flaws occur in the interior of plastic used for automobiles according to a Poisson distribution with a mean of 0.02 flaw per panel.
(a) If 50 panels are inspected, what is the probability that there are no flaws?
(b) What is the expected number of panels that need to be inspected before a flaw is found?
(c) If 50 panels are inspected, what is the probability that the number of panels that have one or more flaws is less than or equal to 2?

MIND-EXPANDING EXERCISES

3-130. Derive the mean and variance of a hypergeometric random variable (difficult exercise).

3-131. Show that the function $f(x)$ in Example 3-5 satisfies the properties of a probability mass function by summing the infinite series.

3-132. Derive the formula for the mean and standard deviation of a discrete uniform random variable over the range of integers $a, a + 1, \ldots, b$.

3-133. A company performs inspection on shipments from suppliers in order to defect nonconforming products. Assume a lot contains 1000 items and 1% are nonconforming. What sample size is needed so that the probability of choosing at least one nonconforming item in the sample is at least 0.90? Assume the binomial approximation to the hypergeometric distribution is adequate.

3-134. A company performs inspection on shipments from suppliers in order to detect nonconforming products. The company's policy is to use a sample size that is always 10% of the lot size. Comment on the effectiveness of this policy as a general rule for all sizes of lots.

3-135. Surface flaws in automobile exterior panels follow a Poisson distribution with a mean of 0.1 flaws per panel. If 100 panels are checked, what is the probability that fewer than five panels have any flaws?

3-136. A large bakery can produce rolls in lots of either 0, 1000, 2000, or 3000 per day. The production cost per item is $0.10. The demand varies randomly according to the following distribution:

demand for rolls	0	1000	2000	3000
probability of demand	0.3	0.2	0.3	0.2

Every roll for which there is a demand is sold for $0.30. Every roll for which there is no demand is sold in a secondary market for $0.05. How many rolls should the bakery produce each day to maximize the mean profit?

3-137. A manufacturer stocks components obtained from a supplier. Suppose that 2% of the components are defective and that the defective components occur independently. How many components must the manufacturer have in stock so that the probability that 100 orders can be completed without reordering components is at least 0.95?

IMPORTANT TERMS AND CONCEPTS

In the E-book, click on any term or concept below to go to that subject.

Bernoulli trial
Binomial distribution
Cumulative probability distribution function-discrete random. variable
Discrete uniform distribution

Expected value of a function of a random variable
Finite population correction factor
Geometric distribution
Hypergeometric distribution
Lack of memory property-discrete random variable

Mean-discrete random variable
Mean-function of a discrete random variable
Negative binomial distribution
Poisson distribution
Poisson process
Probability distribution-discrete random variable

Probability mass function
Standard deviation-discrete random variable
Variance-discrete random variable

Continuous Random Variables and Probability Distributions

LEARNING OBJECTIVES

After careful study of this chapter you should be able to do the following:

1. Determine probabilities from probability density functions.

2. Determine probabilities from cumulative distribution functions and cumulative distribution functions from probability density functions, and the reverse.

3. Calculate means and variances for continuous random variables.

4. Understand the assumptions for each of the continuous probability distributions presented.

5. Select an appropriate continuous probability distribution to calculate probabilities in specific applications.

6. Calculate probabilities, determine means and variances for each of the continuous probability distributions presented.

7. Standardize normal random variables.

8. Use the table for the cumulative distribution function of a standard normal distribution to calculate probabilities.

9. Approximate probabilities for some binomial and Poisson distributions.

CD MATERIAL

10. Use continuity corrections to improve the normal approximation to those binomial and Poisson distributions.

Answers for most odd numbered exercises are at the end of the book. Answers to exercises whose numbers are surrounded by a box can be accessed in the e-Text by clicking on the box. Complete worked solutions to certain exercises are also available in the e-Text. These are indicated in the Answers to Selected Exercises section by a box around the exercise number. Exercises are also available for some of the text sections that appear on CD only. These exercises may be found within the e-Text immediately following the section they accompany.

4-1 CONTINUOUS RANDOM VARIABLES

Previously, we discussed the measurement of the current in a thin copper wire. We noted that the results might differ slightly in day-to-day replications because of small variations in variables that are not controlled in our experiment—changes in ambient temperatures, small impurities in the chemical composition of the wire, current source drifts, and so forth.

Another example is the selection of one part from a day's production and very accurately measuring a dimensional length. In practice, there can be small variations in the actual measured lengths due to many causes, such as vibrations, temperature fluctuations, operator differences, calibrations, cutting tool wear, bearing wear, and raw material changes. Even the measurement procedure can produce variations in the final results.

In these types of experiments, the measurement of interest—current in a copper wire experiment, length of a machined part—can be represented by a random variable. It is reasonable to model the range of possible values of the random variable by an interval (finite or infinite) of real numbers. For example, for the length of a machined part, our model enables the measurement from the experiment to result in any value within an interval of real numbers. Because the range is any value in an interval, the model provides for any precision in length measurements. However, because the number of possible values of the random variable X is uncountably infinite, X has a distinctly different distribution from the discrete random variables studied previously. The range of X includes all values in an interval of real numbers; that is, the range of X can be thought of as a continuum.

A number of continuous distributions frequently arise in applications. These distributions are described, and example computations of probabilities, means, and variances are provided in the remaining sections of this chapter.

4-2 PROBABILITY DISTRIBUTIONS AND PROBABILITY DENSITY FUNCTIONS

Density functions are commonly used in engineering to describe physical systems. For example, consider the density of a loading on a long, thin beam as shown in Fig. 4-1. For any point x along the beam, the density can be described by a function (in grams/cm). Intervals with large loadings correspond to large values for the function. The total loading between points a and b is determined as the integral of the density function from a to b. This integral is the area

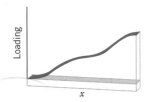

Figure 4-1   Density function of a loading on a long, thin beam.

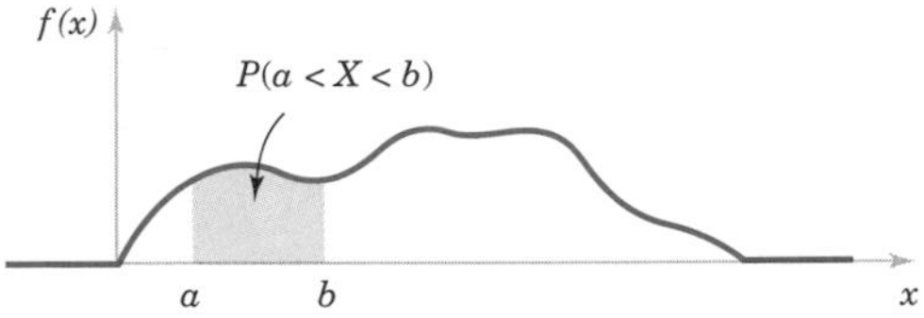

Figure 4-2   Probability determined from the area under $f(x)$.

under the density function over this interval, and it can be loosely interpreted as the sum of all the loadings over this interval.

Similarly, a **probability density function** $f(x)$ can be used to describe the probability distribution of a **continuous random variable** X. If an interval is likely to contain a value for X, its probability is large and it corresponds to large values for $f(x)$. The probability that X is between a and b is determined as the integral of $f(x)$ from a to b. See Fig. 4-2.

Definition

> For a continuous random variable X, a **probability density function** is a function such that
>
> (1) $f(x) \geq 0$
>
> (2) $\displaystyle\int_{-\infty}^{\infty} f(x)\, dx = 1$
>
> (3) $P(a \leq X \leq b) = \displaystyle\int_{a}^{b} f(x)\, dx = $ area under $f(x)$ from a to b
>
> for any a and b (4-1)

A probability density function provides a simple description of the probabilities associated with a random variable. As long as $f(x)$ is nonnegative and $\int_{-\infty}^{\infty} f(x)\, dx = 1$, $0 \leq P(a < X < b) \leq 1$ so that the probabilities are properly restricted. A probability density function is zero for x values that cannot occur and it is assumed to be zero wherever it is not specifically defined.

A **histogram** is an approximation to a probability density function. See Fig. 4-3. For each interval of the histogram, the area of the bar equals the relative frequency (proportion) of the measurements in the interval. The relative frequency is an estimate of the probability that a measurement falls in the interval. Similarly, the area under $f(x)$ over any interval equals the true probability that a measurement falls in the interval.

The important point is that $f(x)$ **is used to calculate an area** that represents the probability that X assumes a value in $[a, b]$. For the current measurement example, the probability that X results in [14 mA, 15 mA] is the integral of the probability density function of X over this interval. The probability that X results in [14.5 mA, 14.6 mA] is the integral of

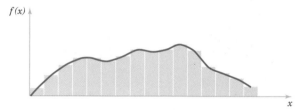

Figure 4-3   Histogram approximates a probability density function.

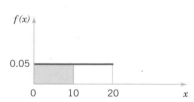

Figure 4-4   Probability density function for Example 4-1.

the same function, $f(x)$, over the smaller interval. By appropriate choice of the shape of $f(x)$, we can represent the probabilities associated with any continuous random variable X. The shape of $f(x)$ determines how the probability that X assumes a value in [14.5 mA, 14.6 mA] compares to the probability of any other interval of equal or different length.

For the density function of a loading on a long thin beam, because every point has zero width, the loading at any point is zero. Similarly, for a continuous random variable X and *any* value x.

$$P(X = x) = 0$$

Based on this result, it might appear that our model of a continuous random variable is useless. However, in practice, when a particular current measurement is observed, such as 14.47 milliamperes, this result can be interpreted as the rounded value of a current measurement that is actually in a range such as $14.465 \leq x \leq 14.475$. Therefore, the probability that the rounded value 14.47 is observed as the value for X is the probability that X assumes a value in the interval [14.465, 14.475], which is not zero. Similarly, because each point has zero probability, one need not distinguish between inequalities such as $<$ or $\leq$ for continuous random variables.

If X is a **continuous random variable**, for any x_1 and x_2,

$$P(x_1 \leq X \leq x_2) = P(x_1 < X \leq x_2) = P(x_1 \leq X < x_2) = P(x_1 < X < x_2) \quad (4\text{-}2)$$

EXAMPLE 4-1   Let the continuous random variable X denote the current measured in a thin copper wire in milliamperes. Assume that the range of X is [0, 20 mA], and assume that the probability density function of X is $f(x) = 0.05$ for $0 \leq x \leq 20$. What is the probability that a current measurement is less than 10 milliamperes?

The probability density function is shown in Fig. 4-4. It is assumed that $f(x) = 0$ wherever it is not specifically defined. The probability requested is indicated by the shaded area in Fig. 4-4.

$$P(X < 10) = \int_0^{10} f(x)\, dx = \int_0^{10} 0.05\, dx = 0.5$$

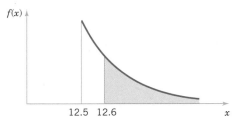

Figure 4-5 Probability density function for Example 4-2.

As another example,

$$P(5 < X < 20) = \int_{5}^{20} f(x)\,dx = 0.75$$

EXAMPLE 4-2   Let the continuous random variable X denote the diameter of a hole drilled in a sheet metal component. The target diameter is 12.5 millimeters. Most random disturbances to the process result in larger diameters. Historical data show that the distribution of X can be modeled by a probability density function $f(x) = 20e^{-20(x-12.5)}, x \geq 12.5$.

If a part with a diameter larger than 12.60 millimeters is scrapped, what proportion of parts is scrapped? The density function and the requested probability are shown in Fig. 4-5. A part is scrapped if $X > 12.60$. Now,

$$P(X > 12.60) = \int_{12.6}^{\infty} f(x)\,dx = \int_{12.6}^{\infty} 20e^{-20(x-12.5)}\,dx = -e^{-20(x-12.5)}\Big|_{12.6}^{\infty} = 0.135$$

What proportion of parts is between 12.5 and 12.6 millimeters? Now,

$$P(12.5 < X < 12.6) = \int_{12.5}^{12.6} f(x)\,dx = -e^{-20(x-12.5)}\Big|_{12.5}^{12.6} = 0.865$$

Because the total area under $f(x)$ equals 1, we can also calculate $P(12.5 < X < 12.6) = 1 - P(X > 12.6) = 1 - 0.135 = 0.865$.

EXERCISES FOR SECTION 4-2

4-1. Suppose that $f(x) = e^{-x}$ for $0 < x$. Determine the following probabilities:
(a) $P(1 < X)$ (b) $P(1 < X < 2.5)$
(c) $P(X = 3)$ (d) $P(X < 4)$
(e) $P(3 \leq X)$

4-2. Suppose that $f(x) = e^{-x}$ for $0 < x$.
(a) Determine x such that $P(x < X) = 0.10$.
(b) Determine x such that $P(X \leq x) = 0.10$.

4-3. Suppose that $f(x) = x/8$ for $3 < x < 5$. Determine the following probabilities:
(a) $P(X < 4)$ (b) $P(X > 3.5)$
(c) $P(4 < X < 5)$ (d) $P(X < 4.5)$
(e) $P(X < 3.5$ or $X > 4.5)$

4-4. Suppose that $f(x) = e^{-(x-4)}$ for $4 < x$. Determine the following probabilities:
(a) $P(1 < X)$ (b) $P(2 \leq X < 5)$

(c) $P(5 < X)$ (d) $P(8 < X < 12)$

(e) Determine x such that $P(X < x) = 0.90$.

4-5. Suppose that $f(x) = 1.5x^2$ for $-1 < x < 1$. Determine the following probabilities:

(a) $P(0 < X)$ (b) $P(0.5 < X)$

(c) $P(-0.5 \le X \le 0.5)$ (d) $P(X < -2)$

(e) $P(X < 0 \text{ or } X > -0.5)$

(f) Determine x such that $P(x < X) = 0.05$.

4-6. The probability density function of the time to failure of an electronic component in a copier (in hours) is $f(x) = \dfrac{e^{-x/1000}}{1000}$ for $x > 0$. Determine the probability that

(a) A component lasts more than 3000 hours before failure.

(b) A component fails in the interval from 1000 to 2000 hours.

(c) A component fails before 1000 hours.

(d) Determine the number of hours at which 10% of all components have failed.

4-7. The probability density function of the net weight in pounds of a packaged chemical herbicide is $f(x) = 2.0$ for $49.75 < x < 50.25$ pounds.

(a) Determine the probability that a package weighs more than 50 pounds.

(b) How much chemical is contained in 90% of all packages?

4-8. The probability density function of the length of a hinge for fastening a door is $f(x) = 1.25$ for $74.6 < x < 75.4$ millimeters. Determine the following:

(a) $P(X < 74.8)$

(b) $P(X < 74.8 \text{ or } X > 75.2)$

(c) If the specifications for this process are from 74.7 to 75.3 millimeters, what proportion of hinges meets specifications?

4-9. The probability density function of the length of a metal rod is $f(x) = 2$ for $2.3 < x < 2.8$ meters.

(a) If the specifications for this process are from 2.25 to 2.75 meters, what proportion of the bars fail to meet the specifications?

(b) Assume that the probability density function is $f(x) = 2$ for an interval of length 0.5 meters. Over what value should the density be centered to achieve the greatest proportion of bars within specifications?

4-10. If X is a continuous random variable, argue that $P(x_1 \le X \le x_2) = P(x_1 < X \le x_2) = P(x_1 \le X < x_2) = P(x_1 < X < x_2)$.

4-3   CUMULATIVE DISTRIBUTION FUNCTIONS

An alternative method to describe the distribution of a discrete random variable can also be used for continuous random variables.

Definition

> The **cumulative distribution function** of a continuous random variable X is
>
> $$F(x) = P(X \le x) = \int_{-\infty}^{x} f(u)\, du \qquad (4\text{-}3)$$
>
> for $-\infty < x < \infty$.

Extending the definition of $f(x)$ to the entire real line enables us to define the cumulative distribution function for all real numbers. The following example illustrates the definition.

EXAMPLE 4-3   For the copper current measurement in Example 4-1, the cumulative distribution function of the random variable X consists of three expressions. If $x < 0$, $f(x) = 0$. Therefore,

$$F(x) = 0, \quad \text{for} \quad x < 0$$

and

$$F(x) = \int_0^x f(u)\, du = 0.05x, \quad \text{for} \quad 0 \le x < 20$$

Finally,

$$F(x) = \int_0^x f(u)\, du = 1, \quad \text{for} \quad 20 \le x$$

Therefore,

$$F(x) = \begin{cases} 0 & x < 0 \\ 0.05x & 0 \le x < 20 \\ 1 & 20 \le x \end{cases}$$

The plot of $F(x)$ is shown in Fig. 4-6.

Notice that in the definition of $F(x)$ any $<$ can be changed to $\le$ and vice versa. That is, $F(x)$ can be defined as either $0.05x$ or 0 at the end-point $x = 0$, and $F(x)$ can be defined as either $0.05x$ or 1 at the end-point $x = 20$. In other words, $F(x)$ is a continuous function. For a discrete random variable, $F(x)$ is not a continuous function. Sometimes, a continuous random variable is defined as one that has a continuous cumulative distribution function.

EXAMPLE 4-4

For the drilling operation in Example 4-2, $F(x)$ consists of two expressions.

$$F(x) = 0 \quad \text{for} \quad x < 12.5$$

and for $12.5 \le x$

$$F(x) = \int_{12.5}^x 20e^{-20(u-12.5)}\, du$$
$$= 1 - e^{-20(x-12.5)}$$

Therefore,

$$F(x) = \begin{cases} 0 & x < 12.5 \\ 1 - e^{-20(x-12.5)} & 12.5 \le x \end{cases}$$

Figure 4-7 displays a graph of $F(x)$.

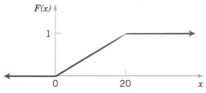

Figure 4-6 Cumulative distribution function for Example 4-3.

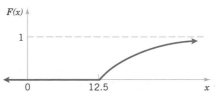

Figure 4-7 Cumulative distribution function for Example 4-4.

The probability density function of a continuous random variable can be determined from the cumulative distribution function by differentiating. Recall that the fundamental theorem of calculus states that

$$\frac{d}{dx} \int_{-\infty}^{x} f(u) \, du = f(x)$$

Then, given $F(x)$

$$f(x) = \frac{dF(x)}{dx}$$

as long as the derivative exists.

EXAMPLE 4-5   The time until a chemical reaction is complete (in milliseconds) is approximated by the cumulative distribution function

$$F(x) = \begin{cases} 0 & x < 0 \\ 1 - e^{-0.01x} & 0 \le x \end{cases}$$

Determine the probability density function of X. What proportion of reactions is complete within 200 milliseconds? Using the result that the probability density function is the derivative of the $F(x)$, we obtain

$$f(x) = \begin{cases} 0 & x < 0 \\ 0.01e^{-0.01x} & 0 \le x \end{cases}$$

The probability that a reaction completes within 200 milliseconds is

$$P(X < 200) = F(200) = 1 - e^{-2} = 0.8647.$$

EXERCISES FOR SECTION 4-3

4-11. Suppose the cumulative distribution function of the random variable X is

$$F(x) = \begin{cases} 0 & x < 0 \\ 0.2x & 0 \le x < 5 \\ 1 & 5 \le x \end{cases}$$

Determine the following:
(a) $P(X < 2.8)$ (b) $P(X > 1.5)$
(c) $P(X < -2)$ (d) $P(X > 6)$

4-12. Suppose the cumulative distribution function of the random variable X is

$$F(x) = \begin{cases} 0 & x < -2 \\ 0.25x + 0.5 & -2 \le x < 2 \\ 1 & 2 \le x \end{cases}$$

Determine the following:
(a) $P(X < 1.8)$ (b) $P(X > -1.5)$
(c) $P(X < -2)$ (d) $P(-1 < X < 1)$

4-13. Determine the cumulative distribution function for the distribution in Exercise 4-1.

4-14. Determine the cumulative distribution function for the distribution in Exercise 4-3.

4-15. Determine the cumulative distribution function for the distribution in Exercise 4-4.

4-16. Determine the cumulative distribution function for the distribution in Exercise 4-6. Use the cumulative distribution function to determine the probability that a component lasts more than 3000 hours before failure.

4-17. Determine the cumulative distribution function for the distribution in Exercise 4-8. Use the cumulative distribution function to determine the probability that a length exceeds 75 millimeters.

Determine the probability density function for each of the following cumulative distribution functions.

4-18. $F(x) = 1 - e^{-2x}$ $x > 0$

4-19.

$$F(x) = \begin{cases} 0 & x < 0 \\ 0.2x & 0 \le x < 4 \\ 0.04x + 0.64 & 4 \le x < 9 \\ 1 & 9 \le x \end{cases}$$

4-20.

$$F(x) = \begin{cases} 0 & x < -2 \\ 0.25x + 0.5 & -2 \le x < 1 \\ 0.5x + 0.25 & 1 \le x < 1.5 \\ 1 & 1.5 \le x \end{cases}$$

4-21. The gap width is an important property of a magnetic recording head. In coded units, if the width is a continuous random variable over the range from $0 < x < 2$ with $f(x) = 0.5x$, determine the cumulative distribution function of the gap width.

4-4   MEAN AND VARIANCE OF A CONTINUOUS RANDOM VARIABLE

The mean and variance of a continuous random variable are defined similarly to a discrete random variable. Integration replaces summation in the definitions. If a probability density function is viewed as a loading on a beam as in Fig. 4-1, the mean is the balance point.

Definition

Suppose X is a continuous random variable with probability density function $f(x)$. The **mean** or **expected value** of X, denoted as μ or $E(X)$, is

$$\mu = E(X) = \int_{-\infty}^{\infty} xf(x)\,dx \tag{4-4}$$

The **variance** of X, denoted as $V(X)$ or σ^2, is

$$\sigma^2 = V(X) = \int_{-\infty}^{\infty} (x - \mu)^2 f(x)\,dx = \int_{-\infty}^{\infty} x^2 f(x)\,dx - \mu^2$$

The **standard deviation** of X is $\sigma = \sqrt{\sigma^2}$.

The equivalence of the two formulas for variance can be derived as one, as was done for discrete random variables.

EXAMPLE 4-6   For the copper current measurement in Example 4-1, the mean of X is

$$E(X) = \int_{0}^{20} xf(x)\,dx = 0.05x^2/2 \Big|_{0}^{20} = 10$$

The variance of X is

$$V(X) = \int_{0}^{20} (x - 10)^2 f(x)\,dx = 0.05(x - 10)^3/3 \Big|_{0}^{20} = 33.33$$

The expected value of a function $h(X)$ of a continuous random variable is defined similarly to a function of a discrete random variable.

Expected Value of a Function of a Continuous Random Variable

If X is a continuous random variable with probability density function $f(x)$,

$$E[h(X)] = \int_{-\infty}^{\infty} h(x)f(x)\,dx \qquad (4\text{-}5)$$

EXAMPLE 4-7

In Example 4-1, X is the current measured in milliamperes. What is the expected value of the squared current? Now, $h(X) = X^2$. Therefore,

$$E[h(X)] = \int_{-\infty}^{\infty} x^2 f(x)\,dx = \int_{0}^{20} 0.05x^2\,dx = 0.05\,\frac{x^3}{3}\,\bigg|_{0}^{20} = 133.33$$

In the previous example, the expected value of X^2 does not equal $E(X)$ squared. However, in the special case that $h(X) = aX + b$ for any constants a and b, $E[h(X)] = aE(X) + b$. This can be shown from the properties of integrals.

EXAMPLE 4-8

For the drilling operation in Example 4-2, the mean of X is

$$E(X) = \int_{12.5}^{\infty} xf(x)\,dx = \int_{12.5}^{\infty} x\,20e^{-20(x-12.5)}\,dx$$

Integration by parts can be used to show that

$$E(X) = -xe^{-20(x-12.5)} - \frac{e^{-20(x-12.5)}}{20}\,\bigg|_{12.5}^{\infty} = 12.5 + 0.05 = 12.55$$

The variance of X is

$$V(X) = \int_{12.5}^{\infty} (x - 12.55)^2 f(x)\,dx$$

Although more difficult, integration by parts can be used two times to show that $V(X) = 0.0025$.

EXERCISES FOR SECTION 4-4

4-22.   Suppose $f(x) = 0.25$ for $0 < x < 4$. Determine the mean and variance of X.

4-23.   Suppose $f(x) = 0.125x$ for $0 < x < 4$. Determine the mean and variance of X.

4-24.   Suppose $f(x) = 1.5x^2$ for $-1 < x < 1$. Determine the mean and variance of X.

4-25.   Suppose that $f(x) = x/8$ for $3 < x < 5$. Determine the mean and variance for x.

4-26.   Determine the mean and variance of the weight of packages in Exercise 4.7.

4-27.   The thickness of a conductive coating in micrometers has a density function of $600x^{-2}$ for $100\ \mu\text{m} < x < 120\ \mu\text{m}$.

(a) Determine the mean and variance of the coating thickness.

(b) If the coating costs $0.50 per micrometer of thickness on each part, what is the average cost of the coating per part?

4-28. Suppose that contamination particle size (in micrometers) can be modeled as $f(x) = 2x^{-3}$ for $1 < x$. Determine the mean of X.

4-29. Integration by parts is required. The probability density function for the diameter of a drilled hole in millimeters is $10e^{-10(x-5)}$ for $x > 5$ mm. Although the target diameter is 5 millimeters, vibrations, tool wear, and other nuisances produce diameters larger than 5 millimeters.

(a) Determine the mean and variance of the diameter of the holes.

(b) Determine the probability that a diameter exceeds 5.1 millimeters.

4-30. Suppose the probability density function of the length of computer cables is $f(x) = 0.1$ from 1200 to 1210 millimeters.

(a) Determine the mean and standard deviation of the cable length.

(b) If the length specifications are $1195 < x < 1205$ millimeters, what proportion of cables are within specifications?

4-5 CONTINUOUS UNIFORM DISTRIBUTION

The simplest continuous distribution is analogous to its discrete counterpart.

Definition

A continuous random variable X with probability density function

$$f(x) = 1/(b - a), \qquad a \leq x \leq b \tag{4-6}$$

is a **continuous uniform random variable**.

The probability density function of a continuous uniform random variable is shown in Fig. 4-8. The mean of the continuous uniform random variable X is

$$E(X) = \int_a^b \frac{x}{b - a}\, dx = \left. \frac{0.5x^2}{b - a} \right|_a^b = \frac{(a + b)}{2}$$

The variance of X is

$$V(X) = \int_a^b \frac{\left(x - \left(\dfrac{a + b}{2}\right)\right)^2}{b - a}\, dx = \left. \frac{\left(x - \dfrac{a + b}{2}\right)^3}{3(b - a)} \right|_a^b = \frac{(b - a)^2}{12}$$

These results are summarized as follows.

If X is a continuous uniform random variable over $a \leq x \leq b$,

$$\mu = E(X) = \frac{(a + b)}{2} \quad \text{and} \quad \sigma^2 = V(X) = \frac{(b - a)^2}{12} \tag{4-7}$$

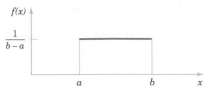

Figure 4-8   Continuous uniform probability density function.

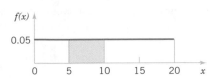

Figure 4-9   Probability for Example 4-9.

EXAMPLE 4-9   Let the continuous random variable X denote the current measured in a thin copper wire in milliamperes. Assume that the range of X is [0, 20 mA], and assume that the probability density function of X is $f(x) = 0.05, 0 \leq x \leq 20$.

What is the probability that a measurement of current is between 5 and 10 milliamperes? The requested probability is shown as the shaded area in Fig. 4-9.

$$P(5 < X < 10) = \int_5^{10} f(x)\, dx$$

$$= 5(0.05) = 0.25$$

The mean and variance formulas can be applied with $a = 0$ and $b = 20$. Therefore,

$$E(X) = 10\text{ mA} \quad \text{and} \quad V(X) = 20^2/12 = 33.33\text{ mA}^2$$

Consequently, the standard deviation of X is 5.77 mA.

The cumulative distribution function of a continuous uniform random variable is obtained by integration. If $a < x < b$,

$$F(x) = \int_a^x 1/(b - a)\, du = x/(b - a) - a/(b - a)$$

Therefore, the complete description of the cumulative distribution function of a continuous uniform random variable is

$$F(x) = \begin{cases} 0 & x < a \\ (x - a)/(b - a) & a \leq x < b \\ 1 & b \leq x \end{cases}$$

An example of $F(x)$ for a continuous uniform random variable is shown in Fig. 4-6.

EXERCISES FOR SECTION 4-5

4-31.   Suppose X has a continuous uniform distribution over the interval [1.5, 5.5].
(a) Determine the mean, variance, and standard deviation of X.
(b) What is $P(X < 2.5)$?

4-32.   Suppose X has a continuous uniform distribution over the interval $[-1, 1]$.

(a) Determine the mean, variance, and standard deviation of X.
(b) Determine the value for x such that $P(-x < X < x) = 0.90$.

4-33.   The net weight in pounds of a packaged chemical herbicide is uniform for $49.75 < x < 50.25$ pounds.
(a) Determine the mean and variance of the weight of packages.

(b) Determine the cumulative distribution function of the weight of packages.

(c) Determine $P(X < 50.1)$.

4-34. The thickness of a flange on an aircraft component is uniformly distributed between 0.95 and 1.05 millimeters.

(a) Determine the cumulative distribution function of flange thickness.

(b) Determine the proportion of flanges that exceeds 1.02 millimeters.

(c) What thickness is exceeded by 90% of the flanges?

(d) Determine the mean and variance of flange thickness.

4-35. Suppose the time it takes a data collection operator to fill out an electronic form for a database is uniformly between 1.5 and 2.2 minutes.

(a) What is the mean and variance of the time it takes an operator to fill out the form?

(b) What is the probability that it will take less than two minutes to fill out the form?

(c) Determine the cumulative distribution function of the time it takes to fill out the form.

4-36. The probability density function of the time it takes a hematology cell counter to complete a test on a blood sample is $f(x) = 0.04$ for $50 < x < 75$ seconds.

(a) What percentage of tests require more than 70 seconds to complete.

(b) What percentage of tests require less than one minute to complete.

(c) Determine the mean and variance of the time to complete a test on a sample.

4-37. The thickness of photoresist applied to wafers in semiconductor manufacturing at a particular location on the wafer is uniformly distributed between 0.2050 and 0.2150 micrometers.

(a) Determine the cumulative distribution function of photoresist thickness.

(b) Determine the proportion of wafers that exceeds 0.2125 micrometers in photoresist thickness.

(c) What thickness is exceeded by 10% of the wafers?

(d) Determine the mean and variance of photoresist thickness.

4-38. The probability density function of the time required to complete an assembly operation is $f(x) = 0.1$ for $30 < x < 40$ seconds.

(a) Determine the proportion of assemblies that requires more than 35 seconds to complete.

(b) What time is exceeded by 90% of the assemblies?

(c) Determine the mean and variance of time of assembly.

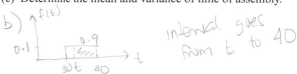

4-6   NORMAL DISTRIBUTION

Undoubtedly, the most widely used model for the distribution of a random variable is a **normal distribution.** Whenever a random experiment is replicated, the random variable that equals the average (or total) result over the replicates tends to have a normal distribution as the number of replicates becomes large. De Moivre presented this fundamental result, known as the **central limit theorem,** in 1733. Unfortunately, his work was lost for some time, and Gauss independently developed a normal distribution nearly 100 years later. Although De Moivre was later credited with the derivation, a normal distribution is also referred to as a **Gaussian** distribution.

When do we average (or total) results? Almost always. For example, an automotive engineer may plan a study to average pull-off force measurements from several connectors. If we assume that each measurement results from a replicate of a random experiment, the normal distribution can be used to make approximate conclusions about this average. These conclusions are the primary topics in the subsequent chapters of this book.

Furthermore, sometimes the central limit theorem is less obvious. For example, assume that the deviation (or error) in the length of a machined part is the sum of a large number of infinitesimal effects, such as temperature and humidity drifts, vibrations, cutting angle variations, cutting tool wear, bearing wear, rotational speed variations, mounting and fixturing variations, variations in numerous raw material characteristics, and variation in levels of contamination. If the component errors are independent and equally likely to be positive or negative, the total error can be shown to have an approximate normal distribution. Furthermore, the normal distribution arises in the study of numerous basic physical phenomena. For example, the physicist Maxwell developed a normal distribution from simple assumptions regarding the velocities of molecules.

The theoretical basis of a normal distribution is mentioned to justify the somewhat complex form of the probability density function. Our objective now is to calculate probabilities for a normal random variable. The central limit theorem will be stated more carefully later.

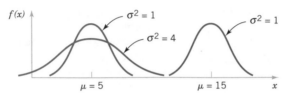

Figure 4-10   Normal probability density functions for selected values of the parameters μ and σ^2.

Random variables with different means and variances can be modeled by normal probability density functions with appropriate choices of the center and width of the curve. The value of $E(X) = \mu$ determines the center of the probability density function and the value of $V(X) = \sigma^2$ determines the width. Figure 4-10 illustrates several normal probability density functions with selected values of μ and σ^2. Each has the characteristic symmetric bell-shaped curve, but the centers and dispersions differ. The following definition provides the formula for normal probability density functions.

Definition

A random variable X with probability density function

$$f(x) = \frac{1}{\sqrt{2\pi}\sigma} e^{\frac{-(x-\mu)^2}{2\sigma^2}} \qquad -\infty < x < \infty \qquad (4\text{-}8)$$

is a **normal random variable** with parameters μ, where $-\infty < \mu < \infty$, and $\sigma > 0$. Also,

$$E(X) = \mu \quad \text{and} \quad V(X) = \sigma^2 \qquad (4\text{-}9)$$

and the notation $N(\mu, \sigma^2)$ is used to denote the distribution. The mean and variance of X are shown to equal μ and σ^2, respectively, at the end of this Section 5-6.

EXAMPLE 4-10   Assume that the current measurements in a strip of wire follow a normal distribution with a mean of 10 milliamperes and a variance of 4 (milliamperes)^2. What is the probability that a measurement exceeds 13 milliamperes?

Let X denote the current in milliamperes. The requested probability can be represented as $P(X > 13)$. This probability is shown as the shaded area under the normal probability density function in Fig. 4-11. Unfortunately, there is no closed-form expression for the integral of a normal probability density function, and probabilities based on the normal distribution are typically found numerically or from a table (that we will later introduce).

Some useful results concerning a normal distribution are summarized below and in Fig. 4-12. For any normal random variable,

$$P(\mu - \sigma < X < \mu + \sigma) = 0.6827$$
$$P(\mu - 2\sigma < X < \mu + 2\sigma) = 0.9545$$
$$P(\mu - 3\sigma < X < \mu + 3\sigma) = 0.9973$$

Also, from the symmetry of $f(x)$, $P(X > \mu) = P(X < \mu) = 0.5$. Because $f(x)$ is positive for all x, this model assigns some probability to each interval of the real line. However, the

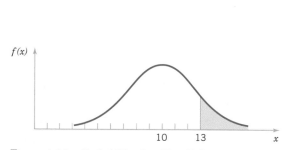

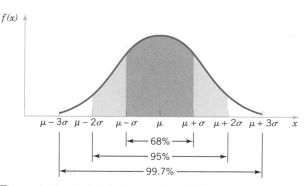

Figure 4-11   Probability that $X > 13$ for a normal random variable with $\mu = 10$ and $\sigma^2 = 4$.

Figure 4-12   Probabilities associated with a normal distribution.

probability density function decreases as x moves farther from μ. Consequently, the probability that a measurement falls far from μ is small, and at some distance from μ the probability of an interval can be approximated as zero.

The area under a normal probability density function beyond 3σ from the mean is quite small. This fact is convenient for quick, rough sketches of a normal probability density function. The sketches help us determine probabilities. Because more than 0.9973 of the probability of a normal distribution is within the interval $(\mu - 3\sigma, \mu + 3\sigma)$, 6σ is often referred to as the **width** of a normal distribution. Advanced integration methods can be used to show that the area under the normal probability density function from $-\infty < x < \infty$ is 1.

Definition

> A normal random variable with
>
> $$\mu = 0 \quad \text{and} \quad \sigma^2 = 1$$
>
> is called a **standard normal random variable** and is denoted as Z.
>
> The cumulative distribution function of a standard normal random variable is denoted as
>
> $$\Phi(z) = P(Z \le z)$$

Appendix Table II provides cumulative probability values for $\Phi(z)$, for a standard normal random variable. Cumulative distribution functions for normal random variables are also widely available in computer packages. They can be used in the same manner as Appendix Table II to obtain probabilities for these random variables. The use of Table II is illustrated by the following example.

EXAMPLE 4-11   Assume Z is a standard normal random variable. Appendix Table II provides probabilities of the form $P(Z \le z)$. The use of Table II to find $P(Z \le 1.5)$ is illustrated in Fig. 4-13. Read down the z column to the row that equals 1.5. The probability is read from the adjacent column, labeled 0.00, to be 0.93319.

The column headings refer to the hundredth's digit of the value of z in $P(Z \le z)$. For example, $P(Z \le 1.53)$ is found by reading down the z column to the row 1.5 and then selecting the probability from the column labeled 0.03 to be 0.93699.

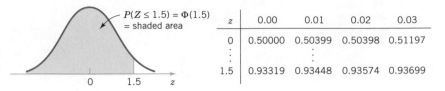

Figure 4-13 Standard normal probability density function.

Probabilities that are not of the form $P(Z \leq z)$ are found by using the basic rules of probability and the symmetry of the normal distribution along with Appendix Table II. The following examples illustrate the method.

EXAMPLE 4-12   The following calculations are shown pictorially in Fig. 4-14. In practice, a probability is often rounded to one or two significant digits.

(1) $P(Z > 1.26) = 1 - P(Z \leq 1.26) = 1 - 0.89616 = 0.10384$

(2) $P(Z < -0.86) = 0.19490$.

(3) $P(Z > -1.37) = P(Z < 1.37) = 0.91465$

(4) $P(-1.25 < Z < 0.37)$. This probability can be found from the difference of two areas, $P(Z < 0.37) - P(Z < -1.25)$. Now,

$$P(Z < 0.37) = 0.64431 \quad \text{and} \quad P(Z < -1.25) = 0.10565$$

Therefore,

$$P(-1.25 < Z < 0.37) = 0.64431 - 0.10565 = 0.53866$$

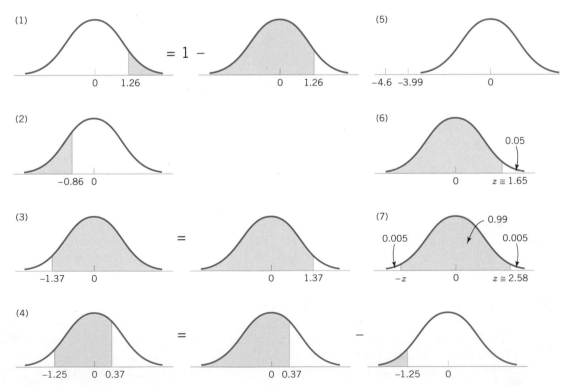

Figure 4-14 Graphical displays for standard normal distributions.

(5) $P(Z \leq -4.6)$ cannot be found exactly from Appendix Table II. However, the last entry in the table can be used to find that $P(Z \leq -3.99) = 0.00003$. Because $P(Z \leq -4.6) < P(Z \leq -3.99)$, $P(Z \leq -4.6)$ is nearly zero.

(6) Find the value z such that $P(Z > z) = 0.05$. This probability expression can be written as $P(Z \leq z) = 0.95$. Now, Table II is used in reverse. We search through the probabilities to find the value that corresponds to 0.95. The solution is illustrated in Fig. 4-14. We do not find 0.95 exactly; the nearest value is 0.95053, corresponding to $z = 1.65$.

(7) Find the value of z such that $P(-z < Z < z) = 0.99$. Because of the symmetry of the normal distribution, if the area of the shaded region in Fig. 4-14(7) is to equal 0.99, the area in each tail of the distribution must equal 0.005. Therefore, the value for z corresponds to a probability of 0.995 in Table II. The nearest probability in Table II is 0.99506, when $z = 2.58$.

The preceding examples show how to calculate probabilities for standard normal random variables. To use the same approach for an arbitrary normal random variable would require a separate table for every possible pair of values for μ and σ. Fortunately, all normal probability distributions are related algebraically, and Appendix Table II can be used to find the probabilities associated with an arbitrary normal random variable by first using a simple transformation.

If X is a normal random variable with $E(X) = \mu$ and $V(X) = \sigma^2$, the random variable

$$Z = \frac{X - \mu}{\sigma} \qquad (4\text{-}10)$$

is a normal random variable with $E(Z) = 0$ and $V(Z) = 1$. That is, Z is a standard normal random variable.

Creating a new random variable by this transformation is referred to as **standardizing**. The random variable Z represents the distance of X from its mean in terms of standard deviations. It is the key step to calculate a probability for an arbitrary normal random variable.

EXAMPLE 4-13 Suppose the current measurements in a strip of wire are assumed to follow a normal distribution with a mean of 10 milliamperes and a variance of 4 (milliamperes)^2. What is the probability that a measurement will exceed 13 milliamperes?

Let X denote the current in milliamperes. The requested probability can be represented as $P(X > 13)$. Let $Z = (X - 10)/2$. The relationship between the several values of X and the transformed values of Z are shown in Fig. 4-15. We note that $X > 13$ corresponds to $Z > 1.5$. Therefore, from Appendix Table II,

$$P(X > 13) = P(Z > 1.5) = 1 - P(Z \leq 1.5) = 1 - 0.93319 = 0.06681$$

Rather than using Fig. 4-15, the probability can be found from the inequality $X > 13$. That is,

$$P(X > 13) = P\left(\frac{(X - 10)}{2} > \frac{(13 - 10)}{2}\right) = P(Z > 1.5) = 0.06681$$

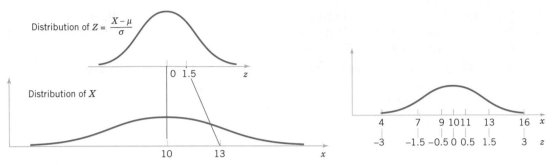

Figure 4-15   Standardizing a normal random variable.

In the preceding example, the value 13 is transformed to 1.5 by standardizing, and 1.5 is often referred to as the **z-value** associated with a probability. The following summarizes the calculation of probabilities derived from normal random variables.

Suppose X is a normal random variable with mean μ and variance σ^2. Then,

$$P(X \le x) = P\left(\frac{X - \mu}{\sigma} \le \frac{x - \mu}{\sigma}\right) = P(Z \le z) \qquad (4\text{-}11)$$

where Z is a **standard normal random variable**, and $z = \dfrac{(x - \mu)}{\sigma}$ is the **z-value** obtained by **standardizing** X.

The probability is obtained by entering Appendix Table II with $z = (x - \mu)/\sigma$.

EXAMPLE 4-14   Continuing the previous example, what is the probability that a current measurement is between 9 and 11 milliamperes? From Fig. 4-15, or by proceeding algebraically, we have

$$P(9 < X < 11) = P((9 - 10)/2 < (X - 10)/2 < (11 - 10)/2)$$
$$= P(-0.5 < Z < 0.5) = P(Z < 0.5) - P(Z < -0.5)$$
$$= 0.69146 - 0.30854 = 0.38292$$

Determine the value for which the probability that a current measurement is below this value is 0.98. The requested value is shown graphically in Fig. 4-16. We need the value of x such that $P(X < x) = 0.98$. By standardizing, this probability expression can be written as

$$P(X < x) = P((X - 10)/2 < (x - 10)/2)$$
$$= P(Z < (x - 10)/2)$$
$$= 0.98$$

Appendix Table II is used to find the z-value such that $P(Z < z) = 0.98$. The nearest probability from Table II results in

$$P(Z < 2.05) = 0.97982$$

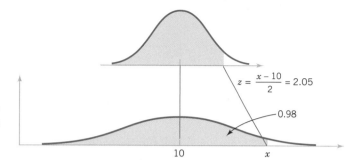

Figure 4-16 Determining the value of x to meet a specified probability.

$$z = \frac{x - 10}{2} = 2.05$$

0.98

10 $\qquad$ x

Therefore, $(x - 10)/2 = 2.05$, and the standardizing transformation is used in reverse to solve for x. The result is

$$x = 2(2.05) + 10 = 14.1 \text{ milliamperes}$$

EXAMPLE 4-15 Assume that in the detection of a digital signal the background noise follows a normal distribution with a mean of 0 volt and standard deviation of 0.45 volt. The system assumes a digital 1 has been transmitted when the voltage exceeds 0.9. What is the probability of detecting a digital 1 when none was sent?

Let the random variable N denote the voltage of noise. The requested probability is

$$P(N > 0.9) = P\left(\frac{N}{0.45} > \frac{0.9}{0.45}\right) = P(Z > 2) = 1 - 0.97725 = 0.02275$$

This probability can be described as the probability of a false detection.

Determine symmetric bounds about 0 that include 99% of all noise readings. The question requires us to find x such that $P(-x < N < x) = 0.99$. A graph is shown in Fig. 4-17. Now,

$$P(-x < N < x) = P(-x/0.45 < N/0.45 < x/0.45)$$
$$= P(-x/0.45 < Z < x/0.45) = 0.99$$

From Appendix Table II

$$P(-2.58 < Z < 2.58) = 0.99$$

Standardized distribution of $\frac{N}{0.45}$

Distribution of N

Figure 4-17 Determining the value of x to meet a specified probability.

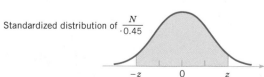

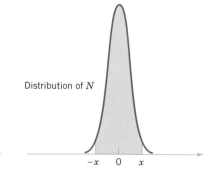

Therefore,

$$x/0.45 = 2.58$$

and

$$x = 2.58(0.45) = 1.16$$

Suppose a digital 1 is represented as a shift in the mean of the noise distribution to 1.8 volts. What is the probability that a digital 1 is not detected? Let the random variable S denote the voltage when a digital 1 is transmitted. Then,

$$P(S < 0.9) = P\left(\frac{S - 1.8}{0.45} < \frac{0.9 - 1.8}{0.45}\right) = P(Z < -2) = 0.02275$$

This probability can be interpreted as the probability of a missed signal.

EXAMPLE 4-16   The diameter of a shaft in an optical storage drive is normally distributed with mean 0.2508 inch and standard deviation 0.0005 inch. The specifications on the shaft are 0.2500 ± 0.0015 inch. What proportion of shafts conforms to specifications?

Let X denote the shaft diameter in inches. The requested probability is shown in Fig. 4-18 and

$$P(0.2485 < X < 0.2515) = P\left(\frac{0.2485 - 0.2508}{0.0005} < Z < \frac{0.2515 - 0.2508}{0.0005}\right)$$
$$= P(-4.6 < Z < 1.4) = P(Z < 1.4) - P(Z < -4.6)$$
$$= 0.91924 - 0.0000 = 0.91924$$

Most of the nonconforming shafts are too large, because the process mean is located very near to the upper specification limit. If the process is centered so that the process mean is equal to the target value of 0.2500,

$$P(0.2485 < X < 0.2515) = P\left(\frac{0.2485 - 0.2500}{0.0005} < Z < \frac{0.2515 - 0.2500}{0.0005}\right)$$

$$= P(-3 < Z < 3)$$
$$= P(Z < 3) - P(Z < -3)$$
$$= 0.99865 - 0.00135$$
$$= 0.9973$$

By recentering the process, the yield is increased to approximately 99.73%.

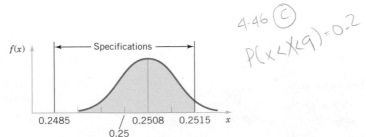

Figure 4-18
Distribution for
Example 4-16.

Mean and Variance of the Normal Distribution (CD Only)

EXERCISES FOR SECTION 4-6

4-39. Use Appendix Table II to determine the following probabilities for the standard normal random variable Z:
(a) $P(Z < 1.32)$ (b) $P(Z < 3.0)$
(c) $P(Z > 1.45)$ (d) $P(Z > -2.15)$
(e) $P(-2.34 < Z < 1.76)$

4-40. Use Appendix Table II to determine the following probabilities for the standard normal random variable Z:
(a) $P(-1 < Z < 1)$ (b) $P(-2 < Z < 2)$
(c) $P(-3 < Z < 3)$ (d) $P(Z > 3)$
(e) $P(0 < Z < 1)$

4-41. Assume Z has a standard normal distribution. Use Appendix Table II to determine the value for z that solves each of the following:
(a) $P(Z < z) = 0.9$ (b) $P(Z < z) = 0.5$
(c) $P(Z > z) = 0.1$ (d) $P(Z > z) = 0.9$
(e) $P(-1.24 < Z < z) = 0.8$

4-42. Assume Z has a standard normal distribution. Use Appendix Table II to determine the value for z that solves each of the following:
(a) $P(-z < Z < z) = 0.95$ (b) $P(-z < Z < z) = 0.99$
(c) $P(-z < Z < z) = 0.68$ (d) $P(-z < Z < z) = 0.9973$

4-43. Assume X is normally distributed with a mean of 10 and a standard deviation of 2. Determine the following:
(a) $P(X < 13)$ (b) $P(X > 9)$
(c) $P(6 < X < 14)$ (d) $P(2 < X < 4)$
(e) $P(-2 < X < 8)$

4-44. Assume X is normally distributed with a mean of 10 and a standard deviation of 2. Determine the value for x that solves each of the following:
(a) $P(X > x) = 0.5$
(b) $P(X > x) = 0.95$
(c) $P(x < X < 10) = 0.2$
(d) $P(-x < X - 10 < x) = 0.95$
(e) $P(-x < X - 10 < x) = 0.99$

4-45. Assume X is normally distributed with a mean of 5 and a standard deviation of 4. Determine the following:
(a) $P(X < 11)$ (b) $P(X > 0)$
(c) $P(3 < X < 7)$ (d) $P(-2 < X < 9)$
(e) $P(2 < X < 8)$

4-46. Assume X is normally distributed with a mean of 5 and a standard deviation of 4. Determine the value for x that solves each of the following:
(a) $P(X > x) = 0.5$ (b) $P(X > x) = 0.95$
(c) $P(x < X < 9) = 0.2$ (d) $P(3 < X < x) = 0.95$
(e) $P(-x < X - 5 < x) = 0.99$

4-47. The compressive strength of samples of cement can be modeled by a normal distribution with a mean of 6000 kilograms per square centimeter and a standard deviation of 100 kilograms per square centimeter.

(a) What is the probability that a sample's strength is less than 6250 Kg/cm^2?
(b) What is the probability that a sample's strength is between 5800 and 5900 Kg/cm^2?
(c) What strength is exceeded by 95% of the samples?

4-48. The tensile strength of paper is modeled by a normal distribution with a mean of 35 pounds per square inch and a standard deviation of 2 pounds per square inch.
(a) What is the probability that the strength of a sample is less than 40 lb/in^2?
(b) If the specifications require the tensile strength to exceed 30 lb/in^2, what proportion of the samples is scrapped?

4-49. The line width of for semiconductor manufacturing is assumed to be normally distributed with a mean of 0.5 micrometer and a standard deviation of 0.05 micrometer.
(a) What is the probability that a line width is greater than 0.62 micrometer?
(b) What is the probability that a line width is between 0.47 and 0.63 micrometer?
(c) The line width of 90% of samples is below what value?

4-50. The fill volume of an automated filling machine used for filling cans of carbonated beverage is normally distributed with a mean of 12.4 fluid ounces and a standard deviation of 0.1 fluid ounce.
(a) What is the probability a fill volume is less than 12 fluid ounces?
(b) If all cans less than 12.1 or greater than 12.6 ounces are scrapped, what proportion of cans is scrapped?
(c) Determine specifications that are symmetric about the mean that include 99% of all cans.

4-51. The time it takes a cell to divide (called mitosis) is normally distributed with an average time of one hour and a standard deviation of 5 minutes.
(a) What is the probability that a cell divides in less than 45 minutes?
(b) What is the probability that it takes a cell more than 65 minutes to divide?
(c) What is the time that it takes approximately 99% of all cells to complete mitosis?

4-52. In the previous exercise, suppose that the mean of the filling operation can be adjusted easily, but the standard deviation remains at 0.1 ounce.
(a) At what value should the mean be set so that 99.9% of all cans exceed 12 ounces?
(b) At what value should the mean be set so that 99.9% of all cans exceed 12 ounces if the standard deviation can be reduced to 0.05 fluid ounce?

4-53. The reaction time of a driver to visual stimulus is normally distributed with a mean of 0.4 seconds and a standard deviation of 0.05 seconds.

(a) What is the probability that a reaction requires more than 0.5 seconds?

(b) What is the probability that a reaction requires between 0.4 and 0.5 seconds?

(c) What is the reaction time that is exceeded 90% of the time?

4-54. The speed of a file transfer from a server on campus to a personal computer at a student's home on a weekday evening is normally distributed with a mean of 60 kilobits per second and a standard deviation of 4 kilobits per second.

(a) What is the probability that the file will transfer at a speed of 70 kilobits per second or more?

(b) What is the probability that the file will transfer at a speed of less than 58 kilobits per second?

(c) If the file is 1 megabyte, what is the average time it will take to transfer the file? (Assume eight bits per byte.)

4-55. The length of an injection-molded plastic case that holds magnetic tape is normally distributed with a length of 90.2 millimeters and a standard deviation of 0.1 millimeter.

(a) What is the probability that a part is longer than 90.3 millimeters or shorter than 89.7 millimeters?

(b) What should the process mean be set at to obtain the greatest number of parts between 89.7 and 90.3 millimeters?

(c) If parts that are not between 89.7 and 90.3 millimeters are scrapped, what is the yield for the process mean that you selected in part (b)?

4-56. In the previous exercise assume that the process is centered so that the mean is 90 millimeters and the standard deviation is 0.1 millimeter. Suppose that 10 cases are measured, and they are assumed to be independent.

(a) What is the probability that all 10 cases are between 89.7 and 90.3 millimeters?

(b) What is the expected number of the 10 cases that are between 89.7 and 90.3 millimeters?

4-57. The sick-leave time of employees in a firm in a month is normally distributed with a mean of 100 hours and a standard deviation of 20 hours.

(a) What is the probability that the sick-leave time for next month will be between 50 and 80 hours?

(b) How much time should be budgeted for sick leave if the budgeted amount should be exceeded with a probability of only 10%?

4-58. The life of a semiconductor laser at a constant power is normally distributed with a mean of 7000 hours and a standard deviation of 600 hours.

(a) What is the probability that a laser fails before 5000 hours?

(b) What is the life in hours that 95% of the lasers exceed?

(c) If three lasers are used in a product and they are assumed to fail independently, what is the probability that all three are still operating after 7000 hours?

4-59. The diameter of the dot produced by a printer is normally distributed with a mean diameter of 0.002 inch and a standard deviation of 0.0004 inch.

(a) What is the probability that the diameter of a dot exceeds 0.0026 inch?

(b) What is the probability that a diameter is between 0.0014 and 0.0026 inch?

(c) What standard deviation of diameters is needed so that the probability in part (b) is 0.995?

4-60. The weight of a sophisticated running shoe is normally distributed with a mean of 12 ounces and a standard deviation of 0.5 ounce.

(a) What is the probability that a shoe weighs more than 13 ounces?

(b) What must the standard deviation of weight be in order for the company to state that 99.9% of its shoes are less than 13 ounces?

(c) If the standard deviation remains at 0.5 ounce, what must the mean weight be in order for the company to state that 99.9% of its shoes are less than 13 ounces?

4-7 NORMAL APPROXIMATION TO THE BINOMIAL AND POISSON DISTRIBUTIONS

We began our section on the normal distribution with the central limit theorem and the normal distribution as an approximation to a random variable with a large number of trials. Consequently, it should not be a surprise to learn that the normal distribution can be used to approximate binomial probabilities for cases in which n is large. The following example illustrates that for many physical systems the binomial model is appropriate with an extremely large value for n. In these cases, it is difficult to calculate probabilities by using the binomial distribution. Fortunately, the normal approximation is most effective in these cases. An illustration is provided in Fig. 4-19. The area of each bar equals the binomial probability of x. Notice that the area of bars can be approximated by areas under the normal density function.

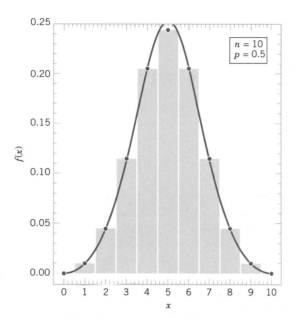

Figure 4-19 Normal approximation to the binomial distribution.

EXAMPLE 4-17   In a digital communication channel, assume that the number of bits received in error can be modeled by a binomial random variable, and assume that the probability that a bit is received in error is 1×10^{-5}. If 16 million bits are transmitted, what is the probability that more than 150 errors occur?

Let the random variable X denote the number of errors. Then X is a binomial random variable and

$$P(X > 150) = 1 - P(x \le 150) = 1 - \sum_{x=0}^{150} \binom{16,000,000}{x} (10^{-5})^x (1 - 10^{-5})^{16,000,000-x}$$

Clearly, the probability in Example 4-17 is difficult to compute. Fortunately, the normal distribution can be used to provide an excellent approximation in this example.

Normal Approximation to the Binomial Distribution

If X is a binomial random variable,

$$Z = \frac{X - np}{\sqrt{np(1 - p)}} \qquad (4\text{-}12)$$

is approximately a standard normal random variable. The approximation is good for

$$np > 5 \quad \text{and} \quad n(1 - p) > 5$$

Recall that for a binomial variable X, $E(X) = np$ and $V(X) = np(1 - p)$. Consequently, the expression in Equation 4-12 is nothing more than the formula for standardizing the random variable X. Probabilities involving X can be approximated by using a standard normal distribution. The approximation is good when n is large relative to p.

EXAMPLE 4-18   The digital communication problem in the previous example is solved as follows:

$$P(X > 150) = P\left(\frac{X - 160}{\sqrt{160(1 - 10^{-5})}} > \frac{150 - 160}{\sqrt{160(1 - 10^{-5})}}\right)$$
$$= P(Z > -0.79) = P(Z < 0.79) = 0.785$$

Because $np = (16 \times 10^6)(1 \times 10^{-5}) = 160$ and $n(1 - p)$ is much larger, the approximation is expected to work well in this case.

EXAMPLE 4-19   Again consider the transmission of bits in Example 4-18. To judge how well the normal approximation works, assume only $n = 50$ bits are to be transmitted and that the probability of an error is $p = 0.1$. The exact probability that 2 or less errors occur is

$$P(X \le 2) = \binom{50}{0} 0.9^{50} + \binom{50}{1} 0.1(0.9^{49}) + \binom{50}{2} 0.1^2(0.9^{48}) = 0.112$$

Based on the normal approximation

$$P(X \le 2) = P\left(\frac{X - 5}{2.12} < \frac{2 - 5}{2.12}\right) = P(Z < -1.42) = 0.08$$

Even for a sample as small as 50 bits, the normal approximation is reasonable.

If np or $n(1 - p)$ is small, the binomial distribution is quite skewed and the symmetric normal distribution is not a good approximation. Two cases are illustrated in Fig. 4-20. However, a correction factor can be used that will further improve the approximation. This factor is called a **continuity correction** and it is discussed in Section 4-8 on the CD.

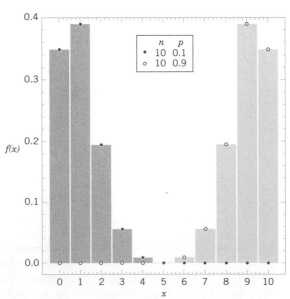

Figure 4-20 Binomial distribution is not symmetrical if p is near 0 or 1.

hypergometric	$\approx$	binomial	$\approx$	normal
distribution	$\dfrac{n}{N} < 0.1$	distribution	$np > 5$ $n(1 - p) > 5$	distribution

Figure 4-21   Conditions for approximating hypergeometric and binomial probabilities.

Recall that the binomial distribution is a satisfactory approximation to the hypergeometric distribution when n, the sample size, is small relative to N, the size of the population from which the sample is selected. A rule of thumb is that the binomial approximation is effective if $n/N < 0.1$. Recall that for a hypergeometric distribution p is defined as $p = K/N$. That is, p is interpreted as the number of successes in the population. Therefore, the normal distribution can provide an effective approximation of hypergeometric probabilities when $n/N < 0.1$, $np > 5$ and $n(1 - p) > 5$. Figure 4-21 provides a summary of these guidelines.

Recall that the Poisson distribution was developed as the limit of a binomial distribution as the number of trials increased to infinity. Consequently, it should not be surprising to find that the normal distribution can also be used to approximate probabilities of a Poisson random variable.

Normal Approximation to the Poisson Distribution

If X is a Poisson random variable with $E(X) = \lambda$ and $V(X) = \lambda$,

$$Z = \frac{X - \lambda}{\sqrt{\lambda}} \tag{4-13}$$

is approximately a standard normal random variable. The approximation is good for

$$\lambda > 5$$

EXAMPLE 4-20

Assume that the number of asbestos particles in a squared meter of dust on a surface follows a Poisson distribution with a mean of 1000. If a squared meter of dust is analyzed, what is the probability that less than 950 particles are found?

This probability can be expressed exactly as

$$P(X \le 950) = \sum_{x=0}^{950} \frac{e^{-1000} x^{1000}}{x!}$$

The computational difficulty is clear. The probability can be approximated as

$$P(X \le x) = P\left(Z \le \frac{950 - 1000}{\sqrt{1000}}\right) = P(Z \le -1.58) = 0.057$$

EXERCISES FOR SECTION 4-7

4-61. Suppose that X is a binomial random variable with $n = 200$ and $p = 0.4$.

(a) Approximate the probability that X is less than or equal to 70.

(b) Approximate the probability that X is greater than 70 and less than 90.

4-62. Suppose that X is a binomial random variable with $n = 100$ and $p = 0.1$.

(a) Compute the exact probability that X is less than 4.

(b) Approximate the probability that X is less than 4 and compare to the result in part (a).

(c) Approximate the probability that $8 < X < 12$.

4-63. The manufacturing of semiconductor chips produces 2% defective chips. Assume the chips are independent and that a lot contains 1000 chips.
(a) Approximate the probability that more than 25 chips are defective.
(b) Approximate the probability that between 20 and 30 chips are defective.

4-64. A supplier ships a lot of 1000 electrical connectors. A sample of 25 is selected at random, without replacement. Assume the lot contains 100 defective connectors.
(a) Using a binomial approximation, what is the probability that there are no defective connectors in the sample?
(b) Use the normal approximation to answer the result in part (a). Is the approximation satisfactory?
(c) Redo parts (a) and (b) assuming the lot size is 500. Is the normal approximation to the probability that there are no defective connectors in the sample satisfactory in this case?

4-65. An electronic office product contains 5000 electronic components. Assume that the probability that each component operates without failure during the useful life of the product is 0.999, and assume that the components fail independently. Approximate the probability that 10 or more of the original 5000 components fail during the useful life of the product.

4-66. Suppose that the number of asbestos particles in a sample of 1 squared centimeter of dust is a Poisson random variable with a mean of 1000. What is the probability that 10 squared centimeters of dust contains more than 10,000 particles?

4-67. A corporate Web site contains errors on 50 of 1000 pages. If 100 pages are sampled randomly, without replace-

ment, approximate the probability that at least 1 of the pages in error are in the sample.

4-68. Hits to a high-volume Web site are assumed to follow a Poisson distribution with a mean of 10,000 per day. Approximate each of the following:
(a) The probability of more than 20,000 hits in a day
(b) The probability of less than 9900 hits in a day
(c) The value such that the probability that the number of hits in a day exceed the value is 0.01

4-69. Continuation of Exercise 4-68.
(a) Approximate the expected number of days in a year (365 days) that exceed 10,200 hits.
(b) Approximate the probability that over a year (365 days) more than 15 days each have more than 10,200 hits.

4-70. The percentage of people exposed to a bacteria who become ill is 20%. Assume that people are independent. Assume that 1000 people are exposed to the bacteria. Approximate each of the following:
(a) The probability that more than 225 become ill
(b) The probability that between 175 and 225 become ill
(c) The value such that the probability that the number of people that become ill exceeds the value is 0.01

4-71. A high-volume printer produces minor print-quality errors on a test pattern of 1000 pages of text according to a Poisson distribution with a mean of 0.4 per page.
(a) Why are the number of errors on each page independent random variables?
(b) What is the mean number of pages with errors (one or more)?
(c) Approximate the probability that more than 350 pages contain errors (one or more).

4-8 CONTINUITY CORRECTION TO IMPROVE THE APPROXIMATION (CD ONLY)

4-9 EXPONENTIAL DISTRIBUTION

The discussion of the Poisson distribution defined a random variable to be the number of flaws along a length of copper wire. The distance between flaws is another random variable that is often of interest. Let the random variable X denote the length from any starting point on the wire until a flaw is detected.

As you might expect, the distribution of X can be obtained from knowledge of the distribution of the number of flaws. The key to the relationship is the following concept. The distance to the first flaw exceeds 3 millimeters if and only if there are no flaws within a length of 3 millimeters—simple, but sufficient for an analysis of the distribution of X.

In general, let the random variable N denote the number of flaws in x millimeters of wire. If the mean number of flaws is λ per millimeter, N has a Poisson distribution with mean λx. We assume that the wire is longer than the value of x. Now,

$$P(X > x) = P(N = 0) = \frac{e^{-\lambda x}(\lambda x)^0}{0!} = e^{-\lambda x}$$

Therefore,

$$F(x) = P(X \le x) = 1 - e^{-\lambda x}, \quad x \ge 0$$

is the cumulative distribution function of X. By differentiating $F(x)$, the probability density function of X is calculated to be

$$f(x) = \lambda e^{-\lambda x}, \quad x \ge 0$$

The derivation of the distribution of X depends only on the assumption that the flaws in the wire follow a **Poisson process.** Also, the starting point for measuring X doesn't matter because the probability of the number of flaws in an interval of a Poisson process depends only on the length of the interval, not on the location. For any Poisson process, the following general result applies.

Definition

> The random variable X that equals the distance between successive counts of a Poisson process with mean $\lambda > 0$ is an **exponential random variable** with parameter λ. The probability density function of X is
>
> $$f(x) = \lambda e^{-\lambda x} \quad \text{for} \quad 0 \le x < \infty \qquad (4\text{-}14)$$

The exponential distribution obtains its name from the exponential function in the probability density function. Plots of the exponential distribution for selected values of λ are shown in Fig. 4-22. For any value of λ, the exponential distribution is quite skewed. The following results are easily obtained and are left as an exercise.

> If the random variable X has an exponential distribution with parameter λ,
>
> $$\mu = E(X) = \frac{1}{\lambda} \quad \text{and} \quad \sigma^2 = V(X) = \frac{1}{\lambda^2} \qquad (4\text{-}15)$$

It is important to **use consistent units** in the calculation of probabilities, means, and variances involving exponential random variables. The following example illustrates unit conversions.

EXAMPLE 4-21

In a large corporate computer network, user log-ons to the system can be modeled as a Poisson process with a mean of 25 log-ons per hour. What is the probability that there are no log-ons in an interval of 6 minutes?

Let X denote the time in hours from the start of the interval until the first log-on. Then, X has an exponential distribution with $\lambda = 25$ log-ons per hour. We are interested in the probability that X exceeds 6 minutes. Because λ is given in log-ons per hour, we express all time units in hours. That is, 6 minutes $= 0.1$ hour. The probability requested is shown as the shaded area under the probability density function in Fig. 4-23. Therefore,

$$P(X > 0.1) = \int_{0.1}^{\infty} 25 e^{-25x} \, dx = e^{-25(0.1)} = 0.082$$

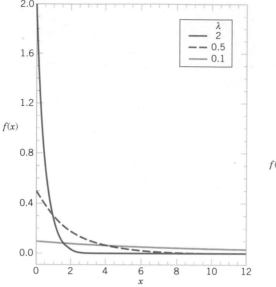

Figure 4-22   Probability density function of exponential random variables for selected values of λ.

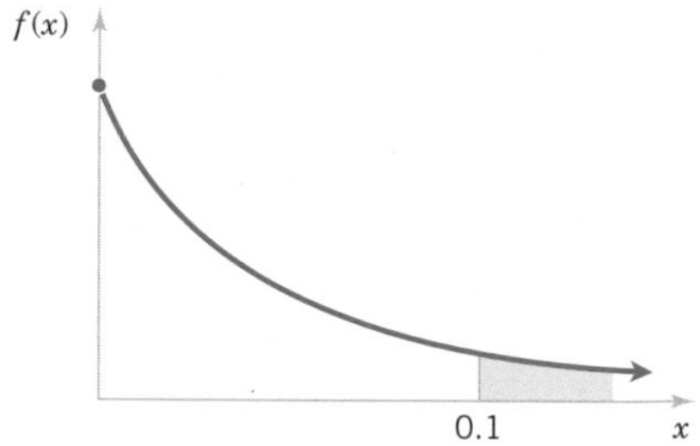

Figure 4-23   Probability for the exponential distribution in Example 4-21.

Also, the cumulative distribution function can be used to obtain the same result as follows:

$$P(X > 0.1) = 1 - F(0.1) = e^{-25(0.1)}$$

An identical answer is obtained by expressing the mean number of log-ons as 0.417 log-ons per minute and computing the probability that the time until the next log-on exceeds 6 minutes. Try it.

What is the probability that the time until the next log-on is between 2 and 3 minutes? Upon converting all units to hours,

$$P(0.033 < X < 0.05) = \int_{0.033}^{0.05} 25e^{-25x}\, dx = \left. -e^{-25x} \right|_{0.033}^{0.05} = 0.152$$

An alternative solution is

$$P(0.033 < X < 0.05) = F(0.05) - F(0.033) = 0.152$$

Determine the interval of time such that the probability that no log-on occurs in the interval is 0.90. The question asks for the length of time x such that $P(X > x) = 0.90$. Now,

$$P(X > x) = e^{-25x} = 0.90$$

Take the (natural) log of both sides to obtain $-25x = \ln(0.90) = -0.1054$. Therefore,

$$x = 0.00421 \text{ hour} = 0.25 \text{ minute}$$

Furthermore, the mean time until the next log-on is

$$\mu = 1/25 = 0.04 \text{ hour} = 2.4 \text{ minutes}$$

The standard deviation of the time until the next log-on is

$$\sigma = 1/25 \text{ hours} = 2.4 \text{ minutes}$$

In the previous example, the probability that there are no log-ons in a 6-minute interval is 0.082 regardless of the starting time of the interval. A Poisson process assumes that events occur uniformly throughout the interval of observation; that is, there is no clustering of events. If the log-ons are well modeled by a Poisson process, the probability that the first log-on after noon occurs after 12:06 P.M. is the same as the probability that the first log-on after 3:00 P.M. occurs after 3:06 P.M. And if someone logs on at 2:22 P.M., the probability the next log-on occurs after 2:28 P.M. is still 0.082.

Our starting point for observing the system does not matter. However, if there are high-use periods during the day, such as right after 8:00 A.M., followed by a period of low use, a Poisson process is not an appropriate model for log-ons and the distribution is not appropriate for computing probabilities. It might be reasonable to model each of the high- and low-use periods by a separate Poisson process, employing a larger value for λ during the high-use periods and a smaller value otherwise. Then, an exponential distribution with the corresponding value of λ can be used to calculate log-on probabilities for the high- and low-use periods.

Lack of Memory Property

An even more interesting property of an exponential random variable is concerned with conditional probabilities.

EXAMPLE 4-22   Let X denote the time between detections of a particle with a geiger counter and assume that X has an exponential distribution with $\lambda = 1.4$ minutes. The probability that we detect a particle within 30 seconds of starting the counter is

$$P(X < 0.5 \text{ minute}) = F(0.5) = 1 - e^{-0.5/1.4} = 0.30$$

In this calculation, all units are converted to minutes. Now, suppose we turn on the geiger counter and wait 3 minutes without detecting a particle. What is the probability that a particle is detected in the next 30 seconds?

Because we have already been waiting for 3 minutes, we feel that we are "due." That is, the probability of a detection in the next 30 seconds should be greater than 0.3. However, for an exponential distribution, this is not true. The requested probability can be expressed as the conditional probability that $P(X < 3.5 \mid X > 3)$. From the definition of conditional probability,

$$P(X < 3.5 \mid X > 3) = P(3 < X < 3.5)/P(X > 3)$$

where

$$P(3 < X < 3.5) = F(3.5) - F(3) = [1 - e^{-3.5/1.4}] - [1 - e^{-3/1.4}] = 0.0035$$

and

$$P(X > 3) = 1 - F(3) = e^{-3/1.4} = 0.117$$

Therefore,

$$P(X < 3.5 \mid X > 3) = 0.035/0.117 = 0.30$$

After waiting for 3 minutes without a detection, the probability of a detection in the next 30 seconds is the same as the probability of a detection in the 30 seconds immediately after starting the counter. The fact that you have waited 3 minutes without a detection does not change the probability of a detection in the next 30 seconds.

Example 4-22 illustrates the **lack of memory property** of an exponential random variable and a general statement of the property follows. In fact, the exponential distribution is the only continuous distribution with this property.

Lack of Memory Property

For an exponential random variable X,

$$P(X < t_1 + t_2 \mid X > t_1) = P(X < t_2) \qquad (4\text{-}16)$$

Figure 4-24 graphically illustrates the lack of memory property. The area of region A divided by the total area under the probability density function $(A + B + C + D = 1)$ equals $P(X < t_2)$. The area of region C divided by the area $C + D$ equals $P(X < t_1 + t_2 \mid X > t_1)$. The lack of memory property implies that the proportion of the total area that is in A equals the proportion of the area in C and D that is in C. The mathematical verification of the lack of memory property is left as a mind-expanding exercise.

The lack of memory property is not that surprising when you consider the development of a Poisson process. In that development, we assumed that an interval could be partitioned into small intervals that were independent. These subintervals are similar to independent Bernoulli trials that comprise a binomial process; knowledge of previous results does not affect the probabilities of events in future subintervals. An exponential random variable is the continuous analog of a geometric random variable, and they share a similar lack of memory property.

The exponential distribution is often used in reliability studies as the model for the time until failure of a device. For example, the lifetime of a semiconductor chip might be modeled as an exponential random variable with a mean of 40,000 hours. The lack of

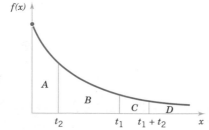

Figure 4-24   Lack of memory property of an exponential distribution.

memory property of the exponential distribution implies that the device does not wear out. That is, regardless of how long the device has been operating, the probability of a failure in the next 1000 hours is the same as the probability of a failure in the first 1000 hours of operation. The lifetime L of a device with failures caused by random shocks might be appropriately modeled as an exponential random variable. However, the lifetime L of a device that suffers slow mechanical wear, such as bearing wear, is better modeled by a distribution such that $P(L < t + \Delta t \mid L > t)$ increases with t. Distributions such as the Weibull distribution are often used, in practice, to model the failure time of this type of device. The Weibull distribution is presented in a later section.

EXERCISES FOR SECTION 4-9

4-72. Suppose X has an exponential distribution with $\lambda = 2$. Determine the following:
(a) $P(X \leq 0)$ (b) $P(X \geq 2)$
(c) $P(X \leq 1)$ (d) $P(1 < X < 2)$
(e) Find the value of x such that $P(X < x) = 0.05$.

4-73. Suppose X has an exponential distribution with mean equal to 10. Determine the following:
(a) $P(X > 10)$
(b) $P(X > 20)$
(c) $P(X > 30)$
(d) Find the value of x such that $P(X < x) = 0.95$.

4-74. Suppose the counts recorded by a geiger counter follow a Poisson process with an average of two counts per minute.
(a) What is the probability that there are no counts in a 30-second interval?
(b) What is the probability that the first count occurs in less than 10 seconds?
(c) What is the probability that the first count occurs between 1 and 2 minutes after start-up?

4-75. Suppose that the log-ons to a computer network follow a Poisson process with an average of 3 counts per minute.
(a) What is the mean time between counts?
(b) What is the standard deviation of the time between counts?
(c) Determine x such that the probability that at least one count occurs before time x minutes is 0.95.

4-76. The time to failure (in hours) for a laser in a cytometry machine is modeled by an exponential distribution with $\lambda = 0.00004$.
(a) What is the probability that the laser will last at least 20,000 hours?
(b) What is the probability that the laser will last at most 30,000 hours?
(c) What is the probability that the laser will last between 20,000 and 30,000 hours?

4-77. The time between calls to a plumbing supply business is exponentially distributed with a mean time between calls of 15 minutes.
(a) What is the probability that there are no calls within a 30-minute interval?

(b) What is the probability that at least one call arrives within a 10-minute interval?
(c) What is the probability that the first call arrives within 5 and 10 minutes after opening?
(d) Determine the length of an interval of time such that the probability of at least one call in the interval is 0.90.

4-78. The life of automobile voltage regulators has an exponential distribution with a mean life of six years. You purchase an automobile that is six years old, with a working voltage regulator, and plan to own it for six years.
(a) What is the probability that the voltage regulator fails during your ownership?
(b) If your regulator fails after you own the automobile three years and it is replaced, what is the mean time until the next failure?

4-79. The time to failure (in hours) of fans in a personal computer can be modeled by an exponential distribution with $\lambda = 0.0003$.
(a) What proportion of the fans will last at least 10,000 hours?
(b) What proportion of the fans will last at most 7000 hours?

4-80. The time between the arrival of electronic messages at your computer is exponentially distributed with a mean of two hours.
(a) What is the probability that you do not receive a message during a two-hour period?
(b) If you have not had a message in the last four hours, what is the probability that you do not receive a message in the next two hours?
(c) What is the expected time between your fifth and sixth messages?

4-81. The time between arrivals of taxis at a busy intersection is exponentially distributed with a mean of 10 minutes.
(a) What is the probability that you wait longer than one hour for a taxi?
(b) Suppose you have already been waiting for one hour for a taxi, what is the probability that one arrives within the next 10 minutes?

4-82. Continuation of Exercise 4-81.
(a) Determine x such that the probability that you wait more than x minutes is 0.10.

(b) Determine x such that the probability that you wait less than x minutes is 0.90.

(c) Determine x such that the probability that you wait less than x minutes is 0.50.

4-83. The distance between major cracks in a highway follows an exponential distribution with a mean of 5 miles.

(a) What is the probability that there are no major cracks in a 10-mile stretch of the highway?

(b) What is the probability that there are two major cracks in a 10-mile stretch of the highway?

(c) What is the standard deviation of the distance between major cracks?

4-84. Continuation of Exercise 4-83.

(a) What is the probability that the first major crack occurs between 12 and 15 miles of the start of inspection?

(b) What is the probability that there are no major cracks in two separate 5-mile stretches of the highway?

(c) Given that there are no cracks in the first 5 miles inspected, what is the probability that there are no major cracks in the next 10 miles inspected?

4-85. The lifetime of a mechanical assembly in a vibration test is exponentially distributed with a mean of 400 hours.

(a) What is the probability that an assembly on test fails in less than 100 hours?

(b) What is the probability that an assembly operates for more than 500 hours before failure?

(c) If an assembly has been on test for 400 hours without a failure, what is the probability of a failure in the next 100 hours?

4-86. Continuation of Exercise 4-85.

(a) If 10 assemblies are tested, what is the probability that at least one fails in less than 100 hours? Assume that the assemblies fail independently.

(b) If 10 assemblies are tested, what is the probability that all have failed by 800 hours? Assume the assemblies fail independently.

4-87. When a bus service reduces fares, a particular trip from New York City to Albany, New York, is very popular. A small bus can carry four passengers. The time between calls for tickets is exponentially distributed with a mean of 30 minutes. Assume that each call orders one ticket. What is the probability that the bus is filled in less than 3 hours from the time of the fare reduction?

4-88. The time between arrivals of small aircraft at a county airport is exponentially distributed with a mean of one hour. What is the probability that more than three aircraft arrive within an hour?

4-89. Continuation of Exercise 4-88.

(a) If 30 separate one-hour intervals are chosen, what is the probability that no interval contains more than three arrivals?

(b) Determine the length of an interval of time (in hours) such that the probability that no arrivals occur during the interval is 0.10.

4-90. The time between calls to a corporate office is exponentially distributed with a mean of 10 minutes.

(a) What is the probability that there are more than three calls in one-half hour?

(b) What is the probability that there are no calls within one-half hour?

(c) Determine x such that the probability that there are no calls within x hours is 0.01.

4-91. Continuation of Exercise 4-90.

(a) What is the probability that there are no calls within a two-hour interval?

(b) If four nonoverlapping one-half hour intervals are selected, what is the probability that none of these intervals contains any call?

(c) Explain the relationship between the results in part (a) and (b).

4-92. If the random variable X has an exponential distribution with mean θ, determine the following:

(a) $P(X > \theta)$ (b) $P(X > 2\theta)$

(c) $P(X > 3\theta)$

(d) How do the results depend on θ?

4-93. Assume that the flaws along a magnetic tape follow a Poisson distribution with a mean of 0.2 flaw per meter. Let X denote the distance between two successive flaws.

(a) What is the mean of X?

(b) What is the probability that there are no flaws in 10 consecutive meters of tape?

(c) Does your answer to part (b) change if the 10 meters are not consecutive?

(d) How many meters of tape need to be inspected so that the probability that at least one flaw is found is 90%?

4-94. Continuation of Exercise 4-93. (More difficult questions.)

(a) What is the probability that the first time the distance between two flaws exceeds 8 meters is at the fifth flaw?

(b) What is the mean number of flaws before a distance between two flaws exceeds 8 meters?

4-95. Derive the formula for the mean and variance of an exponential random variable.

4-10   ERLANG AND GAMMA DISTRIBUTIONS

4-10.1   Erlang Distribution

An exponential random variable describes the length until the first count is obtained in a Poisson process. A generalization of the exponential distribution is the length until r counts occur in a Poisson process. The random variable that equals the interval length until r counts occur in a Poisson process has an **Erlang random variable.**

EXAMPLE 4-23

The failures of the central processor units of large computer systems are often modeled as a Poisson process. Typically, failures are not caused by components wearing out, but by more random failures of the large number of semiconductor circuits in the units. Assume that the units that fail are immediately repaired, and assume that the mean number of failures per hour is 0.0001. Let X denote the time until four failures occur in a system. Determine the probability that X exceeds 40,000 hours.

Let the random variable N denote the number of failures in 40,000 hours of operation. The time until four failures occur exceeds 40,000 hours if and only if the number of failures in 40,000 hours is three or less. Therefore,

$$P(X > 40,000) = P(N \le 3)$$

The assumption that the failures follow a Poisson process implies that N has a Poisson distribution with

$$E(N) = 40,000(0.0001) = 4 \text{ failures per 40,000 hours}$$

Therefore,

$$P(X > 40,000) = P(N \le 3) = \sum_{k=0}^{3} \frac{e^{-4}4^k}{k!} = 0.433$$

The cumulative distribution function of a general Erlang random variable X can be obtained from $P(X \le x) = 1 - P(X > x)$, and $P(X > x)$ can be determined as in the previous example. Then, the probability density function of X can be obtained by differentiating the cumulative distribution function and using a great deal of algebraic simplification. The details are left as an exercise. In general, we can obtain the following result.

Definition

> The random variable X that equals the interval length until r counts occur in a Poisson process with mean $\lambda > 0$ has an **Erlang random variable** with parameters λ and r. The probability density function of X is
>
> $$f(x) = \frac{\lambda^r x^{r-1} e^{-\lambda x}}{(r-1)!}, \quad \text{for } x > 0 \text{ and } r = 1, 2, \dots \qquad (4\text{-}17)$$

Sketches of the Erlang probability density function for several values of r and λ are shown in Fig. 4-25. Clearly, an Erlang random variable with $r = 1$ is an exponential random variable. Probabilities involving Erlang random variables are often determined by computing a summation of Poisson random variables as in Example 4-23. The probability density function of an Erlang random variable can be used to determine probabilities; however, integrating by parts is often necessary. As was the case for the exponential distribution, one must be careful to define the random variable and the parameter in consistent units.

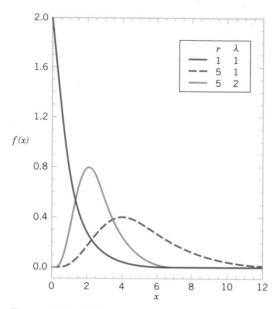

Figure 4-25   Erlang probability density functions
for selected values of r and λ.

EXAMPLE 4-24

An alternative approach to computing the probability requested in Example 4-24 is to integrate the probability density function of X. That is,

$$P(X > 40{,}000) = \int\limits_{40{,}000}^{\infty} f(x)\,dx = \int\limits_{40{,}000}^{\infty} \frac{\lambda^r x^{r-1} e^{-\lambda x}}{(r-1)!}\,dx$$

where $r = 4$ and $\lambda = 0.0001$. Integration by parts can be used to verify the result obtained previously.

An Erlang random variable can be thought of as the continuous analog of a negative binomial random variable. A negative binomial random variable can be expressed as the sum of r geometric random variables. Similarly, an Erlang random variable can be represented as the sum of r exponential random variables. Using this conclusion, we can obtain the following plausible result. Sums of random variables are studied in Chapter 5.

If X is an Erlang random variable with parameters λ and r,

$$\mu = E(X) = r/\lambda \quad \text{and} \quad \sigma^2 = V(X) = r/\lambda^2 \qquad (4\text{-}18)$$

4-9.2   Gamma Distribution

The Erlang distribution is a special case of the **gamma distribution.** If the parameter r of an Erlang random variable is not an integer, but $r > 0$, the random variable has a gamma distribution. However, in the Erlang density function, the parameter r appears as r factorial.

Therefore, to define a gamma random variable, we require a generalization of the factorial function.

Definition

> The **gamma function** is
>
> $$\Gamma(r) = \int_0^\infty x^{r-1} e^{-x} \, dx, \quad \text{for } r > 0 \qquad (4\text{-}19)$$

It can be shown that the integral in the definition of $\Gamma(r)$ is finite. Furthermore, by using integration by parts it can be shown that

$$\Gamma(r) = (r - 1)\Gamma(r - 1)$$

This result is left as an exercise. Therefore, if r is a positive integer (as in the Erlang distribution),

$$\Gamma(r) = (r - 1)!$$

Also, $\Gamma(1) = 0! = 1$ and it can be shown that $\Gamma(1/2) = \pi^{1/2}$. The gamma function can be interpreted as a generalization to noninteger values of r of the term $(r - 1)!$ that is used in the Erlang probability density function.

Now the gamma probability density function can be stated.

Definition

> The random variable X with probability density function
>
> $$f(x) = \frac{\lambda^r x^{r-1} e^{-\lambda x}}{\Gamma(r)}, \quad \text{for } x > 0 \qquad (4\text{-}20)$$
>
> has a **gamma random variable** with parameters $\lambda > 0$ and $r > 0$. If r is an integer, X has an Erlang distribution.

Sketches of the gamma distribution for several values of λ and r are shown in Fig. 4-26. It can be shown that $f(x)$ satisfies the properties of a probability density function, and the following result can be obtained. Repeated integration by parts can be used, but the details are lengthy.

> If X is a **gamma random variable** with parameters λ and r,
>
> $$\mu = E(X) = r/\lambda \quad \text{and} \quad \sigma^2 = V(X) = r/\lambda^2 \qquad (4\text{-}21)$$

Although the gamma distribution is not frequently used as a model for a physical system, the special case of the Erlang distribution is very useful for modeling random experiments. The exercises provide illustrations. Furthermore, the **chi-squared distribution** is a special case of

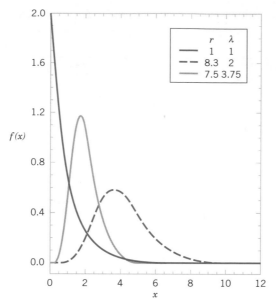

Figure 4-26   Gamma probability density functions for selected values of λ and r.

the gamma distribution in which $\lambda = 1/2$ and r equals one of the values $1/2, 1, 3/2, 2, \ldots$. This distribution is used extensively in interval estimation and tests of hypotheses that are discussed in subsequent chapters.

EXERCISES FOR SECTION 4-10

4-96.   Calls to a telephone system follow a Poisson distribution with a mean of five calls per minute.
(a) What is the name applied to the distribution and parameter values of the time until the tenth call?
(b) What is the mean time until the tenth call?
(c) What is the mean time between the ninth and tenth calls?

4-97.   Continuation of Exercise 4-96.
(a) What is the probability that exactly four calls occur within one minute?
(b) If 10 separate one-minute intervals are chosen, what is the probability that all intervals contain more than two calls?

4-98.   Raw materials are studied for contamination. Suppose that the number of particles of contamination per pound of material is a Poisson random variable with a mean of 0.01 particle per pound.
(a) What is the expected number of pounds of material required to obtain 15 particles of contamination?
(b) What is the standard deviation of the pounds of materials required to obtain 15 particles of contamination?

4-99.   The time between failures of a laser in a cytogenics machine is exponentially distributed with a mean of 25,000 hours.
(a) What is the expected time until the second failure?

(b) What is the probability that the time until the third failure exceeds 50,000 hours?

4-100.   In a data communication system, several messages that arrive at a node are bundled into a packet before they are transmitted over the network. Assume the messages arrive at the node according to a Poisson process with $\tau = 30$ messages per minute. Five messages are used to form a packet.
(a) What is the mean time until a packet is formed, that is, until five messages arrived at the node?
(b) What is the standard deviation of the time until a packet is formed?
(c) What is the probability that a packet is formed in less than 10 seconds?
(d) What is the probability that a packet is formed in less than 5 seconds?

4-101.   Errors caused by contamination on optical disks occur at the rate of one error every 10^5 bits. Assume the errors follow a Poisson distribution.
(a) What is the mean number of bits until five errors occur?
(b) What is the standard deviation of the number of bits until five errors occur?

(c) The error-correcting code might be ineffective if there are three or more errors within 10^5 bits. What is the probability of this event?

4-102. Calls to the help line of a large computer distributor follow a Possion distribution with a mean of 20 calls per minute.
(a) What is the mean time until the one-hundredth call?
(b) What is the mean time between call numbers 50 and 80?
(c) What is the probability that three or more calls occur within 15 seconds?

4-103. The time between arrivals of customers at an automatic teller machine is an exponential random variable with a mean of 5 minutes.
(a) What is the probability that more than three customers arrive in 10 minutes?
(b) What is the probability that the time until the fifth customer arrives is less than 15 minutes?

4-104. The time between process problems in a manufacturing line is exponentially distributed with a mean of 30 days.
(a) What is the expected time until the fourth problem?
(b) What is the probability that the time until the fourth problem exceeds 120 days?

4-105. Use the properties of the gamma function to evaluate the following:
(a) $\Gamma(6)$ (b) $\Gamma(5/2)$
(c) $\Gamma(9/2)$

4-106. Use integration by parts to show that $\Gamma(r) = (r - 1)\Gamma(r - 1)$.

4-107. Show that the gamma density function $f(x, \lambda, r)$ integrates to 1.

4-108. Use the result for the gamma distribution to determine the mean and variance of a chi-square distribution with $r = 7/2$.

4-11 WEIBULL DISTRIBUTION

As mentioned previously, the Weibull distribution is often used to model the time until failure of many different physical systems. The parameters in the distribution provide a great deal of flexibility to model systems in which the number of failures increases with time (bearing wear), decreases with time (some semiconductors), or remains constant with time (failures caused by external shocks to the system).

Definition

The random variable X with probability density function

$$f(x) = \frac{\beta}{\delta}\left(\frac{x}{\delta}\right)^{\beta-1} \exp\left[-\left(\frac{x}{\delta}\right)^{\beta}\right], \quad \text{for } x > 0 \quad (4\text{-}22)$$

is a **Weibull random variable** with scale parameter $\delta > 0$ and shape parameter $\beta > 0$.

The flexibility of the Weibull distribution is illustrated by the graphs of selected probability density functions in Fig. 4-27. By inspecting the probability density function, it is seen that when $\beta = 1$, the Weibull distribution is identical to the exponential distribution.

The cumulative distribution function is often used to compute probabilities. The following result can be obtained.

If X has a Weibull distribution with parameters δ and β, then the cumulative distribution function of X is

$$F(x) = 1 - e^{-\left(\frac{x}{\delta}\right)^{\beta}} \quad (4\text{-}23)$$

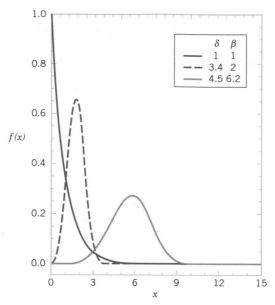

Figure 4-27   Weibull probability density functions for selected values of δ and β.

Also, the following result can be obtained.

If X has a Weibull distribution with parameters δ and β,

$$\mu = E(x) = \delta\Gamma\left(1 + \frac{1}{\beta}\right) \quad \text{and} \quad \sigma^2 = V(x) = \delta^2\Gamma\left(1 + \frac{2}{\beta}\right) - \delta^2\left[\Gamma\left(1 + \frac{1}{\beta}\right)\right]^2$$

$$(4\text{-}24)$$

EXAMPLE 4-25

The time to failure (in hours) of a bearing in a mechanical shaft is satisfactorily modeled as a Weibull random variable with β = 1/2, and δ = 5000 hours. Determine the mean time until failure.

From the expression for the mean,

$$E(X) = 5000\Gamma[1 + (1/0.5)] = 5000\Gamma[3] = 5000 \times 2! = 10{,}000 \text{ hours}$$

Determine the probability that a bearing lasts at least 6000 hours. Now

$$P(x > 6000) = 1 - F(6000) = \exp{-\left[\left(\frac{6000}{5000}\right)^{1/2}\right]} = e^{-1.095} = 0.334$$

Consequently, only 33.4% of all bearings last at least 6000 hours.

EXERCISES FOR SECTION 4-11

4-109. Suppose that X has a Weibull distribution with $\beta = 0.2$ and $\delta = 100$ hours. Determine the mean and variance of X.

4-110. Suppose that X has a Weibull distribution $\beta = 0.2$ and $\delta = 100$ hours. Determine the following:
(a) $P(X < 10,000)$ (b) $P(X > 5000)$

4-111. Assume that the life of a roller bearing follows a Weibull distribution with parameters $\beta = 2$ and $\delta = 10,000$ hours.
(a) Determine the probability that a bearing lasts at least 8000 hours.
(b) Determine the mean time until failure of a bearing.
(c) If 10 bearings are in use and failures occur independently, what is the probability that all 10 bearings last at least 8000 hours?

4-112. The life (in hours) of a computer processing unit (CPU) is modeled by a Weibull distribution with parameters $\beta = 3$ and $\delta = 900$ hours.
(a) Determine the mean life of the CPU.
(b) Determine the variance of the life of the CPU.
(c) What is the probability that the CPU fails before 500 hours?

4-113. Assume the life of a packaged magnetic disk exposed to corrosive gases has a Weibull distribution with $\beta = 0.5$ and the mean life is 600 hours.
(a) Determine the probability that a packaged disk lasts at least 500 hours.
(b) Determine the probability that a packaged disk fails before 400 hours.

4-114. The life of a recirculating pump follows a Weibull distribution with parameters $\beta = 2$, and $\delta = 700$ hours.
(a) Determine the mean life of a pump.
(b) Determine the variance of the life of a pump.
(c) What is the probability that a pump will last longer than its mean?

4-115. The life (in hours) of a magnetic resonance imagining machine (MRI) is modeled by a Weibull distribution with parameters $\beta = 2$ and $\delta = 500$ hours.
(a) Determine the mean life of the MRI.
(b) Determine the variance of the life of the MRI.
(c) What is the probability that the MRI fails before 250 hours?

4-116. If X is a Weibull random variable with $\beta = 1$, and $\delta = 1000$, what is another name for the distribution of X and what is the mean of X?

4-12 LOGNORMAL DISTRIBUTION

Variables in a system sometimes follow an exponential relationship as $x = exp(w)$. If the exponent is a random variable, say W, $X = \exp(W)$ is a random variable and the distribution of X is of interest. An important special case occurs when W has a normal distribution. In that case, the distribution of X is called a **lognormal distribution.** The name follows from the transformation $\ln(X) = W$. That is, the natural logarithm of X is normally distributed.

Probabilities for X are obtained from the transformation to W, but we need to recognize that the range of X is $(0, \infty)$. Suppose that W is normally distributed with mean θ and variance ω^2; then the cumulative distribution function for X is

$$F(x) = P[X \le x] = P[\exp(W) \le x] = P[W \le \ln(x)]$$
$$= P\left[Z \le \frac{\ln(x) - \theta}{\omega}\right] = \Phi\left[\frac{\ln(x) - \theta}{\omega}\right]$$

for $x > 0$, where Z is a standard normal random variable. Therefore, Appendix Table II can be used to determine the probability. Also, $F(x) = 0$, for $x \le 0$.

The probability density function of X can be obtained from the derivative of $F(x)$. This derivative is applied to the last term in the expression for $F(x)$, the integral of the standard normal density function. Furthermore, from the probability density function, the mean and variance of X can be derived. The details are omitted, but a summary of results follows.

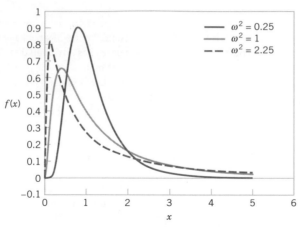

Figure 4-28   Lognormal probability density functions with $\theta = 0$ for selected values of ω^2.

Let W have a normal distribution mean θ and variance ω^2; then $X = \exp(W)$ is a **lognormal random variable** with probability density function

$$f(x) = \frac{1}{x\omega \sqrt{2\pi}} \; \exp\left[-\frac{(\ln x - \theta)^2}{2\omega^2}\right] \qquad 0 < x < \infty$$

The mean and variance of X are

$$E(X) = e^{\theta + \omega^2/2} \qquad \text{and} \qquad V(X) = e^{2\theta + \omega^2}\left(e^{\omega^2} - 1\right) \qquad (4\text{-}25)$$

The parameters of a lognormal distribution are θ and ω^2, but care is needed to interpret that these are the mean and variance of the normal random variable W. The mean and variance of X are the functions of these parameters shown in (4-25). Figure 4-28 illustrates lognormal distributions for selected values of the parameters.

    The lifetime of a product that degrades over time is often modeled by a lognormal random variable. For example, this is a common distribution for the lifetime of a semiconductor laser. A Weibull distribution can also be used in this type of application, and with an appropriate choice for parameters, it can approximate a selected lognormal distribution. However, a lognormal distribution is derived from a simple exponential function of a normal random variable, so it is easy to understand and easy to evaluate probabilities.

EXAMPLE 4-26

The lifetime of a semiconductor laser has a lognormal distribution with $\theta = 10$ hours and $\omega = 1.5$ hours. What is the probability the lifetime exceeds 10,000 hours?

    From the cumulative distribution function for X

$$P(X > 10{,}000) = 1 - P[\exp(W) \le 10{,}000] = 1 - P[W \le \ln(10{,}000)]$$

$$= \Phi\left(\frac{\ln(10{,}000) - 10}{1.5}\right) = 1 - \Phi(-0.52) = 1 - 0.30 = 0.70$$

What lifetime is exceeded by 99% of lasers? The question is to determine x such that $P(X > x) = 0.99$. Therefore,

$$P(X > x) = P[\exp(W) > x] = P[W > \ln(x)] = 1 - \Phi\left(\frac{\ln(x) - 10}{1.5}\right) = 0.99$$

From Appendix Table II, $1 - \Phi(z) = 0.99$ when $z = -2.33$. Therefore,

$$\frac{\ln(x) - 10}{1.5} = -2.33 \quad \text{and} \quad x = \exp(6.505) = 668.48 \text{ hours.}$$

Determine the mean and standard deviation of lifetime. Now,

$$E(X) = e^{\theta + \omega^2/2} = \exp(10 + 1.125) = 67,846.3$$
$$V(X) = e^{2\theta + \omega^2}(e^{\omega^2} - 1) = \exp(20 + 2.25)[\exp(2.25) - 1] = 39,070,059,886.6$$

so the standard deviation of X is 197,661.5 hours. Notice that the standard deviation of lifetime is large relative to the mean.

EXERCISES FOR SECTION 4-12

4-117. Suppose that X has a lognormal distribution with parameters $\theta = 5$ and $\omega^2 = 9$. Determine the following:
(a) $P(X < 13,300)$
(b) The value for x such that $P(X \le x) = 0.95$
(c) The mean and variance of X

4-118. Suppose that X has a lognormal distribution with parameters $\theta = -2$ and $\omega^2 = 9$. Determine the following:
(a) $P(500 < X < 1000)$
(b) The value for x such that $P(X < x) = 0.1$
(c) The mean and variance of X

4-119. Suppose that X has a lognormal distribution with parameters $\theta = 2$ and $\omega^2 = 4$. Determine the following:
(a) $P(X < 500)$
(b) The conditional probability that $X < 1500$ given that $X > 1000$
(c) What does the difference between the probabilities in parts (a) and (b) imply about lifetimes of lognormal random variables?

4-120. The length of time (in seconds) that a user views a page on a Web site before moving to another page is a lognormal random variable with parameters $\theta = 0.5$ and $\omega^2 = 1$.
(a) What is the probability that a page is viewed for more than 10 seconds?
(b) What is the length of time that 50% of users view the page?
(c) What is the mean and standard deviation of the time until a user moves from the page?

4-121. Suppose that X has a lognormal distribution and that the mean and variance of X are 100 and 85,000, respectively.

Determine the parameters θ and ω^2 of the lognormal distribution. (*Hint:* define $x = \exp(\theta)$ and $y = \exp(\omega^2)$ and write two equations in terms of x and y.)

4-122. The lifetime of a semiconductor laser has a lognormal distribution, and it is known that the mean and standard deviation of lifetime are 10,000 and 20,000, respectively.
(a) Calculate the parameters of the lognormal distribution
(b) Determine the probability that a lifetime exceeds 10,000 hours
(c) Determine the lifetime that is exceeded by 90% of lasers

4-123. Derive the probability density function of a lognormal random variable from the derivative of the cumulative distribution function.

Supplemental Exercises

4-124. Suppose that $f(x) = 0.5x - 1$ for $2 < x < 4$. Determine the following:
(a) $P(X < 2.5)$
(b) $P(X > 3)$
(c) $P(2.5 < X < 3.5)$

4-125. Continuation of Exercise 4-124. Determine the cumulative distribution function of the random variable.

4-126. Continuation of Exercise 4-124. Determine the mean and variance of the random variable.

4-127. The time between calls is exponentially distributed with a mean time between calls of 10 minutes.
(a) What is the probability that the time until the first call is less than 5 minutes?
(b) What is the probability that the time until the first call is between 5 and 15 minutes?
(c) Determine the length of an interval of time such that the probability of at least one call in the interval is 0.90.

4-128. Continuation of Exercise 4-127.
(a) If there has not been a call in 10 minutes, what is the probability that the time until the next call is less than 5 minutes?
(b) What is the probability that there are no calls in the intervals from 10:00 to 10:05, from 11:30 to 11:35, and from 2:00 to 2:05?

4-129. Continuation of Exercise 4-127.
(a) What is the probability that the time until the third call is greater than 30 minutes?
(b) What is the mean time until the fifth call?

4-130. The CPU of a personal computer has a lifetime that is exponentially distributed with a mean lifetime of six years. You have owned this CPU for three years. What is the probability that the CPU fails in the next three years?

4-131. Continuation of Exercise 4-130. Assume that your corporation has owned 10 CPUs for three years, and assume that the CPUs fail independently. What is the probability that at least one fails within the next three years?

4-132. Suppose that X has a lognormal distribution with parameters $\theta = 0$ and $\omega^2 = 4$. Determine the following:
(a) $P(10 < X < 50)$
(b) The value for x such that $P(X < x) = 0.05$
(c) The mean and variance of X

4-133. Suppose that X has a lognormal distribution and that the mean and variance of X are 50 and 4000, respectively. Determine the following:
(a) The parameters θ and ω^2 of the lognormal distribution
(b) The probability that X is less than 150

4-134. Asbestos fibers in a dust sample are identified by an electron microscope after sample preparation. Suppose that the number of fibers is a Poisson random variable and the mean number of fibers per squared centimeter of surface dust is 100. A sample of 800 square centimeters of dust is analyzed. Assume a particular grid cell under the microscope represents 1/160,000 of the sample.
(a) What is the probability that at least one fiber is visible in the grid cell?
(b) What is the mean of the number of grid cells that need to be viewed to observe 10 that contain fibers?
(c) What is the standard deviation of the number of grid cells that need to be viewed to observe 10 that contain fibers?

4-135. Without an automated irrigation system, the height of plants two weeks after germination is normally distributed with a mean of 2.5 centimeters and a standard deviation of 0.5 centimeters.

(a) What is the probability that a plant's height is greater than 2.25 centimeters?
(b) What is the probability that a plant's height is between 2.0 and 3.0 centimeters?
(c) What height is exceeded by 90% of the plants?

4-136. Continuation of Exercise 4-135. With an automated irrigation system, a plant grows to a height of 3.5 centimeters two weeks after germination.
(a) What is the probability of obtaining a plant of this height or greater from the distribution of heights in Exercise 4-135.
(b) Do you think the automated irrigation system increases the plant height at two weeks after germination?

4-137. The thickness of a laminated covering for a wood surface is normally distributed with a mean of 5 millimeters and a standard deviation of 0.2 millimeter.
(a) What is the probability that a covering thickness is greater than 5.5 millimeters?
(b) If the specifications require the thickness to be between 4.5 and 5.5 millimeters, what proportion of coverings do not meet specifications?
(c) The covering thickness of 95% of samples is below what value?

4-138. The diameter of the dot produced by a printer is normally distributed with a mean diameter of 0.002 inch. Suppose that the specifications require the dot diameter to be between 0.0014 and 0.0026 inch. If the probability that a dot meets specifications is to be 0.9973, what standard deviation is needed?

4-139. Continuation of Exercise 4-138. Assume that the standard deviation of the size of a dot is 0.0004 inch. If the probability that a dot meets specifications is to be 0.9973, what specifications are needed? Assume that the specifications are to be chosen symmetrically around the mean of 0.002.

4-140. The life of a semiconductor laser at a constant power is normally distributed with a mean of 7000 hours and a standard deviation of 600 hours.
(a) What is the probability that a laser fails before 5,800 hours?
(b) What is the life in hours that 90% of the lasers exceed?

4-141. Continuation of Exercise 4-140. What should the mean life equal in order for 99% of the lasers to exceed 10,000 hours before failure?

4-142. Continuation of Exercise 4-140. A product contains three lasers, and the product fails if any of the lasers fails. Assume the lasers fail independently. What should the mean life equal in order for 99% of the products to exceed 10,000 hours before failure?

4-143. Continuation of Exercise 140. Rework parts (a) and (b). Assume that the lifetime is an exponential random variable with the same mean.

4-144. Continuation of Exercise 4-140. Rework parts (a) and (b). Assume that the lifetime is a lognormal random variable with the same mean and standard deviation.

4-145. A square inch of carpeting contains 50 carpet fibers. The probability of a damaged fiber is 0.0001. Assume the damaged fibers occur independently.
(a) Approximate the probability of one or more damaged fibers in 1 square yard of carpeting.
(b) Approximate the probability of four or more damaged fibers in 1 square yard of carpeting.

4-146. An airline makes 200 reservations for a flight that holds 185 passengers. The probability that a passenger arrives for the flight is 0.9 and the passengers are assumed to be independent.
(a) Approximate the probability that all the passengers that arrive can be seated.
(b) Approximate the probability that there are empty seats.
(c) Approximate the number of reservations that the airline should make so that the probability that everyone who arrives can be seated is 0.95. [*Hint*: Successively try values for the number of reservations.]

MIND-EXPANDING EXERCISES

4-147. The steps in this exercise lead to the probability density function of an Erlang random variable X with parameters λ and r, $f(x) = \lambda^r x^{r-1} e^{-\lambda x}/(r-1)!$, $x > 0$, $r = 1, 2, \ldots$.
(a) Use the Poisson distribution to express $P(X > x)$.
(b) Use the result from part (a) to determine the cumulative distribution function of X.
(c) Differentiate the cumulative distribution function in part (b) and simplify to obtain the probability density function of X.

4-148. A bearing assembly contains 10 bearings. The bearing diameters are assumed to be independent and normally distributed with a mean of 1.5 millimeters and a standard deviation of 0.025 millimeter. What is the probability that the maximum diameter bearing in the assembly exceeds 1.6 millimeters?

4-149. Let the random variable X denote a measurement from a manufactured product. Suppose the target value for the measurement is m. For example, X could denote a dimensional length, and the target might be 10 millimeters. The **quality loss** of the process producing the product is defined to be the expected value of $k(X - m)^2$, where k is a constant that relates a deviation from target to a loss measured in dollars.
(a) Suppose X is a continuous random variable with $E(X) = m$ and $V(X) = \sigma^2$. What is the quality loss of the process?
(b) Suppose X is a continuous random variable with $E(X) = \mu$ and $V(X) = \sigma^2$. What is the quality loss of the process?

4-150. The lifetime of an electronic amplifier is modeled as an exponential random variable. If 10% of the amplifiers have a mean of 20,000 hours and the remaining amplifiers have a mean of 50,000 hours, what proportion of the amplifiers fail before 60,000 hours?

4-151. Lack of Memory Property. Show that for an exponential random variable X, $P(X < t_1 + t_2 \mid X > t_1) = P(X < t_2)$

4-152. A process is said to be of **six-sigma quality** if the process mean is at least six standard deviations from the nearest specification. Assume a normally distributed measurement.
(a) If a process mean is centered between the upper and lower specifications at a distance of six standard deviations from each, what is the probability that a product does not meet specifications? Using the result that 0.000001 equals one part per million, express the answer in parts per million.
(b) Because it is difficult to maintain a process mean centered between the specifications, the probability of a product not meeting specifications is often calculated after assuming the process shifts. If the process mean positioned as in part (a) shifts upward by 1.5 standard deviations, what is the probability that a product does not meet specifications? Express the answer in parts per million.
(c) Rework part (a). Assume that the process mean is at a distance of three standard deviations.
(d) Rework part (b). Assume that the process mean is at a distance of three standard deviations and then shifts upward by 1.5 standard deviations.
(e) Compare the results in parts (b) and (d) and comment.

IMPORTANT TERMS AND CONCEPTS

In the E-book, click on any term or concept below to go to that subject.

Chi-squared distribution

Continuous uniform distribution

Cumulative probability distribution function-continuous random variable

Erlang distribution

Exponential distribution

Gamma distribution

Lack of memory property-continuous random variable

Lognormal distribution

Mean-continuous random variable

Mean-function of a continuous random variable

Normal approximation to binomial and Poisson probabilities

Normal distribution

Probability density function

Probability distribution-continuous random variable

Standard deviation-continuous random variable

Standard normal distribution

Standardizing

Variance-continuous random variable

Weibull distribution

CD MATERIAL

Continuity correction

5

Joint Probability
Distributions

CHAPTER OUTLINE

LEARNING OBJECTIVES

After careful study of this chapter you should be able to do the following:

1. Use joint probability mass functions and joint probability density functions to calculate probabilities

2. Calculate marginal and conditional probability distributions from joint probability distributions

3. Use the multinomial distribution to determine probabilities

4. Interpret and calculate covariances and correlations between random variables

5. Understand properties of a bivariate normal distribution and be able to draw contour plots for the probability density function

6. Calculate means and variance for linear combinations of random variables and calculate probabilities for linear combinations of normally distributed random variables

CD MATERIAL

7. **Determine the distribution of a function of one or more random variables**

8. **Calculate moment generating functions and use them to determine moments for random variables and use the uniqueness property to determine the distribution of a random variable**

9. **Provide bounds on probabilities for arbitrary distributions based on Chebyshev's inequality**

Answers for most odd numbered exercises are at the end of the book. Answers to exercises whose numbers are surrounded by a box can be accessed in the e-Text by clicking on the box. Complete worked solutions to certain exercises are also available in the e-Text. These are indicated in the Answers to Selected Exercises section by a box around the exercise number. Exercises are also available for the text sections that appear on CD only. These exercises may be found within the e-Text immediately following the section they accompany.

In Chapters 3 and 4 we studied probability distributions for a single random variable. However, it is often useful to have more than one random variable defined in a random experiment. For example, in the classification of transmitted and received signals, each signal can be classified as high, medium, or low quality. We might define the random variable X to be the number of high-quality signals received and the random variable Y to be the number of low-quality signals received. In another example, the continuous random variable X can denote the length of one dimension of an injection-molded part, and the continuous random variable Y might denote the length of another dimension. We might be interested in probabilities that can be expressed in terms of both X and Y. For example, if the specifications for X and Y are (2.95 to 3.05) and (7.60 to 7.80) millimeters, respectively, we might be interested in the probability that a part satisfies both specifications; that is, $P(2.95 < X < 3.05 \text{ and } 7.60 < Y < 7.80)$.

In general, if X and Y are two random variables, the probability distribution that defines their simultaneous behavior is called a **joint probability distribution.** In this chapter, we investigate some important properties of these joint distributions.

5-1   TWO DISCRETE RANDOM VARIABLES

5-1.1   Joint Probability Distributions

For simplicity, we begin by considering random experiments in which only two random variables are studied. In later sections, we generalize the presentation to the joint probability distribution of more than two random variables.

EXAMPLE 5-1   In the development of a new receiver for the transmission of digital information, each received bit is rated as *acceptable, suspect,* or *unacceptable,* depending on the quality of the received signal, with probabilities 0.9, 0.08, and 0.02, respectively. Assume that the ratings of each bit are independent.

In the first four bits transmitted, let

X denote the number of acceptable bits

Y denote the number of suspect bits

Then, the distribution of X is binomial with $n = 4$ and $p = 0.9$, and the distribution of Y is binomial with $n = 4$ and $p = 0.08$. However, because only four bits are being rated, the possible values of X and Y are restricted to the points shown in the graph in Fig. 5-1. Although the possible values of X are 0, 1, 2, 3, or 4, if $y = 3$, $x = 0$ or 1. By specifying the probability of each of the points in Fig. 5-1, we specify the joint probability distribution of X and Y. Similarly to an individual random variable, we define the range of the random variables (X, Y) to be the set of points (x, y) in two-dimensional space for which the probability that $X = x$ and $Y = y$ is positive.

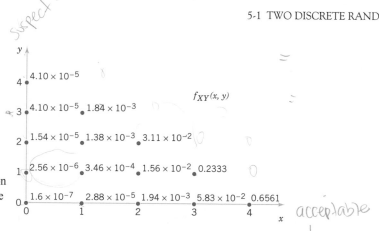

Figure 5-1 Joint probability distribution of X and Y in Example 5-1.

If X and Y are discrete random variables, the joint probability distribution of X and Y is a description of the set of points (x, y) in the range of (X, Y) along with the probability of each point. The joint probability distribution of two random variables is sometimes referred to as the **bivariate probability distribution** or **bivariate distribution** of the random variables. One way to describe the joint probability distribution of two discrete random variables is through a joint probability mass function. Also, $P(X = x$ and $Y = y)$ is usually written as $P(X = x, Y = y)$.

Definition

> The **joint probability mass function** of the discrete random variables X and Y, denoted as $f_{XY}(x, y)$, satisfies
>
> (1) $f_{XY}(x, y) \geq 0$
>
> (2) $\sum_x \sum_y f_{XY}(x, y) = 1$
>
> (3) $f_{XY}(x, y) = P(X = x, Y = y)$ (5-1)

Subscripts are used to indicate the random variables in the bivariate probability distribution. Just as the probability mass function of a single random variable X is assumed to be zero at all values outside the range of X, so the joint probability mass function of X and Y is assumed to be zero at values for which a probability is not specified.

EXAMPLE 5-2

Probabilities for each point in Fig. 5-1 are determined as follows. For example, $P(X = 2, Y = 1)$ is the probability that exactly two acceptable bits and exactly one suspect bit are received among the four bits transferred. Let a, s, and u denote acceptable, suspect, and unacceptable bits, respectively. By the assumption of independence,

$$P(aasu) = 0.9(0.9)(0.08)(0.02) = 0.0013$$

The number of possible sequences consisting of two a's, one s, and one u is shown in the CD material for Chapter 2:

$$\frac{4!}{2!1!1!} = 12$$

Therefore,

$$P(aasu) = 12(0.0013) = 0.0156$$

and

$$f_{XY}(2, 1) = P(X = 2, Y = 1) = 0.0156$$

The probabilities for all points in Fig. 5-1 are shown next to the point and the figure describes the joint probability distribution of X and Y.

5-1.2 Marginal Probability Distributions

If more than one random variable is defined in a random experiment, it is important to distinguish between the joint probability distribution of X and Y and the probability distribution of each variable individually. The individual probability distribution of a random variable is referred to as its **marginal probability distribution.** In Example 5-1, we mentioned that the marginal probability distribution of X is binomial with $n = 4$ and $p = 0.9$ and the marginal probability distribution of Y is binomial with $n = 4$ and $p = 0.08$.

In general, the marginal probability distribution of X can be determined from the joint probability distribution of X and other random variables. For example, to determine $P(X = x)$, we sum $P(X = x, Y = y)$ over all points in the range of (X, Y) for which $X = x$. Subscripts on the probability mass functions distinguish between the random variables.

EXAMPLE 5-3   The joint probability distribution of X and Y in Fig. 5-1 can be used to find the marginal probability distribution of X. For example,

$$P(X = 3) = P(X = 3, Y = 0) + P(X = 3, Y = 1)$$
$$= 0.0583 + 0.2333 = 0.292$$

As expected, this probability matches the result obtained from the binomial probability distribution for X; that is, $P(X = 3) = \binom{4}{3}0.9^3 0.1^1 = 0.292$. The marginal probability distribution for X is found by summing the probabilities in each column, whereas the marginal probability distribution for Y is found by summing the probabilities in each row. The results are shown in Fig. 5-2.

Although the marginal probability distribution of X in the previous example can be determined directly from the description of the experiment, in some problems the marginal probability distribution is determined from the joint probability distribution.

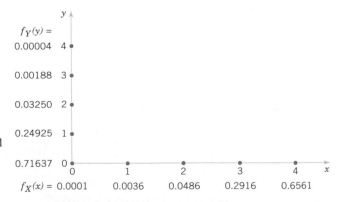

Figure 5-2 Marginal probability distributions of X and Y from Fig. 5-1.

Definition

> If X and Y are discrete random variables with joint probability mass function $f_{XY}(x, y)$, then the **marginal probability mass functions** of X and Y are
>
> $$f_X(x) = P(X = x) = \sum_{R_x} f_{XY}(x, y) \quad \text{and} \quad f_Y(y) = P(Y = y) = \sum_{R_y} f_{XY}(x, y)$$
>
> $$(5\text{-}2)$$
>
> where R_x denotes the set of all points in the range of (X, Y) for which $X = x$ and R_y denotes the set of all points in the range of (X, Y) for which $Y = y$

Given a joint probability mass function for random variables X and Y, $E(X)$ and $V(X)$ can be obtained directly from the joint probability distribution of X and Y or by first calculating the marginal probability distribution of X and then determining $E(X)$ and $V(X)$ by the usual method. This is shown in the following equation.

Mean and Variance from Joint Distribution

> If the marginal probability distribution of X has the probability mass function $f_X(x)$, then
>
> $$E(X) = \mu_X = \sum_x x f_X(x) = \sum_x x \left(\sum_{R_x} f_{XY}(x, y) \right) = \sum_x \sum_{R_x} x f_{XY}(x, y)$$
>
> $$= \sum_R x f_{XY}(x, y) \qquad (5\text{-}3)$$
>
> and
>
> $$V(X) = \sigma_X^2 = \sum_x (x - \mu_X)^2 f_X(x) = \sum_x (x - \mu_X)^2 \sum_{R_x} f_{XY}(x, y)$$
>
> $$= \sum_x \sum_{R_x} (x - \mu_X)^2 f_{XY}(x, y) = \sum_R (x - \mu_X)^2 f_{XY}(x, y)$$
>
> where R_x denotes the set of all points in the range of (X, Y) for which $X = x$ and R denotes the set of all points in the range of (X, Y)

EXAMPLE 5-4

In Example 5-1, $E(X)$ can be found as

$$\begin{aligned}
E(X) &= 0[f_{XY}(0, 0) + f_{XY}(0, 1) + f_{XY}(0, 2) + f_{XY}(0, 3) + f_{XY}(0, 4)] \\
&\quad + 1[f_{XY}(1, 0) + f_{XY}(1, 1) + f_{XY}(1, 2) + f_{XY}(1, 3)] \\
&\quad + 2[f_{XY}(2, 0) + f_{XY}(2, 1) + f_{XY}(2, 2)] \\
&\quad + 3[f_{XY}(3, 0) + f_{XY}(3, 1)] \\
&\quad + 4[f_{XY}(4, 0)] \\
&= 0[0.0001] + 1[0.0036] + 2[0.0486] + 3[0.02916] + 4[0.6561] = 3.6
\end{aligned}$$

Alternatively, because the marginal probability distribution of X is binomial,

$$E(X) = np = 4(0.9) = 3.6$$

The calculation using the joint probability distribution can be used to determine $E(X)$ even in cases in which the marginal probability distribution of X is not known. As practice, you can use the joint probability distribution to verify that $E(Y) = 0.32$ in Example 5-1.

Also,

$$V(X) = np(1 - p) = 4(0.9)(1 - 0.9) = 0.36$$

Verify that the same result can be obtained from the joint probability distribution of X and Y.

5-1.3 Conditional Probability Distributions

When two random variables are defined in a random experiment, knowledge of one can change the probabilities that we associate with the values of the other. Recall that in Example 5-1, X denotes the number of acceptable bits and Y denotes the number of suspect bits received by a receiver. Because only four bits are transmitted, if $X = 4$, Y must equal 0. Using the notation for conditional probabilities from Chapter 2, we can write this result as $P(Y = 0 | X = 4) = 1$. If $X = 3$, Y can only equal 0 or 1. Consequently, the random variables X and Y can be considered to be dependent. Knowledge of the value obtained for X changes the probabilities associated with the values of Y.

Recall that the definition of conditional probability for events A and B is $P(B|A) = P(A \cap B)/P(A)$. This definition can be applied with the event A defined to be $X = x$ and event B defined to be $Y = y$.

EXAMPLE 5-5   For Example 5-1, X and Y denote the number of acceptable and suspect bits received, respectively. The remaining bits are unacceptable.

$$P(Y = 0 | X = 3) = P(X = 3, Y = 0)/P(X = 3)$$
$$= f_{XY}(3, 0)/f_X(3) = 0.05832/0.2916 = 0.200$$

The probability that $Y = 1$ given that $X = 3$ is

$$P(Y = 1 | X = 3) = P(X = 3, Y = 1)/P(X = 3)$$
$$= f_{XY}(3, 1)/f_X(3) = 0.2333/0.2916 = 0.800$$

Given that $X = 3$, the only possible values for Y are 0 and 1. Notice that $P(Y = 0 | X = 3) + P(Y = 1 | X = 3) = 1$. The values 0 and 1 for Y along with the probabilities 0.200 and 0.800 define the conditional probability distribution of Y given that $X = 3$.

Example 5-5 illustrates that the conditional probabilities that $Y = y$ given that $X = x$ can be thought of as a new probability distribution. The following definition generalizes these ideas.

Definition

Given discrete random variables X and Y with joint probability mass function $f_{XY}(x, y)$ the **conditional probability mass function** of Y given $X = x$ is

$$f_{Y|x}(y) = f_{XY}(x, y)/f_X(x) \qquad \text{for } f_X(x) > 0 \tag{5-4}$$

The function $f_{Y|x}(y)$ is used to find the probabilities of the possible values for Y given that $X = x$. That is, it is the probability mass function for the possible values of Y given that $X = x$. More precisely, let R_x denote the set of all points in the range of (X, Y) for which $X = x$. The conditional probability mass function provides the conditional probabilities for the values of Y in the set R_x.

Because a conditional probability mass function $f_{Y|x}(y)$ is a probability mass function for all y in R_x, the following properties are satisfied:

(1) $f_{Y|x}(y) \geq 0$

(2) $\sum_{R_x} f_{Y|x}(y) = 1$

(3) $P(Y = y | X = x) = f_{Y|x}(y)$

$\qquad\qquad\qquad\qquad\qquad\qquad\qquad\qquad\qquad$ (5-5)

EXAMPLE 5-6   For the joint probability distribution in Fig. 5-1, $f_{Y|x}(y)$ is found by dividing each $f_{XY}(x, y)$ by $f_X(x)$. Here, $f_X(x)$ is simply the sum of the probabilities in each column of Fig. 5-1. The function $f_{Y|x}(y)$ is shown in Fig. 5-3. In Fig. 5-3, each column sums to one because it is a probability distribution.

Properties of random variables can be extended to a conditional probability distribution of Y given $X = x$. The usual formulas for mean and variance can be applied to a conditional probability mass function.

Definition

Let R_x denote the set of all points in the range of (X, Y) for which $X = x$. The **conditional mean** of Y given $X = x$, denoted as $E(Y|x)$ or $\mu_{Y|x}$, is

$$E(Y|x) = \sum_{R_x} y f_{Y|x}(y) \qquad\qquad\qquad (5-6)$$

and the **conditional variance** of Y given $X = x$, denoted as $V(Y|x)$ or $\sigma^2_{Y|x}$, is

$$V(Y|x) = \sum_{R_x} (y - \mu_{Y|x})^2 f_{Y|x}(y) = \sum_{R_x} y^2 f_{Y|x}(y) - \mu^2_{Y|x}$$

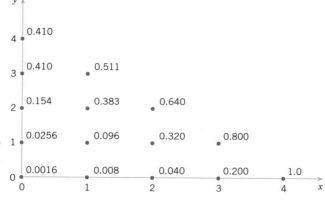

Figure 5-3
Conditional probability distributions of Y given $X = x$, $f_{Y|x}(y)$ in Example 5-6.

EXAMPLE 5-7

For the random variables in Example 5-1, the conditional mean of Y given $X = 2$ is obtained from the conditional distribution in Fig. 5-3:

$$E(Y|2) = \mu_{Y|2} = 0(0.040) + 1(0.320) + 2(0.640) = 1.6$$

The conditional mean is interpreted as the expected number of acceptable bits given that two of the four bits transmitted are suspect. The conditional variance of Y given $X = 2$ is

$$V(Y|2) = (0 - \mu_{Y|2})^2(0.040) + (1 - \mu_{Y|2})^2(0.320) + (2 - \mu_{Y|2})^2(0.640) = 0.32$$

5-1.4   Independence

In some random experiments, knowledge of the values of X does not change any of the probabilities associated with the values for Y.

EXAMPLE 5-8

In a plastic molding operation, each part is classified as to whether it conforms to color and length specifications. Define the random variable X and Y as

$$X = \begin{cases} 1 & \text{if the part conforms to color specifications} \\ 0 & \text{otherwise} \end{cases}$$

$$Y = \begin{cases} 1 & \text{if the part conforms to length specifications} \\ 0 & \text{otherwise} \end{cases}$$

Assume the joint probability distribution of X and Y is defined by $f_{XY}(x, y)$ in Fig. 5-4(a). The marginal probability distributions of X and Y are also shown in Fig. 5-4(a). Note that $f_{XY}(x, y) = f_X(x) f_Y(y)$. The conditional probability mass function $f_{Y|x}(y)$ is shown in Fig. 5-4(b). Notice that for any $x, f_{Y|x}(y) = f_Y(y)$. That is, knowledge of whether or not the part meets color specifications does not change the probability that it meets length specifications.

By analogy with independent events, we define two random variables to be **independent** whenever $f_{XY}(x, y) = f_X(x) f_Y(y)$ for all x and y. Notice that independence implies that $f_{XY}(x, y) = f_X(x) f_Y(y)$ for *all* x and y. If we find one pair of x and y in which the equality fails, X and Y are not independent. If two random variables are independent, then

$$f_{Y|x}(y) = \frac{f_{XY}(x, y)}{f_X(x)} = \frac{f_X(x) f_Y(y)}{f_X(x)} = f_Y(y)$$

With similar calculations, the following equivalent statements can be shown.

Figure 5-4   (a) Joint and marginal probability distributions of X and Y in Example 5-8. (b) Conditional probability distribution of Y given $X = x$ in Example 5-8.

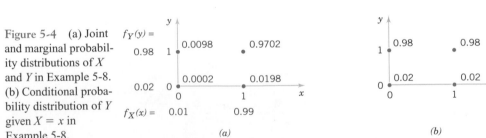

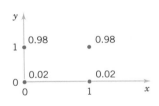

For discrete random variables X and Y, if any one of the following properties is true, the others are also true, and X and Y are **independent**.

(1) $f_{XY}(x, y) = f_X(x) f_Y(y)$ for all x and y

(2) $f_{Y|x}(y) = f_Y(y)$ for all x and y with $f_X(x) > 0$

(3) $f_{X|y}(x) = f_X(x)$ for all x and y with $f_Y(y) > 0$

(4) $P(X \in A, Y \in B) = P(X \in A)P(Y \in B)$ for any sets A and B in the range of X and Y, respectively.   (5-7)

Rectangular Range for (X, Y)!

If the set of points in two-dimensional space that receive positive probability under $f_{XY}(x, y)$ does not form a rectangle, X and Y are not independent because knowledge of X can restrict the range of values of Y that receive positive probability. In Example 5-1 knowledge that $X = 3$ implies that Y can equal only 0 or 1. Consequently, the marginal probability distribution of Y does not equal the conditional probability distribution $f_{Y|3}(y)$ for $X = 3$. Using this idea, we know immediately that the random variables X and Y with joint probability mass function in Fig. 5-1 are not independent. If the set of points in two-dimensional space that receives positive probability under $f_{XY}(x, y)$ forms a rectangle, independence is possible but not demonstrated. One of the conditions in Equation 5-7 must still be verified.

Rather than verifying independence from a joint probability distribution, knowledge of the random experiment is often used to assume that two random variables are independent. Then, the joint probability mass function of X and Y is computed from the product of the marginal probability mass functions.

EXAMPLE 5-9

In a large shipment of parts, 1% of the parts do not conform to specifications. The supplier inspects a random sample of 30 parts, and the random variable X denotes the number of parts in the sample that do not conform to specifications. The purchaser inspects another random sample of 20 parts, and the random variable Y denotes the number of parts in this sample that do not conform to specifications. What is the probability that $X \leq 1$ and $Y \leq 1$?

Although the samples are typically selected without replacement, if the shipment is large, relative to the sample sizes being used, approximate probabilities can be computed by assuming the sampling is with replacement and that X and Y are independent. With this assumption, the marginal probability distribution of X is binomial with $n = 30$ and $p = 0.01$, and the marginal probability distribution of Y is binomial with $n = 20$ and $p = 0.01$.

If independence between X and Y were not assumed, the solution would have to proceed as follows:

$$
\begin{aligned}
P(X \leq 1, Y \leq 1) &= P(X = 0, Y = 0) + P(X = 1, Y = 0) \\
&\quad + P(X = 0, Y = 1) + P(X = 1, Y = 1) \\
&= f_{XY}(0, 0) + f_{XY}(1, 0) + f_{XY}(0, 1) + f_{XY}(1, 1)
\end{aligned}
$$

However, with independence, property (4) of Equation 5-7 can be used as

$$
P(X \leq 1, Y \leq 1) = P(X \leq 1) P(Y \leq 1)
$$

and the binomial distributions for X and Y can be used to determine these probabilities as $P(X \leq 1) = 0.9639$ and $P(Y \leq 1) = 0.9831$. Therefore, $P(X \leq 1, Y \leq 1) = 0.948$.

Consequently, the probability that the shipment is accepted for use in manufacturing is 0.948 even if 1% of the parts do not conform to specifications. If the supplier and the purchaser change their policies so that the shipment is acceptable only if zero nonconforming parts are found in the sample, the probability that the shipment is accepted for production is still quite high. That is,

$$P(X = 0, Y = 0) = P(X = 0)P(Y = 0) = 0.605$$

This example shows that inspection is not an effective means of achieving quality.

EXERCISES FOR SECTION 5-1

5-1. Show that the following function satisfies the properties of a joint probability mass function.

x	y	$f_{XY}(x, y)$
1	1	1/4
1.5	2	1/8
1.5	3	1/4
2.5	4	1/4
3	5	1/8

5-2. Continuation of Exercise 5-1. Determine the following probabilities:
(a) $P(X < 2.5, Y < 3)$ (b) $P(X < 2.5)$
(c) $P(Y < 3)$ (d) $P(X > 1.8, Y > 4.7)$

5-3. Continuation of Exercise 5-1. Determine $E(X)$ and $E(Y)$.

5-4. Continuation of Exercise 5-1. Determine
(a) The marginal probability distribution of the random variable X.
(b) The conditional probability distribution of Y given that $X = 1.5$.
(c) The conditional probability distribution of X given that $Y = 2$.
(d) $E(Y|X = 1.5)$
(e) Are X and Y independent?

5-5. Determine the value of c that makes the function $f(x, y) = c(x + y)$ a joint probability mass function over the nine points with $x = 1, 2, 3$ and $y = 1, 2, 3$.

5-6. Continuation of Exercise 5-5. Determine the following probabilities:
(a) $P(X = 1, Y < 4)$ (b) $P(X = 1)$
(c) $P(Y = 2)$ (d) $P(X < 2, Y < 2)$

5-7. Continuation of Exercise 5-5. Determine $E(X)$, $E(Y)$, $V(X)$, and $V(Y)$.

5-8. Continuation of Exercise 5-5. Determine
(a) The marginal probability distribution of the random variable X.

(b) The conditional probability distribution of Y given that $X = 1$.
(c) The conditional probability distribution of X given that $Y = 2$.
(d) $E(Y|X = 1)$
(e) Are X and Y independent?

5-9. Show that the following function satisfies the properties of a joint probability mass function.

x	y	$f_{XY}(x, y)$
-1	-2	1/8
-0.5	-1	1/4
0.5	1	1/2
1	2	1/8

5-10. Continuation of Exercise 5-9. Determine the following probabilities:
(a) $P(X < 0.5, Y < 1.5)$ (b) $P(X < 0.5)$
(c) $P(Y < 1.5)$ (d) $P(X > 0.25, Y < 4.5)$

5-11. Continuation of Exercise 5-9. Determine $E(X)$ and $E(Y)$.

5-12. Continuation of Exercise 5-9. Determine
(a) The marginal probability distribution of the random variable X.
(b) The conditional probability distribution of Y given that $X = 1$.
(c) The conditional probability distribution of X given that $Y = 1$.
(d) $E(X|y = 1)$
(e) Are X and Y independent?

5-13. Four electronic printers are selected from a large lot of damaged printers. Each printer is inspected and classified as containing either a major or a minor defect. Let the random variables X and Y denote the number of printers with major and minor defects, respectively. Determine the range of the joint probability distribution of X and Y.

5-14. In the transmission of digital information, the probability that a bit has high, moderate, and low distortion is 0.01, 0.04, and 0.95, respectively. Suppose that three bits are transmitted and that the amount of distortion of each bit is assumed to be independent. Let X and Y denote the number of bits with high and moderate distortion out of the three, respectively. Determine

(a) $f_{XY}(x, y)$ (b) $f_X(x)$
(c) $E(X)$ (d) $f_{Y|1}(y)$
(e) $E(Y|X = 1)$ (f) Are X and Y independent?

5-15. A small-business Web site contains 100 pages and 60%, 30%, and 10% of the pages contain low, moderate, and high graphic content, respectively. A sample of four pages is selected without replacement, and X and Y denote the number of pages with moderate and high graphics output in the sample. Determine

(a) $f_{XY}(x, y)$ (b) $f_X(x)$

(c) $E(X)$
(d) $f_{Y|3}(y)$
(e) $E(Y|X = 3)$
(f) $V(Y|X = 3)$
(g) Are X and Y independent?

5-16. A manufacturing company employs two inspecting devices to sample a fraction of their output for quality control purposes. The first inspection monitor is able to accurately detect 99.3% of the defective items it receives, whereas the second is able to do so in 99.7% of the cases. Assume that four defective items are produced and sent out for inspection. Let X and Y denote the number of items that will be identified as defective by inspecting devices 1 and 2, respectively. Assume the devices are independent. Determine

(a) $f_{XY}(x, y)$ (b) $f_X(x)$
(c) $E(X)$ (d) $f_{Y|2}(y)$
(e) $E(Y|X = 2)$ (f) $V(Y|X = 2)$
(g) Are X and Y independent?

5-2 MULTIPLE DISCRETE RANDOM VARIABLES

5-2.1 Joint Probability Distributions

EXAMPLE 5-10

In some cases, more than two random variables are defined in a random experiment, and the concepts presented earlier in the chapter can easily be extended. The notation can be cumbersome and if doubts arise, it is helpful to refer to the equivalent concept for two random variables. Suppose that the quality of each bit received in Example 5-1 is categorized even more finely into one of the four classes, excellent, good, fair, or poor, denoted by E, G, F, and P, respectively. Also, let the random variables X_1, X_2, X_3, and X_4 denote the number of bits that are E, G, F, and P, respectively, in a transmission of 20 bits. In this example, we are interested in the joint probability distribution of four random variables. Because each of the 20 bits is categorized into one of the four classes, only values for x_1, x_2, x_3, and x_4 such that $x_1 + x_2 + x_3 + x_4 = 20$ receive positive probability in the probability distribution.

In general, given discrete random variables $X_1, X_2, X_3, \ldots, X_p$, the joint probability distribution of $X_1, X_2, X_3, \ldots, X_p$ is a description of the set of points $(x_1, x_2, x_3, \ldots, x_p)$ in the range of $X_1, X_2, X_3, \ldots, X_p$, along with the probability of each point. A joint probability mass function is a simple extension of a bivariate probability mass function.

Definition

The **joint probability mass function** of $X_1, X_2, \ldots, X_p$ is

$$f_{X_1 X_2 \ldots X_p}(x_1, x_2, \ldots, x_p) = P(X_1 = x_1, X_2 = x_2, \ldots, X_p = x_p) \qquad (5\text{-}8)$$

for all points $(x_1, x_2, \ldots, x_p)$ in the range of $X_1, X_2, \ldots, X_p$.

A marginal probability distribution is a simple extension of the result for two random variables.

Definition

> If $X_1, X_2, X_3, \ldots, X_p$ are discrete random variables with joint probability mass function $f_{X_1 X_2 \ldots X_p}(x_1, x_2, \ldots, x_p)$, the **marginal probability mass function** of any X_i is
>
> $$f_{X_i}(x_i) = P(X_i = x_i) = \sum_{R_{x_i}} f_{X_1 X_2 \ldots X_p}(x_1, x_2, \ldots, x_p) \tag{5-9}$$
>
> where R_{x_i} denotes the set of points in the range of $(X_1, X_2, \ldots, X_p)$ for which $X_i = x_i$.

EXAMPLE 5-11

Points that have positive probability in the joint probability distribution of three random variables X_1, X_2, X_3 are shown in Fig. 5-5. The range is the nonnegative integers with $x_1 + x_2 + x_3 = 3$. The marginal probability distribution of X_2 is found as follows.

$$P(X_2 = 0) = f_{X_1 X_2 X_3}(3, 0, 0) + f_{X_1 X_2 X_3}(0, 0, 3) + f_{X_1 X_2 X_3}(1, 0, 2) + f_{X_1 X_2 X_3}(2, 0, 1)$$
$$P(X_2 = 1) = f_{X_1 X_2 X_3}(2, 1, 0) + f_{X_1 X_2 X_3}(0, 1, 2) + f_{X_1 X_2 X_3}(1, 1, 1)$$
$$P(X_2 = 2) = f_{X_1 X_2 X_3}(1, 2, 0) + f_{X_1 X_2 X_3}(0, 2, 1)$$
$$P(X_2 = 3) = f_{X_1 X_2 X_3}(0, 3, 0)$$

Furthermore, $E(X_i)$ and $V(X_i)$ for $i = 1, 2, \ldots, p$ can be determined from the marginal probability distribution of X_i or from the joint probability distribution of $X_1, X_2, \ldots, X_p$ as follows.

Mean and Variance from Joint Distribution

> $$E(X_i) = \sum_R x_i f_{X_1 X_2 \ldots X_p}(x_1, x_2, \ldots, x_p)$$
>
> and
>
> $$V(X_i) = \sum_R (x_i - \mu_{X_i})^2 f_{X_1 X_2 \ldots X_p}(x_1, x_2, \ldots, x_p) \tag{5-10}$$
>
> where R is the set of all points in the range of $X_1, X_2, \ldots, X_p$.

With several random variables, we might be interested in the probability distribution of some subset of the collection of variables. The probability distribution of $X_1, X_2, \ldots, X_k, k < p$ can be obtained from the joint probability distribution of $X_1, X_2, \ldots, X_p$ as follows.

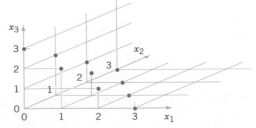

Figure 5-5   Joint probability distribution of X_1, X_2, and X_3.

Distribution of a Subset of Random Variables

If $X_1, X_2, X_3, \ldots, X_p$ are discrete random variables with joint probability mass function $f_{X_1 X_2 \ldots X_p}(x_1, x_2, \ldots, x_p)$, the joint **probability mass function** of $X_1, X_2, \ldots, X_k$, $k < p$, is

$$f_{X_1 X_2 \ldots X_k}(x_1, x_2, \ldots, x_k) = P(X_1 = x_1, X_2 = x_2, \ldots, X_k = x_k)$$
$$= \sum_{R_{x_1 x_2 \ldots x_k}} P(X_1 = x_1, X_2 = x_2, \ldots, X_k = x_k) \qquad (5\text{-}11)$$

where $R_{x_1 x_2 \ldots x_k}$ denotes the set of all points in the range of $X_1, X_2, \ldots, X_p$ for which $X_1 = x_1, X_2 = x_2, \ldots, X_k = x_k$.

That is, $P(X_1 = x_1, X_2 = x_2, \ldots, X_k = x_k)$ is the sum of the probabilities over all points in the range of $X_1, X_2, X_3, \ldots, X_p$ for which $X_1 = x_1, X_2 = x_2, \ldots,$ and $X_k = x_k$. An example is presented in the next section. Any k random variables can be used in the definition. The first k simplifies the notation.

Conditional Probability Distributions

Conditional probability distributions can be developed for multiple discrete random variables by an extension of the ideas used for two discrete random variables. For example, the conditional joint probability mass function of X_1, X_2, X_3 given X_4, X_5 is

$$f_{X_1 X_2 X_3 | x_4 x_5}(x_1, x_2, x_3) = \frac{f_{X_1 X_2 X_3 X_4 X_5}(x_1, x_2, x_3, x_4, x_5)}{f_{X_4 X_5}(x_4, x_5)}$$

for $f_{X_4 X_5}(x_4, x_5) > 0$. The conditional joint probability mass function of X_1, X_2, X_3 given X_4, X_5 provides the conditional probabilities at all points in the range of X_1, X_2, X_3, X_4, X_5 for which $X_4 = x_4$ and $X_5 = x_5$.

The concept of independence can be extended to multiple discrete random variables.

Definition

Discrete variables $X_1, X_2, \ldots, X_p$ are **independent** if and only if

$$f_{X_1 X_2 \ldots X_p}(x_1, x_2, \ldots, x_p) = f_{X_1}(x_1) f_{X_2}(x_2) \ldots f_{X_p}(x_p) \qquad (5\text{-}12)$$

for *all* $x_1, x_2, \ldots, x_p$.

Similar to the result for bivariate random variables, independence implies that Equation 5-12 holds for *all* $x_1, x_2, \ldots, x_p$. If we find one point for which the equality fails, $X_1, X_2, \ldots, X_p$ are not independent. It can be shown that if $X_1, X_2, \ldots, X_p$ are independent,

$$P(X_1 \in A_1, X_2 \in A_2, \ldots, X_p \in A_p) = P(X_1 \in A_1) P(X_2 \in A_2) \ldots P(X_p \in A_p)$$

for *any* sets $A_1, A_2, \ldots, A_p$.

5-2.2 Multinomial Probability Distribution

A joint probability distribution for multiple discrete random variables that is quite useful is an extension of the binomial. The random experiment that generates the probability distribution consists of a series of independent trials. However, the results from each trial can be categorized into one of k classes.

EXAMPLE 5-12   We might be interested in a probability such as the following. Of the 20 bits received, what is the probability that 14 are excellent, 3 are good, 2 are fair, and 1 is poor? Assume that the classifications of individual bits are independent events and that the probabilities of E, G, F, and P are 0.6, 0.3, 0.08, and 0.02, respectively. One sequence of 20 bits that produces the specified numbers of bits in each class can be represented as

$$EEEEEEEEEEEEEEGGGFFP$$

Using independence, we find that the probability of this sequence is

$$P(EEEEEEEEEEEEEEGGGFFP) = 0.6^{14}0.3^{3}0.08^{2}0.02^{1} = 2.708 \times 10^{-9}$$

Clearly, all sequences that consist of the same numbers of E's, G's, F's, and P's have the same probability. Consequently, the requested probability can be found by multiplying 2.708×10^{-9} by the number of sequences with 14 E's, three G's, two F's, and one P. The number of sequences is found from the CD material for Chapter 2 to be

$$\frac{20!}{14!3!2!1!} = 2325600$$

Therefore, the requested probability is

$$P(14E\text{'s, three }G\text{'s, two }F\text{'s, and one }P) = 2325600(2.708 \times 10^{-9}) = 0.0063$$

Example 5-12 leads to the following generalization of a binomial experiment and a binomial distribution.

Multinomial Distribution

Suppose a random experiment consists of a series of n trials. Assume that

(1)   The result of each trial is classified into one of k classes.

(2)   The probability of a trial generating a result in class 1, class 2, $\ldots$, class k is constant over the trials and equal to $p_1, p_2, \ldots, p_k$, respectively.

(3)   The trials are independent.

The random variables $X_1, X_2, \ldots, X_k$ that denote the number of trials that result in class 1, class 2, $\ldots$, class k, respectively, have a **multinomial distribution** and the joint probability mass function is

$$P(X_1 = x_1, X_2 = x_2, \ldots, X_k = x_k) = \frac{n!}{x_1!x_2!\cdots x_k!} p_1^{x_1} p_2^{x_2} \cdots p_k^{x_k} \qquad (5\text{-}13)$$

for $x_1 + x_2 + \cdots + x_k = n$ and $p_1 + p_2 + \cdots + p_k = 1$.

The multinomial distribution is considered a multivariable extension of the binomial distribution.

EXAMPLE 5-13

In Example 5-12, let the random variables X_1, X_2, X_3, and X_4 denote the number of bits that are E, G, F, and P, respectively, in a transmission of 20 bits. The probability that 12 of the bits received are E, 6 are G, 2 are F, and 0 are P is

$$P(X_1 = 12, X_2 = 6, X_3 = 2, X_4 = 0) = \frac{20!}{12!6!2!0!}\ 0.6^{12}0.3^{6}0.08^{2}0.02^{0} = 0.0358$$

Each trial in a multinomial random experiment can be regarded as either generating or not generating a result in class i, for each $i = 1, 2, \ldots, k$. Because the random variable X_i is the number of trials that result in class i, X_i has a binomial distribution.

> If $X_1, X_2, \ldots, X_k$ have a multinomial distribution, the marginal probability distribution of X_i is binomial with
>
> $$E(X_i) = np_i \quad \text{and} \quad V(X_i) = np_i(1 - p_i) \qquad (5\text{-}14)$$

EXAMPLE 5-14

In Example 5-13, the marginal probability distribution of X_2 is binomial with $n = 20$ and $p = 0.3$. Furthermore, the joint marginal probability distribution of X_2 and X_3 is found as follows. The $P(X_2 = x_2, X_3 = x_3)$ is the probability that exactly x_2 trials result in G and that x_3 result in F. The remaining $n - x_2 - x_3$ trials must result in either E or P. Consequently, we can consider each trial in the experiment to result in one of three classes, $\{G\}$, $\{F\}$, or $\{E, P\}$, with probabilities 0.3, 0.08, and $0.6 + 0.02 = 0.62$, respectively. With these new classes, we can consider the trials to comprise a new multinomial experiment. Therefore,

$$f_{X_2 X_3}(x_2, x_3) = P(X_2 = x_2, X_3 = x_3)$$

$$= \frac{n!}{x_2! x_3!\ (n - x_2 - x_3)!}\ (0.3)^{x_2}(0.08)^{x_3}(0.62)^{n - x_2 - x_3}$$

The joint probability distribution of other sets of variables can be found similarly.

EXERCISES FOR SECTION 5-2

5-17. Suppose the random variables X, Y, and Z have the following joint probability distribution

x	y	z	$f(x, y, z)$
1	1	1	0.05
1	1	2	0.10
1	2	1	0.15
1	2	2	0.20
2	1	1	0.20
2	1	2	0.15
2	2	1	0.10
2	2	2	0.05

Determine the following:
(a) $P(X = 2)$ (b) $P(X = 1, Y = 2)$
(c) $P(Z < 1.5)$ (d) $P(X = 1 \text{ or } Z = 2)$
(e) $E(X)$

5-18. Continuation of Exercise 5-17. Determine the following:
(a) $P(X = 1 | Y = 1)$ (b) $P(X = 1, Y = 1 | Z = 2)$
(c) $P(X = 1 | Y = 1, Z = 2)$

5-19. Continuation of Exercise 5-17. Determine the conditional probability distribution of X given that $Y = 1$ and $Z = 2$.

5-20. Based on the number of voids, a ferrite slab is classified as either high, medium, or low. Historically, 5% of the slabs are classified as high, 85% as medium, and 10% as low.

A sample of 20 slabs is selected for testing. Let X, Y, and Z denote the number of slabs that are independently classified as high, medium, and low, respectively.

(a) What is the name and the values of the parameters of the joint probability distribution of X, Y, and Z?

(b) What is the range of the joint probability distribution of X, Y, Z?

(c) What is the name and the values of the parameters of the marginal probability distribution of X?

(d) Determine $E(X)$ and $V(X)$.

5-21. Continuation of Exercise 5-20. Determine the following:

(a) $P(X = 1, Y = 17, Z = 3)$

(b) $P(X \le 1, Y = 17, Z = 3)$

(c) $P(X \le 1)$

(d) $E(X)$

5-22. Continuation of Exercise 5-20. Determine the following:

(a) $P(X = 2, Z = 3 \mid Y = 17)$ (b) $P(X = 2 \mid Y = 17)$

(c) $E(X \mid Y = 17)$

5-23. An order of 15 printers contains four with a graphics-enhancement feature, five with extra memory, and six with both features. Four printers are selected at random, without replacement, from this set. Let the random variables X, Y, and Z denote the number of printers in the sample with graphics enhancement only, extra memory only, and both, respectively.

(a) Describe the range of the joint probability distribution of X, Y, and Z.

(b) Is the probability distribution of X, Y, and Z multinomial? Why or why not?

5-24. Continuation of Exercise 5-23. Determine the conditional probability distribution of X given that $Y = 2$.

5-25. Continuation of Exercise 5-23. Determine the following:

(a) $P(X = 1, Y = 2, Z = 1)$ (b) $P(X = 1, Y = 1)$

(c) $E(X)$ and $V(X)$

5-26. Continuation of Exercise 5-23. Determine the following:

(a) $P(X = 1, Y = 2 \mid Z = 1)$ (b) $P(X = 2 \mid Y = 2)$

(c) The conditional probability distribution of X given that $Y = 0$ and $Z = 3$.

5-27. Four electronic ovens that were dropped during shipment are inspected and classified as containing either a major, a minor, or no defect. In the past, 60% of dropped ovens had a major defect, 30% had a minor defect, and 10% had no defect. Assume that the defects on the four ovens occur independently.

(a) Is the probability distribution of the count of ovens in each category multinomial? Why or why not?

(b) What is the probability that, of the four dropped ovens, two have a major defect and two have a minor defect?

(c) What is the probability that no oven has a defect?

5-28. Continuation of Exercise 5-27. Determine the following:

(a) The joint probability mass function of the number of ovens with a major defect and the number with a minor defect.

(b) The expected number of ovens with a major defect.

(c) The expected number of ovens with a minor defect.

5-29. Continuation of Exercise 5-27. Determine the following:

(a) The conditional probability that two ovens have major defects given that two ovens have minor defects

(b) The conditional probability that three ovens have major defects given that two ovens have minor defects

(c) The conditional probability distribution of the number of ovens with major defects given that two ovens have minor defects

(d) The conditional mean of the number of ovens with major defects given that two ovens have minor defects

5-30. In the transmission of digital information, the probability that a bit has high, moderate, or low distortion is 0.01, 0.04, and 0.95, respectively. Suppose that three bits are transmitted and that the amount of distortion of each bit is assumed to be independent.

(a) What is the probability that two bits have high distortion and one has moderate distortion?

(b) What is the probability that all three bits have low distortion?

5-31. Continuation of Exercise 5-30. Let X and Y denote the number of bits with high and moderate distortion out of the three transmitted, respectively. Determine the following:

(a) The probability distribution, mean and variance of X.

(b) The conditional probability distribution, conditional mean and conditional variance of X given that $Y = 2$.

5-32. A marketing company performed a risk analysis for a manufacturer of synthetic fibers and concluded that new competitors present no risk 13% of the time (due mostly to the diversity of fibers manufactured), moderate risk 72% of the time (some overlapping of products), and very high risk (competitor manufactures the exact same products) 15% of the time. It is known that 12 international companies are planning to open new facilities for the manufacture of synthetic fibers within the next three years. Assume the companies are independent. Let X, Y, and Z denote the number of new competitors that will pose no, moderate, and very high risk for the interested company, respectively.

(a) What is the range of the joint probability distribution of X, Y, and Z?

(b) Determine $P(X = 1, Y = 3, Z = 1)$

(c) Determine $P(Z \le 2)$

5-33. Continuation of Exercise 5-32. Determine the following:

(a) $P(Z = 2 \mid Y = 1, X = 10)$ (b) $P(Z \le 1 \mid X = 10)$

(c) $P(Y \le 1, Z \le 1 \mid X = 10)$ (d) $E(Z \mid X = 10)$

5-3   TWO CONTINUOUS RANDOM VARIABLES

5-3.1   Joint Probability Distributions

Our presentation of the joint probability distribution of two continuous random variables is similar to our discussion of two discrete random variables. As an example, let the continuous random variable X denote the length of one dimenson of an injection-molded part, and let the continuous random variable Y denote the length of another dimension. The sample space of the random experiment consists of points in two dimensions.

We can study each random variable separately. However, because the two random variables are measurements from the same part, small disturbances in the injection-molding process, such as pressure and temperature variations, might be more likely to generate values for X and Y in specific regions of two-dimensional space. For example, a small pressure increase might generate parts such that both X and Y are greater than their respective targets and a small pressure decrease might generate parts such that X and Y are both less than their respective targets. Therefore, based on pressure variations, we expect that the probability of a part with X much greater than its target and Y much less than its target is small. Knowledge of the joint probability distribution of X and Y provides information that is not obvious from the marginal probability distributions.

The joint probability distribution of two continuous random variables X and Y can be specified by providing a method for calculating the probability that X and Y assume a value in any region R of two-dimensional space. Analogous to the probability density function of a single continuous random variable, a **joint probability density function** can be defined over two-dimensional space. The double integral of $f_{XY}(x, y)$ over a region R provides the probability that (X, Y) assumes a value in R. This integral can be interpreted as the volume under the surface $f_{XY}(x, y)$ over the region R.

A joint probability density function for X and Y is shown in Fig. 5-6. The probability that (X, Y) assumes a value in the region R equals the volume of the shaded region in Fig. 5-6. In this manner, a joint probability density function is used to determine probabilities for X and Y.

Definition

> A **joint probability density function** for the continuous random variables X and Y, denoted as $f_{XY}(x, y)$, satisfies the following properties:
>
> (1) $f_{XY}(x, y) \geq 0$ for all x, y
>
> (2) $\displaystyle\int_{-\infty}^{\infty} \int_{-\infty}^{\infty} f_{XY}(x, y)\, dx\, dy = 1$
>
> (3)   For any region R of two-dimensional space
>
> $$P([X, Y] \in R) = \iint_{R} f_{XY}(x, y)\, dx\, dy \qquad (5\text{-}15)$$

Typically, $f_{XY}(x, y)$ is defined over all of two-dimensional space by assuming that $f_{XY}(x, y) = 0$ for all points for which $f_{XY}(x, y)$ is not specified.

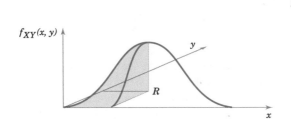

Probability that (X, Y) is in the region R is determined by the volume of $f_{XY}(x, y)$ over the region R.

Figure 5-6   Joint probability density function for random variables X and Y.

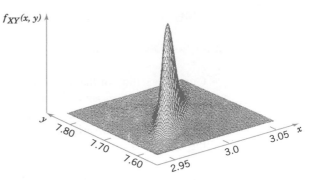

Figure 5-7   Joint probability density function for the lengths of different dimensions of an injection-molded part.

At the start of this chapter, the lengths of different dimensions of an injection-molded part were presented as an example of two random variables. Each length might be modeled by a normal distribution. However, because the measurements are from the same part, the random variables are typically not independent. A probability distribution for two normal random variables that are not independent is important in many applications and it is presented later in this chapter. If the specifications for X and Y are 2.95 to 3.05 and 7.60 to 7.80 millimeters, respectively, we might be interested in the probability that a part satisfies both specifications; that is, $P(2.95 < X < 3.05, 7.60 < Y < 7.80)$. Suppose that $f_{XY}(x, y)$ is shown in Fig. 5-7. The required probability is the volume of $f_{XY}(x, y)$ within the specifications. Often a probability such as this must be determined from a numerical integration.

EXAMPLE 5-15   Let the random variable X denote the time until a computer server connects to your machine (in milliseconds), and let Y denote the time until the server authorizes you as a valid user (in milliseconds). Each of these random variables measures the wait from a common starting time and $X < Y$. Assume that the joint probability density function for X and Y is

$$f_{XY}(x, y) = 6 \times 10^{-6} \exp(-0.001x - 0.002y) \quad \text{for } x < y$$

Reasonable assumptions can be used to develop such a distribution, but for now, our focus is only on the joint probability density function.

The region with nonzero probability is shaded in Fig. 5-8. The property that this joint probability density function integrates to 1 can be verified by the integral of $f_{XY}(x, y)$ over this region as follows:

$$\int_{-\infty}^{\infty} \int_{-\infty}^{\infty} f_{XY}(x, y) \, dy \, dx = \int_{0}^{\infty} \left(\int_{x}^{\infty} 6 \times 10^{-6} e^{-0.001x - 0.002y} \, dy \right) dx$$

$$= 6 \times 10^{-6} \int_{0}^{\infty} \left(\int_{x}^{\infty} e^{-0.002y} \, dy \right) e^{-0.001x} \, dx$$

$$= 6 \times 10^{-6} \int_{0}^{\infty} \left(\frac{e^{-0.002x}}{0.002} \right) e^{-0.001x} \, dx$$

$$= 0.003 \left(\int_{0}^{\infty} e^{-0.003x} \, dx \right) = 0.003 \left(\frac{1}{0.003} \right) = 1$$

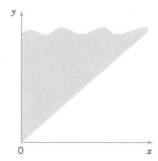

Figure 5-8   The joint probability density function of X and Y is nonzero over the shaded region.

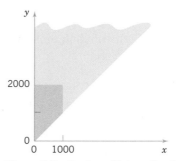

Figure 5-9   Region of integration for the probability that $X < 1000$ and $Y < 2000$ is darkly shaded.

The probability that $X < 1000$ and $Y < 2000$ is determined as the integral over the darkly shaded region in Fig. 5-9.

$$P(X \leq 1000, Y \leq 2000) = \int_0^{1000} \int_x^{2000} f_{XY}(x, y) \, dy \, dx$$

$$= 6 \times 10^{-6} \int_0^{1000} \left(\int_x^{2000} e^{-0.002y} \, dy \right) e^{-0.001x} \, dx$$

$$= 6 \times 10^{-6} \int_0^{1000} \left(\frac{e^{-0.002x} - e^{-4}}{0.002} \right) e^{-0.001x} \, dx$$

$$= 0.003 \int_0^{1000} e^{-0.003x} - e^{-4} e^{-0.001x} \, dx$$

$$= 0.003 \left[\left(\frac{1 - e^{-3}}{0.003} \right) - e^{-4} \left(\frac{1 - e^{-1}}{0.001} \right) \right]$$

$$= 0.003 (316.738 - 11.578) = 0.915$$

5-3.2   Marginal Probability Distributions

Similar to joint discrete random variables, we can find the marginal probability distributions of X and Y from the joint probability distribution.

Definition

> If the joint probability density function of continuous random variables X and Y is $f_{XY}(x, y)$, the **marginal probability density functions of X and Y** are
>
> $$f_X(x) = \int_{R_x} f_{XY}(x, y) \, dy \quad \text{and} \quad f_Y(y) = \int_{R_y} f_{XY}(x, y) \, dx \qquad (5\text{-}16)$$
>
> where R_x denotes the set of all points in the range of (X, Y) for which $X = x$ and R_y denotes the set of all points in the range of (X, Y) for which $Y = y$

A probability involving only one random variable, say, for example, $P(a < X < b)$, can be found from the marginal probability distribution of X or from the joint probability distribution of X and Y. For example, $P(a < X < b)$ equals $P(a < X < b, -\infty < Y < \infty)$. Therefore,

$$P(a < X < b) = \int_a^b \int_{R_x} f_{XY}(x, y)\, dy\, dx = \int_a^b \left(\int_{R_x} f_{XY}(x, y)\, dy \right) dx = \int_a^b f_X(x)\, dx$$

Similarly, $E(X)$ and $V(X)$ can be obtained directly from the joint probability distribution of X and Y or by first calculating the marginal probability distribution of X. The details, shown in the following equations, are similar to those used for discrete random variables.

Mean and Variance from Joint Distribution

$$E(X) = \mu_X = \int_{-\infty}^{\infty} x f_X(x)\, dx = \int_{-\infty}^{\infty} x \left[\int_{R_x} f_{XY}(x, y)\, dy \right] dx$$

$$= \iint_R x f_{XY}(x, y)\, dx\, dy \tag{5-17}$$

and

$$V(X) = \sigma_X^2 \int_{-\infty}^{\infty} (x - \mu_X)^2 f_X(x)\, dx = \int_{-\infty}^{\infty} (x - \mu_X)^2 \left[\int_{R_x} f_{XY}(x, y)\, dy \right] dx$$

$$= \iint_R (x - \mu_X)^2 f_{XY}(x, y)\, dx\, dy$$

where R_X denotes the set of all points in the range of (X, Y) for which $X = x$ and R_Y denotes the set of all points in the range of (X, Y)

EXAMPLE 5-16

For the random variables that denote times in Example 5-15, calculate the probability that Y exceeds 2000 milliseconds.

This probability is determined as the integral of $f_{XY}(x, y)$ over the darkly shaded region in Fig. 5-10. The region is partitioned into two parts and different limits of integration are determined for each part.

$$P(Y > 2000) = \int_0^{2000} \left(\int_{2000}^{\infty} 6 \times 10^{-6} e^{-0.001x - 0.002y}\, dy \right) dx$$

$$+ \int_{2000}^{\infty} \left(\int_x^{\infty} 6 \times 10^{-6} e^{-0.001x - 0.002y}\, dy \right) dx$$

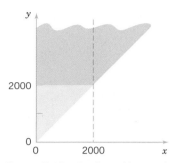

Figure 5-10 Region of integration for the probability that $Y < 2000$ is darkly shaded and it is partitioned into two regions with $x < 2000$ and $x > 2000$.

The first integral is

$$6 \times 10^{-6} \int_0^{2000} \left(\frac{e^{-0.002y}}{-0.002} \bigg|_{2000}^{\infty} \right) e^{-0.001x} \, dx = \frac{6 \times 10^{-6}}{0.002} e^{-4} \int_0^{2000} e^{-0.001x} \, dx$$

$$= \frac{6 \times 10^{-6}}{0.002} e^{-4} \left(\frac{1 - e^{-2}}{0.001} \right) = 0.0475$$

The second integral is

$$6 \times 10^{-6} \int_{2000}^{\infty} \left(\frac{e^{-0.002y}}{-0.002} \bigg|_x^{\infty} \right) e^{-0.001x} \, dx = \frac{6 \times 10^{-6}}{0.002} \int_{2000}^{\infty} e^{-0.003x} \, dx$$

$$= \frac{6 \times 10^{-6}}{0.002} \left(\frac{e^{-6}}{0.003} \right) = 0.0025$$

Therefore,

$$P(Y > 2000) = 0.0475 + 0.0025 = 0.05.$$

Alternatively, the probability can be calculated from the marginal probability distribution of Y as follows. For $y > 0$

$$f_Y(y) = \int_0^y 6 \times 10^{-6} e^{-0.001x - 0.002y} \, dx = 6 \times 10^{-6} e^{-0.002y} \int_0^y e^{-0.001x} \, dx$$

$$= 6 \times 10^{-6} e^{-0.002y} \left(\frac{e^{-0.001x}}{-0.001} \bigg|_0^y \right) = 6 \times 10^{-6} e^{-0.002y} \left(\frac{1 - e^{-0.001y}}{0.001} \right)$$

$$= 6 \times 10^{-3} e^{-0.002y} (1 - e^{-0.001y}) \quad \text{for } y > 0$$

We have obtained the marginal probability density function of Y. Now,

$$P(Y > 2000) = 6 \times 10^{-3} \int_{2000}^{\infty} e^{-0.002y}(1 - e^{-0.001y})\, dy$$

$$= 6 \times 10^{-3}\left[\left(\frac{e^{-0.002y}}{-0.002}\Big|_{2000}^{\infty}\right) - \left(\frac{e^{-0.003y}}{-0.003}\Big|_{2000}^{\infty}\right)\right]$$

$$= 6 \times 10^{-3}\left[\frac{e^{-4}}{0.002} - \frac{e^{-6}}{0.003}\right] = 0.05$$

5-3.3 Conditional Probability Distributions

Analogous to discrete random variables, we can define the conditional probability distribution of Y given $X = x$.

Definition

> Given continuous random variables X and Y with joint probability density function $f_{XY}(x, y)$, the **conditional probability density function** of Y given $X = x$ is
>
> $$f_{Y|x}(y) = \frac{f_{XY}(x, y)}{f_X(x)} \quad \text{for} \quad f_X(x) > 0 \qquad (5\text{-}18)$$

The function $f_{Y|x}(y)$ is used to find the probabilities of the possible values for Y given that $X = x$. Let R_x denote the set of all points in the range of (X, Y) for which $X = x$. The conditional probability density function provides the conditional probabilities for the values of Y in the set R_x.

> Because the conditional probability density function $f_{Y|x}(y)$ is a probability density function for all y in R_x, the following properties are satisfied:
>
> (1) $f_{Y|x}(y) \geq 0$
>
> (2) $\int_{R_x} f_{Y|x}(y)\, dy = 1$
>
> (3) $P(Y \in B \,|\, X = x) = \int_{B} f_{Y|x}(y)\, dy$ for any set B in the range of Y
>
> $$\qquad (5\text{-}19)$$

It is important to state the region in which a joint, marginal, or conditional probability density function is not zero. The following example illustrates this.

EXAMPLE 5-17

For the random variables that denote times in Example 5-15, determine the conditional probability density function for Y given that $X = x$.

First the marginal density function of x is determined. For $x > 0$

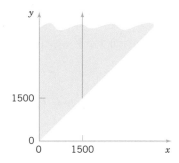

Figure 5-11 The conditional probability density function for Y, given that $x = 1500$, is nonzero over the solid line.

$$f_X(x) = \int_x^\infty 6 \times 10^{-6} e^{-0.001x - 0.002y} \, dy = 6 \times 10^{-6} e^{-0.001x} \left(\frac{e^{-0.002y}}{-0.002} \Big|_x^\infty \right)$$

$$= 6 \times 10^{-6} e^{-0.001x} \left(\frac{e^{-0.002x}}{0.002} \right) = 0.003 e^{-0.003x} \quad \text{for} \quad x > 0$$

This is an exponential distribution with $\lambda = 0.003$. Now, for $0 < x$ and $x < y$ the conditional probability density function is

$$f_{Y|x}(y) = f_{XY}(x,y)/f_x(x) = \frac{6 \times 10^{-6} e^{-0.001x - 0.002y}}{0.003 e^{-0.003x}}$$

$$= 0.002 e^{0.002x - 0.002y} \quad \text{for } 0 < x \quad \text{and} \quad x < y$$

The conditional probability density function of Y, given that $x = 1500$, is nonzero on the solid line in Fig. 5-11.

Determine the probability that Y exceeds 2000, given that $x = 1500$. That is, determine $P(Y > 2000 \mid x = 1500)$. The conditional probability density function is integrated as follows:

$$P(Y > 2000 | x = 1500) = \int_{2000}^\infty f_{Y|1500}(y) \, dy = \int_{2000}^\infty 0.002 e^{0.002(1500) - 0.002y} \, dy$$

$$= 0.002 e^3 \left(\frac{e^{-0.002y}}{-0.002} \Big|_{2000}^\infty \right) = 0.002 e^3 \left(\frac{e^{-4}}{0.002} \right) = 0.368$$

Definition

Let R_x denote the set of all points in the range of (X, Y) for which $X = x$. The **conditional mean** of Y given $X = x$, denoted as $E(Y|x)$ or $\mu_{Y|x}$, is

$$E(Y \mid x) = \int_{R_x} y f_{Y|x}(y) \, dy$$

and the **conditional variance** of Y given $X = x$, denoted as $V(Y|x)$ or $\sigma^2_{Y|x}$, is

$$V(Y \mid x) = \int_{R_x} (y - \mu_{Y|x})^2 f_{Y|x}(y) \, dy = \int_{R_x} y^2 f_{Y|x}(y) \, dy - \mu^2_{Y|x} \qquad (5\text{-}20)$$

EXAMPLE 5-18   For the random variables that denote times in Example 5-15, determine the conditional mean for Y given that $x = 1500$.

The conditional probability density function for Y was determined in Example 5-17. Because $f_{Y|1500}(y)$ is nonzero for $y > 1500$,

$$E(Y|x = 1500) = \int_{1500}^{\infty} y(0.002e^{0.002(1500) - 0.002y})\, dy = 0.002e^3 \int_{1500}^{\infty} ye^{-0.002y}\, dy$$

Integrate by parts as follows:

$$\int_{1500}^{\infty} ye^{-0.002y}\, dy = y\frac{e^{-0.002y}}{-0.002}\Big|_{1500}^{\infty} - \int_{1500}^{\infty}\left(\frac{e^{-0.002y}}{-0.002}\right) dy$$

$$= \frac{1500}{0.002}e^{-3} - \left(\frac{e^{-0.002y}}{(-0.002)(-0.002)}\Big|_{1500}^{\infty}\right)$$

$$= \frac{1500}{0.002}e^{-3} + \frac{e^{-3}}{(0.002)(0.002)} = \frac{e^{-3}}{0.002}(2000)$$

With the constant $0.002e^3$ reapplied

$$E(Y|x = 1500) = 2000$$

5-3.4   Independence

The definition of independence for continuous random variables is similar to the definition for discrete random variables. If $f_{XY}(x, y) = f_X(x)f_Y(y)$ for all x and y, X and Y are **independent.** Independence implies that $f_{XY}(x, y) = f_X(x)f_Y(y)$ for *all* x and y. If we find one pair of x and y in which the equality fails, X and Y are not independent.

Definition

> For continuous random variables X and Y, if any one of the following properties is true, the others are also true, and X and Y are said to be **independent.**
>
> (1) $f_{XY}(x, y) = f_X(x)f_Y(y)$ for all x and y
> (2) $f_{Y|x}(y) = f_Y(y)$ for all x and y with $f_X(x) > 0$
> (3) $f_{X|y}(x) = f_X(x)$ for all x and y with $f_Y(y) > 0$
> (4) $P(X \in A, Y \in B) = P(X \in A)P(Y \in B)$ for any sets A and B in the range of X and Y, respectively.   (5-21)

EXAMPLE 5-19   For the joint distribution of times in Example 5-15, the

- Marginal distribution of Y was determined in Example 5-16.
- Conditional distribution of Y given $X = x$ was determined in Example 5-17.

Because the marginal and conditional probability densities are not the same for all values of x, property (2) of Equation 5-20 implies that the random variables are not independent. The fact that these variables are not independent can be determined quickly by noticing that the range of (X, Y), shown in Fig. 5-8, is not rectangular. Consequently, knowledge of X changes the interval of values for Y that receives nonzero probability.

EXAMPLE 5-20   Suppose that Example 5-15 is modified so that the joint probability density function of X and Y is $f_{XY}(x, y) = 2 \times 10^{-6}e^{-0.001x - 0.002y}$ for $x \geq 0$ and $y \geq 0$. Show that X and Y are independent and determine $P(X > 1000, Y < 1000)$.
The marginal probability density function of X is

$$f_X(x) = \int_0^\infty 2 \times 10^{-6}e^{-0.001x - 0.002y}\,dy = 0.001\,e^{-0.001x} \quad \text{for } x > 0$$

The marginal probability density function of y is

$$f_Y(y) = \int_0^\infty 2 \times 10^{-6}e^{-0.001x - 0.002y}\,dx = 0.002\,e^{-0.002y} \quad \text{for } y > 0$$

Therefore, $f_{XY}(x, y) = f_X(x)f_Y(y)$ for all x and y and X and Y are independent.
To determine the probability requested, property (4) of Equation 5-21 and the fact that each random variable has an exponential distribution can be applied.

$$P(X > 1000, Y < 1000) = P(X > 1000)P(Y < 1000) = e^{-1}(1 - e^{-2}) = 0.318$$

Often, based on knowledge of the system under study, random variables are assumed to be independent. Then, probabilities involving both variables can be determined from the marginal probability distributions.

EXAMPLE 5-21   Let the random variables X and Y denote the lengths of two dimensions of a machined part, respectively. Assume that X and Y are independent random variables, and further assume that the distribution of X is normal with mean 10.5 millimeters and variance 0.0025 (millimeter)^2 and that the distribution of Y is normal with mean 3.2 millimeters and variance 0.0036 (millimeter)^2. Determine the probability that $10.4 < X < 10.6$ and $3.15 < Y < 3.25$.
Because X and Y are independent,

$$P(10.4 < X < 10.6, 3.15 < Y < 3.25) = P(10.4 < X < 10.6)P(3.15 < Y < 3.25)$$

$$= P\left(\frac{10.4 - 10.5}{0.05} < Z < \frac{10.6 - 10.5}{0.05}\right) P\left(\frac{3.15 - 3.2}{0.06} < Z < \frac{3.25 - 3.2}{0.06}\right)$$

$$= P(-2 < Z < 2)P(-0.833 < Z < 0.833) = 0.566$$

where Z denotes a standard normal random variable.

EXERCISES FOR SECTION 5-3

5-34.   Determine the value of c such that the function $f(x, y) = cxy$ for $0 < x < 3$ and $0 < y < 3$ satisfies the properties of a joint probability density function.

5-35.   Continuation of Exercise 5-34. Determine the following:

(a) $P(X < 2, Y < 3)$
(b) $P(X < 2.5)$
(c) $P(1 < Y < 2.5)$
(d) $P(X > 1.8, 1 < Y < 2.5)$
(e) $E(X)$
(f) $P(X < 0, Y < 4)$

5-36. Continuation of Exercise 5-34. Determine the following:
(a) Marginal probability distribution of the random variable X
(b) Conditional probability distribution of Y given that $X = 1.5$
(c) $E(Y|X) = 1.5)$
(d) $P(Y < 2|X = 1.5)$
(e) Conditional probability distribution of X given that $Y = 2$

5-37. Determine the value of c that makes the function $f(x, y) = c(x + y)$ a joint probability density function over the range $0 < x < 3$ and $x < y < x + 2$.

5-38. Continuation of Exercise 5-37. Determine the following:
(a) $P(X < 1, Y < 2)$ (b) $P(1 < X < 2)$
(c) $P(Y > 1)$ (d) $P(X < 2, Y < 2)$
(e) $E(X)$

5-39. Continuation of Exercise 5-37. Determine the following:
(a) Marginal probability distribution of X
(b) Conditional probability distribution of Y given that $X = 1$
(c) $E(Y|X = 1)$
(d) $P(Y > 2|X = 1)$
(e) Conditional probability distribution of X given that $Y = 2$

5-40. Determine the value of c that makes the function $f(x, y) = cxy$ a joint probability density function over the range $0 < x < 3$ and $0 < y < x$.

5-41. Continuation of Exercise 5-40. Determine the following:
(a) $P(X < 1, Y < 2)$ (b) $P(1 < X < 2)$
(c) $P(Y > 1)$ (d) $P(X < 2, Y < 2)$
(e) $E(X)$ (f) $E(Y)$

5-42. Continuation of Exercise 5-40. Determine the following:
(a) Marginal probability distribution of X
(b) Conditional probability distribution of Y given $X = 1$
(c) $E(Y|X = 1)$
(d) $P(Y > 2|X = 1)$
(e) Conditional probability distribution of X given $Y = 2$

5-43. Determine the value of c that makes the function $f(x, y) = ce^{-2x - 3y}$ a joint probability density function over the range $0 < x$ and $0 < y < x$.

5-44. Continuation of Exercise 5-43. Determine the following:
(a) $P(X < 1, Y < 2)$ (b) $P(1 < X < 2)$
(c) $P(Y > 3)$ (d) $P(X < 2, Y < 2)$
(e) $E(X)$ (f) $E(Y)$

5-45. Continuation of Exercise 5-43. Determine the following:
(a) Marginal probability distribution of X
(b) Conditional probability distribution of Y given $X = 1$
(c) $E(Y|X = 1)$
(d) Conditional probability distribution of X given $Y = 2$

5-46. Determine the value of c that makes the function $f(x, y) = ce^{-2x - 3y}$ a joint probability density function over the range $0 < x$ and $x < y$.

5-47. Continuation of Exercise 5-46. Determine the following:
(a) $P(X < 1, Y < 2)$ (b) $P(1 < X < 2)$
(c) $P(Y > 3)$ (d) $P(X < 2, Y < 2)$
(e) $E(X)$ (f) $E(Y)$

5-48. Continuation of Exercise 5-46. Determine the following:
(a) Marginal probability distribution of X
(b) Conditional probability distribution of Y given $X = 1$
(c) $E(Y|X = 1)$
(d) $P(Y < 2|X = 1)$
(e) Conditional probability distribution of X given $Y = 2$

5-49. Two methods of measuring surface smoothness are used to evaluate a paper product. The measurements are recorded as deviations from the nominal surface smoothness in coded units. The joint probability distribution of the two measurements is a uniform distribution over the region $0 < x < 4$, $0 < y$, and $x - 1 < y < x + 1$. That is, $f_{XY}(x, y) = c$ for x and y in the region. Determine the value for c such that $f_{XY}(x, y)$ is a joint probability density function.

5-50. Continuation of Exercise 5-49. Determine the following:
(a) $P(X < 0.5, Y < 0.5)$ (b) $P(X < 0.5)$
(c) $E(X)$ (d) $E(Y)$

5-51. Continuation of Exercise 5-49. Determine the following:
(a) Marginal probability distribution of X
(b) Conditional probability distribution of Y given $X = 1$
(c) $E(Y|X = 1)$
(d) $P(Y < 0.5|X = 1)$

5-52. The time between surface finish problems in a galvanizing process is exponentially distributed with a mean of 40 hours. A single plant operates three galvanizing lines that are assumed to operate independently.
(a) What is the probability that none of the lines experiences a surface finish problem in 40 hours of operation?
(b) What is the probability that all three lines experience a surface finish problem between 20 and 40 hours of operation?
(c) Why is the joint probability density function not needed to answer the previous questions?

5-53. A popular clothing manufacturer receives Internet orders via two different routing systems. The time between orders for each routing system in a typical day is known to be exponentially distributed with a mean of 3.2 minutes. Both systems operate independently.
(a) What is the probability that no orders will be received in a 5 minute period? In a 10 minute period?
(b) What is the probability that both systems receive two orders between 10 and 15 minutes after the site is officially open for business?

(c) Why is the joint probability distribution not needed to answer the previous questions?

5-54.   The conditional probability distribution of Y given $X = x$ is $f_{Y|x}(y) = xe^{-xy}$ for $y > 0$ and the marginal probability distribution of X is a continuous uniform distribution over 0 to 10.

(a) Graph $f_{Y|X}(y) = xe^{-xy}$ for $y > 0$ for several values of x. Determine
(b) $P(Y < 2 | X = 2)$ (c) $E(Y | X = 2)$
(d) $E(Y | X = x)$ (e) $f_{XY}(x, y)$
(f) $f_Y(y)$

5-4   MULTIPLE CONTINUOUS RANDOM VARIABLES

As for discrete random variables, in some cases, more than two continuous random variables are defined in a random experiment.

EXAMPLE 5-22   Many dimensions of a machined part are routinely measured during production. Let the random variables, X_1, X_2, X_3, and X_4 denote the lengths of four dimensions of a part. Then, at least four random variables are of interest in this study.

The joint probability distribution of continuous random variables, $X_1, X_2, X_3 \ldots, X_p$ can be specified by providing a method of calculating the probability that $X_1, X_2, X_3, \ldots, X_p$ assume a value in a region R of p-dimensional space. A joint probability density function $f_{X_1 X_2 \ldots X_p}(x_1, x_2, \ldots, x_p)$ is used to determine the probability that $(X_1, X_2, X_3, \ldots, X_p)$ assume a value in a region R by the multiple integral of $f_{X_1 X_2 \ldots X_p}(x_1, x_2, \ldots, x_p)$ over the region R.

Definition

A **joint probability density function** for the continuous random variables $X_1, X_2, X_3, \ldots, X_p$, denoted as $f_{X_1 X_2 \ldots X_p}(x_1, x_2, \ldots, x_p)$, satisfies the following properties:

(1) $f_{X_1 X_2 \ldots X_p}(x_1, x_2, \ldots, x_p) \geq 0$

(2) $\displaystyle \int_{-\infty}^{\infty} \int_{-\infty}^{\infty} \cdots \int_{-\infty}^{\infty} f_{X_1 X_2 \ldots X_p}(x_1, x_2, \ldots, x_p) \, dx_1 \, dx_2 \ldots dx_p = 1$

(3)   For any region B of p-dimensional space

$$P[(X_1, X_2, \ldots, X_p) \in B] = \int \int_B \cdots \int f_{X_1 X_2 \ldots X_p}(x_1, x_2, \ldots, x_p) \, dx_1 \, dx_2 \ldots dx_p \quad (5\text{-}22)$$

Typically, $f_{X_1 X_2 \ldots X_p}(x_1, x_2, \ldots, x_p)$ is defined over all of p-dimensional space by assuming that $f_{X_1 X_2 \ldots X_p}(x_1, x_2, \ldots, x_p) = 0$ for all points for which $f_{X_1 X_2 \ldots X_p}(x_1, x_2, \ldots, x_p)$ is not specified.

EXAMPLE 5-23   In an electronic assembly, let the random variables X_1, X_2, X_3, X_4 denote the lifetimes of four components in hours. Suppose that the joint probability density function of these variables is

$$f_{X_1 X_2 X_3 X_4}(x_1, x_2, x_3, x_4) = 9 \times 10^{-2} e^{-0.001x_1 - 0.002x_2 - 0.0015x_3 - 0.003x_4}$$
$$\text{for } x_1 \geq 0, x_2 \geq 0, x_3 \geq 0, x_4 \geq 0$$

What is the probability that the device operates for more than 1000 hours without any failures?

The requested probability is $P(X_1 > 1000, X_2 > 1000, X_3 > 1000, X_4 > 1000)$, which equals the multiple integral of $f_{X_1 X_2 X_3 X_4}(x_1, x_2, x_3, x_4)$ over the region $x_1 > 1000$, $x_2 > 1000$, $x_3 > 1000$, $x_4 > 1000$. The joint probability density function can be written as a product of exponential functions, and each integral is the simple integral of an exponential function. Therefore,

$$P(X_1 > 1000, X_2 > 1000, X_3 > 1000, X_4 > 1000) = e^{-1-2-1.5-3} = 0.00055$$

Suppose that the joint probability density function of several continuous random variables is a constant, say c over a region R (and zero elsewhere). In this special case,

$$\int_{-\infty}^{\infty} \int_{-\infty}^{\infty} \cdots \int_{-\infty}^{\infty} f_{X_1 X_2 \ldots X_p}(x_1, x_2, \ldots, x_p) \, dx_1 \, dx_2 \ldots dx_p = c \times (\text{volume of region } R) = 1$$

by property (2) of Equation 5-22. Therefore, $c = 1/\text{volume}(R)$. Furthermore, by property (3) of Equation 5-22.

$$P[(X_1, X_2, \ldots, X_p) \in B]$$

$$= \int \int \cdots \int_B f_{X_1 X_2 \ldots X_p}(x_1, x_2, \ldots, x_p) \, dx_1 \, dx_2 \ldots dx_p = c \times \text{volume } (B \cap R)$$

$$= \frac{\text{volume } (B \cap R)}{\text{volume } (R)}$$

When the joint probability density function is constant, the probability that the random variables assume a value in the region B is just the ratio of the volume of the region $B \cap R$ to the volume of the region R for which the probability is positive.

EXAMPLE 5-24

Suppose the joint probability density function of the continuous random variables X and Y is constant over the region $x^2 + y^2 \le 4$. Determine the probability that $X^2 + Y^2 \le 1$.

The region that receives positive probability is a circle of radius 2. Therefore, the area of this region is 4π. The area of the region $x^2 + y^2 \le 1$ is π. Consequently, the requested probability is $\pi/(4\pi) = 1/4$.

Definition

> If the joint probability density function of continuous random variables $X_1, X_2, \ldots, X_p$ is $f_{X_1 X_2 \ldots X_p}(x_1, x_2 \ldots, x_p)$ the **marginal probability density function** of X_i is
>
> $$f_{X_i}(x_i) = \int \int \cdots \int_{R_{x_i}} f_{X_1 X_2 \ldots X_p}(x_1, x_2, \ldots, x_p) \, dx_1 \, dx_2 \ldots dx_{i-1} \, dx_{i+1} \ldots dx_p \qquad (5\text{-}23)$$
>
> where R_{x_i} denotes the set of all points in the range of $X_1, X_2, \ldots, X_p$ for which $X_i = x_i$.

As for two random variables, a probability involving only one random variable, say, for example $P(a < X_i < b)$, can be determined from the marginal probability distribution of X_i or from the joint probability distribution of $X_1, X_2, \ldots, X_p$. That is,

$$P(a < X_i < b) = P(-\infty < X_1 < \infty, \ldots, -\infty < X_{i-1} < \infty, a < X_i < b,$$
$$-\infty < X_{i+1} < \infty, \ldots, -\infty < X_p < \infty)$$

Furthermore, $E(X_i)$ and $V(X_i)$, for $i = 1, 2, \ldots, p$, can be determined from the marginal probability distribution of X_i or from the joint probability distribution of $X_1, X_2, \ldots, X_p$ as follows.

Mean and Variance from Joint Distribution

$$E(X_i) = \int_{-\infty}^{\infty} \int_{-\infty}^{\infty} \cdots \int_{-\infty}^{\infty} x_i f_{X_1 X_2 \ldots X_p}(x_1, x_2, \ldots, x_p) \, dx_1 \, dx_2 \ldots dx_p$$

and   (5-24)

$$V(X_i) = \int_{-\infty}^{\infty} \int_{-\infty}^{\infty} \cdots \int_{-\infty}^{\infty} (x_i - \mu_{X_i})^2 f_{X_1 X_2 \ldots X_p}(x_1, x_2, \ldots, x_p) \, dx_1 \, dx_2 \ldots dx_p$$

The probability distribution of a subset of variables such as $X_1, X_2, \ldots, X_k, k < p$, can be obtained from the joint probability distribution of $X_1, X_2, X_3, \ldots, X_p$ as follows.

Distribution of a Subset of Random Variables

If the joint probability density function of continuous random variables $X_1, X_2, \ldots, X_p$ is $f_{X_1 X_2 \ldots X_p}(x_1, x_2, \ldots, x_p)$, the **probability density function** of $X_1, X_2, \ldots, X_k, k < p$, is

$$f_{X_1 X_2 \ldots X_k}(x_1, x_2, \ldots, x_k)$$
$$= \int \int_{R_{x_1 x_2 \ldots x_k}} \cdots \int f_{X_1 X_2 \ldots X_p}(x_1, x_2, \ldots, x_p) \, dx_{k+1} \, dx_{k+2} \ldots dx_p \qquad (5\text{-}25)$$

where $R_{x_1 x_2 \ldots x_k}$ denotes the set of all points in the range of $X_1, X_2, \ldots, X_k$ for which $X_1 = x_1, X_2 = x_2, \ldots, X_k = x_k$.

Conditional Probability Distribution

Conditional probability distributions can be developed for multiple continuous random variables by an extension of the ideas used for two continuous random variables.

$$f_{X_1 X_2 X_3 | x_4 x_5}(x_1, x_2, x_3) = \frac{f_{X_1 X_2 X_3 X_4 X_5}(x_1, x_2, x_3, x_4, x_5)}{f_{X_4 X_5}(x_4, x_5)}$$

for $f_{X_4 X_5}(x_4, x_5) > 0$.

The concept of independence can be extended to multiple continuous random variables.

Definition

> Continuous random variables $X_1, X_2, \ldots, X_p$ are **independent** if and only if
>
> $$f_{X_1 X_2 \ldots X_p}(x_1, x_2 \ldots, x_p) = f_{X_1}(x_1) f_{X_2}(x_2) \ldots f_{X_p}(x_p) \quad \text{for all } x_1, x_2, \ldots, x_p \qquad (5\text{-}26)$$

Similar to the result for only two random variables, independence implies that Equation 5-26 holds for *all* $x_1, x_2, \ldots, x_p$. If we find one point for which the equality fails, $X_1, X_2, \ldots, X_p$ are not independent. It is left as an exercise to show that if $X_1, X_2, \ldots, X_p$ are independent,

$$P(X_1 \in A_1, X_2 \in A_2, \ldots, X_p \in A_p) = P(X_1 \in A_1)P(X_2 \in A_2) \ldots P(X_p \in A_p)$$

for *any* regions $A_1, A_2, \ldots, A_p$ in the range of $X_1, X_2, \ldots, X_p$, respectively.

EXAMPLE 5-25   In Chapter 3, we showed that a negative binomial random variable with parameters p and r can be represented as a sum of r geometric random variables $X_1, X_2, \ldots, X_r$. Each geometric random variable represents the additional trials required to obtain the next success. Because the trials in a binomial experiment are independent, $X_1, X_2, \ldots, X_r$ are independent random variables.

EXAMPLE 5-26   Suppose X_1, X_2, and X_3 represent the thickness in micrometers of a substrate, an active layer, and a coating layer of a chemical product. Assume that X_1, X_2, and X_3 are independent and normally distributed with $\mu_1 = 10000, \mu_2 = 1000, \mu_3 = 80, \sigma_1 = 250, \sigma_2 = 20$, and $\sigma_3 = 4$, respectively. The specifications for the thickness of the substrate, active layer, and coating layer are $9200 < x_1 < 10800, 950 < x_2 < 1050$, and $75 < x_3 < 85$, respectively. What proportion of chemical products meets all thickness specifications? Which one of the three thicknesses has the least probability of meeting specifications?

The requested probability is $P(9200 < X_1 < 10800, 950 < X_2 < 1050, 75 < X_3 < 85)$. Because the random variables are independent,

$$P(9200 < X_1 < 10800, 950 < X_2 < 1050, 75 < X_3 < 85)$$
$$= P(9200 < X_1 < 10800)P(950 < X_2 < 1050)P(75 < X_3 < 85)$$

After standardizing, the above equals

$$P(-3.2 < Z < 3.2)P(-2.5 < Z < 2.5)P(-1.25 < Z < 1.25)$$

where Z is a standard normal random variable. From the table of the standard normal distribution, the above equals

$$(0.99862)(0.98758)(0.78870) = 0.7778$$

The thickness of the coating layer has the least probability of meeting specifications. Consequently, a priority should be to reduce variability in this part of the process.

EXERCISES FOR SECTION 5–4

5-55. Suppose the random variables X, Y, and Z have the joint probability density function $f(x, y, z) = 8xyz$ for $0 < x < 1$, $0 < y < 1$, and $0 < z < 1$. Determine the following:
(a) $P(X < 0.5)$ (b) $P(X < 0.5, Y < 0.5)$
(c) $P(Z < 2)$ (d) $P(X < 0.5$ or $Z < 2)$
(e) $E(X)$

5-56. Continuation of Exercise 5-55. Determine the following:
(a) $P(X < 0.5 | Y = 0.5)$
(b) $P(X < 0.5, Y < 0.5 | Z = 0.8)$

5-57. Continuation of Exercise 5-55. Determine the following:
(a) Conditional probability distribution of X given that $Y = 0.5$ and $Z = 0.8$
(b) $P(X < 0.5 | Y = 0.5, Z = 0.8)$

5-58. Suppose the random variables X, Y, and Z have the joint probability density function $f_{XYZ}(x, y, z) = c$ over the cylinder $x^2 + y^2 < 4$ and $0 < z < 4$. Determine the following.
(a) The constant c so that $f_{XYZ}(x, y, z)$ is a probability density function
(b) $P(X^2 + Y^2 < 2)$
(c) $P(Z < 2)$
(d) $E(X)$

5-59. Continuation of Exercise 5-58. Determine the following:
(a) $P(X < 1 | Y = 1)$ (b) $P(X^2 + Y^2 < 1 | Z = 1)$

5-60. Continuation of Exercise 5-58. Determine the conditional probability distribution of Z given that $X = 1$ and $Y = 1$.

5-61. Determine the value of c that makes $f_{XYZ}(x, y, z) = c$ a joint probability density function over the region $x > 0$, $y > 0, z > 0$, and $x + y + z < 1$.

5-62. Continuation of Exercise 5-61. Determine the following:
(a) $P(X < 0.5, Y < 0.5, Z < 0.5)$
(b) $P(X < 0.5, Y < 0.5)$
(c) $P(X < 0.5)$
(d) $E(X)$

5-63. Continuation of Exercise 5-61. Determine the following:
(a) Marginal distribution of X
(b) Joint distribution of X and Y

(c) Conditional probability distribution of X given that $Y = 0.5$ and $Z = 0.5$
(d) Conditional probability distribution of X given that $Y = 0.5$

5-64. The yield in pounds from a day's production is normally distributed with a mean of 1500 pounds and standard deviation of 100 pounds. Assume that the yields on different days are independent random variables.
(a) What is the probability that the production yield exceeds 1400 pounds on each of five days next week?
(b) What is the probability that the production yield exceeds 1400 pounds on at least four of the five days next week?

5-65. The weights of adobe bricks used for construction are normally distributed with a mean of 3 pounds and a standard deviation of 0.25 pound. Assume that the weights of the bricks are independent and that a random sample of 20 bricks is selected.
(a) What is the probability that all the bricks in the sample exceed 2.75 pounds?
(b) What is the probability that the heaviest brick in the sample exceeds 3.75 pounds?

5-66. A manufacturer of electroluminescent lamps knows that the amount of luminescent ink deposited on one of its products is normally distributed with a mean of 1.2 grams and a standard deviation of 0.03 grams. Any lamp with less than 1.14 grams of luminescent ink will fail to meet customer's specifications. A random sample of 25 lamps is collected and the mass of luminescent ink on each is measured.
(a) What is the probability that at least 1 lamp fails to meet specifications?
(b) What is the probability that 5 lamps or fewer fail to meet specifications?
(c) What is the probability that all lamps conform to specifications?
(d) Why is the joint probability distribution of the 25 lamps not needed to answer the previous questions?

5-5 COVARIANCE AND CORRELATION

When two or more random variables are defined on a probability space, it is useful to describe how they vary together; that is, it is useful to measure the relationship between the variables. A common measure of the relationship between two random variables is the **covariance.** To define the covariance, we need to describe the expected value of a function of two random variables $h(X, Y)$. The definition simply extends that used for a function of a single random variable.

Definition

$$E[h(X, Y)] = \begin{cases} \displaystyle\sum_R \sum h(x, y) f_{XY}(x, y) & X, Y \text{ discrete} \\[2em] \displaystyle\iint_R h(x, y) f_{XY}(x, y) \, dx \, dy & X, Y \text{ continuous} \end{cases} \qquad (5\text{-}27)$$

That is, $E[h(X, Y)]$ can be thought of as the weighted average of $h(x, y)$ for each point in the range of (X, Y). The value of $E[h(X, Y)]$ represents the average value of $h(X, Y)$ that is expected in a long sequence of repeated trials of the random experiment.

EXAMPLE 5-27   For the joint probability distribution of the two random variables in Fig. 5-12, calculate $E[(X - \mu_X)(Y - \mu_Y)]$.

The result is obtained by multiplying $x - \mu_X$ times $y - \mu_Y$, times $f_{XY}(x, y)$ for each point in the range of (X, Y). First, μ_X and μ_Y are determined from Equation 5-3 as

$$\mu_X = 1 \times 0.3 + 3 \times 0.7 = 2.4$$

and

$$\mu_Y = 1 \times 0.3 + 2 \times 0.4 + 3 \times 0.3 = 2.0$$

Therefore,

$$\begin{aligned} E[(X - \mu_X)(Y - \mu_Y)] = {}& (1 - 2.4)(1 - 2.0) \times 0.1 \\ & + (1 - 2.4)(2 - 2.0) \times 0.2 + (3 - 2.4)(1 - 2.0) \times 0.2 \\ & + (3 - 2.4)(2 - 2.0) \times 0.2 + (3 - 2.4)(3 - 2.0) \times 0.3 = 0.2 \end{aligned}$$

The covariance is defined for both continuous and discrete random variables by the same formula.

Definition

The **covariance** between the random variables X and Y, denoted as $\text{cov}(X, Y)$ or σ_{XY}, is

$$\sigma_{XY} = E[(X - \mu_X)(Y - \mu_Y)] = E(XY) - \mu_X \mu_Y \qquad (5\text{-}28)$$

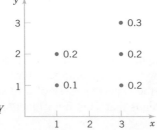

Figure 5-12   Joint distribution of X and Y for Example 5-27.

If the points in the joint probability distribution of X and Y that receive positive probability tend to fall along a line of positive (or negative) slope, σ_{XY} is positive (or negative). If the points tend to fall along a line of positive slope, X tends to be greater than μ_X when Y is greater than μ_Y. Therefore, the product of the two terms $x - \mu_X$ and $y - \mu_Y$ tends to be positive. However, if the points tend to fall along a line of negative slope, $x - \mu_X$ tends to be positive when $y - \mu_Y$ is negative, and vice versa. Therefore, the product of $x - \mu_X$ and $y - \mu_Y$ tends to be negative. In this sense, the covariance between X and Y describes the variation between the two random variables. Figure 5-13 shows examples of pairs of random variables with positive, negative, and zero covariance.

Covariance is a measure of **linear relationship** between the random variables. If the relationship between the random variables is nonlinear, the covariance might not be sensitive to the relationship. This is illustrated in Fig. 5-13(d). The only points with nonzero probability are the points on the circle. There is an identifiable relationship between the variables. Still, the covariance is zero.

The equality of the two expressions for covariance in Equation 5-28 is shown for continuous random variables as follows. By writing the expectations as integrals,

$$
E[(Y - \mu_Y)(X - \mu_X)] = \int\limits_{-\infty}^{\infty} \int\limits_{-\infty}^{\infty} (x - \mu_X)(y - \mu_Y) f_{XY}(x, y)\, dx\, dy
$$

$$
= \int\limits_{-\infty}^{\infty} \int\limits_{-\infty}^{\infty} [xy - \mu_X y - x\mu_Y + \mu_X \mu_Y] f_{XY}(x, y)\, dx\, dy
$$

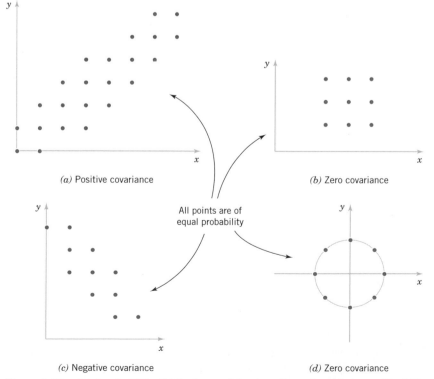

(a) Positive covariance

(b) Zero covariance

All points are of equal probability

(c) Negative covariance

(d) Zero covariance

Figure 5-13   Joint probability distributions and the sign of covariance between X and Y.

Now

$$\int_{-\infty}^{\infty} \int_{-\infty}^{\infty} \mu_X y \, f_{XY}(x, y) \, dx \, dy = \mu_X \left[\int_{-\infty}^{\infty} \int_{-\infty}^{\infty} y f_{XY}(x, y) \, dx \, dy \right] = \mu_X \mu_Y$$

Therefore,

$$E[(X - \mu_X)(Y - \mu_Y)] = \int_{-\infty}^{\infty} \int_{-\infty}^{\infty} xy f_{XY}(x, y) \, dx \, dy - \mu_X \mu_Y - \mu_X \mu_Y + \mu_X \mu_Y$$

$$= \int_{-\infty}^{\infty} \int_{-\infty}^{\infty} xy f_{XY}(x, y) \, dx \, dy - \mu_X \mu_Y = E(XY) - \mu_X \mu_Y$$

EXAMPLE 5-28   In Example 5-1, the random variables X and Y are the number of acceptable and suspect bits among four bits received during a digital communication, respectively. Is the covariance between X and Y positive or negative?

Because X and Y are the number of acceptable and suspect bits out of the four received, $X + Y \leq 4$. If X is near 4, Y must be near 0. Therefore, X and Y have a negative covariance. This can be verified from the joint probability distribution in Fig. 5-1.

There is another measure of the relationship between two random variables that is often easier to interpret than the covariance.

Definition

> The **correlation** between random variables X and Y, denoted as ρ_{XY}, is
>
> $$\rho_{XY} = \frac{\text{cov}(X, Y)}{\sqrt{V(X)V(Y)}} = \frac{\sigma_{XY}}{\sigma_X \sigma_Y} \qquad (5\text{-}29)$$

Because $\sigma_X > 0$ and $\sigma_Y > 0$, if the covariance between X and Y is positive, negative, or zero, the correlation between X and Y is positive, negative, or zero, respectively. The following result can be shown.

> For any two random variables X and Y
>
> $$-1 \leq \rho_{XY} \leq +1 \qquad (5\text{-}30)$$

The correlation just scales the covariance by the standard deviation of each variable. Consequently, the correlation is a dimensionless quantity that can be used to compare the linear relationships between pairs of variables in different units.

If the points in the joint probability distribution of X and Y that receive positive probability tend to fall along a line of positive (or negative) slope, ρ_{XY} is near $+1$ (or -1). If ρ_{XY} equals $+1$ or -1, it can be shown that the points in the joint probability distribution that receive positive probability fall exactly along a straight line. Two random variables with nonzero correlation are said to be **correlated.** Similar to covariance, the correlation is a measure of the **linear relationship** between random variables.

EXAMPLE 5-29 For the discrete random variables X and Y with the joint distribution shown in Fig. 5-14, determine σ_{XY} and ρ_{XY}.

The calculations for $E(XY)$, $E(X)$, and $V(X)$ are as follows.

$$E(XY) = 0 \times 0 \times 0.2 + 1 \times 1 \times 0.1 + 1 \times 2 \times 0.1 + 2 \times 1 \times 0.1$$
$$+ 2 \times 2 \times 0.1 + 3 \times 3 \times 0.4 = 4.5$$
$$E(X) = 0 \times 0.2 + 1 \times 0.2 + 2 \times 0.2 + 3 \times 0.4 = 1.8$$
$$V(X) = (0 - 1.8)^2 \times 0.2 + (1 - 1.8)^2 \times 0.2 + (2 - 1.8)^2 \times 0.2$$
$$+ (3 - 1.8)^2 \times 0.4 = 1.36$$

Because the marginal probability distribution of Y is the same as for X, $E(Y) = 1.8$ and $V(Y) = 1.36$. Consequently,

$$\sigma_{XY} = E(XY) - E(X)E(Y) = 4.5 - (1.8)(1.8) = 1.26$$

Furthermore,

$$\rho_{XY} = \frac{\sigma_{XY}}{\sigma_X \sigma_Y} = \frac{1.26}{(\sqrt{1.36})(\sqrt{1.36})} = 0.926$$

EXAMPLE 5-30 Suppose that the random variable X has the following distribution: $P(X = 1) = 0.2$, $P(X = 2) = 0.6$, $P(X = 3) = 0.2$. Let $Y = 2X + 5$. That is, $P(Y = 7) = 0.2$, $P(Y = 9) = 0.6$, $P(Y = 11) = 0.2$. Determine the correlation between X and Y. Refer to Fig. 5-15.

Because X and Y are linearly related, $\rho = 1$. This can be verified by direct calculations: Try it.

For independent random variables, we do not expect any relationship in their joint probability distribution. The following result is left as an exercise.

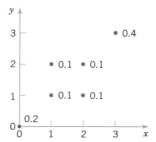

Figure 5-14   Joint distribution for Example 5-29.

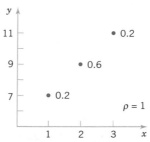

Figure 5-15   Joint distribution for Example 5-30.

> If X and Y are independent random variables,
>
> $$\sigma_{XY} = \rho_{XY} = 0 \qquad (5\text{-}31)$$

EXAMPLE 5-31

For the two random variables in Fig. 5-16, show that $\sigma_{XY} = 0$.

The two random variables in this example are continuous random variables. In this case $E(XY)$ is defined as the double integral over the range of (X, Y). That is,

$$E(XY) = \int_0^4 \int_0^2 xy f_{XY}(x, y)\,dx\,dy = \frac{1}{16}\int_0^4 \left[\int_0^2 x^2 y^2\,dx\right]dy = \frac{1}{16}\int_0^4 y^2 \left[x^3/3 \Big|_0^2\right]$$

$$= \frac{1}{16}\int_0^4 y^2 [8/3]\,dy = \frac{1}{6}\left[y^3/3 \Big|_0^4\right] = \frac{1}{6}[64/3] = 32/9$$

Also,

$$E(X) = \int_0^4 \int_0^2 x f_{XY}(x, y)\,dx\,dy = \frac{1}{16}\int_0^4 \left[\int_0^2 x^2\,dx\right]dy = \frac{1}{16}\int_0^4 \left[x^3/3 \Big|_0^2\right]dy$$

$$= \frac{1}{16}\left[y^2/2 \Big|_0^4\right][8/3] = \frac{1}{6}[16/2] = 4/3$$

$$E(Y) = \int_0^4 \int_0^2 y f_{XY}(x, y)\,dx\,dy = \frac{1}{16}\int_0^4 y^2 \left[\int_0^2 x\,dx\right]dy = \frac{1}{16}\int_0^4 y^2 \left[x^2/2 \Big|_0^2\right]dy$$

$$= \frac{2}{16}\left[y^3/3 \Big|_0^4\right] = \frac{1}{8}[64/3] = 8/3$$

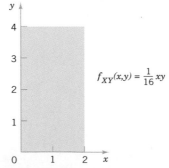

Figure 5-16   Random variables with zero covariance from Example 5-31.

Thus,

$$E(XY) - E(X)E(Y) = 32/9 - (4/3)(8/3) = 0$$

It can be shown that these two random variables are independent. You can check that $f_{XY}(x, y) = f_X(x)f_Y(y)$ for all x and y.

However, if the correlation between two random variables is zero, we *cannot* immediately conclude that the random variables are independent. Figure 5-13(d) provides an example.

EXERCISES FOR SECTION 5-5

5-67. Determine the covariance and correlation for the following joint probability distribution:

0.703

x	1	1	2	4
y	3	4	5	6
$f_{XY}(x, y)$	1/8	1/4	1/2	1/8

5-68. Determine the covariance and correlation for the following joint probability distribution:

x	-1	-0.5	0.5	1
y	-2	-1	1	2
$f_{XY}(x, y)$	1/8	1/4	1/2	1/8

5-69. Determine the value for c and the covariance and correlation for the joint probability mass function $f_{XY}(x, y) = c(x + y)$ for $x = 1, 2, 3$ and $y = 1, 2, 3$.

5-70. Determine the covariance and correlation for the joint probability distribution shown in Fig. 5-4(a) and described in Example 5-8.

5-71. Determine the covariance and correlation for X_1 and X_2 in the joint distribution of the multinomial random variables X_1, X_2 and X_3 in with $p_1 = p_2 = p_3 = \frac{1}{3}$ and $n = 3$. What can you conclude about the sign of the correlation between two random variables in a multinomial distribution?

5-72. Determine the value for c and the covariance and correlation for the joint probability density function $f_{XY}(x, y) = cxy$ over the range $0 < x < 3$ and $0 < y < x$.

5-73. Determine the value for c and the covariance and correlation for the joint probability density function $f_{XY}(x, y) = c$ over the range $0 < x < 5$, $0 < y$, and $x - 1 < y < x + 1$.

5-74. Determine the covariance and correlation for the joint probability density function $f_{XY}(x, y) = 6 \times 10^{-6}e^{-0.001x - 0.002y}$ over the range $0 < x$ and $x < y$ from Example 5-15.

5-75. Determine the covariance and correlation for the joint probability density function $f_{XY}(x, y) = e^{-x-y}$ over the range $0 < x$ and $0 < y$.

5-76. Suppose that the correlation between X and Y is ρ. For constants a, b, c, and d, what is the correlation between the random variables $U = aX + b$ and $V = cY + d$?

5-77. The joint probability distribution is

x	-1	0	0	1
y	0	-1	1	0
$f_{XY}(x, y)$	1/4	1/4	1/4	1/4

Show that the correlation between X and Y is zero, but X and Y are not independent.

5-78. Suppose X and Y are independent continuous random variables. Show that $\sigma_{XY} = 0$.

5-6 BIVARIATE NORMAL DISTRIBUTION

An extension of a normal distribution to two random variables is an important bivariate probability distribution.

EXAMPLE 5-32 At the start of this chapter, the length of different dimensions of an injection-molded part was presented as an example of two random variables. Each length might be modeled by a normal distribution. However, because the measurements are from the same part, the random variables are typically not independent. A probability distribution for two normal random variables that are not independent is important in many applications. As stated at the start of the chapter, if the specifications for X and Y are 2.95 to 3.05 and 7.60 to 7.80 millimeters, respectively, we might be interested in the probability that a part satisfies both specifications; that is, $P(2.95 < X < 3.05, 7.60 < Y < 7.80)$.

Definition

> The probability density function of a **bivariate normal distribution** is
>
> $$f_{XY}(x, y; \sigma_X, \sigma_Y, \mu_X, \mu_Y, \rho) = \frac{1}{2\pi\sigma_X\sigma_Y\sqrt{1 - \rho^2}} \exp\left\{\frac{-1}{2(1 - \rho^2)}\left[\frac{(x - \mu_X)^2}{\sigma_X^2}\right.\right.$$
> $$\left.\left. - \frac{2\rho(x - \mu_X)(y - \mu_Y)}{\sigma_X\sigma_Y} + \frac{(y - \mu_Y)^2}{\sigma_Y^2}\right]\right\} \qquad (5\text{-}32)$$
>
> for $-\infty < x < \infty$ and $-\infty < y < \infty$, with parameters $\sigma_X > 0$, $\sigma_Y > 0$, $-\infty < \mu_X < \infty$, $-\infty < \mu_Y < \infty$, and $-1 < \rho < 1$.

The result that $f_{XY}(x, y; \sigma_X, \sigma_Y, \mu_X, \mu_Y, \rho)$ integrates to 1 is left as an exercise. Also, the bivariate normal probability density function is positive over the entire plane of real numbers.

Two examples of bivariate normal distributions are illustrated in Fig. 5-17 along with corresponding contour plots. Each curve on the contour plots is a set of points for which the probability density function is constant. As seen in the contour plots, the bivariate normal probability density function is constant on ellipses in the (x, y) plane. (We can consider a circle to be a special case of an ellipse.) The center of each ellipse is at the point (μ_X, μ_Y). If $\rho > 0$ ($\rho < 0$), the major axis of each ellipse has positive (negative) slope, respectively. If $\rho = 0$, the major axis of the ellipse is aligned with either the x or y coordinate axis.

EXAMPLE 5-33   The joint probability density function $f_{XY}(x, y) = \frac{1}{\sqrt{2\pi}} e^{-0.5(x^2 + y^2)}$ is a special case of a bivariate normal distribution with $\sigma_X = 1$, $\sigma_Y = 1$, $\mu_X = 0$, $\mu_Y = 0$, and $\rho = 0$. This probability density function is illustrated in Fig. 5-18. Notice that the contour plot consists of concentric circles about the origin.

By completing the square in the exponent, the following results can be shown. The details are left as an exercise.

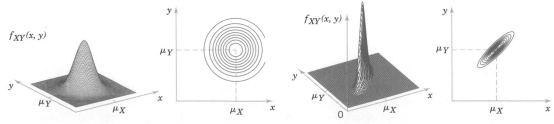

Figure 5-17   Examples of bivariate normal distributions.

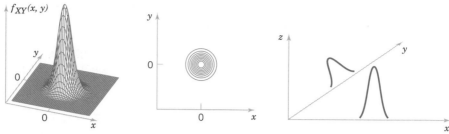

Figure 5-18   Bivariate normal probability density function with $\sigma_X = 1$, $\sigma_Y = 1$, $\rho = 0$, $\mu_X = 0$, and $\mu_Y = 0$.

Figure 5-19   Marginal probability density functions of a bivariate normal distribution.

Marginal Distributions of Bivariate Normal Random Variables

If X and Y have a bivariate normal distribution with joint probability density $f_{XY}(x, y;$ $\sigma_X, \sigma_Y, \mu_X, \mu_Y, \rho)$, the **marginal probability distributions** of X and Y are normal with means μ_X and μ_Y and standard deviations σ_X and σ_Y, respectively.   (5-33)

Figure 5-19 illustrates that the marginal probability distributions of X and Y are normal. Furthermore, as the notation suggests, ρ represents the correlation between X and Y. The following result is left as an exercise.

If X and Y have a bivariate normal distribution with joint probability density function $f_{XY}(x, y; \sigma_X, \sigma_Y, \mu_X, \mu_Y, \rho)$, the correlation between X and Y is ρ.   (5-34)

The contour plots in Fig. 5-17 illustrate that as ρ moves from zero (left graph) to 0.9 (right graph), the ellipses narrow around the major axis. The probability is more concentrated about a line in the (x, y) plane and graphically displays greater correlation between the variables. If $\rho = -1$ or $+1$, all the probability is concentrated on a line in the (x, y) plane. That is, the probability that X and Y assume a value that is not on the line is zero. In this case, the bivariate normal probability density is not defined.

In general, zero correlation does not imply independence. But in the special case that X and Y have a bivariate normal distribution, if $\rho = 0$, X and Y are independent. The details are left as an exercise.

If X and Y have a bivariate normal distribution with $\rho = 0$, X and Y are independent.   (5-35)

An important use of the bivariate normal distribution is to calculate probabilities involving two correlated normal random variables.

EXAMPLE 5-34   Suppose that the X and Y dimensions of an injection-molded part have a bivariate normal distribution with $\sigma_X = 0.04$, $\sigma_Y = 0.08$. $\mu_X = 3.00$. $\mu_Y = 7.70$, and $\rho = 0.8$. Then, the probability that a part satisfies both specifications is

$$P(2.95 < X < 3.05, 7.60 < Y < 7.80)$$

This probability can be obtained by integrating $f_{XY}(x, y; \sigma_X, \sigma_Y, \mu_X \mu_Y, \rho)$ over the region $2.95 < x < 3.05$ and $7.60 < y < 7.80$, as shown in Fig. 5-7. Unfortunately, there is often no closed-form solution to probabilities involving bivariate normal distributions. In this case, the integration must be done numerically.

EXERCISES FOR SECTION 5-6

5-79.   Let X and Y represent concentration and viscosity of a chemical product. Suppose X and Y have a bivariate normal distribution with $\sigma_X = 4$, $\sigma_Y = 1$, $\mu_X = 2$, and $\mu_Y = 1$. Draw a rough contour plot of the joint probability density function for each of the following values for ρ:

(a) $\rho = 0$ (b) $\rho = 0.8$

(c) $\rho = -0.8$

5-80.   Let X and Y represent two dimensions of an injection molded part. Suppose X and Y have a bivariate normal distribution with $\sigma_X = 0.04$, $\sigma_Y = 0.08$, $\mu_X = 3.00$, $\mu_Y = 7.70$, and $\rho_Y = 0$. Determine $P(2.95 < X < 3.05, 7.60 < Y < 7.80)$.

5-81.   In the manufacture of electroluminescent lamps, several different layers of ink are deposited onto a plastic substrate. The thickness of these layers is critical if specifications regarding the final color and intensity of light of the lamp are to be met. Let X and Y denote the thickness of two different layers of ink. It is known that X is normally distributed with a mean of 0.1 millimeter and a standard deviation of 0.00031 millimeter, and Y is also normally distributed with a mean of 0.23 millimeter and a standard deviation of 0.00017 millimeter. The value of ρ for these variables is equal to zero. Specifications call for a lamp to have a thickness of the ink corresponding to X in the range of 0.099535 to 0.100465 millimeters and Y in the range of 0.22966 to 0.23034 millimeters. What is the probability that a randomly selected lamp will conform to specifications?

5-82.   Suppose that X and Y have a bivariate normal distribution with joint probability density function $f_{XY}(x, y; \sigma_X, \sigma_Y, \mu_X, \mu_Y, \rho)$.

(a) Show that the conditional distribution of Y, given that $X = x$ is normal.

(b) Determine $E(Y|X = x)$.

(c) Determine $V(Y|X = x)$.

5-83.   If X and Y have a bivariate normal distribution with $\rho = 0$, show that X and Y are independent.

5-84.   Show that the probability density function $f_{XY}(x, y; \sigma_X, \sigma_Y, \mu_X, \mu_Y, \rho)$ of a bivariate normal distribution integrates to one. [*Hint:* Complete the square in the exponent and use the fact that the integral of a normal probability density function for a single variable is 1.]

5-85.   If X and Y have a bivariate normal distribution with joint probability density $f_{XY}(x, y; \sigma_X, \sigma_Y, \mu_X, \mu_Y, \rho)$, show that the marginal probability distribution of X is normal with mean μ_X and standard deviation σ_X. [*Hint:* Complete the square in the exponent and use the fact that the integral of a normal probability density function for a single variable is 1.]

5-86.   If X and Y have a bivariate normal distribution with joint probability density $f_{XY}(x, y; \sigma_X, \sigma_Y, \mu_X, \mu_Y, \rho)$, show that the correlation between X and Y is ρ. [*Hint:* Complete the square in the exponent].

5-7   LINEAR COMBINATIONS OF RANDOM VARIABLES

A random variable is sometimes defined as a function of one or more random variables. The CD material presents methods to determine the distributions of general functions of random variables. Furthermore, moment-generating functions are introduced on the CD and used to determine the distribution of a sum of random variables. In this section, results for linear functions are highlighted because of their importance in the remainder of the book. References are made to the CD material as needed. For example, if the random variables X_1 and X_2 denote the length and width, respectively, of a manufactured part, $Y = 2X_1 + 2X_2$ is a random variable that represents the perimeter of the part. As another example, recall that the negative binomial random variable was represented as the sum of several geometric random variables.

In this section, we develop results for random variables that are linear combinations of random variables.

Definition

> Given random variables $X_1, X_2, \ldots, X_p$ and constants $c_1, c_2, \ldots, c_p$,
>
> $$Y = c_1 X_1 + c_2 X_2 + \cdots + c_p X_p \qquad (5\text{-}36)$$
>
> is a **linear combination** of $X_1, X_2, \ldots, X_p$.

Now, $E(Y)$ can be found from the joint probability distribution of $X_1, X_2, \ldots, X_p$ as follows. Assume $X_1, X_2, \ldots, X_p$ are continuous random variables. An analogous calculation can be used for discrete random variables.

$$
\begin{aligned}
E(Y) = {}& \int_{-\infty}^{\infty} \int_{-\infty}^{\infty} \cdots \int_{-\infty}^{\infty} (c_1 x_1 + c_2 x_2 + \cdots + c_p x_p) f_{X_1 X_2 \ldots X_p}(x_1, x_2, \ldots, x_p) \, dx_1 \, dx_2 \ldots dx_p \\[6pt]
= {}& c_1 \int_{-\infty}^{\infty} \int_{-\infty}^{\infty} \cdots \int_{-\infty}^{\infty} x_1 f_{X_1 X_2 \ldots X_p}(x_1, x_2, \ldots, x_p) \, dx_1 \, dx_2 \ldots dx_p \\[6pt]
& + c_2 \int_{-\infty}^{\infty} \int_{-\infty}^{\infty} \cdots \int_{-\infty}^{\infty} x_2 f_{X_1 X_2 \ldots X_p}(x_1, x_2, \ldots, x_p) \, dx_1 \, dx_2 \ldots dx_p + , \ldots, \\[6pt]
& + c_p \int_{-\infty}^{\infty} \int_{-\infty}^{\infty} \cdots \int_{-\infty}^{\infty} x_p f_{X_1 X_2 \ldots X_p}(x_1, x_2, \ldots, x_p) \, dx_1 \, dx_2 \ldots dx_p
\end{aligned}
$$

By using Equation 5-24 for each of the terms in this expression, we obtain the following.

Mean of a Linear Combination

> If $Y = c_1 X_1 + c_2 X_2 + \cdots + c_p X_p$,
>
> $$E(Y) = c_1 E(X_1) + c_2 E(X_2) + \cdots + c_p E(X_p) \qquad (5\text{-}37)$$

Furthermore, it is left as an exercise to show the following.

Variance of a
Linear
Combination

If $X_1, X_2, \ldots, X_p$ are random variables, and $Y = c_1 X_1 + c_2 X_2 + \cdots + c_p X_p$, then in general

$$V(Y) = c_1^2 V(X_1) + c_2^2 V(X_2) + \cdots + c_p^2 V(X_p) + 2 \sum \sum_{i<j} c_i c_j \, \text{cov}(X_i, X_j) \quad (5\text{-}38)$$

If $X_1, X_2, \ldots, X_p$ are **independent,**

$$V(Y) = c_1^2 V(X_1) + c_2^2 V(X_2) + \cdots + c_p^2 V(X_p) \quad (5\text{-}39)$$

Note that the result for the variance in Equation 5-39 requires the random variables to be independent. To see why the independence is important, consider the following simple example. Let X_1 denote any random variable and define $X_2 = -X_1$. Clearly, X_1 and X_2 are not independent. In fact, $\rho_{XY} = -1$. Now, $Y = X_1 + X_2$ is 0 with probability 1. Therefore, $V(Y) = 0$, regardless of the variances of X_1 and X_2.

EXAMPLE 5-35   In Chapter 3, we found that if Y is a negative binomial random variable with parameters p and r, $Y = X_1 + X_2 + \cdots + X_r$, where each X_i is a geometric random variable with parameter p and they are independent. Therefore, $E(X_i) = 1/p$ and $E(X_i) = (1 - p)/p^2$. From Equation 5-37, $E(Y) = r/p$ and from Equation 5-39, $V(Y) = r(1 - p)/p^2$.

An approach similar to the one applied in the above example can be used to verify the formulas for the mean and variance of an Erlang random variable in Chapter 4.

EXAMPLE 5-36   Suppose the random variables X_1 and X_2 denote the length and width, respectively, of a manufactured part. Assume $E(X_1) = 2$ centimeters with standard deviation 0.1 centimeter and $E(X_2) = 5$ centimeters with standard deviation 0.2 centimeter. Also, assume that the covariance between X_1 and X_2 is -0.005. Then, $Y = 2X_1 + 2X_2$ is a random variable that represents the perimeter of the part. From Equation 5-36,

$$E(Y) = 2(2) + 2(5) = 14 \text{ centimeters}$$

and from Equation 5-38

$$V(Y) = 2^2(0.1^2) + 2^2(0.2^2) + 2 \times 2 \times 2(-0.005)$$
$$= 0.04 + 0.16 - 0.04 = 0.16 \text{ centimeters squared}$$

Therefore, the standard deviation of Y is $0.16^{1/2} = 0.4$ centimeters.

The particular linear combination that represents the average of p random variables, with identical means and variances, is used quite often in the subsequent chapters. We highlight the results for this special case.

Mean and Variance of an Average

If $\overline{X} = (X_1 + X_2 + \cdots + X_p)/p$ with $E(X_i) = \mu$ for $i = 1, 2, \ldots, p$

$$E(\overline{X}) = \mu \qquad (5\text{-}40a)$$

if $X_1, X_2, \ldots, X_p$ are also independent with $V(X_i) = \sigma^2$ for $i = 1, 2, \ldots, p$,

$$V(\overline{X}) = \frac{\sigma^2}{p} \qquad (5\text{-}40b)$$

The conclusion for $V(\overline{X})$ is obtained as follows. Using Equation 5-39, with $c_i = 1/p$ and $V(X_i) = \sigma^2$, yields

$$V(\overline{X}) = \underbrace{(1/p)^2\sigma^2 + \cdots + (1/p)^2\sigma^2}_{p \text{ terms}} = \sigma^2/p$$

Another useful result concerning linear combinations of random variables is a **reproductive property** that holds for independent, normal random variables.

Reproductive Property of the Normal Distribution

If $X_1, X_2, \ldots, X_p$ are independent, normal random variables with $E(X_i) = \mu_i$ and $V(X_i) = \sigma_i^2$, for $i = 1, 2, \ldots, p$,

$$Y = c_1X_1 + c_2X_2 + \cdots + c_pX_p$$

is a normal random variable with

$$E(Y) = c_1\mu_1 + c_2\mu_2 + \cdots + c_p\mu_p$$

and

$$V(Y) = c_1^2\sigma_1^2 + c_2^2\sigma_2^2 + \cdots + c_p^2\sigma_p^2 \qquad (5\text{-}41)$$

The mean and variance of Y follow from Equations 5-37 and 5-39. The fact that Y has a normal distribution can be obtained from moment-generating functions discussed in Section 5-9 in the CD material.

EXAMPLE 5-37 Let the random variables X_1 and X_2 denote the length and width, respectively, of a manufactured part. Assume that X_1 is normal with $E(X_1) = 2$ centimeters and standard deviation 0.1 centimeter and that X_2 is normal with $E(X_2) = 5$ centimeters and standard deviation 0.2 centimeter. Also, assume that X_1 and X_2 are independent. Determine the probability that the perimeter exceeds 14.5 centimeters.

Then, $Y = 2X_1 + 2X_2$ is a normal random variable that represents the perimeter of the part. We obtain, $E(Y) = 14$ centimeters and the variance of Y is

$$V(Y) = 4 \times 0.1^2 + 4 \times 0.2^2 = 0.0416$$

Now,

$$P(Y > 14.5) = P[(Y - \mu_Y)/\sigma_Y > (14.5 - 14)/\sqrt{0.0416}]$$
$$= P(Z > 1.12) = 0.13$$

EXAMPLE 5-38   Soft-drink cans are filled by an automated filling machine. The mean fill volume is 12.1 fluid ounces, and the standard deviation is 0.1 fluid ounce. Assume that the fill volumes of the cans are independent, normal random variables. What is the probability that the average volume of 10 cans selected from this process is less than 12 fluid ounces?

Let $X_1, X_2, \ldots, X_{10}$ denote the fill volumes of the 10 cans. The average fill volume (denoted as $\overline{X}$) is a normal random variable with

$$E(\overline{X}) = 12.1 \quad \text{and} \quad V(\overline{X}) = \frac{0.1^2}{10} = 0.001$$

Consequently,

$$P(\overline{X} < 12) = P\left[\frac{\overline{X} - \mu_{\overline{X}}}{\sigma_{\overline{X}}} < \frac{12 - 12.1}{\sqrt{0.001}} \right]$$
$$= P(Z < -3.16) = 0.00079$$

EXERCISES FOR SECTION 5-7

5-87. If X and Y are independent, normal random variables with $E(X) = 0$, $V(X) = 4$, $E(Y) = 10$, and $V(Y) = 9$. Determine the following:
(a) $E(2X + 3Y)$ (b) $V(2X + 3Y)$
(c) $P(2X + 3Y < 30)$ (d) $P(2X + 3Y < 40)$

5-88. Suppose that the random variable X represents the length of a punched part in centimeters. Let Y be the length of the part in millimeters. If $E(X) = 5$ and $V(X) = 0.25$, what are the mean and variance of Y?

5-89. A plastic casing for a magnetic disk is composed of two halves. The thickness of each half is normally distributed with a mean of 2 millimeters and a standard deviation of 0.1 millimeter and the halves are independent.
(a) Determine the mean and standard deviation of the total thickness of the two halves.
(b) What is the probability that the total thickness exceeds 4.3 millimeters?

5-90. In the manufacture of electroluminescent lamps, several different layers of ink are deposited onto a plastic substrate. The thickness of these layers is critical if specifications regarding the final color and intensity of light of the lamp are to be met. Let X and Y denote the thickness of two different layers of ink. It is known that X is normally distributed with a mean of 0.1 millimeter and a standard deviation of 0.00031 millimeter and Y is also normally distributed with a mean of 0.23 millimeter and a standard deviation of 0.00017 millimeter. Assume that these variables are independent.
(a) If a particular lamp is made up of these two inks only, what is the probability that the total ink thickness is less than 0.2337 millimeter?
(b) A lamp with a total ink thickness exceeding 0.2405 millimeters lacks the uniformity of color demanded by the customer. Find the probability that a randomly selected lamp fails to meet customer specifications.

5-91. The width of a casing for a door is normally distributed with a mean of 24 inches and a standard deviation of 1/8 inch. The width of a door is normally distributed with a mean of 23 and 7/8 inches and a standard deviation of 1/16 inch. Assume independence.
(a) Determine the mean and standard deviation of the difference between the width of the casing and the width of the door.

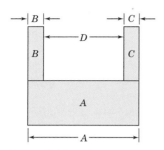

Figure 5-20 Figure for the
U-shaped component.

(b) What is the probability that the width of the casing minus the width of the door exceeds 1/4 inch?

(c) What is the probability that the door does not fit in the casing?

5-92. A U-shaped component is to be formed from the three parts A, B, and C. The picture is shown in Fig. 5-20. The length of A is normally distributed with a mean of 10 millimeters and a standard deviation of 0.1 millimeter. The thickness of parts B and C is normally distributed with a mean of 2 millimeters and a standard deviation of 0.05 millimeter. Assume all dimensions are independent.

(a) Determine the mean and standard deviation of the length of the gap D.

(b) What is the probability that the gap D is less than 5.9 millimeters?

5-93. Soft-drink cans are filled by an automated filling machine and the standard deviation is 0.5 fluid ounce. Assume that the fill volumes of the cans are independent, normal random variables.

(a) What is the standard deviation of the average fill volume of 100 cans?

(b) If the mean fill volume is 12.1 ounces, what is the probability that the average fill volume of the 100 cans is below 12 fluid ounces?

(c) What should the mean fill volume equal so that the probability that the average of 100 cans is below 12 fluid ounces is 0.005?

(d) If the mean fill volume is 12.1 fluid ounces, what should the standard deviation of fill volume equal so that the probability that the average of 100 cans is below 12 fluid ounces is 0.005?

(e) Determine the number of cans that need to be measured such that the probability that the average fill volume is less than 12 fluid ounces is 0.01.

5-94. The photoresist thickness in semiconductor manufacturing has a mean of 10 micrometers and a standard deviation of 1 micrometer. Assume that the thickness is normally distributed and that the thicknesses of different wafers are independent.

(a) Determine the probability that the average thickness of 10 wafers is either greater than 11 or less than 9 micrometers.

(b) Determine the number of wafers that needs to be measured such that the probability that the average thickness exceeds 11 micrometers is 0.01.

(c) If the mean thickness is 10 micrometers, what should the standard deviation of thickness equal so that the probability that the average of 10 wafers is either greater than 11 or less than 9 micrometers is 0.001?

5-95. Assume that the weights of individuals are independent and normally distributed with a mean of 160 pounds and a standard deviation of 30 pounds. Suppose that 25 people squeeze into an elevator that is designed to hold 4300 pounds.

(a) What is the probability that the load (total weight) exceeds the design limit?

(b) What design limit is exceeded by 25 occupants with probability 0.0001?

5-8 FUNCTIONS OF RANDOM VARIABLES (CD ONLY)

5-9 MOMENT GENERATING FUNCTION (CD ONLY)

5-10 CHEBYSHEV'S INEQUALITY (CD ONLY)

Supplemental Exercises

5-96. Show that the following function satisfies the properties of a joint probability mass function:

x	y	$f(x, y)$
0	0	1/4
0	1	1/8
1	0	1/8
1	1	1/4
2	2	1/4

5-97. Continuation of Exercise 5-96. Determine the following probabilities:

(a) $P(X < 0.5, Y < 1.5)$ (b) $P(X \le 1)$

(c) $P(X < 1.5)$ (d) $P(X > 0.5, Y < 1.5)$

(e) Determine $E(X)$, $E(Y)$, $V(X)$, and $V(Y)$.

5-98. Continuation of Exercise 5-96. Determine the following:

(a) Marginal probability distribution of the random variable X

(b) Conditional probability distribution of Y given that $X = 1$

(c) $E(Y|X = 1)$

(d) Are X and Y independent? Why or why not?

(e) Calculate the correlation between X and Y.

5-99. The percentage of people given an antirheumatoid medication who suffer severe, moderate, or minor side effects are 10, 20, and 70%, respectively. Assume that people react independently and that 20 people are given the medication. Determine the following:

(a) The probability that 2, 4, and 14 people will suffer severe, moderate, or minor side effects, respectively

(b) The probability that no one will suffer severe side effects

(c) The mean and variance of the number of people that will suffer severe side effects

(d) What is the conditional probability distribution of the number of people who suffer severe side effects given that 19 suffer minor side effects?

(e) What is the conditional mean of the number of people who suffer severe side effects given that 19 suffer minor side effects?

5-100. The backoff torque required to remove bolts in a steel plate is rated as high, moderate, or low. Historically, the probability of a high, moderate, or low rating is 0.6, 0.3, or 0.1, respectively. Suppose that 20 bolts are evaluated and that the torque ratings are independent.

(a) What is the probability that 12, 6, and 2 bolts are rated as high, moderate, and low, respectively?

(b) What is the marginal distribution of the number of bolts rated low?

(c) What is the expected number of bolts rated low?

(d) What is the probability that the number of bolts rated low is greater than two?

5-101. Continuation of Exercise 5-100

(a) What is the conditional distribution of the number of bolts rated low given that 16 bolts are rated high?

(b) What is the conditional expected number of bolts rated low given that 16 bolts are rated high?

(c) Are the numbers of bolts rated high and low independent random variables?

5-102. To evaluate the technical support from a computer manufacturer, the number of rings before a call is answered by a service representative is tracked. Historically, 70% of the calls are answered in two rings or less, 25% are answered in three or four rings, and the remaining calls require five rings or more. Suppose you call this manufacturer 10 times and assume that the calls are independent.

(a) What is the probability that eight calls are answered in two rings or less, one call is answered in three or four rings, and one call requires five rings or more?

(b) What is the probability that all 10 calls are answered in four rings or less?

(c) What is the expected number of calls answered in four rings or less?

5-103. Continuation of Exercise 5-102

(a) What is the conditional distribution of the number of calls requiring five rings or more given that eight calls are answered in two rings or less?

(b) What is the conditional expected number of calls requiring five rings or more given that eight calls are answered in two rings or less?

(c) Are the number of calls answered in two rings or less and the number of calls requiring five rings or more independent random variables?

5-104. Determine the value of c such that the function $f(x, y) = cx^2y$ for $0 < x < 3$ and $0 < y < 2$ satisfies the properties of a joint probability density function.

5-105. Continuation of Exercise 5-104. Determine the following:

(a) $P(X < 1, Y < 1)$ (b) $P(X < 2.5)$
(c) $P(1 < Y < 2.5)$ (d) $P(X > 2, 1 < Y < 1.5)$
(e) $E(X)$ (f) $E(Y)$

5-106. Continuation of Exercise 5-104.

(a) Determine the marginal probability distribution of the random variable X.

(b) Determine the conditional probability distribution of Y given that $X = 1$.

(c) Determine the conditional probability distribution of X given that $Y = 1$.

5-107. The joint distribution of the continuous random variables X, Y, and Z is constant over the region $x^2 + y^2 \le 1$, $0 < z < 4$.

(a) Determine $P(X^2 + Y^2 \le 0.5)$

(b) Determine $P(X^2 + Y^2 \le 0.5, Z < 2)$

(c) What is the joint conditional probability density function of X and Y given that $Z = 1$?

(d) What is the marginal probability density function of X?

5-108. Continuation of Exercise 5-107.

(a) Determine the conditional mean of Z given that $X = 0$ and $Y = 0$.

(b) In general, determine the conditional mean of Z given that $X = x$ and $Y = y$.

5-109. Suppose that X and Y are independent, continuous uniform random variables for $0 < x < 1$ and $0 < y < 1$. Use the joint probability density function to determine the probability that $|X - Y| < 0.5$.

5-110. The lifetimes of six major components in a copier are independent exponential random variables with means of 8000, 10,000, 10,000, 20,000, 20,000, and 25,000 hours, respectively.

(a) What is the probability that the lifetimes of all the components exceed 5000 hours?

(b) What is the probability that at least one component lifetime exceeds 25,000 hours?

5-111. Contamination problems in semiconductor manufacturing can result in a functional defect, a minor defect, or no defect in the final product. Suppose that 20, 50, and 30% of the contamination problems result in functional, minor, and no defects, respectively. Assume that the effects of 10 contamination problems are independent.

(a) What is the probability that the 10 contamination problems result in two functional defects and five minor defects?

(b) What is the distribution of the number of contamination problems that result in no defects?

(c) What is the expected number of contamination problems that result in no defects?

5-112. The weight of adobe bricks for construction is normally distributed with a mean of 3 pounds and a standard deviation of 0.25 pound. Assume that the weights of the bricks are independent and that a random sample of 25 bricks is chosen.

(a) What is the probability that the mean weight of the sample is less than 2.95 pounds?

(b) What value will the mean weight exceed with probability 0.99?

5-113. The length and width of panels used for interior doors (in inches) are denoted as X and Y, respectively. Suppose that X and Y are independent, continuous uniform random variables for $17.75 < x < 18.25$ and $4.75 < y < 5.25$, respectively.

(a) By integrating the joint probability density function over the appropriate region, determine the probability that the area of a panel exceeds 90 squared inches.

(b) What is the probability that the perimeter of a panel exceeds 46 inches?

5-114. The weight of a small candy is normally distributed with a mean of 0.1 ounce and a standard deviation of 0.01 ounce. Suppose that 16 candies are placed in a package and that the weights are independent.

(a) What are the mean and variance of package net weight?

(b) What is the probability that the net weight of a package is less than 1.6 ounces?

(c) If 17 candies are placed in each package, what is the probability that the net weight of a package is less than 1.6 ounces?

5-115. The time for an automated system in a warehouse to locate a part is normally distributed with a mean of 45 seconds and a standard deviation of 30 seconds. Suppose that independent requests are made for 10 parts.

(a) What is the probability that the average time to locate 10 parts exceeds 60 seconds?

(b) What is the probability that the total time to locate 10 parts exceeds 600 seconds?

5-116. A mechanical assembly used in an automobile engine contains four major components. The weights of the components are independent and normally distributed with the following means and standard deviations (in ounces):

Component	Mean	Standard Deviation
Left case	4	0.4
Right case	5.5	0.5
Bearing assembly	10	0.2
Bolt assembly	8	0.5

(a) What is the probability that the weight of an assembly exceeds 29.5 ounces?

(b) What is the probability that the mean weight of eight independent assemblies exceeds 29 ounces?

5-117. Suppose X and Y have a bivariate normal distribution with $\sigma_X = 4$, $\sigma_Y = 1$, $\mu_X = 4$, $\mu_Y = 4$, and $\rho = -0.2$. Draw a rough contour plot of the joint probability density function.

5-118. If $f_{XY}(x, y) = \dfrac{1}{1.2\pi} \exp\left\{\dfrac{-1}{0.72}\left[(x - 1)^2 \right.\right.$
$$\left.\left. - 1.6(x - 1)(y - 2) + (y - 2)^2\right]\right\}$$
determine $E(X)$, $E(Y)$, $V(X)$, $V(Y)$, and ρ by recorganizing the parameters in the joint probability density function.

5-119. The permeability of a membrane used as a moisture barrier in a biological application depends on the thickness of two integrated layers. The layers are normally distributed with means of 0.5 and 1 millimeters, respectively. The standard deviations of layer thickness are 0.1 and 0.2 millimeters, respectively. The correlation between layers is 0.7.

(a) Determine the mean and variance of the total thickness of the two layers.

(b) What is the probability that the total thickness is less than 1 millimeter?

(c) Let X_1 and X_2 denote the thickness of layers 1 and 2, respectively. A measure of performance of the membrane is a function $2X_1 + 3X_2$ of the thickness. Determine the mean and variance of this performance measure.

5-120. The permeability of a membrane used as a moisture barrier in a biological application depends on the thickness of three integrated layers. Layers 1, 2, and 3 are normally distributed with means of 0.5, 1, and 1.5 millimeters, respectively. The standard deviations of layer thickness are 0.1, 0.2, and 0.3, respectively. Also, the correlation between layers 1 and 2 is 0.7, between layers 2 and 3 is 0.5, and between layers 1 and 3 is 0.3.

(a) Determine the mean and variance of the total thickness of the three layers.

(b) What is the probability that the total thickness is less than 1.5 millimeters?

5-121. A small company is to decide what investments to use for cash generated from operations. Each investment has a mean and standard deviation associated with the percentage gain. The first security has a mean percentage gain of 5% with a standard deviation of 2%, and the second security provides the same mean of 5% with a standard deviation of 4%. The securities have a correlation of -0.5, so there is a negative correlation between the percentage returns. If the company invests two million dollars with half in each security, what is the mean and standard deviation of the percentage return? Compare the standard deviation of this strategy to one that invests the two million dollars into the first security only.

MIND-EXPANDING EXERCISES

5-122. Show that if $X_1, X_2, \ldots, X_p$ are independent, continuous random variables, $P(X_1 \in A_1, X_2 \in A_2, \ldots, X_p \in A_p) = P(X_1 \in A_1)P(X_2 \in A_2)\ldots P(X_p \in A_p)$ for any regions $A_1, A_2, \ldots, A_p$ in the range of $X_1, X_2, \ldots, X_p$ respectively.

5-123. Show that if $X_1, X_2, \ldots, X_p$ are independent random variables and $Y = c_1X_1 + c_2X_2 + \cdots + c_pX_p$,

$$V(Y) = c_1^2 V(X_1) + c_2^2 V(X_2) + \cdots + c_p^2 V(X_p)$$

You can assume that the random variables are continuous.

5-124. Suppose that the joint probability function of the continuous random variables X and Y is constant on the rectangle $0 < x < a, 0 < y < b$. Show that X and Y are independent.

5-125. Suppose that the range of the continuous variables X and Y is $0 < x < a$ and $0 < y < b$. Also suppose that the joint probability density function $f_{XY}(x, y) = g(x)h(y)$, where $g(x)$ is a function only of x and $h(y)$ is a function only of y. Show that X and Y are independent.

IMPORTANT TERMS AND CONCEPTS

In the E-book, click on any term or concept below to go to that subject.

Bivariate normal distribution

Conditional mean

Conditional probability density function

Conditional probability mass function

Conditional variance

Contour plots

Correlation

Covariance

Independence

Joint probability density function

Joint probability mass function

Linear combinations of random variables

Marginal probability distribution

Multinomial distribution

Reproductive property of the normal distribution

CD MATERIAL

Convolution

Functions of random variables

Jacobian of a transformation

Moment generating function

Uniqueness property of moment generating function

Chebyshev's inequality

Random Sampling and Data Description

LEARNING OBJECTIVES

After careful study of this chapter you should be able to do the following:

1. Compute and interpret the sample mean, sample variance, sample standard deviation, sample median, and sample range

2. Explain the concepts of sample mean, sample variance, population mean, and population variance

3. Construct and interpret visual data displays, including the stem-and-leaf display, the histogram, and the box plot

4. Explain the concept of random sampling

5. Construct and interpret normal probability plots

6. Explain how to use box plots and other data displays to visually compare two or more samples of data

7. Know how to use simple time series plots to visually display the important features of time-oriented data.

CD MATERIAL

8. Interpret probability plots for distributions other than normal.

Answers for most odd numbered exercises are at the end of the book. Answers to exercises whose numbers are surrounded by a box can be accessed in the e-Text by clicking on the box. Complete worked solutions to certain exercises are also available in the e-Text. These are indicated in the Answers to Selected Exercises section by a box around the exercise number. Exercises are also available for some of the text sections that appear on CD only. These exercises may be found within the e-Text immediately following the section they accompany.

6-1   DATA SUMMARY AND DISPLAY

Well-constructed data summaries and displays are essential to good statistical thinking, because they can focus the engineer on important features of the data or provide insight about the type of model that should be used in solving the problem. The computer has become an important tool in the presentation and analysis of data. While many statistical techniques require only a hand-held calculator, much time and effort may be required by this approach, and a computer will perform the tasks much more efficiently.

Most statistical analysis is done using a prewritten library of statistical programs. The user enters the data and then selects the types of analysis and output displays that are of interest. Statistical software packages are available for both mainframe machines and personal computers. We will present examples of output from Minitab (one of the most widely-used PC packages), throughout the book. We will not discuss the hands-on use of Minitab for entering and editing data or using commands. This information is found in the software documentation.

We often find it useful to describe data features **numerically.** For example, we can characterize the location or central tendency in the data by the ordinary arithmetic average or mean. Because we almost always think of our data as a sample, we will refer to the arithmetic mean as the **sample mean.**

Definition

> If the n observations in a sample are denoted by $x_1, x_2, \ldots, x_n$, the **sample mean** is
>
> $$\bar{x} = \frac{x_1 + x_2 + \cdots + x_n}{n} = \frac{\sum\limits_{i=1}^{n} x_i}{n} \qquad (6\text{-}1)$$

EXAMPLE 6-1

Let's consider the eight observations collected from the prototype engine connectors from Chapter 1. The eight observations are $x_1 = 12.6$, $x_2 = 12.9$, $x_3 = 13.4$, $x_4 = 12.3$, $x_5 = 13.6$, $x_6 = 13.5$, $x_7 = 12.6$, and $x_8 = 13.1$. The sample mean is

$$\bar{x} = \frac{x_1 + x_2 + \cdots + x_n}{n} = \frac{\sum\limits_{i=1}^{8} x_i}{8} = \frac{12.6 + 12.9 + \cdots + 13.1}{8}$$

$$= \frac{104}{8} = 13.0 \text{ pounds}$$

A physical interpretation of the sample mean as a measure of location is shown in the dot diagram of the pull-off force data. See Figure 6-1. Notice that the sample mean $\bar{x} = 13.0$ can be thought of as a "balance point." That is, if each observation represents 1 pound of mass placed at the point on the x-axis, a fulcrum located at $\bar{x}$ would exactly balance this system of weights.

The sample mean is the average value of all the observations in the data set. Usually, these data are a **sample** of observations that have been selected from some larger **population** of observations. Here the population might consist of all the connectors that will be manufactured and sold to customers. Recall that this type of population is called a **conceptual** or

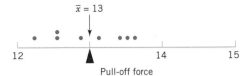

Figure 6-1   The
sample mean as a
balance point for a
system of weights.

hypothetical population, because it does not physically exist. Sometimes there is an actual physical population, such as a lot of silicon wafers produced in a semiconductor factory.

In previous chapters we have introduced the mean of a probability distribution, denoted μ. If we think of a probability distribution as a **model** for the population, one way to think of the mean is as the average of all the measurements in the population. For a finite population with N equally-likely observations, the probability mass function is $f(x_i) = 1/N$ and the mean is

$$\mu = \sum_{i=1}^{n} x_i f(x_i) = \frac{\sum_{i=1}^{N} x_i}{N} \tag{6-2}$$

The sample mean, $\bar{x}$, is a reasonable estimate of the population mean, μ. Therefore, the engineer designing the connector using a 3/32-inch wall thickness would conclude, on the basis of the data, that an estimate of the mean pull-off force is 13.0 pounds.

Although the sample mean is useful, it does not convey all of the information about a sample of data. The variability or scatter in the data may be described by the **sample variance** or the **sample standard deviation.**

Definition

> If $x_1, x_2, \ldots, x_n$ is a sample of n observations, the **sample variance** is
>
> $$s^2 = \frac{\sum_{i=1}^{n} (x_i - \bar{x})^2}{n - 1} \tag{6-3}$$
>
> The **sample standard deviation,** s, is the positive square root of the sample variance.

The units of measurements for the sample variance are the square of the original units of the variable. Thus, if x is measured in pounds, the units for the sample variance are (pounds)^2. The standard deviation has the desirable property of measuring variability in the original units of the variable of interest, x.

How Does the Sample Variance Measure Variability?

To see how the sample variance measures dispersion or variability, refer to Fig. 6-2, which shows the deviations $x_i - \bar{x}$ for the connector pull-off force data. The greater the amount of variability in the pull-off force data, the larger in absolute magnitude some of the deviations $x_i - \bar{x}$ will be. Since the deviations $x_i - \bar{x}$ always sum to zero, we must use a measure of variability that changes the negative deviations to nonnegative quantities. Squaring the deviations is the approach used in the sample variance. Consequently, if s^2 is small, there is relatively little variability in the data, but if s^2 is large, the variability is relatively large.

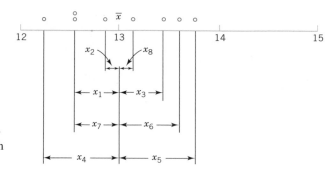

Figure 6-2   How the sample variance measures variability through the deviations $x_i - \bar{x}$.

EXAMPLE 6-2

Table 6-1 displays the quantities needed for calculating the sample variance and sample standard deviation for the pull-off force data. These data are plotted in Fig. 6-2. The numerator of s^2 is

$$\sum_{i=1}^{8} (x_i - \bar{x})^2 = 1.60$$

so the sample variance is

$$s^2 = \frac{1.60}{8 - 1} = \frac{1.60}{7} = 0.2286 \ (\text{pounds})^2$$

and the sample standard deviation is

$$s = \sqrt{0.2286} = 0.48 \ \text{pounds}$$

Computation of s^2

The computation of s^2 requires calculation of $\bar{x}$, n subtractions, and n squaring and adding operations. If the original observations or the deviations $x_i - \bar{x}$ are not integers, the deviations $x_i - \bar{x}$ may be tedious to work with, and several decimals may have to be carried to ensure

Table 6-1   Calculation of Terms for the Sample Variance and Sample Standard Deviation

i	x_i	$x_i - \bar{x}$	$(x_i - \bar{x})^2$
1	12.6	−0.4	0.16
2	12.9	−0.1	0.01
3	13.4	0.4	0.16
4	12.3	−0.7	0.49
5	13.6	0.6	0.36
6	13.5	0.5	0.25
7	12.6	−0.4	0.16
8	13.1	0.1	0.01
	104.0	0.0	1.60

numerical accuracy. A more efficient computational formula for the sample variance is obtained as follows:

$$s^2 = \frac{\sum_{i=1}^{n}(x_i - \bar{x})^2}{n-1} = \frac{\sum_{i=1}^{n}(x_i^2 + \bar{x}^2 + 2\bar{x}x_i)}{n-1} = \frac{\sum_{i=1}^{n}x_i^2 + n\bar{x}^2 - 2\bar{x}\sum_{i=1}^{n}x_i}{n-1}$$

and since $\bar{x} = (1/n)\sum_{i=1}^{n} x_i$, this last equation reduces to

$$s^2 = \frac{\sum_{i=1}^{n}x_i^2 - \frac{\left(\sum_{i=1}^{n}x_i\right)^2}{n}}{n-1} \tag{6-4}$$

Note that Equation 6-4 requires squaring each individual x_i, then squaring the sum of the x_i, subtracting $(\sum x_i)^2/n$ from $\sum x_i^2$, and finally dividing by $n-1$. Sometimes this is called the shortcut method for calculating s^2 (or s).

EXAMPLE 6-3   We will calculate the sample variance and standard deviation using the shortcut method, Equation 6-4. The formula gives

$$s^2 = \frac{\sum_{i=1}^{n}x_i^2 - \frac{\left(\sum_{i=1}^{n}x_i\right)^2}{n}}{n-1} = \frac{1353.6 - \frac{(104)^2}{8}}{7} = \frac{1.60}{7} = 0.2286 \text{ (pounds)}^2$$

and

$$s = \sqrt{0.2286} = 0.48 \text{ pounds}$$

These results agree exactly with those obtained previously.

Analogous to the sample variance s^2, the variability in the population is defined by the **population variance** (σ^2). As in earlier chapters, the positive square root of σ^2, or σ, will denote the **population standard deviation.** When the population is finite and consists of N equally-likely values, we may define the population variance as

$$\sigma^2 = \frac{\sum_{i=1}^{N}(x_i - \mu)^2}{N} \tag{6-5}$$

We observed previously that the sample mean could be used as an estimate of the population mean. Similarly, the sample variance is an estimate of the population variance. In Chapter 7, we will discuss **estimation of parameters** more formally.

Note that the divisor for the sample variance is the sample size minus one $(n-1)$, while for the population variance it is the population size N. If we knew the true value of the population mean μ, we could find the *sample* variance as the average squared deviation of the sample observations about μ. In practice, the value of μ is almost never known, and so the sum of the squared deviations about the sample average $\bar{x}$ must be used instead. However, the observations x_i tend to be closer to their average, $\bar{x}$, than to the population mean, μ. Therefore, to compensate for this we use $n - 1$ as the divisor rather than n. If we used n as the divisor in the sample variance, we would obtain a measure of variability that is, on the average, consistently smaller than the true population variance σ^2.

Another way to think about this is to consider the sample variance s^2 as being based on $n - 1$ **degrees of freedom**. The term *degrees of freedom* results from the fact that the n deviations $x_1 - \bar{x}, x_2 - \bar{x}, \ldots, x_n - \bar{x}$ always sum to zero, and so specifying the values of any $n - 1$ of these quantities automatically determines the remaining one. This was illustrated in Table 6-1. Thus, only $n - 1$ of the n deviations, $x_i - \bar{x}$, are freely determined.

In addition to the sample variance and sample standard deviation, the **sample range,** or the difference between the largest and smallest observations, is a useful measure of variability. The sample range is defined as follows.

Definition

> If the n observations in a sample are denoted by $x_1, x_2, \ldots, x_n$, the **sample range** is
>
> $$r = \max(x_i) - \min(x_i) \tag{6-6}$$

For the pull-off force data, the sample range is $r = 13.6 - 12.3 = 1.3$. Generally, as the variability in sample data increases, the sample range increases.

The sample range is easy to calculate, but it ignores all of the information in the sample data between the largest and smallest values. For example, the two samples 1, 3, 5, 8, and 9 and 1, 5, 5, 5, and 9, both have the same range ($r = 8$). However, the standard deviation of the first sample is $s_1 = 3.35$, while the standard deviation of the second sample is $s_2 = 2.83$. The variability is actually less in the second sample.

Sometimes, when the sample size is small, say $n < 8$ or 10, the information loss associated with the range is not too serious. For example, the range is used widely in statistical quality control where sample sizes of 4 or 5 are fairly common. We will discuss some of these applications in Chapter 16.

EXERCISES FOR SECTION 6-1

6-1. Eight measurements were made on the inside diameter of forged piston rings used in an automobile engine. The data (in millimeters) are 74.001, 74.003, 74.015, 74.000, 74.005, 74.002, 74.005, and 74.004. Calculate the sample mean and sample standard deviation, construct a dot diagram, and comment on the data.

6-2. In *Applied Life Data Analysis* (Wiley, 1982), Wayne Nelson presents the breakdown time of an insulating fluid between electrodes at 34 kV. The times, in minutes, are as follows: 0.19, 0.78, 0.96, 1.31, 2.78, 3.16, 4.15, 4.67, 4.85, 6.50, 7.35, 8.01, 8.27, 12.06, 31.75, 32.52, 33.91, 36.71, and 72.89. Calculate the sample mean and sample standard deviation.

6-3. The January 1990 issue of *Arizona Trend* contains a supplement describing the 12 "best" golf courses in the state. The yardages (lengths) of these courses are as follows: 6981, 7099, 6930, 6992, 7518, 7100, 6935, 7518, 7013, 6800, 7041, and 6890. Calculate the sample mean and sample standard deviation. Construct a dot diagram of the data.

6-4. An article in the *Journal of Structural Engineering* (Vol. 115, 1989) describes an experiment to test the yield strength of circular tubes with caps welded to the ends. The first yields (in kN) are 96, 96, 102, 102, 102, 104, 104, 108, 126, 126, 128, 128, 140, 156, 160, 160, 164, and 170. Calculate the sample mean and sample standard deviation. Construct a dot diagram of the data.

6-5. An article in *Human Factors* (June 1989) presented data on visual accommodation (a function of eye movement) when recognizing a speckle pattern on a high-resolution CRT screen. The data are as follows: 36.45, 67.90, 38.77, 42.18, 26.72, 50.77, 39.30, and 49.71. Calculate the sample mean and sample standard deviation. Construct a dot diagram of the data.

6-6.   The following data are direct solar intensity measurements (watts/m^2) on different days at a location in southern Spain: 562, 869, 708, 775, 775, 704, 809, 856, 655, 806, 878, 909, 918, 558, 768, 870, 918, 940, 946, 661, 820, 898, 935, 952, 957, 693, 835, 905, 939, 955, 960, 498, 653, 730, and 753. Calculate the sample mean and sample standard deviation.

6-7.   The April 22, 1991 issue of *Aviation Week and Space Technology* reports that during Operation Desert Storm, U.S. Air Force F-117A pilots flew 1270 combat sorties for a total of 6905 hours. What is the mean duration of an F-117A mission during this operation? Why is the parameter you have calculated a population mean?

6-8.   Preventing fatigue crack propagation in aircraft structures is an important element of aircraft safety. An engineering study to investigate fatigue crack in $n = 9$ cyclically loaded wing boxes reported the following crack lengths (in mm): 2.13, 2.96, 3.02, 1.82, 1.15, 1.37, 2.04, 2.47, 2.60.
(a) Calculate the sample mean.
(b) Calculate the sample variance and sample standard deviation.
(c) Prepare a dot diagram of the data.

6-9.   Consider the solar intensity data in Exercise 6-6. Prepare a dot diagram of this data. Indicate where the sample mean falls on this diagram. Give a practical interpretation of the sample mean.

6-10.   Exercise 6-5 describes data from an article in *Human Factors* on visual accommodation from an experiment involving a high-resolution CRT screen.
(a) Construct a dot diagram of this data.
(b) Data from a second experiment using a low-resolution screen were also reported in the article. They are 8.85,

35.80, 26.53, 64.63, 9.00, 15.38, 8.14, and 8.24. Prepare a dot diagram for this second sample and compare it to the one for the first sample. What can you conclude about CRT resolution in this situation?

6-11.   The pH of a solution is measured eight times by one operator using the same instrument. She obtains the following data: 7.15, 7.20, 7.18, 7.19, 7.21, 7.20, 7.16, and 7.18.
(a) Calculate the sample mean.
(b) Calculate the sample variance and sample standard deviation.
(c) What are the major sources of variability in this experiment?

6-12.   An article in the *Journal of Aircraft* (1988) describes the computation of drag coefficients for the NASA 0012 airfoil. Different computational algorithms were used at $M_\infty = 0.7$ with the following results (drag coefficients are in units of drag counts; that is, one count is equivalent to a drag coefficient of 0.0001): 79, 100, 74, 83, 81, 85, 82, 80, and 84. Compute the sample mean, sample variance, and sample standard deviation, and construct a dot diagram.

6-13.   The following data are the joint temperatures of the O-rings (°F) for each test firing or actual launch of the space shuttle rocket motor (from *Presidential Commission on the Space Shuttle Challenger Accident*, Vol. 1, pp. 129–131): 84, 49, 61, 40, 83, 67, 45, 66, 70, 69, 80, 58, 68, 60, 67, 72, 73, 70, 57, 63, 70, 78, 52, 67, 53, 67, 75, 61, 70, 81, 76, 79, 75, 76, 58, 31.
(a) Compute the sample mean and sample standard deviation.
(b) Construct a dot diagram of the temperature data.
(c) Set aside the smallest observation (31°F) and recompute the quantities in part (a). Comment on your findings. How "different" are the other temperatures from this last value?

6-2   RANDOM SAMPLING

In most statistics problems, we work with a sample of observations selected from the population that we are interested in studying. Figure 6-3 illustrates the relationship between the population and the sample. We have informally discussed these concepts before; however, we now give the formal definitions of some of these terms.

Definition

> A **population** consists of the totality of the observations with which we are concerned.

In any particular problem, the population may be small, large but finite, or infinite. The number of observations in the population is called the **size** of the population. For example, the number of underfilled bottles produced on one day by a soft-drink company is a population of finite size. The observations obtained by measuring the carbon monoxide level every day is a population of infinite size. We often use a **probability distribution** as a **model** for a population. For example, a structural engineer might consider the population of tensile strengths of a

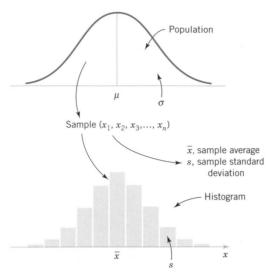

Figure 6-3   Relationship between a population and a sample.

chassis structural element to be normally distributed with mean μ and variance σ^2. We could refer to this as a **normal population** or a normally distributed population.

In most situations, it is impossible or impractical to observe the entire population. For example, we could not test the tensile strength of all the chassis structural elements because it would be too time consuming and expensive. Furthermore, some (perhaps many) of these structural elements do not yet exist at the time a decision is to be made, so to a large extent, we must view the population as **conceptual.** Therefore, we depend on a subset of observations from the population to help make decisions about the population.

Definition

> A **sample** is a subset of observations selected from a population.

For statistical methods to be valid, the sample must be representative of the population. It is often tempting to select the observations that are most convenient as the sample or to exercise judgment in sample selection. These procedures can frequently introduce **bias** into the sample, and as a result the parameter of interest will be consistently underestimated (or overestimated) by such a sample. Furthermore, the behavior of a judgment sample cannot be statistically described. To avoid these difficulties, it is desirable to select a **random sample** as the result of some chance mechanism. Consequently, the selection of a sample is a random experiment and each observation in the sample is the observed value of a random variable. The observations in the population determine the probability distribution of the random variable.

To define a random sample, let X be a random variable that represents the result of one selection of an observation from the population. Let $f(x)$ denote the probability density function of X. Suppose that each observation in the sample is obtained independently, under unchanging conditions. That is, the observations for the sample are obtained by observing X independently under unchanging conditions, say, n times. Let X_i denote the random variable that represents the ith replicate. Then, $X_1, X_2, \ldots, X_n$ is a random sample and the numerical values obtained are denoted as $x_1, x_2, \ldots, x_n$. The random variables in a random sample are independent with the same probability distribution $f(x)$ because of the identical conditions under which each observation is obtained. That is, the marginal probability density function of $X_1, X_2, \ldots, X_n$ is

$f(x_1), f(x_2), \ldots, f(x_n)$, respectively, and by independence the joint probability density function of the random sample is $f_{X_1 X_2 \ldots X_n}(x_1, x_2, \ldots, x_n) = f(x_1)f(x_2) \ldots f(x_n)$.

Definition

> The random variables $X_1, X_2, \ldots, X_n$ are a random sample of size n if (a) the X_i's are independent random variables, and (b) every X_i has the same probability distribution.

To illustrate this definition, suppose that we are investigating the effective service life of an electronic component used in a cardiac pacemaker and that component life is normally distributed. Then we would expect each of the observations on component life $X_1, X_2, \ldots, X_n$ in a random sample of n components to be independent random variables with exactly the same normal distribution. After the data are collected, the numerical values of the observed lifetimes are denoted as $x_1, x_2, \ldots, x_n$.

The primary purpose in taking a random sample is to obtain information about the unknown population parameters. Suppose, for example, that we wish to reach a conclusion about the proportion of people in the United States who prefer a particular brand of soft drink. Let p represent the unknown value of this proportion. It is impractical to question every individual in the population to determine the true value of p. In order to make an inference regarding the true proportion p, a more reasonable procedure would be to select a random sample (of an appropriate size) and use the observed proportion $\hat{p}$ of people in this sample favoring the brand of soft drink.

The sample proportion, $\hat{p}$ is computed by dividing the number of individuals in the sample who prefer the brand of soft drink by the total sample size n. Thus, $\hat{p}$ is a function of the observed values in the random sample. Since many random samples are possible from a population, the value of $\hat{p}$ will vary from sample to sample. That is, $\hat{p}$ is a random variable. Such a random variable is called a **statistic.**

Definition

> A **statistic** is any function of the observations in a random sample.

We have encountered statistics before. For example, if $X, X_2, \ldots, X_n$ is a random sample of size n, the **sample mean** $\overline{X}$, the **sample variance** S^2, and the **sample standard deviation** S are statistics.

Although numerical summary statistics are very useful, **graphical displays** of sample data are a very powerful and extremely useful way to visually examine the data. We now present a few of the techniques that are most relevant to engineering applications of probability and statistics.

6-3 STEM-AND-LEAF DIAGRAMS

The dot diagram is a useful data display for small samples, up to (say) about 20 observations. However, when the number of observations is moderately large, other graphical displays may be more useful.

For example, consider the data in Table 6-2. These data are the compressive strengths in pounds per square inch (psi) of 80 specimens of a new aluminum-lithium alloy undergoing evaluation as a possible material for aircraft structural elements. The data were recorded in the order of testing, and in this format they do not convey much information about compressive strength. Questions such as "What percent of the specimens fail below 120 psi?" are not easy to answer.

Table 6-2   Compressive Strength (in psi) of 80 Aluminum-Lithium Alloy Specimens

105	221	183	186	121	181	180	143
97	154	153	174	120	168	167	141
245	228	174	199	181	158	176	110
163	131	154	115	160	208	158	133
207	180	190	193	194	133	156	123
134	178	76	167	184	135	229	146
218	157	101	171	165	172	158	169
199	151	142	163	145	171	148	158
160	175	149	87	160	237	150	135
196	201	200	176	150	170	118	149

Because there are many observations, constructing a dot diagram of these data would be relatively inefficient; more effective displays are available for large data sets.

A **stem-and-leaf diagram** is a good way to obtain an informative visual display of a data set $x_1, x_2, \ldots, x_n$, where each number x_i consists of at least two digits. To construct a stem-and-leaf diagram, use the following steps.

Steps for Constructing a Stem-and-Leaf Diagram

(1)   Divide each number x_i into two parts: a **stem,** consisting of one or more of the leading digits and a **leaf,** consisting of the remaining digit.

(2)   List the stem values in a vertical column.

(3)   Record the leaf for each observation beside its stem.

(4)   Write the units for stems and leaves on the display.

To illustrate, if the data consist of percent defective information between 0 and 100 on lots of semiconductor wafers, we can divide the value 76 into the stem 7 and the leaf 6. In general, we should choose relatively few stems in comparison with the number of observations. It is usually best to choose between 5 and 20 stems.

EXAMPLE 6-4

To illustrate the construction of a stem-and-leaf diagram, consider the alloy compressive strength data in Table 6-2. We will select as stem values the numbers $7, 8, 9, \ldots, 24$. The resulting stem-and-leaf diagram is presented in Fig. 6-4. The last column in the diagram is a frequency count of the number of leaves associated with each stem. Inspection of this display immediately reveals that most of the compressive strengths lie between 110 and 200 psi and that a central value is somewhere between 150 and 160 psi. Furthermore, the strengths are distributed approximately symmetrically about the central value. The stem-and-leaf diagram enables us to determine quickly some important features of the data that were not immediately obvious in the original display in Table 6-2.

In some data sets, it may be desirable to provide more classes or stems. One way to do this would be to modify the original stems as follows: Divide the stem 5 (say) into two new stems, 5L and 5U. The stem 5L has leaves 0, 1, 2, 3, and 4, and stem 5U has leaves 5, 6, 7, 8, and 9. This will double the number of original stems. We could increase the number of original stems by four by defining five new stems: 5z with leaves 0 and 1, 5t (for twos and three) with leaves 2 and 3, 5f (for fours and fives) with leaves 4 and 5, 5s (for six and seven) with leaves 6 and 7, and 5e with leaves 8 and 9.

Stem	Leaf	Frequency
7	6	1
8	7	1
9	7	1
10	5 1	2
11	5 8 0	3
12	1 0 3	3
13	4 1 3 5 3 5	6
14	2 9 5 8 3 1 6 9	8
15	4 7 1 3 4 0 8 8 6 8 0 8	12
16	3 0 7 3 0 5 0 8 7 9	10
17	8 5 4 4 1 6 2 1 0 6	10
18	0 3 6 1 4 1 0	7
19	9 6 0 9 3 4	6
20	7 1 0 8	4
21	8	1
22	1 8 9	3
23	7	1
24	5	1

Figure 6-4 Stem-and-leaf diagram for the compressive strength data in Table 6-2.

Stem : Tens and hundreds digits (psi); Leaf: Ones digits (psi)

EXAMPLE 6-5

Figure 6-5 illustrates the stem-and-leaf diagram for 25 observations on batch yields from a chemical process. In Fig. 6-5(a) we have used 6, 7, 8, and 9 as the stems. This results in too few stems, and the stem-and-leaf diagram does not provide much information about the data. In Fig. 6-5(b) we have divided each stem into two parts, resulting in a display that more

Stem	Leaf
6	1 3 4 5 5 6
7	0 1 1 3 5 7 8 8 9
8	1 3 4 4 7 8 8
9	2 3 5

(a)

Stem	Leaf
6L	1 3 4
6U	5 5 6
7L	0 1 1 3
7U	5 7 8 8 9
8L	1 3 4 4
8U	7 8 8
9L	2 3
9U	5

(b)

Stem	Leaf
6z	1
6t	3
6f	4 5 5
6s	6
6e	
7z	0 1 1
7t	3
7f	5
7s	7
7e	8 8 9
8z	1
8t	3
8f	4 4
8s	7
8e	8 8
9z	
9t	2 3
9f	5
9s	
9e	

(c)

Figure 6-5 Stem-and-leaf displays for Example 6-5. Stem: Tens digits. Leaf: Ones digits.

Character Stem-and-Leaf Display

Stem-and-leaf of Strength
N = 80   Leaf Unit = 1.0

1	7	6
2	8	7
3	9	7
5	10	1 5
8	11	0 5 8
11	12	0 1 3
17	13	1 3 3 4 5 5
25	14	1 2 3 5 6 8 9 9
37	15	0 0 1 3 4 4 6 7 8 8 8 8
(10)	16	0 0 0 3 3 5 7 7 8 9
33	17	0 1 1 2 4 4 5 6 6 8
23	18	0 0 1 1 3 4 6
16	19	0 3 4 6 9 9
10	20	0 1 7 8
6	21	8
5	22	1 8 9
2	23	7
1	24	5

Figure 6-6   A stem-and-leaf diagram from Minitab.

adequately displays the data. Figure 6-5(c) illustrates a stem-and-leaf display with each stem divided into five parts. There are too many stems in this plot, resulting in a display that does not tell us much about the shape of the data.

Figure 6-6 shows a stem-and-leaf display of the compressive strength data in Table 6-2 produced by Minitab. The software uses the same stems as in Fig. 6-4. Note also that the computer orders the leaves from smallest to largest on each stem. This form of the plot is usually called an **ordered stem-and-leaf diagram.** This is not usually done when the plot is constructed manually because it can be time consuming. The computer adds a column to the left of the stems that provides a count of the observations at and above each stem in the upper half of the display and a count of the observations at and below each stem in the lower half of the display. At the middle stem of 16, the column indicates the number of observations at this stem.

The ordered stem-and-leaf display makes it relatively easy to find data features such as percentiles, quartiles, and the median. The sample **median** is a measure of central tendency that divides the data into two equal parts, half below the median and half above. If the number of observations is even, the median is halfway between the two central values. From Fig. 6-6 we find the 40th and 41st values of strength as 160 and 163, so the median is $(160 + 163)/2 = 161.5$. If the number of observations is odd, the median is the central value. The sample **mode** is the most frequently occurring data value. Figure 6-6 indicates that the mode is 158; this value occurs four times, and no other value occurs as frequently in the sample.

We can also divide data into more than two parts. When an ordered set of data is divided into four equal parts, the division points are called **quartiles.** The *first* or *lower quartile*, q_1, is a value that has approximately 25% of the observations below it and approximately 75% of the observations above. The *second quartile*, q_2, has approximately 50% of the observations below its value. The second quartile is exactly equal to the median. The *third* or *upper quartile*, q_3, has approximately 75% of the observations below its value. As in the case of the median, the quartiles may not be unique. The compressive strength data in Fig. 6-6 contains $n = 80$ observations. Minitab software calculates the first and third quartiles as the $(n + 1)/4$

Table 6-3 Summary Statistics for the Compressive Strength Data from Minitab

Variable	N	Mean	Median	StDev	SE Mean
	80	162.66	161.50	33.77	3.78
	Min	Max	Q1	Q3	
	76.00	245.00	143.50	181.00	

and $3(n + 1)/4$ ordered observations and interpolates as needed. For example, $(80 + 1)/4 = 20.25$ and $3(80 + 1)/4 = 60.75$. Therefore, Minitab interpolates between the 20th and 21st ordered observation to obtain $q_1 = 143.50$ and between the 60th and 61st observation to obtain $q_3 = 181.00$. In general, the $100k$th **percentile** is a data value such that approximately $100k\%$ of the observations are at or below this value and approximately $100(1 - k)\%$ of them are above it. Finally, we may use the **interquartile range,** defined as IQR $= q_3 - q_1$, as a measure of variability. The interquartile range is less sensitive to the extreme values in the sample than is the ordinary sample range.

Many statistics software packages provide data summaries that include these quantities. The output obtained for the compressive strength data in Table 6-2 from Minitab is shown in Table 6-3.

EXERCISES FOR SECTION 6-3

6-14. An article in *Technometrics* (Vol. 19, 1977, p. 425) presents the following data on the motor fuel octane ratings of several blends of gasoline:

88.5	98.8	89.6	92.2	92.7	88.4	87.5	90.9
94.7	88.3	90.4	83.4	87.9	92.6	87.8	89.9
84.3	90.4	91.6	91.0	93.0	93.7	88.3	91.8
90.1	91.2	90.7	88.2	94.4	96.5	89.2	89.7
89.0	90.6	88.6	88.5	90.4	84.3	92.3	92.2
89.8	92.2	88.3	93.3	91.2	93.2	88.9	
91.6	87.7	94.2	87.4	86.7	88.6	89.8	
90.3	91.1	85.3	91.1	94.2	88.7	92.7	
90.0	86.7	90.1	90.5	90.8	92.7	93.3	
91.5	93.4	89.3	100.3	90.1	89.3	86.7	
89.9	96.1	91.1	87.6	91.8	91.0	91.0	

Construct a stem-and-leaf display for these data.

6-15. The following data are the numbers of cycles to failure of aluminum test coupons subjected to repeated alternating stress at 21,000 psi, 18 cycles per second:

1115	865	1015	885	1594	1000	1416	1501
1310	2130	845	1223	2023	1820	1560	1238
1540	1421	1674	375	1315	1940	1055	990
1502	1109	1016	2265	1269	1120	1764	1468
1258	1481	1102	1910	1260	910	1330	1512
1315	1567	1605	1018	1888	1730	1608	1750
1085	1883	706	1452	1782	1102	1535	1642
798	1203	2215	1890	1522	1578	1781	
1020	1270	785	2100	1792	758	1750	

(a) Construct a stem-and-leaf display for these data.

(b) Does it appear likely that a coupon will "survive" beyond 2000 cycles? Justify your answer.

6-16. The percentage of cotton in material used to manufacture men's shirts follows. Construct a stem-and-leaf display for the data.

34.2	37.8	33.6	32.6	33.8	35.8	34.7	34.6
33.1	36.6	34.7	33.1	34.2	37.6	33.6	33.6
34.5	35.4	35.0	34.6	33.4	37.3	32.5	34.1
35.6	34.6	35.4	35.9	34.7	34.6	34.1	34.7
36.3	33.8	36.2	34.7	34.6	35.5	35.1	35.7
35.1	37.1	36.8	33.6	35.2	32.8	36.8	36.8
34.7	34.0	35.1	32.9	35.0	32.1	37.9	34.3
33.6	34.1	35.3	33.5	34.9	34.5	36.4	32.7

6-17. The following data represent the yield on 90 consecutive batches of ceramic substrate to which a metal coating has been applied by a vapor-deposition process. Construct a stem-and-leaf display for these data.

94.1	86.1	95.3	84.9	88.8	84.6	94.4	84.1
93.2	90.4	94.1	78.3	86.4	83.6	96.1	83.7
90.6	89.1	97.8	89.6	85.1	85.4	98.0	82.9
91.4	87.3	93.1	90.3	84.0	89.7	85.4	87.3
88.2	84.1	86.4	93.1	93.7	87.6	86.6	86.4
86.1	90.1	87.6	94.6	87.7	85.1	91.7	84.5
95.1	95.2	94.1	96.3	90.6	89.6	87.5	
90.0	86.1	92.1	94.7	89.4	90.0	84.2	
92.4	94.3	96.4	91.1	88.6	90.1	85.1	
87.3	93.2	88.2	92.4	84.1	94.3	90.5	
86.6	86.7	86.4	90.6	82.6	97.3	95.6	
91.2	83.0	85.0	89.1	83.1	96.8	88.3	

6-18.   Find the median and the quartiles for the motor fuel octane data in Exercise 6-14.

6-19.   Find the median and the quartiles for the failure data in Exercise 6-15.

6-20.   Find the median, mode, and sample average of the data in Exercise 6-16. Explain how these three measures of location describe different features in the data.

6-21.   Find the median and the quartiles for the yield data in Exercise 6-17.

6-22.   The female students in an undergraduate engineering core course at ASU self-reported their heights to the nearest inch. The data are

62 64 66 67 65 68 61 65 67 65 64 63 67
68 64 66 68 69 65 67 62 66 68 67 66 65
69 65 70 65 67 68 65 63 64 67 67

(a) Calculate the sample mean and standard deviation of height.
(b) Construct a stem-and-leaf diagram for the height data and comment on any important features that you notice.
(c) What is the median height of this group of female engineering students?

6-23.   The shear strengths of 100 spot welds in a titanium alloy follow. Construct a stem-and-leaf diagram for the weld strength data and comment on any important features that you notice.

5408 5431 5475 5442 5376 5388 5459 5422 5416 5435
5420 5429 5401 5446 5487 5416 5382 5357 5388 5457
5407 5469 5416 5377 5454 5375 5409 5459 5445 5429
5463 5408 5481 5453 5422 5354 5421 5406 5444 5466
5399 5391 5477 5447 5329 5473 5423 5441 5412 5384
5445 5436 5454 5453 5428 5418 5465 5427 5421 5396
5381 5425 5388 5388 5378 5481 5387 5440 5482 5406
5401 5411 5399 5431 5440 5413 5406 5342 5452 5420
5458 5485 5431 5416 5431 5390 5399 5435 5387 5462
5383 5401 5407 5385 5440 5422 5448 5366 5430 5418

6-24.   An important quality characteristic of water is the concentration of suspended solid material. Following are 60 measurements on suspended solids from a certain lake. Construct a stem-and-leaf diagram for this data and comment on any important features that you notice. Compute the sample mean, sample standard deviation, and the sample median.

42.4 65.7 29.8 58.7 52.1 55.8 57.0 68.7 67.3 67.3
54.3 54.0 73.1 81.3 59.9 56.9 62.2 69.9 66.9 59.0
56.3 43.3 57.4 45.3 80.1 49.7 42.8 42.4 59.6 65.8
61.4 64.0 64.2 72.6 72.5 46.1 53.1 56.1 67.2 70.7
42.6 77.4 54.7 57.1 77.3 39.3 76.4 59.3 51.1 73.8
61.4 73.1 77.3 48.5 89.8 50.7 52.0 59.6 66.1 31.6

6-25.   The United States Golf Association tests golf balls to ensure that they conform to the rules of golf. Balls are tested for weight, diameter, roundness, and overall distance. The overall distance test is conducted by hitting balls with a driver swung by a mechanical device nicknamed "Iron Byron" after the legendary great Byron Nelson, whose swing the machine is said to emulate. Following are 100 distances (in yards) achieved by a particular brand of golf ball in the overall distance test. Construct a stem-and-leaf diagram for this data and comment on any important features that you notice. Compute the sample mean, sample standard deviation, and the sample median.

261.3 259.4 265.7 270.6 274.2 261.4 254.5 283.7
258.1 270.5 255.1 268.9 267.4 253.6 234.3 263.2
254.2 270.7 233.7 263.5 244.5 251.8 259.5 257.5
257.7 272.6 253.7 262.2 252.0 280.3 274.9 233.7
237.9 274.0 264.5 244.8 264.0 268.3 272.1 260.2
255.8 260.7 245.5 279.6 237.8 278.5 273.3 263.7
241.4 260.6 280.3 272.7 261.0 260.0 279.3 252.1
244.3 272.2 248.3 278.7 236.0 271.2 279.8 245.6
241.2 251.1 267.0 273.4 247.7 254.8 272.8 270.5
254.4 232.1 271.5 242.9 273.6 256.1 251.6
256.8 273.0 240.8 276.6 264.5 264.5 226.8
255.3 266.6 250.2 255.8 285.3 255.4 240.5
255.0 273.2 251.4 276.1 277.8 266.8 268.5

6-26.   A semiconductor manufacturer produces devices used as central processing units in personal computers. The speed of the device (in megahertz) is important because it determines the price that the manufacturer can charge for the devices. The following table contains measurements on 120 devices. Construct a stem-and-leaf diagram for this data and comment on any important features that you notice. Compute the sample mean, sample standard deviation, and the sample median. What percentage of the devices has a speed exceeding 700 megahertz?

680 669 719 699 670 710 722 663 658 634 720 690
677 669 700 718 690 681 702 696 692 690 694 660
649 675 701 721 683 735 688 763 672 698 659 704
681 679 691 683 705 746 706 649 668 672 690 724
652 720 660 695 701 724 668 698 668 660 680 739
717 727 653 637 660 693 679 682 724 642 704 695
704 652 664 702 661 720 695 670 656 718 660 648
683 723 710 680 684 705 681 748 697 703 660 722
662 644 683 695 678 674 656 667 683 691 680 685
681 715 665 676 665 675 655 659 720 675 697 663

6-27.   A group of wine enthusiasts taste-tested a pinot noir wine from Oregon. The evaluation was to grade the wine on a 0 to 100 point scale. The results follow:

94 90 92 91 91 86 89 91 91 90
90 93 87 90 91 92 89 86 89 90
88 95 91 88 89 92 87 89 95 92
85 91 85 89 88 84 85 90 90 83

(a) Construct a stem-and-leaf diagram for this data and comment on any important features that you notice.

(b) Compute the sample mean, sample standard deviation, and the sample median.

(c) A wine rated above 90 is considered truly exceptional. What proportion of the taste-tasters considered this particular pinot noir truly exceptional?

6-28. In their book *Introduction to Linear Regression Analysis* (3rd edition, Wiley, 2001) Montgomery, Peck, and Vining present measurements on $NbOCl_3$ concentration from a tube-flow reactor experiment. The data, in gram $-$ mole per liter $\times 10^{-3}$, are as follows:

450	450	473	507	457	452	453	1215	1256
1145	1085	1066	1111	1364	1254	1396	1575	1617
1733	2753	3186	3227	3469	1911	2588	2635	2725

(a) Construct a stem-and-leaf diagram for this data and comment on any important features that you notice.

(b) Compute the sample mean, sample standard deviation, and the sample median.

6-29. **A Comparative Stem-and-Leaf Diagram.** In Exercise 6-22, we presented height data that was self-reported by female undergraduate engineering students in a core course at ASU. In the same class, the male students self-reported their heights as follows:

69 67 69 70 65 68 69 70 71 69 66 67 69 75 68 67 68
69 70 71 72 68 69 69 70 71 68 72 69 69 68 69 73 70
73 68 69 71 67 68 65 68 68 69 70 74 71 69 70 69

(a) Construct a comparative stem-and-leaf diagram by listing the stems in the center of the display and then placing the female leaves on the left and the male leaves on the right.

(b) Comment on any important features that you notice in this display.

6-4 FREQUENCY DISTRIBUTIONS AND HISTOGRAMS

A frequency distribution is a more compact summary of data than a stem-and-leaf diagram. To construct a frequency distribution, we must divide the range of the data into intervals, which are usually called **class intervals, cells,** or **bins.** If possible, the bins should be of equal width in order to enhance the visual information in the frequency distribution. Some judgment must be used in selecting the number of bins so that a reasonable display can be developed. The number of bins depends on the number of observations and the amount of scatter or dispersion in the data. A frequency distribution that uses either too few or too many bins will not be informative. We usually find that between 5 and 20 bins is satisfactory in most cases and that the number of bins should increase with n. Choosing the **number of bins** approximately equal to the square root of the number of observations often works well in practice.

A frequency distribution for the comprehensive strength data in Table 6-2 is shown in Table 6-4. Since the data set contains 80 observations, and since $\sqrt{80} \approx 9$, we suspect that about eight to nine bins will provide a satisfactory frequency distribution. The largest and smallest data values are 245 and 76, respectively, so the bins must cover a range of at least $245 - 76 = 169$ units on the psi scale. If we want the lower limit for the first bin to begin slightly below the smallest data value and the upper limit for the last bin to be slightly above the largest data value, we might start the frequency distribution at 70 and end it at 250. This is an interval or range of 180 psi units. Nine bins, each of width 20 psi, give a reasonable frequency distribution, so the frequency distribution in Table 6-4 is based on nine bins.

The second row of Table 6-4 contains a **relative frequency distribution.** The relative frequencies are found by dividing the observed frequency in each bin by the total number of

Table 6-4 Frequency Distribution for the Compressive Strength Data in Table 6-2

Class	$70 \leq x < 90$	$90 \leq x < 110$	$110 \leq x < 130$	$130 \leq x < 150$	$150 \leq x < 170$	$170 \leq x < 190$	$190 \leq x < 210$	$210 \leq x < 230$	$230 \leq x < 250$
Frequency	2	3	6	14	22	17	10	4	2
Relative frequency	0.0250	0.0375	0.0750	0.1750	0.2750	0.2125	0.1250	0.0500	0.0250
Cumulative relative frequency	0.0250	0.0625	0.1375	0.3125	0.5875	0.8000	0.9250	0.9750	1.0000

observations. The last row of Table 6-4 expresses the relative frequencies on a cumulative basis. Frequency distributions are often easier to interpret than tables of data. For example, from Table 6-4 it is very easy to see that most of the specimens have compressive strengths between 130 and 190 psi and that 97.5 percent of the specimens fail below 230 psi.

The **histogram** is a visual display of the frequency distribution. The stages for constructing a histogram follow.

Constructing a Histogram (Equal Bin Widths)

> (1) Label the bin (class interval) boundaries on a horizontal scale.
> (2) Mark and label the vertical scale with the frequencies or the relative frequencies.
> (3) Above each bin, draw a rectangle where height is equal to the frequency (or relative frequency) corresponding to that bin.

Figure 6-7 is the histogram for the compression strength data. The histogram, like the **stem-and-leaf diagram,** provides a visual impression of the shape of the distribution of the measurements and information about the central tendency and scatter or dispersion in the data. Notice the symmetric, bell-shaped distribution of the strength measurements in Fig. 6-7. This display often gives insight about possible choices of probability distribution to use as a model for the population. For example, here we would likely conclude that the **normal distribution** is a reasonable model for the population of compression strength measurements.

Sometimes a histogram with **unequal bin widths** will be employed. For example, if the data have several extreme observations or outliers, using a few equal-width bins will result in nearly all observations falling in just of few of the bins. Using many equal-width bins will result in many bins with zero frequency. A better choice is to use shorter intervals in the region where most of the data falls and a few wide intervals near the extreme observations. When the bins are of unequal width, the rectangle's **area** (not its height) should be proportional to the bin frequency. This implies that the rectangle height should be

$$\text{Rectangle height} = \frac{\text{bin frequency}}{\text{bin width}}$$

In passing from either the original data or stem-and-leaf diagram to a frequency distribution or histogram, we have lost some information because we no longer have the individual observations. However, this information loss is often small compared with the conciseness and ease of interpretation gained in using the frequency distribution and histogram.

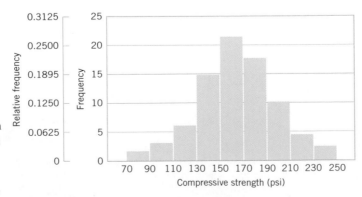

Figure 6-7   Histogram of compressive strength for 80 aluminum-lithium alloy specimens.

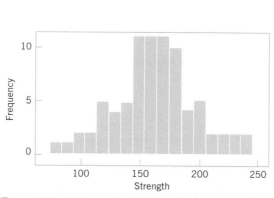

Figure 6-8 A histogram of the compressive strength data from Minitab with 17 bins.

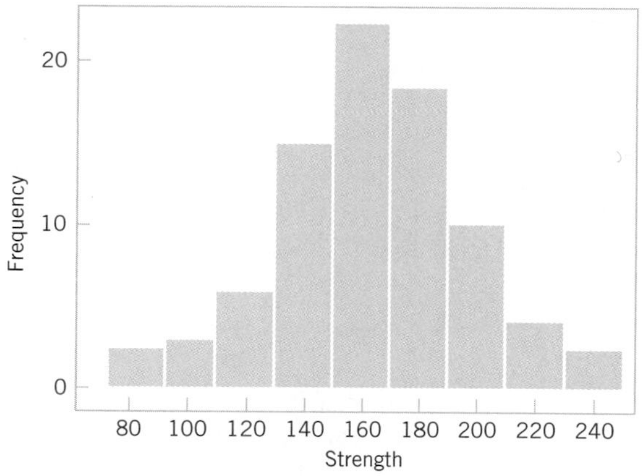

Figure 6-9 A histogram of the compressive strength data from Minitab with nine bins.

Figure 6-8 shows a histogram of the compressive strength data from Minitab. The "default" settings were used in this histogram, leading to 17 bins. We have noted that histograms may be relatively sensitive to the number of bins and their width. For small data sets, histograms may change dramatically in appearance if the number and/or width of the bins changes. Histograms are more stable for larger data sets, preferably of size 75 to 100 or more. Figure 6-9 shows the Minitab histogram for the compressive strength data with nine bins. This is similar to the original histogram shown in Fig. 6-7. Since the number of observations is moderately large ($n = 80$), the choice of the number of bins is not especially important, and both Figs. 6-8 and 6-9 convey similar information.

Figure 6-10 shows a variation of the histogram available in Minitab, the **cumulative frequency plot.** In this plot, the height of each bar is the total number of observations that are less than or equal to the upper limit of the bin. Cumulative distributions are also useful in data interpretation; for example, we can read directly from Fig. 6-10 that there are approximately 70 observations less than or equal to 200 psi.

When the sample size is large, the histogram can provide a reasonably reliable indicator of the general **shape** of the distribution or population of measurements from which the sample was drawn. Figure 6-11 presents three cases. The median is denoted as $\tilde{x}$. Generally, if the data are symmetric, as in Fig. 6-11(*b*), the mean and median coincide. If, in addition, the data have only one mode (we say the data are *unimodal*), the mean, median, and mode all coincide. If the data are *skewed* (asymmetric, with a long tail to one side), as in Fig. 6-11(*a*) and (*c*), the mean, median, and mode do not coincide. Usually, we find that mode < median < mean if the

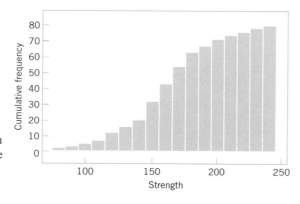

Figure 6-10 A cumulative distribution plot of the compressive strength data from Minitab.

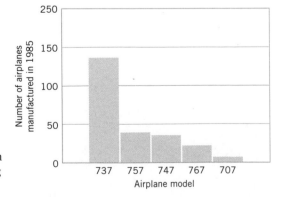

Figure 6-11
Histograms for symmetric and skewed distributions.

Negative or left skew
(a)

Symmetric
(b)

Positive or right skew
(c)

distribution is skewed to the right, whereas mode $>$ median $>$ mean if the distribution is skewed to the left.

Frequency distributions and histograms can also be used with qualitative or categorical data. In some applications there will be a natural ordering of the categories (such as freshman, sophomore, junior, and senior), whereas in others the order of the categories will be arbitrary (such as male and female). When using categorical data, the bins should have equal width.

EXAMPLE 6-6

Figure 6-12 presents the production of transport aircraft by the Boeing Company in 1985. Notice that the 737 was the most popular model, followed by the 757, 747, 767, and 707.

A chart of occurrences by category (in which the categories are ordered by the number of occurrences) is sometimes referred to as a **Pareto chart.** See Exercise 6-41.

In this section we have concentrated on descriptive methods for the situation in which each observation in a data set is a single number or belongs to one category. In many cases, we work with data in which each observation consists of several measurements. For example, in a gasoline mileage study, each observation might consist of a measurement of miles per gallon, the size of the engine in the vehicle, engine horsepower, vehicle weight, and vehicle length. This is an example of **multivariate data.** In later chapters, we will discuss analyzing this type of data.

EXERCISES FOR SECTION 6-4

 6-30. Construct a frequency distribution and histogram for the motor fuel octane data from Exercise 6-14. Use eight bins.

 6-31. Construct a frequency distribution and histogram using the failure data from Exercise 6-15.

6-32. Construct a frequency distribution and histogram for the cotton content data in Exercise 6-16.

6-33. Construct a frequency distribution and histogram for the yield data in Exercise 6-17.

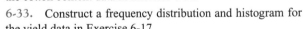

Figure 6-12
Airplane production in 1985. (*Source:* Boeing Company.)

6-34.   Construct a frequency distribution and histogram with 16 bins for the motor fuel octane data in Exercise 6-14. Compare its shape with that of the histogram with eight bins from Exercise 6-30. Do both histograms display similar information?

6-35.   Construct a histogram for the female student height data in Exercise 6-22.

6-36.   Construct a histogram with 10 bins for the spot weld shear strength data in Exercise 6-23. Comment on the shape of the histogram. Does it convey the same information as the stem-and-leaf display?

6-37.   Construct a histogram for the water quality data in Exercise 6-24. Comment on the shape of the histogram. Does it convey the same information as the stem-and-leaf display?

6-38.   Construct a histogram with 10 bins for the overall distance data in Exercise 6-25. Comment on the shape of the histogram. Does it convey the same information as the stem-and-leaf display?

6-39.   Construct a histogram for the semiconductor speed data in Exercise 6-26. Comment on the shape of the his-

togram. Does it convey the same information as the stem-and-leaf display?

6-40.   Construct a histogram for the pinot noir wine rating data in Exercise 6-27. Comment on the shape of the histogram. Does it convey the same information as the stem-and-leaf display?

6-41.   **The Pareto Chart.** An important variation of a histogram for categorical data is the Pareto chart. This chart is widely used in quality improvement efforts, and the categories usually represent different types of defects, failure modes, or product/process problems. The categories are ordered so that the category with the largest frequency is on the left, followed by the category with the second largest frequency and so forth. These charts are named after the Italian economist V. Pareto, and they usually exhibit "Pareto's law"; that is, most of the defects can be accounted for by only a few categories. Suppose that the following information on structural defects in automobile doors is obtained: dents, 4; pits, 4; parts assembled out of sequence, 6; parts undertrimmed, 21; missing holes/slots, 8; parts not lubricated, 5; parts out of contour, 30; and parts not deburred, 3. Construct and interpret a Pareto chart.

6-5   BOX PLOTS

The stem-and-leaf display and the histogram provide general visual impressions about a data set, while numerical quantities such as $\bar{x}$ or s provide information about only one feature of the data. The **box plot** is a graphical display that simultaneously describes several important features of a data set, such as center, spread, departure from symmetry, and identification of unusual observations or outliers.

A box plot displays the three quartiles, the minimum, and the maximum of the data on a rectangular box, aligned either horizontally or vertically. The box encloses the interquartile range with the left (or lower) edge at the first quartile, q_1, and the right (or upper) edge at the third quartile, q_3. A line is drawn through the box at the second quartile (which is the 50th percentile or the median), $q_2 = \bar{x}$. A line, or **whisker,** extends from each end of the box. The lower whisker is a line from the first quartile to the smallest data point within 1.5 interquartile ranges from the first quartile. The upper whisker is a line from the third quartile to the largest data point within 1.5 interquartile ranges from the third quartile. Data farther from the box than the whiskers are plotted as individual points. A point beyond a whisker, but less than 3 interquartile ranges from the box edge, is called an **outlier.** A point more than 3 interquartile ranges from the box edge is called an **extreme outlier.** See Fig. 6-13. Occasionally, different symbols, such as open and filled circles, are used to identify the two types of outliers. Sometimes box plots are called *box-and-whisker plots.*

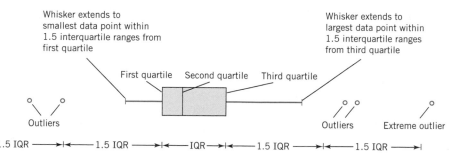

Figure 6-13   Description of a box plot.

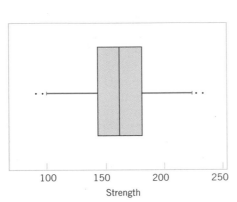

Figure 6-14   Box plot for compressive strength data in Table 6-2.

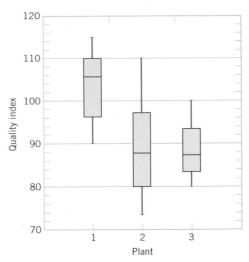

Figure 6-15   Comparative box plots of a quality index at three plants.

Figure 6-14 presents the box plot from Minitab for the alloy compressive strength data shown in Table 6-2. This box plot indicates that the distribution of compressive strengths is fairly symmetric around the central value, because the left and right whiskers and the lengths of the left and right boxes around the median are about the same. There are also two mild outliers on either end of the data.

Box plots are very useful in graphical comparisons among data sets, because they have high visual impact and are easy to understand. For example, Fig. 6-15 shows the comparative box plots for a manufacturing quality index on semiconductor devices at three manufacturing plants. Inspection of this display reveals that there is too much variability at plant 2 and that plants 2 and 3 need to raise their quality index performance.

EXERCISES FOR SECTION 6-5

6-42. Exercise 6-13 presented the joint temperatures of the O-rings (°F) for each test firing or actual launch of the space shuttle rocket motor. In that exercise you were asked to find the sample mean and sample standard deviation of temperature.
(a) Find the upper and lower quartiles of temperature.
(b) Find the median.
(c) Set aside the smallest observation (31°F) and recompute the quantities in parts (a) and (b). Comment on your findings. How "different" are the other temperatures from this smallest value?
(d) Construct a box plot of the data and comment on the possible presence of outliers.

6-43. An article in the *Transactions of the Institution of Chemical Engineers* (Vol. 34, 1956, pp. 280–293) reported data from an experiment investigating the effect of several process variables on the vapor phase oxidation of naphthalene. A sample of the percentage mole conversion of naphthalene to maleic anhydride follows: 4.2, 4.7, 4.7, 5.0, 3.8, 3.6, 3.0, 5.1, 3.1, 3.8, 4.8, 4.0, 5.2, 4.3, 2.8, 2.0, 2.8, 3.3, 4.8, 5.0.
(a) Calculate the sample mean.
(b) Calculate the sample variance and sample standard deviation.
(c) Construct a box plot of the data.

6-44. The "cold start ignition time" of an automobile engine is being investigated by a gasoline manufacturer. The following times (in seconds) were obtained for a test vehicle: 1.75, 1.92, 2.62, 2.35, 3.09, 3.15, 2.53, 1.91.
(a) Calculate the sample mean and sample standard deviation.
(b) Construct a box plot of the data.

6-45. The nine measurements that follow are furnace temperatures recorded on successive batches in a semiconductor manufacturing process (units are °F): 953, 950, 948, 955, 951, 949, 957, 954, 955.

(a) Calculate the sample mean, sample variance, and standard deviation.

(b) Find the median. How much could the largest temperature measurement increase without changing the median value?

(c) Construct a box plot of the data.

6-46. Exercise 6-12 presents drag coefficients for the NASA 0012 airfoil. You were asked to calculate the sample mean, sample variance, and sample standard deviation of those coefficients.

(a) Find the upper and lower quartiles of the drag coefficients.

(b) Construct a box plot of the data.

(c) Set aside the largest observation (100) and rework parts a and b. Comment on your findings.

6-47. The following data are the temperatures of effluent at discharge from a sewage treatment facility on consecutive days:

43	47	51	48	52	50	46	49
45	52	46	51	44	49	46	51
49	45	44	50	48	50	49	50

(a) Calculate the sample mean and median.

(b) Calculate the sample variance and sample standard deviation.

(c) Construct a box plot of the data and comment on the information in this display.

6-48. Reconsider the golf course yardage data in Exercise 6-3. Construct a box plot of the yardages and write an interpretation of the plot.

6-49. Reconsider the motor fuel octane rating data in Exercise 6-14. Construct a box plot of the yardages and write an interpretation of the plot. How does the box plot compare in interpretive value to the original stem-and-leaf diagram in Exercise 6-14?

6-50. Reconsider the spot weld shear strength data in Exercise 6-23. Construct a box plot of the strengths and write an interpretation of the plot. How does the box plot compare in interpretive value to the original stem-and-leaf diagram in Exercise 6-23?

6-51. Reconsider the female engineering student height data in Exercise 6-22. Construct a box plot of the heights and write an interpretation of the plot. How does the box plot compare in interpretive value to the original stem-and-leaf diagram in Exercise 6-22?

6-52. Reconsider the water quality data in Exercise 6-24. Construct a box plot of the concentrations and write an interpretation of the plot. How does the box plot compare in interpretive value to the original stem-and-leaf diagram in Exercise 6-24?

6-53. Reconsider the golf ball overall distance data in Exercise 6-25. Construct a box plot of the yardage distance and write an interpretation of the plot. How does the box plot compare in interpretive value to the original stem-and-leaf diagram in Exercise 6-25?

6-54. Reconsider the wine rating data in Exercise 6-27. Construct a box plot of the wine ratings and write an interpretation of the plot. How does the box plot compare in interpretive value to the original stem-and-leaf diagram in Exercise 6-27?

6-55. Use the data on heights of female and male engineering students from Exercises 6-22 and 6-29 to construct comparative box plots. Write an interpretation of the information that you see in these plots.

6-56. In Exercise 6-44, data was presented on the cold start ignition time of a particular gasoline used in a test vehicle. A second formulation of the gasoline was tested in the same vehicle, with the following times (in seconds): 1.83, 1.99, 3.13, 3.29, 2.65, 2.87, 3.40, 2.46, 1.89, and 3.35. Use this new data along with the cold start times reported in Exercise 6-44 to construct comparative box plots. Write an interpretation of the information that you see in these plots.

6-6 TIME SEQUENCE PLOTS

The graphical displays that we have considered thus far such as histograms, stem-and-leaf plots, and box plots are very useful visual methods for showing the variability in data. However, we noted in Section 1-2.2 that time is an important factor that contributes to variability in data, and those graphical methods do not take this into account. A **time series** or **time sequence** is a data set in which the observations are recorded in the order in which they occur. A **time series plot** is a graph in which the vertical axis denotes the observed value of the variable (say x) and the horizontal axis denotes the time (which could be minutes, days, years, etc.) When measurements are plotted as a time series, we often see trends, cycles, or other broad features of the data that could not be seen otherwise.

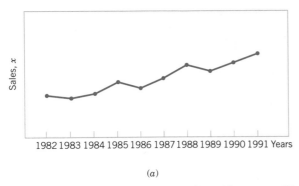

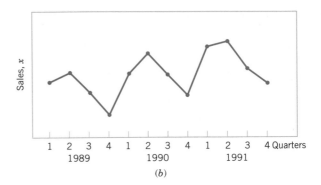

(a)

(b)

Figure 6-16   Company sales by year (a) and by quarter (b).

For example, consider Fig. 6-16(a), which presents a time series plot of the annual sales of a company for the last 10 years. The general impression from this display is that sales show an upward **trend.** There is some variability about this trend, with some years' sales increasing over those of the last year and some years' sales decreasing. Figure 6-16(b) shows the last three years of sales reported by quarter. This plot clearly shows that the annual sales in this business exhibit a **cyclic** variability by quarter, with the first- and second-quarter sales being generally greater than sales during the third and fourth quarters.

Sometimes it can be very helpful to combine a time series plot with some of the other graphical displays that we have considered previously. J. Stuart Hunter (*The American Statistician*, Vol. 42, 1988, p. 54) has suggested combining the stem-and-leaf plot with a time series plot to form a **digidot plot.**

Figure 6-17 shows a digidot plot for the observations on compressive strength from Table 6-2, assuming that these observations are recorded in the order in which they

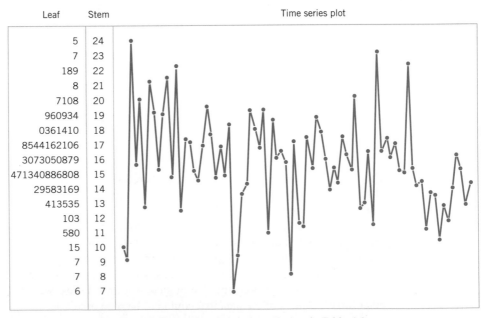

Figure 6-17   A digidot plot of the compressive strength data in Table 6-2.

Leaf	Stem	Time series plot
8	9e	
6	9s	
45	9f	
2333	9t	
0010000	9z	
99998	8e	
66677	8s	
45	8f	
23	8t	
1	8z	

Figure 6-18 A digi-dot plot of chemical process concentration readings, observed hourly.

occurred. This plot effectively displays the overall variability in the compressive strength data and simultaneously shows the variability in these measurements over time. The general impression is that compressive strength varies around the mean value of 162.67, and there is no strong obvious pattern in this variability over time.

The digidot plot in Fig. 6-18 tells a different story. This plot summarizes 30 observations on concentration of the output product from a chemical process, where the observations are recorded at one-hour time intervals. This plot indicates that during the first 20 hours of operation this process produced concentrations generally above 85 grams per liter, but that following sample 20, something may have occurred in the process that results in lower concentrations. If this variability in output product concentration can be reduced, operation of this process can be improved.

EXERCISES FOR SECTION 6-6

6-57. The College of Engineering and Applied Science at Arizona State University had a VAX computer system. Response times for 20 consecutive jobs were recorded and are as follows: (read across)

5.3	10.1	5.9	12.2	11.2	12.4	9.2
5.0	5.8	7.2	8.5	7.3	3.9	10.5
9.5	6.2	10.0	4.7	6.4	8.1	

Construct and interpret a time series plot of these data.

6-58. The following data are the viscosity measurements for a chemical product observed hourly (read down, then left to right).

47.9	48.6	48.0	48.1	43.0	43.2
47.9	48.8	47.5	48.0	42.9	43.6
48.6	48.1	48.6	48.3	43.6	43.2
48.0	48.3	48.0	43.2	43.3	43.5
48.4	47.2	47.9	43.0	43.0	43.0
48.1	48.9	48.3	43.5	42.8	
48.0	48.6	48.5	43.1	43.1	

(a) Construct and interpret either a digidot plot or a separate stem-and-leaf and time series plot of these data.

(b) Specifications on product viscosity are at 48 ± 2. What conclusions can you make about process performance?

6-59. The pull-off force for a connector is measured in a laboratory test. Data for 40 test specimens follow (read down, then left to right).

241	203	201	251	236	190
258	195	195	238	245	175
237	249	255	210	209	178
210	220	245	198	212	175
194	194	235	199	185	190
225	245	220	183	187	
248	209	249	213	218	

(a) Construct a time series plot of the data.
(b) Construct and interpret either a digidot plot or a stem-and-leaf plot of the data.

6-60. In their book *Time Series Analysis, Forecasting, and Control* (Prentice Hall, 1994), G. E. P. Box, G. M. Jenkins, and G. C. Reinsel present chemical process concentration readings made every two hours. Some of these data follow (read down, then left to right).

Table 6-5   United Kingdom Passenger Airline Miles Flown

Month	1964	1965	1966	1967	1968	1969	1970
Jan.	7.269	8.350	8.186	8.334	8.639	9.491	10.840
Feb.	6.775	7.829	7.444	7.899	8.772	8.919	10.436
Mar.	7.819	8.829	8.484	9.994	10.894	11.607	13.589
Apr.	8.371	9.948	9.864	10.078	10.455	8.852	13.402
May	9.069	10.638	10.252	10.801	11.179	12.537	13.103
June	10.248	11.253	12.282	12.953	10.588	14.759	14.933
July	11.030	11.424	11.637	12.222	10.794	13.667	14.147
Aug.	10.882	11.391	11.577	12.246	12.770	13.731	14.057
Sept.	10.333	10.665	12.417	13.281	13.812	15.110	16.234
Oct.	9.109	9.396	9.637	10.366	10.857	12.185	12.389
Nov.	7.685	7.775	8.094	8.730	9.290	10.645	11.594
Dec.	7.682	7.933	9.280	9.614	10.925	12.161	12.772

17.0	16.7	17.1	17.5	17.6	41	10	16	8	62	94
16.6	17.4	17.4	18.1	17.5	21	8	7	13	98	96
16.3	17.2	17.4	17.5	16.5	16	2	4	57	124	77
16.1	17.4	17.5	17.4	17.8	6	0	2	122	96	59
17.1	17.4	17.4	17.4	17.3	4	1	8	138	66	44
16.9	17.0	17.6	17.1	17.3	7	5	17	103	64	47
16.8	17.3	17.4	17.6	17.1	14	12	36	86	54	30
17.4	17.2	17.3	17.7	17.4	34	14	50	63	39	16
17.1	17.4	17.0	17.4	16.9	45	35	62	37	21	7
17.0	16.8	17.8	17.8	17.3	43	46	67	24	7	37
					48	41	71	11	4	74
					42	30	48	15	23	
					28	24	28	40	55	

Construct and interpret either a digidot plot or a stem-and-leaf plot of these data.

6-61.   The 100 annual Wolfer sunspot numbers from 1770 to 1869 follow. (For an interesting analysis and interpretation of these numbers, see the book by Box, Jenkins, and Reinsel referenced in Exercise 6-60. Their analysis requires some advanced knowledge of statistics and statistical model building.) (read down, then left to right)
(a) Construct a time series plot of these data.
(b) Construct and interpret either a digidot plot or a stem-and-leaf plot of these data.

101	31	154	38	83	90
82	7	125	23	132	67
66	20	85	10	131	60
35	92	68	24	118	47

6-62.   In their book *Forecasting and Time Series Analysis*, 2nd edition (McGraw-Hill, 1990), D. C. Montgomery, L. A. Johnson, and J. S. Gardiner analyze the data in Table 6-5, which are the monthly total passenger airline miles flown in the United Kingdom, 1964–1970 (in millions of miles).
(a) Draw a time series plot of the data and comment on any features of the data that are apparent.
(b) Construct and interpret either a digidot plot or a stem-and-leaf plot of these data.

6-7   PROBABILITY PLOTS

How do we know if a particular probability distribution is a reasonable model for data? Sometimes, this is an important question because many of the statistical techniques presented in subsequent chapters are based on an assumption that the population distribution is of a specific type. Thus, we can think of determining whether data come from a specific probability distribution as **verifying assumptions.** In other cases, the form of the distribution can give insight into the underlying physical mechanism generating the data. For example, in reliability engineering, verifying that time-to-failure data come from an exponential distribution identifies the **failure mechanism** in the sense that the failure rate is constant with respect to time.

Some of the visual displays we have used earlier, such as the histogram, can provide insight about the form of the underlying distribution. However, histograms are usually not really reliable indicators of the distribution form unless the sample size is very large. **Probability plotting** is a graphical method for determining whether sample data conform to a hypothesized distribution based on a subjective visual examination of the data. The general procedure is very simple and can be performed quickly. It is also more reliable than the histogram for small to moderate size samples. Probability plotting typically uses special graph paper, known as **probability paper,** that has been designed for the hypothesized distribution. Probability paper is widely available for the normal, lognormal, Weibull, and various chi-square and gamma distributions. We focus primarily on normal probability plots because many statistical techniques are appropriate only when the population is (at least approximately) normal.

To construct a probability plot, the observations in the sample are first ranked from smallest to largest. That is, the sample $x_1, x_2, \ldots, x_n$ is arranged as $x_{(1)}, x_{(2)}, \ldots, x_{(n)}$, where $x_{(1)}$ is the smallest observation, $x_{(2)}$ is the second smallest observation, and so forth, with $x_{(n)}$ the largest. The ordered observations $x_{(j)}$ are then plotted against their observed cumulative frequency $(j - 0.5)/n$ on the appropriate probability paper. If the hypothesized distribution adequately describes the data, the plotted points will fall approximately along a straight line; if the plotted points deviate significantly from a straight line, the hypothesized model is not appropriate. Usually, the determination of whether or not the data plot as a straight line is subjective. The procedure is illustrated in the following example.

EXAMPLE 6-7   Ten observations on the effective service life in minutes of batteries used in a portable personal computer are as follows: 176, 191, 214, 220, 205, 192, 201, 190, 183, 185. We hypothesize that battery life is adequately modeled by a normal distribution. To use probability plotting to investigate this hypothesis, first arrange the observations in ascending order and calculate their cumulative frequencies $(j - 0.5)/10$ as shown in Table 6-6.

The pairs of values $x_{(j)}$ and $(j - 0.5)/10$ are now plotted on normal probability paper. This plot is shown in Fig. 6-19. Most normal probability paper plots $100(j - 0.5)/n$ on the left vertical scale and $100[1 - (j - 0.5)/n]$ on the right vertical scale, with the variable value plotted on the horizontal scale. A straight line, chosen subjectively, has been drawn through the plotted points. In drawing the straight line, you should be influenced more by the points near the middle of the plot than by the extreme points. A good rule of thumb is to draw the line approximately between the 25th and 75th percentile points. This is how the line in Fig. 6-19 was determined. In assessing the "closeness" of the points to the straight line, imagine a "fat pencil" lying along the line. If all the points are covered by this imaginary pencil, a normal distribution adequately describes the data. Since the points in Fig. 6-19 would pass the "fat pencil" test, we conclude that the normal distribution is an appropriate model.

A **normal probability plot** can also be constructed on ordinary graph paper by plotting the standardized normal scores z_j against $x_{(j)}$, where the standardized normal scores satisfy

$$\frac{j - 0.5}{n} = P(Z \le z_j) = \Phi(z_j)$$

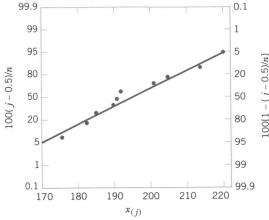

Figure 6-19   Normal probability plot for battery life.

Table 6-6   Calculation for Constructing a Normal Probability Plot

j	$x_{(j)}$	$(j - 0.5)/10$	z_j
1	176	0.05	−1.64
2	183	0.15	−1.04
3	185	0.25	−0.67
4	190	0.35	−0.39
5	191	0.45	−0.13
6	192	0.55	0.13
7	201	0.65	0.39
8	205	0.75	0.67
9	214	0.85	1.04
10	220	0.95	1.64

For example, if $(j - 0.5)/n = 0.05$, $\Phi(z_j) = 0.05$ implies that $z_j = -1.64$. To illustrate, consider the data from Example 6-4. In the last column of Table 6-6 we show the standardized normal scores. Figure 6-20 presents the plot of z_j versus $x_{(j)}$. This normal probability plot is equivalent to the one in Fig. 6-19.

We have constructed our probability plots with the probability scale (or the z-scale) on the vertical axis. Some computer packages "flip" the axis and put the probability scale on the horizontal axis.

The normal probability plot can be useful in identifying distributions that are symmetric but that have tails that are "heavier" or "lighter" than the normal. They can also be useful in identifying skewed distributions. When a sample is selected from a light-tailed distribution (such as the uniform distribution), the smallest and largest observations will not be as extreme as would be expected in a sample from a normal distribution. Thus if we consider the straight line drawn through the observations at the center of the normal probability plot, observations on the left side will tend to fall below the line, whereas observations on the right side will tend to fall above the line. This will produce an S-shaped normal probability plot such as shown in

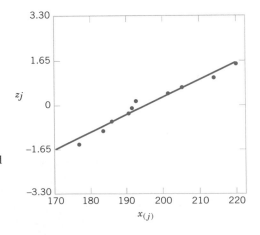

Figure 6-20   Normal probability plot obtained from standardized normal scores.

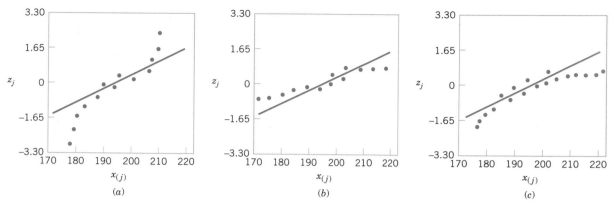

Figure 6-21   Normal probability plots indicating a nonnormal distribution. (a) Light-tailed distribution. (b) Heavy-tailed distribution. (c) A distribution with positive (or right) skew.

Fig. 6-21(a). A heavy-tailed distribution will result in data that also produces an S-shaped normal probability plot, but now the observations on the left will be above the straight line and the observations on the right will lie below the line. See Fig. 6-19(b). A positively skewed distribution will tend to produce a pattern such as shown in Fig. 6-19(c), where points on both ends of the plot tend to fall below the line, giving a curved shape to the plot. This occurs because both the smallest and the largest observations from this type of distribution are larger than expected in a sample from a normal distribution.

Even when the underlying population is exactly normal, the sample data will not plot exactly on a straight line. Some judgment and experience are required to evaluate the plot. Generally, if the sample size is $n < 30$, there can be a lot of deviation from linearity in normal plots, so in these cases only a very severe departure from linearity should be interpreted as a strong indication of nonnormality. As n increases, the linear pattern will tend to become stronger, and the normal probability plot will be easier to interpret and more reliable as an indicator of the form of the distribution.

EXERCISES FOR SECTION 6-7

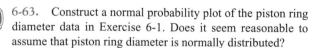 **6-63.**   Construct a normal probability plot of the piston ring diameter data in Exercise 6-1. Does it seem reasonable to assume that piston ring diameter is normally distributed?

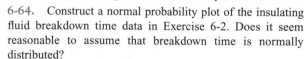 **6-64.**   Construct a normal probability plot of the insulating fluid breakdown time data in Exercise 6-2. Does it seem reasonable to assume that breakdown time is normally distributed?

6-65.   Construct a normal probability plot of the visual accommodation data in Exercise 6-5. Does it seem reasonable to assume that visual accommodation is normally distributed?

6-66.   Construct a normal probability plot of the O-ring joint temperature data in Exercise 6-13. Does it seem reasonable to assume that O-ring joint temperature is normally distributed? Discuss any interesting features that you see on the plot.

 6-67.   Construct a normal probability plot of the octane rating data in Exercise 6-14. Does it seem reasonable to assume that octane rating is normally distributed?

 6-68.   Construct a normal probability plot of the cycles to failure data in Exercise 6-15. Does it seem reasonable to assume that cycles to failure is normally distributed?

 6-69.   Construct a normal probability plot of the wine quality rating data in Exercise 6-27. Does it seem reasonable to assume that this variable is normally distributed?

 6-70.   Construct a normal probability plot of the suspended solids concentration data in Exercise 6-24. Does it seem reasonable to assume that the concentration of suspended solids in water from this particular lake is normally distributed?

6-71.   Construct two normal probability plots for the height data in Exercises 6-22 and 6-29. Plot the data for female and male students on the same axes. Does height seem to be normally distributed for either group of students? If both populations have the same variance, the two normal probability plots should have identical slopes. What conclusions would you draw about the heights of the two groups of students from visual examination of the normal probability plots?

6-72.   It is possible to obtain a "quick and dirty" estimate of the mean of a normal distribution from the fiftieth percentile value on a normal probability plot. Provide an argument why this is so. It is also possible to obtain an estimate of the standard deviation of a normal distribution by subtracting the sixty-fourth percentile value from the fiftieth percentile value. Provide an argument why this is so.

6-8   MORE ABOUT PROBABILITY PLOTTING (CD ONLY)

Supplemental Exercises

6-73.   The concentration of a solution is measured six times by one operator using the same instrument. She obtains the following data: 63.2, 67.1, 65.8, 64.0, 65.1, and 65.3 (grams per liter).
(a) Calculate the sample mean. Suppose that the desirable value for this solution has been specified to be 65.0 grams per liter. Do you think that the sample mean value computed here is close enough to the target value to accept the solution as conforming to target? Explain your reasoning.
(b) Calculate the sample variance and sample standard deviation.
(c) Suppose that in measuring the concentration, the operator must set up an apparatus and use a reagent material. What do you think the major sources of variability are in this experiment? Why is it desirable to have a small variance of these measurements?

6-74.   A sample of six resistors yielded the following resistances (ohms): $x_1 = 45$, $x_2 = 38$, $x_3 = 47$, $x_4 = 41$, $x_5 = 35$, and $x_6 = 43$.
(a) Compute the sample variance and sample standard deviation.
(b) Subtract 35 from each of the original resistance measurements and compute s^2 and s. Compare your results with those obtained in part (a) and explain your findings.
(c) If the resistances were 450, 380, 470, 410, 350, and 430 ohms, could you use the results of previous parts of this problem to find s^2 and s?

6-75.   Consider the following two samples:

Sample 1: 10, 9, 8, 7, 8, 6, 10, 6

Sample 2: 10, 6, 10, 6, 8, 10, 8, 6

(a) Calculate the sample range for both samples. Would you conclude that both samples exhibit the same variability? Explain.
(b) Calculate the sample standard deviations for both samples. Do these quantities indicate that both samples have the same variability? Explain.
(c) Write a short statement contrasting the sample range versus the sample standard deviation as a measure of variability.

6-76.   An article in *Quality Engineering* (Vol. 4, 1992, pp. 487–495) presents viscosity data from a batch chemical process. A sample of these data follows:

13.3	14.3	14.9	15.2	15.8	14.2	16.0	14.0
14.5	16.1	13.7	15.2	13.7	16.9	14.9	14.4
15.3	13.1	15.2	15.9	15.1	14.9	13.6	13.7
15.3	15.5	14.5	16.5	13.4	15.2	15.3	13.8
14.3	12.6	15.3	14.8	14.1	14.4	14.3	15.6
14.8	14.6	15.6	15.1	14.8	15.2	15.6	14.5
15.2	14.3	15.8	17.0	14.3	14.6	16.1	12.8
14.5	15.4	13.3	14.9	14.3	16.4	13.9	16.1
14.6	15.2	14.1	14.8	16.4	14.2	15.2	16.6
14.1	16.8	15.4	14.0	16.9	15.7	14.4	15.6

(a) Reading down and left to right, draw a time series plot of all the data and comment on any features of the data that are revealed by this plot.
(b) Consider the notion that the first 40 observations were generated from a specific process, whereas the last 40 observations were generated from a different process. Does the plot indicate that the two processes generate similar results?
(c) Compute the sample mean and sample variance of the first 40 observations; then compute these values for the second 40 observations. Do these quantities indicate that both processes yield the same mean level? The same variability? Explain.

6-77.   Reconsider the data from Exercise 6-76. Prepare comparative box plots for two groups of observations: the first 40 and the last 40. Comment on the information in the box plots.

6-78.   The data shown in Table 6-7 are monthly champagne sales in France (1962-1969) in thousands of bottles.
(a) Construct a time series plot of the data and comment on any features of the data that are revealed by this plot.
(b) Speculate on how you would use a graphical procedure to forecast monthly champagne sales for the year 1970.

6-79.   A manufacturer of coil springs is interested in implementing a quality control system to monitor his production process. As part of this quality system, it is decided to record the number of nonconforming coil springs in each production

Table 6-7 Champagne Sales in France

Month	1962	1963	1964	1965	1966	1967	1968	1969
Jan.	2.851	2.541	3.113	5.375	3.633	4.016	2.639	3.934
Feb.	2.672	2.475	3.006	3.088	4.292	3.957	2.899	3.162
Mar.	2.755	3.031	4.047	3.718	4.154	4.510	3.370	4.286
Apr.	2.721	3.266	3.523	4.514	4.121	4.276	3.740	4.676
May	2.946	3.776	3.937	4.520	4.647	4.968	2.927	5.010
June	3.036	3.230	3.986	4.539	4.753	4.677	3.986	4.874
July	2.282	3.028	3.260	3.663	3.965	3.523	4.217	4.633
Aug.	2.212	1.759	1.573	1.643	1.723	1.821	1.738	1.659
Sept.	2.922	3.595	3.528	4.739	5.048	5.222	5.221	5.591
Oct.	4.301	4.474	5.211	5.428	6.922	6.873	6.424	6.981
Nov.	5.764	6.838	7.614	8.314	9.858	10.803	9.842	9.851
Dec.	7.132	8.357	9.254	10.651	11.331	13.916	13.076	12.670

batch of size 50. During 40 days of production, 40 batches of data were collected as follows:

Read data across.

9	12	6	9	7	14	12	4	6	7
8	5	9	7	8	11	3	6	7	7
11	4	4	8	7	5	6	4	5	8
19	19	18	12	11	17	15	17	13	13

(a) Construct a stem-and-leaf plot of the data.
(b) Find the sample average and standard deviation.
(c) Construct a time series plot of the data. Is there evidence that there was an increase or decrease in the average number of nonconforming springs made during the 40 days? Explain.

6-80. A communication channel is being monitored by recording the number of errors in a string of 1000 bits. Data for 20 of these strings follow:

Read data across.

| 3 | 1 | 0 | 1 | 3 | 2 | 4 | 1 | 3 | 1 |
| 1 | 1 | 2 | 3 | 3 | 2 | 0 | 2 | 0 | 1 |

(a) Construct a stem-and-leaf plot of the data.
(b) Find the sample average and standard deviation.
(c) Construct a time series plot of the data. Is there evidence that there was an increase or decrease in the number of errors in a string? Explain.

6-81. Reconsider the data in Exercise 6-76. Construct normal probability plots for two groups of the data: the first 40 and the last 40 observations. Construct both plots on the same axes. What tentative conclusions can you draw?

6-82. Construct a normal probability plot of the effluent discharge temperature data from Exercise 6-47. Based on the plot, what tentative conclusions can you draw?

6-83. Construct normal probability plots of the cold start ignition time data presented in Exercises 6-44 and 6-56.

Construct a separate plot for each gasoline formulation, but arrange the plots on the same axes. What tentative conclusions can you draw?

6-84. **Transformations.** In some data sets, a transformation by some mathematical function applied to the original data, such as $\sqrt{y}$ or log y, can result in data that are simpler to work with statistically than the original data. To illustrate the effect of a transformation, consider the following data, which represent cycles to failure for a yarn product: 675, 3650, 175, 1150, 290, 2000, 100, 375.

(a) Construct a normal probability plot and comment on the shape of the data distribution.
(b) Transform the data using logarithms; that is, let y^* (new value) = log y (old value). Construct a normal probability plot of the transformed data and comment on the effect of the transformation.

6-85. In 1879, A. A. Michelson made 100 determinations of the velocity of light in air using a modification of a method proposed by the French physicist Foucault. He made the measurements in five trials of 20 measurements each. The observations (in kilometers per second) follow. Each value has 299,000 substracted from it.

Trial 1

850	900	930	950	980
1000	930	760	1000	960
740	1070	850	980	880
980	650	810	1000	960

Trial 2

960	960	880	850	900
830	810	880	800	760
940	940	800	880	840
790	880	830	790	800

Trial 3

880	880	720	620	970
880	850	840	850	840
880	860	720	860	950
910	870	840	840	840

Trial 4

890	810	800	760	750
910	890	880	840	850
810	820	770	740	760
920	860	720	850	780

Trial 5

890	780	760	790	820
870	810	810	950	810
840	810	810	810	850
870	740	940	800	870

The currently accepted true velocity of light in a vacuum is 299, 792.5 kilometers per second. Stigler (1977, *The Annals of Statistics*) reports that the "true" value for comparison to these measurements is 734.5. Construct comparative box plots of these measurements. Does it seem that all five trials are con-sistent with respect to the variability of the measurements? Are all five trials centered on the same value? How does each group of trials compare to the true value? Could there have been "startup" effects in the experiment that Michelson performed? Could there have been bias in the measuring instrument?

6-86. In 1789, Henry Cavendish estimated the density of the earth by using a torsion balance. His 29 measurements follow, expressed as a multiple of the density of water.

5.50	5.30	5.47	5.10	5.29	5.65
5.55	5.61	5.75	5.63	5.27	5.44
5.57	5.36	4.88	5.86	5.34	5.39
5.34	5.53	5.29	4.07	5.85	5.46
5.42	5.79	5.62	5.58	5.26	

(a) Calculate the sample mean, sample standard deviation, and median of the Cavendish density data.

(b) Construct a normal probability plot of the data. Comment on the plot. Does there seem to be a "low" outlier in the data?

(c) Would the sample median be a better estimate of the density of the earth than the sample mean? Why?

MIND-EXPANDING EXERCISES

6-87. Consider the airfoil data in Exercise 6-12. Subtract 30 from each value and then multiply the resulting quantities by 10. Now compute s^2 for the new data. How is this quantity related to s^2 for the *original* data? Explain why.

6-88. Consider the quantity $\sum_{i=1}^{n} (x_i - a)^2$. For what value of a is this quantity minimized?

6-89 Using the results of Exercise 6-88, which of the two quantities $\sum_{i=1}^{n} (x_i - \bar{x})^2$ and $\sum_{i=1}^{n} (x_i - \mu)^2$ will be smaller, provided that $\bar{x} \neq \mu$?

6-90. **Coding the Data.** Let $y_i = a + bx_i$, $i = 1, 2, \ldots, n$, where a and b are nonzero constants. Find the relationship between $\bar{x}$ and $\bar{y}$, and between s_x and s_y.

6-91. A sample of temperature measurements in a furnace yielded a sample average (°F) of 835.00 and a sample standard deviation of 10.5. Using the results from Exercise 6-90, what are the sample average and sample standard deviations expressed in °C?

6-92. Consider the sample $x_1, x_2, \ldots, x_n$ with sample mean $\bar{x}$ and sample standard deviation s. Let $z_i = (x_i - \bar{x})/s$, $i = 1, 2, \ldots, n$. What are the values of the sample mean and sample standard deviation of the z_i?

6-93. An experiment to investigate the survival time in hours of an electronic component consists of placing the parts in a test cell and running them for 100 hours under elevated temperature conditions. (This is called an "accelerated" life test.) Eight components were tested with the following resulting failure times:

75, 63, 100$^+$, 36, 51, 45, 80, 90

The observation 100$^+$ indicates that the unit still functioned at 100 hours. Is there any meaningful measure of location that can be calculated for these data? What is its numerical value?

6-94. Suppose that we have a sample $x_1, x_2, \ldots, x_n$ and we have calculated $\bar{x}_n$ and s_n^2 for the sample. Now an $(n + 1)$st observation becomes available. Let $\bar{x}_{n+1}$ and s_{n+1}^2 be the sample mean and sample variance for the sample using all $n + 1$ observations.
(a) Show how $\bar{x}_{n+1}$ can be computed using $\bar{x}_n$ and x_{n+1}.
(b) Show that $ns_{n+1}^2 = (n - 1)s_n^2 + \dfrac{n(x_{n+1} - \bar{x}_n)^2}{n + 1}$
(c) Use the results of parts (a) and (b) to calculate the new sample average and standard deviation for the data of Exercise 6-22, when the new observation is $x_{38} = 64$.

6-95. **The Trimmed Mean.** Suppose that the data are arranged in increasing order, $T\%$ of the observations are removed from each end and the sample mean of the remaining numbers is calculated. The resulting quantity is called a *trimmed mean*. The trimmed mean generally lies between the sample mean $\bar{x}$ and the sample median $\tilde{x}$. Why?
(a) Calculate the 10% trimmed mean for the yield data in Exercise 6-17.
(b) Calculate the 20% trimmed mean for the yield data in Exercise 6-17 and compare it with the quantity found in part (a).
(c) Compare the values calculated in parts (a) and (b) with the sample mean and median for the yield data. Is there much difference in these quantities? Why?

6-96. **The Trimmed Mean.** Suppose that the sample size n is such that the quantity $nT/100$ is not an integer. Develop a procedure for obtaining a trimmed mean in this case.

IMPORTANT TERMS AND CONCEPTS

In the E-book, click on any term or concept below to go to that subject.

Box plot
Frequency distribution and histogram
Median, quartiles and percentiles

Normal probability plot
Population mean
Population standard deviation
Population variance
Random sample

Sample mean
Sample standard deviation
Sample variance
Stem-and-leaf diagram
Time series plots

CD MATERIAL
Exponential probability plot
Goodness of fit
Weibull probability plot

Point Estimation of Parameters

CHAPTER OUTLINE

LEARNING OBJECTIVES

After careful study of this chapter you should be able to do the following:

1. Explain the general concepts of estimating the parameters of a population or a probability distribution
2. Explain important properties of point estimators, including bias, variance, and mean square error
3. Know how to construct point estimators using the method of moments and the method of maximum likelihood
4. Know how to compute and explain the precision with which a parameter is estimated
5. Understand the central limit theorem
6. Explain the important role of the normal distribution as a sampling distribution

CD MATERIAL

7. Use bootstrapping to find the standard error of a point estimate
8. Know how to construct a point estimator using the Bayesian approach

Answers for most odd numbered exercises are at the end of the book. Answers to exercises whose numbers are surrounded by a box can be accessed in the e-Text by clicking on the box. Complete worked solutions to certain exercises are also available in the e-Text. These are indicated in the Answers to Selected Exercises section by a box around the exercise number. Exercises are also available for some of the text sections that appear on CD only. These exercises may be found within the e-Text immediately following the section they accompany.

7-1 INTRODUCTION

The field of statistical inference consists of those methods used to make decisions or to draw conclusions about a **population.** These methods utilize the information contained in a **sample** from the population in drawing conclusions. This chapter begins our study of the statistical methods used for inference and decision making.

Statistical inference may be divided into two major areas: **parameter estimation** and **hypothesis testing.** As an example of a parameter estimation problem, suppose that a structural engineer is analyzing the tensile strength of a component used in an automobile chassis. Since variability in tensile strength is naturally present between the individual components because of differences in raw material batches, manufacturing processes, and measurement procedures (for example), the engineer is interested in estimating the mean tensile strength of the components. In practice, the engineer will use sample data to compute a number that is in some sense a reasonable value (or guess) of the true mean. This number is called a **point estimate.** We will see that it is possible to establish the precision of the estimate.

Now consider a situation in which two different reaction temperatures can be used in a chemical process, say t_1 and t_2. The engineer conjectures that t_1 results in higher yields than does t_2. Statistical hypothesis testing is a framework for solving problems of this type. In this case, the hypothesis would be that the mean yield using temperature t_1 is greater than the mean yield using temperature t_2. Notice that there is no emphasis on estimating yields; instead, the focus is on drawing conclusions about a stated hypothesis.

Suppose that we want to obtain a point estimate of a population parameter. We know that before the data is collected, the observations are considered to be random variables, say $X_1, X_2, \ldots, X_n$. Therefore, any function of the observation, or any **statistic,** is also a random variable. For example, the sample mean $\overline{X}$ and the sample variance S^2 are statistics and they are also random variables.

Since a statistic is a random variable, it has a probability distribution. We call the probability distribution of a statistic a **sampling distribution.** The notion of a sampling distribution is very important and will be discussed and illustrated later in the chapter.

When discussing inference problems, it is convenient to have a general symbol to represent the parameter of interest. We will use the Greek symbol θ (theta) to represent the parameter. The objective of point estimation is to select a single number, based on sample data, that is the most plausible value for θ. A numerical value of a sample statistic will be used as the point estimate.

In general, if X is a random variable with probability distribution $f(x)$, characterized by the unknown parameter θ, and if $X_1, X_2, \ldots, X_n$ is a random sample of size n from X, the statistic $\hat{\Theta} = h(X_1, X_2, \ldots, X_n)$ is called a **point estimator** of θ. Note that $\hat{\Theta}$ is a random variable because it is a function of random variables. After the sample has been selected, $\hat{\Theta}$ takes on a particular numerical value $\hat{\theta}$ called the **point estimate** of θ.

Definition

A **point estimate** of some population parameter θ is a single numerical value $\hat{\theta}$ of a statistic $\hat{\Theta}$. The statistic $\hat{\Theta}$ is called the **point estimator**.

As an example, suppose that the random variable X is normally distributed with an unknown mean μ. The sample mean is a point estimator of the unknown population mean μ. That is, $\hat{\mu} = \overline{X}$. After the sample has been selected, the numerical value $\overline{x}$ is the point estimate of μ. Thus, if $x_1 = 25$, $x_2 = 30$, $x_3 = 29$, and $x_4 = 31$, the point estimate of μ is

$$\overline{x} = \frac{25 + 30 + 29 + 31}{4} = 28.75$$

Similarly, if the population variance σ^2 is also unknown, a point estimator for σ^2 is the sample variance S^2, and the numerical value $s^2 = 6.9$ calculated from the sample data is called the point estimate of σ^2.

Estimation problems occur frequently in engineering. We often need to estimate

- The mean μ of a single population
- The variance σ^2 (or standard deviation σ) of a single population
- The proportion p of items in a population that belong to a class of interest
- The difference in means of two populations, $\mu_1 - \mu_2$
- The difference in two population proportions, $p_1 - p_2$

Reasonable point estimates of these parameters are as follows:

- For μ, the estimate is $\hat{\mu} = \overline{x}$, the sample mean.
- For σ^2, the estimate is $\hat{\sigma}^2 = s^2$, the sample variance.
- For p, the estimate is $\hat{p} = x/n$, the sample proportion, where x is the number of items in a random sample of size n that belong to the class of interest.
- For $\mu_1 - \mu_2$, the estimate is $\hat{\mu}_1 - \hat{\mu}_2 = \overline{x}_1 - \overline{x}_2$, the difference between the sample means of two independent random samples.
- For $p_1 - p_2$, the estimate is $\hat{p}_1 - \hat{p}_2$, the difference between two sample proportions computed from two independent random samples.

We may have several different choices for the point estimator of a parameter. For example, if we wish to estimate the mean of a population, we might consider the sample mean, the sample median, or perhaps the average of the smallest and largest observations in the sample as point estimators. In order to decide which point estimator of a particular parameter is the best one to use, we need to examine their statistical properties and develop some criteria for comparing estimators.

7-2   GENERAL CONCEPTS OF POINT ESTIMATION

7-2.1   Unbiased Estimators

An estimator should be "close" in some sense to the true value of the unknown parameter. Formally, we say that $\hat{\Theta}$ is an unbiased estimator of θ if the expected value of $\hat{\Theta}$ is equal to θ. This is equivalent to saying that the mean of the probability distribution of $\hat{\Theta}$ (or the mean of the sampling distribution of $\hat{\Theta}$) is equal to θ.

Definition

> The point estimator $\hat{\Theta}$ is an **unbiased estimator** for the parameter θ if
>
> $$E(\hat{\Theta}) = \theta \tag{7-1}$$
>
> If the estimator is not unbiased, then the difference
>
> $$E(\hat{\Theta}) - \theta \tag{7-2}$$
>
> is called the **bias** of the estimator $\hat{\Theta}$.

When an estimator is unbiased, the bias is zero; that is, $E(\hat{\Theta}) - \theta = 0$.

EXAMPLE 7-1

Suppose that X is a random variable with mean μ and variance σ^2. Let $X_1, X_2, \ldots, X_n$ be a random sample of size n from the population represented by X. Show that the sample mean $\overline{X}$ and sample variance S^2 are unbiased estimators of μ and σ^2, respectively.

First consider the sample mean. In Equation 5.40a in Chapter 5, we showed that $E(\overline{X}) = \mu$. Therefore, the sample mean $\overline{X}$ is an unbiased estimator of the population mean μ.

Now consider the sample variance. We have

$$E(S^2) = E\left[\frac{\sum_{i=1}^{n}(X_i - \overline{X})^2}{n-1}\right] = \frac{1}{n-1} E \sum_{i=1}^{n}(X_i - \overline{X})^2$$

$$= \frac{1}{n-1} E \sum_{i-1}^{n}(X_i^2 + \overline{X}^2 - 2\overline{X}X_i) = \frac{1}{n-1} E\left(\sum_{i=1}^{n} X_i^2 - n\overline{X}^2\right)$$

$$= \frac{1}{n-1}\left[\sum_{i=1}^{n} E(X_i^2) - nE(\overline{X}^2)\right]$$

The last equality follows from Equation 5-37 in Chapter 5. However, since $E(X_i^2) = \mu^2 + \sigma^2$ and $E(\overline{X}^2) = \mu^2 + \sigma^2/n$, we have

$$E(S^2) = \frac{1}{n-1}\left[\sum_{i=1}^{n}(\mu^2 + \sigma^2) - n(\mu^2 + \sigma^2/n)\right]$$

$$= \frac{1}{n-1}(n\mu^2 + n\sigma^2 - n\mu^2 - \sigma^2)$$

$$= \sigma^2$$

Therefore, the sample variance S^2 is an unbiased estimator of the population variance σ^2.

Although S^2 is unbiased for σ^2, S is a biased estimator of σ. For large samples, the bias is very small. However, there are good reasons for using S as an estimator of σ in samples from normal distributions, as we will see in the next three chapters when are discuss confidence intervals and hypothesis testing.

Sometimes there are several unbiased estimators of the sample population parameter. For example, suppose we take a random sample of size $n = 10$ from a normal population and obtain the data $x_1 = 12.8$, $x_2 = 9.4$, $x_3 = 8.7$, $x_4 = 11.6$, $x_5 = 13.1$, $x_6 = 9.8$, $x_7 = 14.1$, $x_8 = 8.5$, $x_9 = 12.1$, $x_{10} = 10.3$. Now the sample mean is

$$\bar{x} = \frac{12.8 + 9.4 + 8.7 + 11.6 + 13.1 + 9.8 + 14.1 + 8.5 + 12.1 + 10.3}{10}$$
$$= 11.04$$

the sample median is

$$\tilde{x} = \frac{10.3 + 11.6}{2} = 10.95$$

and a 10% trimmed mean (obtained by discarding the smallest and largest 10% of the sample before averaging) is

$$\bar{x}_{tr(10)} = \frac{8.7 + 9.4 + 9.8 + 10.3 + 11.6 + 12.1 + 12.8 + 13.1}{8}$$
$$= 10.98$$

We can show that all of these are unbiased estimates of μ. Since there is not a unique unbiased estimator, we cannot rely on the property of unbiasedness alone to select our estimator. We need a method to select among unbiased estimators. We suggest a method in Section 7-2.3.

7-2.2   Proof That S is a Biased Estimator of σ (CD Only)

7-2.3   Variance of a Point Estimator

Suppose that $\hat{\Theta}_1$ and $\hat{\Theta}_2$ are unbiased estimators of θ. This indicates that the distribution of each estimator is centered at the true value of θ. However, the variance of these distributions may be different. Figure 7-1 illustrates the situation. Since $\hat{\Theta}_1$ has a smaller variance than $\hat{\Theta}_2$, the estimator $\hat{\Theta}_1$ is more likely to produce an estimate close to the true value θ. A logical principle of estimation, when selecting among several estimators, is to choose the estimator that has minimum variance.

Definition

> If we consider all unbiased estimators of θ, the one with the smallest variance is called the **minimum variance unbiased estimator** (MVUE).

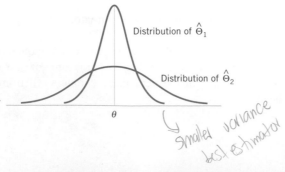

Figure 7-1   The sampling distributions of two unbiased estimators $\hat{\Theta}_1$ and $\hat{\Theta}_2$.

Distribution of $\hat{\Theta}_1$

Distribution of $\hat{\Theta}_2$

In a sense, the MVUE is most likely among all unbiased estimators to produce an estimate $\hat{\theta}$ that is close to the true value of θ. It has been possible to develop methodology to identify the MVUE in many practical situations. While this methodology is beyond the scope of this book, we give one very important result concerning the normal distribution.

Theorem 7-1

> If $X_1, X_2, \ldots, X_n$ is a random sample of size n from a normal distribution with mean μ and variance σ^2, the sample mean $\overline{X}$ is the MVUE for μ.

In situations in which we do not know whether an MVUE exists, we could still use a minimum variance principle to choose among competing estimators. Suppose, for example, we wish to estimate the mean of a population (not necessarily a *normal* population). We have a random sample of n observations $X_1, X_2, \ldots, X_n$ and we wish to compare two possible estimators for μ: the sample mean $\overline{X}$ and a single observation from the sample, say, X_i. Note that both $\overline{X}$ and X_i are unbiased estimators of μ; for the sample mean, we have $V(\overline{X}) = \sigma^2/n$ from Equation 5-40b and the variance of any observation is $V(X_i) = \sigma^2$. Since $V(\overline{X}) < V(X_i)$ for sample sizes $n \geq 2$, we would conclude that the sample mean is a better estimator of μ than a single observation X_i.

7-2.4 Standard Error: Reporting a Point Estimate

When the numerical value or point estimate of a parameter is reported, it is usually desirable to give some idea of the precision of estimation. The measure of precision usually employed is the standard error of the estimator that has been used.

Definition

> The **standard error** of an estimator $\hat{\Theta}$ is its standard deviation, given by $\sigma_{\hat{\Theta}} = \sqrt{V(\hat{\Theta})}$. If the standard error involves unknown parameters that can be estimated, substitution of those values into $\sigma_{\hat{\Theta}}$ produces an **estimated standard error**, denoted by $\hat{\sigma}_{\hat{\Theta}}$.

Sometimes the estimated standard error is denoted by $s_{\hat{\Theta}}$ or $se(\hat{\Theta})$.

Suppose we are sampling from a normal distribution with mean μ and variance σ^2. Now the distribution of $\overline{X}$ is normal with mean μ and variance σ^2/n, so the standard error of $\overline{X}$ is

$$\sigma_{\overline{X}} = \frac{\sigma}{\sqrt{n}}$$

If we did not know σ but substituted the sample standard deviation S into the above equation, the estimated standard error of $\overline{X}$ would be

$$\hat{\sigma}_{\overline{X}} = \frac{S}{\sqrt{n}}$$

When the estimator follows a normal distribution, as in the above situation, we can be reasonably confident that the true value of the parameter lies within two standard errors of the estimate. Since many point estimators are normally distributed (or approximately so) for large *n*, this is a very useful result. Even in cases in which the point estimator is not normally distributed, we can state that so long as the estimator is unbiased, the estimate of the parameter will deviate from the true value by as much as four standard errors at most 6 percent of the time. Thus a very conservative statement is that the true value of the parameter differs from the point estimate by at most four standard errors. See Chebyshev's inequality in the CD only material.

EXAMPLE 7-2

An article in the *Journal of Heat Transfer* (Trans. ASME, Sec. C, 96, 1974, p. 59) described a new method of measuring the thermal conductivity of Armco iron. Using a temperature of 100°F and a power input of 550 watts, the following 10 measurements of thermal conductivity (in Btu/hr-ft-°F) were obtained:

$$41.60, 41.48, 42.34, 41.95, 41.86,$$
$$42.18, 41.72, 42.26, 41.81, 42.04$$

A point estimate of the mean thermal conductivity at 100°F and 550 watts is the sample mean or

$$\bar{x} = 41.924 \text{ Btu/hr-ft-°F}$$

The standard error of the sample mean is $\sigma_{\bar{X}} = \sigma/\sqrt{n}$, and since σ is unknown, we may replace it by the sample standard deviation $s = 0.284$ to obtain the estimated standard error of $\bar{X}$ as

$$\hat{\sigma}_{\bar{X}} = \frac{s}{\sqrt{n}} = \frac{0.284}{\sqrt{10}} = 0.0898$$

Notice that the standard error is about 0.2 percent of the sample mean, implying that we have obtained a relatively precise point estimate of thermal conductivity. If we can assume that thermal conductivity is normally distributed, 2 times the standard error is $2\hat{\sigma}_{\bar{X}} = 2(0.0898) = 0.1796$, and we are highly confident that the true mean thermal conductivity is with the interval 41.924 ± 0.1756, or between 41.744 and 42.104.

7-2.5 Bootstrap Estimate of the Standard Error (CD Only)

7-2.6 Mean Square Error of an Estimator

Sometimes it is necessary to use a biased estimator. In such cases, the mean square error of the estimator can be important. The **mean square error** of an estimator $\hat{\Theta}$ is the expected squared difference between $\hat{\Theta}$ and θ.

Definition

> The **mean square error** of an estimator $\hat{\Theta}$ of the parameter θ is defined as
>
> $$\text{MSE}(\hat{\Theta}) = E(\hat{\Theta} - \theta)^2 \qquad (7\text{-}3)$$

The mean square error can be rewritten as follows:

$$\text{MSE}(\hat{\Theta}) = E[\hat{\Theta} - E(\hat{\Theta})]^2 + [\theta - E(\hat{\Theta})]^2$$
$$= V(\hat{\Theta}) + (\text{bias})^2$$

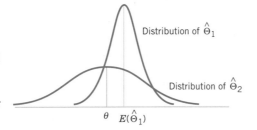

Figure 7-2   A biased estimator $\hat{\Theta}_1$ that has smaller variance than the unbiased estimator $\hat{\Theta}_2$.

That is, the mean square error of $\hat{\Theta}$ is equal to the variance of the estimator plus the squared bias. If $\hat{\Theta}$ is an unbiased estimator of θ, the mean square error of $\hat{\Theta}$ is equal to the variance of $\hat{\Theta}$.

The mean square error is an important criterion for comparing two estimators. Let $\hat{\Theta}_1$ and $\hat{\Theta}_2$ be two estimators of the parameter θ, and let MSE $(\hat{\Theta}_1)$ and MSE $(\hat{\Theta}_2)$ be the mean square errors of $\hat{\Theta}_1$ and $\hat{\Theta}_2$. Then the **relative efficiency** of $\hat{\Theta}_2$ to $\hat{\Theta}_1$ is defined as

$$\frac{\text{MSE}(\hat{\Theta}_1)}{\text{MSE}(\hat{\Theta}_2)} \tag{7-4}$$

If this relative efficiency is less than 1, we would conclude that $\hat{\Theta}_1$ is a more efficient estimator of θ than $\hat{\Theta}_2$, in the sense that it has a smaller mean square error.

Sometimes we find that biased estimators are preferable to unbiased estimators because they have smaller mean square error. That is, we may be able to reduce the variance of the estimator considerably by introducing a relatively small amount of bias. As long as the reduction in variance is greater than the squared bias, an improved estimator from a mean square error viewpoint will result. For example, Fig. 7-2 shows the probability distribution of a biased estimator $\hat{\Theta}_1$ that has a smaller variance than the unbiased estimator $\hat{\Theta}_2$. An estimate based on $\hat{\Theta}_1$ would more likely be close to the true value of θ than would an estimate based on $\hat{\Theta}_2$. Linear regression analysis (Chapters 11 and 12) is an area in which biased estimators are occasionally used.

An estimator $\hat{\Theta}$ that has a mean square error that is less than or equal to the mean square error of any other estimator, for all values of the parameter θ, is called an **optimal** estimator of θ. Optimal estimators rarely exist.

EXERCISES FOR SECTION 7-2

7-1. Suppose we have a random sample of size $2n$ from a population denoted by X, and $E(X) = \mu$ and $V(X) = \sigma^2$. Let

$$\overline{X}_1 = \frac{1}{2n} \sum_{i=1}^{2n} X_i \quad \text{and} \quad \overline{X}_2 = \frac{1}{n} \sum_{i=1}^{n} X_i$$

be two estimators of μ. Which is the better estimator of μ? Explain your choice.

7-2. Let $X_1, X_2, \ldots, X_7$ denote a random sample from a population having mean μ and variance σ^2. Consider the following estimators of μ:

$$\hat{\Theta}_1 = \frac{X_1 + X_2 + \cdots + X_7}{7}$$

$$\hat{\Theta}_2 = \frac{2X_1 - X_6 + X_4}{2}$$

(a) Is either estimator unbiased?

(b) Which estimator is best? In what sense is it best?

7-3. Suppose that $\hat{\Theta}_1$ and $\hat{\Theta}_2$ are unbiased estimators of the parameter θ. We know that $V(\hat{\Theta}_1) = 10$ and $V(\hat{\Theta}_2) = 4$. Which estimator is best and in what sense is it best?

7-4. Calculate the relative efficiency of the two estimators in Exercise 7-2.

7-5. Calculate the relative efficiency of the two estimators in Exercise 7-3.

7-6. Suppose that $\hat{\Theta}_1$ and $\hat{\Theta}_2$ are estimators of the parameter θ. We know that $E(\hat{\Theta}_1) = \theta$, $E(\hat{\Theta}_2) = \theta/2$, $V(\hat{\Theta}_1) = 10$, $V(\hat{\Theta}_2) = 4$. Which estimator is best? In what sense is it best?

7-7. Suppose that $\hat{\Theta}_1$, $\hat{\Theta}_2$, and $\hat{\Theta}_3$ are estimators of θ. We know that $E(\hat{\Theta}_1) = E(\hat{\Theta}_2) = \theta$, $E(\hat{\Theta}_3) \neq \theta$, $V(\hat{\Theta}_1) = 12$, $V(\hat{\Theta}_2) = 10$, and $E(\hat{\Theta}_3 - \theta)^2 = 6$. Compare these three estimators. Which do you prefer? Why?

7-8. Let three random samples of sizes $n_1 = 20$, $n_2 = 10$, and $n_3 = 8$ be taken from a population with mean μ and variance σ^2. Let S_1^2, S_2^2, and S_3^2 be the sample variances. Show that $S^2 = (20S_1^2 + 10S_2^2 + 8S_3^2)/38$ is an unbiased estimator of σ^2.

7-9. (a) Show that $\sum_{i=1}^{n} (X_i - \overline{X})^2/n$ is a biased estimator of σ^2.

(b) Find the amount of bias in the estimator.

(c) What happens to the bias as the sample size n increases?

7-10. Let $X_1, X_2, \ldots, X_n$ be a random sample of size n from a population with mean μ and variance σ^2.

(a) Show that $\overline{X}^2$ is a biased estimator for μ^2.

(b) Find the amount of bias in this estimator.

(c) What happens to the bias as the sample size n increases?

7-11. Data on pull-off force (pounds) for connectors used in an automobile engine application are as follows: 79.3, 75.1, 78.2, 74.1, 73.9, 75.0, 77.6, 77.3, 73.8, 74.6, 75.5, 74.0, 74.7, 75.9, 72.9, 73.8, 74.2, 78.1, 75.4, 76.3, 75.3, 76.2, 74.9, 78.0, 75.1, 76.8.

(a) Calculate a point estimate of the mean pull-off force of all connectors in the population. State which estimator you used and why.

(b) Calculate a point estimate of the pull-off force value that separates the weakest 50% of the connectors in the population from the strongest 50%.

(c) Calculate point estimates of the population variance and the population standard deviation.

(d) Calculate the standard error of the point estimate found in part (a). Provide an interpretation of the standard error.

(e) Calculate a point estimate of the proportion of all connectors in the population whose pull-off force is less than 73 pounds.

7-12. Data on oxide thickness of semiconductors are as follows: 425, 431, 416, 419, 421, 436, 418, 410, 431, 433, 423, 426, 410, 435, 436, 428, 411, 426, 409, 437, 422, 428, 413, 416.

(a) Calculate a point estimate of the mean oxide thickness for all wafers in the population.

(b) Calculate a point estimate of the standard deviation of oxide thickness for all wafers in the population.

(c) Calculate the standard error of the point estimate from part (a).

(d) Calculate a point estimate of the median oxide thickness for all wafers in the population.

(e) Calculate a point estimate of the proportion of wafers in the population that have oxide thickness greater than 430 angstrom.

7-13. Consider the continuous random variable with probability density function

$$f(x) = \frac{1}{2}(1 + \theta x), \quad -1 \le x \le 1, \; -1 \le \theta \le 1$$

(a) Find $E(X)$

(b) Suppose that $X_1, X_2, \ldots, X_n$ is a random sample from this distribution. Find an unbiased estimator of the parameter θ.

7-14. Suppose that X is the number of observed "successes" in a sample of n observations where p is the probability of success on each observation.

(a) Show that $\hat{P} = X/n$ is an unbiased estimator of p.

(b) Show that the standard error of $\hat{P}$ is $\sqrt{p(1-p)/n}$. How would you estimate the standard error?

7-15. $\overline{X}_1$ and S_1^2 are the sample mean and sample variance from a population with mean μ_1 and variance σ_1^2. Similarly, $\overline{X}_2$ and S_2^2 are the sample mean and sample variance from a second independent population with mean μ_2 and variance σ_2^2. The sample sizes are n_1 and n_2, respectively.

(a) Show that $\overline{X}_1 - \overline{X}_2$ is an unbiased estimator of $\mu_1 - \mu_2$.

(b) Find the standard error of $\overline{X}_1 - \overline{X}_2$. How could you estimate the standard error?

7-16. Continuation of Exercise 7-15. Suppose that both populations have the same variance; that is, $\sigma_1^2 = \sigma_2^2 = \sigma_2$. Show that

$$S_p^2 = \frac{(n_1 - 1)S_1^2 + (n_2 - 1)S_2^2}{n_1 + n_2 - 2}$$

is an unbiased estimator of σ^2.

7-17. Two different plasma etchers in a semiconductor factory have the same mean etch rate μ. However, machine 1 is newer than machine 2 and consequently has smaller variability in etch rate. We know that the variance of etch rate for machine 1 is σ_1^2 and for machine 2 it is $\sigma_2^2 = a\sigma_1^2$. Suppose that we have n_1 independent observations on etch rate from machine 1 and n_2 independent observations on etch rate from machine 2.

(a) Show that $\hat{\mu} = \alpha \overline{X}_1 + (1 - \alpha) \overline{X}_2$ is an unbiased estimator of μ for any value of α between 0 and 1.

(b) Find the standard error of the point estimate of μ in part (a).

(c) What value of α would minimize the standard error of the point estimate of μ?

(d) Suppose that $a = 4$ and $n_1 = 2n_2$. What value of α would you select to minimize the standard error of the point estimate of μ. How "bad" would it be to arbitrarily choose $\alpha = 0.5$ in this case?

7-18. Of n_1 randomly selected engineering students at ASU, X_1 owned an HP calculator, and of n_2 randomly selected engineering students at Virginia Tech X_2 owned an HP calculator. Let p_1 and p_2 be the probability that randomly selected ASU and Va. Tech engineering students, respectively, own HP calculators.

(a) Show that an unbiased estimate for $p_1 - p_2$ is $(X_1/n_1) - (X_2/n_2)$.

(b) What is the standard error of the point estimate in part (a)?

(c) How would you compute an estimate of the standard error found in part (b)?

(d) Suppose that $n_1 = 200$, $X_1 = 150$, $n_2 = 250$, and $X_2 = 185$. Use the results of part (a) to compute an estimate of $p_1 - p_2$.

(e) Use the results in parts (b) through (d) to compute an estimate of the standard error of the estimate.

7-3   METHODS OF POINT ESTIMATION

The definitions of unbiasness and other properties of estimators do not provide any guidance about how good estimators can be obtained. In this section, we discuss two methods for obtaining point estimators: the method of moments and the method of maximum likelihood. Maximum likelihood estimates are generally preferable to moment estimators because they have better efficiency properties. However, moment estimators are sometimes easier to compute. Both methods can produce unbiased point estimators.

7-3.1   Method of Moments

The general idea behind the method of moments is to equate **population moments,** which are defined in terms of expected values, to the corresponding **sample moments.** The population moments will be functions of the unknown parameters. Then these equations are solved to yield estimators of the unknown parameters.

Definition

> Let $X_1, X_2, \ldots, X_n$ be a random sample from the probability distribution $f(x)$, where $f(x)$ can be a discrete probability mass function or a continuous probability density function. The kth **population moment** (or **distribution moment**) is $E(X^k)$, $k = 1, 2, \ldots$. The corresponding kth **sample moment** is $(1/n) \sum_{i=1}^{n} X_i^k$, $k = 1, 2, \ldots$.

To illustrate, the first population moment is $E(X) = \mu$, and the first sample moment is $(1/n) \sum_{i=1}^{n} X_i = \overline{X}$. Thus by equating the population and sample moments, we find that $\hat{\mu} = \overline{X}$. That is, the sample mean is the **moment estimator** of the population mean. In the general case, the population moments will be functions of the unknown parameters of the distribution, say, $\theta_1, \theta_2, \ldots, \theta_m$.

Definition

> Let $X_1, X_2, \ldots, X_n$ be a random sample from either a probability mass function or probability density function with m unknown parameters $\theta_1, \theta_2, \ldots, \theta_m$. The **moment estimators** $\hat{\Theta}_1, \hat{\Theta}_2, \ldots, \hat{\Theta}_m$ are found by equating the first m population moments to the first m sample moments and solving the resulting equations for the unknown parameters.

EXAMPLE 7-3   Suppose that $X_1, X_2, \ldots, X_n$ is a random sample from an exponential distribution with parameter λ. Now there is only one parameter to estimate, so we must equate $E(X)$ to $\overline{X}$. For the exponential, $E(X) = 1/\lambda$. Therefore $E(X) = \overline{X}$ results in $1/\lambda = \overline{X}$, so $\hat{\lambda} = 1/\overline{X}$ is the moment estimator of λ.

As an example, suppose that the time to failure of an electronic module used in an automobile engine controller is tested at an elevated temperature to accelerate the failure mechanism.

The time to failure is exponentially distributed. Eight units are randomly selected and tested, resulting in the following failure time (in hours): $x_1 = 11.96$, $x_2 = 5.03$, $x_3 = 67.40$, $x_4 = 16.07$, $x_5 = 31.50$, $x_6 = 7.73$, $x_7 = 11.10$, and $x_8 = 22.38$. Because $\bar{x} = 21.65$, the moment estimate of λ is $\lambda = 1/\bar{x} = 1/21.65 = 0.0462$.

EXAMPLE 7-4

Suppose that $X_1, X_2, \ldots, X_n$ is a random sample from a normal distribution with parameters μ and σ^2. For the normal distribution $E(X) = \mu$ and $E(X^2) = \mu^2 + \sigma^2$. Equating $E(X)$ to $\bar{X}$ and $E(X^2)$ to $\frac{1}{n}\sum_{i=1}^{n} X_i^2$ gives

$$\mu = \bar{X}, \qquad \mu^2 + \sigma^2 = \frac{1}{n}\sum_{i=1}^{n} X_i^2$$

Solving these equations gives the moment estimators

$$\hat{\mu} = \bar{X}, \qquad \hat{\sigma}^2 = \frac{\sum_{i=1}^{n} X_i^2 - \left(\frac{1}{n}\sum_{i=1}^{n} X_i\right)^2}{n} = \frac{\sum_{i=1}^{n}(X_i - \bar{X})^2}{n}$$

Notice that the moment estimator of σ^2 is not an unbiased estimator.

EXAMPLE 7-5

Suppose that $X_1, X_2, \ldots, X_n$ is a random sample from a gamma distribution with parameters r and λ. For the gamma distribution $E(X) = r/\lambda$ and $E(X^2) = r(r + 1)/\lambda^2$. The moment estimators are found by solving

$$r/\lambda = \bar{X}, \quad r(r + 1)/\lambda^2 = \frac{1}{n}\sum_{i=1}^{n} X_i^2$$

The resulting estimators are

$$\hat{r} = \frac{\bar{X}^2}{(1/n)\sum_{i=1}^{n} X_i^2 - \bar{X}^2} \qquad \hat{\lambda} = \frac{\bar{X}}{(1/n)\sum_{i=1}^{n} X_i^2 - \bar{X}^2}$$

To illustrate, consider the time to failure data introduced following Example 7-3. For this data, $\bar{x} = 21.65$ and $\sum_{i=1}^{8} x_i^2 = 6639.40$, so the moment estimates are

$$\hat{r} = \frac{(21.65)^2}{(1/8)6639.40 - (21.65)^2} = 1.30, \qquad \hat{\lambda} = \frac{21.65}{(1/8)6639.40 - (21.65)^2} = 0.0599$$

When $r = 1$, the gamma reduces to the exponential distribution. Because $\hat{r}$ slightly exceeds unity, it is quite possible that either the gamma or the exponential distribution would provide a reasonable model for the data.

7-3.2 Method of Maximum Likelihood

One of the best methods of obtaining a point estimator of a parameter is the method of maximum likelihood. This technique was developed in the 1920s by a famous British statistician, Sir R. A. Fisher. As the name implies, the estimator will be the value of the parameter that maximizes the **likelihood function.**

Definition

> Suppose that X is a random variable with probability distribution $f(x; \theta)$, where θ is a single unknown parameter. Let $x_1, x_2, \ldots, x_n$ be the observed values in a random sample of size n. Then the **likelihood function** of the sample is
>
> $$L(\theta) = f(x_1; \theta) \cdot f(x_2; \theta) \cdot \cdots \cdot f(x_n; \theta) \tag{7-5}$$
>
> Note that the likelihood function is now a function of only the unknown parameter θ. The **maximum likelihood estimator** of θ is the value of θ that maximizes the likelihood function $L(\theta)$.

In the case of a discrete random variable, the interpretation of the likelihood function is clear. The likelihood function of the sample $L(\theta)$ is just the probability

$$P(X_1 = x_1, X_2 = x_2, \ldots, X_n = x_n)$$

That is, $L(\theta)$ is just the probability of obtaining the sample values $x_1, x_2, \ldots, x_n$. Therefore, in the discrete case, the maximum likelihood estimator is an estimator that maximizes the probability of occurrence of the sample values.

EXAMPLE 7-6

Let X be a Bernoulli random variable. The probability mass function is

$$f(x; p) = \begin{cases} p^x(1-p)^{1-x}, & x = 0, 1 \\ 0, & \text{otherwise} \end{cases}$$

where p is the parameter to be estimated. The likelihood function of a random sample of size n is

$$L(p) = p^{x_1}(1-p)^{1-x_1} p^{x_2}(1-p)^{1-x_2} \cdots p^{x_n}(1-p)^{1-x_n}$$
$$= \prod_{i=1}^{n} p^{x_i}(1-p)^{1-x_i} = p^{\sum_{i=1}^{n} x_i}(1-p)^{n-\sum_{i=1}^{n} x_i}$$

We observe that if $\hat{p}$ maximizes $L(p)$, $\hat{p}$ also maximizes $\ln L(p)$. Therefore,

$$\ln L(p) = \left(\sum_{i=1}^{n} x_i\right) \ln p + \left(n - \sum_{i=1}^{n} x_i\right) \ln(1-p)$$

Now

$$\frac{d \ln L(p)}{dp} = \frac{\sum_{i=1}^{n} x_i}{p} - \frac{\left(n - \sum_{i=1}^{n} x_i\right)}{1-p}$$

Equating this to zero and solving for p yields $\hat{p} = (1/n) \sum_{i=1}^{n} x_i$. Therefore, the maximum likelihood estimator of p is

$$\hat{P} = \frac{1}{n} \sum_{i=1}^{n} X_i$$

Suppose that this estimator was applied to the following situation: n items are selected at random from a production line, and each item is judged as either defective (in which case we set $x_i = 1$) or nondefective (in which case we set $x_i = 0$). Then $\sum_{i=1}^{n} x_i$ is the number of defective units in the sample, and $\hat{p}$ is the **sample proportion defective.** The parameter p is the **population proportion defective;** and it seems intuitively quite reasonable to use $\hat{p}$ as an estimate of p.

Although the interpretation of the likelihood function given above is confined to the discrete random variable case, the method of maximum likelihood can easily be extended to a continuous distribution. We now give two examples of maximum likelihood estimation for continuous distributions.

EXAMPLE 7-7

Let X be normally distributed with unknown μ and known variance σ^2. The likelihood function of a random sample of size n, say $X_1, X_2, \ldots, X_n$, is

$$L(\mu) = \prod_{i=1}^{n} \frac{1}{\sigma\sqrt{2\pi}} \, e^{-(x_i-\mu)^2/(2\sigma^2)} = \frac{1}{(2\pi\sigma^2)^{n/2}} \, e^{-(1/2\sigma^2)\sum_{i=1}^{n}(x_i-\mu)^2}$$

Now

$$\ln L(\mu) = -(n/2)\ln(2\pi\sigma^2) - (2\sigma^2)^{-1} \sum_{i=1}^{n} (x_i - \mu)^2$$

and

$$\frac{d \ln L(\mu)}{d\mu} = (\sigma^2)^{-1} \sum_{i=1}^{n} (x_i - \mu)$$

Equating this last result to zero and solving for μ yields

$$\hat{\mu} = \frac{\sum\limits_{i=1}^{n} X_i}{n} = \overline{X}$$

Thus the sample mean is the maximum likelihood estimator of μ. Notice that this is identical to the moment estimator.

EXAMPLE 7-8

Let X be exponentially distributed with parameter λ. The likelihood function of a random sample of size n, say $X_1, X_2, \ldots, X_n$, is

$$L(\lambda) = \prod_{i=1}^{n} \lambda e^{-\lambda x_i} = \lambda^n e^{-\lambda \sum_{i=1}^{n} x_i}$$

The log likelihood is

$$\ln L(\lambda) = n \ln \lambda - \lambda \sum_{i=1}^{n} x_i$$

Now

$$\frac{d \ln L(\lambda)}{d\lambda} = \frac{n}{\lambda} - \sum_{i=1}^{n} x_i$$

and upon equating this last result to zero we obtain

$$\hat{\lambda} = n/\sum_{i=1}^{n} X_i = 1/\overline{X}$$

Thus the maximum likelihood estimator of λ is the reciprocal of the sample mean. Notice that this is the same as the moment estimator.

It is easy to illustrate graphically just how the method of maximum likelihood works. Figure 7-3(a) plots the log of the likelihood function for the exponential parameter from Example 7-8, using the $n = 8$ observations on failure time given following Example 7-3. We found that the estimate of λ was $\hat{\lambda} = 0.0462$. From Example 7-8, we know that this is a maximum likelihood estimate. Figure 7-3(a) shows clearly that the log likelihood function is maximized at a value of λ that is approximately equal to 0.0462. Notice that the log likelihood function is relatively flat in the region of the maximum. This implies that the parameter is not estimated very precisely. If the parameter were estimated precisely, the log likelihood function would be very peaked at the maximum value. The sample size here is relatively small, and this has led to the imprecision in estimation. This is illustrated in Fig. 7-3(b) where we have plotted the difference in log likelihoods for the maximum value, assuming that the sample sizes were $n = 8$, 20, and 40 but that the sample average time to failure remained constant at $\overline{x} = 21.65$. Notice how much steeper the log likelihood is for $n = 20$ in comparsion to $n = 8$, and for $n = 40$ in comparison to both smaller sample sizes.

The method of maximum likelihood can be used in situations where there are several unknown parameters, say, $\theta_1, \theta_2, \ldots, \theta_k$ to estimate. In such cases, the likelihood function is a function of the k unknown parameters $\theta_1, \theta_2, \ldots, \theta_k$, and the maximum likelihood estimators $\{\hat{\Theta}_i\}$ would be found by equating the k partial derivatives $\partial L(\theta_1, \theta_2, \ldots, \theta_k)/\partial\theta_i$, $i = 1, 2, \ldots, k$ to zero and solving the resulting system of equations.

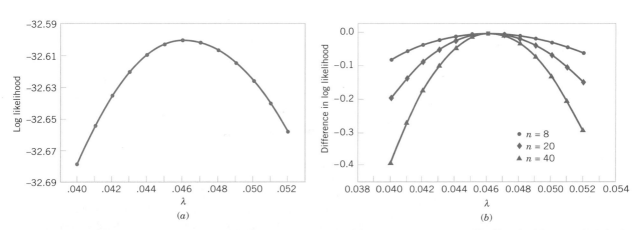

Figure 7-3   Log likelihood for the exponential distribution, using the failure time data. (a) Log likelihood with $n = 8$ (original data). (b) Log likelihood if $n = 8$, 20, and 40.

EXAMPLE 7-9   Let X be normally distributed with mean μ and variance σ^2, where both μ and σ^2 are unknown. The likelihood function for a random sample of size n is

$$L(\mu, \sigma^2) = \prod_{i=1}^{n} \frac{1}{\sigma\sqrt{2\pi}} \, e^{-(x_i - \mu)^2/(2\sigma^2)} = \frac{1}{(2\pi\sigma^2)^{n/2}} \, e^{-(1/2\sigma^2)\sum_{i=1}^{n}(x_i - \mu)^2}$$

and

$$\ln L(\mu, \sigma^2) = -\frac{n}{2} \ln(2\pi\sigma^2) - \frac{1}{2\sigma^2} \sum_{i=1}^{n} (x_i - \mu)^2$$

Now

$$\frac{\partial \ln L(\mu, \sigma^2)}{\partial \mu} = \frac{1}{\sigma^2} \sum_{i=1}^{n} (x_i - \mu) = 0$$

$$\frac{\partial \ln L(\mu, \sigma^2)}{\partial (\sigma^2)} = -\frac{n}{2\sigma^2} + \frac{1}{2\sigma^4} \sum_{i=1}^{n} (x_i - \mu)^2 = 0$$

The solutions to the above equation yield the maximum likelihood estimators

$$\hat{\mu} = \overline{X} \qquad \hat{\sigma}^2 = \frac{1}{n} \sum_{i=1}^{n} (X_i - \overline{X})^2$$

Once again, the maximum likelihood estimators are equal to the moment estimators.

Properties of the Maximum Likelihood Estimator

The method of maximum likelihood is often the estimation method that mathematical statisticians prefer, because it is usually easy to use and produces estimators with good statistical properties. We summarize these properties as follows.

Properties of the Maximum Likelihood Estimator

> Under very general and not restrictive conditions, when the sample size n is large and if $\hat{\Theta}$ is the maximum likelihood estimator of the parameter θ,
>
> (1) $\hat{\Theta}$ is an approximately unbiased estimator for θ $[E(\hat{\Theta}) \simeq \theta]$,
>
> (2) the variance of $\hat{\Theta}$ is nearly as small as the variance that could be obtained with any other estimator, and
>
> (3) $\hat{\Theta}$ has an approximate normal distribution.

Properties 1 and 2 essentially state that the maximum likelihood estimator is approximately an MVUE. This is a very desirable result and, coupled with the fact that it is fairly easy to obtain in many situations and has an asymptotic normal distribution (the "asymptotic" means "when n is large"), explains why the maximum likelihood estimation technique is widely used. To use maximum likelihood estimation, remember that the distribution of the population must be either known or assumed.

To illustrate the "large-sample" or asymptotic nature of the above properties, consider the maximum likelihood estimator for σ^2, the variance of the normal distribution, in Example 7-9. It is easy to show that

$$E(\hat{\sigma}^2) = \frac{n-1}{n}\sigma^2$$

The bias is

$$E(\hat{\sigma}^2) - \sigma^2 = \frac{n-1}{n}\sigma^2 - \sigma^2 = \frac{-\sigma^2}{n}$$

Because the bias is negative, $\hat{\sigma}^2$ tends to underestimate the true variance σ^2. Note that the bias approaches zero as n increases. Therefore, $\hat{\sigma}^2$ is an asymptotically unbiased estimator for σ^2.

We now give another very important and useful property of maximum likelihood estimators.

The Invariance Property

Let $\hat{\Theta}_1, \hat{\Theta}_2, \ldots, \hat{\Theta}_k$ be the maximum likelihood estimators of the parameters $\theta_1, \theta_2, \ldots, \theta_k$. Then the maximum likelihood estimator of any function $h(\theta_1, \theta_2, \ldots, \theta_k)$ of these parameters is the same function $h(\hat{\Theta}_1, \hat{\Theta}_2, \ldots, \hat{\Theta}_k)$ of the estimators $\hat{\Theta}_1, \hat{\Theta}_2, \ldots, \hat{\Theta}_k$.

EXAMPLE 7-10

In the normal distribution case, the maximum likelihood estimators of μ and σ^2 were $\hat{\mu} = \overline{X}$ and $\hat{\sigma}^2 = \sum_{i=1}^{n}(X_i - \overline{X})^2/n$. To obtain the maximum likelihood estimator of the function $h(\mu, \sigma^2) = \sqrt{\sigma^2} = \sigma$, substitute the estimators $\hat{\mu}$ and $\hat{\sigma}^2$ into the function h, which yields

$$\hat{\sigma} = \sqrt{\hat{\sigma}^2} = \left[\frac{1}{n}\sum_{i=1}^{n}(X_i - \overline{X})^2\right]^{1/2}$$

Thus, the maximum likelihood estimator of the standard deviation σ is *not* the sample standard deviation S.

Complications in Using Maximum Likelihood Estimation

While the method of maximum likelihood is an excellent technique, sometimes complications arise in its use. For example, it is not always easy to maximize the likelihood function because the equation(s) obtained from $dL(\theta)/d\theta = 0$ may be difficult to solve. Furthermore, it may not always be possible to use calculus methods directly to determine the maximum of $L(\theta)$. These points are illustrated in the following two examples.

EXAMPLE 7-11

Let X be uniformly distributed on the interval 0 to a. Since the density function is $f(x) = 1/a$ for $0 \leq x \leq a$ and zero otherwise, the likelihood function of a random sample of size n is

$$L(a) = \prod_{i=1}^{n}\frac{1}{a} = \frac{1}{a^n}$$

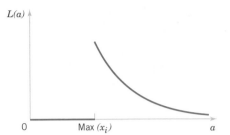

Figure 7-4 The like-
lihood function for the
uniform distribution in
Example 7-11.

if $0 \leq x_1 \leq a, 0 \leq x_2 \leq a, \ldots, 0 \leq x_n \leq a$. Note that the slope of this function is not zero anywhere. That is, as long as $\max(x_i) \leq a$, the likelihood is $1/a^n$, which is positive, but when $a < \max(x_i)$, the likelihood goes to zero, as illustrated in Fig. 7-4. Therefore, calculus methods cannot be used directly because the maximum value of the likelihood function occurs at a point of discontinuity. However, since $d/da(a^{-n}) = -n/a^{n+1}$ is less than zero for all values of $a > 0$, a^{-n} is a decreasing function of a. This implies that the maximum of the likelihood function $L(a)$ occurs at the lower boundary point. The figure clearly shows that we could maximize $L(a)$ by setting $\hat{a}$ equal to the smallest value that it could logically take on, which is $\max(x_i)$. Clearly, a cannot be smaller than the largest sample observation, so setting $\hat{a}$ equal to the largest sample value is reasonable.

EXAMPLE 7-12   Let $X_1, X_2, \ldots, X_n$ be a random sample from the gamma distribution. The log of the likelihood function is

$$\ln L(r, \lambda) = \ln \left(\prod_{i=1}^{n} \frac{\lambda^r X_i^{r-1} e^{-\lambda x_i}}{\Gamma(r)} \right)$$

$$= nr \ln(\lambda) + (r - 1) \sum_{i=1}^{n} \ln(x_i) - n \ln[\Gamma(r)] - \lambda \sum_{i=1}^{n} x_i$$

The derivatives of the log likelihood are

$$\frac{\partial \ln L(r, \lambda)}{\partial r} = n \ln(\lambda) + \sum_{i=1}^{n} \ln(x_i) - n \frac{\Gamma'(r)}{\Gamma(r)}$$

$$\frac{\partial \ln L(r, \lambda)}{\partial \lambda} = \frac{nr}{\lambda} - \sum_{i=1}^{n} x_i$$

When the derivatives are equated to zero, we obtain the equations that must be solved to find the maximum likelihood estimators of r and λ:

$$\hat{\lambda} = \frac{\hat{r}}{\bar{x}}$$

$$n \ln(\hat{\lambda}) + \sum_{i=1}^{n} \ln(x_i) = n \frac{\Gamma'(\hat{r})}{\Gamma(\hat{r})}$$

There is no closed form solution to these equations.

Figure 7-5 shows a graph of the log likelihood for the gamma distribution using the $n = 8$ observations on failure time introduced previously. Figure 7-5(a) shows the **log likelihood**

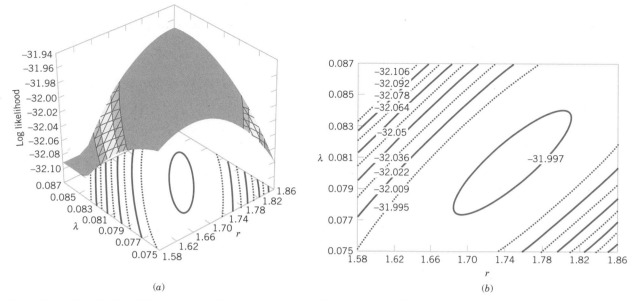

Figure 7-5   Log likelihood for the gamma distribution using the failure time data. (a) Log likelihood surface. (b) Contour plot.

surface as a function of r and λ, and Figure 7-5(b) is a **contour plot.** These plots reveal that the log likelihood is maximized at approximately $\hat{r} = 1.75$ and $\hat{\lambda} = 0.08$. Many statistics computer programs use numerical techniques to solve for the maximum likelihood estimates when no simple solution exists.

7-3.3   Bayesian Estimation of Parameters (CD Only)

EXERCISES FOR SECTION 7-3

7-19.   Consider the Poisson distribution

$$f(x) = \frac{e^{-\lambda}\lambda^x}{x!}, \qquad x = 0, 1, 2, \ldots$$

Find the maximum likelihood estimator of λ, based on a random sample of size n.

7-20.   Consider the shifted exponential distribution

$$f(x) = \lambda e^{-\lambda(x-\theta)}, \qquad x \geq \theta$$

When $\theta = 0$, this density reduces to the usual exponential distribution. When $\theta > 0$, there is only positive probability to the right of θ.

(a)   Find the maximum likelihood estimator of λ and θ, based on a random sample of size n.

(b)   Describe a practical situation in which one would suspect that the shifted exponential distribution is a plausible model.

7-21.   Let X be a geometric random variable with parameter p. Find the maximum likelihood estimator of p, based on a random sample of size n.

7-22.   Let X be a random variable with the following probability distribution:

$$f(x) = \begin{cases} (\theta + 1)x^{\theta}, & 0 \leq x \leq 1 \\ 0 & , \quad \text{otherwise} \end{cases}$$

Find the maximum likelihood estimator of θ, based on a random sample of size n.

7-23.   Consider the Weibull distribution

$$f(x) = \begin{cases} \dfrac{\beta}{\delta}\left(\dfrac{x}{\delta}\right)^{\beta-1} e^{-\left(\frac{x}{\delta}\right)^{\beta}}, & 0 < x \\ 0 & , \quad \text{otherwise} \end{cases}$$

(a)   Find the likelihood function based on a random sample of size n. Find the log likelihood.

(b) Show that the log likelihood is maximized by solving the equations

$$\beta = \left[\frac{\sum\limits_{i=1}^{n} x_i^\beta \ln(x_i)}{\sum\limits_{i=1}^{n} x_i^\beta} - \frac{\sum\limits_{i=1}^{n} \ln(x_i)}{n} \right]^{-1}$$

$$\delta = \left[\frac{\sum\limits_{i=1}^{n} x_i^\beta}{n} \right]^{1/\beta}$$

(c) What complications are involved in solving the two equations in part (b)?

7-24. Consider the probability distribution in Exercise 7-22. Find the moment estimator of θ.

7-25. Let $X_1, X_2, \ldots, X_n$ be uniformly distributed on the interval 0 to a. Show that the moment estimator of a is $\hat{a} = 2\overline{X}$. Is this an unbiased estimator? Discuss the reasonableness of this estimator.

7-26. Let $X_1, X_2, \ldots, X_n$ be uniformly distributed on the interval 0 to a. Recall that the maximum likelihood estimator of a is $\hat{a} = \max(X_i)$.
(a) Argue intuitively why $\hat{a}$ cannot be an unbiased estimator for a.
(b) Suppose that $E(\hat{a}) = na/(n + 1)$. Is it reasonable that $\hat{a}$ consistently underestimates a? Show that the bias in the estimator approaches zero as n gets large.
(c) Propose an unbiased estimator for a.
(d) Let $Y = \max(X_i)$. Use the fact that $Y \le y$ if and only if each $X_i \le y$ to derive the cumulative distribution function of Y. Then show that the probability density function of Y is

$$f(y) = \begin{cases} \dfrac{ny^{n-1}}{a^n}, & 0 \le y \le a \\ 0, & \text{otherwise} \end{cases}$$

Use this result to show that the maximum likelihood estimator for a is biased.

7-27. For the continuous distribution of the interval 0 to a, we have two unbiased estimators for a: the moment estimator $\hat{a}_1 = 2\overline{X}$ and $\hat{a}_2 = [(n + 1)/n] \max(X_i)$, where $\max(X_i)$ is the largest observation in a random sample of size n (see Exercise 7-26). It can be shown that $V(\hat{a}_1) = a^2/(3n)$ and that

$V(\hat{a}_2) = a^2/[n(n + 2)]$. Show that if $n > 1$, $\hat{a}_2$ is a better estimator than $\hat{a}$. In what sense is it a better estimator of a?

7-28. Consider the probability density function

$$f(x) = \frac{1}{\theta^2} xe^{-x/\theta}, \qquad 0 \le x < \infty, \quad 0 < \theta < \infty$$

Find the maximum likelihood estimator for θ.

7-29. The Rayleigh distribution has probability density function

$$f(x) = \frac{x}{\theta} e^{-x^2/2\theta}, \qquad x > 0, \qquad 0 < \theta < \infty$$

(a) It can be shown that $E(X^2) = 2\theta$. Use this information to construct an unbiased estimator for θ.
(b) Find the maximum likelihood estimator of θ. Compare your answer to part (a).
(c) Use the invariance property of the maximum likelihood estimator to find the maximum likelihood estimator of the median of the Raleigh distribution.

7-30. Consider the probability density function

$$f(x) = c(1 + \theta x), \quad -1 \le x \le 1$$

(a) Find the value of the constant c.
(b) What is the moment estimator for θ?
(c) Show that $\hat{\theta} = 3\overline{X}$ is an unbiased estimator for θ.
(d) Find the maximum likelihood estimator for θ.

7-31. Reconsider the oxide thickness data in Exercise 7-12 and suppose that it is reasonable to assume that oxide thickness is normally distributed.
(a) Use the results of Example 7-9 to compute the maximum likelihood estimates of μ and σ^2.
(b) Graph the likelihood function in the vicinity of $\hat{\mu}$ and $\hat{\sigma}^2$, the maximum likelihood estimates, and comment on its shape.

7-32. Continuation of Exercise 7-31. Suppose that for the situation of Exercise 7-12, the sample size was larger ($n = 40$) but the maximum likelihood estimates were numerically equal to the values obtained in Exercise 7-31. Graph the likelihood function for $n = 40$, compare it to the one from Exercise 7-31 (b), and comment on the effect of the larger sample size.

7-4   SAMPLING DISTRIBUTIONS

Statistical inference is concerned with making **decisions** about a population based on the information contained in a random sample from that population. For instance, we may be interested in the mean fill volume of a can of soft drink. The mean fill volume in the population is required to be 300 milliliters. An engineer takes a random sample of 25 cans and computes the sample average fill volume to be $\bar{x} = 298$ milliliters. The engineer will probably decide that the population mean is $\mu = 300$ milliliters, even though the sample mean was 298 milliliters because he or she knows that the sample mean is a reasonable estimate of μ and that a sample mean of 298 milliliters is very likely to occur, even if the true population mean is $\mu = 300$ milliliters. In fact, if the true mean is 300 milliliters, tests of 25 cans made repeatedly, perhaps every five minutes, would produce values of $\bar{x}$ that vary both above and below $\mu = 300$ milliliters.

The sample mean is a statistic; that is, it is a random variable that depends on the results obtained in each particular sample. Since a statistic is a random variable, it has a probability distribution.

Definition

> The probability distribution of a statistic is called a **sampling distribution.**

For example, the probability distribution of $\overline{X}$ is called the **sampling distribution of the mean.**

The sampling distribution of a statistic depends on the distribution of the population, the size of the sample, and the method of sample selection. The next section presents perhaps the most important sampling distribution. Other sampling distributions and their applications will be illustrated extensively in the following two chapters.

7-5   SAMPLING DISTRIBUTIONS OF MEANS

Consider determining the sampling distribution of the sample mean $\overline{X}$. Suppose that a random sample of size n is taken from a normal population with mean μ and variance σ^2. Now each observation in this sample, say, $X_1, X_2, \ldots, X_n$, is a normally and independently distributed random variable with mean μ and variance σ^2. Then by the reproductive property of the normal distribution, Equation 5-41 in Chapter 5, we conclude that the sample mean

$$\overline{X} = \frac{X_1 + X_2 + \cdots + X_n}{n}$$

has a normal distribution with mean

$$\mu_{\overline{X}} = \frac{\mu + \mu + \cdots + \mu}{n} = \mu$$

and variance

$$\sigma_{\overline{X}}^2 = \frac{\sigma^2 + \sigma^2 + \cdots + \sigma^2}{n^2} = \frac{\sigma^2}{n}$$

If we are sampling from a population that has an unknown probability distribution, the sampling distribution of the sample mean will still be approximately normal with mean μ and variance σ^2/n, if the sample size n is large. This is one of the most useful theorems in statistics, called the **central limit theorem.** The statement is as follows:

Theorem 7-2:
The Central
Limit Theorem

> If $X_1, X_2, \ldots, X_n$ is a random sample of size n taken from a population (either finite or infinite) with mean μ and finite variance σ^2, and if $\overline{X}$ is the sample mean, the limiting form of the distribution of
>
> $$Z = \frac{\overline{X} - \mu}{\sigma/\sqrt{n}} \qquad (7\text{-}6)$$
>
> as $n \to \infty$, is the standard normal distribution.

The normal approximation for $\overline{X}$ depends on the sample size n. Figure 7-6(a) shows the distribution obtained for throws of a single, six-sided true die. The probabilities are equal (1/6) for all the values obtained, 1, 2, 3, 4, 5, or 6. Figure 7-6(b) shows the distribution of the average score obtained when tossing two dice, and Fig. 7-6(c), 7-6(d), and 7-6(e) show the distributions of average scores obtained when tossing three, five, and ten dice, respectively. Notice that, while the population (one die) is relatively far from normal, the distribution of averages is approximated reasonably well by the normal distribution for sample sizes as small as five. (The dice throw distributions are discrete, however, while the normal is continuous). Although the central limit theorem will work well for small samples ($n = 4, 5$) in most cases, particularly where the population is continuous, unimodal, and symmetric, larger samples will be required in other situations, depending on the shape of the population. In many cases of practical interest, if $n \geq 30$, the normal approximation will be satisfactory regardless of the

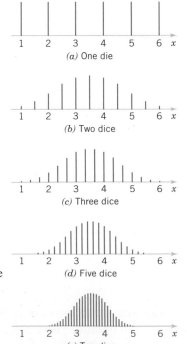

Figure 7-6
Distributions of average scores from throwing dice. [Adapted with permission from Box, Hunter, and Hunter (1978).]

shape of the population. If $n < 30$, the central limit theorem will work if the distribution of the population is not severely nonnormal.

EXAMPLE 7-13

An electronics company manufactures resistors that have a mean resistance of 100 ohms and a standard deviation of 10 ohms. The distribution of resistance is normal. Find the probability that a random sample of $n = 25$ resistors will have an average resistance less than 95 ohms.

Note that the sampling distribution of $\overline{X}$ is normal, with mean $\mu_{\overline{X}} = 100$ ohms and a standard deviation of

$$\sigma_{\overline{X}} = \frac{\sigma}{\sqrt{n}} = \frac{10}{\sqrt{25}} = 2$$

Therefore, the desired probability corresponds to the shaded area in Fig. 7-7. Standardizing the point $\overline{X} = 95$ in Fig. 7-7, we find that

$$z = \frac{95 - 100}{2} = -2.5$$

and therefore,

$$P(\overline{X} < 95) = P(Z < -2.5)$$
$$= 0.0062$$

The following example makes use of the central limit theorem.

EXAMPLE 7-14

Suppose that a random variable X has a continuous uniform distribution

$$f(x) = \begin{cases} 1/2, & 4 \leq x \leq 6 \\ 0, & \text{otherwise} \end{cases}$$

Find the distribution of the sample mean of a random sample of size $n = 40$.

The mean and variance of X are $\mu = 5$ and $\sigma^2 = (6 - 4)^2/12 = 1/3$. The central limit theorem indicates that the distribution of $\overline{X}$ is approximately normal with mean $\mu_{\overline{X}} = 5$ and variance $\sigma_{\overline{X}}^2 = \sigma^2/n = 1/[3(40)] = 1/120$. The distributions of X and $\overline{X}$ are shown in Fig. 7-8.

Now consider the case in which we have two independent populations. Let the first population have mean μ_1 and variance σ_1^2 and the second population have mean μ_2 and variance σ_2^2. Suppose that both populations are normally distributed. Then, using the fact that linear

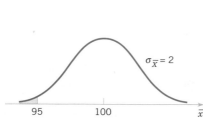

Figure 7-7   Probability distribution for Example 7-13.

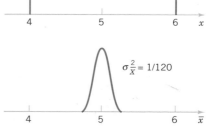

Figure 7-8   The distributions of X and $\overline{X}$ for Example 7-14.

combinations of independent normal random variables follow a normal distribution (see Equation 5-41), we can say that the sampling distribution of $\overline{X}_1 - \overline{X}_2$ is normal with mean

$$\mu_{\overline{X}_1 - \overline{X}_2} = \mu_{\overline{X}_1} - \mu_{\overline{X}_2} = \mu_1 - \mu_2 \tag{7-7}$$

and variance

$$\sigma^2_{\overline{X}_1 - \overline{X}_2} = \sigma^2_{\overline{X}_1} + \sigma^2_{\overline{X}_2} = \frac{\sigma^2_1}{n_1} + \frac{\sigma^2_2}{n_2} \tag{7-8}$$

If the two populations are not normally distributed and if both sample sizes n_1 and n_2 are greater than 30, we may use the central limit theorem and assume that $\overline{X}_1$ and $\overline{X}_2$ follow approximately independent normal distributions. Therefore, the sampling distribution of $\overline{X}_1 - \overline{X}_2$ is approximately normal with mean and variance given by Equations 7-7 and 7-8, respectively. If either n_1 or n_2 is less than 30, the sampling distribution of $\overline{X}_1 - \overline{X}_2$ will still be approximately normal with mean and variance given by Equations 7-7 and 7-8, provided that the population from which the small sample is taken is not dramatically different from the normal. We may summarize this with the following definition.

Definition

> If we have two independent populations with means μ_1 and μ_2 and variances σ^2_1 and σ^2_2 and if $\overline{X}_1$ and $\overline{X}_2$ are the sample means of two independent random samples of sizes n_1 and n_2 from these populations, then the sampling distribution of
>
> $$Z = \frac{\overline{X}_1 - \overline{X}_2 - (\mu_1 - \mu_2)}{\sqrt{\sigma^2_1/n_1 + \sigma^2_2/n_2}} \tag{7-9}$$
>
> is approximately standard normal, if the conditions of the central limit theorem apply. If the two populations are normal, the sampling distribution of Z is exactly standard normal.

EXAMPLE 7-15

The effective life of a component used in a jet-turbine aircraft engine is a random variable with mean 5000 hours and standard deviation 40 hours. The distribution of effective life is fairly close to a normal distribution. The engine manufacturer introduces an improvement into the manufacturing process for this component that increases the mean life to 5050 hours and decreases the standard deviation to 30 hours. Suppose that a random sample of $n_1 = 16$ components is selected from the "old" process and a random sample of $n_2 = 25$ components is selected from the "improved" process. What is the probability that the difference in the two sample means $\overline{X}_2 - \overline{X}_1$ is at least 25 hours? Assume that the old and improved processes can be regarded as independent populations.

To solve this problem, we first note that the distribution of $\overline{X}_1$ is normal with mean $\mu_1 = 5000$ hours and standard deviation $\sigma_1/\sqrt{n_1} = 40/\sqrt{16} = 10$ hours, and the distribution of $\overline{X}_2$ is normal with mean $\mu_2 = 5050$ hours and standard deviation $\sigma_2/\sqrt{n_2} = 30/\sqrt{25} = 6$ hours. Now the distribution of $\overline{X}_2 - \overline{X}_1$ is normal with mean $\mu_2 - \mu_1 = 5050 - 5000 = 50$ hours and variance $\sigma^2_2/n_2 + \sigma^2_1/n_1 = (6)^2 + (10)^2 = 136$ hours^2. This sampling distribution is shown in Fig. 7-9. The probability that $\overline{X}_2 - \overline{X}_1 \geq 25$ is the shaded portion of the normal distribution in this figure.

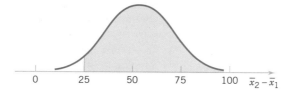

Figure 7-9   The sampling distribution of $\bar{X}_2 - \bar{X}_1$ in Example 7-15.

Corresponding to the value $\bar{x}_2 - \bar{x}_1 = 25$ in Fig. 7-9, we find that

$$z = \frac{25 - 50}{\sqrt{136}} = -2.14$$

and we find that

$$P(\bar{X}_2 - \bar{X}_1 \geq 25) = P(Z \geq -2.14)$$
$$= 0.9838$$

EXERCISES FOR SECTION 7-5

7-33.   PVC pipe is manufactured with a mean diameter of 1.01 inch and a standard deviation of 0.003 inch. Find the probability that a random sample of $n = 9$ sections of pipe will have a sample mean diameter greater than 1.009 inch and less than 1.012 inch.

7-34.   Suppose that samples of size $n = 25$ are selected at random from a normal population with mean 100 and standard deviation 10. What is the probability that the sample mean falls in the interval from $\mu_{\bar{X}} - 1.8\sigma_{\bar{X}}$ to $\mu_{\bar{X}} + 1.0\sigma_{\bar{X}}$?

7-35.   A synthetic fiber used in manufacturing carpet has tensile strength that is normally distributed with mean 75.5 psi and standard deviation 3.5 psi. Find the probability that a random sample of $n = 6$ fiber specimens will have sample mean tensile strength that exceeds 75.75 psi.

7-36.   Consider the synthetic fiber in the previous exercise. How is the standard deviation of the sample mean changed when the sample size is increased from $n = 6$ to $n = 49$?

7-37.   The compressive strength of concrete is normally distributed with $\mu = 2500$ psi and $\sigma = 50$ psi. Find the probability that a random sample of $n = 5$ specimens will have a sample mean diameter that falls in the interval from 2499 psi to 2510 psi.

7-38.   Consider the concrete specimens in the previous example. What is the standard error of the sample mean?

7-39.   A normal population has mean 100 and variance 25. How large must the random sample be if we want the standard error of the sample average to be 1.5?

7-40.   Suppose that the random variable X has the continuous uniform distribution

$$f(x) = \begin{cases} 1, & 0 \leq x \leq 1 \\ 0, & \text{otherwise} \end{cases}$$

Suppose that a random sample of $n = 12$ observations is selected from this distribution. What is the probability distribution of $\bar{X} - 6$? Find the mean and variance of this quantity.

7-41.   Suppose that X has a discrete uniform distribution

$$f(x) = \begin{cases} ^1/_3, & x = 1, 2, 3 \\ 0, & \text{otherwise} \end{cases}$$

A random sample of $n = 36$ is selected from this population. Find the probability that the sample mean is greater than 2.1 but less than 2.5, assuming that the sample mean would be measured to the nearest tenth.

7-42.   The amount of time that a customer spends waiting at an airport check-in counter is a random variable with mean 8.2 minutes and standard deviation 1.5 minutes. Suppose that a random sample of $n = 49$ customers is observed. Find the probability that the average time waiting in line for these customers is
(a) Less than 10 minutes
(b) Between 5 and 10 minutes
(c) Less than 6 minutes

7-43.   A random sample of size $n_1 = 16$ is selected from a normal population with a mean of 75 and a standard deviation of 8. A second random sample of size $n_2 = 9$ is taken from another normal population with mean 70 and standard deviation 12. Let $\bar{X}_1$ and $\bar{X}_2$ be the two sample means. Find
(a) The probability that $\bar{X}_1 - \bar{X}_2$ exceeds 4
(b) The probability that $3.5 \leq \bar{X}_1 - \bar{X}_2 \leq 5.5$

7-44.   A consumer electronics company is comparing the brightness of two different types of picture tubes for use in its television sets. Tube type A has mean brightness of 100 and standard deviation of 16, while tube type B has unknown mean brightness, but the standard deviation is assumed to be identical to that for type A. A random sample of $n = 25$ tubes of each type is selected, and $\bar{X}_B - \bar{X}_A$ is computed. If μ_B equals or exceeds μ_A, the manufacturer would like to adopt type B for use. The observed difference is $\bar{x}_B - \bar{x}_A = 3.5$. What decision would you make, and why?

7-45. The elasticity of a polymer is affected by the concentration of a reactant. When low concentration is used, the true mean elasticity is 55, and when high concentration is used the mean elasticity is 60. The standard deviation of elasticity is 4, regardless of concentration. If two random samples of size 16 are taken, find the probability that $\bar{X}_{\text{high}} - \bar{X}_{\text{low}} \geq 2$.

Supplemental Exercises

7-46. Suppose that a random variable is normally distributed with mean μ and variance σ^2, and we draw a random sample of five observations from this distribution. What is the joint probability density function of the sample?

7-47. Transistors have a life that is exponentially distributed with parameter λ. A random sample of n transistors is taken. What is the joint probability density function of the sample?

7-48. Suppose that X is uniformly distributed on the interval from 0 to 1. Consider a random sample of size 4 from X. What is the joint probability density function of the sample?

7-49. A procurement specialist has purchased 25 resistors from vendor 1 and 30 resistors from vendor 2. Let $X_{1,1}$, $X_{1,2}, \ldots, X_{1,25}$ represent the vendor 1 observed resistances, which are assumed to be normally and independently distributed with mean 100 ohms and standard deviation 1.5 ohms. Similarly, let $X_{2,1}, X_{2,2}, \ldots, X_{2,30}$ represent the vendor 2 observed resistances, which are assumed to be normally and independently distributed with mean 105 ohms and standard deviation of 2.0 ohms. What is the sampling distribution of $\bar{X}_1 - \bar{X}_2$?

7-50. Consider the resistor problem in Exercise 7-49. What is the standard error of $\bar{X}_1 - \bar{X}_2$?

7-51. A random sample of 36 observations has been drawn from a normal distribution with mean 50 and standard deviation 12. Find the probability that the sample mean is in the interval $47 \leq \bar{X} \leq 53$.

7-52. Is the assumption of normality important in Exercise 7-51? Why?

7-53. A random sample of $n = 9$ structural elements is tested for compressive strength. We know that the true mean compressive strength $\mu = 5500$ psi and the standard deviation is $\sigma = 100$ psi. Find the probability that the sample mean compressive strength exceeds 4985 psi.

7-54. A normal population has a known mean 50 and known variance $\sigma^2 = 2$. A random sample of $n = 16$ is selected from this population, and the sample mean is $\bar{x} = 52$. How unusual is this result?

7-55. A random sample of size $n = 16$ is taken from a normal population with $\mu = 40$ and $\sigma^2 = 5$. Find the probability that the sample mean is less than or equal to 37.

7-56. A manufacturer of semiconductor devices takes a random sample of 100 chips and tests them, classifying each chip as defective or nondefective. Let $X_i = 0$ if the chip is nondefective and $X_i = 1$ if the chip is defective. The sample fraction defective is

$$\hat{P} = \frac{X_1 + X_2 + \cdots + X_{100}}{100}$$

What is the sampling distribution of the random variable $\hat{P}$?

7-57. Let X be a random variable with mean μ and variance σ^2. Given two independent random samples of sizes n_1 and n_2, with sample means $\bar{X}_1$ and $\bar{X}_2$, show that

$$\bar{X} = a\bar{X}_1 + (1 - a)\bar{X}_2, \quad 0 < a < 1$$

is an unbiased estimator for μ. If $\bar{X}_1$ and $\bar{X}_2$ are independent, find the value of a that minimizes the standard error of $\bar{X}$.

7-58. A random variable x has probability density function

$$f(x) = \frac{1}{2\theta^3} x^2 e^{-x/\theta}, \quad 0 < x < \infty, \quad 0 < \theta < \infty$$

Find the maximum likelihood estimator for θ.

7-59. Let $f(x) = \theta x^{\theta - 1}$, $0 < \theta < \infty$, and $0 < x < 1$. Show that $\hat{\Theta} = -n/(\ln \prod_{i=1}^{n} X_i)$ is the maximum likelihood estimator for θ.

7-60. Let $f(x) = (1/\theta)x^{(1-\theta)/\theta}$, $0 < x < 1$, and $0 < \theta < \infty$. Show that $\hat{\Theta} = -(1/n) \sum_{i=1}^{n} \ln(X_i)$ is the maximum likelihood estimator for θ and that $\hat{\Theta}$ is an unbiased estimator for θ.

MIND-EXPANDING EXERCISES

7-61. A lot consists of N transistors, and of these M ($M \le N$) are defective. We randomly select two transistors without replacement from this lot and determine whether they are defective or nondefective. The random variable

$$
X_i = \begin{cases} 1, & \text{if the } i\text{th transistor} \\ & \text{is nondefective} \\ 0, & \text{if the } i\text{th transistor} \\ & \text{is defective} \end{cases} \quad i = 1, 2
$$

Determine the joint probability function for X_1 and X_2. What are the marginal probability functions for X_1 and X_2? Are X_1 and X_2 independent random variables?

7-62. When the sample standard deviation is based on a random sample of size n from a normal population, it can be shown that S is a biased estimator for σ. Specifically,

$$
E(S) = \sigma \sqrt{2/(n-1)} \; \Gamma(n/2)/\Gamma[(n-1)/2]
$$

(a) Use this result to obtain an unbiased estimator for σ of the form $c_n S$, when the constant c_n depends on the sample size n.

(b) Find the value of c_n for $n = 10$ and $n = 25$. Generally, how well does S perform as an estimator of σ for large n with respect to bias?

7-63. A collection of n randomly selected parts is measured twice by an operator using a gauge. Let X_i and Y_i denote the two measured values for the two parts. Assume that these two random variables are independent and normally distributed and that both have mean μ_i and variance σ^2. Consider the new random variable $Z_i = X_i - Y_i$.

(a) Show that the maximum likelihood estimator of σ^2 is

$$
\hat{\sigma}^2 = \frac{1}{2n} \sum_{i=1}^{n} (Y_i - X_i)^2 .
$$

(b) Is $\hat{\sigma}^2$ and unbiased estimator of σ^2?

7-64. **Consistent Estimator.** Another way to measure the closeness of an estimator $\hat{\Theta}$ to the parameter θ is in terms of consistency. If $\hat{\Theta}_n$ is an estimator of θ based on a random sample of n observations, $\hat{\Theta}_n$ is consistent for θ if

$$
\lim_{n \to \infty} P(|\hat{\Theta}_n - \theta| < \epsilon) = 1
$$

Thus, consistency is a large-sample property, describing the limiting behavior of $\hat{\Theta}_n$ as n tends to infinity. It is usually difficult to prove consistency using the above definition, although it can be done from other approaches. To illustrate, show that $\overline{X}$ is a consistent estimator of μ (when $\sigma^2 < \infty$) by using Chebyshev's inequality. See Section 5-10 (CD Only).

7-65. **Order Statistics.** Let $X_1, X_2, \ldots, X_n$ be a random sample of size n from X, a random variable having distribution function $F(x)$. Rank the elements in order of increasing numerical magnitude, resulting in $X_{(1)}$, $X_{(2)}, \ldots, X_{(n)}$, where $X_{(1)}$ is the smallest sample element ($X_{(1)} = \min\{X_1, X_2, \ldots, X_n\}$) and $X_{(n)}$ is the largest sample element ($X_{(n)} = \max\{X_1, X_2, \ldots, X_n\}$). $X_{(i)}$ is called the ith order statistic. Often the distribution of some of the order statistics is of interest, particularly the minimum and maximum sample values. $X_{(1)}$ and $X_{(n)}$, respectively. Prove that the cumulative distribution functions of these two order statistics, denoted respectively by $F_{X_{(1)}}(t)$ and $F_{X_{(n)}}(t)$ are

$$
F_{X_{(1)}}(t) = 1 - [1 - F(t)]^n
$$
$$
F_{X_{(n)}}(t) = [F(t)]^n
$$

Prove that if X is continuous with probability density function $f(x)$, the probability distributions of $X_{(1)}$ and $X_{(n)}$ are

$$
f_{X_{(1)}}(t) = n[1 - F(t)]^{n-1} f(t)
$$
$$
f_{X_{(n)}}(t) = n[F(t)]^{n-1} f(t)
$$

7-66. Continuation of Exercise 7-65. Let $X_1, X_2, \ldots,$ X_n be a random sample of a Bernoulli random variable with parameter p. Show that

$$
P(X_{(n)} = 1) = 1 - (1 - p)^n
$$
$$
P(X_{(1)} = 0) = 1 - p^n
$$

Use the results of Exercise 7-65.

7-67. Continuation of Exercise 7-65. Let $X_1, X_2, \ldots,$ X_n be a random sample of a normal random variable with mean μ and variance σ^2. Using the results of Exercise 7-65, derive the probability density functions of $X_{(1)}$ and $X_{(n)}$.

MIND-EXPANDING EXERCISES

7-68. Continuation of Exercise 7-65. Let $X_1, X_2, \ldots,$ X_n be a random sample of an exponential random variable of parameter λ. Derive the cumulative distribution functions and probability density functions for $X_{(1)}$ and $X_{(n)}$. Use the result of Exercise 7-65.

7-69. Let $X_1, X_2, \ldots, X_n$ be a random sample of a continuous random variable with cumulative distribution function $F(x)$. Find

$$E[F(X_{(n)})]$$

and

$$E[F(X_{(1)})]$$

7-70. Let X be a random variable with mean μ and variance σ^2, and let $X_1, X_2, \ldots, X_n$ be a random sample of size n from X. Show that the statistic $V = k \sum_{i=1}^{n-1} (X_{i+1} - X_i)^2$ is an unbiased estimator for σ^2 for an appropriate choice for the constant k. Find this value for k.

7-71. When the population has a normal distribution, the estimator

$$\hat{\sigma} = \text{median} (|X_1 - \overline{X}|, |X_2 - \overline{X}|,$$
$$\cdots, |X_n - \overline{X}|)/0.6745$$

is sometimes used to estimate the population standard deviation. This estimator is more robust to outliers than the usual sample standard deviation and usually does not differ much from S when there are no unusual observations.

(a) Calculate $\hat{\sigma}$ and S for the data 10, 12, 9, 14, 18, 15, and 16.

(b) Replace the first observation in the sample (10) with 50 and recalculate both S and $\hat{\sigma}$.

7-72.   Censored Data. A common problem in industry is life testing of components and systems. In this problem, we will assume that lifetime has an exponential distribution with parameter λ, so $\hat{\mu} = 1/\hat{\lambda} = \overline{X}$ is an unbiased estimate of μ. When n components are tested until failure and the data $X_1, X_2, \ldots, X_n$ represent actual lifetimes, we have a complete sample, and $\overline{X}$ is indeed an unbiased estimator of μ. However, in many situations, the components are only left under test until $r < n$ failures have occurred. Let Y_1 be the time of the first failure, Y_2 be the time of the second failure, $\ldots,$ and Y_r be the time of the last failure. This type of test results in **censored data.** There are $n - r$ units still running when the test is terminated. The total accumulated test time at termination is

$$T_r = \sum_{i=1}^{r} Y_i + (n - r)Y_r$$

(a) Show that $\hat{\mu} = T_r/r$ is an unbiased estimator for μ. [*Hint:* You will need to use the memoryless property of the exponential distribution and the results of Exercise 7-68 for the distribution of the minimum of a sample from an exponential distribution with parameter λ.]

(b) It can be shown that $V(T_r/r) = 1/(\lambda^2 r)$. How does this compare to $V(\overline{X})$ in the uncensored experiment?

IMPORTANT TERMS AND CONCEPTS

In the E-book, click on any term or concept below to go to that subject.

Bias in parameter estimation
Central limit theorem
Estimator versus estimate
Likelihood function
Maximum likelihood estimator

Mean square error of an estimator
Minimum variance unbiased estimator
Moment estimator
Normal distribution as the sampling distribution of a sample mean
Normal distribution as the sampling distribution of the differ-

ence in two sample means
Parameter estimation
Point estimator
Population or distribution moments
Sample moments
Sampling distribution
Standard error and estimated standard error of an estimator

Statistic
Statistical inference
Unbiased estimator

CD MATERIAL
Bayes estimator
Bootstrap
Posterior distribution
Prior distribution

Statistical Intervals
for a Single Sample

LEARNING OBJECTIVES

After careful study of this chapter, you should be able to do the following:

1. Construct confidence intervals on the mean of a normal distribution, using either the normal distribution or the t distribution method

2. Construct confidence intervals on the variance and standard deviation of a normal distribution

3. Construct confidence intervals on a population proportion

4. Construct prediction intervals for a future observation

5. Construct a tolerance interval for a normal population

6. Explain the three types of interval estimates: confidence intervals, prediction intervals, and tolerance intervals

7. Use the general method for constructing a confidence interval

CD MATERIAL

8. Use the bootstrap technique to construct a confidence interval

---

Answers for many odd numbered exercises are at the end of the book. Answers to exercises whose numbers are surrounded by a box can be accessed in the e-Text by clicking on the box. Complete worked solutions to certain exercises are also available in the e-Text. These are indicated in the Answers to Selected Exercises section by a box around the exercise number. Exercises are also available for some of the text sections that appear on CD only. These exercises may be found within the e-Text immediately following the section they accompany.

8-1   INTRODUCTION

In the previous chapter we illustrated how a parameter can be estimated from sample data. However, it is important to understand how good is the estimate obtained. For example, suppose that we estimate the mean viscosity of a chemical product to be $\hat{\mu} = \bar{x} = 1000$. Now because of sampling variability, it is almost never the case that $\mu = \bar{x}$. The point estimate says nothing about how close $\hat{\mu}$ is to μ. Is the process mean likely to be between 900 and 1100? Or is it likely to be between 990 and 1010? The answer to these questions affects our decisions regarding this process. Bounds that represent an interval of plausible values for a parameter are an example of an interval estimate. Surprisingly, it is easy to determine such intervals in many cases, and the same data that provided the point estimate are typically used.

An interval estimate for a population parameter is called a **confidence interval.** We cannot be certain that the interval contains the true, unknown population parameter—we only use a sample from the full population to compute the point estimate and the interval. However, the confidence interval is constructed so that we have high confidence that it does contain the unknown population parameter. Confidence intervals are widely used in engineering and the sciences.

A **tolerance interval** is another important type of interval estimate. For example, the chemical product viscosity data might be assumed to be normally distributed. We might like to calculate limits that bound 95% of the viscosity values. For a normal distribution, we know that 95% of the distribution is in the interval

$$\mu - 1.96\sigma, \mu + 1.96\sigma \tag{8-1}$$

However, this is not a useful tolerance interval because the parameters μ and σ are unknown. Point estimates such as $\bar{x}$ and s can be used in Equation 8-1 for μ and σ. However, we need to account for the potential error in each point estimate to form a tolerance interval for the distribution. The result is an interval of the form

$$\bar{x} - ks, \bar{x} + ks \tag{8-2}$$

where k is an appropriate constant (that is larger than 1.96 to account for the estimation error). As for a confidence interval, it is not certain that Equation 8-2 bounds 95% of the distribution, but the interval is constructed so that we have high confidence that it does. Tolerance intervals are widely used and, as we will subsequently see, they are easy to calculate for normal distributions.

Confidence and tolerance intervals bound unknown elements of a distribution. In this chapter you will learn to appreciate the value of these intervals. A **prediction interval** provides bounds on one (or more) future observations from the population. For example, a prediction interval could be used to bound a single, new measurement of viscosity—another useful interval. With a large sample size, the prediction interval for normally distributed data tends to the tolerance interval in Equation 8-1, but for more modest sample sizes the prediction and tolerance intervals are different.

Keep the purpose of the three types of interval estimates clear:

- A confidence interval bounds population or distribution parameters (such as the mean viscosity).

- A tolerance interval bounds a selected proportion of a distribution.

- A prediction interval bounds future observations from the population or distribution.

8-2   CONFIDENCE INTERVAL ON THE MEAN OF A NORMAL DISTRIBUTION, VARIANCE KNOWN

The basic ideas of a confidence interval (CI) are most easily understood by initially considering a simple situation. Suppose that we have a normal population with unknown mean μ and known variance σ^2. This is a somewhat unrealistic scenario because typically we know the distribution mean before we know the variance. However, in subsequent sections we will present confidence intervals for more general situations.

8-2.1   Development of the Confidence Interval and its Basic Properties

Suppose that $X_1, X_2, \ldots, X_n$ is a random sample from a normal distribution with unknown mean μ and known variance σ^2. From the results of Chapter 5 we know that the sample mean $\overline{X}$ is normally distributed with mean μ and variance σ^2/n. We may **standardize** $\overline{X}$ by subtracting the mean and dividing by the standard deviation, which results in the variable

$$Z = \frac{\overline{X} - \mu}{\sigma/\sqrt{n}} \tag{8-3}$$

Now Z has a standard normal distribution.

A **confidence interval** estimate for μ is an interval of the form $l \leq \mu \leq u$, where the end-points l and u are computed from the sample data. Because different samples will produce different values of l and u, these end-points are values of random variables L and U, respectively. Suppose that we can determine values of L and U such that the following probability statement is true:

$$P\{L \leq \mu \leq U\} = 1 - \alpha \tag{8-4}$$

where $0 \leq \alpha \leq 1$. There is a probability of $1 - \alpha$ of selecting a sample for which the CI will contain the true value of μ. Once we have selected the sample, so that $X_1 = x_1, X_2 = x_2, \ldots, X_n = x_n$, and computed l and u, the resulting **confidence interval** for μ is

$$l \leq \mu \leq u \tag{8-5}$$

The end-points or bounds l and u are called the **lower-** and **upper-confidence limits,** respectively, and $1 - \alpha$ is called the **confidence coefficient.**

In our problem situation, because $Z = (\bar{X} - \mu)/(\sigma/\sqrt{n})$ has a standard normal distribution, we may write

$$P\left\{-z_{\alpha/2} \leq \frac{\bar{X} - \mu}{\sigma/\sqrt{n}} \leq z_{\alpha/2}\right\} = 1 - \alpha$$

Now manipulate the quantities inside the brackets by (1) multiplying through by $\sigma/\sqrt{n}$, (2) subtracting $\bar{X}$ from each term, and (3) multiplying through by -1. This results in

$$P\left\{\bar{X} - z_{\alpha/2}\frac{\sigma}{\sqrt{n}} \leq \mu \leq \bar{X} + z_{\alpha/2}\frac{\sigma}{\sqrt{n}}\right\} = 1 - \alpha \tag{8-6}$$

From consideration of Equation 8-4, the lower and upper limits of the inequalities in Equation 8-6 are the lower- and upper-confidence limits L and U, respectively. This leads to the following definition.

Definition

> If $\bar{x}$ is the sample mean of a random sample of size n from a normal population with known variance σ^2, a $100(1 - \alpha)\%$ CI on μ is given by
>
> $$\bar{x} - z_{\alpha/2}\sigma/\sqrt{n} \leq \mu \leq \bar{x} + z_{\alpha/2}\sigma/\sqrt{n} \tag{8-7}$$
>
> where $z_{\alpha/2}$ is the upper $100\alpha/2$ percentage point of the standard normal distribution.

EXAMPLE 8-1

ASTM Standard E23 defines standard test methods for notched bar impact testing of metallic materials. The Charpy V-notch (CVN) technique measures impact energy and is often used to determine whether or not a material experiences a ductile-to-brittle transition with decreasing temperature. Ten measurements of impact energy (J) on specimens of A238 steel cut at 60°C are as follows: 64.1, 64.7, 64.5, 64.6, 64.5, 64.3, 64.6, 64.8, 64.2, and 64.3. Assume that impact energy is normally distributed with $\sigma = 1J$. We want to find a 95% CI for μ, the mean impact energy. The required quantities are $z_{\alpha/2} = z_{0.025} = 1.96$, $n = 10$, $\sigma = 1$, and $\bar{x} = 64.46$. The resulting 95% CI is found from Equation 8-7 as follows:

$$\bar{x} - z_{\alpha/2}\frac{\sigma}{\sqrt{n}} \leq \mu \leq \bar{x} + z_{\alpha/2}\frac{\sigma}{\sqrt{n}}$$

$$64.46 - 1.96\frac{1}{\sqrt{10}} \leq \mu \leq 64.46 + 1.96\frac{1}{\sqrt{10}}$$

$$63.84 \leq \mu \leq 65.08$$

That is, based on the sample data, a range of highly plausible vaules for mean impact energy for A238 steel at 60°C is $63.84J \leq \mu \leq 65.08J$.

Interpreting a Confidence Interval

How does one interpret a confidence interval? In the impact energy estimation problem in Example 8-1 the 95% CI is $63.84 \leq \mu \leq 65.08$, so it is tempting to conclude that μ is within this interval with probability 0.95. However, with a little reflection, it's easy to see that this cannot be correct; the true value of μ is unknown and the statement $63.84 \leq \mu \leq 65.08$ is either correct (true with probability 1) or incorrect (false with probability 1). The correct interpretation lies in the realization that a CI is **a random interval** because in the probability statement defining the end-points of the interval (Equation 8-4), L and U are random variables. Consequently, the correct interpretation of a $100(1 - \alpha)$% CI depends on the relative frequency view of probability. Specifically, if an infinite number of random samples are collected and a $100(1 - \alpha)$% confidence interval for μ is computed from each sample, $100(1 - \alpha)$% of these intervals will contain the true value of μ.

The situation is illustrated in Fig. 8-1, which shows several $100(1 - \alpha)$% confidence intervals for the mean μ of a normal distribution. The dots at the center of the intervals indicate the point estimate of μ (that is, $\bar{x}$). Notice that one of the intervals fails to contain the true value of μ. If this were a 95% confidence interval, in the long run only 5% of the intervals would fail to contain μ.

Now in practice, we obtain only one random sample and calculate one confidence interval. Since this interval either will or will not contain the true value of μ, it is not reasonable to attach a probability level to this specific event. The appropriate statement is the observed interval $[l, u]$ brackets the true value of μ with **confidence** $100(1 - \alpha)$. This statement has a frequency interpretation; that is, we don't know if the statement is true for this specific sample, but the **method** used to obtain the interval $[l, u]$ yields correct statements $100(1 - \alpha)$% of the time.

Confidence Level and Precision of Estimation

Notice in Example 8-1 that our choice of the 95% level of confidence was essentially arbitrary. What would have happened if we had chosen a higher level of confidence, say, 99%? In fact, doesn't it seem reasonable that we would want the higher level of confidence? At $\alpha = 0.01$, we find $z_{\alpha/2} = z_{0.01/2} = z_{0.005} = 2.58$, while for $\alpha = 0.05$, $z_{0.025} = 1.96$. Thus, the **length** of the 95% confidence interval is

$$2(1.96\sigma/\sqrt{n}) = 3.92\sigma/\sqrt{n}$$

whereas the length of the 99% CI is

$$2(2.58\sigma/\sqrt{n}) = 5.16\sigma/\sqrt{n}$$

Thus, the 99% CI is longer than the 95% CI. This is why we have a higher level of confidence in the 99% confidence interval. Generally, for a fixed sample size n and standard deviation σ, the higher the confidence level, the longer the resulting CI.

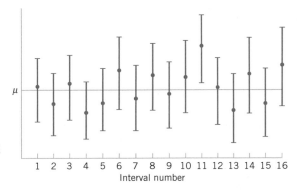

Figure 8-1 Repeated construction of a confidence interval for μ.

Interval number

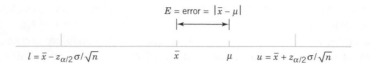

Figure 8-2   Error in
estimating μ with $\bar{x}$.

The length of a confidence interval is a measure of the **precision** of estimation. From the preceeding discussion, we see that precision is inversely related to the confidence level. It is desirable to obtain a confidence interval that is short enough for decision-making purposes and that also has adequate confidence. One way to achieve this is by choosing the sample size n to be large enough to give a CI of specified length or precision with prescribed confidence.

8-2.2   Choice of Sample Size

The precision of the confidence interval in Equation 8-7 is $2z_{\alpha/2}\sigma/\sqrt{n}$. This means that in using $\bar{x}$ to estimate μ, the error $E = |\bar{x} - \mu|$ is less than or equal to $z_{\alpha/2}\sigma/\sqrt{n}$ with confidence $100(1 - \alpha)$. This is shown graphically in Fig. 8-2. In situations where the sample size can be controlled, we can choose n so that we are $100(1 - \alpha)$ percent confident that the error in estimating μ is less than a specified bound on the error E. The appropriate sample size is found by choosing n such that $z_{\alpha/2}\sigma/\sqrt{n} = E$. Solving this equation gives the following formula for n.

Definition

> If $\bar{x}$ is used as an estimate of μ, we can be $100(1 - \alpha)\%$ confident that the error $|\bar{x} - \mu|$ will not exceed a specified amount E when the sample size is
>
> $$n = \left(\frac{z_{\alpha/2}\sigma}{E}\right)^2 \tag{8-8}$$

If the right-hand side of Equation 8-8 is not an integer, it must be rounded up. This will ensure that the level of confidence does not fall below $100(1 - \alpha)\%$. Notice that $2E$ is the length of the resulting confidence interval.

EXAMPLE 8-2   To illustrate the use of this procedure, consider the CVN test described in Example 8-1, and suppose that we wanted to determine how many specimens must be tested to ensure that the 95% CI on μ for A238 steel cut at 60°C has a length of at most 1.0 J. Since the bound on error in estimation E is one-half of the length of the CI, to determine n we use Equation 8-8 with $E = 0.5$, $\sigma = 1$, and $z_{\alpha/2} = 0.025$. The required sample size is 16

$$n = \left(\frac{z_{\alpha/2}\sigma}{E}\right)^2 = \left[\frac{(1.96)1}{0.5}\right]^2 = 15.37$$

and because n must be an integer, the required sample size is $n = 16$.

Notice the general relationship between sample size, desired length of the confidence interval $2E$, confidence level $100(1 - \alpha)$, and standard deviation σ:

- As the desired length of the interval $2E$ decreases, the required sample size n increases for a fixed value of σ and specified confidence.

- As σ increases, the required sample size n increases for a fixed desired length $2E$ and specified confidence.

- As the level of confidence increases, the required sample size n increases for fixed desired length $2E$ and standard deviation σ.

8-2.3 One-Sided Confidence Bounds

The confidence interval in Equation 8-7 gives both a lower confidence bound and an upper confidence bound for μ. Thus it provides a two-sided CI. It is also possible to obtain one-sided confidence bounds for μ by setting either $l = -\infty$ or $u = \infty$ and replacing $z_{\alpha/2}$ by z_{α}.

Definition

> A $100(1 - \alpha)\%$ **upper-confidence bound** for μ is
>
> $$\mu \leq u = \bar{x} + z_{\alpha}\sigma/\sqrt{n} \tag{8-9}$$
>
> and a $100(1 - \alpha)\%$ **lower-confidence bound** for μ is
>
> $$\bar{x} - z_{\alpha}\sigma/\sqrt{n} = l \leq \mu \tag{8-10}$$

8-2.4 General Method to Derive a Confidence Interval

It is easy to give a general method for finding a confidence interval for an unknown parameter θ. Let $X_1, X_2, \ldots, X_n$ be a random sample of n observations. Suppose we can find a statistic $g(X_1, X_2, \ldots, X_n; \theta)$ with the following properties:

1. $g(X_1, X_2, \ldots, X_n; \theta)$ depends on both the sample and θ.
2. The probability distribution of $g(X_1, X_2, \ldots, X_n; \theta)$ does not depend on θ or any other unknown parameter.

In the case considered in this section, the parameter $\theta = \mu$. The random variable $g(X_1, X_2, \ldots, X_n; \mu) = (\bar{X} - \mu)/(\sigma/\sqrt{n})$ and satisfies both conditions above; it depends on the sample and on μ, and it has a standard normal distribution since σ is known. Now one must find constants C_L and C_U so that

$$P[C_L \leq g(X_1, X_2, \ldots, X_n; \theta) \leq C_U] = 1 - \alpha \tag{8-11}$$

Because of property 2, C_L and C_U do not depend on θ. In our example, $C_L = -z_{\alpha/2}$ and $C_U = z_{\alpha/2}$. Finally, you must manipulate the inequalities in the probability statement so that

$$P[L(X_1, X_2, \ldots, X_n) \leq \theta \leq U(X_1, X_2, \ldots, X_n)] = 1 - \alpha \tag{8-12}$$

This gives $L(X_1, X_2, \ldots, X_n)$ and $U(X_1, X_2, \ldots, X_n)$ as the lower and upper confidence limits defining the $100(1 - \alpha)\%$ confidence interval for θ. The quantity $g(X_1, X_2, \ldots, X_n; \theta)$ is often called a "pivotal quantity" because we pivot on this quantity in Equation 8-11 to produce Equation 8-12. In our example, we manipulated the pivotal quantity $(\bar{X} - \mu)/(\sigma/\sqrt{n})$ to obtain $L(X_1, X_2, \ldots, X_n) = \bar{X} - z_{\alpha/2}\sigma/\sqrt{n}$ and $U(X_1, X_2, \ldots, X_n) = \bar{X} + z_{\alpha/2}\sigma/\sqrt{n}$.

8-2.5   A Large-Sample Confidence Interval for μ

We have assumed that the population distribution is normal with unknown mean and known standard deviation σ. We now present a **large-sample CI** and μ that does not require these assumptions. Let $X_1, X_2, \ldots, X_n$ be a random sample from a population with unknown mean μ and variance σ^2. Now if the sample size n is large, the central limit theorem implies that $\overline{X}$ has approximately a normal distribution with mean μ and variance σ^2/n. Therefore $Z = (\overline{X} - \mu)/(\sigma/\sqrt{n})$ has approximately a standard normal distribution. This ratio could be used as a pivotal quantity and manipulated as in Section 8-2.1 to produce an approximate CI for μ. However, the standard deviation σ is unknown. It turns out that when n is large, replacing σ by the sample standard deviation S has little effect on the distribution of Z. This leads to the following useful result.

Definition

When n is large, the quantity

$$\frac{\overline{X} - \mu}{S/\sqrt{n}}$$

has an approximate standard normal distribution. Consequently,

$$\overline{x} - z_{\alpha/2}\frac{s}{\sqrt{n}} \le \mu \le \overline{x} + z_{\alpha/2}\frac{s}{\sqrt{n}} \tag{8-13}$$

is a **large sample confidence interval** for μ, with confidence level of approximately $100(1 - \alpha)\%$.

Equation 8-13 holds regardless of the shape of the population distribution. Generally n should be at least 40 to use this result reliably. The central limit theorem generally holds for $n \ge 30$, but the larger sample size is recommended here because replacing σ by S in Z results in additional variability.

EXAMPLE 8-3

An article in the 1993 volume of the *Transactions of the American Fisheries Society* reports the results of a study to investigate the mercury contamination in largemouth bass. A sample of fish was selected from 53 Florida lakes and mercury concentration in the muscle tissue was measured (ppm). The mercury concentration values are

1.230	0.490	0.490	1.080	0.590	0.280	0.180	0.100	0.940
1.330	0.190	1.160	0.980	0.340	0.340	0.190	0.210	0.400
0.040	0.830	0.050	0.630	0.340	0.750	0.040	0.860	0.430
0.044	0.810	0.150	0.560	0.840	0.870	0.490	0.520	0.250
1.200	0.710	0.190	0.410	0.500	0.560	1.100	0.650	0.270
0.270	0.500	0.770	0.730	0.340	0.170	0.160	0.270	

The summary statistics from Minitab are displayed below:

Descriptive Statistics: Concentration

Variable	N	Mean	Median	TrMean	StDev	SE Mean
Concentration	53	0.5250	0.4900	0.5094	0.3486	0.0479

Variable	Minimum	Maximum	Q1	Q3
Concentration	0.0400	1.3300	0.2300	0.7900

Figure 8-3(a) and (b) presents the histogram and normal probability plot of the mercury concentration data. Both plots indicate that the distribution of mercury concentration is not normal and is positively skewed. We want to find an approximate 95% CI on μ. Because $n > 40$, the assumption of normality is not necessary to use Equation 8-13. The required quantities are $n = 53$, $\bar{x} = 0.5250$, $s = 0.3486$, and $z_{0.025} = 1.96$. The approximate 95% CI on μ is

$$\bar{x} - z_{0.025} \frac{s}{\sqrt{n}} \leq \mu \leq x + z_{0.025} \frac{s}{\sqrt{n}}$$

$$0.5250 - 1.96 \frac{0.3486}{\sqrt{53}} \leq \mu \leq 0.5250 + 1.96 \frac{0.3486}{\sqrt{53}}$$

$$0.4311 \leq \mu \leq 0.6189$$

This interval is fairly wide because there is a lot of variability in the mercury concentration measurements.

A General Large Sample Confidence Interval

The large-sample confidence interval for μ in Equation 8-13 is a special case of a more general result. Suppose that θ is a parameter of a probability distribution and let $\hat{\Theta}$ be an estimator of θ. If $\hat{\Theta}$ (1) has an approximate normal distribution, (2) is approximately unbiased

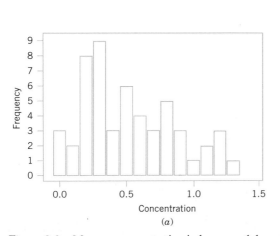

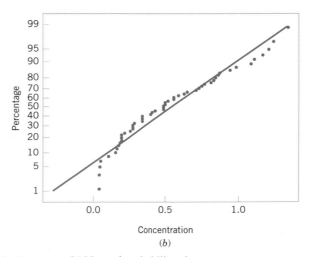

Figure 8-3   Mercury concentration in largemouth bass (a) Histogram. (b) Normal probability plot.

for θ, and (3) has standard deviation $\sigma_{\hat{\Theta}}$ that can be estimated from the sample data, then the quantity $(\hat{\Theta} - \theta)/\sigma_{\hat{\Theta}}$ has an approximate standard normal distribution. Then a **large-sample approximate CI for θ** is given by

$$\hat{\theta} - z_{\alpha/2}\sigma_{\hat{\Theta}} \le \theta \le \hat{\theta} + z_{\alpha/2}\sigma_{\hat{\Theta}} \tag{8-14}$$

Maximum likelihood estimators usually satisfy the three conditions listed above, so Equation 8-14 is often used when $\hat{\Theta}$ is the maximum likelihood estimator of θ. Finally, note that Equation 8-14 can be used even when $\sigma_{\hat{\Theta}}$ is a function of other unknown parameters (or of θ). Essentially, all one does is to use the sample data to compute estimates of the unknown parameters and substitute those estimates into the expression for $\sigma_{\hat{\Theta}}$.

8-2.6 Bootstrap Confidence Intervals (CD Only)

EXERCISES FOR SECTION 8-2

8-1. For a normal population with known variance σ^2, answer the following questions:
(a) What is the confidence level for the interval $\bar{x} - 2.14\sigma/\sqrt{n} \le \mu \le \bar{x} + 2.14\sigma/\sqrt{n}$?
(b) What is the confidence level for the interval $\bar{x} - 2.49\sigma/\sqrt{n} \le \mu \le \bar{x} + 2.49\sigma/\sqrt{n}$?
(c) What is the confidence level for the interval $\bar{x} - 1.85\sigma/\sqrt{n} \le \mu \le \bar{x} + 1.85\sigma/\sqrt{n}$?

8-2. For a normal population with known variance σ^2:
(a) What value of $z_{\alpha/2}$ in Equation 8-7 gives 98% confidence?
(b) What value of $z_{\alpha/2}$ in Equation 8-7 gives 80% confidence?
(c) What value of $z_{\alpha/2}$ in Equation 8-7 gives 75% confidence?

8-3. Consider the one-sided confidence interval expressions, Equations 8-9 and 8-10.
(a) What value of z_α would result in a 90% CI?   96.76
(b) What value of z_α would result in a 95% CI?   98 72
(c) What value of z_α would result in a 99% CI?   93 56

8-4. A confidence interval estimate is desired for the gain in a circuit on a semiconductor device. Assume that gain is normally distributed with standard deviation $\sigma = 20$.
(a) Find a 95% CI for μ when $n = 10$ and $\bar{x} = 1000$.
(b) Find a 95% CI for μ when $n = 25$ and $\bar{x} = 1000$.
(c) Find a 99% CI for μ when $n = 10$ and $\bar{x} = 1000$.
(d) Find a 99% CI for μ when $n = 25$ and $\bar{x} = 1000$.

8-5. Consider the gain estimation problem in Exercise 8-4. How large must n be if the length of the 95% CI is to be 40?

8-6. Following are two confidence interval estimates of the mean μ of the cycles to failure of an automotive door latch mechanism (the test was conducted at an elevated stress level to accelerate the failure).

$3124.9 \le \mu \le 3215.7 \qquad 3110.5 \le \mu \le 3230.1$

(a) What is the value of the sample mean cycles to failure?
(b) The confidence level for one of these CIs is 95% and the confidence level for the other is 99%. Both CIs are calculated from the same sample data. Which is the 95% CI? Explain why.

8-7. $n = 100$ random samples of water from a fresh water lake were taken and the calcium concentration (milligrams per liter) measured. A 95% CI on the mean calcium concentration is $0.49 \le \mu \le 0.82$.
(a) Would a 99% CI calculated from the same sample data been longer or shorter?
(b) Consider the following statement: There is a 95% chance that μ is between 0.49 and 0.82. Is this statement correct? Explain your answer.
(c) Consider the following statement: If $n = 100$ random samples of water from the lake were taken and the 95% CI on μ computed, and this process was repeated 1000 times, 950 of the CIs will contain the true value of μ. Is this statement correct? Explain your answer.

8-8. The breaking strength of yarn used in manufacturing drapery material is required to be at least 100 psi. Past experience has indicated that breaking strength is normally distributed and that $\sigma = 2$ psi. A random sample of nine specimens is tested, and the average breaking strength is found to be 98 psi. Find a 95% two-sided confidence interval on the true mean breaking strength.

8-9. The yield of a chemical process is being studied. From previous experience yield is known to be normally distributed and $\sigma = 3$. The past five days of plant operation have resulted in the following percent yields: 91.6, 88.75, 90.8, 89.95, and 91.3. Find a 95% two-sided confidence interval on the true mean yield.

8-10. The diameter of holes for cable harness is known to have a normal distribution with $\sigma = 0.01$ inch. A random sample of size 10 yields an average diameter of 1.5045 inch. Find a 99% two-sided confidence interval on the mean hole diameter.

8-11. A manufacturer produces piston rings for an automobile engine. It is known that ring diameter is normally distributed with $\sigma = 0.001$ millimeters. A random sample of 15 rings has a mean diameter of $\bar{x} = 74.036$ millimeters.

(a) Construct a 99% two-sided confidence interval on the mean piston ring diameter.

(b) Construct a 95% lower-confidence bound on the mean piston ring diameter.

8-12. The life in hours of a 75-watt light bulb is known to be normally distributed with $\sigma = 25$ hours. A random sample of 20 bulbs has a mean life of $\bar{x} = 1014$ hours.

(a) Construct a 95% two-sided confidence interval on the mean life.

(b) Construct a 95% lower-confidence bound on the mean life.

8-13. A civil engineer is analyzing the compressive strength of concrete. Compressive strength is normally distributed with $\sigma^2 = 1000(\text{psi})^2$. A random sample of 12 specimens has a mean compressive strength of $\bar{x} = 3250$ psi.

(a) Construct a 95% two-sided confidence interval on mean compressive strength.

(b) Construct a 99% two-sided confidence interval on mean compressive strength. Compare the width of this confidence interval with the width of the one found in part (a).

8-14. Suppose that in Exercise 8-12 we wanted to be 95% confident that the error in estimating the mean life is less than five hours. What sample size should be used?

8-15. Suppose that in Exercise 8-12 we wanted the total width of the two-sided confidence interval on mean life to be six hours at 95% confidence. What sample size should be used?

8-16. Suppose that in Exercise 8-13 it is desired to estimate the compressive strength with an error that is less than 15 psi at 99% confidence. What sample size is required?

8-17. By how much must the sample size n be increased if the length of the CI on μ in Equation 8-7 is to be halved?

8-18. If the sample size n is doubled, by how much is the length of the CI on μ in Equation 8-7 reduced? What happens to the length of the interval if the sample size is increased by a factor of four?

8-3 CONFIDENCE INTERVAL ON THE MEAN OF A NORMAL DISTRIBUTION, VARIANCE UNKNOWN

When we are constructing confidence intervals on the mean μ of a normal population when σ^2 is known, we can use the procedure in Section 8-2.1. This CI is also approximately valid (because of the central limit theorem) regardless of whether or not the underlying population is normal, so long as n is reasonably large ($n \geq 40$, say). As noted in Section 8-2.5, we can even handle the case of unknown variance for the large-sample-size situation. However, when the sample is small and σ^2 is unknown, we must make an assumption about the form of the underlying distribution to obtain a valid CI procedure. A reasonable assumption in many cases is that the underlying distribution is **normal.**

Many populations encountered in practice are well approximated by the normal distribution, so this assumption will lead to confidence interval procedures of wide applicability. In fact, moderate departure from normality will have little effect on validity. When the assumption is unreasonable, an alternate is to use the nonparametric procedures in Chapter 15 that are valid for any underlying distribution.

Suppose that the population of interest has a normal distribution with unknown mean μ and unknown variance σ^2. Assume that a random sample of size n, say $X_1, X_2, \ldots, X_n$, is available, and let $\bar{X}$ and S^2 be the sample mean and variance, respectively.

We wish to construct a two-sided CI on μ. If the variance σ^2 is known, we know that $Z = (\bar{X} - \mu)/(\sigma/\sqrt{n})$ has a standard normal distribution. When σ^2 is unknown, a logical procedure is to replace σ with the sample standard deviation S. The random variable Z now becomes $T = (\bar{X} - \mu)/(S/\sqrt{n})$. A logical question is what effect does replacing σ by S have on the distribution of the random variable T? If n is large, the answer to this question is "very little," and we can proceed to use the confidence interval based on the normal distribution from

Section 8-2.5. However, n is usually small in most engineering problems, and in this situation a different distribution must be employed to construct the CI.

8-3.1   The t Distribution

Definition

> Let $X_1, X_2, \ldots, X_n$ be a random sample from a normal distribution with unknown mean μ and unknown variance σ^2. The random variable
>
> $$T = \frac{\bar{X} - \mu}{S/\sqrt{n}} \tag{8-15}$$
>
> has a t distribution with $n - 1$ degrees of freedom.

The t probability density function is

$$f(x) = \frac{\Gamma[(k + 1)/2]}{\sqrt{\pi k}\,\Gamma(k/2)} \cdot \frac{1}{[(x^2/k) + 1]^{(k+1)/2}} \qquad -\infty < x < \infty \tag{8-16}$$

where k is the number of degrees of freedom. The mean and variance of the t distribution are zero and $k/(k - 2)$ (for $k > 2$), respectively.

Several t distributions are shown in Fig. 8-4. The general appearance of the t distribution is similar to the standard normal distribution in that both distributions are symmetric and unimodal, and the maximum ordinate value is reached when the mean $\mu = 0$. However, the t distribution has heavier tails than the normal; that is, it has more probability in the tails than the normal distribution. As the number of degrees of freedom $k \to \infty$, the limiting form of the t distribution is the standard normal distribution. Generally, the number of degrees of freedom for t are the number of degrees of freedom associated with the estimated standard deviation.

Appendix Table IV provides **percentage points** of the t distribution. We will let $t_{\alpha,k}$ be the value of the random variable T with k degrees of freedom above which we find an area (or probability) α. Thus, $t_{\alpha,k}$ is an upper-tail 100α percentage point of the t distribution with k degrees of freedom. This percentage point is shown in Fig. 8-5. In the Appendix Table IV the α values are the column headings, and the degrees of freedom are listed in the left column. To

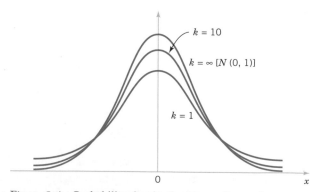

Figure 8-4   Probability density functions of several t distributions.

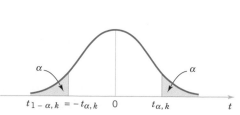

Figure 8-5   Percentage points of the t distribution.

illustrate the use of the table, note that the t-value with 10 degrees of freedom having an area of 0.05 to the right is $t_{0.05,10} = 1.812$. That is,

$$P(T_{10} > t_{0.05,10}) = P(T_{10} > 1.812) = 0.05$$

Since the t distribution is symmetric about zero, we have $t_{1-\alpha} = -t_\alpha$; that is, the t-value having an area of $1 - \alpha$ to the right (and therefore an area of α to the left) is equal to the negative of the t-value that has area α in the right tail of the distribution. Therefore, $t_{0.95,10} = -t_{0.05,10} = -1.812$. Finally, because t_∞ is the standard normal distribution, the familiar z_α values appear in the last row of Appendix Table IV.

8-3.2   Development of the t Distribution (CD Only)

8-3.3   The t Confidence Interval on μ

It is easy to find a $100(1 - \alpha)$ percent confidence interval on the mean of a normal distribution with unknown variance by proceeding essentially as we did in Section 8-2.1. We know that the distribution of $T = (\overline{X} - \mu)/(S/\sqrt{n})$ is t with $n - 1$ degrees of freedom. Letting $t_{\alpha/2,n-1}$ be the upper $100\alpha/2$ percentage point of the t distribution with $n - 1$ degrees of freedom, we may write:

$$P(-t_{\alpha/2,n-1} \le T \le t_{\alpha/2,n-1}) = 1 - \alpha$$

or

$$P\left(-t_{\alpha/2,n-1} \le \frac{\overline{X} - \mu}{S/\sqrt{n}} \le t_{\alpha/2,n-1}\right) = 1 - \alpha$$

Rearranging this last equation yields

$$P(\overline{X} - t_{\alpha/2,n-1}S/\sqrt{n} \le \mu \le \overline{X} + t_{\alpha/2,n-1}S/\sqrt{n}) = 1 - \alpha \qquad (8\text{-}17)$$

This leads to the following definition of the $100(1 - \alpha)$ percent two-sided confidence interval on μ.

Definition

> If $\overline{x}$ and s are the mean and standard deviation of a random sample from a normal distribution with unknown variance σ^2, a **$100(1 - \alpha)$ percent confidence interval on μ** is given by
>
> $$\overline{x} - t_{\alpha/2,n-1}s/\sqrt{n} \le \mu \le \overline{x} + t_{\alpha/2,n-1}s/\sqrt{n} \qquad (8\text{-}18)$$
>
> where $t_{\alpha/2,n-1}$ is the upper $100\alpha/2$ percentage point of the t distribution with $n - 1$ degrees of freedom.

One-sided confidence bounds on the mean of a normal distribution are also of interest and are easy to find. Simply use only the appropriate lower or upper confidence limit from Equation 8-18 and replace $t_{\alpha/2,n-1}$ by $t_{\alpha,n-1}$.

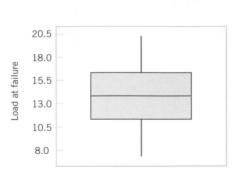

Figure 8-6   Box and whisker plot for the load at failure data in Example 8-4.

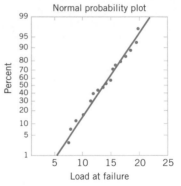

Figure 8-7   Normal probability plot of the load at failure data from Example 8-4.

EXAMPLE 8-4

An article in the journal *Materials Engineering* (1989, Vol. II, No. 4, pp. 275–281) describes the results of tensile adhesion tests on 22 U-700 alloy specimens. The load at specimen failure is as follows (in megapascals):

19.8	10.1	14.9	7.5	15.4	15.4
15.4	18.5	7.9	12.7	11.9	11.4
11.4	14.1	17.6	16.7	15.8	
19.5	8.8	13.6	11.9	11.4	

The sample mean is $\bar{x} = 13.71$, and the sample standard deviation is $s = 3.55$. Figures 8-6 and 8-7 show a box plot and a normal probability plot of the tensile adhesion test data, respectively. These displays provide good support for the assumption that the population is normally distributed. We want to find a 95% CI on μ. Since $n = 22$, we have $n - 1 = 21$ degrees of freedom for t, so $t_{0.025,21} = 2.080$. The resulting CI is

$$\bar{x} - t_{\alpha/2,n-1}s/\sqrt{n} \le \mu \le \bar{x} + t_{\alpha/2,n-1}s/\sqrt{n}$$

$$13.71 - 2.080(3.55)/\sqrt{22} \le \mu \le 13.71 + 2.080(3.55)/\sqrt{22}$$

$$13.71 - 1.57 \le \mu \le 13.71 + 1.57$$

$$12.14 \le \mu \le 15.28$$

The CI is fairly wide because there is a lot of variability in the tensile adhesion test measurements.

It is not as easy to select a sample size n to obtain a specified length (or precision of estimation) for this CI as it was in the known-σ case because the length of the interval involves s (which is unknown before the data are collected), n, and $t_{\alpha/2,n-1}$. Note that the t-percentile depends on the sample size n. Consequently, an appropriate n can only be obtained through trial and error. The results of this will, of course, also depend on the reliability of our prior "guess" for σ.

EXERCISES FOR SECTION 8-3

8-19. Find the values of the following percentiles: $t_{0.025,15}$, $t_{0.05,10}$, $t_{0.10,20}$, $t_{0.005,25}$, and $t_{0.001,30}$.

8-20. Determine the t-percentile that is required to construct each of the following two-sided confidence intervals:
(a) Confidence level = 95%, degrees of freedom = 12
(b) Confidence level = 95%, degrees of freedom = 24
(c) Confidence level = 99%, degrees of freedom = 13
(d) Confidence level = 99.9%, degrees of freedom = 15

8-21. Determine the t-percentile that is required to construct each of the following one-sided confidence intervals:
(a) Confidence level = 95%, degrees of freedom = 14
(b) Confidence level = 99%, degrees of freedom = 19
(c) Confidence level = 99.9%, degrees of freedom = 24

8-22. A research engineer for a tire manufacturer is investigating tire life for a new rubber compound and has built 16 tires and tested them to end-of-life in a road test. The sample mean and standard deviation are 60,139.7 and 3645.94 kilometers. Find a 95% confidence interval on mean tire life.

8-23. An Izod impact test was performed on 20 specimens of PVC pipe. The sample mean is $\bar{x} = 1.25$ and the sample standard deviation is $s = 0.25$. Find a 99% lower confidence bound on Izod impact strength.

8-24. The brightness of a television picture tube can be evaluated by measuring the amount of current required to achieve a particular brightness level. A sample of 10 tubes results in $\bar{x} = 317.2$ and $s = 15.7$. Find (in microamps) a 99% confidence interval on mean current required. State any necessary assumptions about the underlying distribution of the data.

8-25. A particular brand of diet margarine was analyzed to determine the level of polyunsaturated fatty acid (in percentages). A sample of six packages resulted in the following data: 16.8, 17.2, 17.4, 16.9, 16.5, 17.1.

(a) Is there evidence to support the assumption that the level of polyunsaturated fatty acid is normally distributed?

(b) Find a 99% confidence interval on the mean μ. Provide a practical interpretation of this interval.

8-26. The compressive strength of concrete is being tested by a civil engineer. He tests 12 specimens and obtains the following data.

2216	2237	2249	2204
2225	2301	2281	2263
2318	2255	2275	2295

(a) Is there evidence to support the assumption that compressive strength is normally distributed? Does this data set support your point of view? Include a graphical display in your answer.

(b) Construct a 95% two-sided confidence interval on the mean strength.

(c) Construct a 95% lower-confidence bound on the mean strength.

8-27. A machine produces metal rods used in an automobile suspension system. A random sample of 15 rods is selected, and the diameter is measured. The resulting data (in millimeters) are as follows:

8.24	8.25	8.20	8.23	8.24
8.21	8.26	8.26	8.20	8.25
8.23	8.23	8.19	8.28	8.24

(a) Check the assumption of normality for rod diameter.

(b) Find a 95% two-sided confidence interval on mean rod diameter.

8-28. Rework Exercise 8-27 to compute a 95% lower confidence bound on rod diameter. Compare this bound with the lower limit of the two-sided confidence limit from Exercise 8-27. Discuss why they are different.

8-29. The wall thickness of 25 glass 2-liter bottles was measured by a quality-control engineer. The sample mean was $\bar{x} = 4.05$ millimeters, and the sample standard deviation was $s = 0.08$ millimeter. Find a 95% lower confidence bound for mean wall thickness. Interpret the interval you have obtained.

8-30. An article in *Nuclear Engineering International* (February 1988, p. 33) describes several characteristics of fuel rods used in a reactor owned by an electric utility in Norway. Measurements on the percentage of enrichment of 12 rods were reported as follows:

2.94	3.00	2.90	2.75	3.00	2.95
2.90	2.75	2.95	2.82	2.81	3.05

(a) Use a normal probability plot to check the normality assumption.

(b) Find a 99% two-sided confidence interval on the mean percentage of enrichment. Are you comfortable with the statement that the mean percentage of enrichment is 2.95 percent? Why?

8-31. A postmix beverage machine is adjusted to release a certain amount of syrup into a chamber where it is mixed with carbonated water. A random sample of 25 beverages was found to have a mean syrup content of $\bar{x} = 1.10$ fluid ounces and a standard deviation of $s = 0.015$ fluid ounces. Find a 95% CI on the mean volume of syrup dispensed.

8-32. An article in the *Journal of Composite Materials* (December 1989, Vol 23, p. 1200) describes the effect of delamination on the natural frequency of beams made from composite laminates. Five such delaminated beams were subjected to loads, and the resulting frequencies were as follows (in hertz):

230.66, 233.05, 232.58, 229.48, 232.58

Find a 90% two-sided confidence interval on mean natural frequency. Is there evidence to support the assumption of normality in the population?

8-4 CONFIDENCE INTERVAL ON THE VARIANCE AND STANDARD DEVIATION OF A NORMAL POPULATION

Sometimes confidence intervals on the population variance or standard deviation are needed. When the population is modeled by a normal distribution, the tests and intervals described in this section are applicable. The following result provides the basis of constructing these confidence intervals.

Definition

Let $X_1, X_2, \ldots, X_n$ be a random sample from a normal distribution with mean μ and variance σ^2, and let S^2 be the sample variance. Then the random variable

$$X^2 = \frac{(n-1)\,S^2}{\sigma^2} \tag{8-19}$$

has a chi-square (χ^2) distribution with $n-1$ degrees of freedom.

The probability density function of a χ^2 random variable is

$$f(x) = \frac{1}{2^{k/2}\Gamma(k/2)}\, x^{(k/2)-1} e^{-x/2} \qquad x > 0 \tag{8-20}$$

where k is the number of degrees of freedom. The mean and variance of the χ^2 distribution are k and $2k$, respectively. Several chi-square distributions are shown in Fig. 8-8. Note that the chi-square random variable is nonnegative and that the probability distribution is skewed to the right. However, as k increases, the distribution becomes more symmetric. As $k \to \infty$, the limiting form of the chi-square distribution is the normal distribution.

The **percentage points** of the χ^2 distribution are given in Table III of the Appendix. Define $\chi^2_{\alpha,k}$ as the percentage point or value of the chi-square random variable with k degrees of freedom such that the probability that X^2 exceeds this value is α. That is,

$$P(X^2 > \chi^2_{\alpha,k}) = \int_{\chi^2_{\alpha,k}}^{\infty} f(u)\,du = \alpha$$

This probability is shown as the shaded area in Fig. 8-9(a). To illustrate the use of Table III, note that the areas α are the column headings and the degrees of freedom k are given in the left column. Therefore, the value with 10 degrees of freedom having an area (probability) of 0.05 to the right is $\chi^2_{0.05,10} = 18.31$. This value is often called an **upper** 5% point of chi-square with

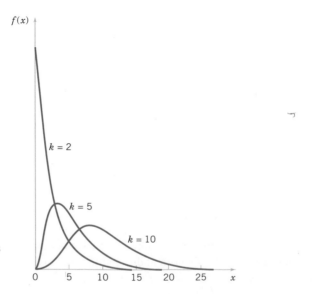

Figure 8-8   Probability density functions of several χ^2 distributions.

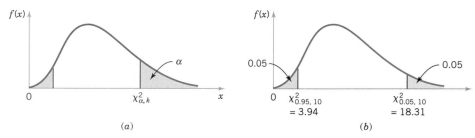

Figure 8-9   Percentage point of the χ^2 distribution. (a) The percentage point $\chi^2_{\alpha,k}$. (b) The upper percentage point $\chi^2_{0.05,10} = 18.31$ and the lower percentage point $\chi^2_{0.95,10} = 3.94$.

10 degrees of freedom. We may write this as a probability statement as follows:

$$P(X^2 > \chi^2_{0.05,10}) = P(X^2 > 18.31) = 0.05$$

Conversely, a **lower** 5% point of chi-square with 10 degrees of freedom would be $\chi^2_{0.95,10} = 3.94$ (from Appendix Table III). Both of these percentage points are shown in Figure 8-9(b).

The construction of the $100(1-\alpha)\%$ CI for σ^2 is straightforward. Because

$$X^2 = \frac{(n-1)S^2}{\sigma^2}$$

is chi-square with $n-1$ degrees of freedom, we may write

$$P(\chi^2_{1-\alpha/2,n-1} \le X^2 \le \chi^2_{\alpha/2,n-1}) = 1 - \alpha$$

so that

$$P\left(\chi^2_{1-\alpha/2,n-1} \le \frac{(n-1)S^2}{\sigma^2} \le \chi^2_{\alpha/2,n-1}\right) = 1 - \alpha$$

This last equation can be rearranged as

$$P\left(\frac{(n-1)S^2}{\chi^2_{\alpha/2,n-1}} \le \sigma^2 \le \frac{(n-1)S^2}{\chi^2_{1-\alpha/2,n-1}}\right) = 1 - \alpha$$

This leads to the following definition of the confidence interval for σ^2.

Definition

If s^2 is the sample variance from a random sample of n observations from a normal distribution with unknown variance σ^2, then **a $100(1-\alpha)\%$ confidence interval on σ^2 is**

$$\frac{(n-1)s^2}{\chi^2_{\alpha/2,n-1}} \le \sigma^2 \le \frac{(n-1)s^2}{\chi^2_{1-\alpha/2,n-1}} \tag{8-21}$$

where $\chi^2_{\alpha/2,n-1}$ and $\chi^2_{1-\alpha/2,n-1}$ are the upper and lower $100\alpha/2$ percentage points of the chi-square distribution with $n-1$ degrees of freedom, respectively. A **confidence interval for σ** has lower and upper limits that are the square roots of the corresponding limits in Equation 8-21.

It is also possible to find a $100(1 - \alpha)\%$ lower confidence bound or upper confidence bound on σ^2.

The $100(1 - \alpha)\%$ lower and upper confidence bounds on σ^2 are

$$\frac{(n - 1)s^2}{\chi^2_{\alpha,n-1}} \leq \sigma^2 \quad \text{and} \quad \sigma^2 \leq \frac{(n - 1)s^2}{\chi^2_{1-\alpha,n-1}} \tag{8-22}$$

respectively.

EXAMPLE 8-5

An automatic filling machine is used to fill bottles with liquid detergent. A random sample of 20 bottles results in a sample variance of fill volume of $s^2 = 0.0153$ (fluid ounces)^2. If the variance of fill volume is too large, an unacceptable proportion of bottles will be under- or overfilled. We will assume that the fill volume is approximately normally distributed. A 95% upper-confidence interval is found from Equation 8-22 as follows:

$$\sigma^2 \leq \frac{(n - 1)s^2}{\chi^2_{0.95,19}}$$

or

$$\sigma^2 \leq \frac{(19)0.0153}{10.117} = 0.0287 \text{ (fluid ounce)}^2$$

This last expression may be converted into a confidence interval on the standard deviation σ by taking the square root of both sides, resulting in

$$\sigma \leq 0.17$$

Therefore, at the 95% level of confidence, the data indicate that the process standard deviation could be as large as 0.17 fluid ounce.

EXERCISES FOR SECTION 8-4

8-33. Determine the values of the following percentiles: $\chi^2_{0.05,10}$, $\chi^2_{0.025,15}$, $\chi^2_{0.01,12}$, $\chi^2_{0.95,20}$, $\chi^2_{0.99,18}$, $\chi^2_{0.995,16}$, and $\chi^2_{0.005,25}$.

8-34. Determine the χ^2 percentile that is required to construct each of the following CIs:
(a) Confidence level = 95%, degrees of freedom = 24, one-sided (upper)
(b) Confidence level = 99%, degrees of freedom = 9, one-sided (lower)
(c) Confidence level = 90%, degrees of freedom = 19, two-sided.

8-35. A rivet is to be inserted into a hole. A random sample of $n = 15$ parts is selected, and the hole diameter is measured.

The sample standard deviation of the hole diameter measurements is $s = 0.008$ millimeters. Construct a 99% lower confidence bound for σ^2.

8-36. The sugar content of the syrup in canned peaches is normally distributed. A random sample of $n = 10$ cans yields a sample standard deviation of $s = 4.8$ milligrams. Find a 95% two-sided confidence interval for σ.

8-37. Consider the tire life data in Exercise 8-22. Find a 95% lower confidence bound for σ^2.

8-38. Consider the Izod impact test data in Exercise 8-23. Find a 99% two-sided confidence interval for σ^2.

8-39.   The percentage of titanium in an alloy used in aerospace castings is measured in 51 randomly selected parts. The sample standard deviation is $s = 0.37$. Construct a 95% two-sided confidence interval for σ.

8-40.   Consider the hole diameter data in Exercise 8-35. Construct a 99% two-sided confidence interval for σ.

8-41.   Consider the sugar content data in Exercise 8-36. Find a 90% lower confidence bound for σ.

8-5   A LARGE-SAMPLE CONFIDENCE INTERVAL FOR A POPULATION PROPORTION

It is often necessary to construct confidence intervals on a population proportion. For example, suppose that a random sample of size n has been taken from a large (possibly infinite) population and that $X(\leq n)$ observations in this sample belong to a class of interest. Then $\hat{P} = X/n$ is a point estimator of the proportion of the population p that belongs to this class. Note that n and p are the parameters of a binomial distribution. Furthermore, from Chapter 4 we know that the sampling distribution of $\hat{P}$ is approximately normal with mean p and variance $p(1 - p)/n$, if p is not too close to either 0 or 1 and if n is relatively large. Typically, to apply this approximation we require that np and $n(1 - p)$ be greater than or equal to 5. We will make use of the normal approximation in this section.

Definition

> If n is large, the distribution of
>
> $$Z = \frac{X - np}{\sqrt{np(1 - p)}} = \frac{\hat{P} - p}{\sqrt{\dfrac{p(1 - p)}{n}}}$$
>
> is approximately standard normal.

To construct the confidence interval on p, note that

$$P(-z_{\alpha/2} \leq Z \leq z_{\alpha/2}) \simeq 1 - \alpha$$

so

$$P\left(-z_{\alpha/2} \leq \frac{\hat{P} - p}{\sqrt{\dfrac{p(1 - p)}{n}}} \leq z_{\alpha/2}\right) \simeq 1 - \alpha$$

This may be rearranged as

$$P\left(\hat{P} - z_{\alpha/2}\sqrt{\frac{p(1 - p)}{n}} \leq p \leq \hat{P} + z_{\alpha/2}\sqrt{\frac{p(1 - p)}{n}}\right) \simeq 1 - \alpha \qquad (8\text{-}23)$$

The quantity $\sqrt{p(1 - p)/n}$ in Equation 8-23 is called the **standard error of the point estimator** $\hat{P}$. Unfortunately, the upper and lower limits of the confidence interval obtained from

Equation 8-23 contain the unknown parameter p. However, as suggested at the end of Section 8-2.5, a satisfactory solution is to replace p by $\hat{P}$ in the standard error, which results in

$$P\left(\hat{P} - z_{\alpha/2}\sqrt{\frac{\hat{P}(1-\hat{P})}{n}} \leq p \leq \hat{P} + z_{\alpha/2}\sqrt{\frac{\hat{P}(1-\hat{P})}{n}}\right) \simeq 1 - \alpha \qquad (8\text{-}24)$$

This leads to the approximate $100(1 - \alpha)\%$ confidence interval on p.

Definition

If $\hat{p}$ is the proportion of observations in a random sample of size n that belongs to a class of interest, an approximate $100(1 - \alpha)\%$ confidence interval on the proportion p of the population that belongs to this class is

$$\hat{p} - z_{\alpha/2}\sqrt{\frac{\hat{p}(1-\hat{p})}{n}} \leq p \leq \hat{p} + z_{\alpha/2}\sqrt{\frac{\hat{p}(1-\hat{p})}{n}} \qquad (8\text{-}25)$$

where $z_{\alpha/2}$ is the upper $\alpha/2$ percentage point of the standard normal distribution.

This procedure depends on the adequacy of the normal approximation to the binomial. To be reasonably conservative, this requires that np and $n(1 - p)$ be greater than or equal to 5. In situations where this approximation is inappropriate, particularly in cases where n is small, other methods must be used. Tables of the binomial distribution could be used to obtain a confidence interval for p. However, we could also use numerical methods based on the binomial probability mass function that are implemented in computer programs.

EXAMPLE 8-6

In a random sample of 85 automobile engine crankshaft bearings, 10 have a surface finish that is rougher than the specifications allow. Therefore, a point estimate of the proportion of bearings in the population that exceeds the roughness specification is $\hat{p} = x/n = 10/85 = 0.12$. A 95% two-sided confidence interval for p is computed from Equation 8-25 as

$$\hat{p} - z_{0.025}\sqrt{\frac{\hat{p}(1-\hat{p})}{n}} \leq p \leq \hat{p} + z_{0.025}\sqrt{\frac{\hat{p}(1-\hat{p})}{n}}$$

or

$$0.12 - 1.96\sqrt{\frac{0.12(0.88)}{85}} \leq p \leq 0.12 + 1.96\sqrt{\frac{0.12(0.88)}{85}}$$

which simplifies to

$$0.05 \leq p \leq 0.19$$

Choice of Sample Size

Since $\hat{P}$ is the point estimator of p, we can define the error in estimating p by $\hat{P}$ as $E = |p - \hat{P}|$. Note that we are approximately $100(1 - \alpha)\%$ confident that this error is less than $z_{\alpha/2}\sqrt{p(1 - p)/n}$. For instance, in Example 8-6, we are 95% confident that the sample proportion $\hat{p} = 0.12$ differs from the true proportion p by an amount not exceeding 0.07.

In situations where the sample size can be selected, we may choose n to be $100(1 - \alpha)\%$ confident that the error is less than some specified value E. If we set $E = z_{\alpha/2}\sqrt{p(1 - p)/n}$ and solve for n, the appropriate sample size is

$$n = \left(\frac{z_{\alpha/2}}{E}\right)^2 p(1 - p) \qquad (8\text{-}26)$$

An estimate of p is required to use Equation 8-26. If an estimate $\hat{p}$ from a previous sample is available, it can be substituted for p in Equation 8-26, or perhaps a subjective estimate can be made. If these alternatives are unsatisfactory, a preliminary sample can be taken, $\hat{p}$ computed, and then Equation 8-26 used to determine how many additional observations are required to estimate p with the desired accuracy. Another approach to choosing n uses the fact that the sample size from Equation 8-26 will always be a maximum for $p = 0.5$ [that is, $p(1 - p) \leq 0.25$ with equality for $p = 0.5$], and this can be used to obtain an upper bound on n. In other words, we are at least $100(1 - \alpha)\%$ confident that the error in estimating p by $\hat{p}$ is less than E if the sample size is

$$n = \left(\frac{z_{\alpha/2}}{E}\right)^2 (0.25) \qquad (8\text{-}27)$$

EXAMPLE 8-7   Consider the situation in Example 8-6. How large a sample is required if we want to be 95% confident that the error in using $\hat{p}$ to estimate p is less than 0.05? Using $\hat{p} = 0.12$ as an initial estimate of p, we find from Equation 8-26 that the required sample size is

$$n = \left(\frac{z_{0.025}}{E}\right)^2 \hat{p}(1 - \hat{p}) = \left(\frac{1.96}{0.05}\right)^2 0.12(0.88) \cong 163$$

If we wanted to be *at least* 95% confident that our estimate $\hat{p}$ of the true proportion p was within 0.05 regardless of the value of p, we would use Equation 8-27 to find the sample size

$$n = \left(\frac{z_{0.025}}{E}\right)^2 (0.25) = \left(\frac{1.96}{0.05}\right)^2 (0.25) \cong 385$$

Notice that if we have information concerning the value of p, either from a preliminary sample or from past experience, we could use a smaller sample while maintaining both the desired precision of estimation and the level of confidence.

One-Sided Confidence Bounds

We may find approximate one-sided confidence bounds on p by a simple modification of Equation 8-25.

The approximate $100(1 - \alpha)\%$ lower and upper confidence bounds are

$$\hat{p} - z_\alpha \sqrt{\frac{\hat{p}(1 - \hat{p})}{n}} \leq p \quad \text{and} \quad p \leq \hat{p} + z_\alpha \sqrt{\frac{\hat{p}(1 - \hat{p})}{n}} \qquad (8\text{-}28)$$

respectively.

EXERCISES FOR SECTION 8-5

8-42.   Of 1000 randomly selected cases of lung cancer, 823 resulted in death within 10 years. Construct a 95% two-sided confidence interval on the death rate from lung cancer.

8-43.   How large a sample would be required in Exercise 8-42 to be at least 95% confident that the error in estimating the 10-year death rate from lung cancer is less than 0.03?

8-44.   A random sample of 50 suspension helmets used by motorcycle riders and automobile race-car drivers was subjected to an impact test, and on 18 of these helmets some damage was observed.

(a) Find a 95% two-sided confidence interval on the true proportion of helmets of this type that would show damage from this test.

(b) Using the point estimate of p obtained from the preliminary sample of 50 helmets, how many helmets must be tested to be 95% confident that the error in estimating the true value of p is less than 0.02?

(c) How large must the sample be if we wish to be at least 95% confident that the error in estimating p is less than 0.02, regardless of the true value of p?

8-45.   The Arizona Department of Transportation wishes to survey state residents to determine what proportion of the population would like to increase statewide highway speed limits to 75 mph from 65 mph. How many residents do they need to survey if they want to be at least 99% confident that the sample proportion is within 0.05 of the true proportion?

8-46.   A manufacturer of electronic calculators is interested in estimating the fraction of defective units produced. A random sample of 800 calculators contains 10 defectives. Compute a 99% upper-confidence bound on the fraction defective.

8-47.   A study is to be conducted of the percentage of homeowners who own at least two television sets. How large a sample is required if we wish to be 99% confident that the error in estimating this quantity is less than 0.017?

8-48.   The fraction of defective integrated circuits produced in a photolithography process is being studied. A random sample of 300 circuits is tested, revealing 13 defectives. Find a 95% two-sided CI on the fraction of defective circuits produced by this particular tool.

8-6   A PREDICTION INTERVAL FOR A FUTURE OBSERVATION

In some problem situations, we may be interested in **predicting** a future observation of a variable. This is a different problem than estimating the mean of that variable, so a confidence interval is not appropriate. In this section we show how to obtain a $100(1 - \alpha)\%$ **prediction interval** on a future value of a normal random variable.

Suppose that $X_1, X_2, \ldots, X_n$ is a random sample from a normal population. We wish to predict the value X_{n+1}, a single **future** observation. A point prediction of X_{n+1} is $\bar{X}$, the sample mean. The prediction error is $X_{n+1} - \bar{X}$. The expected value of the prediction error is

$$E(X_{n+1} - \bar{X}) = \mu - \mu = 0$$

and the variance of the prediction error is

$$V(X_{n+1} - \bar{X}) = \sigma^2 + \frac{\sigma^2}{n} = \sigma^2\left(1 + \frac{1}{n}\right)$$

because the future observation, X_{n+1} is independent of the mean of the current sample $\bar{X}$. The prediction error $X_{n+1} - \bar{X}$ is normally distributed. Therefore

$$Z = \frac{X_{n+1} - \bar{X}}{\sigma\sqrt{1 + \frac{1}{n}}}$$

has a standard normal distribution. Replacing σ with S results in

$$T = \frac{X_{n+1} - \overline{X}}{S\sqrt{1 + \dfrac{1}{n}}}$$

which has a t distribution with $n - 1$ degrees of freedom. Manipulating T as we have done previously in the development of a CI leads to a prediction interval on the future observation X_{n+1}.

Definition

> **A $100(1 - \alpha)\%$ prediction interval on a single future observation from a normal distribution** is given by
>
> $$\overline{x} - t_{\alpha/2, n-1}\, s\sqrt{1 + \frac{1}{n}} \leq X_{n+1} \leq \overline{x} + t_{\alpha/2, n-1}\, s\sqrt{1 + \frac{1}{n}} \qquad (8\text{-}29)$$

The prediction interval for X_{n+1} will always be longer than the confidence interval for μ because there is more variability associated with the prediction error than with the error of estimation. This is easy to see because the prediction error is the difference between two random variables $(X_{n+1} - \overline{X})$, and the estimation error in the CI is the difference between one random variable and a constant $(\overline{X} - \mu)$. As n gets larger $(n \to \infty)$, the length of the CI decreases to zero, essentially becoming the single value μ, but the length of the prediction interval approaches $2z_{\alpha/2}\sigma$. So as n increases, the uncertainty in estimating μ goes to zero, although there will always be uncertainty about the future value X_{n+1} even when there is no need to estimate any of the distribution parameters.

EXAMPLE 8-8

Reconsider the tensile adhesion tests on specimens of U-700 alloy described in Example 8-4. The load at failure for $n = 22$ specimens was observed, and we found that $\overline{x} = 13.71$ and $s = 3.55$. The 95% confidence interval on μ was $12.14 \leq \mu \leq 15.28$. We plan to test a twenty-third specimen. A 95% prediction interval on the load at failure for this specimen is

$$\overline{x} - t_{\alpha/2, n-1}\, s\sqrt{1 + \frac{1}{n}} \leq X_{n+1} \leq \overline{x} + t_{\alpha/2, n-1}\, s\sqrt{1 + \frac{1}{n}}$$

$$13.71 - (2.080)3.55\sqrt{1 + \frac{1}{22}} \leq X_{23} \leq 13.71 + (2.080)3.55\sqrt{1 + \frac{1}{22}}$$

$$6.16 \leq X_{23} \leq 21.26$$

Notice that the prediction interval is considerably longer than the CI.

EXERCISES FOR SECTION 8-6

8-49. Consider the tire-testing data described in Exercise 8-22. Compute a 95% prediction interval on the life of the next tire of this type tested under conditions that are similar to those employed in the original test. Compare the length of the prediction interval with the length of the 95% CI on the population mean.

8-50. Consider the Izod impact test described in Exercise 8-23. Compute a 99% prediction interval on the impact strength of the next specimen of PVC pipe tested. Compare the length of the prediction interval with the length of the 99% CI on the population mean.

8-51. Consider the television tube brightness test described in Exercise 8-24. Compute a 99% prediction interval on the brightness of the next tube tested. Compare the length of the prediction interval with the length of the 99% CI on the population mean.

8-52. Consider the margarine test described in Exercise 8-25. Compute a 99% prediction interval on the polyunsaturated fatty acid in the next package of margarine that is tested. Compare the length of the prediction interval with the length of the 99% CI on the population mean.

8-53. Consider the test on the compressive strength of concrete described in Exercise 8-26. Compute a 90% prediction interval on the next specimen of concrete tested.

8-54. Consider the suspension rod diameter measurements described in Exercise 8-27. Compute a 95% prediction interval on the diameter of the next rod tested. Compare the length of the prediction interval with the length of the 95% CI on the population mean.

8-55. Consider the bottle wall thickness measurements described in Exercise 8-29. Compute a 90% prediction interval on the wall thickness of the next bottle tested.

8-56. How would you obtain a one-sided prediction bound on a future observation? Apply this procedure to obtain a 95% one-sided prediction bound on the wall thickness of the next bottle for the situation described in Exercise 8-29.

8-57. Consider the fuel rod enrichment data described in Exercise 8-30. Compute a 90% CI prediction interval on the enrichment of the next rod tested. Compare the length of the prediction interval with the length of the 90% CI on the population mean.

8-58. Consider the syrup dispensing measurements described in Exercise 8-31. Compute a 95% prediction interval on the syrup volume in the next beverage dispensed. Compare the length of the prediction interval with the length of the 99% CI on the population mean.

8-59. Consider the natural frequency of beams described in Exercise 8-32. Compute a 90% prediction interval on the diameter of the natural frequency of the next beam of this type that will be tested. Compare the length of the prediction interval with the length of the 90% CI on the population mean.

8-7   TOLERANCE INTERVALS FOR A NORMAL DISTRIBUTION

Consider a population of semiconductor processors. Suppose that the speed of these processors has a normal distribution with mean $\mu = 600$ megahertz and standard deviation $\sigma = 30$ megahertz. Then the interval from $600 - 1.96(30) = 541.2$ to $600 + 1.96(30) = 658.8$ megahertz captures the speed of 95% of the processors in this population because the interval from -1.96 to 1.96 captures 95% of the area under the standard normal curve. The interval from $\mu - z_{\alpha/2}\sigma$ to $\mu + z_{\alpha/2}\sigma$ is called a **tolerance interval.**

If μ and σ are unknown, we can use the data from a random sample of size n to compute $\bar{x}$ and s, and then form the interval $(\bar{x} - 1.96s, \bar{x} + 1.96s)$. However, because of sampling variability in $\bar{x}$ and s, it is likely that this interval will contain less than 95% of the values in the population. The solution to this problem is to replace 1.96 by some value that will make the proportion of the distribution contained in the interval 95% with some level of confidence. Fortunately, it is easy to do this.

Definition

> A **tolerance interval** for capturing at least γ% of the values in a normal distribution with confidence level $100(1 - \alpha)$% is
>
> $$\bar{x} - ks, \qquad \bar{x} + ks$$
>
> where k is a tolerance interval factor found in Appendix Table XI. Values are given for $\gamma = 90\%$, 95%, and 95% and for 95% and 99% confidence.

One-sided tolerance bounds can also be computed. The tolerance factors for these bounds are also given in Appendix Table XI.

EXAMPLE 8-9

Let's reconsider the tensile adhesion tests originally described in Example 8-4. The load at failure for $n = 22$ specimens was observed, and we found that $\bar{x} = 31.71$ and $s = 3.55$. We want to find a tolerance interval for the load at failure that includes 90% of the values in the population with 95% confidence. From Appendix Table XI the tolerance factor k for $n = 22$, $\gamma = 0.90$, and 95% confidence is $k = 2.264$. The desired tolerance interval is

$$(\bar{x} - ks, \bar{x} + ks) \quad \text{or} \quad [31.71 - (2.264)3.55, 31.71 + (2.264)3.55]$$

which reduces to (23.67, 39.75). We can be 95% confident that at least 90% of the values of load at failure for this particular alloy lie between 23.67 and 39.75 megapascals.

From Appendix Table XI, we note that as $n \rightarrow \infty$, the value of k goes to the z-value associated with the desired level of containment for the normal distribution. For example, if we want 90% of the population to fall in the two-sided tolerance interval, k approaches $z_{0.05} = 1.645$ as $n \rightarrow \infty$. Note that as $n \rightarrow \infty$, a $100(1 - \alpha)\%$ prediction interval on a future value approaches a tolerance interval that contains $100(1 - \alpha)\%$ of the distribution.

EXERCISES FOR SECTION 8-7

8-60.   Compute a 95% tolerance interval on the life of the tires described in Exercise 8-22, that has confidence level 95%. Compare the length of the tolerance interval with the length of the 95% CI on the population mean. Which interval is shorter? Discuss the difference in interpretation of these two intervals.

8-61.   Consider the Izod impact test described in Exercise 8-23. Compute a 99% tolerance interval on the impact strength of PVC pipe that has confidence level 90%. Compare the length of the tolerance interval with the length of the 99% CI on the population mean. Which interval is shorter? Discuss the difference in interpretation of these two intervals.

8-62.   Compute a 99% tolerance interval on the brightness of the television tubes in Exercise 8-24 that has confidence level 95%. Compare the length of the prediction interval with the length of the 99% CI on the population mean. Which interval is shorter? Discuss the difference in interpretation of these two intervals.

8-63.   Consider the margarine test described in Exercise 8-25. Compute a 99% tolerance interval on the polyunsaturated fatty acid in this particular type of margarine that has confidence level 95%. Compare the length of the prediction interval with the length of the 99% CI on the population mean. Which interval is shorter? Discuss the difference in interpretation of these two intervals.

8-64.   Compute a 90% tolerance interval on the compressive strength of the concrete described in Exercise 8-26 that has 90% confidence.

8-65.   Compute a 95% tolerance interval on the diameter of the rods described in Exercise 8-27 that has 90% confidence. Compare the length of the prediction interval with the length of the 95% CI on the population mean. Which interval is shorter? Discuss the difference in interpretation of these two intervals.

8-66.   Consider the bottle wall thickness measurements described in Exercise 8-29. Compute a 90% tolerance interval on bottle wall thickness that has confidence level 90%.

8-67.   Consider the bottle wall thickness measurements described in Exercise 8-29. Compute a 90% lower tolerance bound on bottle wall thickness that has confidence level 90%. Why would a lower tolerance bound likely be of interest here?

8-68.   Consider the fuel rod enrichment data described in Exercise 8-30. Compute a 99% tolerance interval on rod enrichment that has confidence level 95%. Compare the length of the prediction interval with the length of the 95% CI on the population mean.

8-69.   Compute a 95% tolerance interval on the syrup volume described in Exercise 8-31 that has confidence level 90%. Compare the length of the prediction interval with the length of the 95% CI on the population mean.

Supplemental Exercises

8-70.   Consider the confidence interval for μ with known standard deviation σ:

$$\bar{x} - z_{\alpha_1}\sigma/\sqrt{n} \le \mu \le \bar{x} + z_{\alpha_2}\sigma/\sqrt{n}$$

where $\alpha_1 + \alpha_2 = \alpha$. Let $\alpha = 0.05$ and find the interval for $\alpha_1 = \alpha_2 = \alpha/2 = 0.025$. Now find the interval for the case $\alpha_1 = 0.01$ and $\alpha_2 = 0.04$. Which interval is shorter? Is there any advantage to a "symmetric" confidence interval?

8-71.   A normal population has a known mean 50 and unknown variance.
(a) A random sample of $n = 16$ is selected from this population, and the sample results are $\bar{x} = 52$ and $s = 8$. How unusual are these results? That is, what is the probability of observing a sample average as large as 52 (or larger) if the known, underlying mean is actually 50?
(b) A random sample of $n = 30$ is selected from this population, and the sample results are $\bar{x} = 52$ and $s = 8$. How unusual are these results?
(c) A random sample of $n = 100$ is selected from this population, and the sample results are $\bar{x} = 52$ and $s = 8$. How unusual are these results?
(d) Compare your answers to parts (a)–(c) and explain why they are the same or differ.

8-72.   A normal population has known mean $\mu = 50$ and variance $\sigma^2 = 5$. What is the approximate probability that the sample variance is greater than or equal to 7.44? less than or equal to 2.56?
(a) For a random sample of $n = 16$.
(b) For a random sample of $n = 30$.
(c) For a random sample of $n = 71$.
(d) Compare your answers to parts (a)–(c) for the approximate probability that the sample variance is greater than or equal to 7.44. Explain why this tail probability is increasing or decreasing with increased sample size.
(e) Compare your answers to parts (a)–(c) for the approximate probability that the sample variance is less than or equal to 2.56. Explain why this tail probability is increasing or decreasing with increased sample size.

8-73.   An article in the *Journal of Sports Science* (1987, Vol. 5, pp. 261–271) presents the results of an investigation of the hemoglobin level of Canadian Olympic ice hockey players. The data reported are as follows (in g/dl):

15.3	16.0	14.4	16.2	16.2
14.9	15.7	15.3	14.6	15.7
16.0	15.0	15.7	16.2	14.7
14.8	14.6	15.6	14.5	15.2

(a) Given the following probability plot of the data, what is a logical assumption about the underlying distribution of the data?

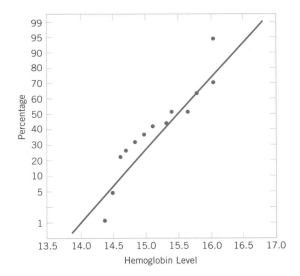

(b) Explain why this check of the distribution underlying the sample data is important if we want to construct a confidence interval on the mean.
(c) Based on this sample data, a 95% confidence interval for the mean is (15.04, 15.62). Is it reasonable to infer that the true mean could be 14.5? Explain your answer.
(d) Explain why this check of the distribution underlying the sample data is important if we want to construct a confidence interval on the variance.
(e) Based on this sample data, a 95% confidence interval for the variance is (0.22, 0.82). Is it reasonable to infer that the true variance could be 0.35? Explain your answer.
(f) Is it reasonable to use these confidence intervals to draw an inference about the mean and variance of hemoglobin levels
    (i) of Canadian doctors? Explain your answer.
    (ii) of Canadian children ages 6–12? Explain your answer.

8-74.   The article "Mix Design for Optimal Strength Development of Fly Ash Concrete" (*Cement and Concrete Research*, 1989, Vol. 19, No. 4, pp. 634–640) investigates the compressive strength of concrete when mixed with fly ash (a mixture of silica, alumina, iron, magnesium oxide, and other ingredients). The compressive strength for nine samples in dry conditions on the twenty-eighth day are as follows (in megapascals):

40.2	30.4	28.9	30.5	22.4
25.8	18.4	14.2	15.3	

(a) Given the following probability plot of the data, what is a logical assumption about the underlying distribution of the data?

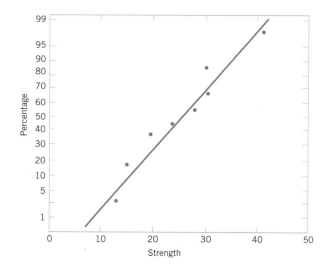

(b) Find a 99% lower one-sided confidence interval on mean compressive strength. Provide a practical interpretation of this interval.

(c) Find a 98% two-sided confidence interval on mean compressive strength. Provide a practical interpretation of this interval and explain why the lower end-point of the interval is or is not the same as in part (b).

(d) Find a 99% upper one-sided confidence interval on the variance of compressive strength. Provide a practical interpretation of this interval.

(e) Find a 98% two-sided confidence interval on the variance of compression strength. Provide a practical interpretation of this interval and explain why the upper end-point of the interval is or is not the same as in part (d).

(f) Suppose that it was discovered that the largest observation 40.2 was misrecorded and should actually be 20.4. Now the sample mean $\bar{x} = 23$ and the sample variance $s^2 = 36.9$. Use these new values and repeat parts (c) and (e). Compare the original computed intervals and the newly computed intervals with the corrected observation value. How does this mistake affect the values of the sample mean, sample variance, and the width of the two-sided confidence intervals?

(g) Suppose, instead, that it was discovered that the largest observation 40.2 is correct, but that the observation 25.8 is incorrect and should actually be 24.8. Now the sample mean $\bar{x} = 25$ and the sample standard deviation $s^2 = 8.41$. Use these new values and repeat parts (c) and (e). Compare the original computed intervals and the newly computed intervals with the corrected observation value. How does this mistake affect the values of the sample mean, sample variance, and the width of the two-sided confidence intervals?

(h) Use the results from parts (f) and (g) to explain the effect of mistakenly recorded values on sample estimates. Comment on the effect when the mistaken values are near the sample mean and when they are not.

8-75. An operating system for a personal computer has been studied extensively, and it is known that the standard deviation of the response time following a particular command is $\sigma = 8$ milliseconds. A new version of the operating system is installed, and we wish to estimate the mean response time for the new system to ensure that a 95% confidence interval for μ has length at most 5 milliseconds.
(a) If we can assume that response time is normally distributed and that $\sigma = 8$ for the new system, what sample size would you recommend?
(b) Suppose that we are told by the vendor that the standard deviation of the response time of the new system is smaller, say $\sigma = 6$; give the sample size that you recommend and comment on the effect the smaller standard deviation has on this calculation.

8-76. Consider the hemoglobin data in Exercise 8-73. Find the following:
(a) An interval that contains 95% of the hemoglobin values with 90% confidence.
(b) An interval that contains 99% of the hemoglobin values with 90% confidence.

8-77. Consider the compressive strength of concrete data from Exercise 8-74. Find a 95% prediction interval on the next sample that will be tested.

8-78. The maker of a shampoo knows that customers like this product to have a lot of foam. Ten sample bottles of the product are selected at random and the foam heights observed are as follows (in millimeters): 210, 215, 194, 195, 211, 201, 198, 204, 208, and 196.
(a) Is there evidence to support the assumption that foam height is normally distributed?
(b) Find a 95% CI on the mean foam height.
(c) Find a 95% prediction interval on the next bottle of shampoo that will be tested.
(d) Find an interval that contains 95% of the shampoo foam heights with 99% confidence.
(e) Explain the difference in the intervals computed in parts (b), (c), and (d).

8-79. During the 1999 and 2000 baseball seasons, there was much speculation that the unusually large number of home runs that were hit was due at least in part to a livelier ball. One way to test the "liveliness" of a baseball is to launch the ball at a vertical surface with a known velocity V_L and measure the ratio of the outgoing velocity V_O of the ball to V_L. The ratio $R = V_O/V_L$ is called the coefficient of restitution. Following are measurements of the coefficient of restitution for 40 randomly selected baseballs. The balls were thrown from a pitching machine at an oak surface.

0.6248	0.6237	0.6118	0.6159	0.6298	0.6192
0.6520	0.6368	0.6220	0.6151	0.6121	0.6548
0.6226	0.6280	0.6096	0.6300	0.6107	0.6392
0.6230	0.6131	0.6223	0.6297	0.6435	0.5978

0.6351	0.6275	0.6261	0.6262	0.6262	0.6314
0.6128	0.6403	0.6521	0.6049	0.6170	
0.6134	0.6310	0.6065	0.6214	0.6141	

(a) Is there evidence to support the assumption that the coefficient of restitution is normally distributed?

(b) Find a 99% CI on the mean coefficient of restitution.

(c) Find a 99% prediction interval on the coefficient of restitution for the next baseball that will be tested.

(d) Find an interval that will contain 99% of the values of the coefficient of restitution with 95% confidence.

(e) Explain the difference in the three intervals computed in parts (b), (c), and (d).

8-80.   Consider the baseball coefficient of restitution data in Exercise 8-79. Suppose that any baseball that has a coefficient of restitution that exceeds 0.635 is considered too lively. Based on the available data, what proportion of the baseballs in the sampled population are too lively? Find a 95% lower confidence bound on this proportion.

8-81.   An article in the *ASCE Journal of Energy Engineering* ("Overview of Reservoir Release Improvements at 20 TVA Dams," Vol. 125, April 1999, pp. 1–17) presents data on dissolved oxygen concentrations in streams below 20 dams in the Tennessee Valley Authority system. The observations are (in milligrams per liter): 5.0, 3.4, 3.9, 1.3, 0.2, 0.9, 2.7, 3.7, 3.8, 4.1, 1.0, 1.0, 0.8, 0.4, 3.8, 4.5, 5.3, 6.1, 6.9, and 6.5.

(a) Is there evidence to support the assumption that the dissolved oxygen concentration is normally distributed?

(b) Find a 95% CI on the mean dissolved oxygen concentration.

(c) Find a 95% prediction interval on the dissolved oxygen concentration for the next stream in the system that will be tested.

(d) Find an interval that will contain 95% of the values of the dissolved oxygen concentration with 99% confidence.

(e) Explain the difference in the three intervals computed in parts (b), (c), and (d).

8-82.   The tar content in 30 samples of cigar tobacco follows:

1.542	1.585	1.532	1.466	1.499	1.611
1.622	1.466	1.546	1.494	1.548	1.626
1.440	1.608	1.520	1.478	1.542	1.511
1.459	1.533	1.532	1.523	1.397	1.487
1.598	1.498	1.600	1.504	1.545	1.558

(a) Is there evidence to support the assumption that the tar content is normally distributed?

(b) Find a 99% CI on the mean tar content.

(c) Find a 99% prediction interval on the tar content for the next observation that will be taken on this particular type of tobacco.

(d) Find an interval that will contain 99% of the values of the tar content with 95% confidence.

(e) Explain the difference in the three intervals computed in parts (b), (c), and (d).

8-83.   A manufacturer of electronic calculators takes a random sample of 1200 calculators and finds that there are eight defective units.

(a) Construct a 95% confidence interval on the population proportion.

(b) Is there evidence to support a claim that the fraction of defective units produced is 1% or less?

8-84.   An article in *The Engineer* ("Redesign for Suspect Wiring," June 1990) reported the results of an investigation into wiring errors on commercial transport aircraft that may produce faulty information to the flight crew. Such a wiring error may have been responsible for the crash of a British Midland Airways aircraft in January 1989 by causing the pilot to shut down the wrong engine. Of 1600 randomly selected aircraft, eight were found to have wiring errors that could display incorrect information to the flight crew.

(a) Find a 99% confidence interval on the proportion of aircraft that have such wiring errors.

(b) Suppose we use the information in this example to provide a preliminary estimate of p. How large a sample would be required to produce an estimate of p that we are 99% confident differs from the true value by at most 0.008?

(c) Suppose we did not have a preliminary estimate of p. How large a sample would be required if we wanted to be at least 99% confident that the sample proportion differs from the true proportion by at most 0.008 regardless of the true value of p?

(d) Comment on the usefulness of preliminary information in computing the needed sample size.

8-85.   An article in *Engineering Horizons* (Spring 1990, p. 26) reported that 117 of 484 new engineering graduates were planning to continue studying for an advanced degree. Consider this as a random sample of the 1990 graduating class.

(a) Find a 90% confidence interval on the proportion of such graduates planning to continue their education.

(b) Find a 95% confidence interval on the proportion of such graduates planning to continue their education.

(c) Compare your answers to parts (a) and (b) and explain why they are the same or different.

(d) Could you use either of these confidence intervals to determine whether the proportion is actually 0.25? Explain your answer. *Hint:* Use the normal approximation to the binomial.

MIND-EXPANDING EXERCISES

8-86. An electrical component has a time-to-failure (or lifetime) distribution that is exponential with parameter λ, so the mean lifetime is $\mu = 1/\lambda$. Suppose that a sample of n of these components is put on test, and let X_i be the observed lifetime of component i. The test continues only until the rth unit fails, where $r < n$. This results in a **censored** life test. Let X_1 denote the time at which the first failure occurred, X_2 denote the time at which the second failure occurred, and so on. Then the total lifetime that has been accumulated at test termination is

$$T_r = \sum_{i=1}^{r} X_i + (n - r)X_r$$

We have previously shown in Exercise 7-72 that T_r/r is an unbiased estimator for μ.

(a) It can be shown that $2\lambda T_r$ has a chi-square distribution with $2r$ degrees of freedom. Use this fact to develop a $100(1 - \alpha)\%$ confidence interval for mean lifetime $\mu = 1/\lambda$.

(b) Suppose 20 units were put on test, and the test terminated after 10 failures occurred. The failure times (in hours) are 15, 18, 19, 20, 21, 21, 22, 27, 28, 29. Find a 95% confidence interval on mean lifetime.

8-87. Consider a two-sided confidence interval for the mean μ when σ is known;

$$\bar{x} - z_{\alpha_1}\sigma/\sqrt{n} \le \mu \le \bar{x} + z_{\alpha_2}\sigma/\sqrt{n}$$

where $\alpha_1 + \alpha_2 = \alpha$. If $\alpha_1 = \alpha_2 = \alpha/2$, we have the usual $100(1 - \alpha)\%$ confidence interval for μ. In the above, when $\alpha_1 \ne \alpha_2$, the interval is not symmetric about μ. The length of the interval is $L = \sigma(z_{\alpha_1} + z_{\alpha_2})/\sqrt{n}$. Prove that the length of the interval L is minimized when $\alpha_1 = \alpha_2 = \alpha/2$. *Hint:* Remember that $\Phi(z_a) = 1 - \alpha$, so $\Phi^{-1}(1 - \alpha) = z_\alpha$, and the relationship between the derivative of a function $y = f(x)$ and the inverse $x = f^{-1}(y)$ is $(d/dy)f^{-1}(y) = 1/[(d/dx)f(x)]$.

8-88. It is possible to construct a **nonparametric tolerance interval** that is based on the extreme values in a random sample of size n from any continuous population. If p is the minimum proportion of the population contained between the smallest and largest sample observations with confidence $1 - \alpha$, it can be shown that

$$np^{n-1} - (n - 1)p^n = \alpha$$

and n is approximately

$$n = \frac{1}{2} + \left(\frac{1 + p}{1 - p}\right)\left(\frac{\chi_{\alpha,4}^2}{4}\right)$$

(a) In order to be 95% confident that at least 90% of the population will be included between the extreme values of the sample, what sample size will be required?

(b) A random sample of 10 transistors gave the following measurements on saturation current (in milliamps): 10.25, 10.41, 10.30, 10.26, 10.19, 10.37, 10.29, 10.34, 10.23, 10.38. Find the limits that contain a proportion p of the saturation current measurements at 95% confidence. What is the proportion p contained by these limits?

8-89. Suppose that X_1, X_2, ... , X_n is a random sample from a continuous probability distribution with median $\tilde{\mu}$.

(a) Show that

$$P\{\min(X_i) < \tilde{\mu} < \max(X_i)\}$$
$$= 1 - \left(\frac{1}{2}\right)^{n-1}$$

Hint: The complement of the event $[\min(X_i) < \tilde{\mu} < \max(X_i)]$ is $[\max(X_i) \le \tilde{\mu}] \cup [\min(X_i) \le \tilde{\mu}]$, but $\max(X_i) \le \tilde{\mu}$ if and only if $X_i \le \tilde{\mu}$ for all i.]

(b) Write down a $100(1 - \alpha)\%$ confidence interval for the median $\tilde{\mu}$, where

$$\alpha = \left(\frac{1}{2}\right)^{n-1}.$$

8-90. Students in the industrial statistics lab at ASU calculate a lot of confidence intervals on μ. Suppose all these CIs are independent of each other. Consider the next one thousand 95% confidence intervals that will be calculated. How many of these CIs do you expect to capture the true value of μ? What is the probability that between 930 and 970 of these intervals contain the true value of μ?

IMPORTANT TERMS AND CONCEPTS

In the E-book, click on any term or concept below to go to that subject.

Confidence coefficient

Confidence interval

Confidence interval for a population proportion

Chi-squared distribution

Confidence intervals for the mean of a normal distribution

Confidence interval for the variance of a normal distribution

Confidence level

Error in estimation

Large sample confidence intervals

One-sided confidence bounds

Precision of parameter estimation

Prediction interval

Tolerance interval

Two-sided confidence interval

t distribution

CD MATERIAL

Boostrap samples

Percentile method for boostrap confidence intervals

Tests of Hypotheses for a Single Sample

LEARNING OBJECTIVES

After careful study of this chapter, you should be able to do the following:
1. Structure engineering decision-making problems as hypothesis tests
2. Test hypotheses on the mean of a normal distribution using either a Z-test or a *t*-test procedure
3. Test hypotheses on the variance or standard deviation of a normal distribution
4. Test hypotheses on a population proportion
5. Use the P-value approach for making decisions in hypotheses tests
6. Compute power, type II error probability, and make sample size selection decisions for tests on means, variances, and proportions
7. Explain and use the relationship between confidence intervals and hypothesis tests
8. Use the chi-square goodness of fit test to check distributional assumptions
9. Use contingency table tests

CD MATERIAL
10. Appreciate the likelihood ratio approach to construction of test statistics
11. Conduct small sample tests on a population proportion

Answers for many odd numbered exercises are at the end of the book. Answers to exercises whose numbers are surrounded by a box can be accessed in the e-Text by clicking on the box. Complete worked solutions to certain exercises are also available in the e-Text. These are indicated in the Answers to Selected Exercises section by a box around the exercise number. Exercises are also available for some of the text sections that appear on CD only. These exercises may be found within the e-Text immediately following the section they accompany.

9-1 HYPOTHESIS TESTING

9-1.1 Statistical Hypotheses

In the previous chapter we illustrated how to construct a confidence interval estimate of a parameter from sample data. However, many problems in engineering require that we decide whether to accept or reject a statement about some parameter. The statement is called a **hypothesis,** and the decision-making procedure about the hypothesis is called **hypothesis testing.** This is one of the most useful aspects of statistical inference, since many types of decision-making problems, tests, or experiments in the engineering world can be formulated as hypothesis-testing problems. Furthermore, as we will see, there is a very close connection between hypothesis testing and confidence intervals.

Statistical hypothesis testing and confidence interval estimation of parameters are the fundamental methods used at the data analysis stage of a **comparative experiment,** in which the engineer is interested, for example, in comparing the mean of a population to a specified value. These simple comparative experiments are frequently encountered in practice and provide a good foundation for the more complex experimental design problems that we will discuss in Chapters 13 and 14. In this chapter we discuss comparative experiments involving a single population, and our focus is on testing hypotheses concerning the parameters of the population.

We now give a formal definition of a statistical hypothesis.

Definition

> A **statistical hypothesis** is a statement about the parameters of one or more populations.

Since we use probability distributions to represent populations, a statistical hypothesis may also be thought of as a statement about the probability distribution of a random variable. The hypothesis will usually involve one or more parameters of this distribution.

For example, suppose that we are interested in the burning rate of a solid propellant used to power aircrew escape systems. Now burning rate is a random variable that can be described by a probability distribution. Suppose that our interest focuses on the mean burning rate (a parameter of this distribution). Specifically, we are interested in deciding whether or not the mean burning rate is 50 centimeters per second. We may express this formally as

two sided
$$\begin{cases} H_0: \mu = 50 \text{ centimeters per second} \quad \textit{Null hypothesis} \\ H_1: \mu \neq 50 \text{ centimeters per second} \quad \textit{alternative hypothesis} \end{cases} \tag{9-1}$$

The statement $H_0: \mu = 50$ centimeters per second in Equation 9-1 is called the **null hypothesis,** and the statement $H_1: \mu \neq 50$ centimeters per second is called the **alternative hypothesis.** Since the alternative hypothesis specifies values of μ that could be either greater or less than 50 centimeters per second, it is called a **two-sided alternative hypothesis.** In some situations, we may wish to formulate a **one-sided alternative hypothesis,** as in

one sided
$$\begin{aligned} H_0: \mu = 50 \text{ centimeters per second} \qquad\qquad H_0: \mu = 50 \text{ centimeters per second} \\ \text{or} \qquad\qquad\qquad\qquad\qquad\qquad (9\text{-}2) \\ H_1: \mu < 50 \text{ centimeters per second} \qquad\qquad H_1: \mu > 50 \text{ centimeters per second} \end{aligned}$$

It is important to remember that hypotheses are always statements about the population or distribution under study, not statements about the sample. The value of the population parameter specified in the null hypothesis (50 centimeters per second in the above example) is usually determined in one of three ways. First, it may result from past experience or knowledge of the process, or even from previous tests or experiments. The objective of hypothesis testing then is usually to determine whether the parameter value has changed. Second, this value may be determined from some theory or model regarding the process under study. Here the objective of hypothesis testing is to verify the theory or model. A third situation arises when the value of the population parameter results from external considerations, such as design or engineering specifications, or from contractual obligations. In this situation, the usual objective of hypothesis testing is conformance testing.

A procedure leading to a decision about a particular hypothesis is called a **test of a hypothesis.** Hypothesis-testing procedures rely on using the information in a random sample from the population of interest. If this information is consistent with the hypothesis, we will conclude that the hypothesis is true; however, if this information is inconsistent with the hypothesis, we will conclude that the hypothesis is false. We emphasize that the truth or falsity of a particular hypothesis can never be known with certainty, unless we can examine the entire population. This is usually impossible in most practical situations. Therefore, a hypothesis-testing procedure should be developed with the probability of reaching a wrong conclusion in mind.

The structure of hypothesis-testing problems is identical in all the applications that we will consider. The null hypothesis is the hypothesis we wish to test. Rejection of the null hypothesis always leads to accepting the alternative hypothesis. In our treatment of hypothesis testing, the null hypothesis will always be stated so that it specifies an exact value of the parameter (as in the statement $H_0: \mu = 50$ centimeters per second in Equation 9-1). The alternate hypothesis will allow the parameter to take on several values (as in the statement $H_1: \mu \neq 50$ centimeters per second in Equation 9-1). Testing the hypothesis involves taking a random sample, computing a **test statistic** from the sample data, and then using the test statistic to make a decision about the null hypothesis.

9-1.2 Tests of Statistical Hypotheses

To illustrate the general concepts, consider the propellant burning rate problem introduced earlier. The null hypothesis is that the mean burning rate is 50 centimeters per second, and the alternate is that it is not equal to 50 centimeters per second. That is, we wish to test

$$H_0: \mu = 50 \text{ centimeters per second}$$
$$H_1: \mu \neq 50 \text{ centimeters per second}$$

Suppose that a sample of $n = 10$ specimens is tested and that the sample mean burning rate $\bar{x}$ is observed. The sample mean is an estimate of the true population mean μ. A value of the sample mean $\bar{x}$ that falls close to the hypothesized value of $\mu = 50$ centimeters per second is evidence that the true mean μ is really 50 centimeters per second; that is, such evidence supports the null hypothesis H_0. On the other hand, a sample mean that is considerably different from 50 centimeters per second is evidence in support of the alternative hypothesis H_1. Thus, the sample mean is the test statistic in this case.

The sample mean can take on many different values. Suppose that if $48.5 \leq \bar{x} \leq 51.5$, we will not reject the null hypothesis $H_0: \mu = 50$, and if either $\bar{x} < 48.5$ or $\bar{x} > 51.5$, we will reject the null hypothesis in favor of the alternative hypothesis $H_1: \mu \neq 50$. This is illustrated in Fig. 9-1. The values of $\bar{x}$ that are less than 48.5 and greater than 51.5 constitute the **critical region** for the test, while all values that are in the interval $48.5 \leq \bar{x} \leq 51.5$ form a region for which we will fail to reject the null hypothesis. By convention, this is usually called the **acceptance region**. The boundaries between the critical regions and the acceptance region are called the **critical values**. In our example the critical values are 48.5 and 51.5. It is customary to state conclusions relative to the null hypothesis H_0. Therefore, we reject H_0 in favor of H_1 if the test statistic falls in the critical region and fail to reject H_0 otherwise.

This decision procedure can lead to either of two wrong conclusions. For example, the true mean burning rate of the propellant could be equal to 50 centimeters per second. However, for the randomly selected propellant specimens that are tested, we could observe a value of the test statistic $\bar{x}$ that falls into the critical region. We would then reject the null hypothesis H_0 in favor of the alternate H_1 when, in fact, H_0 is really true. This type of wrong conclusion is called a **type I error.**

Definition

> Rejecting the null hypothesis H_0 when it is true is defined as a **type I error.**

Now suppose that the true mean burning rate is different from 50 centimeters per second, yet the sample mean $\bar{x}$ falls in the acceptance region. In this case we would fail to reject H_0 when it is false. This type of wrong conclusion is called a **type II error.**

Definition

> Failing to reject the null hypothesis when it is false is defined as a **type II error.**

Thus, in testing any statistical hypothesis, four different situations determine whether the final decision is correct or in error. These situations are presented in Table 9-1.

Reject H_0	Fail to Reject H_0	Reject H_0
$\mu \neq 50$ cm/s	$\mu = 50$ cm/s	$\mu \neq 50$ cm/s

Figure 9-1 Decision criteria for testing H_0: $\mu = 50$ centimeters per second versus H_1: $\mu \neq 50$ centimeters per second.

Table 9-1 Decisions in Hypothesis Testing

Decision	H_0 Is True	H_0 Is False
Fail to reject H_0	no error	type II error
Reject H_0	type I error	no error

Because our decision is based on random variables, probabilities can be associated with the type I and type II errors in Table 9-1. The probability of making a type I error is denoted by the Greek letter α. That is,

$$\alpha = P(\text{type I error}) = P(\text{reject } H_0 \text{ when } H_0 \text{ is true}) \tag{9-3}$$

Sometimes the type I error probability is called the **significance level,** or the **α-error,** or the size of the test. In the propellant burning rate example, a type I error will occur when either $\bar{x} > 51.5$ or $\bar{x} < 48.5$ when the true mean burning rate is $\mu = 50$ centimeters per second. Suppose that the standard deviation of burning rate is $\sigma = 2.5$ centimeters per second and that the burning rate has a distribution for which the conditions of the central limit theorem apply, so the distribution of the sample mean is approximately normal with mean $\mu = 50$ and standard deviation $\sigma/\sqrt{n} = 2.5/\sqrt{10} = 0.79$. The probability of making a type I error (or the significance level of our test) is equal to the sum of the areas that have been shaded in the tails of the normal distribution in Fig. 9-2. We may find this probability as

$$\alpha = P(\bar{X} < 48.5 \text{ when } \mu = 50) + P(\bar{X} > 51.5 \text{ when } \mu = 50)$$

The z-values that correspond to the critical values 48.5 and 51.5 are

$$z_1 = \frac{48.5 - 50}{0.79} = -1.90 \quad \text{and} \quad z_2 = \frac{51.5 - 50}{0.79} = 1.90$$

Therefore

$$\alpha = P(Z < -1.90) + P(Z > 1.90) = 0.028717 + 0.028717 = 0.057434$$

This implies that 5.76% of all random samples would lead to rejection of the hypothesis H_0: $\mu = 50$ centimeters per second when the true mean burning rate is really 50 centimeters per second.

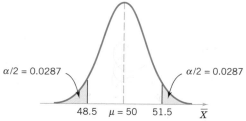

Figure 9-2 The critical region for H_0: $\mu = 50$ versus H_1: $\mu \neq 50$ and $n = 10$.

From inspection of Fig. 9-2, notice that we can reduce α by widening the acceptance region. For example, if we make the critical values 48 and 52, the value of α is

$$\alpha = P\left(Z < \frac{48 - 50}{0.79}\right) + P\left(Z > \frac{52 - 50}{0.79}\right) = P(Z < -2.53) + P(Z > 2.53)$$
$$= 0.0057 + 0.0057 = 0.0114$$

We could also reduce α by increasing the sample size. If $n = 16$, $\sigma/\sqrt{n} = 2.5/\sqrt{16} = 0.625$, and using the original critical region from Fig. 9-1, we find

$$z_1 = \frac{48.5 - 50}{0.625} = -2.40 \quad \text{and} \quad z_2 = \frac{51.5 - 50}{0.625} = 2.40$$

Therefore

$$\alpha = P(Z < -2.40) + P(Z > 2.40) = 0.0082 + 0.0082 = 0.0164$$

In evaluating a hypothesis-testing procedure, it is also important to examine the probability of a **type II error**, which we will denote by β. That is,

$$\beta = P(\text{type II error}) = P(\text{fail to reject } H_0 \text{ when } H_0 \text{ is false}) \qquad (9\text{-}4)$$

To calculate β (sometimes called the **β-error**), we must have a specific alternative hypothesis; that is, we must have a particular value of μ. For example, suppose that it is important to reject the null hypothesis H_0: $\mu = 50$ whenever the mean burning rate μ is greater than 52 centimeters per second or less than 48 centimeters per second. We could calculate the probability of a type II error β for the values $\mu = 52$ and $\mu = 48$ and use this result to tell us something about how the test procedure would perform. Specifically, how will the test procedure work if we wish to detect, that is, reject H_0, for a mean value of $\mu = 52$ or $\mu = 48$? Because of symmetry, it is necessary only to evaluate one of the two cases—say, find the probability of accepting the null hypothesis H_0: $\mu = 50$ centimeters per second when the true mean is $\mu = 52$ centimeters per second.

Figure 9-3 will help us calculate the probability of type II error β. The normal distribution on the left in Fig. 9-3 is the distribution of the test statistic $\overline{X}$ when the null hypothesis H_0: $\mu = 50$ is true (this is what is meant by the expression "under H_0: $\mu = 50$"), and the normal distribution on the right is the distribution of $\overline{X}$ when the alternative hypothesis is true and the value of the mean is 52 (or "under H_1: $\mu = 52$"). Now a type II error will be committed if the sample mean $\overline{X}$ falls between 48.5 and 51.5 (the critical region boundaries) when $\mu = 52$. As seen in Fig. 9-3, this is just the probability that $48.5 \leq \overline{X} \leq 51.5$ when the true mean is $\mu = 52$, or the shaded area under the normal distribution on the right. Therefore, referring to Fig. 9-3, we find that

$$\beta = P(48.5 \leq \overline{X} \leq 51.5 \text{ when } \mu = 52)$$

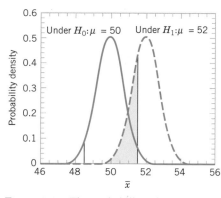

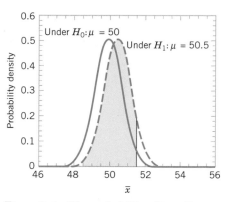

Figure 9-3   The probability of type II error when $\mu = 52$ and $n = 10$.

Figure 9-4   The probability of type II error when $\mu = 50.5$ and $n = 10$.

The z-values corresponding to 48.5 and 51.5 when $\mu = 52$ are

$$z_1 = \frac{48.5 - 52}{0.79} = -4.43 \qquad \text{and} \qquad z_2 = \frac{51.5 - 52}{0.79} = -0.63$$

Therefore

$$\beta = P(-4.43 \le Z \le -0.63) = P(Z \le -0.63) - P(Z \le -4.43)$$
$$= 0.2643 - 0.0000 = 0.2643$$

Thus, if we are testing H_0: $\mu = 50$ against H_1: $\mu \ne 50$ with $n = 10$, and the true value of the mean is $\mu = 52$, the probability that we will fail to reject the false null hypothesis is 0.2643. By symmetry, if the true value of the mean is $\mu = 48$, the value of β will also be 0.2643.

The probability of making a type II error β increases rapidly as the true value of μ approaches the hypothesized value. For example, see Fig. 9-4, where the true value of the mean is $\mu = 50.5$ and the hypothesized value is H_0: $\mu = 50$. The true value of μ is very close to 50, and the value for β is

$$\beta = P(48.5 \le \overline{X} \le 51.5 \text{ when } \mu = 50.5)$$

As shown in Fig. 9-4, the z-values corresponding to 48.5 and 51.5 when $\mu = 50.5$ are

$$z_1 = \frac{48.5 - 50.5}{0.79} = -2.53 \qquad \text{and} \qquad z_2 = \frac{51.5 - 50.5}{0.79} = 1.27$$

Therefore

$$\beta = P(-2.53 \le Z \le 1.27) = P(Z \le 1.27) - P(Z \le -2.53)$$
$$= 0.8980 - 0.0057 = 0.8923$$

Thus, the type II error probability is much higher for the case where the true mean is 50.5 centimeters per second than for the case where the mean is 52 centimeters per second. Of course,

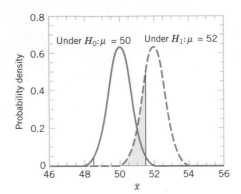

Figure 9-5 The probability of type II error when $\mu = 52$ and $n = 16$.

in many practical situations we would not be as concerned with making a type II error if the mean were "close" to the hypothesized value. We would be much more interested in detecting large differences between the true mean and the value specified in the null hypothesis.

The type II error probability also depends on the sample size n. Suppose that the null hypothesis is H_0: $\mu = 50$ centimeters per second and that the true value of the mean is $\mu = 52$. If the sample size is increased from $n = 10$ to $n = 16$, the situation of Fig. 9-5 results. The normal distribution on the left is the distribution of $\overline{X}$ when the mean $\mu = 50$, and the normal distribution on the right is the distribution of $\overline{X}$ when $\mu = 52$. As shown in Fig. 9-5, the type II error probability is

$$\beta = P(48.5 \leq \overline{X} \leq 51.5 \text{ when } \mu = 52)$$

When $n = 16$, the standard deviation of $\overline{X}$ is $\sigma/\sqrt{n} = 2.5/\sqrt{16} = 0.625$, and the z-values corresponding to 48.5 and 51.5 when $\mu = 52$ are

$$z_1 = \frac{48.5 - 52}{0.625} = -5.60 \quad \text{and} \quad z_2 = \frac{51.5 - 52}{0.625} = -0.80$$

Therefore

$$\beta = P(-5.60 \leq Z \leq -0.80) = P(Z \leq -0.80) - P(Z \leq -5.60)$$
$$= 0.2119 - 0.0000 = 0.2119$$

Recall that when $n = 10$ and $\mu = 52$, we found that $\beta = 0.2643$; therefore, increasing the sample size results in a decrease in the probability of type II error.

The results from this section and a few other similar calculations are summarized in the following table:

Acceptance Region	Sample Size	α	β at $\mu = 52$	β at $\mu = 50.5$
$48.5 < \bar{x} < 51.5$	10	0.0576	0.2643	0.8923
$48 < \bar{x} < 52$	10	0.0114	0.5000	0.9705
$48.5 < \bar{x} < 51.5$	16	0.0164	0.2119	0.9445
$48 < \bar{x} < 52$	16	0.0014	0.5000	0.9918

The results in boxes were not calculated in the text but can easily be verified by the reader. This display and the discussion above reveal four important points:

1. The size of the critical region, and consequently the probability of a type I error α, can always be reduced by appropriate selection of the critical values.

2. Type I and type II errors are related. A decrease in the probability of one type of error always results in an increase in the probability of the other, provided that the sample size n does not change.

3. An increase in sample size will generally reduce both α and β, provided that the critical values are held constant.

4. When the null hypothesis is false, β increases as the true value of the parameter approaches the value hypothesized in the null hypothesis. The value of β decreases as the difference between the true mean and the hypothesized value increases.

Generally, the analyst controls the type I error probability α when he or she selects the critical values. Thus, it is usually easy for the analyst to set the type I error probability at (or near) any desired value. Since the analyst can directly control the probability of wrongly rejecting H_0, we always think of rejection of the null hypothesis H_0 as a **strong conclusion.**

On the other hand, the probability of type II error β is not a constant, but depends on the true value of the parameter. It also depends on the sample size that we have selected. Because the type II error probability β is a function of both the sample size and the extent to which the null hypothesis H_0 is false, it is customary to think of the decision to accept H_0 as a **weak conclusion,** unless we know that β is acceptably small. Therefore, rather than saying we "accept H_0", we prefer the terminology "fail to reject H_0". Failing to reject H_0 implies that we have not found sufficient evidence to reject H_0, that is, to make a strong statement. Failing to reject H_0 does not necessarily mean that there is a high probability that H_0 is true. It may simply mean that more data are required to reach a strong conclusion. This can have important implications for the formulation of hypotheses.

An important concept that we will make use of is the **power** of a statistical test.

Definition

> The **power** of a statistical test is the probability of rejecting the null hypothesis H_0 when the alternative hypothesis is true.

The power is computed as $1 - \beta$, and **power** can be interpreted as **the probability of correctly rejecting a false null hypothesis.** We often compare statistical tests by comparing their power properties. For example, consider the propellant burning rate problem when we are testing $H_0: \mu = 50$ centimeters per second against $H_1: \mu \neq 50$ centimeters per second. Suppose that the true value of the mean is $\mu = 52$. When $n = 10$, we found that $\beta = 0.2643$, so the power of this test is $1 - \beta = 1 - 0.2643 = 0.7357$ when $\mu = 52$.

Power is a very descriptive and concise measure of the **sensitivity** of a statistical test, where by sensitivity we mean the ability of the test to detect differences. In this case, the sensitivity of the test for detecting the difference between a mean burning rate of 50 centimeters per second and 52 centimeters per second is 0.7357. That is, if the true mean is really 52 centimeters per second, this test will correctly reject $H_0: \mu = 50$ and "detect" this difference 73.57% of the time. If this value of power is judged to be too low, the analyst can increase either α or the sample size n.

9-1.3 One-Sided and Two-Sided Hypotheses

A test of any hypothesis such as

$$H_0: \mu = \mu_0$$
$$H_1: \mu \neq \mu_0$$

is called a **two-sided** test, because it is important to detect differences from the hypothesized value of the mean μ_0 that lie on either side of μ_0. In such a test, the critical region is split into two parts, with (usually) equal probability placed in each tail of the distribution of the test statistic.

Many hypothesis-testing problems naturally involve a **one-sided** alternative hypothesis, such as

$$H_0: \mu = \mu_0 \qquad \qquad H_0: \mu = \mu_0$$
$$\text{or}$$
$$H_1: \mu > \mu_0 \qquad \qquad H_1: \mu < \mu_0$$

If the alternative hypothesis is $H_1: \mu > \mu_0$, the critical region should lie in the upper tail of the distribution of the test statistic, whereas if the alternative hypothesis is $H_1: \mu < \mu_0$, the critical region should lie in the lower tail of the distribution. Consequently, these tests are sometimes called **one-tailed** tests. The location of the critical region for one-sided tests is usually easy to determine. Simply visualize the behavior of the test statistic if the null hypothesis is true and place the critical region in the appropriate end or tail of the distribution. Generally, the inequality in the alternative hypothesis "points" in the direction of the critical region.

In constructing hypotheses, we will always state the null hypothesis as an equality so that the probability of type I error α can be controlled at a specific value. The alternative hypothesis might be either one-sided or two-sided, depending on the conclusion to be drawn if H_0 is rejected. If the objective is to make a claim involving statements such as greater than, less than, superior to, exceeds, at least, and so forth, a one-sided alternative is appropriate. If no direction is implied by the claim, or if the claim not equal to is to be made, a two-sided alternative should be used.

EXAMPLE 9-1

Consider the propellant burning rate problem. Suppose that if the burning rate is less than 50 centimeters per second, we wish to show this with a strong conclusion. The hypotheses should be stated as

$$H_0: \mu = 50 \text{ centimeters per second}$$
$$H_1: \mu < 50 \text{ centimeters per second}$$

Here the critical region lies in the lower tail of the distribution of $\overline{X}$. Since the rejection of H_0 is always a strong conclusion, this statement of the hypotheses will produce the desired outcome if H_0 is rejected. Notice that, although the null hypothesis is stated with an equal sign, it is understood to include any value of μ not specified by the alternative hypothesis. Therefore, failing to reject H_0 does not mean that $\mu = 50$ centimeters per second exactly, but only that we do not have strong evidence in support of H_1.

In some real-world problems where one-sided test procedures are indicated, it is occasionally difficult to choose an appropriate formulation of the alternative hypothesis. For example, suppose that a soft-drink beverage bottler purchases 10-ounce bottles from a glass company. The bottler wants to be sure that the bottles meet the specification on mean internal pressure or bursting strength, which for 10-ounce bottles is a minimum strength of 200 psi. The bottler has decided to formulate the decision procedure for a specific lot of bottles as a hypothesis testing problem. There are two possible formulations for this problem, either

$$H_0: \mu = 200 \text{ psi}$$
$$H_1: \mu > 200 \text{ psi} \tag{9-5}$$

or

$$H_0: \mu = 200 \text{ psi}$$
$$H_1: \mu < 200 \text{ psi} \tag{9-6}$$

Consider the formulation in Equation 9-5. If the null hypothesis is rejected, the bottles will be judged satisfactory; if H_0 is not rejected, the implication is that the bottles do not conform to specifications and should not be used. Because rejecting H_0 is a strong conclusion, this formulation forces the bottle manufacturer to "demonstrate" that the mean bursting strength of the bottles exceeds the specification. Now consider the formulation in Equation 9-6. In this situation, the bottles will be judged satisfactory unless H_0 is rejected. That is, we conclude that the bottles are satisfactory unless there is strong evidence to the contrary.

Which formulation is correct, the one of Equation 9-5 or Equation 9-6? The answer is it depends. For Equation 9-5, there is some probability that H_0 will not be rejected (i.e., we would decide that the bottles are not satisfactory), even though the true mean is slightly greater than 200 psi. This formulation implies that we want the bottle manufacturer to demonstrate that the product meets or exceeds our specifications. Such a formulation could be appropriate if the manufacturer has experienced difficulty in meeting specifications in the past or if product safety considerations force us to hold tightly to the 200 psi specification. On the other hand, for the formulation of Equation 9-6 there is some probability that H_0 will be accepted and the bottles judged satisfactory, even though the true mean is slightly less than 200 psi. We would conclude that the bottles are unsatisfactory only when there is strong evidence that the mean does not exceed 200 psi, that is, when $H_0: \mu = 200$ psi is rejected. This formulation assumes that we are relatively happy with the bottle manufacturer's past performance and that small deviations from the specification of $\mu \geq 200$ psi are not harmful.

In formulating one-sided alternative hypotheses, we should remember that rejecting H_0 is always a strong conclusion. Consequently, we should put the statement about which it is important to make a strong conclusion in the alternative hypothesis. In real-world problems, this will often depend on our point of view and experience with the situation.

9-1.4   General Procedure for Hypothesis Tests

This chapter develops hypothesis-testing procedures for many practical problems. Use of the following sequence of steps in applying hypothesis-testing methodology is recommended.

1. From the problem context, identify the parameter of interest.
2. State the null hypothesis, H_0.
3. Specify an appropriate alternative hypothesis, H_1.
4. Choose a significance level α.

5. Determine an appropriate test statistic.

6. State the rejection region for the statistic.

7. Compute any necessary sample quantities, substitute these into the equation for the test statistic, and compute that value.

8. Decide whether or not H_0 should be rejected and report that in the problem context.

Steps 1–4 should be completed prior to examination of the sample data. This sequence of steps will be illustrated in subsequent sections.

EXERCISES FOR SECTION 9-1

9-1. In each of the following situations, state whether it is a correctly stated hypothesis testing problem and why.
(a) $H_0: \mu = 25, H_1: \mu \neq 25$ 2 sided
(b) $H_0: \sigma > 10, H_1: \sigma = 10$ 'cuz $H_0 =$
(c) $H_0: \bar{x} = 50, H_1: \bar{x} \neq 50$ 'cuz $\bar{x}$ is stat not charac
(d) $H_0: p = 0.1, H_1: p = 0.5$ 'cuz $H_0 = H_1$
(e) $H_0: s = 30, H_1: s > 30$ 'cuz isn't a pop parameter

9-2. A textile fiber manufacturer is investigating a new drapery yarn, which the company claims has a mean thread elongation of 12 kilograms with a standard deviation of 0.5 kilograms. The company wishes to test the hypothesis $H_0: \mu = 12$ against $H_1: \mu < 12$, using a random sample of four specimens.
(a) What is the type I error probability if the critical region is defined as $\bar{x} < 11.5$ kilograms?
(b) Find β for the case where the true mean elongation is 11.25 kilograms.

9-3. Repeat Exercise 9-2 using a sample size of $n = 16$ and the same critical region.

9-4. In Exercise 9-2, find the boundary of the critical region if the type I error probability is specified to be $\alpha = 0.01$.

9-5. In Exercise 9-2, find the boundary of the critical region if the type I error probability is specified to be 0.05.

9-6. The heat evolved in calories per gram of a cement mixture is approximately normally distributed. The mean is thought to be 100 and the standard deviation is 2. We wish to test $H_0: \mu = 100$ versus $H_1: \mu \neq 100$ with a sample of $n = 9$ specimens.
(a) If the acceptance region is defined as $98.5 \leq \bar{x} \leq 101.5$, find the type I error probability α.
(b) Find β for the case where the true mean heat evolved is 103.
(c) Find β for the case where the true mean heat evolved is 105. This value of β is smaller than the one found in part (b) above. Why?

9-7. Repeat Exercise 9-6 using a sample size of $n = 5$ and the same acceptance region.

9-8. A consumer products company is formulating a new shampoo and is interested in foam height (in millimeters). Foam height is approximately normally distributed and has a standard deviation of 20 millimeters. The company wishes to test $H_0: \mu = 175$ millimeters versus $H_1: \mu > 175$ millimeters, using the results of $n = 10$ samples.
(a) Find the type I error probability α if the critical region is $\bar{x} > 185$.
(b) What is the probability of type II error if the true mean foam height is 195 millimeters?

9-9. In Exercise 9-8, suppose that the sample data result in $\bar{x} = 190$ millimeters.
(a) What conclusion would you reach?
(b) How "unusual" is the sample value $\bar{x} = 190$ millimeters if the true mean is really 175 millimeters? That is, what is the probability that you would observe a sample average as large as 190 millimeters (or larger), if the true mean foam height was really 175 millimeters?

9-10. Repeat Exercise 9-8 assuming that the sample size is $n = 16$ and the boundary of the critical region is the same.

9-11. Consider Exercise 9-8, and suppose that the sample size is increased to $n = 16$.
(a) Where would the boundary of the critical region be placed if the type I error probability were to remain equal to the value that it took on when $n = 10$?
(b) Using $n = 16$ and the new critical region found in part (a), find the type II error probability β if the true mean foam height is 195 millimeters.
(c) Compare the value of β obtained in part (b) with the value from Exercise 9-8 (b). What conclusions can you draw?

9-12. A manufacturer is interested in the output voltage of a power supply used in a PC. Output voltage is assumed to be normally distributed, with standard deviation 0.25 Volts, and the manufacturer wishes to test $H_0: \mu = 5$ Volts against $H_1: \mu \neq 5$ Volts, using $n = 8$ units.
(a) The acceptance region is $4.85 \leq \bar{x} \leq 5.15$. Find the value of α.
(b) Find the power of the test for detecting a true mean output voltage of 5.1 Volts.

9-13. Rework Exercise 9-12 when the sample size is 16 and the boundaries of the acceptance region do not change.

9-14. Consider Exercise 9-12, and suppose that the manufacturer wants the type I error probability for the test to be $\alpha = 0.05$. Where should the acceptance region be located?

9-15. If we plot the probability of accepting H_0: $\mu = \mu_0$ versus various values of μ and connect the points with a smooth curve, we obtain the **operating characteristic curve** (or the **OC curve**) of the test procedure. These curves are used extensively in industrial applications of hypothesis testing to display the sensitivity and relative performance of the test. When the true mean is really equal to μ_0, the probability of accepting H_0 is $1 - \alpha$. Construct an OC curve for Exercise 9-8, using values of the true mean μ of 178, 181, 184, 187, 190, 193, 196, and 199.

9-16. Convert the OC curve in Exercise 9-15 into a plot of the **power function** of the test.

9-17. A random sample of 500 registered voters in Phoenix is asked if they favor the use of oxygenated fuels year-round to reduce air pollution. If more than 400 voters respond positively, we will conclude that at least 60% of the voters favor the use of these fuels.
(a) Find the probability of type I error if exactly 60% of the voters favor the use of these fuels.
(b) What is the type II error probability β if 75% of the voters favor this action?
  Hint: use the normal approximation to the binomial.

9-18. The proportion of residents in Phoenix favoring the building of toll roads to complete the freeway system is believed to be $p = 0.3$. If a random sample of 10 residents shows that 1 or fewer favor this proposal, we will conclude that $p < 0.3$.
(a) Find the probability of type I error if the true proportion is $p = 0.3$.
(b) Find the probability of committing a type II error with this procedure if $p = 0.2$.
(c) What is the power of this procedure if the true proportion is $p = 0.2$?

9-19. The proportion of adults living in Tempe, Arizona, who are college graduates is estimated to be $p = 0.4$. To test this hypothesis, a random sample of 15 Tempe adults is selected. If the number of college graduates is between 4 and 8, the hypothesis will be accepted; otherwise, we will conclude that $p \neq 0.4$.
(a) Find the type I error probability for this procedure, assuming that $p = 0.4$.
(b) Find the probability of committing a type II error if the true proportion is really $p = 0.2$.

9-2 TESTS ON THE MEAN OF A NORMAL DISTRIBUTION, VARIANCE KNOWN

In this section, we consider hypothesis testing about the mean μ of a single, normal population where the variance of the population σ^2 is known. We will assume that a random sample X_1, $X_2, \ldots, X_n$ has been taken from the population. Based on our previous discussion, the sample mean $\overline{X}$ is an **unbiased point estimator** of μ with variance σ^2/n.

9-2.1 Hypothesis Tests on the Mean

Suppose that we wish to test the hypotheses

$$H_0: \mu = \mu_0$$
$$H_1: \mu \neq \mu_0 \tag{9-7}$$

where μ_0 is a specified constant. We have a random sample $X_1, X_2, \ldots, X_n$ from a normal population. Since $\overline{X}$ has a normal distribution (i.e., the **sampling distribution** of $\overline{X}$ is normal) with mean μ_0 and standard deviation $\sigma/\sqrt{n}$ if the null hypothesis is true, we could construct a critical region based on the computed value of the sample mean $\overline{X}$, as in Section 9-1.2.

It is usually more convenient to *standardize* the sample mean and use a test statistic based on the standard normal distribution. That is, the test procedure for H_0: $\mu = \mu_0$ uses the **test statistic**

$$Z_0 = \frac{\overline{X} - \mu_0}{\sigma/\sqrt{n}} \tag{9-8}$$

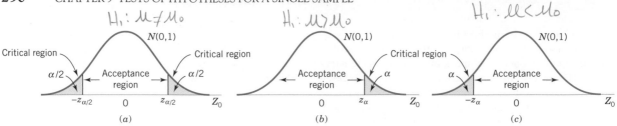

Figure 9-6   The distribution of Z_0 when H_0: $\mu = \mu_0$ is true, with critical region for (a) the two-sided alternative H_1: $\mu \neq \mu_0$, (b) the one-sided alternative H_1: $\mu > \mu_0$, and (c) the one-sided alternative H_1: $\mu < \mu_0$.

If the null hypothesis H_0: $\mu = \mu_0$ is true, $E(\overline{X}) = \mu_0$, and it follows that the distribution of Z_0 is the standard normal distribution [denoted $N(0, 1)$]. Consequently, if H_0: $\mu = \mu_0$ is true, the probability is $1 - \alpha$ that the test statistic Z_0 falls between $-z_{\alpha/2}$ and $z_{\alpha/2}$, where $z_{\alpha/2}$ is the $100\alpha/2$ percentage point of the standard normal distribution. The regions associated with $z_{\alpha/2}$ and $-z_{\alpha/2}$ are illustrated in Fig. 9-6(a). Note that the probability is α that the test statistic Z_0 will fall in the region $Z_0 > z_{\alpha/2}$ or $Z_0 < -z_{\alpha/2}$ when H_0: $\mu = \mu_0$ is true. Clearly, a sample producing a value of the test statistic that falls in the tails of the distribution of Z_0 would be unusual if H_0: $\mu = \mu_0$ is true; therefore, it is an indication that H_0 is false. Thus, we should reject H_0 if the observed value of the test statistic z_0 is either

$$z_0 > z_{\alpha/2} \quad \text{or} \quad z_0 < -z_{\alpha/2} \tag{9-9}$$

and we should fail to reject H_0 if

$$-z_{\alpha/2} \leq z_0 \leq z_{\alpha/2} \quad \text{two sided} \tag{9-10}$$

The inequalities in Equation 9-10 defines the **acceptance region** for H_0, and the two inequalities in Equation 9-9 define the **critical region** or **rejection region.** The type I error probability for this test procedure is α.

It is easier to understand the critical region and the test procedure, in general, when the test statistic is Z_0 rather than $\overline{X}$. However, the same critical region can always be written in terms of the computed value of the sample mean $\bar{x}$. A procedure identical to the above is as follows:

$$\text{Reject } H_0: \mu = \mu_0 \text{ if either } \bar{x} > a \text{ or } \bar{x} < b$$

where

$$a = \mu_0 + z_{\alpha/2}\sigma/\sqrt{n} \quad \text{and} \quad b = \mu_0 - z_{\alpha/2}\sigma/\sqrt{n}$$

EXAMPLE 9-2   Aircrew escape systems are powered by a solid propellant. The burning rate of this propellant is an important product characteristic. Specifications require that the mean burning rate must be 50 centimeters per second. We know that the standard deviation of burning rate is $\sigma = 2$ centimeters per second. The experimenter decides to specify a type I error probability or significance level of $\alpha = 0.05$ and selects a random sample of $n = 25$ and obtains a sample average burning rate of $\bar{x} = 51.3$ centimeters per second. What conclusions should be drawn?

We may solve this problem by following the eight-step procedure outlined in Section 9-1.4. This results in

1. The parameter of interest is μ, the mean burning rate.
2. H_0: $\mu = 50$ centimeters per second
3. H_1: $\mu \neq 50$ centimeters per second
4. $\alpha = 0.05$
5. The test statistic is

$$z_0 = \frac{\bar{x} - \mu_0}{\sigma/\sqrt{n}}$$

6. Reject H_0 if $z_0 > 1.96$ or if $z_0 < -1.96$. Note that this results from step 4, where we specified $\alpha = 0.05$, and so the boundaries of the critical region are at $z_{0.025} = 1.96$ and $-z_{0.025} = -1.96$.
7. Computations: Since $\bar{x} = 51.3$ and $\sigma = 2$,

$$z_0 = \frac{51.3 - 50}{2/\sqrt{25}} = 3.25$$

8. Conclusion: Since $z_0 = 3.25 > 1.96$, we reject H_0: $\mu = 50$ at the 0.05 level of significance. Stated more completely, we conclude that the mean burning rate differs from 50 centimeters per second, based on a sample of 25 measurements. In fact, there is strong evidence that the mean burning rate exceeds 50 centimeters per second.

We may also develop procedures for testing hypotheses on the mean μ where the alternative hypothesis is one-sided. Suppose that we specify the hypotheses as

$$\begin{aligned} H_0: \mu &= \mu_0 \\ H_1: \mu &> \mu_0 \end{aligned} \tag{9-11}$$

In defining the critical region for this test, we observe that a negative value of the test statistic Z_0 would never lead us to conclude that H_0: $\mu = \mu_0$ is false. Therefore, we would place the critical region in the **upper tail** of the standard normal distribution and reject H_0 if the computed value of z_0 is too large. That is, we would reject H_0 if

$$z_0 > z_\alpha \tag{9-12}$$

as shown in Figure 9-6(b). Similarly, to test

$$\begin{aligned} H_0: \mu &= \mu_0 \\ H_1: \mu &< \mu_0 \end{aligned} \tag{9-13}$$

we would calculate the test statistic Z_0 and reject H_0 if the value of z_0 is too small. That is, the critical region is in the **lower tail** of the standard normal distribution as shown in Figure 9-6(c), and we reject H_0 if

$$z_0 < -z_\alpha \tag{9-14}$$

9-2.2 P-Values in Hypothesis Tests

One way to report the results of a hypothesis test is to state that the null hypothesis was or was not rejected at a specified α-value or level of significance. For example, in the propellant problem above, we can say that H_0: $\mu = 50$ was rejected at the 0.05 level of significance. This statement of conclusions is often inadequate because it gives the decision maker no idea about whether the computed value of the test statistic was just barely in the rejection region or whether it was very far into this region. Furthermore, stating the results this way imposes the predefined level of significance on other users of the information. This approach may be unsatisfactory because some decision makers might be uncomfortable with the risks implied by $\alpha = 0.05$.

To avoid these difficulties the **P-value approach** has been adopted widely in practice. The *P*-value is the probability that the test statistic will take on a value that is at least as extreme as the observed value of the statistic when the null hypothesis H_0 is true. Thus, a *P*-value conveys much information about the weight of evidence against H_0, and so a decision maker can draw a conclusion at *any* specified level of significance. We now give a formal definition of a *P*-value.

Definition

> The *P*-value is the smallest level of significance that would lead to rejection of the null hypothesis H_0 with the given data.

It is customary to call the test statistic (and the data) significant when the null hypothesis H_0 is rejected; therefore, we may think of the *P*-value as the smallest level α at which the data are significant. Once the *P*-value is known, the decision maker can determine how significant the data are without the data analyst formally imposing a preselected level of significance.

For the foregoing normal distribution tests it is relatively easy to compute the *P*-value. If z_0 is the computed value of the test statistic, the *P*-value is

$$P = \begin{cases} 2[1 - \Phi(|z_0|)] & \text{for a two-tailed test: } H_0: \mu = \mu_0 \qquad H_1: \mu \neq \mu_0 \\ 1 - \Phi(z_0) & \text{for a upper-tailed test: } H_0: \mu = \mu_0 \qquad H_1: \mu > \mu_0 \\ \Phi(z_0) & \text{for a lower-tailed test: } H_0: \mu = \mu_0 \qquad H_1: \mu < \mu_0 \end{cases} \qquad (9\text{-}15)$$

Here, $\Phi(z)$ is the standard normal cumulative distribution function defined in Chapter 4. Recall that $\Phi(z) = P(Z \leq z)$, where Z is $N(0, 1)$. To illustrate this, consider the propellant problem in Example 9-2. The computed value of the test statistic is $z_0 = 3.25$ and since the alternative hypothesis is two-tailed, the *P*-value is

$$P\text{-value} = 2[1 - \Phi(3.25)] = 0.0012$$

Thus, H_0: $\mu = 50$ would be rejected at any level of significance $\alpha \geq P\text{-value} = 0.0012$. For example, H_0 would be rejected if $\alpha = 0.01$, but it would not be rejected if $\alpha = 0.001$.

It is not always easy to compute the exact *P*-value for a test. However, most modern computer programs for statistical analysis report *P*-values, and they can be obtained on some hand-held calculators. We will also show how to approximate the *P*-value. Finally, if the

P-value approach is used, step 6 of the hypothesis-testing procedure can be modified. Specifically, it is not necessary to state explicitly the critical region.

9-2.3 Connection between Hypothesis Tests and Confidence Intervals

There is a close relationship between the test of a hypothesis about any parameter, say θ, and the confidence interval for θ. If $[l, u]$ is a $100(1 - \alpha)\%$ confidence interval for the parameter θ, the test of size α of the hypothesis

$$H_0: \theta = \theta_0$$
$$H_1: \theta \neq \theta_0$$

will lead to rejection of H_0 if and only if θ_0 is **not** in the $100(1 - \alpha)\%$ CI $[l, u]$. As an illustration, consider the escape system propellant problem discussed above. The null hypothesis $H_0: \mu = 50$ was rejected, using $\alpha = 0.05$. The 95% two-sided CI on μ can be calculated using Equation 8-7. This CI is $50.52 \leq \mu \leq 52.08$. Because the value $\mu_0 = 50$ is not included in this interval, the null hypothesis $H_0: \mu = 50$ is rejected.

Although hypothesis tests and CIs are equivalent procedures insofar as decision making or **inference** about μ is concerned, each provides somewhat different insights. For instance, the confidence interval provides a range of likely values for μ at a stated confidence level, whereas hypothesis testing is an easy framework for displaying the **risk levels** such as the *P*-value associated with a specific decision. We will continue to illustrate the connection between the two procedures throughout the text.

9-2.4 Type II Error and Choice of Sample Size

In testing hypotheses, the analyst directly selects the type I error probability. However, the probability of type II error β depends on the choice of sample size. In this section, we will show how to calculate the probability of type II error β. We will also show how to select the sample size to obtain a specified value of β.

Finding the Probability of Type II Error β
Consider the two-sided hypothesis

$$H_0: \mu = \mu_0$$
$$H_1: \mu \neq \mu_0$$

Suppose that the null hypothesis is false and that the true value of the mean is $\mu = \mu_0 + \delta$, say, where $\delta > 0$. The test statistic Z_0 is

$$Z_0 = \frac{\overline{X} - \mu_0}{\sigma/\sqrt{n}} = \frac{\overline{X} - (\mu_0 + \delta)}{\sigma/\sqrt{n}} + \frac{\delta\sqrt{n}}{\sigma}$$

Therefore, the distribution of Z_0 when H_1 is true is

$$Z_0 \sim N\left(\frac{\delta\sqrt{n}}{\sigma}, 1\right) \tag{9-16}$$

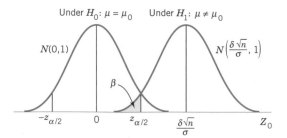

Figure 9-7   The distribution of Z_0 under H_0 and H_1.

The distribution of the test statistic Z_0 under both the null hypothesis H_0 and the alternate hypothesis H_1 is shown in Fig. 9-7. From examining this figure, we note that if H_1 is true, a type II error will be made only if $-z_{\alpha/2} \leq Z_0 \leq z_{\alpha/2}$ where $Z_0 \sim N(\delta\sqrt{n}/\sigma, 1)$. That is, the probability of the type II error β is the probability that Z_0 falls between $-z_{\alpha/2}$ and $z_{\alpha/2}$ *given that H_1 is true*. This probability is shown as the shaded portion of Fig. 9-7. Expressed mathematically, this probability is

$$\beta = \Phi\left(z_{\alpha/2} - \frac{\delta\sqrt{n}}{\sigma}\right) - \Phi\left(-z_{\alpha/2} - \frac{\delta\sqrt{n}}{\sigma}\right) \tag{9-17}$$

where $\Phi(z)$ denotes the probability to the left of z in the standard normal distribution. Note that Equation 9-17 was obtained by evaluating the probability that Z_0 falls in the interval $[-z_{\alpha/2}, z_{\alpha/2}]$ when H_1 is true. Furthermore, note that Equation 9-17 also holds if $\delta < 0$, due to the symmetry of the normal distribution. It is also possible to derive an equation similar to Equation 9-17 for a one-sided alternative hypothesis.

Sample Size Formulas

One may easily obtain formulas that determine the appropriate sample size to obtain a particular value of β for a given δ and α. For the two-sided alternative hypothesis, we know from Equation 9-17 that

$$\beta = \Phi\left(z_{\alpha/2} - \frac{\delta\sqrt{n}}{\sigma}\right) - \Phi\left(-z_{\alpha/2} - \frac{\delta\sqrt{n}}{\sigma}\right)$$

or if $\delta > 0$,

$$\beta \simeq \Phi\left(z_{\alpha/2} - \frac{\delta\sqrt{n}}{\sigma}\right) \tag{9-18}$$

since $\Phi(-z_{\alpha/2} - \delta\sqrt{n}/\sigma) \simeq 0$ when δ is positive. Let z_β be the 100β upper percentile of the standard normal distribution. Then, $\beta = \Phi(-z_\beta)$. From Equation 9-18

$$-z_\beta \simeq z_{\alpha/2} - \frac{\delta\sqrt{n}}{\sigma}$$

or

$$n \simeq \frac{(z_{\alpha/2} + z_\beta)^2 \sigma^2}{\delta^2} \qquad \text{two sided.} \qquad (9\text{-}19)$$

where

$$\delta = \mu - \mu_0$$

This approximation is good when $\Phi(-z_{\alpha/2} - \delta\sqrt{n}/\sigma)$ is small compared to β. For either of the one-sided alternative hypotheses the sample size required to produce a specified type II error with probability β given δ and α is

$$n = \frac{(z_\alpha + z_\beta)^2 \sigma^2}{\delta^2} \qquad \text{one sided} \qquad (9\text{-}20)$$

where

$$\delta = \mu - \mu_0$$

EXAMPLE 9-3

Consider the rocket propellant problem of Example 9-2. Suppose that the analyst wishes to design the test so that if the true mean burning rate differs from 50 centimeters per second by as much as 1 centimeter per second, the test will detect this (i.e., reject H_0: $\mu = 50$) with a high probability, say 0.90. Now, we note that $\sigma = 2$, $\delta = 51 - 50 = 1$, $\alpha = 0.05$, and $\beta = 0.10$. Since $z_{\alpha/2} = z_{0.025} = 1.96$ and $z_\beta = z_{0.10} = 1.28$, the sample size required to detect this departure from H_0: $\mu = 50$ is found by Equation 9-19 as

μ = true mean
μ_0 = mean

$$n \simeq \frac{(z_{\alpha/2} + z_\beta)^2 \sigma^2}{\delta^2} = \frac{(1.96 + 1.28)^2 2^2}{(1)^2} \simeq 42$$

The approximation is good here, since $\Phi(-z_{\alpha/2} - \delta\sqrt{n}/\sigma) = \Phi(-1.96 - (1)\sqrt{42}/2) = \Phi(-5.20) \simeq 0$, which is small relative to β.

Using Operating Characteristic Curves
When performing sample size or type II error calculations, it is sometimes more convenient to use the **operating characteristic curves** in Appendix Charts VI*a* and VI*b*. These curves plot β as calculated from Equation 9-17 against a parameter d for various sample sizes n. Curves are provided for both $\alpha = 0.05$ and $\alpha = 0.01$. The parameter d is defined as

$$d = \frac{|\mu - \mu_0|}{\sigma} = \frac{|\delta|}{\sigma} \qquad (9\text{-}21)$$

so one set of operating characteristic curves can be used for all problems regardless of the values of μ_0 and σ. From examining the operating characteristic curves or Equation 9-17 and Fig. 9-7, we note that

1. The further the true value of the mean μ is from μ_0, the smaller the probability of type II error β for a given n and α. That is, we see that for a specified sample size and α, large differences in the mean are easier to detect than small ones.

2. For a given δ and α, the probability of type II error β decreases as n increases. That is, to detect a specified difference δ in the mean, we may make the test more powerful by increasing the sample size.

EXAMPLE 9-4

Consider the propellant problem in Example 9-2. Suppose that the analyst is concerned about the probability of type II error if the true mean burning rate is $\mu = 51$ centimeters per second. We may use the operating characteristic curves to find β. Note that $\delta = 51 - 50 = 1$, $n = 25$, $\sigma = 2$, and $\alpha = 0.05$. Then using Equation 9-21 gives

$$d = \frac{|\mu - \mu_0|}{\sigma} = \frac{|\delta|}{\sigma} = \frac{1}{2}$$

and from Appendix Chart VIa, with $n = 25$, we find that $\beta = 0.30$. That is, if the true mean burning rate is $\mu = 51$ centimeters per second, there is approximately a 30% chance that this will not be detected by the test with $n = 25$.

EXAMPLE 9-5

Once again, consider the propellant problem in Example 9-2. Suppose that the analyst would like to design the test so that if the true mean burning rate differs from 50 centimeters per second by as much as 1 centimeter per second, the test will detect this (i.e., reject H_0: $\mu = 50$) with a high probability, say, 0.90. This is exactly the same requirement as in Example 9-3, where we used Equation 9-19 to find the required sample size to be $n = 42$. The operating characteristic curves can also be used to find the sample size for this test. Since $d = |\mu - \mu_0|/\sigma = 1/2$, $\alpha = 0.05$, and $\beta = 0.10$, we find from Appendix Chart VIa that the required sample size is approximately $n = 40$. This closely agrees with the sample size calculated from Equation 9-19.

In general, the operating characteristic curves involve three parameters: β, d, and n. Given any two of these parameters, the value of the third can be determined. There are two typical applications of these curves:

1. For a given n and d, find β (as illustrated in Example 9-3). This kind of problem is often encountered when the analyst is concerned about the sensitivity of an experiment already performed, or when sample size is restricted by economic or other factors.

2. For a given β and d, find n. This was illustrated in Example 9-4. This kind of problem is usually encountered when the analyst has the opportunity to select the sample size at the outset of the experiment.

Operating characteristic curves are given in Appendix Charts VIc and VId for the one-sided alternatives. If the alternative hypothesis is either H_1: $\mu > \mu_0$ or H_1: $\mu < \mu_0$, the abscissa scale on these charts is

$$d = \frac{|\mu - \mu_0|}{\sigma} \tag{9-22}$$

Using the Computer

Many statistics software packages will calculate sample sizes and type II error probabilities. To illustrate, here are some computations from Minitab for the propellant burning rate problem.

Power and Sample Size

1-Sample Z Test
Testing mean = null (versus not = null)
Calculating power for mean = null + difference
Alpha = 0.05 Sigma = 2

Difference	Sample Size	Target Power	Actual Power
1	43	0.9000	0.9064

Power and Sample Size

1-Sample Z Test
Testing mean = null (versus not = null)
Calculating power for mean = null + difference
Alpha = 0.05 Sigma = 2

Difference	Sample Size	Target Power	Actual Power
1	28	0.7500	0.7536

Power and Sample Size

1-Sample Z Test
Testing mean = null (versus not = null)
Calculating power for mean = null + difference
Alpha = 0.05 Sigma = 2

Difference	Sample Size	Power
1	25	0.7054

In the first part of the boxed display, we asked Minitab to work Example 9-3, that is, to find the sample size n that would allow detection of a difference from $\mu_0 = 50$ of 1 centimeter per second with power of 0.9 and $\alpha = 0.05$. The answer, $n = 43$, agrees closely with the calculated value from Equation 9-19 in Example 9-3, which was $n = 42$. The difference is due to Minitab using a value of z_β that has more than two decimal places. The second part of the computer output relaxes the power requirement to 0.75. Note that the effect is to reduce the required sample size to $n = 28$. The third part of the output is the solution to Example 9-4, where we wish to determine the type II error probability of (β) or the power = $1 - \beta$ for the sample size $n = 25$. Note that Minitab computes the power to be 0.7054, which agrees closely with the answer obtained from the O.C. curve in Example 9-4. Generally, however, the computer calculations will be more accurate than visually reading values from an O.C. curve.

9-2.5 Large-Sample Test

We have developed the test procedure for the null hypothesis H_0: $\mu = \mu_0$ assuming that the population is normally distributed and that σ^2 is known. In many if not most practical situations σ^2

will be unknown. Furthermore, we may not be certain that the population is well modeled by a normal distribution. In these situations if n is large (say $n > 40$) the sample standard deviation s can be substituted for σ in the test procedures with little effect. Thus, while we have given a test for the mean of a normal distribution with known σ^2, it can be easily converted into a **large-sample test procedure for unknown σ^2** that is valid regardless of the form of the distribution of the population. This large-sample test relies on the central limit theorem just as the large-sample confidence interval on μ that was presented in the previous chapter did. Exact treatment of the case where the population is normal, σ^2 is unknown, and n is small involves use of the t distribution and will be deferred until Section 9-3.

9-2.6   Some Practical Comments on Hypothesis Tests

The Eight-Step Procedure

In Section 9-1.4 we described an eight-step procedure for statistical hypothesis testing. This procedure was illustrated in Example 9-2 and will be encountered many times in both this chapter and Chapter 10. In practice, such a formal and (seemingly) rigid procedure is not always necessary. Generally, once the experimenter (or decision maker) has decided on the question of interest and has determined the *design of the experiment* (that is, how the data are to be collected, how the measurements are to be made, and how many observations are required), only three steps are really required:

1. Specify the test statistic to be used (such as Z_0).
2. Specify the location of the critical region (two-tailed, upper-tailed, or lower-tailed).
3. Specify the criteria for rejection (typically, the value of α, or the P-value at which rejection should occur).

These steps are often completed almost simultaneously in solving real-world problems, although we emphasize that it is important to think carefully about each step. That is why we present and use the eight-step process: it seems to reinforce the essentials of the correct approach. While you may not use it every time in solving real problems, it is a helpful framework when you are first learning about hypothesis testing.

Statistical versus Practical Significance

We noted previously that reporting the results of a hypothesis test in terms of a P-value is very useful because it conveys more information than just the simple statement "reject H_0" or "fail to reject H_0". That is, rejection of H_0 at the 0.05 level of significance is much more meaningful if the value of the test statistic is well into the critical region, greatly exceeding the 5% critical value, than if it barely exceeds that value.

Even a very small P-value can be difficult to interpret from a practical viewpoint when we are making decisions because, while a small P-value indicates **statistical significance** in the sense that H_0 should be rejected in favor of H_1, the actual departure from H_0 that has been detected may have little (if any) **practical significance** (engineers like to say "engineering significance"). This is particularly true when the sample size n is large.

For example, consider the propellant burning rate problem of Example 9-3 where we are testing H_0: $\mu = 50$ centimeters per second versus H_1: $\mu \neq 50$ centimeters per second with $\sigma = 2$. If we suppose that the mean rate is really 50.5 centimeters per second, this is not a serious departure from H_0: $\mu = 50$ centimeters per second in the sense that if the mean really is 50.5 centimeters per second there is no practical observable effect on the performance of the aircrew escape system. In other words, concluding that $\mu = 50$ centimeters per second when it is really 50.5 centimeters per second is an inexpensive error and has no practical significance. For a reasonably large sample size, a true value of $\mu = 50.5$ will lead to a sample $\bar{x}$ that is close to 50.5 centimeters per second, and we would not want this value of $\bar{x}$ from the sample to result in rejection of H_0. The following display shows the P-value for testing H_0: $\mu = 50$ when we observe $\bar{x} = 50.5$ centimeters per second and the power of the test at $\alpha = 0.05$ when the true mean is 50.5 for various sample sizes n:

Sample Size n	P-value When $\bar{x} = 50.5$	Power (at $\alpha = 0.05$) When True $\mu = 50.5$
10	0.4295	0.1241
25	0.2113	0.2396
50	0.0767	0.4239
100	0.0124	0.7054
400	5.73×10^{-7}	0.9988
1000	2.57×10^{-15}	1.0000

The P-value column in this display indicates that for large sample sizes the observed sample value of $\bar{x} = 50.5$ would strongly suggest that H_0: $\mu = 50$ should be rejected, even though the observed sample results imply that from a practical viewpoint the true mean does not differ much at all from the hypothesized value $\mu_0 = 50$. The power column indicates that if we test a hypothesis at a fixed significance level α and even if there is little practical difference between the true mean and the hypothesized value, a large sample size will almost always lead to rejection of H_0. The moral of this demonstration is clear:

> Be careful when interpreting the results from hypothesis testing when the sample size is large, because any small departure from the hypothesized value μ_0 will probably be detected, even when the difference is of little or no practical significance.

EXERCISES FOR SECTION 9-2

9-20. The mean water temperature downstream from a power plant cooling tower discharge pipe should be no more than 100°F. Past experience has indicated that the standard deviation of temperature is 2°F. The water temperature is measured on nine randomly chosen days, and the average temperature is found to be 98°F.
(a) Should the water temperature be judged acceptable with $\alpha = 0.05$?
(b) What is the P-value for this test?
(c) What is the probability of accepting the null hypothesis at $\alpha = 0.05$ if the water has a true mean temperature of 104 °F?

9-21. Reconsider the chemical process yield data from Exercise 8-9. Recall that $\sigma = 3$, yield is normally distributed and that $n = 5$ observations on yield are 91.6%, 88.75%, 90.8%, 89.95%, and 91.3%. Use $\alpha = 0.05$.
(a) Is there evidence that the mean yield is not 90%?
(b) What is the P-value for this test?
(c) What sample size would be required to detect a true mean yield of 85% with probability 0.95?

(d) What is the type II error probability if the true mean yield is 92%?
(e) Compare the decision you made in part (c) with the 95% CI on mean yield that you constructed in Exercise 8-7.

9-22. A manufacturer produces crankshafts for an automobile engine. The wear of the crankshaft after 100,000 miles (0.0001 inch) is of interest because it is likely to have an impact on warranty claims. A random sample of $n = 15$ shafts is tested and $\bar{x} = 2.78$. It is known that $\sigma = 0.9$ and that wear is normally distributed.
(a) Test H_0: $\mu = 3$ versus H_0: $\mu \neq 3$ using $\alpha = 0.05$.
(b) What is the power of this test if $\mu = 3.25$?
(c) What sample size would be required to detect a true mean of 3.75 if we wanted the power to be at least 0.9?

9-23. A melting point test of $n = 10$ samples of a binder used in manufacturing a rocket propellant resulted in $\bar{x} = 154.2°$F. Assume that melting point is normally distributed with $\sigma = 1.5°$F.
(a) Test H_0: $\mu = 155$ versus H_0: $\mu \neq 155$ using $\alpha = 0.01$.
(b) What is the P-value for this test?

(c) What is the β-error if the true mean is $\mu = 150$?

(d) What value of n would be required if we want $\beta < 0.1$ when $\mu = 150$? Assume that $\alpha = 0.01$.

9-24. The life in hours of a battery is known to be approximately normally distributed, with standard deviation $\sigma = 1.25$ hours. A random sample of 10 batteries has a mean life of $\bar{x} = 40.5$ hours.

(a) Is there evidence to support the claim that battery life exceeds 40 hours? Use $\alpha = 0.05$.

(b) What is the P-value for the test in part (a)?

(c) What is the β-error for the test in part (a) if the true mean life is 42 hours?

(d) What sample size would be required to ensure that β does not exceed 0.10 if the true mean life is 44 hours?

(e) Explain how you could answer the question in part (a) by calculating an appropriate confidence bound on life.

9-25. An engineer who is studying the tensile strength of a steel alloy intended for use in golf club shafts knows that tensile strength is approximately normally distributed with $\sigma = 60$ psi. A random sample of 12 specimens has a mean tensile strength of $\bar{x} = 3250$ psi.

(a) Test the hypothesis that mean strength is 3500 psi. Use $\alpha = 0.01$.

(b) What is the smallest level of significance at which you would be willing to reject the null hypothesis?

(c) Explain how you could answer the question in part (a) with a two-sided confidence interval on mean tensile strength.

9-26. Suppose that in Exercise 9-25 we wanted to reject the null hypothesis with probability at least 0.8 if mean strength $\mu = 3500$. What sample size should be used?

9-27. Supercavitation is a propulsion technology for undersea vehicles that can greatly increase their speed. It occurs above approximately 50 meters per second, when pressure drops sufficiently to allow the water to dissociate into water vapor, forming a gas bubble behind the vehicle. When the gas bubble completely encloses the vehicle, supercavitation is said to occur. Eight tests were conducted on a scale model of an undersea vehicle in a towing basin with the average observed speed $\bar{x} = 102.2$ meters per second. Assume that speed is normally distributed with known standard deviation $\sigma = 4$ meters per second.

(a) Test the hypotheses $H_0: \mu = 100$ versus $H_1: \mu < 100$ using $\alpha = 0.05$.

(b) Compute the power of the test if the true mean speed is as low as 95 meters per second.

(c) What sample size would be required to detect a true mean speed as low as 95 meters per second if we wanted the power of the test to be at least 0.85?

(d) Explain how the question in part (a) could be answered by constructing a one-sided confidence bound on the mean speed.

9-28. A bearing used in an automotive application is suppose to have a nominal inside diameter of 1.5 inches. A random sample of 25 bearings is selected and the average inside diameter of these bearings is 1.4975 inches. Bearing diameter is known to be normally distributed with standard deviation $\sigma = 0.01$ inch.

(a) Test the hypotheses $H_0: \mu = 1.5$ versus $H_1: \mu \neq 1.5$ using $\alpha = 0.01$.

(b) Compute the power of the test if the true mean diameter is 1.495 inches.

(c) What sample size would be required to detect a true mean diameter as low as 1.495 inches if we wanted the power of the test to be at least 0.9?

(d) Explain how the question in part (a) could be answered by constructing a two-sided confidence interval on the mean diameter.

9-29. Medical researchers have developed a new artificial heart constructed primarily of titanium and plastic. The heart will last and operate almost indefinitely once it is implanted in the patient's body, but the battery pack needs to be recharged about every four hours. A random sample of 50 battery packs is selected and subjected to a life test. The average life of these batteries is 4.05 hours. Assume that battery life is normally distributed with standard deviation $\sigma = 0.2$ hour.

(a) Is there evidence to support the claim that mean battery life exceeds 4 hours? Use $\alpha = 0.05$.

(b) Compute the power of the test if the true mean battery life is 4.5 hours.

(c) What sample size would be required to detect a true mean battery life of 4.5 hours if we wanted the power of the test to be at least 0.9?

(d) Explain how the question in part (a) could be answered by constructing a one-sided confidence bound on the mean life.

9-3 TESTS ON THE MEAN OF A NORMAL DISTRIBUTION, VARIANCE UNKNOWN

9-3.1 Hypothesis Tests on the Mean

We now consider the case of **hypothesis testing** on the mean of a population with **unknown variance** σ^2. The situation is analogous to Section 8-3, where we considered a **confidence interval** on the mean for the same situation. As in that section, the validity of the test procedure we will describe rests on the assumption that the population distribution is at least approximately

normal. The important result upon which the test procedure relies is that if $X_1, X_2, \ldots, X_n$ is a random sample from a normal distribution with mean μ and variance σ^2, the random variable

$$T = \frac{\overline{X} - \mu}{S/\sqrt{n}}$$

has a t distribution with $n - 1$ degrees of freedom. Recall that we used this result in Section 8-3 to devise the t-confidence interval for μ. Now consider testing the hypotheses

$$H_0: \mu = \mu_0$$
$$H_1: \mu \neq \mu_0$$

We will use the **test statistic**

$$T_0 = \frac{\overline{X} - \mu_0}{S/\sqrt{n}} \qquad (9\text{-}23)$$

If the null hypothesis is true, T_0 has a t distribution with $n - 1$ degrees of freedom. When we know the distribution of the test statistic when H_0 is true (this is often called the **reference distribution** or the **null distribution**), we can locate the critical region to control the type I error probability at the desired level. In this case we would use the t percentage points $-t_{\alpha/2,n-1}$ and $t_{\alpha/2,n-1}$ as the boundaries of the critical region so that we would reject $H_0: \mu = \mu_0$ if

$$t_0 > t_{\alpha/2,n-1} \qquad \text{or if} \quad t_0 < -t_{\alpha/2,n-1}$$

where t_0 is the observed value of the test statistic T_0. The test procedure is very similar to the test on the mean with known variance described in Section 9-2, except that T_0 is used as the test statistic instead of Z_0 and the t_{n-1} distribution is used to define the critical region instead of the standard normal distribution. A summary of the test procedures for both two- and one-sided alternative hypotheses follows:

The One-Sample t-Test

Null hypothesis:	$H_0: \mu = \mu_0$
Test statistic:	$T_0 = \dfrac{\overline{X} - \mu_0}{S/\sqrt{n}}$

Alternative hypothesis	Rejection criteria
$H_1: \mu \neq \mu_0$	$t_0 > t_{\alpha/2,n-1}$ or $t_0 < -t_{\alpha/2,n-1}$
$H_1: \mu > \mu_0$	$t_0 > t_{\alpha,n-1}$
$H_1: \mu < \mu_0$	$t_0 < -t_{\alpha,n-1}$

Figure 9-8 shows the location of the critical region for these situations.

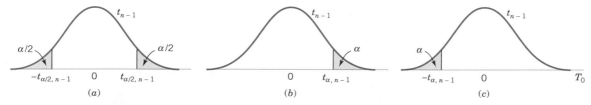

Figure 9-8 The reference distribution for $H_0: \mu = \mu_0$ with critical region for (a) $H_1: \mu \neq \mu_0$, (b) $H_1: \mu > \mu_0$, and (c) $H_1: \mu < \mu_0$.

EXAMPLE 9-6

The increased availability of light materials with high strength has revolutionized the design and manufacture of golf clubs, particularly drivers. Clubs with hollow heads and very thin faces can result in much longer tee shots, especially for players of modest skills. This is due partly to the "spring-like effect" that the thin face imparts to the ball. Firing a golf ball at the head of the club and measuring the ratio of the outgoing velocity of the ball to the incoming velocity can quantify this spring-like effect. The ratio of velocities is called the coefficient of restitution of the club. An experiment was performed in which 15 drivers produced by a particular club maker were selected at random and their coefficients of restitution measured. In the experiment the golf balls were fired from an air cannon so that the incoming velocity and spin rate of the ball could be precisely controlled. It is of interest to determine if there is evidence (with $\alpha = 0.05$) to support a claim that the mean coefficient of restitution exceeds 0.82. The observations follow:

0.8411	0.8191	0.8182	0.8125	0.8750
0.8580	0.8532	0.8483	0.8276	0.7983
0.8042	0.8730	0.8282	0.8359	0.8660

The sample mean and sample standard deviation are $\bar{x} = 0.83725$ and $s = 0.02456$. The normal probability plot of the data in Fig. 9-9 supports the assumption that the coefficient of restitution is normally distributed. Since the objective of the experimenter is to demonstrate that the mean coefficient of restitution exceeds 0.82, a one-sided alternative hypothesis is appropriate.

The solution using the eight-step procedure for hypothesis testing is as follows:

1. The parameter of interest is the mean coefficient of restitution, μ.

2. H_0: $\mu = 0.82$

3. H_1: $\mu > 0.82$. We want to reject H_0 if the mean coefficient of restitution exceeds 0.82.

4. $\alpha = 0.05$

5. The test statistic is

$$t_0 = \frac{\bar{x} - \mu_0}{s/\sqrt{n}}$$

6. Reject H_0 if $t_0 > t_{0.05,14} = 1.761$

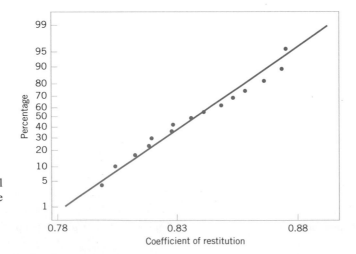

Figure 9-9.   Normal probability plot of the coefficient of restitution data from Example 9-6.

7. Computations: Since $\bar{x} = 0.83725$, $s = 0.02456$, $\mu_0 = 0.82$, and $n = 15$, we have

$$t_0 = \frac{0.83725 - 0.82}{0.02456/\sqrt{15}} = 2.72$$

8. Conclusions: Since $t_0 = 2.72 > 1.761$, we reject H_0 and conclude at the 0.05 level of significance that the mean coefficient of restitution exceeds 0.82.

Minitab will conduct the one-sample t-test. The output from this software package is in the following display:

One-Sample T: COR

Test of mu = 0.82 vs mu > 0.82

Variable	N	Mean	StDev	SE Mean
COR	15	0.83725	0.02456	0.00634

Variable	95.0% Lower Bound	T	P
COR	0.82608	2.72	0.008

Notice that Minitab computes both the test statistic T_0 and a 95% lower confidence bound for the coefficient of restitution. Because the 95% lower confidence bound exceeds 0.82, we would reject the hypothesis that H_0: $\mu = 0.82$ and conclude that the alternative hypothesis H_1: $\mu > 0.82$ is true. Minitab also calculates a P-value for the test statistic T_0. In the next section we explain how this is done.

9-3.2 P-Value for a t-Test

The P-value for a t-test is just the smallest level of significance at which the null hypothesis would be rejected. That is, it is the tail area beyond the value of the test statistic t_0 for a one-sided test or twice this area for a two-sided test. Because the t-table in Appendix Table IV contains only 10 critical values for each t distribution, computation of the exact P-value directly from the table is usually impossible. However, it is easy to find upper and lower bounds on the P-value from this table.

To illustrate, consider the t-test based on 14 degrees of freedom in Example 9-6. The relevant critical values from Appendix Table IV are as follows:

Critical Value:	0.258	0.692	1.345	1.761	2.145	2.624	2.977	3.326	3.787	4.140
Tail Area:	0.40	0.25	0.10	0.05	0.025	0.01	0.005	0.0025	0.001	0.0005

Notice that $t_0 = 2.72$ in Example 9-6, and that this is between two tabulated values, 2.624 and 2.977. Therefore, the P-value must be between 0.01 and 0.005. These are effectively the upper and lower bounds on the P-value.

Example 9-6 is an upper-tailed test. If the test is lower-tailed, just change the sign of t_0 and proceed as above. Remember that for a two-tailed test the level of significance associated with a particular critical value is twice the corresponding tail area in the column heading. This consideration must be taken into account when we compute the bound on the P-value. For example, suppose that $t_0 = 2.72$ for a two-tailed alternate based on 14 degrees of freedom. The value $t_0 > 2.624$ (corresponding to $\alpha = 0.02$) and $t_0 < 2.977$ (corresponding to $\alpha = 0.01$), so the lower and upper bounds on the P-value would be $0.01 < P < 0.02$ for this case.

Finally, most computer programs report P-values along with the computed value of the test statistic. Some hand-held calculators also have this capability. In Example 9-6, Minitab gave the P-value for the value $t_0 = 2.72$ in Example 9-6 as 0.008.

9-3.3   Choice of Sample Size

The type II error probability for tests on the mean of a normal distribution with unknown variance depends on the distribution of the test statistic in Equation 9-23 when the null hypothesis H_0: $\mu = \mu_0$ is false. When the true value of the mean is $\mu = \mu_0 + \delta$, the distribution for T_0 is called the **noncentral t distribution** with $n - 1$ degrees of freedom and noncentrality parameter $\delta \sqrt{n}/\sigma$. Note that if $\delta = 0$, the noncentral t distribution reduces to the usual **central t distribution.** Therefore, the type II error of the two-sided alternative (for example) would be

$$\beta = P\{-t_{\alpha/2,n-1} \le T_0 \le t_{\alpha/2,n-1} | \delta \ne 0\}$$
$$= P\{-t_{\alpha/2,n-1} \le T_0' \le t_{\alpha/2,n-1}\}$$

where T_0' denotes the noncentral t random variable. Finding the type II error probability β for the t-test involves finding the probability contained between two points of the noncentral t distribution. Because the noncentral t-random variable has a messy density function, this integration must be done numerically.

Fortunately, this ugly task has already been done, and the results are summarized in a series of O.C. curves in Appendix Charts VIe, VIf, VIg, and VIh that plot β for the t-test against a parameter d for various sample sizes n. Curves are provided for two-sided alternatives on Charts VIe and VIf. The abscissa scale factor d on these charts is defined as

$$d = \frac{|\mu - \mu_0|}{\sigma} = \frac{|\delta|}{\sigma} \tag{9-24}$$

For the one-sided alternative $\mu > \mu_0$ or $\mu < \mu_0$, we use charts VIg and VIh with

$$d = \frac{|\mu - \mu_0|}{\sigma} = \frac{|\delta|}{\sigma} \tag{9-25}$$

We note that d depends on the unknown parameter σ^2. We can avoid this difficulty in several ways. In some cases, we may use the results of a previous experiment or prior information to make a rough initial estimate of σ^2. If we are interested in evaluating test performance after the data have been collected, we could use the sample variance s^2 to estimate σ^2. If there is no previous experience on which to draw in estimating σ^2, we then define the difference in the mean d that we wish to detect relative to σ. For example, if we wish to detect a small difference in the mean, we might use a value of $d = |\delta|/\sigma \le 1$ (for example), whereas if we are interested in detecting only moderately large differences in the mean, we might select $d = |\delta|/\sigma = 2$ (for example). That is, it is the value of the ratio $|\delta|/\sigma$ that is important in determining sample size, and if it is possible to specify the relative size of the difference in means that we are interested in detecting, then a proper value of d can usually be selected.

EXAMPLE 9-7

Consider the golf club testing problem from Example 9-6. If the mean coefficient of restitution exceeds 0.82 by as much as 0.02, is the sample size $n = 15$ adequate to ensure that H_0: $\mu = 0.82$ will be rejected with probability at least 0.8?

To solve this problem, we will use the sample standard deviation $s = 0.02456$ to estimate σ. Then $d = |\delta|/\sigma = 0.02/0.02456 = 0.81$. By referring to the operating characteristic curves in Appendix Chart VIg (for $\alpha = 0.05$) with $d = 0.81$ and $n = 15$, we find that $\beta = 0.10$, approximately. Thus, the probability of rejecting H_0: $\mu = 0.82$ if the true mean exceeds this by 0.02 is approximately $1 - \beta = 1 - 0.10 = 0.90$, and we conclude that a sample size of $n = 15$ is adequate to provide the desired sensitivity.

Minitab will also perform power and sample size computations for the one-sample *t*-test. Below are several calculations based on the golf club testing problem:

Power and Sample Size

1-Sample t Test
Testing mean = null (versus > null)
Calculating power for mean = null + difference
Alpha = 0.05 Sigma = 0.02456

Difference	Sample Size	Power
0.02	15	0.9117

Power and Sample Size

1-Sample t Test
Testing mean = null (versus > null)
Calculating power for mean = null + difference
Alpha = 0.05 Sigma = 0.02456

Difference	Sample Size	Power
0.01	15	0.4425

Power and Sample Size

1-Sample t Test
Testing mean = null (versus > null)
Calculating power for mean = null + difference
Alpha = 0.05 Sigma = 0.02456

Difference	Sample Size	Target Power	Actual Power
0.01	39	0.8000	0.8029

In the first portion of the computer output, Minitab reproduces the solution to Example 9-7, verifying that a sample size of $n = 15$ is adequate to give power of at least 0.8 if the mean coefficient of restitution exceeds 0.82 by at least 0.02. In the middle section of the output, we asked Minitab to compute the power if the difference in μ and $\mu_0 = 0.82$ we wanted to detect was 0.01. Notice that with $n = 15$, the power drops considerably to 0.4425. The final portion of the output is the sample size required to give a power of at least 0.8 if the difference between μ and μ_0 of interest is actually 0.01. A much larger n is required to detect this smaller difference.

9-3.4 Likelihood Ratio Approach to Development of Test Procedures (CD Only)

EXERCISES FOR SECTION 9-3

9-30. An article in the *ASCE Journal of Energy Engineering* (1999, Vol. 125, pp. 59–75) describes a study of the thermal inertia properties of autoclaved aerated concrete used as a building material. Five samples of the material were tested in a structure, and the average interior temperature (°C) reported was as follows: 23.01, 22.22, 22.04, 22.62, and 22.59.

(a) Test the hypotheses H_0: $\mu = 22.5$ versus H_1: $\mu \neq 22.5$, using $\alpha = 0.05$. Find the P-value.

(b) Is there evidence to support the assumption that interior temperature is normally distributed?

(c) Compute the power of the test if the true mean interior temperature is as high as 22.75.

(d) What sample size would be required to detect a true mean interior temperature as high as 22.75 if we wanted the power of the test to be at least 0.9?

(e) Explain how the question in part (a) could be answered by constructing a two-sided confidence interval on the mean interior temperature.

9-31. A 1992 article in the *Journal of the American Medical Association* ("A Critical Appraisal of 98.6 Degrees F, the Upper Limit of the Normal Body Temperature, and Other Legacies of Carl Reinhold August Wundrlich") reported body temperature, gender, and heart rate for a number of subjects. The body temperatures for 25 female subjects follow: 97.8, 97.2, 97.4, 97.6, 97.8, 97.9, 98.0, 98.0, 98.0, 98.1, 98.2, 98.3, 98.3, 98.4, 98.4, 98.4, 98.5, 98.6, 98.6, 98.7, 98.8, 98.8, 98.9, 98.9, and 99.0.

(a) Test the hypotheses H_0: $\mu = 98.6$ versus H_1: $\mu \neq 98.6$, using $\alpha = 0.05$. Find the P-value.

(b) Compute the power of the test if the true mean female body temperature is as low as 98.0.

(c) What sample size would be required to detect a true mean female body temperature as low as 98.2 if we wanted the power of the test to be at least 0.9?

(d) Explain how the question in part (a) could be answered by constructing a two-sided confidence interval on the mean female body temperature.

(e) Is there evidence to support the assumption that female body temperature is normally distributed?

9-32. Cloud seeding has been studied for many decades as a weather modification procedure (for an interesting study of this subject, see the article in *Technometrics* by Simpson, Alsen, and Eden, "A Bayesian Analysis of a Multiplicative Treatment Effect in Weather Modification", Vol. 17, pp. 161–166). The rainfall in acre-feet from 20 clouds that were selected at random and seeded with silver nitrate follows: 18.0, 30.7, 19.8, 27.1, 22.3, 18.8, 31.8, 23.4, 21.2, 27.9, 31.9, 27.1, 25.0, 24.7, 26.9, 21.8, 29.2, 34.8, 26.7, and 31.6.

(a) Can you support a claim that mean rainfall from seeded clouds exceeds 25 acre-feet? Use $\alpha = 0.01$.

(b) Is there evidence that rainfall is normally distributed?

(c) Compute the power of the test if the true mean rainfall is 27 acre-feet.

(d) What sample size would be required to detect a true mean rainfall of 27.5 acre-feet if we wanted the power of the test to be at least 0.9?

(e) Explain how the question in part (a) could be answered by constructing a one-sided confidence bound on the mean diameter.

9-33. The sodium content of thirty 300-gram boxes of organic corn flakes was determined. The data (in milligrams) are as

follows: 131.15, 130.69, 130.91, 129.54, 129.64, 128.77, 130.72, 128.33, 128.24, 129.65, 130.14, 129.29, 128.71, 129.00, 129.39, 130.42, 129.53, 130.12, 129.78, 130.92, 131.15, 130.69, 130.91, 129.54, 129.64, 128.77, 130.72, 128.33, 128.24, and 129.65.

(a) Can you support a claim that mean sodium content of this brand of cornflakes is 130 milligrams? Use $\alpha = 0.05$.

(b) Is there evidence that sodium content is normally distributed?

(c) Compute the power of the test if the true mean sodium content is 130.5 miligrams.

(d) What sample size would be required to detect a true mean sodium content of 130.1 milligrams if we wanted the power of the test to be at least 0.75?

(e) Explain how the question in part (a) could be answered by constructing a two-sided confidence interval on the mean sodium content.

9-34. Reconsider the tire testing experiment described in Exercise 8-22.

(a) The engineer would like to demonstrate that the mean life of this new tire is in excess of 60,000 kilometers. Formulate and test appropriate hypotheses, and draw conclusions using $\alpha = 0.05$.

(b) Suppose that if the mean life is as long as 61,000 kilometers, the engineer would like to detect this difference with probability at least 0.90. Was the sample size $n = 16$ used in part (a) adequate? Use the sample standard deviation s as an estimate of σ in reaching your decision.

9-35. Reconsider the Izod impact test on PVC pipe described in Exercise 8-23. Suppose that you want to use the data from this experiment to support a claim that the mean impact strength exceeds the ASTM standard (1 foot-pounds per inch). Formulate and test the appropriate hypotheses using $\alpha = 0.05$.

9-36. Reconsider the television tube brightness experiment in Exercise 8-24. Suppose that the design engineer believes that this tube will require 300 microamps of current to produce the desired brightness level. Formulate and test an appropriate hypothesis using $\alpha = 0.05$. Find the P-value for this test. State any necessary assumptions about the underlying distribution of the data.

9-37. Consider the baseball coefficient of restitution data first presented in Exercise 8-79.

(a) Does the data support the claim that the mean coefficient of restitution of baseballs exceeds 0.635? Use $\alpha = 0.05$.

(b) What is the P-value of the test statistic computed in part (a)?

(c) Compute the power of the test if the true mean coefficient of restitution is as high as 0.64.

(d) What sample size would be required to detect a true mean coefficient of restitution as high as 0.64 if we wanted the power of the test to be at least 0.75?

9-38. Consider the dissolved oxygen concentration at TVA dams first presented in Exercise 8-81.

(a) Test the hypotheses H_0: $\mu = 4$ versus H_1: $\mu \neq 4$. Use $\alpha = 0.01$.

(b) What is the *P*-value of the test statistic computed in part (a)?

(c) Compute the power of the test if the true mean dissolved oxygen concentration is as low as 3.

(d) What sample size would be required to detect a true mean dissolved oxygen concentration as low as 2.5 if we wanted the power of the test to be at least 0.9?

9-39. Consider the cigar tar content data first presented in Exercise 8-82.

(a) Can you support a claim that mean tar content exceeds 1.5? Use $\alpha = 0.05$

(b) What is the *P*-value of the test statistic computed in part (a)?

(c) Compute the power of the test if the true mean tar content is 1.6.

(d) What sample size would be required to detect a true mean tar content of 1.6 if we wanted the power of the test to be at least 0.8?

9-40. Exercise 6-22 gave data on the heights of female engineering students at ASU.

(a) Can you support a claim that mean height of female engineering students at ASU is 65 inches? Use $\alpha = 0.05$

(b) What is the *P*-value of the test statistic computed in part (a)?

(c) Compute the power of the test if the true mean height is 62 inches.

(d) What sample size would be required to detect a true mean height of 64 inches if we wanted the power of the test to be at least 0.8?

9-41. Exercise 6-24 presented data on the concentration of suspended solids in lake water.

(a) Test the hypotheses $H_0\colon \mu = 55$ versus $H_1\colon \mu \neq 55$, use $\alpha = 0.05$.

(b) What is the *P*-value of the test statistic computed in part (a)?

(c) Compute the power of the test if the true mean concentration is as low as 50.

(d) What sample size would be required to detect a true mean concentration as low as 50 if we wanted the power of the test to be at least 0.9?

9-42. Exercise 6-25 describes testing golf balls for an overall distance standard.

(a) Can you support a claim that mean distance achieved by this particular golf ball exceeds 280 yards? Use $\alpha = 0.05$.

(b) What is the *P*-value of the test statistic computed in part (a)?

(c) Compute the power of the test if the true mean distance is 290 yards.

(e) What sample size would be required to detect a true mean distance of 290 yards if we wanted the power of the test to be at least 0.8?

9-4 HYPOTHESIS TESTS ON THE VARIANCE AND STANDARD DEVIATION OF A NORMAL POPULATION

Sometimes hypothesis tests on the population variance or standard deviation are needed. When the population is modeled by a normal distribution, the tests and intervals described in this section are applicable.

9-4.1 The Hypothesis Testing Procedures

Suppose that we wish to test the hypothesis that the variance of a normal population σ^2 equals a specified value, say σ_0^2, or equivalently, that the standard deviation σ is equal to σ_0. Let $X_1, X_2, \ldots, X_n$ be a random sample of n observations from this population. To test

$$H_0\colon \sigma^2 = \sigma_0^2$$
$$H_1\colon \sigma^2 \neq \sigma_0^2$$

(9-26)

we will use the **test statistic:**

$$X_0^2 = \frac{(n-1)S^2}{\sigma_0^2}$$

(9-27)

If the null hypothesis $H_0\colon \sigma^2 = \sigma_0^2$ is true, the test statistic X_0^2 defined in Equation 9-27 follows the chi-square distribution with $n - 1$ degrees of freedom. This is the reference

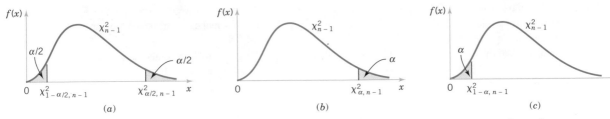

Figure 9-10 Reference distribution for the test of H_0: $\sigma^2 = \sigma_0^2$ with critical region values for (a) H_1: $\sigma^2 \neq \sigma_0^2$, (b) H_1: $\sigma^2 > \sigma_0^2$, and (c) H_1: $\sigma^2 < \sigma_0^2$.

distribution for this test procedure. Therefore, we calculate χ_0^2, the value of the test statistic X_0^2, and the null hypothesis H_0: $\sigma^2 = \sigma_0^2$ would be rejected if

$$\chi_0^2 > \chi_{\alpha/2,n-1}^2 \qquad \text{or if} \quad \chi_0^2 < \chi_{1-\alpha/2,n-1}^2$$

where $\chi_{\alpha/2,n-1}^2$ and $\chi_{1-\alpha/2,n-1}^2$ are the upper and lower $100\alpha/2$ percentage points of the chi-square distribution with $n - 1$ degrees of freedom, respectively. Figure 9-10(a) shows the critical region.

The same test statistic is used for one-sided alternative hypotheses. For the one-sided hypothesis

$$\begin{aligned} H_0&: \sigma^2 = \sigma_0^2 \\ H_1&: \sigma^2 > \sigma_0^2 \end{aligned} \qquad (9\text{-}28)$$

we would reject H_0 if $\chi_0^2 > \chi_{\alpha,n-1}^2$, whereas for the other one-sided hypothesis

$$\begin{aligned} H_0&: \sigma^2 = \sigma_0^2 \\ H_1&: \sigma^2 < \sigma_0^2 \end{aligned} \qquad (9\text{-}29)$$

we would reject H_0 if $\chi_0^2 < \chi_{1-\alpha,n-1}^2$. The one-sided critical regions are shown in Figure 9-10(b) and (c).

EXAMPLE 9-8 An automatic filling machine is used to fill bottles with liquid detergent. A random sample of 20 bottles results in a sample variance of fill volume of $s^2 = 0.0153$ (fluid ounces)^2. If the variance of fill volume exceeds 0.01 (fluid ounces)^2, an unacceptable proportion of bottles will be underfilled or overfilled. Is there evidence in the sample data to suggest that the manufacturer has a problem with underfilled or overfilled bottles? Use $\alpha = 0.05$, and assume that fill volume has a normal distribution.

Using the eight-step procedure results in the following:

1. The parameter of interest is the population variance σ^2.
2. H_0: $\sigma^2 = 0.01$
3. H_1: $\sigma^2 > 0.01$
4. $\alpha = 0.05$
5. The test statistic is

$$\chi_0^2 = \frac{(n-1)s^2}{\sigma_0^2}$$

6. Reject H_0 if $\chi_0^2 > \chi_{0.05,19}^2 = 30.14$.

7. Computations:

$$\chi_0^2 = \frac{19(0.0153)}{0.01} = 29.07$$

8. Conclusions: Since $\chi_0^2 = 29.07 < \chi_{0.05,19}^2 = 30.14$, we conclude that there is no strong evidence that the variance of fill volume exceeds 0.01 (fluid ounces)^2.

Using Appendix Table III, it is easy to place bounds on the P-value of a chi-square test. From inspection of the table, we find that $\chi_{0.10,19}^2 = 27.20$ and $\chi_{0.05,19}^2 = 30.14$. Since $27.20 < 29.07 < 30.14$, we conclude that the P-value for the test in Example 9-8 is in the interval $0.05 < P < 0.10$. The actual P-value is $P = 0.0649$. (This value was obtained from a calculator.)

9-4.2 β-Error and Choice of Sample Size

Operating characteristic curves for the chi-square tests in Section 9-4.1 are provided in Appendix Charts VIi through VIn for $\alpha = 0.05$ and $\alpha = 0.01$. For the two-sided alternative hypothesis of Equation 9-26, Charts VIi and VIj plot β against an abscissa parameter

$$\lambda = \frac{\sigma}{\sigma_0} \qquad (9\text{-}30)$$

for various sample sizes n, where σ denotes the true value of the standard deviation. Charts VIk and VIIl are for the one-sided alternative $H_1: \sigma^2 > \sigma_0^2$, while Charts VI$m$ and VIn are for the other one-sided alternative $H_1: \sigma^2 < \sigma_0^2$. In using these charts, we think of σ as the value of the standard deviation that we want to detect.

These curves can be used to evaluate the β-error (or power) associated with a particular test. Alternatively, they can be used to **design** a test—that is, to determine what sample size is necessary to detect a particular value of σ that differs from the hypothesized value σ_0.

EXAMPLE 9-9

Consider the bottle-filling problem from Example 9-8. If the variance of the filling process exceeds 0.01 (fluid ounces)^2, too many bottles will be underfilled. Thus, the hypothesized value of the standard deviation is $\sigma_0 = 0.10$. Suppose that if the true standard deviation of the filling process exceeds this value by 25%, we would like to detect this with probability at least 0.8. Is the sample size of $n = 20$ adequate?

To solve this problem, note that we require

$$\lambda = \frac{\sigma}{\sigma_0} = \frac{0.125}{0.10} = 1.25$$

This is the abscissa parameter for Chart VIk. From this chart, with $n = 20$ and $\lambda = 1.25$, we find that $\beta \simeq 0.6$. Therefore, there is only about a 40% chance that the null hypothesis will be rejected if the true standard deviation is really as large as $\sigma = 0.125$ fluid ounce.

To reduce the β-error, a larger sample size must be used. From the operating characteristic curve with $\beta = 0.20$ and $\lambda = 1.25$, we find that $n = 75$, approximately. Thus, if we want the test to perform as required above, the sample size must be at least 75 bottles.

EXERCISES FOR SECTION 9-4

9-43. Consider the rivet holes from Exercise 8-35. If the standard deviation of hole diameter exceeds 0.01 millimeters, there is an unacceptably high probability that the rivet will not fit. Recall that $n = 15$ and $s = 0.008$ millimeters.

(a) Is there strong evidence to indicate that the standard deviation of hole diameter exceeds 0.01 millimeters? Use $\alpha = 0.01$. State any necessary assumptions about the underlying distribution of the data.

(b) Find the P-value for this test.

(c) If σ is really as large as 0.0125 millimeters, what sample size will be required to defect this with power of at least 0.8?

9-44. Recall the sugar content of the syrup in canned peaches from Exercise 8-36. Suppose that the variance is thought to be $\sigma^2 = 18$ (milligrams)^2. A random sample of $n = 10$ cans yields a sample standard deviation of $s = 4.8$ milligrams.

(a) Test the hypothesis H_0: $\sigma^2 = 18$ versus H_1: $\sigma^2 \neq 18$ using $\alpha = 0.05$.

(b) What is the P-value for this test?

(c) Discuss how part (a) could be answered by constructing a 95% two-sided confidence interval for σ.

9-45. Consider the tire life data in Exercise 8-22.

(a) Can you conclude, using $\alpha = 0.05$, that the standard deviation of tire life exceeds 200 kilometers? State any necessary assumptions about the underlying distribution of the data.

(b) Find the P-value for this test.

9-46. Consider the Izod impact test data in Exercise 8-23.

(a) Test the hypothesis that $\sigma = 0.10$ against an alternative specifying that $\sigma \neq 0.10$, using $\alpha = 0.01$, and draw a conclusion. State any necessary assumptions about the underlying distribution of the data.

(b) What is the P-value for this test?

(c) Could the question in part (a) have been answered by constructing a 99% two-sided confidence interval for σ^2?

9-47. Reconsider the percentage of titanium in an alloy used in aerospace castings from Exercise 8-39. Recall that $s = 0.37$ and $n = 51$.

(a) Test the hypothesis H_0: $\sigma = 0.25$ versus H_1: $\sigma \neq 0.25$ using $\alpha = 0.05$. State any necessary assumptions about the underlying distribution of the data.

(b) Explain how you could answer the question in part (a) by constructing a 95% two-sided confidence interval for σ.

9-48. Consider the hole diameter data in Exercise 8-35. Suppose that the actual standard deviation of hole diameter exceeds the hypothesized value by 50%. What is the probability that this difference will be detected by the test described in Exercise 9-43?

9-49. Consider the sugar content in Exercise 9-44. Suppose that the true variance is $\sigma^2 = 40$. How large a sample would be required to detect this difference with probability at least 0.90?

9-5 TESTS ON A POPULATION PROPORTION

It is often necessary to test hypotheses on a population proportion. For example, suppose that a random sample of size n has been taken from a large (possibly infinite) population and that $X(\leq n)$ observations in this sample belong to a class of interest. Then $\hat{P} = X/n$ is a point estimator of the proportion of the population p that belongs to this class. Note that n and p are the parameters of a binomial distribution. Furthermore, from Chapter 7 we know that the sampling distribution of $\hat{P}$ is approximately normal with mean p and variance $p(1 - p)/n$, if p is not too close to either 0 or 1 and if n is relatively large. Typically, to apply this approximation we require that np and $n(1 - p)$ be greater than or equal to 5. We will give a large-sample test that makes use of the normal approximation to the binomial distribution.

9-5.1 Large-Sample Tests on a Proportion

In many engineering problems, we are concerned with a random variable that follows the binomial distribution. For example, consider a production process that manufactures items that are classified as either acceptable or defective. It is usually reasonable to model the occurrence of defectives with the binomial distribution, where the binomial parameter p represents the proportion of defective items produced. Consequently, many engineering decision problems include hypothesis testing about p.

We will consider testing

$$H_0: p = p_0$$
$$H_1: p \neq p_0$$

(9-31)

An approximate test based on the normal approximation to the binomial will be given. As noted above, this approximate procedure will be valid as long as p is not extremely close to zero or one, and if the sample size is relatively large. Let X be the number of observations in a random sample of size n that belongs to the class associated with p. Then, if the null hypothesis $H_0: p = p_0$ is true, we have $X \sim N[np_0, np_0(1 - p_0)]$, approximately. To test $H_0: p = p_0$, calculate the **test statistic**

$$Z_0 = \frac{X - np_0}{\sqrt{np_0(1 - p_0)}}$$

(9-32)

and reject $H_0: p = p_0$ if

$$z_0 > z_{\alpha/2} \quad \text{or} \quad z_0 < -z_{\alpha/2}$$

Note that the standard normal distribution is the **reference distribution** for this test statistic. Critical regions for the one-sided alternative hypotheses would be constructed in the usual manner.

EXAMPLE 9-10

A semiconductor manufacturer produces controllers used in automobile engine applications. The customer requires that the process fallout or fraction defective at a critical manufacturing step not exceed 0.05 and that the manufacturer demonstrate process capability at this level of quality using $\alpha = 0.05$. The semiconductor manufacturer takes a random sample of 200 devices and finds that four of them are defective. Can the manufacturer demonstrate process capability for the customer?

We may solve this problem using the eight-step hypothesis-testing procedure as follows:

1. The parameter of interest is the process fraction defective p.
2. $H_0: p = 0.05$
3. $H_1: p < 0.05$
   This formulation of the problem will allow the manufacturer to make a strong claim about process capability if the null hypothesis $H_0: p = 0.05$ is rejected.
4. $\alpha = 0.05$
5. The test statistic is (from Equation 9-32)

   $$z_0 = \frac{x - np_0}{\sqrt{np_0(1 - p_0)}}$$

   where $x = 4$, $n = 200$, and $p_0 = 0.05$.
6. Reject $H_0: p = 0.05$ if $z_0 < -z_{0.05} = -1.645$
7. Computations: The test statistic is

   $$z_0 = \frac{4 - 200(0.05)}{\sqrt{200(0.05)(0.95)}} = -1.95$$

8. Conclusions: Since $z_0 = -1.95 < -z_{0.05} = -1.645$, we reject H_0 and conclude that the process fraction defective p is less than 0.05. The P-value for this value of the test statistic z_0 is $P = 0.0256$, which is less than $\alpha = 0.05$. We conclude that the process is capable.

Another form of the test statistic Z_0 in Equation 9-32 is occasionally encountered. Note that if X is the number of observations in a random sample of size n that belongs to a class of interest, then $\hat{P} = X/n$ is the sample proportion that belongs to that class. Now divide both numerator and denominator of Z_0 in Equation 9-32 by n, giving

$$Z_0 = \frac{X/n - p_0}{\sqrt{p_0(1 - p_0)/n}}$$

or

$$Z_0 = \frac{\hat{P} - p_0}{\sqrt{p_0(1 - p_0)/n}} \tag{9-33}$$

This presents the test statistic in terms of the sample proportion instead of the number of items X in the sample that belongs to the class of interest.

Statistical software packages usually provide the one sample Z-test for a proportion. The Minitab output for Example 9-10 follows.

Test and CI for One Proportion

Test of $p = 0.05$ vs $p < 0.05$

Sample	X	N	Sample p	95.0% Upper Bound	Z-Value	P-Value
1	4	200	0.020000	0.036283	−1.95	0.026

* NOTE * The normal approximation may be inaccurate for small samples.

Notice that both the test statistic (and accompanying P-value) and the 95% one-sided upper confidence bound are displayed. The 95% upper confidence bound is 0.036283, which is less than 0.05. This is consistent with rejection of the null hypothesis H_0: $p = 0.05$.

9-5.2 Small-Sample Tests on a Proportion (CD Only)

9-5.3 Type II Error and Choice of Sample Size

It is possible to obtain closed-form equations for the approximate β-error for the tests in Section 9-5.1. Suppose that p is the true value of the population proportion. The approximate β-error for the two-sided alternative H_1: $p \neq p_0$ is

$$\beta = \Phi\left(\frac{p_0 - p + z_{\alpha/2}\sqrt{p_0(1 - p_0)/n}}{\sqrt{p(1 - p)/n}}\right) - \Phi\left(\frac{p_0 - p - z_{\alpha/2}\sqrt{p_0(1 - p_0)/n}}{\sqrt{p(1 - p)/n}}\right) \tag{9-34}$$

If the alternative is H_1: $p < p_0$,

$$\beta = 1 - \Phi\left(\frac{p_0 - p - z_\alpha\sqrt{p_0(1 - p_0)/n}}{\sqrt{p(1 - p)/n}}\right) \tag{9-35}$$

whereas if the alternative is $H_1: p > p_0$,

$$\beta = \Phi\left(\frac{p_0 - p + z_\alpha\sqrt{p_0(1 - p_0)/n}}{\sqrt{p(1 - p)/n}}\right) \tag{9-36}$$

These equations can be solved to find the approximate sample size n that gives a test of level α that has a specified β risk. The sample size equations are

$$n = \left[\frac{z_{\alpha/2}\sqrt{p_0(1 - p_0)} + z_\beta\sqrt{p(1 - p)}}{p - p_0}\right]^2 \tag{9-37}$$

for the two-sided alternative and

$$n = \left[\frac{z_\alpha\sqrt{p_0(1 - p_0)} + z_\beta\sqrt{p(1 - p)}}{p - p_0}\right]^2 \tag{9-38}$$

for a one-sided alternative.

EXAMPLE 9-11   Consider the semiconductor manufacturer from Example 9-10. Suppose that its process fallout is really $p = 0.03$. What is the β-error for a test of process capability that uses $n = 200$ and $\alpha = 0.05$?

The β-error can be computed using Equation 9-35 as follows:

$$\beta = 1 - \Phi\left[\frac{0.05 - 0.03 - (1.645)\sqrt{0.05(0.95)/200}}{\sqrt{0.03(1 - 0.03)/200}}\right] = 1 - \Phi(-0.44) = 0.67$$

Thus, the probability is about 0.7 that the semiconductor manufacturer will fail to conclude that the process is capable if the true process fraction defective is $p = 0.03$ (3%). That is, the power of the test against this particular alternative is only about 0.3. This appears to be a large β-error (or small power), but the difference between $p = 0.05$ and $p = 0.03$ is fairly small, and the sample size $n = 200$ is not particularly large.

Suppose that the semiconductor manufacturer was willing to accept a β-error as large as 0.10 if the true value of the process fraction defective was $p = 0.03$. If the manufacturer continues to use $\alpha = 0.05$, what sample size would be required?

The required sample size can be computed from Equation 9-38 as follows:

$$n = \left[\frac{1.645\sqrt{0.05(0.95)} + 1.28\sqrt{0.03(0.97)}}{0.03 - 0.05}\right]^2$$
$$\approx 832$$

where we have used $p = 0.03$ in Equation 9-38. Note that $n = 832$ is a very large sample size. However, we are trying to detect a fairly small deviation from the null value $p_0 = 0.05$.

Minitab will also perform power and sample size calculations for the one-sample Z-test on a proportion. Output from Minitab for the engine controllers tested in Example 9-10 follows.

Power and Sample Size

Test for One Proportion
Testing proportion = 0.05 (versus < 0.05)
Alpha = 0.05

Alternative Proportion	Sample Size	Power
3.00E-02	200	0.3287

Power and Sample Size

Test for One Proportion
Testing proportion = 0.05 (versus < 0.05)
Alpha = 0.05

Alternative Proportion	Sample Size	Target Power	Actual Power
3.00E-02	833	0.9000	0.9001

Power and Sample Size

Test for One Proportion
Testing proportion = 0.05 (versus < 0.05)
Alpha = 0.05

Alternative Proportion	Sample Size	Target Power	Actual Power
3.00E-02	561	0.7500	0.7503

The first part of the output shows the power calculation based on the situation described in Example 9-11, where the true proportion is really 0.03. The power calculation from Minitab agrees with the results from Equation 9-35 in Example 9-11. The second part of the output computes the sample size necessary to give a power of 0.9 ($\beta = 0.1$) if $p = 0.03$. Again, the results agree closely with those obtained from Equation 9-38. The final portion of the display shows the sample size that would be required if $p = 0.03$ and the power requirement is relaxed to 0.75. Notice that the sample size of $n = 561$ is still quite large because the difference between $p = 0.05$ and $p = 0.03$ is fairly small.

EXERCISES FOR SECTION 9-5

9-50.   In a random sample of 85 automobile crankshaft bearings, 10 have a surface finish roughness that exceeds the specifications. Does this data present strong evidence that the proportion of crankshaft bearings exhibiting excess surface roughness exceeds 0.10? State and test the appropriate hypotheses using $\alpha = 0.05$.

9-51.   **Continuation of Exercise 9-50.** If it is really the situation that $p = 0.15$, how likely is it that the test procedure in Exercise 9-50 will not reject the null hypothesis? If

$p = 0.15$, how large would the sample size have to be for us to have a probability of correctly rejecting the null hypothesis of 0.9?

9-52.   Reconsider the integrated circuits described in Exercise 8-48.
(a) Use the data to test $H_0: p = 0.05$ versus $H_1: p \neq 0.05$. Use $\alpha = 0.05$.
(b) Find the P-value for the test.

9-53. Consider the defective circuit data in Exercise 8-48.
(a) Do the data support the claim that the fraction of defective units produced is less than 0.05, using $\alpha = 0.05$?
(b) Find the *P*-value for the test.

9-54. An article in *Fortune* (September 21, 1992) claimed that nearly one-half of all engineers continue academic studies beyond the B.S. degree, ultimately receiving either an M.S. or a Ph.D. degree. Data from an article in *Engineering Horizons* (Spring 1990) indicated that 117 of 484 new engineering graduates were planning graduate study.
(a) Are the data from *Engineering Horizons* consistent with the claim reported by *Fortune*? Use $\alpha = 0.05$ in reaching your conclusions.
(b) Find the *P*-value for this test.
(c) Discuss how you could have answered the question in part (a) by constructing a two-sided confidence interval on *p*.

9-55. A manufacturer of interocular lenses is qualifying a new grinding machine and will qualify the machine if the percentage of polished lenses that contain surface defects does not exceed 2%. A random sample of 250 lenses contains six defective lenses.
(a) Formulate and test an appropriate set of hypotheses to determine if the machine can be qualified. Use $\alpha = 0.05$.
(b) Find the *P*-value for the test in part (a).

9-56. A researcher claims that at least 10% of all football helmets have manufacturing flaws that could potentially cause injury to the wearer. A sample of 200 helmets revealed that 16 helmets contained such defects.
(a) Does this finding support the researcher's claim? Use $\alpha = 0.01$.
(b) Find the *P*-value for this test.

9-57. A random sample of 500 registered voters in Phoenix is asked if they favor the use of oxygenated fuels year-round to reduce air pollution. If more than 315 voters respond positively, we will conclude that at least 60% of the voters favor the use of these fuels.
(a) Find the probability of type I error if exactly 60% of the voters favor the use of these fuels.
(b) What is the type II error probability β if 75% of the voters favor this action?

9-58. The advertized claim for batteries for cell phones is set at 48 operating hours, with proper charging procedures. A study of 5000 batteries is carried out and 15 stop operating prior to 48 hours. Do these experimental results support the claim that less than 0.2 percent of the company's batteries will fail during the advertized time period, with proper charging procedures? Use a hypothesis-testing procedure with $\alpha = 0.01$.

9-6   SUMMARY TABLE OF INFERENCE PROCEDURES FOR A SINGLE SAMPLE

The table in the end papers of this book (inside front cover) presents a summary of all the single-sample inference procedures from Chapters 8 and 9. The table contains the null hypothesis statement, the test statistic, the various alternative hypotheses and the criteria for rejecting H_0, and the formulas for constructing the $100(1 - \alpha)\%$ two-sided confidence interval.

9-7   TESTING FOR GOODNESS OF FIT

The hypothesis-testing procedures that we have discussed in previous sections are designed for problems in which the population or probability distribution is known and the hypotheses involve the parameters of the distribution. Another kind of hypothesis is often encountered: we do not know the underlying distribution of the population, and we wish to test the hypothesis that a particular distribution will be satisfactory as a population model. For example, we might wish to test the hypothesis that the population is normal.

We have previously discussed a very useful graphical technique for this problem called **probability plotting** and illustrated how it was applied in the case of a normal distribution. In this section, we describe a formal goodness-of-fit test procedure based on the chi-square distribution.

The test procedure requires a random sample of size n from the population whose probability distribution is unknown. These n observations are arranged in a frequency histogram, having k bins or class intervals. Let O_i be the observed frequency in the ith class interval. From the hypothesized probability distribution, we compute the expected frequency in the ith class interval, denoted E_i. The test statistic is

$$X_0^2 = \sum_{i=1}^{k} \frac{(O_i - E_i)^2}{E_i} \tag{9-39}$$

It can be shown that, if the population follows the hypothesized distribution, X_0^2 has, approximately, a chi-square distribution with $k - p - 1$ degrees of freedom, where p represents the number of parameters of the hypothesized distribution estimated by sample statistics. This approximation improves as n increases. We would reject the hypothesis that the distribution of the population is the hypothesized distribution if the calculated value of the test statistic $X_0^2 > \chi_{\alpha,k-p-1}^2$.

One point to be noted in the application of this test procedure concerns the magnitude of the expected frequencies. If these expected frequencies are too small, the test statistic X_0^2 will not reflect the departure of observed from expected, but only the small magnitude of the expected frequencies. There is no general agreement regarding the minimum value of expected frequencies, but values of 3, 4, and 5 are widely used as minimal. Some writers suggest that an expected frequency could be as small as 1 or 2, so long as most of them exceed 5. Should an expected frequency be too small, it can be combined with the expected frequency in an adjacent class interval. The corresponding observed frequencies would then also be combined, and k would be reduced by 1. Class intervals are not required to be of equal width.

We now give two examples of the test procedure.

EXAMPLE 9-12

A Poisson Distribution

The number of defects in printed circuit boards is hypothesized to follow a Poisson distribution. A random sample of $n = 60$ printed boards has been collected, and the following number of defects observed.

Number of Defects	Observed Frequency
0	32
1	15
2	9
3	4

The mean of the assumed Poisson distribution in this example is unknown and must be estimated from the sample data. The estimate of the mean number of defects per board is the sample average, that is, $(32 \cdot 0 + 15 \cdot 1 + 9 \cdot 2 + 4 \cdot 3)/60 = 0.75$. From the Poisson distribution with parameter 0.75, we may compute p_i, the theoretical, hypothesized probability associated with the ith class interval. Since each class interval corresponds to a particular number of defects, we may find the p_i as follows:

$$p_1 = P(X = 0) = \frac{e^{-0.75}(0.75)^0}{0!} = 0.472$$

$$p_2 = P(X = 1) = \frac{e^{-0.75}(0.75)^1}{1!} = 0.354$$

$$p_3 = P(X = 2) = \frac{e^{-0.75}(0.75)^2}{2!} = 0.133$$

$$p_4 = P(X \geq 3) = 1 - (p_1 + p_2 + p_3) = 0.041$$

The expected frequencies are computed by multiplying the sample size $n = 60$ times the probabilities p_i. That is, $E_i = np_i$. The expected frequencies follow:

Number of Defects	Probability	Expected Frequency
0	0.472	28.32
1	0.354	21.24
2	0.133	7.98
3 (or more)	0.041	2.46

Since the expected frequency in the last cell is less than 3, we combine the last two cells:

Number of Defects	Observed Frequency	Expected Frequency
0	32	28.32
1	15	21.24
2 (or more)	13	10.44

The chi-square test statistic in Equation 9-39 will have $k - p - 1 = 3 - 1 - 1 = 1$ degree of freedom, because the mean of the Poisson distribution was estimated from the data.

The eight-step hypothesis-testing procedure may now be applied, using $\alpha = 0.05$, as follows:

1. The variable of interest is the form of the distribution of defects in printed circuit boards.
2. H_0: The form of the distribution of defects is Poisson.
3. H_1: The form of the distribution of defects is not Poisson.
4. $\alpha = 0.05$
5. The test statistic is

$$\chi_0^2 = \sum_{i=1}^{k} \frac{(o_i - E_i)^2}{E_i}$$

6. Reject H_0 if $\chi_0^2 > \chi_{0.05,1}^2 = 3.84$.

7. Computations:

$$\chi_0^2 = \frac{(32 - 28.32)^2}{28.32} + \frac{(15 - 21.24)^2}{21.24} + \frac{(13 - 10.44)^2}{10.44} = 2.94$$

8. Conclusions: Since $\chi_0^2 = 2.94 < \chi_{0.05,1}^2 = 3.84$, we are unable to reject the null hypothesis that the distribution of defects in printed circuit boards is Poisson. The P-value for the test is $P = 0.0864$. (This value was computed using an HP-48 calculator.)

EXAMPLE 9-13

A Continuous Distribution

A manufacturing engineer is testing a power supply used in a notebook computer and, using $\alpha = 0.05$, wishes to determine whether output voltage is adequately described by a normal distribution. Sample estimates of the mean and standard deviation of $\bar{x} = 5.04$ V and $s = 0.08$ V are obtained from a random sample of $n = 100$ units.

A common practice in constructing the class intervals for the frequency distribution used in the chi-square goodness-of-fit test is to choose the cell boundaries so that the expected frequencies $E_i = np_i$ are equal for all cells. To use this method, we want to choose the cell boundaries $a_0, a_1, \ldots, a_k$ for the k cells so that all the probabilities

$$p_i = P(a_{i-1} \le X \le a_i) = \int_{a_{i-1}}^{a_i} f(x)\, dx$$

are equal. Suppose we decide to use $k = 8$ cells. For the standard normal distribution, the intervals that divide the scale into eight equally likely segments are $[0, 0.32)$, $[0.32, 0.675)$ $[0.675, 1.15)$, $[1.15, \infty)$ and their four "mirror image" intervals on the other side of zero. For each interval $p_i = 1/8 = 0.125$, so the expected cell frequencies are $E_i = np_i = 100(0.125) = 12.5$. The complete table of observed and expected frequencies is as follows:

Class Interval	Observed Frequency o_i	Expected Frequency E_i
$x < 4.948$	12	12.5
$4.948 \le x < 4.986$	14	12.5
$4.986 \le x < 5.014$	12	12.5
$5.014 \le x < 5.040$	13	12.5
$5.040 \le x < 5.066$	12	12.5
$5.066 \le x < 5.094$	11	12.5
$5.094 \le x < 5.132$	12	12.5
$5.132 \le x$	14	12.5
Totals	100	100

The boundary of the first class interval is $\bar{x} - 1.15s = 4.948$. The second class interval is $[\bar{x} - 1.15s, \bar{x} - 0.675s)$ and so forth. We may apply the eight-step hypothesis-testing procedure to this problem.

1. The variable of interest is the form of the distribution of power supply voltage.

2. H_0: The form of the distribution is normal.

3. H_1: The form of the distribution is nonnormal.

4. $\alpha = 0.05$

5. The test statistic is

$$\chi_0^2 = \sum_{i=1}^{k} \frac{(o_i - E_i)^2}{E_i}$$

6. Since two parameters in the normal distribution have been estimated, the chi-square statistic above will have $k - p - 1 = 8 - 2 - 1 = 5$ degrees of freedom. Therefore, we will reject H_0 if $\chi_0^2 > \chi_{0.05,5}^2 = 11.07$.

7. Computations:

$$\chi_0^2 = \sum_{i=1}^{8} \frac{(o_i - E_i)^2}{E_i}$$
$$= \frac{(12 - 12.5)^2}{12.5} + \frac{(14 - 12.5)^2}{12.5} + \cdots + \frac{(14 - 12.5)^2}{12.5}$$
$$= 0.64$$

8. Conclusions: Since $\chi_0^2 = 0.64 < \chi_{0.05,5}^2 = 11.07$, we are unable to reject H_0, and there is no strong evidence to indicate that output voltage is not normally distributed. The P-value for the chi-square statistic $\chi_0^2 = 0.64$ is $P = 0.9861$.

EXERCISES FOR SECTION 9-7

9-59. Consider the following frequency table of observations on the random variable X.

Values	0	1	2	3	4
Observed Frequency	24	30	31	11	4

(a) Based on these 100 observations, is a Poisson distribution with a mean of 1.2 an appropriate model? Perform a goodness-of-fit procedure with $\alpha = 0.05$.
(b) Calculate the P-value for this test.

9-60. Let X denote the number of flaws observed on a large coil of galvanized steel. Seventy-five coils are inspected and the following data were observed for the values of X:

Values	1	2	3	4	5	6	7	8
Observed Frequency	1	11	8	13	11	12	10	9

(a) Does the assumption of the Poisson distribution seem appropriate as a probability model for this data? Use $\alpha = 0.01$.
(b) Calculate the P-value for this test.

9-61. The number of calls arriving at a switchboard from noon to 1 PM during the business days Monday through Friday is monitored for six weeks (i.e., 30 days). Let X be defined as the number of calls during that one-hour period. The relative frequency of calls was recorded and reported as

Value	5	6	8	9	10
Relative Frequency	0.067	0.067	0.100	0.133	0.200
Value	11	12	13	14	15
Relative Frequency	0.133	0.133	0.067	0.033	0.067

(a) Does the assumption of a Poisson distribution seem appropriate as a probability model for this data? Use $\alpha = 0.05$.
(b) Calculate the P-value for this test.

9-62. Consider the following frequency table of observations on the random variable X:

Values	0	1	2	3	4
Frequency	4	21	10	13	2

(a) Based on these 50 observations, is a binomial distribution with $n = 6$ and $p = 0.25$ an appropriate model? Perform a goodness-of-fit procedure with $\alpha = 0.05$.
(b) Calculate the P-value for this test.

9-63. Define X as the number of underfilled bottles from a filling operation in a carton of 24 bottles. Sixty cartons are inspected and the following observations on X are recorded:

Values	0	1	2	3
Frequency	39	23	12	1

(a) Based on these 75 observations, is a binomial distribution an appropriate model? Perform a goodness-of-fit procedure with $\alpha = 0.05$.
(b) Calculate the P-value for this test.

9-64. The number of cars passing eastbound through the intersection of Mill and University Avenues has been tabulated by a group of civil engineering students. They have obtained the data in the adjacent table:

(a) Does the assumption of a Poisson distribution seem appropriate as a probability model for this process? Use $\alpha = 0.05$.
(b) Calculate the P-value for this test.

Vehicles per Minute	Observed Frequency	Vehicles per Minute	Observed Frequency
40	14	53	102
41	24	54	96
42	57	55	90
43	111	56	81
44	194	57	73
45	256	58	64
46	296	59	61
47	378	60	59
48	250	61	50
49	185	62	42
50	171	63	29
51	150	64	18
52	110	65	15

9-8 CONTINGENCY TABLE TESTS

Many times, the n elements of a sample from a population may be classified according to two different criteria. It is then of interest to know whether the two methods of classification are statistically independent; for example, we may consider the population of graduating engineers, and we may wish to determine whether starting salary is independent of academic disciplines. Assume that the first method of classification has r levels and that the second method has c levels. We will let O_{ij} be the observed frequency for level i of the first classification method and level j on the second classification method. The data would, in general, appear as shown in Table 9-2. Such a table is usually called an $r \times c$ **contingency table.**

We are interested in testing the hypothesis that the row-and-column methods of classification are independent. If we reject this hypothesis, we conclude there is some interaction between the two criteria of classification. The exact test procedures are difficult to obtain, but an approximate test statistic is valid for large n. Let p_{ij} be the probability that a randomly selected element falls in the ijth cell, given that the two classifications are independent. Then $p_{ij} = u_i v_j$,

Table 9-2 An $r \times c$ Contingency Table

		Columns			
		1	2	...	c
	1	O_{11}	O_{12}	...	O_{1c}
	2	O_{21}	O_{22}	...	O_{2c}
Rows	⋮	⋮	⋮	⋮	⋮
	r	O_{r1}	O_{r2}	...	O_{rc}

where u_i is the probability that a randomly selected element falls in row class i and v_j is the probability that a randomly selected element falls in column class j. Now, assuming independence, the estimators of u_i and v_j are

$$\hat{u}_i = \frac{1}{n} \sum_{j=1}^{c} O_{ij}$$

$$\hat{v}_j = \frac{1}{n} \sum_{i=1}^{r} O_{ij} \qquad (9\text{-}40)$$

Therefore, the expected frequency of each cell is

$$E_{ij} = n\hat{u}_i\hat{v}_j = \frac{1}{n} \sum_{j=1}^{c} O_{ij} \sum_{i=1}^{r} O_{ij} \qquad (9\text{-}41)$$

Then, for large n, the statistic

$$\chi_0^2 = \sum_{i=1}^{r} \sum_{j=1}^{c} \frac{(O_{ij} - E_{ij})^2}{E_{ij}} \qquad (9\text{-}42)$$

has an approximate chi-square distribution with $(r-1)(c-1)$ degrees of freedom if the null hypothesis is true. Therefore, we would reject the hypothesis of independence if the observed value of the test statistic χ_0^2 exceeded $\chi_{\alpha,(r-1)(c-1)}^2$.

EXAMPLE 9-14   A company has to choose among three pension plans. Management wishes to know whether the preference for plans is independent of job classification and wants to use $\alpha = 0.05$. The opinions of a random sample of 500 employees are shown in Table 9-3.

To find the expected frequencies, we must first compute $\hat{u}_1 = (340/500) = 0.68$, $\hat{u}_2 = (160/500) = 0.32$, $\hat{v}_1 = (200/500) = 0.40$, $\hat{v}_2 = (200/500) = 0.40$, and $\hat{v}_3 = (100/500) = 0.20$. The expected frequencies may now be computed from Equation 9-41. For example, the expected number of salaried workers favoring pension plan 1 is

$$E_{11} = n\hat{u}_1\hat{v}_1 = 500(0.68)(0.40) = 136$$

The expected frequencies are shown in Table 9-4.

The eight-step hypothesis-testing procedure may now be applied to this problem.

1.   The variable of interest is employee preference among pension plans.

2. H_0: Preference is independent of salaried versus hourly job classification.

Table 9-3   Observed Data for Example 9-14

Job Classification	Pension Plan			Totals
	1	2	3	
Salaried workers	160	140	40	340
Hourly workers	40	60	60	160
Totals	200	200	100	500

Table 9-4   Expected Frequencies for Example 9-14

Job Classification	Pension Plan			Totals
	1	2	3	
Salaried workers	136	136	68	340
Hourly workers	64	64	32	160
Totals	200	200	100	500

3. H_1: Preference is not independent of salaried versus hourly job classification.

4. $\alpha = 0.05$

5. The test statistic is

$$\chi_0^2 = \sum_{i=1}^{r} \sum_{j=1}^{c} \frac{(o_{ij} - E_{ij})^2}{E_{ij}}$$

6. Since $r = 2$ and $c = 3$, the degrees of freedom for chi-square are $(r - 1)(c - 1) = (1)(2) = 2$, and we would reject H_0 if $\chi_0^2 > \chi_{0.05,2}^2 = 5.99$.

7. Computations:

$$\chi_0^2 = \sum_{i=1}^{2} \sum_{j=1}^{3} \frac{(o_{ij} - E_{ij})^2}{E_{ij}}$$

$$= \frac{(160 - 136)^2}{136} + \frac{(140 - 136)^2}{136} + \frac{(40 - 68)^2}{68} + \frac{(40 - 64)^2}{64}$$

$$+ \frac{(60 - 64)^2}{64} + \frac{(60 - 32)^2}{32} = 49.63$$

8. Conclusions: Since $\chi_0^2 = 49.63 > \chi_{0.05,2}^2 = 5.99$, we reject the hypothesis of independence and conclude that the preference for pension plans is not independent of job classification. The P-value for $\chi_0^2 = 49.63$ is $P = 1.671 \times 10^{-11}$. (This value was computed using a hand-held calculator.) Further analysis would be necessary to explore the nature of the association between these factors. It might be helpful to examine the table of observed minus expected frequencies.

Using the two-way contingency table to test independence between two variables of classification in a sample from a single population of interest is only one application of contingency table methods. Another common situation occurs when there are r populations of interest and each population is divided into the same c categories. A sample is then taken from the ith population, and the counts are entered in the appropriate columns of the ith row. In this situation we want to investigate whether or not the proportions in the c categories are the same for all populations. The null hypothesis in this problem states that the populations are **homogeneous** with respect to the categories. For example, when there are only two categories, such as success and failure, defective and nondefective, and so on, the test for homogeneity is really a test of the equality of r binomial parameters. Calculation of expected frequencies, determination of degrees of freedom, and computation of the chi-square statistic for the test for homogeneity are identical to the test for independence.

EXERCISES FOR SECTION 9-8

9-65. A company operates four machines three shifts each day. From production records, the following data on the number of breakdowns are collected:

Shift	Machines			
	A	B	C	D
1	41	20	12	16
2	31	11	9	14
3	15	17	16	10

Test the hypothesis (using $\alpha = 0.05$) that breakdowns are independent of the shift. Find the *P*-value for this test.

9-66.   Patients in a hospital are classified as surgical or medical. A record is kept of the number of times patients require nursing service during the night and whether or not these patients are on Medicare. The data are presented here:

	Patient Category	
Medicare	Surgical	Medical
Yes	46	52
No	36	43

Test the hypothesis (using $\alpha = 0.01$) that calls by surgical-medical patients are independent of whether the patients are receiving Medicare. Find the *P*-value for this test.

9-67.   Grades in a statistics course and an operations research course taken simultaneously were as follows for a group of students.

	Operation Research Grade			
Statistics Grade	*A*	*B*	*C*	Other
A	25	6	17	13
B	17	16	15	6
C	18	4	18	10
Other	10	8	11	20

Are the grades in statistics and operations research related? Use $\alpha = 0.01$ in reaching your conclusion. What is the *P*-value for this test?

9-68.   An experiment with artillery shells yields the following data on the characteristics of lateral deflections and ranges. Would you conclude that deflection and range are independent? Use $\alpha = 0.05$. What is the *P*-value for this test?

	Lateral Deflection		
Range (yards)	Left	Normal	Right
0–1,999	6	14	8
2,000–5,999	9	11	4
6,000–11,999	8	17	6

9-69.   A study is being made of the failures of an electronic component. There are four types of failures possible and two mounting positions for the device. The following data have been taken:

	Failure Type			
Mounting Position	*A*	*B*	*C*	*D*
1	22	46	18	9
2	4	17	6	12

Would you conclude that the type of failure is independent of the mounting position? Use $\alpha = 0.01$. Find the *P*-value for this test.

9-70.   A random sample of students is asked their opinions on a proposed core curriculum change. The results are as follows.

	Opinion	
Class	Favoring	Opposing
Freshman	120	80
Sophomore	70	130
Junior	60	70
Senior	40	60

Test the hypothesis that opinion on the change is independent of class standing. Use $\alpha = 0.05$. What is the *P*-value for this test?

Supplemental Exercises

9-71.   A manufacturer of semiconductor devices takes a random sample of size *n* of chips and tests them, classifying each chip as defective or nondefective. Let $X_i = 0$ if the chip is nondefective and $X_i = 1$ if the chip is defective. The sample fraction defective is

$$\hat{p} = \frac{X_1 + X_2 + \cdots + X_n}{n}$$

What are the sampling distribution, the sample mean, and sample variance estimates of $\hat{p}$ when
(a) The sample size is $n = 50$?
(b) The sample size is $n = 80$?
(c) The sample size is $n = 100$?
(d) Compare your answers to parts (a)–(c) and comment on the effect of sample size on the variance of the sampling distribution.

9-72.   Consider the situation of Exercise 9-76. After collecting a sample, we are interested in testing H_0: $p = 0.10$ versus H_1: $p \neq 0.10$ with $\alpha = 0.05$. For each of the following situations, compute the *p*-value for this test:
(a) $n = 50, \hat{p} = 0.095$
(b) $n = 100, \hat{p} = 0.095$
(c) $n = 500, \hat{p} = 0.095$
(d) $n = 1000, \hat{p} = 0.095$
(e) Comment on the effect of sample size on the observed *P*-value of the test.

9-73.   An inspector of flow metering devices used to administer fluid intravenously will perform a hypothesis test to determine whether the mean flow rate is different from the flow rate setting of 200 milliliters per hour. Based on prior information the standard deviation of the flow rate is assumed to be known and equal to 12 milliliters per hour. For each of the following sample sizes, and a fixed $\alpha = 0.05$, find the probability of a type II error if the true mean is 205 milliliters per hour.
(a) $n = 20$
(b) $n = 50$
(c) $n = 100$
(d) Does the probability of a type II error increase or decrease as the sample size increases? Explain your answer.

9-74.   Suppose that in Exercise 9-73, the experimenter had believed that $\sigma = 14$. For each of the following sample sizes, and a fixed $\alpha = 0.05$, find the probability of a type II error if the true mean is 205 milliliters per hour.
(a) $n = 20$
(b) $n = 50$
(c) $n = 100$
(d) Comparing your answers to those in Exercise 9-73, does the probability of a type II error increase or decrease with the increase in standard deviation? Explain your answer.

9-75.   The marketers of shampoo products know that customers like their product to have a lot of foam. A manufacturer of shampoo claims that the foam height of his product exceeds 200 millimeters. It is known from prior experience that the standard deviation of foam height is 8 millimeters. For each of the following sample sizes, and a fixed $\alpha = 0.05$, find the power of the test if the true mean is 204 millimeters.
(a) $n = 20$
(b) $n = 50$
(c) $n = 100$
(d) Does the power of the test increase or decrease as the sample size increases? Explain your answer.

9-76.   Suppose we wish to test the hypothesis H_0: $\mu = 85$ versus the alternative H_1: $\mu > 85$ where $\sigma = 16$. Suppose that the true mean is $\mu = 86$ and that in the practical context of the problem this is not a departure from $\mu_0 = 85$ that has practical significance.
(a) For a test with $\alpha = 0.01$, compute β for the sample sizes $n = 25, 100, 400$, and 2500 assuming that $\mu = 86$.
(b) Suppose the sample average is $\bar{x} = 86$. Find the P-value for the test statistic for the different sample sizes specified in part (a). Would the data be statistically significant at $\alpha = 0.01$?
(c) Comment on the use of a large sample size in this problem.

9-77.   The cooling system in a nuclear submarine consists of an assembly of welded pipes through which a coolant is circulated. Specifications require that weld strength must meet or exceed 150 psi.
(a) Suppose that the design engineers decide to test the hypothesis H_0: $\mu = 150$ versus H_1: $\mu > 150$. Explain

why this choice of alternative hypothesis is better than H_1: $\mu < 150$.
(b) A random sample of 20 welds results in $\bar{x} = 153.7$ psi and $s = 11.3$ psi. What conclusions can you draw about the hypothesis in part (a)? State any necessary assumptions about the underlying distribution of the data.

9-78.   Suppose we are testing H_0: $p = 0.5$ versus H_0: $p \neq 0.5$. Suppose that p is the true value of the population proportion.
(a) Using $\alpha = 0.05$, find the power of the test for $n = 100$, 150, and 300 assuming that $p = 0.6$. Comment on the effect of sample size on the power of the test.
(b) Using $\alpha = 0.01$, find the power of the test for $n = 100$, 150, and 300 assuming that $p = 0.6$. Compare your answers to those from part (a) and comment on the effect of α on the power of the test for different sample sizes.
(c) Using $\alpha = 0.05$, find the power of the test for $n = 100$, assuming $p = 0.08$. Compare your answer to part (a) and comment on the effect of the true value of p on the power of the test for the same sample size and α level.
(d) Using $\alpha = 0.01$, what sample size is required if $p = 0.6$ and we want $\beta = 0.05$? What sample is required if $p = 0.8$ and we want $\beta = 0.05$? Compare the two sample sizes and comment on the effect of the true value of p on sample size required when β is held approximately constant.

9-79.   Consider the television picture tube brightness experiment described in Exercise 8-24.
(a) For the sample size $n = 10$, do the data support the claim that the standard deviation of current is less than 20 microamps?
(b) Suppose instead of $n = 10$, the sample size was 51. Repeat the analysis performed in part (a) using $n = 51$.
(c) Compare your answers and comment on how sample size affects your conclusions drawn in parts (a) and (b).

9-80.   Consider the fatty acid measurements for the diet margarine described in Exercise 8-25.
(a) For the sample size $n = 6$, using a two-sided alternative hypothesis and $\alpha = 0.01$, test H_0: $\sigma^2 = 1.0$.
(b) Suppose instead of $n = 6$, the sample size was $n = 51$. Repeat the analysis performed in part (a) using $n = 51$.
(c) Compare your answers and comment on how sample size affects your conclusions drawn in parts (a) and (b).

9-81.   A manufacturer of precision measuring instruments claims that the standard deviation in the use of the instruments is at most 0.00002 millimeter. An analyst, who is unaware of the claim, uses the instrument eight times and obtains a sample standard deviation of 0.00001 millimeter.
(a) Confirm using a test procedure and an α level of 0.01 that there is insufficient evidence to support the claim that the standard deviation of the instruments is at most 0.00002. State any necessary assumptions about the underlying distribution of the data.

(b) Explain why the sample standard deviation, $s = 0.00001$, is less than 0.00002, yet the statistical test procedure results do not support the claim.

9-82. A biotechnology company produces a therapeutic drug whose concentration has a standard deviation of 4 grams per liter. A new method of producing this drug has been proposed, although some additional cost is involved. Management will authorize a change in production technique only if the standard deviation of the concentration in the new process is less than 4 grams per liter. The researchers chose $n = 10$ and obtained the following data in grams per liter. Perform the necessary analysis to determine whether a change in production technique should be implemented.

16.628	16.630
16.622	16.631
16.627	16.624
16.623	16.622
16.618	16.626

9-83. Consider the 40 observations collected on the number of nonconforming coil springs in production batches of size 50 given in Exercise 6-79.
(a) Based on the description of the random variable and these 40 observations, is a binomial distribution an appropriate model? Perform a goodness-of-fit procedure with $\alpha = 0.05$.
(b) Calculate the P-value for this test.

9-84. Consider the 20 observations collected on the number of errors in a string of 1000 bits of a communication channel given in Exercise 6-80.
(a) Based on the description of the random variable and these 20 observations, is a binomial distribution an appropriate model? Perform a goodness-of-fit procedure with $\alpha = 0.05$.
(b) Calculate the P-value for this test.

9-85. Consider the spot weld shear strength data in Exercise 6-23. Does the normal distribution seem to be a reasonable model for these data? Perform an appropriate goodness-of-fit test to answer this question.

9-86. Consider the water quality data in Exercise 6-24.
(a) Do these data support the claim that mean concentration of suspended solids does not exceed 50 parts per million? Use $\alpha = 0.05$.
(b) What is the P-value for the test in part (a)?
(c) Does the normal distribution seem to be a reasonable model for these data? Perform an appropriate goodness-of-fit test to answer this question.

9-87. Consider the golf ball overall distance data in Exercise 6-25.
(a) Do these data support the claim that the mean overall distance for this brand of ball does not exceed 270 yards? Use $\alpha = 0.05$.
(b) What is the P-value for the test in part (a)?

(c) Do these data appear to be well modeled by a normal distribution? Use a formal goodness-of-fit test in answering this question.

9-88. Consider the baseball coefficient of restitution data in Exercise 8-79. If the mean coefficient of restitution exceeds 0.635, the population of balls from which the sample has been taken will be too "lively" and considered unacceptable for play.
(a) Formulate an appropriate hypothesis testing procedure to answer this question.
(b) Test these hypotheses using the data in Exercise 8-79 and draw conclusions, using $\alpha = 0.01$.
(c) Find the P-value for this test.
(d) In Exercise 8-79(b), you found a 99% confidence interval on the mean coefficient of restitution. Does this interval, or a one-sided CI, provide additional useful information to the decision maker? Explain why or why not.

9-89. Consider the dissolved oxygen data in Exercise 8-81. Water quality engineers are interested in knowing whether these data support a claim that mean dissolved oxygen concentration is 2.5 milligrams per liter.
(a) Formulate an appropriate hypothesis testing procedure to investigate this claim.
(b) Test these hypotheses, using $\alpha = 0.05$, and the data from Exercise 8-81.
(c) Find the P-value for this test.
(d) In Exercise 8-81(b) you found a 95% CI on the mean dissolved oxygen concentration. Does this interval provide useful additional information beyond that of the hypothesis testing results? Explain your answer.

9-90. The mean pull-off force of an adhesive used in manufacturing a connector for an automotive engine application should be at least 75 pounds. This adhesive will be used unless there is strong evidence that the pull-off force does not meet this requirement. A test of an appropriate hypothesis is to be conducted with sample size $n = 10$ and $\alpha = 0.05$. Assume that the pull-off force is normally distributed, and σ is not known.
(a) If the true standard deviation is $\sigma = 1$, what is the risk that the adhesive will be judged acceptable when the true mean pull-off force is only 73 pounds? Only 72 pounds?
(b) What sample size is required to give a 90% chance of detecting that the true mean is only 72 pounds when $\sigma = 1$?
(c) Rework parts (a) and (b) assuming that $\sigma = 2$. How much impact does increasing the value of σ have on the answers you obtain?

MIND-EXPANDING EXERCISES

9-91. Suppose that we wish to test $H_0: \mu = \mu_0$ versus $H_1: \mu \neq \mu_0$, where the population is normal with known σ. Let $0 < \epsilon < \alpha$, and define the critical region so that we will reject H_0 if $z_0 > z_\epsilon$ or if $z_0 < -z_{\alpha-\epsilon}$, where z_0 is the value of the usual test statistic for these hypotheses.

(a) Show that the probability of type I error for this test is α.

(b) Suppose that the true mean is $\mu_1 = \mu_0 + \delta$. Derive an expression for β for the above test.

9-92. Derive an expression for β for the test on the variance of a normal distribution. Assume that the two-sided alternative is specified.

9-93. When $X_1, X_2, \ldots, X_n$ are independent Poisson random variables, each with parameter λ, and n is large, the sample mean $\overline{X}$ has an approximate normal distribution with mean λ and variance λ/n. Therefore,

$$Z = \frac{\overline{X} - \lambda}{\sqrt{\lambda/n}}$$

has approximately a standard normal distribution. Thus we can test $H_0: \lambda = \lambda_0$ by replacing λ in Z by λ_0. When X_i are Poisson variables, this test is preferable to the large-sample test of Section 9-2.5, which would use $S/\sqrt{n}$ in the denominator, because it is designed just for the Poisson distribution. Suppose that the number of open circuits on a semiconductor wafer has a Poisson distribution. Test data for 500 wafers indicate a total of 1038 opens. Using $\alpha = 0.05$, does this suggest that the mean number of open circuits per wafer exceeds 2.0?

9-94. When $X_1, X_2, \ldots, X_n$ is a random sample from a normal distribution and n is large, the sample standard deviation has approximately a normal distribution with mean σ and variance $\sigma^2/(2n)$. Therefore, a large-sample test for $H_0: \sigma = \sigma_0$ can be based on the statistic

$$Z = \frac{S - \sigma_0}{\sqrt{\sigma_0^2/(2n)}}$$

Use this result to test $H_0: \sigma = 10$ versus $H_1: \sigma < 10$ for the golf ball overall distance data in Exercise 6-25.

9-95. Continuation of Exercise 9-94. Using the results of the previous exercise, find an approximately unbiased estimator of the 95 percentile $\theta = \mu + 1.645\sigma$. From the fact that $\overline{X}$ and S are independent random variables, find the standard error of θ. How would you estimate the standard error?

9-96. Continuation of Exercises 9-94 and 9-95. Consider the golf ball overall distance data in Exercise 6-25. We wish to investigate a claim that the 95 percentile of overall distance does not exceed 285 yards. Construct a test statistic that can be used for testing the appropriate hypotheses. Apply this procedure to the data from Exercise 6-25. What are your conclusions?

9-97. Let $X_1, X_2, \ldots, X_n$ be a sample from an exponential distribution with parameter λ. It can be shown that $2\lambda \sum_{i=1}^{n} X_i$ has a chi-square distribution with $2n$ degrees of freedom. Use this fact to devise a test statistic and critical region for $H_0: \lambda = \lambda_0$ versus the three usual alternatives.

IMPORTANT TERMS AND CONCEPTS

In the E-book, click on any term or concept below to go to that subject.

Connection between hypothesis tests and confidence intervals

Critical region for a test statistic

Null hypothesis

One- and two-sided alternative hypotheses

Operating characteristic curves

Power of the test

P-value

Reference distribution for a test statistic

Sample size determination for hypothesis tests

Significance level of a test

Statistical hypotheses

Statistical versus practical significance

Test for goodness of fit

Test for homogeneity

Test for independence

Test statistic

Type I and type II errors

CD MATERIAL

Likelihood ratio test

10

Statistical Inference for Two Samples

CHAPTER OUTLINE

LEARNING OBJECTIVES

After careful study of this chapter, you should be able to do the following:

1. Structure comparative experiments involving two samples as hypothesis tests

2. Test hypotheses and construct confidence intervals on the difference in means of two normal distributions

327

3. Test hypotheses and construct confidence intervals on the ratio of the variances or standard deviations of two normal distributions

4. Test hypotheses and construct confidence intervals on the difference in two population proportions

5. Use the P-value approach for making decisions in hypotheses tests

6. Compute power, type II error probability, and make sample size decisions for two-sample tests on means, variances, and proportions

7. Explain and use the relationship between confidence intervals and hypothesis tests

CD MATERIAL

8. Use the Fisher-Irwin test to compare two population proportions when the normal approximation to the binomial distribution does not apply

---

Answers for many odd numbered exercises are at the end of the book. Answers to exercises whose numbers are surrounded by a box can be accessed in the e-Text by clicking on the box. Complete worked solutions to certain exercises are also available in the e-Text. These are indicated in the Answers to Selected Exercises section by a box around the exercise number. Exercises are also available for some of the text sections that appear on CD only. These exercises may be found within the e-Text immediately following the section they accompany.

10-1   INTRODUCTION

The previous chapter presented hypothesis tests and confidence intervals for a single population parameter (the mean μ, the variance σ^2, or a proportion p). This chapter extends those results to the case of two independent populations.

The general situation is shown in Fig. 10-1. Population 1 has mean μ_1 and variance σ_1^2, while population 2 has mean μ_2 and variance σ_2^2. Inferences will be based on two random samples of sizes n_1 and n_2, respectively. That is, $X_{11}, X_{12}, \ldots, X_{1n_1}$ is a random sample of n_1 observations from population 1, and $X_{21}, X_{22}, \ldots, X_{2n_2}$ is a random sample of n_2 observations from population 2. Most of the practical applications of the procedures in this chapter arise in the context of **simple comparative experiments** in which the objective is to study the difference in the parameters of the two populations.

10-2   INFERENCE FOR A DIFFERENCE IN MEANS OF TWO NORMAL DISTRIBUTIONS, VARIANCES KNOWN

In this section we consider statistical inferences on the difference in means $\mu_1 - \mu_2$ of two normal distributions, where the variances σ_1^2 and σ_2^2 are known. The assumptions for this section are summarized as follows.

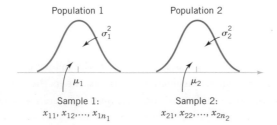

Figure 10-1 Two independent populations.

Assumptions

1. $X_{11}, X_{12}, \ldots, X_{1n_1}$ is a random sample from population 1.
2. $X_{21}, X_{22}, \ldots, X_{2n_2}$ is a random sample from population 2.
3. The two populations represented by X_1 and X_2 are independent.
4. Both populations are normal.

A logical point estimator of $\mu_1 - \mu_2$ is the difference in sample means $\overline{X}_1 - \overline{X}_2$. Based on the properties of expected values

$$E(\overline{X}_1 - \overline{X}_2) = E(\overline{X}_1) - E(\overline{X}_2) = \mu_1 - \mu_2$$

and the variance of $\overline{X}_1 - \overline{X}_2$ is

$$V(\overline{X}_1 - \overline{X}_2) = V(\overline{X}_1) + V(\overline{X}_2) = \frac{\sigma_1^2}{n_1} + \frac{\sigma_2^2}{n_2}$$

Based on the assumptions and the preceding results, we may state the following.

The quantity

$$Z = \frac{\overline{X}_1 - \overline{X}_2 - (\mu_1 - \mu_2)}{\sqrt{\dfrac{\sigma_1^2}{n_1} + \dfrac{\sigma_2^2}{n_2}}} \qquad (10\text{-}1)$$

has a $N(0, 1)$ distribution.

This result will be used to form tests of hypotheses and confidence intervals on $\mu_1 - \mu_2$. Essentially, we may think of $\mu_1 - \mu_2$ as a parameter θ, and its estimator is $\hat{\Theta} = \overline{X}_1 - \overline{X}_2$ with variance $\sigma_{\hat{\Theta}}^2 = \sigma_1^2/n_1 + \sigma_2^2/n_2$. If θ_0 is the null hypothesis value specified for θ, the test statistic will be $(\hat{\Theta} - \theta_0)/\sigma_{\hat{\Theta}}$. Notice how similar this is to the test statistic for a single mean used in Equation 9-8 of Chapter 9.

10-2.1 Hypothesis Tests for a Difference in Means, Variances Known

We now consider hypothesis testing on the difference in the means $\mu_1 - \mu_2$ of two normal populations. Suppose that we are interested in testing that the difference in means $\mu_1 - \mu_2$ is equal to a specified value Δ_0. Thus, the null hypothesis will be stated as $H_0: \mu_1 - \mu_2 = \Delta_0$. Obviously, in many cases, we will specify $\Delta_0 = 0$ so that we are testing the equality of two means (i.e., $H_0: \mu_1 = \mu_2$). The appropriate test statistic would be found by replacing $\mu_1 - \mu_2$ in Equation 10-1 by Δ_0, and this test statistic would have a standard normal distribution under H_0. That is, the standard normal distribution is the **reference distribution** for the test statistic. Suppose that the alternative hypothesis is $H_1: \mu_1 - \mu_2 \neq \Delta_0$. Now, a sample value of $\overline{x}_1 - \overline{x}_2$ that is considerably different from Δ_0 is evidence that H_1 is true. Because Z_0 has the $N(0, 1)$

distribution when H_0 is true, we would take $-z_{\alpha/2}$ and $z_{\alpha/2}$ as the boundaries of the critical region just as we did in the single-sample hypothesis-testing problem of Section 9-2.1. This would give a test with level of significance α. Critical regions for the one-sided alternatives would be located similarly. Formally, we summarize these results below.

Null hypothesis: H_0: $\mu_1 - \mu_2 = \Delta_0$

Test statistic: $$Z_0 = \dfrac{\bar{X}_1 - \bar{X}_2 - \Delta_0}{\sqrt{\dfrac{\sigma_1^2}{n_1} + \dfrac{\sigma_2^2}{n_2}}}$$ (10-2)

Alternative Hypotheses	Rejection Criterion
H_1: $\mu_1 - \mu_2 \neq \Delta_0$	$z_0 > z_{\alpha/2}$ or $z_0 < -z_{\alpha/2}$
H_1: $\mu_1 - \mu_2 > \Delta_0$	$z_0 > z_{\alpha}$
H_1: $\mu_1 - \mu_2 < \Delta_0$	$z_0 < -z_{\alpha}$

EXAMPLE 10-1   A product developer is interested in reducing the drying time of a primer paint. Two formulations of the paint are tested; formulation 1 is the standard chemistry, and formulation 2 has a new drying ingredient that should reduce the drying time. From experience, it is known that the standard deviation of drying time is 8 minutes, and this inherent variability should be unaffected by the addition of the new ingredient. Ten specimens are painted with formulation 1, and another 10 specimens are painted with formulation 2; the 20 specimens are painted in random order. The two sample average drying times are $\bar{x}_1 = 121$ minutes and $\bar{x}_2 = 112$ minutes, respectively. What conclusions can the product developer draw about the effectiveness of the new ingredient, using $\alpha = 0.05$?

We apply the eight-step procedure to this problem as follows:

1. The quantity of interest is the difference in mean drying times, $\mu_1 - \mu_2$, and $\Delta_0 = 0$.
2. H_0: $\mu_1 - \mu_2 = 0$, or H_0: $\mu_1 = \mu_2$.
3. H_1: $\mu_1 > \mu_2$. We want to reject H_0 if the new ingredient reduces mean drying time.
4. $\alpha = 0.05$
5. The test statistic is

$$z_0 = \dfrac{\bar{x}_1 - \bar{x}_2 - 0}{\sqrt{\dfrac{\sigma_1^2}{n_1} + \dfrac{\sigma_2^2}{n_2}}}$$

    where $\sigma_1^2 = \sigma_2^2 = (8)^2 = 64$ and $n_1 = n_2 = 10$.

6. Reject H_0: $\mu_1 = \mu_2$ if $z_0 > 1.645 = z_{0.05}$.
7. Computations: Since $\bar{x}_1 = 121$ minutes and $\bar{x}_2 = 112$ minutes, the test statistic is

$$z_0 = \dfrac{121 - 112}{\sqrt{\dfrac{(8)^2}{10} + \dfrac{(8)^2}{10}}} = 2.52$$

8. Conclusion: Since $z_0 = 2.52 > 1.645$, we reject H_0: $\mu_1 = \mu_2$ at the $\alpha = 0.05$ level and conclude that adding the new ingredient to the paint significantly reduces the drying time. Alternatively, we can find the P-value for this test as

$$P\text{-value} = 1 - \Phi(2.52) = 0.0059$$

Therefore, H_0: $\mu_1 = \mu_2$ would be rejected at any significance level $\alpha \geq 0.0059$.

When the population variances are unknown, the sample variances s_1^2 and s_2^2 can be substituted into the test statistic Equation 10-2 to produce a **large-sample test** for the difference in means. This procedure will also work well when the populations are not necessarily normally distributed. However, both n_1 and n_2 should exceed 40 for this large-sample test to be valid.

10-2.2 Choice of Sample Size

Use of Operating Characteristic Curves

The operating characteristic curves in Appendix Charts VI*a*, VI*b*, VI*c*, and VI*d* may be used to evaluate the type II error probability for the hypotheses in the display (10-2). These curves are also useful in determining sample size. Curves are provided for $\alpha = 0.05$ and $\alpha = 0.01$. For the two-sided alternative hypothesis, the abscissa scale of the operating characteristic curve in charts VI*a* and VI*b* is d, where

$$d = \frac{|\mu_1 - \mu_2 - \Delta_0|}{\sqrt{\sigma_1^2 + \sigma_2^2}} = \frac{|\Delta - \Delta_0|}{\sqrt{\sigma_1^2 + \sigma_2^2}} \qquad (10\text{-}3)$$

and one must choose equal sample sizes, say, $n = n_1 = n_2$. The one-sided alternative hypotheses require the use of Charts VI*c* and VI*d*. For the one-sided alternatives H_1: $\mu_1 - \mu_2 > \Delta_0$ or H_1: $\mu_1 - \mu_2 < \Delta_0$, the abscissa scale is also given by

$$d = \frac{|\mu_1 - \mu_2 - \Delta_0|}{\sqrt{\sigma_1^2 + \sigma_2^2}} = \frac{|\Delta - \Delta_0|}{\sqrt{\sigma_1^2 + \sigma_2^2}}$$

It is not unusual to encounter problems where the costs of collecting data differ substantially between the two populations, or where one population variance is much greater than the other. In those cases, we often use unequal sample sizes. If $n_1 \neq n_2$, the operating characteristic curves may be entered with an *equivalent* value of n computed from

$$n = \frac{\sigma_1^2 + \sigma_2^2}{\sigma_1^2/n_1 + \sigma_2^2/n_2} \qquad (10\text{-}4)$$

If $n_1 \neq n_2$, and their values are fixed in advance, Equation 10-4 is used directly to calculate n, and the operating characteristic curves are entered with a specified d to obtain β. If we are given d and it is necessary to determine n_1 and n_2 to obtain a specified β, say, β^*, we guess at trial values of n_1 and n_2, calculate n in Equation 10-4, and enter the curves with the specified value of d to find β. If $\beta = \beta^*$, the trial values of n_1 and n_2 are satisfactory. If $\beta \neq \beta^*$, adjustments to n_1 and n_2 are made and the process is repeated.

EXAMPLE 10-2

Consider the paint drying time experiment from Example 10-1. If the true difference in mean drying times is as much as 10 minutes, find the sample sizes required to detect this difference with probability at least 0.90.

The appropriate value of the abscissa parameter is (since $\Delta_0 = 0$, and $\Delta = 10$)

$$d = \frac{|\mu_1 - \mu_2|}{\sqrt{\sigma_1^2 + \sigma_2^2}} = \frac{10}{\sqrt{8^2 + 8^2}} = 0.88$$

and since the detection probability or power of the test must be at least 0.9, with $\alpha = 0.05$, we find from Appendix Chart VIc that $n = n_1 = n_2 \simeq 11$.

Sample Size Formulas

It is also possible to obtain formulas for calculating the sample sizes directly. Suppose that the null hypothesis H_0: $\mu_1 - \mu_2 = \Delta_0$ is false and that the true difference in means is $\mu_1 - \mu_2 = \Delta$, where $\Delta > \Delta_0$. One may find formulas for the sample size required to obtain a specific value of the type II error probability β for a given difference in means Δ and level of significance α.

For the two-sided alternative hypothesis with significance level α, the sample size $n_1 = n_2 = n$ required to detect a true difference in means of Δ with power at least $1 - \beta$ is

$$n \simeq \frac{(z_{\alpha/2} + z_\beta)^2(\sigma_1^2 + \sigma_2^2)}{(\Delta - \Delta_0)^2} \qquad (10\text{-}5)$$

This approximation is valid when $\Phi(-z_{\alpha/2} - (\Delta - \Delta_0)\sqrt{n}/\sqrt{\sigma_1^2 + \sigma_2^2})$ is small compared to β.

For a one-sided alternative hypothesis with significance level α, the sample size $n_1 = n_2 = n$ required to detect a true difference in means of $\Delta(\neq\Delta_0)$ with power at least $1 - \beta$ is

$$n = \frac{(z_\alpha + z_\beta)^2(\sigma_1^2 + \sigma_2^2)}{(\Delta - \Delta_0)^2} \qquad (10\text{-}6)$$

The derivation of Equations 10-5 and 10-6 closely follows the single-sample case in Section 9-2.3. For example, to obtain Equation 10-6, we first write the expression for the β-error for the two-sided alternate, which is

$$\beta = \Phi\left(z_{\alpha/2} - \frac{\Delta - \Delta_0}{\sqrt{\dfrac{\sigma_1^2}{n_1} + \dfrac{\sigma_2^2}{n_2}}}\right) - \Phi\left(-z_{\alpha/2} - \frac{\Delta - \Delta_0}{\sqrt{\dfrac{\sigma_1^2}{n_1} + \dfrac{\sigma_2^2}{n_2}}}\right)$$

where Δ is the true difference in means of interest. Then by following a procedure similar to that used to obtain Equation 9-17, the expression for β can be obtained for the case where $n = n_1 = n_2$.

EXAMPLE 10-3   To illustrate the use of these sample size equations, consider the situation described in Example 10-1, and suppose that if the true difference in drying times is as much as 10 minutes, we want to detect this with probability at least 0.90. Under the null hypothesis, $\Delta_0 = 0$. We have a one-sided alternative hypothesis with $\Delta = 10$, $\alpha = 0.05$ (so $z_\alpha = z_{0.05} = 1.645$), and since the power is 0.9, $\beta = 0.10$ (so $z_\beta = z_{0.10} = 1.28$). Therefore we may find the required sample size from Equation 10-6 as follows:

$$n = \frac{(z_\alpha + z_\beta)^2(\sigma_1^2 + \sigma_2^2)}{(\Delta - \Delta_0)^2} = \frac{(1.645 + 1.28)^2[(8)^2 + (8)^2]}{(10 - 0)^2} = 11$$

This is exactly the same as the result obtained from using the O.C. curves.

10-2.3   Identifying Cause and Effect

Engineers and scientists are often interested in comparing two different conditions to determine whether either condition produces a significant effect on the response that is observed. These conditions are sometimes called **treatments.** Example 10-1 illustrates such a situation; the two different treatments are the two paint formulations, and the response is the drying time. The purpose of the study is to determine whether the new formulation results in a significant effect—reducing drying time. In this situation, the product developer (the experimenter) randomly assigned 10 test specimens to one formulation and 10 test specimens to the other formulation. Then the paints were applied to the test specimens in random order until all 20 specimens were painted. This is an example of a **completely randomized experiment.**

When statistical significance is observed in a randomized experiment, the experimenter can be confident in the conclusion that it was the difference in treatments that resulted in the difference in response. That is, we can be confident that a cause-and-effect relationship has been found.

Sometimes the objects to be used in the comparison are not assigned at random to the treatments. For example, the September 1992 issue of *Circulation* (a medical journal published by the American Heart Association) reports a study linking high iron levels in the body with increased risk of heart attack. The study, done in Finland, tracked 1931 men for five years and showed a statistically significant effect of increasing iron levels on the incidence of heart attacks. In this study, the comparison was not performed by randomly selecting a sample of men and then assigning some to a "low iron level" treatment and the others to a "high iron level" treatment. The researchers just tracked the subjects over time. Recall from Chapter 1 that this type of study is called an **observational study.**

It is difficult to identify causality in observational studies, because the observed statistically significant difference in response between the two groups may be due to some other underlying factor (or group of factors) that was not equalized by randomization and not due to the treatments. For example, the difference in heart attack risk could be attributable to the difference in iron levels, or to other underlying factors that form a reasonable explanation for the observed results—such as cholesterol levels or hypertension.

The difficulty of establishing causality from observational studies is also seen in the smoking and health controversy. Numerous studies show that the incidence of lung cancer and other respiratory disorders is higher among smokers than nonsmokers. However, establishing cause and effect here has proven enormously difficult. Many individuals had decided to smoke long before the start of the research studies, and many factors other than smoking could have a role in contracting lung cancer.

10-2.4 Confidence Interval on a Difference in Means, Variances Known

The $100(1 - \alpha)\%$ confidence interval on the difference in two means $\mu_1 - \mu_2$ when the variances are known can be found directly from results given previously in this section. Recall that $X_{11}, X_{12}, \ldots, X_{1n_1}$ is a random sample of n_1 observations from the first population and $X_{21}, X_{22}, \ldots, X_{2n_2}$ is a random sample of n_2 observations from the second population. The difference in sample means $\overline{X}_1 - \overline{X}_2$ is a point estimator of $\mu_1 - \mu_2$, and

$$Z = \frac{\overline{X}_1 - \overline{X}_2 - (\mu_1 - \mu_2)}{\sqrt{\dfrac{\sigma_1^2}{n_1} + \dfrac{\sigma_2^2}{n_2}}}$$

has a standard normal distribution if the two populations are normal or is approximately standard normal if the conditions of the central limit theorem apply, respectively. This implies that $P(-z_{\alpha/2} \leq Z \leq z_{\alpha/2}) = 1 - \alpha$, or

$$P\left[-z_{\alpha/2} \leq \frac{\overline{X}_1 - \overline{X}_2 - (\mu_1 - \mu_2)}{\sqrt{\dfrac{\sigma_1^2}{n_1} + \dfrac{\sigma_2^2}{n_2}}} \leq z_{\alpha/2} \right] = 1 - \alpha$$

This can be rearranged as

$$P\left(\overline{X}_1 - \overline{X}_2 - z_{\alpha/2}\sqrt{\frac{\sigma_1^2}{n_1} + \frac{\sigma_2^2}{n_2}} \leq \mu_1 - \mu_2 \leq \overline{X}_1 - \overline{X}_2 + z_{\alpha/2}\sqrt{\frac{\sigma_1^2}{n_1} + \frac{\sigma_2^2}{n_2}} \right) = 1 - \alpha$$

Therefore, the $100(1 - \alpha)\%$ confidence interval for $\mu_1 - \mu_2$ is defined as follows.

Definition

> If $\bar{x}_1$ and $\bar{x}_2$ are the means of independent random samples of sizes n_1 and n_2 from two independent normal populations with known variances σ_1^2 and σ_2^2, respectively, **a $100(1 - \alpha)\%$ confidence interval for $\mu_1 - \mu_2$ is**
>
> $$\bar{x}_1 - \bar{x}_2 - z_{\alpha/2}\sqrt{\frac{\sigma_1^2}{n_1} + \frac{\sigma_2^2}{n_2}} \leq \mu_1 - \mu_2 \leq \bar{x}_1 - \bar{x}_2 + z_{\alpha/2}\sqrt{\frac{\sigma_1^2}{n_1} + \frac{\sigma_2^2}{n_2}} \qquad (10\text{-}7)$$
>
> where $z_{\alpha/2}$ is the upper $\alpha/2$ percentage point of the standard normal distribution.

The confidence level $1 - \alpha$ is exact when the populations are normal. For nonnormal populations, the confidence level is approximately valid for large sample sizes.

EXAMPLE 10-4

Tensile strength tests were performed on two different grades of aluminum spars used in manufacturing the wing of a commercial transport aircraft. From past experience with the spar manufacturing process and the testing procedure, the standard deviations of tensile strengths are assumed to be known. The data obtained are as follows: $n_1 = 10$, $\bar{x}_1 = 87.6$, $\sigma_1 = 1$, $n_2 = 12$, $\bar{x}_2 = 74.5$, and $\sigma_2 = 1.5$. If μ_1 and μ_2 denote the true mean tensile strengths for the two grades of spars, we may find a 90% confidence interval on the difference in mean strength $\mu_1 - \mu_2$ as follows:

$$\bar{x}_1 - \bar{x}_2 - z_{\alpha/2}\sqrt{\frac{\sigma_1^2}{n_1} + \frac{\sigma_2^2}{n_2}} \leq \mu_1 - \mu_2 \leq \bar{x}_1 - \bar{x}_2 + z_{\alpha/2}\sqrt{\frac{\sigma_1^2}{n_1} + \frac{\sigma_2^2}{n_2}}$$

$$87.6 - 74.5 - 1.645\sqrt{\frac{(1)^2}{10} + \frac{(1.5)^3}{12}} \leq \mu_1 - \mu_2 \leq 87.6 - 74.5 + 1.645\sqrt{\frac{(1^2)}{10} + \frac{(1.5)^2}{12}}$$

Therefore, the 90% confidence interval on the difference in mean tensile strength (in kilograms per square millimeter) is

$$12.22 \leq \mu_1 - \mu_2 \leq 13.98 \text{ (in kilograms per square millimeter)}$$

Notice that the confidence interval does not include zero, implying that the mean strength of aluminum grade 1 (μ_1) exceeds the mean strength of aluminum grade 2 (μ_2). In fact, we can state that we are 90% confident that the mean tensile strength of aluminum grade 1 exceeds that of aluminum grade 2 by between 12.22 and 13.98 kilograms per square millimeter.

Choice of Sample Size

If the standard deviations σ_1 and σ_2 are known (at least approximately) and the two sample sizes n_1 and n_2 are equal ($n_1 = n_2 = n$, say), we can determine the sample size required so that the error in estimating $\mu_1 - \mu_2$ by $\bar{x}_1 - \bar{x}_2$ will be less than E at $100(1 - \alpha)\%$ confidence. The required sample size from each population is

$$n = \left(\frac{z_{\alpha/2}}{E}\right)^2 (\sigma_1^2 + \sigma_2^2) \tag{10-8}$$

Remember to round up if n is not an integer. This will ensure that the level of confidence does not drop below $100(1 - \alpha)\%$.

One-Sided Confidence Bounds

One-sided confidence bounds on $\mu_1 - \mu_2$ may also be obtained. A $100(1 - \alpha)\%$ upper-confidence bound on $\mu_1 - \mu_2$ is

$$\mu_1 - \mu_2 \leq \bar{x}_1 - \bar{x}_2 + z_\alpha\sqrt{\frac{\sigma_1^2}{n_1} + \frac{\sigma_2^2}{n_2}} \tag{10-9}$$

and a $100(1 - \alpha)\%$ lower-confidence bound is

$$\bar{x}_1 - \bar{x}_2 - z_\alpha \sqrt{\frac{\sigma_1^2}{n_1} + \frac{\sigma_2^2}{n_2}} \le \mu_1 - \mu_2 \qquad (10\text{-}10)$$

EXERCISES FOR SECTION 10-2

10-1. Two machines are used for filling plastic bottles with a net volume of 16.0 ounces. The fill volume can be assumed normal, with standard deviation $\sigma_1 = 0.020$ and $\sigma_2 = 0.025$ ounces. A member of the quality engineering staff suspects that both machines fill to the same mean net volume, whether or not this volume is 16.0 ounces. A random sample of 10 bottles is taken from the output of each machine.

Machine 1		Machine 2	
16.03	16.01	16.02	16.03
16.04	15.96	15.97	16.04
16.05	15.98	15.96	16.02
16.05	16.02	16.01	16.01
16.02	15.99	15.99	16.00

(a) Do you think the engineer is correct? Use $\alpha = 0.05$.
(b) What is the P-value for this test?
(c) What is the power of the test in part (a) for a true difference in means of 0.04?
(d) Find a 95% confidence interval on the difference in means. Provide a practical interpretation of this interval.
(e) Assuming equal sample sizes, what sample size should be used to assure that $\beta = 0.05$ if the true difference in means is 0.04? Assume that $\alpha = 0.05$.

10-2. Two types of plastic are suitable for use by an electronics component manufacturer. The breaking strength of this plastic is important. It is known that $\sigma_1 = \sigma_2 = 1.0$ psi. From a random sample of size $n_1 = 10$ and $n_2 = 12$, we obtain $\bar{x}_1 = 162.5$ and $\bar{x}_2 = 155.0$. The company will not adopt plastic 1 unless its mean breaking strength exceeds that of plastic 2 by at least 10 psi. Based on the sample information, should it use plastic 1? Use $\alpha = 0.05$ in reaching a decision.

10-3. Reconsider the situation in Exercise 10-2. Suppose that the true difference in means is really 12 psi. Find the power of the test assuming that $\alpha = 0.05$. If it is really important to detect this difference, are the sample sizes employed in Exercise 10-2 adequate, in your opinion?

10-4. The burning rates of two different solid-fuel propellants used in aircrew escape systems are being studied. It is known that both propellants have approximately the same standard deviation of burning rate; that is $\sigma_1 = \sigma_2 = 3$ centimeters per second. Two random samples of $n_1 = 20$

and $n_2 = 20$ specimens are tested; the sample mean burning rates are $\bar{x}_1 = 18$ centimeters per second and $\bar{x}_2 = 24$ centimeters per second.
(a) Test the hypothesis that both propellants have the same mean burning rate. Use $\alpha = 0.05$.
(b) What is the P-value of the test in part (a)?
(c) What is the β-error of the test in part (a) if the true difference in mean burning rate is 2.5 centimeters per second?
(d) Construct a 95% confidence interval on the difference in means $\mu_1 - \mu_2$. What is the practical meaning of this interval?

10-5. Two machines are used to fill plastic bottles with dishwashing detergent. The standard deviations of fill volume are known to be $\sigma_1 = 0.10$ fluid ounces and $\sigma_2 = 0.15$ fluid ounces for the two machines, respectively. Two random samples of $n_1 = 12$ bottles from machine 1 and $n_2 = 10$ bottles from machine 2 are selected, and the sample mean fill volumes are $\bar{x}_1 = 30.87$ fluid ounces and $\bar{x}_2 = 30.68$ fluid ounces. Assume normality.
(a) Construct a 90% two-sided confidence interval on the mean difference in fill volume. Interpret this interval.
(b) Construct a 95% two-sided confidence interval on the mean difference in fill volume. Compare and comment on the width of this interval to the width of the interval in part (a).
(c) Construct a 95% upper-confidence interval on the mean difference in fill volume. Interpret this interval.

10-6. Reconsider the situation described in Exercise 10-5.
(a) Test the hypothesis that both machines fill to the same mean volume. Use $\alpha = 0.05$.
(b) What is the P-value of the test in part (a)?
(c) If the β-error of the test when the true difference in fill volume is 0.2 fluid ounces should not exceed 0.1, what sample sizes must be used? Use $\alpha = 0.05$.

10-7. Two different formulations of an oxygenated motor fuel are being tested to study their road octane numbers. The variance of road octane number for formulation 1 is $\sigma_1^2 = 1.5$, and for formulation 2 it is $\sigma_2^2 = 1.2$. Two random samples of size $n_1 = 15$ and $n_2 = 20$ are tested, and the mean road octane numbers observed are $\bar{x}_1 = 89.6$ and $\bar{x}_2 = 92.5$. Assume normality.
(a) Construct a 95% two-sided confidence interval on the difference in mean road octane number.
(b) If formulation 2 produces a higher road octane number than formulation 1, the manufacturer would like to detect

it. Formulate and test an appropriate hypothesis, using $\alpha = 0.05$.

(c) What is the P-value for the test you conducted in part (b)?

10-8. Consider the situation described in Exercise 10-4. What sample size would be required in each population if we wanted the error in estimating the difference in mean burning rates to be less than 4 centimeters per second with 99% confidence?

10-9. Consider the road octane test situation described in Exercise 10-7. What sample size would be required in each population if we wanted to be 95% confident that the error in estimating the difference in mean road octane number is less than 1?

$\longrightarrow$ **10-10.** A polymer is manufactured in a batch chemical process. Viscosity measurements are normally made on each batch, and long experience with the process has indicated that the variability in the process is fairly stable with $\sigma = 20$. Fifteen batch viscosity measurements are given as follows: 724, 718, 776, 760, 745, 759, 795, 756, 742, 740, 761, 749, 739, 747, 742. A process change is made which involves switching the type of catalyst used in the process. Following the process change, eight batch viscosity measurements are taken: 735, 775, 729, 755, 783, 760, 738, 780. Assume that process variability is unaffected by the catalyst change. Find a 90% confidence interval on the difference in mean batch viscosity resulting from the process change.

10-11. The concentration of active ingredient in a liquid laundry detergent is thought to be affected by the type of catalyst used in the process. The standard deviation of active concentration is known to be 3 grams per liter, regardless of the catalyst type. Ten observations on concentration are taken with each catalyst, and the data follow:

Catalyst 1: 57.9, 66.2, 65.4, 65.4, 65.2, 62.6, 67.6, 63.7, 67.2, 71.0

Catalyst 2: 66.4, 71.7, 70.3, 69.3, 64.8, 69.6, 68.6, 69.4, 65.3, 68.8

(a) Find a 95% confidence interval on the difference in mean active concentrations for the two catalysts.

(b) Is there any evidence to indicate that the mean active concentrations depend on the choice of catalyst? Base your answer on the results of part (a).

10-12. Consider the polymer batch viscosity data in Exercise 10-10. If the difference in mean batch viscosity is 10 or less, the manufacturer would like to detect it with a high probability.

(a) Formulate and test an appropriate hypothesis using $\alpha = 0.10$. What are your conclusions?

(b) Calculate the P-value for this test.

(c) Compare the results of parts (a) and (b) to the length of the 90% confidence interval obtained in Exercise 10-10 and discuss your findings.

10-13. For the laundry detergent problem in Exercise 10-11, test the hypothesis that the mean active concentrations are the same for both types of catalyst. Use $\alpha = 0.05$. What is the P-value for this test? Compare your answer to that found in part (b) of Exercise 10-11, and comment on why they are same or different.

10-14. Reconsider the laundry detergent problem in Exercise 10-11. Suppose that the true mean difference in active concentration is 5 grams per liter. What is the power of the test to detect this difference if $\alpha = 0.05$? If this difference is really important, do you consider the sample sizes used by the experimenter to be adequate?

10-15. Consider the polymer viscosity data in Exercise 10-10. Does the assumption of normality seem reasonable for both samples?

10-16. Consider the concentration data in Exercise 10-11. Does the assumption of normality seem reasonable?

10-3 INFERENCE FOR THE DIFFERENCE IN MEANS OF TWO NORMAL DISTRIBUTIONS, VARIANCES UNKNOWN

We now extend the results of the previous section to the difference in means of the two distributions in Fig. 10-1 when the variances of both distributions σ_1^2 and σ_2^2 are unknown. If the sample sizes n_1 and n_2 exceed 40, the normal distribution procedures in Section 10-2 could be used. However, when small samples are taken, we will assume that the populations are normally distributed and base our hypotheses tests and confidence intervals on the t distribution. This nicely parallels the case of inference on the mean of a single sample with unknown variance.

10-3.1 Hypotheses Tests for a Difference in Means, Variances Unknown

We now consider tests of hypotheses on the difference in means $\mu_1 - \mu_2$ of two normal distributions where the variances σ_1^2 and σ_2^2 are unknown. A t-statistic will be used to test these hypotheses. As noted above and in Section 9-3, the normality assumption is required to develop the test procedure, but moderate departures from normality do not adversely affect the procedure. Two different situations must be treated. In the first case, we assume that the variances of the two normal distributions are unknown but equal; that is, $\sigma_1^2 = \sigma_2^2 = \sigma^2$. In the second, we assume that σ_1^2 and σ_2^2 are unknown and not necessarily equal.

Case 1: $\sigma_1^2 = \sigma_2^2 = \sigma^2$

Suppose we have two independent normal populations with unknown means μ_1 and μ_2, and unknown but equal variances, $\sigma_1^2 = \sigma_2^2 = \sigma^2$. We wish to test

$$H_0: \mu_1 - \mu_2 = \Delta_0$$
$$H_1: \mu_1 - \mu_2 \neq \Delta_0 \tag{10-11}$$

Let $X_{11}, X_{12}, \ldots, X_{1n_1}$ be a random sample of n_1 observations from the first population and $X_{21}, X_{22}, \ldots, X_{2n_2}$ be a random sample of n_2 observations from the second population. Let $\overline{X}_1, \overline{X}_2, S_1^2$, and S_2^2 be the sample means and sample variances, respectively. Now the expected value of the difference in sample means $\overline{X}_1 - \overline{X}_2$ is $E(\overline{X}_1 - \overline{X}_2) = \mu_1 - \mu_2$, so $\overline{X}_1 - \overline{X}_2$ is an unbiased estimator of the difference in means. The variance of $\overline{X}_1 - \overline{X}_2$ is

$$V(\overline{X}_1 - \overline{X}_2) = \frac{\sigma^2}{n_1} + \frac{\sigma^2}{n_2} = \sigma^2 \left(\frac{1}{n_1} + \frac{1}{n_2} \right)$$

It seems reasonable to combine the two sample variances S_1^2 and S_2^2 to form an estimator of σ^2. The **pooled estimator** of σ^2 is defined as follows.

The **pooled estimator** of σ^2, denoted by S_p^2, is defined by

$$S_p^2 = \frac{(n_1 - 1)S_1^2 + (n_2 - 1)S_2^2}{n_1 + n_2 - 2} \tag{10-12}$$

It is easy to see that the pooled estimator S_p^2 can be written as

$$S_p^2 = \frac{n_1 - 1}{n_1 + n_2 - 2} S_1^2 + \frac{n_2 - 1}{n_1 + n_2 - 2} S_2^2 = wS_1^2 + (1 - w)S_2^2$$

where $0 < w \leq 1$. Thus S_p^2 is a **weighted average** of the two sample variances S_1^2 and S_2^2, where the weights w and $1 - w$ depend on the two sample sizes n_1 and n_2. Obviously, if $n_1 = n_2 = n$, $w = 0.5$ and S_p^2 is just the arithmetic average of S_1^2 and S_2^2. If $n_1 = 10$ and $n_2 = 20$ (say), $w = 0.32$ and $1 - w = 0.68$. The first sample contributes $n_1 - 1$ degrees of freedom to S_p^2 and the second sample contributes $n_2 - 1$ degrees of freedom. Therefore, S_p^2 has $n_1 + n_2 - 2$ degrees of freedom.

Now we know that

$$Z = \frac{\overline{X}_1 - \overline{X}_2 - (\mu_1 - \mu_2)}{\sigma \sqrt{\dfrac{1}{n_1} + \dfrac{1}{n_2}}}$$

has a $N(0, 1)$ distribution. Replacing σ by S_p gives the following.

Given the assumptions of this section, the quantity

$$T = \frac{\bar{X}_1 - \bar{X}_2 - (\mu_1 - \mu_2)}{S_p\sqrt{\dfrac{1}{n_1} + \dfrac{1}{n_2}}}$$ (10-13)

has a t distribution with $n_1 + n_2 - 2$ degrees of freedom.

The use of this information to test the hypotheses in Equation 10-11 is now straightforward: simply replace $\mu_1 - \mu_2$ by Δ_0, and the resulting **test statistic** has a t distribution with $n_1 + n_2 - 2$ degrees of freedom under H_0: $\mu_1 - \mu_2 = \Delta_0$. Therefore, the reference distribution for the test statistic is the t distribution with $n_1 + n_2 - 2$ degrees of freedom. The location of the critical region for both two- and one-sided alternatives parallels those in the one-sample case. Because a pooled estimate of variance is used, the procedure is often called the **pooled t-test.**

Definition:
The Two-Sample or Pooled t-Test*

Null hypothesis: H_0: $\mu_1 - \mu_2 = \Delta_0$

Test statistic: $$T_0 = \frac{\bar{X}_1 - \bar{X}_2 - \Delta_0}{S_p\sqrt{\dfrac{1}{n_1} + \dfrac{1}{n_2}}}$$ (10-14)

Alternative Hypothesis	Rejection Criterion
H_1: $\mu_1 - \mu_2 \neq \Delta_0$	$t_0 > t_{\alpha/2,n_1+n_2-2}$ or $t_0 < -t_{\alpha/2,n_1+n_2-2}$
H_1: $\mu_1 - \mu_2 > \Delta_0$	$t_0 > t_{\alpha,n_1+n_2-2}$
H_1: $\mu_1 - \mu_2 < \Delta_0$	$t_0 < -t_{\alpha,n_1+n_2-2}$

EXAMPLE 10-5   Two catalysts are being analyzed to determine how they affect the mean yield of a chemical process. Specifically, catalyst 1 is currently in use, but catalyst 2 is acceptable. Since catalyst 2 is cheaper, it should be adopted, providing it does not change the process yield. A test is run in the pilot plant and results in the data shown in Table 10-1. Is there any difference between the mean yields? Use $\alpha = 0.05$, and assume equal variances.

The solution using the eight-step hypothesis-testing procedure is as follows:

1.   The parameters of interest are μ_1 and μ_2, the mean process yield using catalysts 1 and 2, respectively, and we want to know if $\mu_1 - \mu_2 = 0$.

2. H_0: $\mu_1 - \mu_2 = 0$, or H_0: $\mu_1 = \mu_2$

---

*While we have given the development of this procedure for the case where the sample sizes could be different, there is an advantage to using equal sample sizes $n_1 = n_2 = n$. When the sample sizes are the same from both populations, the t-test is more robust to the assumption of equal variances. Please see Section 10-3.2 on the CD.

Table 10-1   Catalyst Yield Data, Example 10-5

Observation Number	Catalyst 1	Catalyst 2
1	91.50	89.19
2	94.18	90.95
3	92.18	90.46
4	95.39	93.21
5	91.79	97.19
6	89.07	97.04
7	94.72	91.07
8	89.21	92.75
	$\bar{x}_1 = 92.255$	$\bar{x}_2 = 92.733$
	$s_1 = 2.39$	$s_2 = 2.98$

3. $H_1: \mu_1 \neq \mu_2$

4. $\alpha = 0.05$

5. The test statistic is

$$t_0 = \frac{\bar{x}_1 - \bar{x}_2 - 0}{s_p\sqrt{\dfrac{1}{n_1} + \dfrac{1}{n_2}}}$$

6. Reject H_0 if $t_0 > t_{0.025,14} = 2.145$ or if $t_0 < -t_{0.025,14} = -2.145$.

7. Computations: From Table 10-1 we have $\bar{x}_1 = 92.255$, $s_1 = 2.39$, $n_1 = 8$, $\bar{x}_2 = 92.733$, $s_2 = 2.98$, and $n_2 = 8$. Therefore

$$s_p^2 = \frac{(n_1 - 1)s_1^2 + (n_2 - 1)s_2^2}{n_1 + n_2 - 2} = \frac{(7)(2.39)^2 + 7(2.98)^2}{8 + 8 - 2} = 7.30$$

$$s_p = \sqrt{7.30} = 2.70$$

and

$$t_0 = \frac{\bar{x}_1 - \bar{x}_2}{2.70\sqrt{\dfrac{1}{n_1} + \dfrac{1}{n_2}}} = \frac{92.255 - 92.733}{2.70\sqrt{\dfrac{1}{8} + \dfrac{1}{8}}} = -0.35$$

8. Conclusions: Since $-2.145 < t_0 = -0.35 < 2.145$, the null hypothesis cannot be rejected. That is, at the 0.05 level of significance, we do not have strong evidence to conclude that catalyst 2 results in a mean yield that differs from the mean yield when catalyst 1 is used.

A P-value could also be used for decision making in this example. From Appendix Table IV we find that $t_{0.40,14} = 0.258$ and $t_{0.25,14} = 0.692$. Therefore, since $0.258 < 0.35 < 0.692$, we conclude that lower and upper bounds on the P-value are $0.50 < P < 0.80$. Therefore, since the P-value exceeds $\alpha = 0.05$, the null hypothesis cannot be rejected.

The Minitab two-sample t-test and confidence interval procedure for Example 10-5 follows:

Two-Sample T-Test and CI: Cat 1, Cat 2

Two-sample T for Cat 1 vs Cat 2

	N	Mean	StDev	SE Mean
Cat 1	8	92.26	2.39	0.84
Cat 2	8	92.73	2.99	1.1

Difference = mu Cat 1 − mu Cat 2
Estimate for difference: −0.48
95% CI for difference: (−3.37, 2.42)
T-Test of difference = 0 (vs not =): T-Value = −0.35 P-Value = 0.730 DF = 14
Both use Pooled StDev = 2.70

Notice that the numerical results are essentially the same as the manual computations in Example 10-5. The P-value is reported as $P = 0.73$. The two-sided CI on $\mu_1 - \mu_2$ is also reported. We will give the computing formula for the CI in Section 10-3.3. Figure 10-2 shows the normal probability plot of the two samples of yield data and comparative box plots. The normal probability plots indicate that there is no problem with the normality assumption. Furthermore, both straight lines have similar slopes, providing some verification of the assumption of equal variances. The comparative box plots indicate that there is no obvious difference in the two catalysts, although catalyst 2 has slightly greater sample variability.

Case 2: $\sigma_1^2 \neq \sigma_2^2$

In some situations, we cannot reasonably assume that the unknown variances σ_1^2 and σ_2^2 are equal. There is not an exact t-statistic available for testing H_0: $\mu_1 - \mu_2 = \Delta_0$ in this case. However, if H_0: $\mu_1 - \mu_2 = \Delta_0$ is true, the statistic

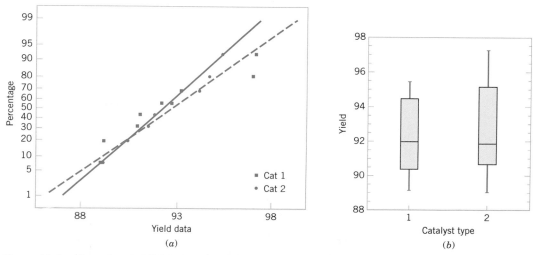

Figure 10-2   Normal probability plot and comparative box plot for the catalyst yield data in Example 10-5. (a) Normal probability plot, (b) Box plots.

$$T_0^* = \frac{\bar{X}_1 - \bar{X}_2 - \Delta_0}{\sqrt{\dfrac{S_1^2}{n_1} + \dfrac{S_2^2}{n_2}}} \qquad (10\text{-}15)$$

is distributed approximately as t with degrees of freedom given by

$$v = \frac{\left(\dfrac{S_1^2}{n_1} + \dfrac{S_2^2}{n_2}\right)^2}{\dfrac{(S_1^2/n_1)^2}{n_1 - 1} + \dfrac{(S_2^2/n_2)^2}{n_2 - 1}} \qquad (10\text{-}16)$$

Therefore, if $\sigma_1^2 \neq \sigma_2^2$, the hypotheses on differences in the means of two normal distributions are tested as in the equal variances case, except that T_0^* is used as the test statistic and $n_1 + n_2 - 2$ is replaced by v in determining the degrees of freedom for the test.

EXAMPLE 10-6

Arsenic concentration in public drinking water supplies is a potential health risk. An article in the *Arizona Republic* (Sunday, May 27, 2001) reported drinking water arsenic concentrations in parts per billion (ppb) for 10 metropolitan Phoenix communities and 10 communities in rural Arizona. The data follow:

Metro Phoenix ($\bar{x}_1 = 12.5$, $s_1 = 7.63$)	Rural Arizona ($\bar{x}_2 = 27.5$, $s_2 = 15.3$)
Phoenix, 3	Rimrock, 48
Chandler, 7	Goodyear, 44
Gilbert, 25	New River, 40
Glendale, 10	Apachie Junction, 38
Mesa, 15	Buckeye, 33
Paradise Valley, 6	Nogales, 21
Peoria, 12	Black Canyon City, 20
Scottsdale, 25	Sedona, 12
Tempe, 15	Payson, 1
Sun City, 7	Casa Grande, 18

We wish to determine it there is any difference in mean arsenic concentrations between metropolitan Phoenix communities and communities in rural Arizona. Figure 10-3 shows a normal probability plot for the two samples of arsenic concentration. The assumption of normality appears quite reasonable, but since the slopes of the two straight lines are very different, it is unlikely that the population variances are the same.

Applying the eight-step procedure gives the following:

1. The parameters of interest are the mean arsenic concentrations for the two geographic regions, say, μ_1 and μ_2, and we are interested in determining whether $\mu_1 - \mu_2 = 0$.

2. H_0: $\mu_1 - \mu_2 = 0$, or H_0: $\mu_1 = \mu_2$

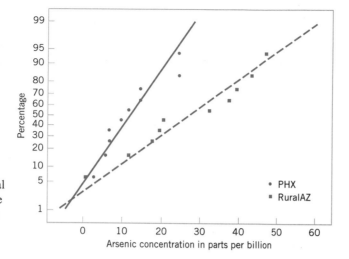

Figure 10-3 Normal probability plot of the arsenic concentration data from Example 10-6.

3. $H_1: \mu_1 \neq \mu_2$

4. $\alpha = 0.05$ (say)

5. The test statistic is

$$t_0^* = \frac{\bar{x}_1 - \bar{x}_2 - 0}{\sqrt{\dfrac{s_1^2}{n_1} + \dfrac{s_2^2}{n_2}}}$$

6. The degrees of freedom on t_0^* are found from Equation 10-16 as

$$\nu = \frac{\left(\dfrac{s_1^2}{n_1} + \dfrac{s_2^2}{n_2}\right)^2}{\dfrac{(s_1^2/n_1)^2}{n_1 - 1} + \dfrac{(s_2^2/n_2)^2}{n_2 - 1}} = \frac{\left[\dfrac{(7.63)^2}{10} + \dfrac{(15.3)^2}{10}\right]^2}{\dfrac{[(7.63)^2/10]^2}{9} + \dfrac{[(15.3)^2/10]^2}{9}} = 13.2 \simeq 13$$

Therefore, using $\alpha = 0.05$, we would reject $H_0: \mu_1 = \mu_2$ if $t_0^* > t_{0.025,13} = 2.160$ or if $t_0^* < -t_{0.025,13} = -2.160$

7. Computations: Using the sample data we find

$$t_0^* = \frac{\bar{x}_1 - \bar{x}_2}{\sqrt{\dfrac{s_1^2}{n_1} + \dfrac{s_2^2}{n_2}}} = \frac{12.5 - 27.5}{\sqrt{\dfrac{(7.63)^2}{10} + \dfrac{(15.3)^2}{10}}} = -2.77$$

8. Conclusions: Because $t_0^* = -2.77 < t_{0.025,13} = -2.160$, we reject the null hypothesis. Therefore, there is evidence to conclude that mean arsenic concentration in the drinking water in rural Arizona is different from the mean arsenic concentration in metropolitan Phoenix drinking water. Furthermore, the mean arsenic concentration is higher in rural Arizona communities. The P-value for this test is approximately $P = 0.016$.

The Minitab output for this example follows:

Two-Sample T-Test and CI: PHX, RuralAZ

Two-sample T for PHX vs RuralAZ

	N	Mean	StDev	SE Mean
PHX	10	12.50	7.63	2.4
RuralAZ	10	27.5	15.3	4.9

Difference = mu PHX − mu RuralAZ
Estimate for difference: −15.00
95% CI for difference: (−26.71, −3.29)
T-Test of difference = 0 (vs not =): T-Value = −2.77   P-Value = 0.016 DF = 13

The numerical results from Minitab exactly match the calculations from Example 10-6. Note that a two-sided 95% CI on $\mu_1 - \mu_2$ is also reported. We will discuss its computation in Section 10-3.4; however, note that the interval does not include zero. Indeed, the upper 95% of confidence limit is −3.29 ppb, well below zero, and the mean observed difference is $\bar{x}_1 - \bar{x}_2 = 12 - 5 - 17.5 = -15$ ppb.

10-3.2   More about the Equal Variance Assumption (CD Only)

10-3.3   Choice of Sample Size

The operating characteristic curves in Appendix Charts VIe, VIf, VIg, and VIh are used to evaluate the type II error for the case where $\sigma_1^2 = \sigma_2^2 = \sigma^2$. Unfortunately, when $\sigma_1^2 \neq \sigma_2^2$, the distribution of T_0^* is unknown if the null hypothesis is false, and no operating characteristic curves are available for this case.

For the two-sided alternative $H_1: \mu_1 - \mu_2 = \Delta \neq \Delta_0$, when $\sigma_1^2 = \sigma_2^2 = \sigma^2$ and $n_1 = n_2 = n$, Charts VIe and VIf are used with

$$d = \frac{|\Delta - \Delta_0|}{2\sigma} \tag{10-17}$$

where Δ is the true difference in means that is of interest. To use these curves, they must be entered with the sample size $n^* = 2n - 1$. For the one-sided alternative hypothesis, we use Charts VIg and VIh and define d and Δ as in Equation 10-17. It is noted that the parameter d is a function of σ, which is unknown. As in the single-sample t-test, we may have to rely on a prior estimate of σ or use a subjective estimate. Alternatively, we could define the differences in the mean that we wish to detect relative to σ.

EXAMPLE 10-7

Consider the catalyst experiment in Example 10-5. Suppose that, if catalyst 2 produces a mean yield that differs from the mean yield of catalyst 1 by 4.0%, we would like to reject the null hypothesis with probability at least 0.85. What sample size is required?

Using $s_p = 2.70$ as a rough estimate of the common standard deviation σ, we have $d = |\Delta|/2\sigma = |4.0|/[(2)(2.70)] = 0.74$. From Appendix Chart VI$e$ with $d = 0.74$ and $\beta = 0.15$, we find $n^* = 20$, approximately. Therefore, since $n^* = 2n - 1$,

$$n = \frac{n^* + 1}{2} = \frac{20 + 1}{2} = 10.5 \approx 11 (\text{say})$$

and we would use sample sizes of $n_1 = n_2 = n = 11$.

Minitab will also perform power and sample size calculations for the two-sample t-test (equal variances). The output from Example 10-7 is as follows:

Power and Sample Size

2-Sample t Test
Testing mean 1 = mean 2 (versus not =)
Calculating power for mean 1 = mean 2 + difference
Alpha = 0.05 Sigma = 2.7

Difference	Sample Size	Target Power	Actual Power
4	10	0.8500	0.8793

The results agree fairly closely with the results obtained from the O.C. curve.

10-3.4 Confidence Interval on the Difference in Means

Case 1: $\sigma_1^2 = \sigma_2^2 = \sigma^2$

To develop the confidence interval for the difference in means $\mu_1 - \mu_2$ when both variances are equal, note that the distribution of the statistic

$$T = \frac{\overline{X}_1 - \overline{X}_2 - (\mu_1 - \mu_2)}{S_p\sqrt{\dfrac{1}{n_1} + \dfrac{1}{n_2}}} \tag{10-18}$$

is the t distribution with $n_1 + n_2 - 2$ degrees of freedom. Therefore $P(-t_{\alpha/2,n_1+n_2-2} \leq T \leq t_{\alpha/2,n_1+n_2-2}) = 1 - \alpha$. Now substituting Equation 10-18 for T and manipulating the quantities inside the probability statement will lead to the $100(1 - \alpha)\%$ confidence interval on $\mu_1 - \mu_2$.

Definition

If $\overline{x}_1, \overline{x}_2, s_1^2$ and s_2^2 are the sample means and variances of two random samples of sizes n_1 and n_2, respectively, from two independent normal populations with unknown but equal variances, then a **$100(1 - \alpha)\%$ confidence interval on the difference in means $\mu_1 - \mu_2$ is**

$$\overline{x}_1 - \overline{x}_2 - t_{\alpha/2,n_1+n_2-2}\, s_p \sqrt{\frac{1}{n_1} + \frac{1}{n_2}}$$

$$\leq \mu_1 - \mu_2 \leq \overline{x}_1 - \overline{x}_2 + t_{\alpha/2,n_1+n_2-2}\, s_p \sqrt{\frac{1}{n_1} + \frac{1}{n_2}} \tag{10-19}$$

where $s_p = \sqrt{[(n_1 - 1)s_1^2 + (n_2 - 1)s_2^2]/(n_1 + n_2 - 2)}$ is the pooled estimate of the common population standard deviation, and $t_{\alpha/2,n_1+n_2-2}$ is the upper $\alpha/2$ percentage point of the t distribution with $n_1 + n_2 - 2$ degrees of freedom.

EXAMPLE 10-8 | An article in the journal *Hazardous Waste and Hazardous Materials* (Vol. 6, 1989) reported the results of an analysis of the weight of calcium in standard cement and cement doped with lead. Reduced levels of calcium would indicate that the hydration mechanism in the cement is blocked and would allow water to attack various locations in the cement structure. Ten samples of standard cement had an average weight percent calcium of $\bar{x}_1 = 90.0$, with a sample standard deviation of $s_1 = 5.0$, while 15 samples of the lead-doped cement had an average weight percent calcium of $\bar{x}_2 = 87.0$, with a sample standard deviation of $s_2 = 4.0$.

We will assume that weight percent calcium is normally distributed and find a 95% confidence interval on the difference in means, $\mu_1 - \mu_2$, for the two types of cement. Furthermore, we will assume that both normal populations have the same standard deviation.

The pooled estimate of the common standard deviation is found using Equation 10-12 as follows:

$$s_p^2 = \frac{(n_1 - 1)s_1^2 + (n_2 - 1)s_2^2}{n_1 + n_2 - 2}$$

$$= \frac{9(5.0)^2 + 14(4.0)^2}{10 + 15 - 2}$$

$$= 19.52$$

Therefore, the pooled standard deviation estimate is $s_p = \sqrt{19.52} = 4.4$. The 95% confidence interval is found using Equation 10-19:

$$\bar{x}_1 - \bar{x}_2 - t_{0.025,23}\, s_p \sqrt{\frac{1}{n_1} + \frac{1}{n_2}} \leq \mu_1 - \mu_2 \leq \bar{x}_1 - \bar{x}_2 + t_{0.025,23}\, s_p \sqrt{\frac{1}{n_1} + \frac{1}{n_2}}$$

or upon substituting the sample values and using $t_{0.025,23} = 2.069$,

$$90.0 - 87.0 - 2.069(4.4)\sqrt{\frac{1}{10} + \frac{1}{15}} \leq \mu_1 - \mu_2$$

$$\leq 90.0 - 87.0 + 2.069(44)\sqrt{\frac{1}{10} + \frac{1}{15}}$$

which reduces to

$$-0.72 \leq \mu_1 - \mu_2 \leq 6.72$$

Notice that the 95% confidence interval includes zero; therefore, at this level of confidence we cannot conclude that there is a difference in the means. Put another way, there is no evidence that doping the cement with lead affected the mean weight percent of calcium; therefore, we cannot claim that the presence of lead affects this aspect of the hydration mechanism at the 95% level of confidence.

Case 2: $\sigma_1^2 \neq \sigma_2^2$

In many situations it is not reasonable to assume that $\sigma_1^2 = \sigma_2^2$. When this assumption is unwarranted, we may still find a $100(1 - \alpha)\%$ confidence interval on $\mu_1 - \mu_2$ using the fact that $T^* = [\bar{X}_1 - \bar{X}_2 - (\mu_1 - \mu_2)]/\sqrt{S_1^2/n_1 + S_2^2/n_2}$ is distributed approximately as t with degrees of freedom v given by Equation 10-16. The CI expression follows.

Definition

> If $\bar{x}_1$, $\bar{x}_2$, s_1^2, and s_2^2 are the means and variances of two random samples of sizes n_1 and n_2, respectively, from two independent normal populations with unknown and unequal variances, an approximate $100(1 - \alpha)\%$ confidence interval on the difference in means $\mu_1 - \mu_2$ is
>
> $$\bar{x}_1 - \bar{x}_2 - t_{\alpha/2,v}\sqrt{\frac{s_1^2}{n_1} + \frac{s_2^2}{n_2}} \leq \mu_1 - \mu_2 \leq \bar{x}_1 - \bar{x}_2 + t_{\alpha/2,v}\sqrt{\frac{s_1^2}{n_1} + \frac{s_2^2}{n_2}} \quad (10\text{-}20)$$
>
> where v is given by Equation 10-16 and $t_{\alpha/2,v}$ is the upper $\alpha/2$ percentage point of the t distribution with v degrees of freedom.

EXERCISES FOR SECTION 10-3

10-17. The diameter of steel rods manufactured on two different extrusion machines is being investigated. Two random samples of sizes $n_1 = 15$ and $n_2 = 17$ are selected, and the sample means and sample variances are $\bar{x}_1 = 8.73$, $s_1^2 = 0.35$, $\bar{x}_2 = 8.68$, and $s_2^2 = 0.40$, respectively. Assume that $\sigma_1^2 = \sigma_2^2$ and that the data are drawn from a normal distribution.

(a) Is there evidence to support the claim that the two machines produce rods with different mean diameters? Use $\alpha = 0.05$ in arriving at this conclusion.

(b) Find the P-value for the t-statistic you calculated in part (a).

(c) Construct a 95% confidence interval for the difference in mean rod diameter. Interpret this interval.

10-18. An article in *Fire Technology* investigated two different foam expanding agents that can be used in the nozzles of fire-fighting spray equipment. A random sample of five observations with an aqueous film-forming foam (AFFF) had a sample mean of 4.7 and a standard deviation of 0.6. A random sample of five observations with alcohol-type concentrates (ATC) had a sample mean of 6.9 and a standard deviation 0.8. Find a 95% confidence interval on the difference in mean foam expansion of these two agents. Can you draw any conclusions about which agent produces the greatest mean foam expansion? Assume that both populations are well represented by normal distributions with the same standard deviations.

10-19. Two catalysts may be used in a batch chemical process. Twelve batches were prepared using catalyst 1, resulting in an average yield of 86 and a sample standard deviation of 3. Fifteen batches were prepared using catalyst 2, and they resulted in an average yield of 89 with a standard deviation of 2. Assume that yield measurements are approximately normally distributed with the same standard deviation.

(a) Is there evidence to support a claim that catalyst 2 produces a higher mean yield than catalyst 1? Use $\alpha = 0.01$.

(b) Find a 95% confidence interval on the difference in mean yields.

10-20. The deflection temperature under load for two different types of plastic pipe is being investigated. Two random samples of 15 pipe specimens are tested, and the deflection temperatures observed are as follows (in °F):

Type 1: 206, 188, 205, 187, 194, 193, 207, 185, 189, 213, 192, 210, 194, 178, 205.

Type 2: 177, 197, 206, 201, 180, 176, 185, 200, 197, 192, 198, 188, 189, 203, 192.

(a) Construct box plots and normal probability plots for the two samples. Do these plots provide support of the assumptions of normality and equal variances? Write a practical interpretation for these plots.

(b) Do the data support the claim that the deflection temperature under load for type 2 pipe exceeds that of type 1? In reaching your conclusions, use $\alpha = 0.05$.

(c) Calculate a P-value for the test in part (b).

(d) Suppose that if the mean deflection temperature for type 2 pipe exceeds that of type 1 by as much as 5°F, it is important to detect this difference with probability at least 0.90. Is the choice of $n_1 = n_2 = 15$ in part (a) of this problem adequate?

10-21. In semiconductor manufacturing, wet chemical etching is often used to remove silicon from the backs of wafers prior to metalization. The etch rate is an important characteristic in this process and known to follow a normal distribution. Two different etching solutions have been compared, using two random samples of 10 wafers for each solution. The observed etch rates are as follows (in mils per minute):

Solution 1		Solution 2	
9.9	10.6	10.2	10.0
9.4	10.3	10.6	10.2
9.3	10.0	10.7	10.7
9.6	10.3	10.4	10.4
10.2	10.1	10.5	10.3

(a) Do the data support the claim that the mean etch rate is the same for both solutions? In reaching your conclusions, use $\alpha = 0.05$ and assume that both population variances are equal.

(b) Calculate a P-value for the test in part (a).

(c) Find a 95% confidence interval on the difference in mean etch rates.

(d) Construct normal probability plots for the two samples. Do these plots provide support for the assumptions of normality and equal variances? Write a practical interpretation for these plots.

10-22. Two suppliers manufacture a plastic gear used in a laser printer. The impact strength of these gears measured in foot-pounds is an important characteristic. A random sample of 10 gears from supplier 1 results in $\bar{x}_1 = 290$ and $s_1 = 12$, while another random sample of 16 gears from the second supplier results in $\bar{x}_2 = 321$ and $s_2 = 22$.

(a) Is there evidence to support the claim that supplier 2 provides gears with higher mean impact strength? Use $\alpha = 0.05$, and assume that both populations are normally distributed but the variances are not equal.

(b) What is the P-value for this test?

(c) Do the data support the claim that the mean impact strength of gears from supplier 2 is at least 25 foot-pounds higher than that of supplier 1? Make the same assumptions as in part (a).

10-23. Reconsider the situation in Exercise 10-22, part (a). Construct a confidence interval estimate for the difference in mean impact strength, and explain how this interval could be used to answer the question posed regarding supplier-to-supplier differences.

10-24. A photoconductor film is manufactured at a nominal thickness of 25 mils. The product engineer wishes to increase the mean speed of the film, and believes that this can be achieved by reducing the thickness of the film to 20 mils. Eight samples of each film thickness are manufactured in a pilot production process, and the film speed (in microjoules per square inch) is measured. For the 25-mil film the sample data result is $\bar{x}_1 = 1.15$ and $s_1 = 0.11$, while for the 20-mil film, the data yield $\bar{x}_2 = 1.06$ and $s_2 = 0.09$. Note that an increase in film speed would lower the value of the observation in microjoules per square inch.

(a) Do the data support the claim that reducing the film thickness increases the mean speed of the film? Use $\alpha = 0.10$ and assume that the two population variances are equal and the underlying population of film speed is normally distributed.

(b) What is the P-value for this test?

(c) Find a 95% confidence interval on the difference in the two means.

10-25. The melting points of two alloys used in formulating solder were investigated by melting 21 samples of each material. The sample mean and standard deviation for alloy 1 was $\bar{x}_1 = 420°F$ and $s_1 = 4°F$, while for alloy 2 they were

$\bar{x}_2 = 426°F$, and $s_2 = 3°F$. Do the sample data support the claim that both alloys have the same melting point? Use $\alpha = 0.05$ and assume that both populations are normally distributed and have the same standard deviation. Find the P-value for the test.

10-26. Referring to the melting point experiment in Exercise 10-25, suppose that the true mean difference in melting points is 3°F. How large a sample would be required to detect this difference using an $\alpha = 0.05$ level test with probability at least 0.9? Use $\sigma_1 = \sigma_2 = 4$ as an initial estimate of the common standard deviation.

10-27. Two companies manufacture a rubber material intended for use in an automotive application. The part will be subjected to abrasive wear in the field application, so we decide to compare the material produced by each company in a test. Twenty-five samples of material from each company are tested in an abrasion test, and the amount of wear after 1000 cycles is observed. For company 1, the sample mean and standard deviation of wear are $\bar{x}_1 = 20$ milligrams/1000 cycles and $s_1 = 2$ milligrams/1000 cycles, while for company 2 we obtain $\bar{x}_2 = 15$ milligrams/1000 cycles and $s_2 = 8$ milligrams/1000 cycles.

(a) Do the data support the claim that the two companies produce material with different mean wear? Use $\alpha = 0.05$, and assume each population is normally distributed but that their variances are not equal.

(b) What is the P-value for this test?

(c) Do the data support a claim that the material from company 1 has higher mean wear than the material from company 2? Use the same assumptions as in part (a).

10-28. The thickness of a plastic film (in mils) on a substrate material is thought to be influenced by the temperature at which the coating is applied. A completely randomized experiment is carried out. Eleven substrates are coated at 125°F, resulting in a sample mean coating thickness of $\bar{x}_1 = 103.5$ and a sample standard deviation of $s_1 = 10.2$. Another 13 substrates are coated at 150°F, for which $\bar{x}_2 = 99.7$ and $s_2 = 20.1$ are observed. It was originally suspected that raising the process temperature would reduce mean coating thickness. Do the data support this claim? Use $\alpha = 0.01$ and assume that the two population standard deviations are not equal. Calculate an approximate P-value for this test.

10-29. Reconsider the coating thickness experiment in Exercise 10-28. How could you have answered the question posed regarding the effect of temperature on coating thickness by using a confidence interval? Explain your answer.

10-30. Reconsider the abrasive wear test in Exercise 10-27. Construct a confidence interval that will address the questions in parts (a) and (c) in that exercise.

10-31. The overall distance traveled by a golf ball is tested by hitting the ball with Iron Byron, a mechanical golfer with a swing that is said to emulate the legendary champion, Byron Nelson. Ten randomly selected balls of two different brands are tested and the overall distance measured. The data follow:

Brand 1: 275, 286, 287, 271, 283, 271, 279, 275, 263, 267

Brand 2: 258, 244, 260, 265, 273, 281, 271, 270, 263, 268

(a) Is there evidence that overall distance is approximately normally distributed? Is an assumption of equal variances justified?

(b) Test the hypothesis that both brands of ball have equal mean overall distance. Use $\alpha = 0.05$.

(c) What is the *P*-value of the test statistic in part (b)?

(d) What is the power of the statistical test in part (b) to detect a true difference in mean overall distance of 5 yards?

(e) What sample size would be required to detect a true difference in mean overall distance of 3 yards with power of approximately 0.75?

(f) Construct a 95% two-sided CI on the mean difference in overall distance between the two brands of golf balls.

10-32. In Example 9-6 we described how the "spring-like effect" in a golf club could be determined by measuring the coefficient of restitution (the ratio of the outbound velocity to the inbound velocity of a golf ball fired at the clubhead). Twelve randomly selected drivers produced by two clubmakers are tested and the coefficient of restitution measured. The data follow:

Club 1: 0.8406, 0.8104, 0.8234, 0.8198, 0.8235, 0.8562, 0.8123, 0.7976, 0.8184, 0.8265, 0.7773, 0.7871

Club 2: 0.8305, 0.7905, 0.8352, 0.8380, 0.8145, 0.8465, 0.8244, 0.8014, 0.8309, 0.8405, 0.8256, 0.8476

(a) Is there evidence that coefficient of restitution is approximately normally distributed? Is an assumption of equal variances justified?

(b) Test the hypothesis that both brands of ball have equal mean coefficient of restitution. Use $\alpha = 0.05$.

(c) What is the *P*-value of the test statistic in part (b)?

(d) What is the power of the statistical test in part (b) to detect a true difference in mean coefficient of restitution of 0.2?

(e) What sample size would be required to detect a true difference in mean coefficient of restitution of 0.1 with power of approximately 0.8?

(f) Construct a 95% two-sided CI on the mean difference in coefficient of restitution between the two brands of golf clubs.

10-4 PAIRED *t*-TEST

A special case of the two-sample *t*-tests of Section 10-3 occurs when the observations on the two populations of interest are collected in **pairs.** Each pair of observations, say (X_{1j}, X_{2j}), is taken under homogeneous conditions, but these conditions may change from one pair to another. For example, suppose that we are interested in comparing two different types of tips for a hardness-testing machine. This machine presses the tip into a metal specimen with a known force. By measuring the depth of the depression caused by the tip, the hardness of the specimen can be determined. If several specimens were selected at random, half tested with tip 1, half tested with tip 2, and the pooled or independent *t*-test in Section 10-3 was applied, the results of the test could be erroneous. The metal specimens could have been cut from bar stock that was produced in different heats, or they might not be homogeneous in some other way that might affect hardness. Then the observed difference between mean hardness readings for the two tip types also includes hardness differences between specimens.

A more powerful experimental procedure is to collect the data in pairs—that is, to make two hardness readings on each specimen, one with each tip. The test procedure would then consist of analyzing the *differences* between hardness readings on each specimen. If there is no difference between tips, the mean of the differences should be zero. This test procedure is called the **paired *t*-test.**

Let $(X_{11}, X_{21}), (X_{12}, X_{22}), \ldots, (X_{1n}, X_{2n})$ be a set of *n* paired observations where we assume that the mean and variance of the population represented by X_1 are μ_1 and σ_1^2, and the mean and variance of the population represented by X_2 are μ_2 and σ_2^2. Define the differences between each pair of observations as $D_j = X_{1j} - X_{2j}, \; j = 1, 2, \ldots, n$. The D_j's are assumed to be normally distributed with mean

$$\mu_D = E(X_1 - X_2) = E(X_1) - E(X_2) = \mu_1 - \mu_2$$

and variance σ_D^2, so testing hypotheses about the difference between μ_1 and μ_2 can be accomplished by performing a one-sample t-test on μ_D. Specifically, testing H_0: $\mu_1 - \mu_2 = \Delta_0$ against H_1: $\mu_1 - \mu_2 \neq \Delta_0$ is equivalent to testing

$$H_0: \mu_D = \Delta_0$$
$$H_1: \mu_D \neq \Delta_0 \tag{10-21}$$

The test statistic is given below.

The Paired t-Test

Null hypothesis: H_0: $\mu_D = \Delta_0$

Test statistic: $T_0 = \dfrac{\overline{D} - \Delta_0}{S_D/\sqrt{n}}$ $\tag{10-22}$

Alternative Hypothesis	Rejection Region
H_1: $\mu_D \neq \Delta_0$	$t_0 > t_{\alpha/2, n-1}$ or $t_0 < -t_{\alpha/2, n-1}$
H_1: $\mu_D > \Delta_0$	$t_0 > t_{\alpha, n-1}$
H_1: $\mu_D < \Delta_0$	$t_0 < -t_{\alpha, n-1}$

In Equation 10-22, $\overline{D}$ is the sample average of the n differences $D_1, D_2, \ldots, D_n$, and S_D is the sample standard deviation of these differences.

EXAMPLE 10-9   An article in the *Journal of Strain Analysis* (1983, Vol. 18, No. 2) compares several methods for predicting the shear strength for steel plate girders. Data for two of these methods, the Karlsruhe and Lehigh procedures, when applied to nine specific girders, are shown in Table 10-2. We wish to determine whether there is any difference (on the average) between the two methods.

The eight-step procedure is applied as follows:

1. The parameter of interest is the difference in mean shear strength between the two methods, say, $\mu_D = \mu_1 - \mu_2 = 0$.
2. H_0: $\mu_D = 0$

Table 10-2   Strength Predictions for Nine Steel Plate Girders (Predicted Load/Observed Load)

Girder	Karlsruhe Method	Lehigh Method	Difference d_j
S1/1	1.186	1.061	0.119
S2/1	1.151	0.992	0.159
S3/1	1.322	1.063	0.259
S4/1	1.339	1.062	0.277
S5/1	1.200	1.065	0.138
S2/1	1.402	1.178	0.224
S2/2	1.365	1.037	0.328
S2/3	1.537	1.086	0.451
S2/4	1.559	1.052	0.507

3. $H_1: \mu_D \neq 0$
4. $\alpha = 0.05$
5. The test statistic is

$$t_0 = \frac{\bar{d}}{s_D/\sqrt{n}}$$

6. Reject H_0 if $t_0 > t_{0.025,8} = 2.306$ or if $t_0 < -t_{0.025,8} = -2.306$.
7. Computations: The sample average and standard deviation of the differences d_j are $\bar{d} = 0.2736$ and $s_D = 0.1356$, so the test statistic is

$$t_0 = \frac{\bar{d}}{s_D/\sqrt{n}} = \frac{0.2736}{0.1356/\sqrt{9}} = 6.05$$

8. Conclusions: Since $t_0 = 6.05 > 2.306$, we conclude that the strength prediction methods yield different results. Specifically, the data indicate that the Karlsruhe method produces, on the average, higher strength predictions than does the Lehigh method. The P-value for $t_0 = 6.05$ is $P = 0.0002$, so the test statistic is well into the critical region.

Paired Versus Unpaired Comparisons

In performing a comparative experiment, the investigator can sometimes choose between the paired experiment and the two-sample (or unpaired) experiment. If n measurements are to be made on each population, the two-sample t-statistic is

$$T_0 = \frac{\bar{X}_1 - \bar{X}_2 - \Delta_0}{S_p\sqrt{\frac{1}{n} + \frac{1}{n}}}$$

which would be compared to t_{2n-2}, and of course, the paired t-statistic is

$$T_0 = \frac{\bar{D} - \Delta_0}{S_D/\sqrt{n}}$$

which is compared to t_{n-1}. Notice that since

$$\bar{D} = \sum_{j=1}^{n} \frac{D_j}{n} = \sum_{j=1}^{n} \frac{(X_{1j} - X_{2j})}{n} = \sum_{j=1}^{n} \frac{X_{1j}}{n} - \sum_{j=1}^{n} \frac{X_{2j}}{n} = \bar{X}_1 - \bar{X}_2$$

the numerators of both statistics are identical. However, the denominator of the two-sample t-test is based on the assumption that X_1 and X_2 are *independent*. In many paired experiments, a strong positive correlation ρ exists between X_1 and X_2. Then it can be shown that

$$\begin{aligned} V(\bar{D}) &= V(\bar{X}_1 - \bar{X}_2 - \Delta_0) \\ &= V(\bar{X}_1) + V(\bar{X}_2) - 2\,\mathrm{cov}(\bar{X}_1, \bar{X}_2) \\ &= \frac{2\sigma^2(1 - \rho)}{n} \end{aligned}$$

assuming that both populations X_1 and X_2 have identical variances σ^2. Furthermore, S_D^2/n estimates the variance of $\overline{D}$. Whenever there is positive correlation within the pairs, the denominator for the paired t-test will be smaller than the denominator of the two-sample t-test. This can cause the two-sample t-test to considerably understate the significance of the data if it is incorrectly applied to paired samples.

Although pairing will often lead to a smaller value of the variance of $\overline{X}_1 - \overline{X}_2$, it does have a disadvantage—namely, the paired t-test leads to a loss of $n - 1$ degrees of freedom in comparison to the two-sample t-test. Generally, we know that increasing the degrees of freedom of a test increases the power against any fixed alternative values of the parameter.

So how do we decide to conduct the experiment? Should we pair the observations or not? Although there is no general answer to this question, we can give some guidelines based on the above discussion.

1. If the experimental units are relatively homogeneous (small σ) and the correlation within pairs is small, the gain in precision attributable to pairing will be offset by the loss of degrees of freedom, so an independent-sample experiment should be used.

2. If the experimental units are relatively heterogeneous (large σ) and there is large positive correlation within pairs, the paired experiment should be used. Typically, this case occurs when the experimental units are the *same* for both treatments; as in Example 10-9, the same girders were used to test the two methods.

Implementing the rules still requires judgment, because σ and ρ are never known precisely. Furthermore, if the number of degrees of freedom is large (say, 40 or 50), the loss of $n - 1$ of them for pairing may not be serious. However, if the number of degrees of freedom is small (say, 10 or 20), losing half of them is potentially serious if not compensated for by increased precision from pairing.

A Confidence Interval for μ_D

To construct the confidence interval for $\mu_D = \mu_1 - \mu_2$, note that

$$T = \frac{\overline{D} - \mu_D}{S_D/\sqrt{n}}$$

follows a t distribution with $n - 1$ degrees of freedom. Then, since $P(-t_{\alpha/2,n-1} \leq T \leq t_{\alpha/2,n-1}) = 1 - \alpha$, we can substitute for T in the above expression and perform the necessary steps to isolate $\mu_D = \mu_1 - \mu_2$ between the inequalities. This leads to the following $100(1 - \alpha)\%$ confidence interval on $\mu_1 - \mu_2$.

Definition

> If $\overline{d}$ and s_D are the sample mean and standard deviation of the difference of n random pairs of normally distributed measurements, a **$100(1 - \alpha)\%$ confidence interval on the difference in means $\mu_D = \mu_1 - \mu_2$** is
>
> $$\overline{d} - t_{\alpha/2,n-1}s_D/\sqrt{n} \leq \mu_D \leq \overline{d} + t_{\alpha/2,n-1}s_D/\sqrt{n} \qquad (10\text{-}23)$$
>
> where $t_{\alpha/2,n-1}$ is the upper $\alpha/2\%$ point of the t-distribution with $n - 1$ degrees of freedom.

Table 10-3   Time in Seconds to Parallel Park Two Automobiles

Subject	Automobile 1(x_{1j})	Automobile 2(x_{2j})	Difference (d_j)
1	37.0	17.8	19.2
2	25.8	20.2	5.6
3	16.2	16.8	−0.6
4	24.2	41.4	−17.2
5	22.0	21.4	0.6
6	33.4	38.4	−5.0
7	23.8	16.8	7.0
8	58.2	32.2	26.0
9	33.6	27.8	5.8
10	24.4	23.2	1.2
11	23.4	29.6	−6.2
12	21.2	20.6	0.6
13	36.2	32.2	4.0
14	29.8	53.8	−24.0

This confidence interval is also valid for the case where $\sigma_1^2 \neq \sigma_2^2$, because s_D^2 estimates $\sigma_D^2 = V(X_1 - X_2)$. Also, for large samples (say, $n \geq 30$ pairs), the explicit assumption of normality is unnecessary because of the central limit theorem.

EXAMPLE 10-10

The journal *Human Factors* (1962, pp. 375-380) reports a study in which $n = 14$ subjects were asked to parallel park two cars having very different wheel bases and turning radii. The time in seconds for each subject was recorded and is given in Table 10-3. From the column of observed differences we calculate $\bar{d} = 1.21$ and $s_D = 12.68$. The 90% confidence interval for $\mu_D = \mu_1 - \mu_2$ is found from Equation 9-24 as follows:

$$\bar{d} - t_{0.05,13}s_D/\sqrt{n} \leq \mu_D \leq \bar{d} + t_{0.05,13}s_D/\sqrt{n}$$

$$1.21 - 1.771(12.68)/\sqrt{14} \leq \mu_D \leq 1.21 + 1.771(12.68)/\sqrt{14}$$

$$-4.79 \leq \mu_D \leq 7.21$$

Notice that the confidence interval on μ_D includes zero. This implies that, at the 90% level of confidence, the data do not support the claim that the two cars have different mean parking times μ_1 and μ_2. That is, the value $\mu_D = \mu_1 - \mu_2 = 0$ is not inconsistent with the observed data.

EXERCISES FOR SECTION 10-4

10-33.   Consider the shear strength experiment described in Example 10-9. Construct a 95% confidence interval on the difference in mean shear strength for the two methods. Is the result you obtained consistent with the findings in Example 10-9? Explain why.

10-34.   Reconsider the shear strength experiment described in Example 10-9. Do each of the individual shear strengths have to be normally distributed for the paired t-test to be appropriate, or is it only the difference in shear strengths that

must be normal? Use a normal probability plot to investigate the normality assumption.

10-35.   Consider the parking data in Example 10-10. Use the paired t-test to investigate the claim that the two types of cars have different levels of difficulty to parallel park. Use $\alpha = 0.10$. Compare your results with the confidence interval constructed in Example 10-10 and comment on why they are the same or different.

10-36.   Reconsider the parking data in Example 10-10. Investigate the assumption that the differences in parking times are normally distributed.

10-37.   The manager of a fleet of automobiles is testing two brands of radial tires and assigns one tire of each brand at random to the two rear wheels of eight cars and runs the cars until the tires wear out. The data (in kilometers) follow. Find a 99% confidence interval on the difference in mean life. Which brand would you prefer, based on this calculation?

Car	Brand 1	Brand 2
1	36,925	34,318
2	45,300	42,280
3	36,240	35,500
4	32,100	31,950
5	37,210	38,015
6	48,360	47,800
7	38,200	37,810
8	33,500	33,215

10-38.   A computer scientist is investigating the usefulness of two different design languages in improving programming tasks. Twelve expert programmers, familiar with both languages, are asked to code a standard function in both languages, and the time (in minutes) is recorded. The data follow:

	Time	
Programmer	Design Language 1	Design Language 2
1	17	18
2	16	14
3	21	19
4	14	11
5	18	23
6	24	21
7	16	10
8	14	13
9	21	19
10	23	24
11	13	15
12	18	20

(a) Find a 95% confidence interval on the difference in mean coding times. Is there any indication that one design language is preferable?

(b) Is the assumption that the difference in coding time is normally distributed reasonable? Show evidence to support your answer.

10-39.   Fifteen adult males between the ages of 35 and 50 participated in a study to evaluate the effect of diet and exercise on blood cholesterol levels. The total cholesterol was measured in each subject initially and then three months after participating in an aerobic exercise program and switching to a low-fat diet. The data are shown in the accompanying table. Do the data support the claim that low-fat diet and aerobic exercise are of value in producing a mean reduction in blood cholesterol levels? Use $\alpha = 0.05$.

Blood Cholesterol Level		
Subject	Before	After
1	265	229
2	240	231
3	258	227
4	295	240
5	251	238
6	245	241
7	287	234
8	314	256
9	260	247
10	279	239
11	283	246
12	240	218
13	238	219
14	225	226
15	247	233

10-40.   An article in the *Journal of Aircraft* (Vol. 23, 1986, pp. 859–864) describes a new equivalent plate analysis method formulation that is capable of modeling aircraft structures such as cranked wing boxes and that produces results similar to the more computationally intensive finite element analysis method. Natural vibration frequencies for the cranked wing box structure are calculated using both methods, and results for the first seven natural frequencies follow:

Freq.	Finite Element Cycle/s	Equivalent Plate, Cycle/s
1	14.58	14.76
2	48.52	49.10
3	97.22	99.99
4	113.99	117.53
5	174.73	181.22
6	212.72	220.14
7	277.38	294.80

(a) Do the data suggest that the two methods prove the same mean value for natural vibration frequency? Use $\alpha = 0.05$.

(b) Find a 95% confidence interval on the mean difference between the two methods.

10-41.   Ten individuals have participated in a diet-modification program to stimulate weight loss. Their weight both before and after participation in the program is shown in the following list. Is there evidence to support the claim that this particular diet-modification program is effective in producing a mean weight reduction? Use $\alpha = 0.05$.

Subject	Before	After
1	195	187
2	213	195
3	247	221
4	201	190
5	187	175
6	210	197
7	215	199
8	246	221
9	294	278
10	310	285

10-42.   Two different analytical tests can be used to determine the impurity level in steel alloys. Eight specimens

are tested using both procedures, and the results are shown in the following tabulation. Is there sufficient evidence to conclude that both tests give the same mean impurity level, using $\alpha = 0.01$?

Specimen	Test 1	Test 2
1	1.2	1.4
2	1.3	1.7
3	1.5	1.5
4	1.4	1.3
5	1.7	2.0
6	1.8	2.1
7	1.4	1.7
8	1.3	1.6

10-43.   Consider the weight-loss data in Exercise 10-41. Is there evidence to support the claim that this particular diet-modification program will result in a mean weight loss of at least 10 pounds? Use $\alpha = 0.05$.

10-44.   Consider the weight-loss experiment in Exercise 10-41. Suppose that, if the diet-modification program results in mean weight loss of at least 10 pounds, it is important to detect this with probability of at least 0.90. Was the use of 10 subjects an adequate sample size? If not, how many subjects should have been used?

10-5   INFERENCES ON THE VARIANCES OF TWO NORMAL POPULATIONS

We now introduce tests and confidence intervals for the two population variances shown in Fig. 10-1. We will assume that both populations are normal. Both the hypothesis-testing and confidence interval procedures are relatively sensitive to the normality assumption.

10-5.1   The F Distribution

Suppose that two independent normal populations are of interest, where the population means and variances, say, μ_1, σ_1^2, μ_2, and σ_2^2, are unknown. We wish to test hypotheses about the equality of the two variances, say, $H_0: \sigma_1^2 = \sigma_2^2$. Assume that two random samples of size n_1 from population 1 and of size n_2 from population 2 are available, and let S_1^2 and S_2^2 be the sample variances. We wish to test the hypotheses

$$H_0: \sigma_1^2 = \sigma_2^2$$
$$H_1: \sigma_1^2 \neq \sigma_2^2 \tag{10-24}$$

The development of a test procedure for these hypotheses requires a new probability distribution, the F distribution. The random variable F is defined to be the ratio of two independent chi-square random variables, each divided by its number of degrees of freedom. That is,

$$F = \frac{W/u}{Y/v} \tag{10-25}$$

where W and Y are independent chi-square random variables with u and v degrees of freedom, respectively. We now formally state the sampling distribution of F.

Definition

Let W and Y be independent chi-square random variables with u and v degrees of freedom, respectively. Then the ratio

$$F = \frac{W/u}{Y/v} \tag{10-26}$$

has the probability density function

$$f(x) = \frac{\Gamma\left(\dfrac{u+v}{2}\right)\left(\dfrac{u}{v}\right)^{u/2} x^{(u/2)-1}}{\Gamma\left(\dfrac{u}{2}\right)\Gamma\left(\dfrac{v}{2}\right)\left[\left(\dfrac{u}{v}\right)x + 1\right]^{(u+v)/2}}, \qquad 0 < x < \infty \tag{10-27}$$

and is said to follow the F distribution with u degrees of freedom in the numerator and v degrees of freedom in the denominator. It is usually abbreviated as $F_{u,v}$.

The mean and variance of the F distribution are $\mu = v/(v-2)$ for $v > 2$, and

$$\sigma^2 = \frac{2v^2(u+v-2)}{u(v-2)^2(v-4)}, \qquad v > 4$$

Two F distributions are shown in Fig. 10-4. The F random variable is nonnegative, and the distribution is skewed to the right. The F distribution looks very similar to the chi-square distribution; however, the two parameters u and v provide extra flexibility regarding shape.

The percentage points of the F distribution are given in Table V of the Appendix. Let $f_{\alpha,u,v}$ be the percentage point of the F distribution, with numerator degrees of freedom u and denominator degrees of freedom v such that the probability that the random variable F exceeds this value is

$$P(F > f_{\alpha,u,v}) = \int_{f_{\alpha,u,v}}^{\infty} f(x)\, dx = \alpha$$

This is illustrated in Fig. 10-5. For example, if $u = 5$ and $v = 10$, we find from Table V of the Appendix that

$$P(F > f_{0.05,5,10}) = P(F_{5,10} > 3.33) = 0.05$$

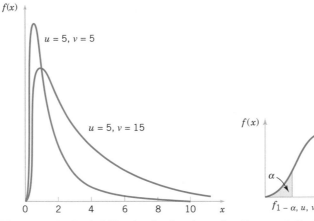

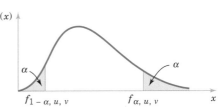

Figure 10-4   Probability density functions of two F distributions.

Figure 10-5   Upper and lower percentage points of the F distribution.

That is, the upper 5 percentage point of $F_{5,10}$ is $f_{0.05,5,10} = 3.33$.

Table V contains only upper-tail percentage points (for selected values of $f_{\alpha,u,v}$ for $\alpha \leq 0.25$) of the F distribution. The lower-tail percentage points $f_{1-\alpha,u,v}$ can be found as follows.

$$f_{1-\alpha,u,v} = \frac{1}{f_{\alpha,v,u}} \qquad (10\text{-}28)$$

For example, to find the lower-tail percentage point $f_{0.95,5,10}$, note that

$$f_{0.95,5,10} = \frac{1}{f_{0.05,10,5}} = \frac{1}{4.74} = 0.211$$

10-5.2   Development of the F Distribution (CD Only)

10-5.3   Hypothesis Tests on the Ratio of Two Variances

A hypothesis-testing procedure for the equality of two variances is based on the following result.

Let $X_{11}, X_{12}, \ldots, X_{1n_1}$ be a random sample from a normal population with mean μ_1 and variance σ_1^2, and let $X_{21}, X_{22}, \ldots, X_{2n_2}$ be a random sample from a second normal population with mean μ_2 and variance σ_2^2. Assume that both normal populations are independent. Let S_1^2 and S_2^2 be the sample variances. Then the ratio

$$F = \frac{S_1^2/\sigma_1^2}{S_2^2/\sigma_2^2}$$

has an F distribution with $n_1 - 1$ numerator degrees of freedom and $n_2 - 1$ denominator degrees of freedom.

This result is based on the fact that $(n_1 - 1)S_1^2/\sigma_1^2$ is a chi-square random variable with $n_1 - 1$ degrees of freedom, that $(n_2 - 1)S_2^2/\sigma_2^2$ is a chi-square random variable with $n_2 - 1$ degrees of freedom, and that the two normal populations are independent. Clearly under the null hypothesis $H_0: \sigma_1^2 = \sigma_2^2$ the ratio $F_0 = S_1^2/S_2^2$ has an F_{n_1-1,n_2-1} distribution. This is the basis of the following test procedure.

Null hypothesis: $H_0: \sigma_1^2 = \sigma_2^2$

Test statistic: $F_0 = \dfrac{S_1^2}{S_2^2}$ (10-29)

Alternative Hypotheses	**Rejection Criterion**
$H_1: \sigma_1^2 \neq \sigma_2^2$	$f_0 > f_{\alpha/2,n_1-1,n_2-1}$ or $f_0 < f_{1-\alpha/2,n_1-1,n_2-1}$
$H_1: \sigma_1^2 > \sigma_2^2$	$f_0 > f_{\alpha,n_1-1,n_2-1}$
$H_1: \sigma_1^2 < \sigma_2^2$	$f_0 < f_{1-\alpha,\,n_1-1,n_2-1}$

EXAMPLE 10-11   Oxide layers on semiconductor wafers are etched in a mixture of gases to achieve the proper thickness. The variability in the thickness of these oxide layers is a critical characteristic of the wafer, and low variability is desirable for subsequent processing steps. Two different mixtures of gases are being studied to determine whether one is superior in reducing the variability of the oxide thickness. Twenty wafers are etched in each gas. The sample standard deviations of oxide thickness are $s_1 = 1.96$ angstroms and $s_2 = 2.13$ angstroms, respectively. Is there any evidence to indicate that either gas is preferable? Use $\alpha = 0.05$.

The eight-step hypothesis-testing procedure may be applied to this problem as follows:

1. The parameters of interest are the variances of oxide thickness σ_1^2 and σ_2^2. We will assume that oxide thickness is a normal random variable for both gas mixtures.

2. $H_0: \sigma_1^2 = \sigma_2^2$

3. $H_1: \sigma_1^2 \neq \sigma_2^2$

4. $\alpha = 0.05$

5. The test statistic is given by Equation 10-29:

$$f_0 = \frac{s_1^2}{s_2^2}$$

6. Since $n_1 = n_2 = 20$, we will reject $H_0: \sigma_1^2 = \sigma_2^2$ if $f_0 > f_{0.025,19,19} = 2.53$ or if $f_0 < f_{0.975,19,19} = 1/f_{0.025,19,19} = 1/2.53 = 0.40$.

7. Computations: Since $s_1^2 = (1.96)^2 = 3.84$ and $s_2^2 = (2.13)^2 = 4.54$, the test statistic is

$$f_0 = \frac{s_1^2}{s_2^2} = \frac{3.84}{4.54} = 0.85$$

8. Conclusions: Since $f_{0.975,19,19} = 0.40 < f_0 = 0.85 < f_{0.025,19,19} = 2.53$, we cannot reject the null hypothesis $H_0: \sigma_1^2 = \sigma_2^2$ at the 0.05 level of significance. Therefore, there is no strong evidence to indicate that either gas results in a smaller variance of oxide thickness.

We may also find a P-value for the F-statistic in Example 10-11. Since $f_{0.50,19,19} = 1.00$, the computed value of the test statistic $f_0 = s_1^2/s_2^2 = 3.84/4.54 = 0.85$ is nearer the lower tail of the F distribution than the upper tail. The probability that an F-random variable with 19 numerator and denominator degrees of freedom is less than 0.85 is 0.3634. Since it is arbitrary which population is identified as "one," we could have computed the test statistic as $f_0 = 4.54/3.84 = 1.18$. The probability that an F-random variable with 19 numerator and denominator degrees of freedom exceeds 1.18 is 0.3610. Therefore, the P-value for the test statistic $f_0 = 0.85$ is the sum of these two probabilities, or $P = 0.3634 + 0.3610 = 0.7244$. Since the P-value exceeds 0.05, the null hypothesis $H_0: \sigma_1^2 = \sigma_2^2$ cannot be rejected. (The probabilities given above were computed using a hand-held calculator.)

10-5.4 β-Error and Choice of Sample Size

Appendix Charts VIo, VIp, VIq, and VIr provide operating characteristic curves for the F-test given in Section 10-5.1 for $\alpha = 0.05$ and $\alpha = 0.01$, assuming that $n_1 = n_2 = n$. Charts VIo and VIp are used with the two-sided alternate hypothesis. They plot β against the abscissa parameter

$$\lambda = \frac{\sigma_1}{\sigma_2} \tag{10-30}$$

for various $n_1 = n_2 = n$. Charts VIq and VIr are used for the one-sided alternative hypotheses.

EXAMPLE 10-12 For the semiconductor wafer oxide etching problem in Example 10-11, suppose that one gas resulted in a standard deviation of oxide thickness that is half the standard deviation of oxide thickness of the other gas. If we wish to detect such a situation with probability at least 0.80, is the sample size $n_1 = n_2 = 20$ adequate?

Note that if one standard deviation is half the other,

$$\lambda = \frac{\sigma_1}{\sigma_2} = 2$$

By referring to Appendix Chart VIo with $n_1 = n_2 = n = 20$ and $\lambda = 2$, we find that $\beta \simeq 0.20$. Therefore, if $\beta = 0.20$, the power of the test (which is the probability that the difference in standard deviations will be detected by the test) is 0.80, and we conclude that the sample sizes $n_1 = n_2 = 20$ are adequate.

10-5.5 Confidence Interval on the Ratio of Two Variances

To find the confidence interval on σ_1^2/σ_2^2, recall that the sampling distribution of

$$F = \frac{S_2^2/\sigma_2^2}{S_1^2/\sigma_1^2}$$

is an F with $n_2 - 1$ and $n_1 - 1$ degrees of freedom. Therefore, $P(f_{1-\alpha/2, n_2-1, n_1-1} \leq F \leq f_{\alpha/2, n_2-1, n_1-1}) = 1 - \alpha$. Substitution for F and manipulation of the inequalities will lead to the $100(1 - \alpha)\%$ confidence interval for σ_1^2/σ_2^2.

Definition

> If s_1^2 and s_2^2 are the sample variances of random samples of sizes n_1 and n_2, respectively, from two independent normal populations with unknown variances σ_1^2 and σ_2^2, then a **100(1 − α)% confidence interval on the ratio σ_1^2/σ_2^2** is
>
> $$\frac{s_1^2}{s_2^2} f_{1-\alpha/2,n_2-1,n_1-1} \le \frac{\sigma_1^2}{\sigma_2^2} \le \frac{s_1^2}{s_2^2} f_{\alpha/2,n_2-1,n_1-1} \tag{10-31}$$
>
> where $f_{\alpha/2,n_2-1,n_1-1}$ and $f_{1-\alpha/2,n_2-1,n_1-1}$ are the upper and lower $\alpha/2$ percentage points of the F distribution with $n_2 - 1$ numerator and $n_1 - 1$ denominator degrees of freedom, respectively. A confidence interval on the ratio of the standard deviations can be obtained by taking square roots in Equation 10-31.

EXAMPLE 10-13

A company manufactures impellers for use in jet-turbine engines. One of the operations involves grinding a particular surface finish on a titanium alloy component. Two different grinding processes can be used, and both processes can produce parts at identical mean surface roughness. The manufacturing engineer would like to select the process having the least variability in surface roughness. A random sample of $n_1 = 11$ parts from the first process results in a sample standard deviation $s_1 = 5.1$ microinches, and a random sample of $n_2 = 16$ parts from the second process results in a sample standard deviation of $s_2 = 4.7$ microinches. We will find a 90% confidence interval on the ratio of the two standard deviations, σ_1/σ_2.

Assuming that the two processes are independent and that surface roughness is normally distributed, we can use Equation 10-31 as follows:

$$\frac{s_1^2}{s_2^2} f_{0.95,15,10} \le \frac{\sigma_1^2}{\sigma_2^2} \le \frac{s_1^2}{s_2^2} f_{0.05,15,10}$$

$$\frac{(5.1)^2}{(4.7)^2} 0.39 \le \frac{\sigma_1^2}{\sigma_2^2} \le \frac{(5.1)^2}{(4.7)^2} 2.85$$

or upon completing the implied calculations and taking square roots,

$$0.678 \le \frac{\sigma_1}{\sigma_2} \le 1.887$$

Notice that we have used Equation 10-28 to find $f_{0.95,15,10} = 1/f_{0.05,10,15} = 1/2.54 = 0.39$. Since this confidence interval includes unity, we cannot claim that the standard deviations of surface roughness for the two processes are different at the 90% level of confidence.

EXERCISES FOR SECTION 10-5

10-45. For an F distribution, find the following:
(a) $f_{0.25,5,10}$ (b) $f_{0.10,24,9}$
(c) $f_{0.05,8,15}$ (d) $f_{0.75,5,10}$
(e) $f_{0.90,24,9}$ (f) $f_{0.95,8,15}$

10-46. For an F distribution, find the following:
(a) $f_{0.25,7,15}$ (b) $f_{0.10,10,12}$

(c) $f_{0.01,20,10}$ (d) $f_{0.75,7,15}$
(e) $f_{0.90,10,12}$ (f) $f_{0.99,20,10}$

10-47. Two chemical companies can supply a raw material. The concentration of a particular element in this material is important. The mean concentration for both suppliers is the same, but we suspect that the variability in concentration may differ between the two companies. The standard deviation of concentration in a random sample of $n_1 = 10$ batches produced by company 1 is $s_1 = 4.7$ grams per liter, while for company 2, a random sample of $n_2 = 16$ batches yields $s_2 = 5.8$ grams per liter. Is there sufficient evidence to conclude that the two population variances differ? Use $\alpha = 0.05$.

10-48.   Consider the etch rate data in Exercise 10-21. Test the hypothesis $H_0: \sigma_1^2 = \sigma_2^2$ against $H_1: \sigma_1^2 \neq \sigma_2^2$ using $\alpha = 0.05$, and draw conclusions.

10-49.   Consider the etch rate data in Exercise 10-21. Suppose that if one population variance is twice as large as the other, we want to detect this with probability at least 0.90 (using $\alpha = 0.05$). Are the sample sizes $n_1 = n_2 = 10$ adequate?

10-50.   Consider the diameter data in Exercise 10-17. Construct the following:
(a) A 90% two-sided confidence interval on σ_1/σ_2.
(b) A 95% two-sided confidence interval on σ_1/σ_2. Comment on the comparison of the width of this interval with the width of the interval in part (a).
(c) A 90% lower-confidence bound on σ_1/σ_2.

10-51.   Consider the foam data in Exercise 10-18. Construct the following:
(a) A 90% two-sided confidence interval on σ_1^2/σ_2^2.
(b) A 95% two-sided confidence interval on σ_1^2/σ_2^2. Comment on the comparison of the width of this interval with the width of the interval in part (a).
(c) A 90% lower-confidence bound on σ_1/σ_2.

10-52.   Consider the film speed data in Exercise 10-24. Test $H_0: \sigma_1^2 = \sigma_2^2$ versus $H_1: \sigma_1^2 \neq \sigma_2^2$ using $\alpha = 0.02$.

10-53.   Consider the gear impact strength data in Exercise 10-22. Is there sufficient evidence to conclude that the variance of impact strength is different for the two suppliers? Use $\alpha = 0.05$.

10-54.   Consider the melting point data in Exercise 10-25. Do the sample data support a claim that both alloys have the same variance of melting point? Use $\alpha = 0.05$ in reaching your conclusion.

10-55.   Exercise 10-28 presented measurements of plastic coating thickness at two different application temperatures. Test $H_0: \sigma_1^2 = \sigma_2^2$ against $H_1: \sigma_1^2 \neq \sigma_2^2$ using $\alpha = 0.01$.

10-56.   A study was performed to determine whether men and women differ in their repeatability in assembling components on printed circuit boards. Random samples of 25 men and 21 women were selected, and each subject assembled the units. The two sample standard deviations of assembly time were $s_{men} = 0.98$ minutes and $s_{women} = 1.02$ minutes. Is there evidence to support the claim that men and women differ in repeatability for this assembly task? Use $\alpha = 0.02$ and state any necessary assumptions about the underlying distribution of the data.

10-57.   Reconsider the assembly repeatability experiment described in Exercise 10-56. Find a 98% confidence interval on the ratio of the two variances. Provide an interpretation of the interval.

10-58.   Reconsider the film speed experiment in Exercise 10-24. Suppose that one population standard deviation is 50% larger than the other. Is the sample size $n_1 = n_2 = 8$ adequate to detect this difference with high probability? Use $\alpha = 0.01$ in answering this question.

10-59.   Reconsider the overall distance data for golf balls in Exercise 10-31. Is there evidence to support the claim that the standard deviation of overall distance is the same for both brands of balls (use $\alpha = 0.05$)? Explain how this question can be answered with a 95% confidence interval on σ_1/σ_2.

10-60.   Reconsider the coefficient of restitution data in Exercise 10-32. Do the data suggest that the standard deviation is the same for both brands of drivers (use $\alpha = 0.05$)? Explain how to answer this question with a confidence interval on σ_1/σ_2.

10-6   INFERENCE ON TWO POPULATION PROPORTIONS

We now consider the case where there are two binomial parameters of interest, say, p_1 and p_2, and we wish to draw inferences about these proportions. We will present large-sample hypothesis testing and confidence interval procedures based on the normal approximation to the binomial.

10-6.1   Large-Sample Test for H_0: $p_1 = p_2$

Suppose that two independent random samples of sizes n_1 and n_2 are taken from two populations, and let X_1 and X_2 represent the number of observations that belong to the class of interest in samples 1 and 2, respectively. Furthermore, suppose that the normal approximation to the binomial is applied to each population, so the estimators of the population proportions

$\hat{P}_1 = X_1/n_1$ and $\hat{P}_2 = X_2/n_2$ have approximate normal distributions. We are interested in testing the hypotheses

$$H_0: p_1 = p_2$$
$$H_1: p_1 \neq p_2$$

The statistic

$$Z = \frac{\hat{P}_1 - \hat{P}_2 - (p_1 - p_2)}{\sqrt{\dfrac{p_1(1 - p_1)}{n_1} + \dfrac{p_2(1 - p_2)}{n_2}}} \qquad (10\text{-}32)$$

is distributed approximately as standard normal and is the basis of a test for $H_0: p_1 = p_2$. Specifically, if the null hypothesis $H_0: p_1 = p_2$ is true, using the fact that $p_1 = p_2 = p$, the random variable

$$Z = \frac{\hat{P}_1 - \hat{P}_2}{\sqrt{p(1 - p)\left(\dfrac{1}{n_1} + \dfrac{1}{n_2}\right)}}$$

is distributed approximately $N(0, 1)$. An estimator of the common parameter p is

$$\hat{P} = \frac{X_1 + X_2}{n_1 + n_2}$$

The **test statistic** for $H_0: p_1 = p_2$ is then

$$Z_0 = \frac{\hat{P}_1 - \hat{P}_2}{\sqrt{\hat{P}(1 - \hat{P})\left(\dfrac{1}{n_1} + \dfrac{1}{n_2}\right)}}$$

This leads to the test procedures described below.

Null hypothesis: $H_0: p_1 = p_2$

Test statistic: $$Z_0 = \frac{\hat{P}_1 - \hat{P}_2}{\sqrt{\hat{P}(1 - \hat{P})\left(\dfrac{1}{n_1} + \dfrac{1}{n_2}\right)}} \qquad (10\text{-}33)$$

Alternative Hypotheses	Rejection Criterion
$H_1: p_1 \neq p_2$	$z_0 > z_{\alpha/2}$ or $z_0 < -z_{\alpha/2}$
$H_1: p_1 > p_2$	$z_0 > z_\alpha$
$H_1: p_1 < p_2$	$z_0 < -z_\alpha$

EXAMPLE 10-14   Extracts of St. John's Wort are widely used to treat depression. An article in the April 18, 2001 issue of the *Journal of the American Medical Association* ("Effectiveness of St. John's Wort on Major Depression: A Randomized Controlled Trial") compared the efficacy of a standard extract of St. John's Wort with a placebo in 200 outpatients diagnosed with major depression. Patients were randomly assigned to two groups; one group received the St. John's Wort, and the other received the placebo. After eight weeks, 19 of the placebo-treated patients showed improvement, whereas 27 of those treated with St. John's Wort improved. Is there any reason to believe that St. John's Wort is effective in treating major depression? Use $\alpha = 0.05$.

The eight-step hypothesis testing procedure leads to the following results:

1. The parameters of interest are p_1 and p_2, the proportion of patients who improve following treatment with St. John's Wort (p_1) or the placebo (p_2).

2. $H_0: p_1 = p_2$

3. $H_1: p_1 \neq p_2$

4. $\alpha = 0.05$

5. The test statistic is

$$z_0 = \frac{\hat{p}_1 - \hat{p}_2}{\sqrt{\hat{p}(1 - \hat{p})\left(\dfrac{1}{n_1} + \dfrac{1}{n_2}\right)}}$$

where $\hat{p}_1 = 27/100 = 0.27$, $\hat{p}_2 = 19/100 = 0.19$, $n_1 = n_2 = 100$, and

$$\hat{p} = \frac{x_1 + x_2}{n_1 + n_2} = \frac{19 + 27}{100 + 100} = 0.23$$

6. Reject $H_0: p_1 = p_2$ if $z_0 > z_{0.025} = 1.96$ or if $z_0 < -z_{0.025} = -1.96$.

7. Computations: The value of the test statistic is

$$z_0 = \frac{0.27 - 0.19}{\sqrt{0.23(0.77)\left(\dfrac{1}{100} + \dfrac{1}{100}\right)}} = 1.35$$

8. Conclusions: Since $z_0 = 1.35$ does not exceed $z_{0.025}$, we cannot reject the null hypothesis. Note that the P-value is $P \simeq 0.177$. There is insufficient evidence to support the claim that St. John's Wort is effective in treating major depression.

The following box shows the Minitab two-sample hypothesis test and CI procedure for proportions. Notice that the 95% CI on $p_1 - p_2$ includes zero. The equation for constructing the CI will be given in Section 10-6.4.

Test and CI for Two Proportions

Sample	X	N	Sample p
1	27	100	0.270000
2	19	100	0.190000

Estimate for p(1) − p(2): 0.08
95% CI for p(1) − p(2): (−0.0361186, 0.196119)
Test for p(1) − p(2) = 0 (vs not = 0): Z = 1.35 P-Value = 0.177

10-6.2   Small-Sample Test for H_0: $p_1 = p_2$ (CD Only)

10-6.3   β-Error and Choice of Sample Size

The computation of the β-error for the large-sample test of H_0: $p_1 = p_2$ is somewhat more involved than in the single-sample case. The problem is that the denominator of the test statistic Z_0 is an estimate of the standard deviation of $\hat{P}_1 - \hat{P}_2$ under the assumption that $p_1 = p_2 = p$. When H_0: $p_1 = p_2$ is false, the standard deviation of $\hat{P}_1 - \hat{P}_2$ is

$$\sigma_{\hat{P}_1 - \hat{P}_2} = \sqrt{\frac{p_1(1 - p_1)}{n_1} + \frac{p_2(1 - p_2)}{n_2}} \tag{10-34}$$

If the alternative hypothesis is two sided, the β-error is

$$\beta = \Phi\left[\frac{z_{\alpha/2}\sqrt{\bar{p}\bar{q}(1/n_1 + 1/n_2)} - (p_1 - p_2)}{\sigma_{\hat{P}_1 - \hat{P}_2}}\right]$$

$$- \Phi\left[\frac{-z_{\alpha/2}\sqrt{\bar{p}\bar{q}(1/n_1 + 1/n_2)} - (p_1 - p_2)}{\sigma_{\hat{P}_1 - \hat{P}_2}}\right] \tag{10-35}$$

where

$$\bar{p} = \frac{n_1 p_1 + n_2 p_2}{n_1 + n_2} \quad \text{and} \quad \bar{q} = \frac{n_1(1 - p_1) + n_2(1 - p_2)}{n_1 + n_2}$$

and $\sigma_{\hat{P}_1 - \hat{P}_2}$ is given by Equation 10-34.

If the alternative hypothesis is H_1: $p_1 > p_2$,

$$\beta = \Phi\left[\frac{z_{\alpha}\sqrt{\bar{p}\bar{q}(1/n_1 + 1/n_2)} - (p_1 - p_2)}{\sigma_{\hat{P}_1 - \hat{P}_2}}\right] \tag{10-36}$$

and if the alternative hypothesis is H_1: $p_1 < p_2$,

$$\beta = 1 - \Phi\left[\frac{-z_{\alpha}\sqrt{\bar{p}\bar{q}(1/n_1 + 1/n_2)} - (p_1 - p_2)}{\sigma_{\hat{P}_1 - \hat{P}_2}}\right] \tag{10-37}$$

For a specified pair of values p_1 and p_2, we can find the sample sizes $n_1 = n_2 = n$ required to give the test of size α that has specified type II error β.

For the two-sided alternative, the common sample size is

$$n = \frac{\left[z_{\alpha/2}\sqrt{(p_1 + p_2)(q_1 + q_2)/2} + z_\beta\sqrt{p_1 q_1 + p_2 q_2}\right]^2}{(p_1 - p_2)^2} \qquad (10\text{-}38)$$

where $q_1 = 1 - p_1$ and $q_2 = 1 - p_2$.

For a one-sided alternative, replace $z_{\alpha/2}$ in Equation 10-38 by z_α.

10-6.4 Confidence Interval for $p_1 - p_2$

The confidence interval for $p_1 - p_2$ can be found directly, since we know that

$$Z = \frac{\hat{P}_1 - \hat{P}_2 - (p_1 - p_2)}{\sqrt{\dfrac{p_1(1 - p_1)}{n_1} + \dfrac{p_2(1 - p_2)}{n_2}}}$$

is a standard normal random variable. Thus $P(-z_{\alpha/2} \leq Z \leq z_{\alpha/2}) \simeq 1 - \alpha$, so we can substitute for Z in this last expression and use an approach similar to the one employed previously to find an approximate $100(1 - \alpha)\%$ two-sided confidence interval for $p_1 - p_2$.

Definition

If $\hat{p}_1$ and $\hat{p}_2$ are the sample proportions of observation in two independent random samples of sizes n_1 and n_2 that belong to a class of interest, **an approximate two-sided $100(1 - \alpha)\%$ confidence interval on the difference in the true proportions $p_1 - p_2$ is**

$$\hat{p}_1 - \hat{p}_2 - z_{\alpha/2}\sqrt{\frac{\hat{p}_1(1 - \hat{p}_1)}{n_1} + \frac{\hat{p}_2(1 - \hat{p}_2)}{n_2}}$$

$$\leq p_1 - p_2 \leq \hat{p}_1 - \hat{p}_2 + z_{\alpha/2}\sqrt{\frac{\hat{p}_1(1 - \hat{p}_1)}{n_1} + \frac{\hat{p}_2(1 - \hat{p}_2)}{n_2}} \qquad (10\text{-}39)$$

where $z_{\alpha/2}$ is the upper $\alpha/2$ percentage point of the standard normal distribution.

EXAMPLE 10-15   Consider the process manufacturing crankshaft bearings described in Example 8-6. Suppose that a modification is made in the surface finishing process and that, subsequently, a second random sample of 85 axle shafts is obtained. The number of defective shafts in this second sample is 8. Therefore, since $n_1 = 85$, $\hat{p}_1 = 0.12$, $n_2 = 85$, and $\hat{p}_2 = 8/85 = 0.09$, we can obtain an approximate 95% confidence interval on the difference in the proportion of defective bearings produced under the two processes from Equation 10-39 as follows:

$$\hat{p}_1 - \hat{p}_2 - z_{0.025} \sqrt{\frac{\hat{p}_1(1 - \hat{p}_1)}{n_1} + \frac{\hat{p}_2(1 - \hat{p}_2)}{n_2}}$$

$$\leq p_1 - p_2 \leq \hat{p}_1 - \hat{p}_2 + z_{0.025} \sqrt{\frac{\hat{p}_1(1 - \hat{p}_1)}{n_1} + \frac{\hat{p}_2(1 - \hat{p}_2)}{n_2}}$$

or

$$0.12 - 0.09 - 1.96 \sqrt{\frac{0.12(0.88)}{85} + \frac{0.09(0.91)}{85}}$$

$$\leq p_1 - p_2 \leq 0.12 - 0.09 + 1.96 \sqrt{\frac{0.12(0.88)}{85} + \frac{0.09(0.91)}{85}}$$

This simplifies to

$$-0.06 \leq p_1 - p_2 \leq 0.12$$

This confidence interval includes zero, so, based on the sample data, it seems unlikely that the changes made in the surface finish process have reduced the proportion of defective crankshaft bearings being produced.

EXERCISES FOR SECTION 10-6

10-61. Two different types of injection-molding machines are used to form plastic parts. A part is considered defective if it has excessive shrinkage or is discolored. Two random samples, each of size 300, are selected, and 15 defective parts are found in the sample from machine 1 while 8 defective parts are found in the sample from machine 2. Is it reasonable to conclude that both machines produce the same fraction of defective parts, using $\alpha = 0.05$? Find the P-value for this test.

10-62. Two different types of polishing solution are being evaluated for possible use in a tumble-polish operation for manufacturing interocular lenses used in the human eye following cataract surgery. Three hundred lenses were tumble-polished using the first polishing solution, and of this number 253 had no polishing-induced defects. Another 300 lenses were tumble-polished using the second polishing solution, and 196 lenses were satisfactory upon completion. Is there any reason to believe that the two polishing solutions differ? Use $\alpha = 0.01$. Discuss how this question could be answered with a confidence interval on $p_1 - p_2$.

10-63. Consider the situation described in Exercise 10-61. Suppose that $p_1 = 0.05$ and $p_2 = 0.01$.
(a) With the sample sizes given here, what is the power of the test for this two-sided alternate?

(b) Determine the sample size needed to detect this difference with a probability of at least 0.9. Use $\alpha = 0.05$.

10-64. Consider the situation described in Exercise 10-61. Suppose that $p_1 = 0.05$ and $p_2 = 0.02$.
(a) With the sample sizes given here, what is the power of the test for this two-sided alternate?
(b) Determine the sample size needed to detect this difference with a probability of at least 0.9. Use $\alpha = 0.05$.

10-65. A random sample of 500 adult residents of Maricopa County found that 385 were in favor of increasing the highway speed limit to 75 mph, while another sample of 400 adult residents of Pima County found that 267 were in favor of the increased speed limit. Do these data indicate that there is a difference in the support for increasing the speed limit between the residents of the two counties? Use $\alpha = 0.05$. What is the P-value for this test?

10-66. Construct a 95% confidence interval on the difference in the two fractions defective for Exercise 10-61.

10-67. Construct a 95% confidence interval on the difference in the two proportions for Exercise 10-65. Provide a practical interpretation of this interval.

10-7 SUMMARY TABLE FOR INFERENCE PROCEDURES FOR TWO SAMPLES

The table in the end papers of the book summarizes all of the two-sample inference procedures given in this chapter. The table contains the null hypothesis statements, the test statistics, the criteria for rejection of the various alternative hypotheses, and the formulas for constructing the $100(1 - \alpha)\%$ confidence intervals.

Supplemental Exercises

10-68. A procurement specialist has purchased 25 resistors from vendor 1 and 35 resistors from vendor 2. Each resistor's resistance is measured with the following results:

Vendor 1

96.8	100.0	100.3	98.5	98.3	98.2
99.6	99.4	99.9	101.1	103.7	97.7
99.7	101.1	97.7	98.6	101.9	101.0
99.4	99.8	99.1	99.6	101.2	98.2
98.6					

Vendor 2

106.8	106.8	104.7	104.7	108.0	102.2
103.2	103.7	106.8	105.1	104.0	106.2
102.6	100.3	104.0	107.0	104.3	105.8
104.0	106.3	102.2	102.8	104.2	103.4
104.6	103.5	106.3	109.2	107.2	105.4
106.4	106.8	104.1	107.1	107.7	

(a) What distributional assumption is needed to test the claim that the variance of resistance of product from vendor 1 is not significantly different from the variance of resistance of product from vendor 2? Perform a graphical procedure to check this assumption.

(b) Perform an appropriate statistical hypothesis-testing procedure to determine whether the procurement specialist can claim that the variance of resistance of product from vendor 1 is significantly different from the variance of resistance of product from vendor 2.

10-69. An article in the *Journal of Materials Engineering* (1989, Vol. 11, No. 4, pp. 275–282) reported the results of an experiment to determine failure mechanisms for plasma-sprayed thermal barrier coatings. The failure stress for one particular coating (NiCrAlZr) under two different test conditions is as follows:

Failure stress ($\times 10^6$ Pa) after nine 1-hour cycles: 19.8, 18.5, 17.6, 16.7, 16.7, 14.8, 15.4, 14.1, 13.6

Failure stress ($\times 10^6$ Pa) after six 1-hour cycles: 14.9, 12.7, 11.9, 11.4, 10.1, 7.9

(a) What assumptions are needed to construct confidence intervals for the difference in mean failure stress under the

two different test conditions? Use normal probability plots of the data to check these assumptions.

(b) Find a 99% confidence interval on the difference in mean failure stress under the two different test conditions.

(c) Using the confidence interval constructed in part (b), does the evidence support the claim that the first test conditions yield higher results, on the average, than the second? Explain your answer.

10-70. Consider Supplemental Exercise 10-69.

(a) Construct a 95% confidence interval on the ratio of the variances, σ_1/σ_2, of failure stress under the two different test conditions.

(b) Use your answer in part (b) to determine whether there is a significant difference in variances of the two different test conditions. Explain your answer.

10-71. A liquid dietary product implies in its advertising that use of the product for one month results in an average weight loss of at least 3 pounds. Eight subjects use the product for one month, and the resulting weight loss data are reported below. Use hypothesis-testing procedures to answer the following questions.

Subject	Initial Weight (lb)	Final Weight (lb)
1	165	161
2	201	195
3	195	192
4	198	193
5	155	150
6	143	141
7	150	146
8	187	183

(a) Do the data support the claim of the producer of the dietary product with the probability of a type I error set to 0.05?

(b) Do the data support the claim of the producer of the dietary product with the probability of a type I error set to 0.01?

(c) In an effort to improve sales, the producer is considering changing its claim from "at least 3 pounds" to "at least 5 pounds." Repeat parts (a) and (b) to test this new claim.

10-72. The breaking strength of yarn supplied by two manufacturers is being investigated. We know from experience with the manufacturers' processes that $\sigma_1 = 5$ psi and $\sigma_2 = 4$ psi. A random sample of 20 test specimens from each manufacturer results in $\bar{x}_1 = 88$ psi and $\bar{x}_2 = 91$ psi, respectively.

(a) Using a 90% confidence interval on the difference in mean breaking strength, comment on whether or not there is evidence to support the claim that manufacturer 2 produces yarn with higher mean breaking strength.

(b) Using a 98% confidence interval on the difference in mean breaking strength, comment on whether or not there is evidence to support the claim that manufacturer 2 produces yarn with higher mean breaking strength.

(c) Comment on why the results from parts (a) and (b) are different or the same. Which would you choose to make your decision and why?

10-73. The Salk polio vaccine experiment in 1954 focused on the effectiveness of the vaccine in combatting paralytic polio. Because it was felt that without a control group of children there would be no sound basis for evaluating the efficacy of the Salk vaccine, the vaccine was administered to one group, and a placebo (visually identical to the vaccine but known to have no effect) was administered to a second group. For ethical reasons, and because it was suspected that knowledge of vaccine administration would affect subsequent diagnoses, the experiment was conducted in a double-blind fashion. That is, neither the subjects nor the administrators knew who received the vaccine and who received the placebo. The actual data for this experiment are as follows:

Placebo group: $n = 201,299$: 110 cases of polio observed

Vaccine group: $n = 200,745$: 33 cases of polio observed

(a) Use a hypothesis-testing procedure to determine if the proportion of children in the two groups who contracted paralytic polio is statistically different. Use a probability of a type I error equal to 0.05.

(b) Repeat part (a) using a probability of a type I error equal to 0.01.

(c) Compare your conclusions from parts (a) and (b) and explain why they are the same or different.

10-74. Consider Supplemental Exercise 10-72. Suppose that prior to collecting the data, you decide that you want the error in estimating $\mu_1 - \mu_2$ by $x_1 - x_2$ to be less than 1.5 psi. Specify the sample size for the following percentage confidence:

(a) 90%

(b) 98%

(c) Comment on the effect of increasing the percentage confidence on the sample size needed.

(d) Repeat parts (a)–(c) with an error of less than 0.75 psi instead of 1.5 psi.

(e) Comment on the effect of decreasing the error on the sample size needed.

10-75. A random sample of 1500 residential telephones in Phoenix in 1990 found that 387 of the numbers were unlisted.

A random sample in the same year of 1200 telephones in Scottsdale found that 310 were unlisted.

(a) Find a 95% confidence interval on the difference in the two proportions and use this confidence interval to determine if there is a statistically significant difference in proportions of unlisted numbers between the two cities.

(b) Find a 90% confidence interval on the difference in the two proportions and use this confidence interval to determine if there is a statistically significant difference in proportions of unlisted numbers between the two cities.

(c) Suppose that all the numbers in the problem description were doubled. That is, 774 residents out of 3000 sampled in Phoenix and 620 residents out of 2400 in Scottsdale had unlisted phone numbers. Repeat parts (a) and (b) and comment on the effect of increasing the sample size without changing the proportions on your results.

10-76. In a random sample of 200 Phoenix residents who drive a domestic car, 165 reported wearing their seat belt regularly, while another sample of 250 Phoenix residents who drive a foreign car revealed 198 who regularly wore their seat belt.

(a) Perform a hypothesis-testing procedure to determine if there is a statistically significant difference in seat belt usage between domestic and foreign car drivers. Set your probability of a type I error to 0.05.

(b) Perform a hypothesis-testing procedure to determine if there is a statistically significant difference in seat belt usage between domestic and foreign car drivers. Set your probability of a type I error to 0.1.

(c) Compare your answers for parts (a) and (b) and explain why they are the same or different.

(d) Suppose that all the numbers in the problem description were doubled. That is, in a random sample of 400 Phoenix residents who drive a domestic car, 330 reported wearing their seat belt regularly, while another sample of 500 Phoenix residents who drive a foreign car revealed 396 who regularly wore their seat belt. Repeat parts (a) and (b) and comment on the effect of increasing the sample size without changing the proportions on your results.

10-77. Consider the previous exercise, which summarized data collected from drivers about their seat belt usage.

(a) Do you think there is a reason not to believe these data? Explain your answer.

(b) Is it reasonable to use the hypothesis-testing results from the previous problem to draw an inference about the difference in proportion of seat belt usage

  (i) of the spouses of these drivers of domestic and foreign cars? Explain your answer.

  (ii) of the children of these drivers of domestic and foreign cars? Explain your answer.

  (iii) of all drivers of domestic and foreign cars? Explain your answer.

  (iv) of all drivers of domestic and foreign trucks? Explain your answer.

10-78. Consider the situation described in Exercise 10-62.

(a) Redefine the parameters of interest to be the proportion of lenses that are unsatisfactory following tumble polishing with polishing fluids 1 or 2. Test the hypothesis that the two polishing solutions give different results using $\alpha = 0.01$.

(b) Compare your answer in part (a) with that for Exercise 10-62. Explain why they are the same or different.

10-79. Consider the situation of Exercise 10-62, and recall that the hypotheses of interest are $H_0: p_1 = p_2$ versus $H_1: p_1 \neq p_2$. We wish to use $\alpha = 0.01$. Suppose that if $p_1 = 0.9$ and $p_2 = 0.6$, we wish to detect this with a high probability, say, at least 0.9. What sample sizes are required to meet this objective?

10-80. A manufacturer of a new pain relief tablet would like to demonstrate that its product works twice as fast as the competitor's product. Specifically, the manufacturer would like to test

$$H_0: \mu_1 = 2\mu_2$$
$$H_1: \mu_1 > 2\mu_2$$

where μ_1 is the mean absorption time of the competitive product and μ_2 is the mean absorption time of the new product. Assuming that the variances σ_1^2 and σ_2^2 are known, develop a procedure for testing this hypothesis.

10-81. Suppose that we are testing $H_0: \mu_1 = \mu_2$ versus $H_1: \mu_1 \neq \mu_2$, and we plan to use equal sample sizes from the two populations. Both populations are assumed to be normal with unknown but equal variances. If we use $\alpha = 0.05$ and if the true mean $\mu_1 = \mu_2 + \sigma$, what sample size must be used for the power of this test to be at least 0.90?

10-82. Consider the fire-fighting foam expanding agents investigated in Exercise 10-18, in which five observations of each agent were recorded. Suppose that, if agent 1 produces a mean expansion that differs from the mean expansion of agent 1 by 1.5, we would like to reject the null hypothesis with probability at least 0.95.

(a) What sample size is required?

(b) Do you think that the original sample size in Exercise 10-18 was appropriate to detect this difference? Explain your answer.

10-83. A fuel-economy study was conducted for two German automobiles, Mercedes and Volkswagen. One vehicle of each brand was selected, and the mileage performance was observed for 10 tanks of fuel in each car. The data are as follows (in miles per gallon):

Mercedes		Volkswagen	
24.7	24.9	41.7	42.8
24.8	24.6	42.3	42.4
24.9	23.9	41.6	39.9
24.7	24.9	39.5	40.8
24.5	24.8	41.9	29.6

(a) Construct a normal probability plot of each of the data sets. Based on these plots, is it reasonable to assume that they are each drawn from a normal population?

(b) Suppose that it was determined that the lowest observation of the Mercedes data was erroneously recorded and should be 24.6. Furthermore, the lowest observation of the Volkswagen data was also mistaken and should be 39.6. Again construct normal probability plots of each of the data sets with the corrected values. Based on these new plots, is it reasonable to assume that they are each drawn from a normal population?

(c) Compare your answers from parts (a) and (b) and comment on the effect of these mistaken observations on the normality assumption.

(d) Using the corrected data from part (b) and a 95% confidence interval, is there evidence to support the claim that the variability in mileage performance is greater for a Volkswagen than for a Mercedes?

10-84. Reconsider the fuel-economy study in Supplemental Exercise 10-83. Rework part (d) of this problem using an appropriate hypothesis-testing procedure. Did you get the same answer as you did originally? Why?

10-85. An experiment was conducted to compare the filling capability of packaging equipment at two different wineries. Ten bottles of pinot noir from Ridgecrest Vineyards were randomly selected and measured, along with 10 bottles of pinot noir from Valley View Vineyards. The data are as follows (fill volume is in milliliters):

Ridgecrest				Valley View			
755	751	752	753	756	754	757	756
753	753	753	754	755	756	756	755
752	751			755	756		

(a) What assumptions are necessary to perform a hypothesis-testing procedure for equality of means of these data? Check these assumptions.

(b) Perform the appropriate hypothesis-testing procedure to determine whether the data support the claim that both wineries will fill bottles to the same mean volume.

10-86. Consider Supplemental Exercise 10-85. Suppose that the true difference in mean fill volume is as much as 2 fluid ounces; did the sample sizes of 10 from each vineyard provide good detection capability when $\alpha = 0.05$? Explain your answer.

10-87. A Rockwell hardness-testing machine presses a tip into a test coupon and uses the depth of the resulting depression to indicate hardness. Two different tips are being compared to determine whether they provide the same Rockwell C-scale hardness readings. Nine coupons are tested, with both tips being tested on each coupon. The data are shown in the accompanying table.

Coupon	Tip 1	Tip 2	Coupon	Tip 1	Tip 2
1	47	46	6	41	41
2	42	40	7	45	46
3	43	45	8	45	46
4	40	41	9	49	48
5	42	43			

(a) State any assumptions necessary to test the claim that both tips produce the same Rockwell C-scale hardness readings. Check those assumptions for which you have the information.

(b) Apply an appropriate statistical method to determine if the data support the claim that the difference in Rockwell C-scale hardness readings of the two tips is significantly different from zero

(c) Suppose that if the two tips differ in mean hardness readings by as much as 1.0, we want the power of the test to be at least 0.9. For an $\alpha = 0.01$, how many coupons should have been used in the test?

10-88. Two different gauges can be used to measure the depth of bath material in a Hall cell used in smelting aluminum. Each gauge is used once in 15 cells by the same operator.

Cell	Gauge 1	Gauge 2	Cell	Gauge 1	Gauge 2
1	46 in.	47 in.	9	52	51
2	50	53	10	47	45
3	47	45	11	49	51
4	53	50	12	45	45
5	49	51	13	47	49
6	48	48	14	46	43
7	53	54	15	50	51
8	56	53			

(a) State any assumptions necessary to test the claim that both gauges produce the same mean bath depth readings. Check those assumptions for which you have the information.

(b) Apply an appropriate statistical procedure to determine if the data support the claim that the two gauges produce different mean bath depth readings.

(c) Suppose that if the two gauges differ in mean bath depth readings by as much as 1.65 inch, we want the power of the test to be at least 0.8. For $\alpha = 0.01$, how many cells should have been used?

10-89. An article in the *Journal of the Environmental Engineering Division* ("Distribution of Toxic Substances in Rivers," 1982, Vol. 108, pp. 639–649) investigates the concentration of several hydrophobic organic substances in the Wolf River in Tennessee. Measurements on hexachlorobenzene (HCB) in nanograms per liter were taken at different depth downstream of an abandoned dump site. Data for two depths follow:

Surface: 3.74, 4.61, 4.00, 4.67, 4.87, 5.12, 4.52, 5.29, 5.74, 5.48

Bottom: 5.44, 6.88, 5.37, 5.44, 5.03, 6.48, 3.89, 5.85, 6.85, 7.16

(a) What assumptions are required to test the claim that mean HCB concentration is the same at both depths? Check those assumptions for which you have the information.

(b) Apply an appropriate procedure to determine if the data support the claim in part a.

(c) Suppose that the true difference in mean concentrations is 2.0 nanograms per liter. For $\alpha = 0.05$, what is the power of a statistical test for $H_0: \mu_1 = \mu_2$ versus $H_1: \mu_1 \neq \mu_2$?

(d) What sample size would be required to detect a difference of 1.0 nanograms per liter at $\alpha = 0.05$ if the power must be at least 0.9?

MIND-EXPANDING EXERCISES

10-90. Three different pesticides can be used to control infestation of grapes. It is suspected that pesticide 3 is more effective than the other two. In a particular vineyard, three different plantings of pinot noir grapes are selected for study. The following results on yield are obtained:

Pesticide	$\bar{x}_i$ (Bushels/ Plant)	s_i	n_i (Number of Plants)
1	4.6	0.7	100
2	5.2	0.6	120
3	6.1	0.8	130

If μ_i is the true mean yield after treatment with the ith pesticide, we are interested in the quantity

$$\mu = \frac{1}{2}(\mu_1 + \mu_2) - \mu_3$$

which measures the difference in mean yields between pesticides 1 and 2 and pesticide 3. If the sample sizes n_i are large, the estimator (say, $\hat{\mu}$) obtained by replacing each individual μ_i by $\bar{X}_i$ is approximately normal.

(a) Find an approximate $100(1 - \alpha)\%$ large-sample confidence interval for μ.

MIND-EXPANDING EXERCISES

(b) Do these data support the claim that pesticide 3 is more effective than the other two? Use $\alpha = 0.05$ in determining your answer.

10-91.   Suppose that we wish to test H_0: $\mu_1 = \mu_2$ versus H_1: $\mu_1 \neq \mu_2$, where σ_1^2 and σ_2^2 are known. The total sample size N is to be determined, and the allocation of observations to the two populations such that $n_1 + n_2 = N$ is to be made on the basis of cost. If the cost of sampling for populations 1 and 2 are C_1 and C_2, respectively, find the minimum cost sample sizes that provide a specified variance for the difference in sample means.

10-92.   Suppose that we wish to test the hypothesis H_0: $\mu_1 = \mu_2$ versus H_1: $\mu_1 \neq \mu_2$, where both variances σ_1^2 and σ_2^2 are known. A total of $n_1 + n_2 = N$ observations can be taken. How should these observations be allocated to the two populations to maximize the probability that H_0 will be rejected if H_1 is true and $\mu_1 - \mu_2 = \Delta \neq 0$?

10-93.   Suppose that we wish to test H_0: $\mu = \mu_0$ versus H_1: $\mu \neq \mu_0$, where the population is normal with known σ. Let $0 < \epsilon < \alpha$, and define the critical region so that we will reject H_0 if $z_0 > z_\epsilon$ or if $z_0 < -z_{\alpha - \epsilon}$, where z_0 is the value of the usual test statistic for these hypotheses.

(a) Show that the probability of type I error for this test is α.

(b) Suppose that the true mean is $\mu_1 = \mu_0 + \Delta$. Derive an expression for β for the above test.

10-94.   Construct a data set for which the paired t-test statistic is very large, indicating that when this analysis is used the two population means are different, but t_0 for the two-sample t-test is very small so that the incorrect analysis would indicate that there is no significant difference between the means.

10-95.   In some situations involving proportions, we are interested in the ratio $\theta = p_1/p_2$ rather than the difference $p_1 - p_2$. Let $\hat{\theta} = \hat{p}_1/\hat{p}_2$. We can show that $\ln(\hat{\theta})$ has an approximate normal distribution with the mean (n/θ) and variance $[(n_1 - x_1)/(n_1 x_1) + (n_2 - x_2)/(n_2 x_2)]^{1/2}$.

(a) Use the information above to derive a large-sample confidence interval for $\ln \theta$.

(b) Show how to find a large-sample CI for θ.

(c) Use the data from the St. John's Wort study in Example 10-14, and find a 95% CI on $\theta = p_1/p_2$. Provide a practical interpretation for this CI.

10-96.   Derive an expression for β for the test of the equality of the variances of two normal distributions. Assume that the two-sided alternative is specified.

IMPORTANT TERMS AND CONCEPTS

In the E-book, click on any term or concept below to go to that subject.

Comparative experiments

Critical region for a test statistic

Identifying cause and effect

Null and alternative hypotheses

One-sided and two-sided alternative hypotheses

Operating characteristic curves

Paired t-test

Pooled t-test

P-value

Reference distribution for a test statistic

Sample size determination for hypothesis tests and confidence intervals

Statistical hypotheses

Test statistic

CD MATERIAL

Fisher-Irwin test on two proportions

11

Simple Linear Regression and Correlation

CHAPTER OUTLINE

LEARNING OBJECTIVES

After careful study of this chapter, you should be able to do the following:

1. Use simple linear regression for building empirical models to engineering and scientific data

2. Understand how the method of least squares is used to estimate the parameters in a linear regression model

3. Analyze residuals to determine if the regression model is an adequate fit to the data or to see if any underlying assumptions are violated

4. Test statistical hypotheses and construct confidence intervals on regression model parameters

5. Use the regression model to make a prediction of a future observation and construct an appropriate prediction interval on the future observation

6. Use simple transformations to achieve a linear regression model

7. Apply the correlation model

CD MATERIAL

8. Conduct a lack-of-fit test in a regression model where there are replicated observations.

Answers for many odd numbered exercises are at the end of the book. Answers to exercises whose numbers are surrounded by a box can be accessed in the e-Text by clicking on the box. Complete worked solutions to certain exercises are also available in the e-Text. These are indicated in the Answers to Selected Exercises section by a box around the exercise number. Exercises are also available for some of the text sections that appear on CD only. These exercises may be found within the e-Text immediately following the section they accompany.

11-1 EMPIRICAL MODELS

Many problems in engineering and science involve exploring the relationships between two or more variables. **Regression analysis** is a statistical technique that is very useful for these types of problems. For example, in a chemical process, suppose that the yield of the product is related to the process-operating temperature. Regression analysis can be used to build a model to predict yield at a given temperature level. This model can also be used for process optimization, such as finding the level of temperature that maximizes yield, or for process control purposes.

As an illustration, consider the data in Table 11-1. In this table y is the purity of oxygen produced in a chemical distillation process, and x is the percentage of hydrocarbons that are present in the main condenser of the distillation unit. Figure 11-1 presents a **scatter diagram**

Table 11-1 Oxygen and Hydrocarbon Levels

Observation Number	Hydrocarbon Level $x\,(\%)$	Purity $y\,(\%)$
1	0.99	90.01
2	1.02	89.05
3	1.15	91.43
4	1.29	93.74
5	1.46	96.73
6	1.36	94.45
7	0.87	87.59
8	1.23	91.77
9	1.55	99.42
10	1.40	93.65
11	1.19	93.54
12	1.15	92.52
13	0.98	90.56
14	1.01	89.54
15	1.11	89.85
16	1.20	90.39
17	1.26	93.25
18	1.32	93.41
19	1.43	94.98
20	0.95	87.33

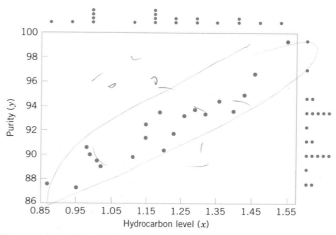

Figure 11-1 Scatter diagram of oxygen purity versus hydrocarbon level from Table 11-1.

of the data in Table 11-1. This is just a graph on which each (x_i, y_i) pair is represented as a point plotted in a two-dimensional coordinate system. This scatter diagram was produced by Minitab, and we selected an option that shows dot diagrams of the x and y variables along the top and right margins of the graph, respectively, making it easy to see the distributions of the individual variables (box plots or histograms could also be selected). Inspection of this scatter diagram indicates that, although no simple curve will pass exactly through all the points, there is a strong indication that the points lie scattered randomly around a straight line. Therefore, it is probably reasonable to assume that the mean of the random variable Y is related to x by the following straight-line relationship:

$$E(Y|x) = \mu_{Y|x} = \beta_0 + \beta_1 x$$

where the slope and intercept of the line are called **regression coefficients.** While the mean of Y is a linear function of x, the actual observed value y does not fall exactly on a straight line. The appropriate way to generalize this to a **probabilistic linear model** is to assume that the expected value of Y is a linear function of x, but that for a fixed value of x the actual value of Y is determined by the mean value function (the linear model) plus a random error term, say,

$$Y = \beta_0 + \beta_1 x + \epsilon \tag{11-1}$$

where ϵ is the random error term. We will call this model the **simple linear regression model,** because it has only one independent variable or **regressor.** Sometimes a model like this will arise from a theoretical relationship. At other times, we will have no theoretical knowledge of the relationship between x and y, and the choice of the model is based on inspection of a scatter diagram, such as we did with the oxygen purity data. We then think of the regression model as an **empirical model.**

To gain more insight into this model, suppose that we can fix the value of x and observe the value of the random variable Y. Now if x is fixed, the random component ϵ on the right-hand side of the model in Equation 11-1 determines the properties of Y. Suppose that the mean and variance of ϵ are 0 and σ^2, respectively. Then

$$E(Y|x) = E(\beta_0 + \beta_1 x + \epsilon) = \beta_0 + \beta_1 x + E(\epsilon) = \beta_0 + \beta_1 x$$

Notice that this is the same relationship that we initially wrote down empirically from inspection of the scatter diagram in Fig. 11-1. The variance of Y given x is

$$V(Y|x) = V(\beta_0 + \beta_1 x + \epsilon) = V(\beta_0 + \beta_1 x) + V(\epsilon) = 0 + \sigma^2 = \sigma^2$$

Thus, the true regression model $\mu_{Y|x} = \beta_0 + \beta_1 x$ is a line of mean values; that is, the height of the regression line at any value of x is just the expected value of Y for that x. The slope, β_1, can be interpreted as the change in the mean of Y for a unit change in x. Furthermore, the variability of Y at a particular value of x is determined by the error variance σ^2. This implies that there is a distribution of Y-values at each x and that the variance of this distribution is the same at each x.

For example, suppose that the true regression model relating oxygen purity to hydrocarbon level is $\mu_{Y|x} = 75 + 15x$, and suppose that the variance is $\sigma^2 = 2$. Figure 11-2 illustrates this situation. Notice that we have used a normal distribution to describe the random variation in ϵ. Since Y is the sum of a constant $\beta_0 + \beta_1 x$ (the mean) and a normally distributed random variable, Y is a normally distributed random variable. The variance σ^2 determines the

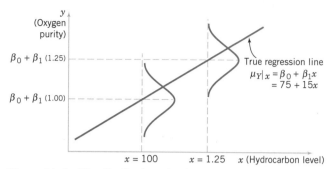

Figure 11-2   The distribution of Y for a given value of x for the oxygen purity-hydrocarbon data.

variability in the observations Y on oxygen purity. Thus, when σ^2 is small, the observed values of Y will fall close to the line, and when σ^2 is large, the observed values of Y may deviate considerably from the line. Because σ^2 is constant, the variability in Y at any value of x is the same.

The regression model describes the relationship between oxygen purity Y and hydrocarbon level x. Thus, for any value of hydrocarbon level, oxygen purity has a normal distribution with mean $75 + 15x$ and variance 2. For example, if $x = 1.25$, Y has mean value $\mu_{Y|x} = 75 + 15(1.25) = 93.75$ and variance 2.

In most real-world problems, the values of the intercept and slope (β_0, β_1) and the error variance σ^2 will not be known, and they must be estimated from sample data. Then this fitted regression equation or model is typically used in prediction of future observations of Y, or for estimating the mean response at a particular level of x. To illustrate, a chemical engineer might be interested in estimating the mean purity of oxygen produced when the hydrocarbon level is $x = 1.25\%$. This chapter discusses such procedures and applications for the simple linear regression model. Chapter 12 will discuss multiple linear regression models that involve more than one regressor.

11-2   SIMPLE LINEAR REGRESSION

The case of **simple linear regression** considers a single **regressor** or **predictor** x and a dependent or **response variable** Y. Suppose that the true relationship between Y and x is a straight line and that the observation Y at each level of x is a random variable. As noted previously, the expected value of Y for each value of x is

$$E(Y|x) = \beta_0 + \beta_1 x$$

where the intercept β_0 and the slope β_1 are unknown regression coefficients. We assume that each observation, Y, can be described by the model

$$Y = \beta_0 + \beta_1 x + \epsilon \tag{11-2}$$

where ϵ is a random error with mean zero and (unknown) variance σ^2. The random errors corresponding to different observations are also assumed to be uncorrelated random variables.

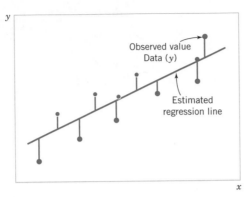

Figure 11-3   Deviations of the data from the estimated regression model.

Suppose that we have n pairs of observations (x_1, y_1), (x_2, y_2), ... (x_n, y_n). Figure 11-3 shows a typical scatter plot of observed data and a candidate for the estimated regression line. The estimates of β_0 and β_1 should result in a line that is (in some sense) a "best fit" to the data. The German scientist Karl Gauss (1777–1855) proposed estimating the parameters β_0 and β_1 in Equation 11-2 to minimize the sum of the squares of the vertical deviations in Fig. 11-3.

We call this criterion for estimating the regression coefficients the **method of least squares.** Using Equation 11-2, we may express the n observations in the sample as

$$y_i = \beta_0 + \beta_1 x_i + \epsilon_i, \qquad i = 1, 2, \ldots, n \tag{11-3}$$

and the sum of the squares of the deviations of the observations from the true regression line is

$$L = \sum_{i=1}^{n} \epsilon_i^2 = \sum_{i=1}^{n} (y_i - \beta_0 - \beta_1 x_i)^2 \tag{11-4}$$

The least squares estimators of β_0 and β_1, say, $\hat{\beta}_0$ and $\hat{\beta}_1$, must satisfy

$$\left. \frac{\partial L}{\partial \beta_0} \right|_{\hat{\beta}_0, \hat{\beta}_1} = -2 \sum_{i=1}^{n} (y_i - \hat{\beta}_0 - \hat{\beta}_1 x_i) = 0$$

$$\left. \frac{\partial L}{\partial \beta_1} \right|_{\hat{\beta}_0, \hat{\beta}_1} = -2 \sum_{i=1}^{n} (y_i - \hat{\beta}_0 - \hat{\beta}_1 x_i) x_i = 0 \tag{11-5}$$

Simplifying these two equations yields

$$n\hat{\beta}_0 + \hat{\beta}_1 \sum_{i=1}^{n} x_i = \sum_{i=1}^{n} y_i$$

$$\hat{\beta}_0 \sum_{i=1}^{n} x_i + \hat{\beta}_1 \sum_{i=1}^{n} x_i^2 = \sum_{i=1}^{n} y_i x_i \tag{11-6}$$

Equations 11-6 are called the **least squares normal equations.** The solution to the normal equations results in the least squares estimators $\hat{\beta}_0$ and $\hat{\beta}_1$.

Definition

> The **least squares estimates** of the intercept and slope in the simple linear regression model are
>
> $$\hat{\beta}_0 = \bar{y} - \hat{\beta}_1\bar{x} \tag{11-7}$$
>
> $$\hat{\beta}_1 = \frac{\displaystyle\sum_{i=1}^{n} y_i x_i - \frac{\left(\displaystyle\sum_{i=1}^{n} y_i\right)\left(\displaystyle\sum_{i=1}^{n} x_i\right)}{n}}{\displaystyle\sum_{i=1}^{n} x_i^2 - \frac{\left(\displaystyle\sum_{i=1}^{n} x_i\right)^2}{n}} \tag{11-8}$$
>
> where $\bar{y} = (1/n)\sum_{i=1}^{n} y_i$ and $\bar{x} = (1/n)\sum_{i=1}^{n} x_i$.

The **fitted** or **estimated regression line** is therefore

$$\hat{y} = \hat{\beta}_0 + \hat{\beta}_1 x \tag{11-9}$$

Note that each pair of observations satisfies the relationship

$$y_i = \hat{\beta}_0 + \hat{\beta}_1 x_i + e_i, \qquad i = 1, 2, \ldots, n$$

where $e_i = y_i - \hat{y}_i$ is called the **residual.** The residual describes the error in the fit of the model to the ith observation y_i. Later in this chapter we will use the residuals to provide information about the adequacy of the fitted model.

Notationally, it is occasionally convenient to give special symbols to the numerator and denominator of Equation 11-8. Given data $(x_1, y_1), (x_2, y_2), \ldots, (x_n, y_n)$, let

$$S_{xx} = \sum_{i=1}^{n}(x_i - \bar{x})^2 = \sum_{i=1}^{n} x_i^2 - \frac{\left(\displaystyle\sum_{i=1}^{n} x_i\right)^2}{n} \tag{11-10}$$

and

$$S_{xy} = \sum_{i=1}^{n} y_i(x_i - \bar{x})^2 = \sum_{i=1}^{n} x_i y_i - \frac{\left(\displaystyle\sum_{i=1}^{n} x_i\right)\left(\displaystyle\sum_{i=1}^{n} y_i\right)}{n} \tag{11-11}$$

EXAMPLE 11-1

We will fit a simple linear regression model to the oxygen purity data in Table 11-1. The following quantities may be computed:

$$n = 20 \quad \sum_{i=1}^{20} x_i = 23.92 \quad \sum_{i=1}^{20} y_i = 1{,}843.21 \quad \bar{x} = 1.1960 \quad \bar{y} = 92.1605$$

$$\sum_{i=1}^{20} y_i^2 = 170{,}044.5321 \quad \sum_{i=1}^{20} x_i^2 = 29.2892 \quad \sum_{i=1}^{20} x_i y_i = 2{,}214.6566$$

$$S_{xx} = \sum_{i=1}^{20} x_i^2 - \frac{\left(\sum_{i=1}^{20} x_i\right)^2}{20} = 29.2892 - \frac{(23.92)^2}{20} = 0.68088$$

and

$$S_{xy} = \sum_{i=1}^{20} x_i y_i - \frac{\left(\sum_{i=1}^{20} x_i\right)\left(\sum_{i=1}^{20} y_i\right)}{20} = 2,214.6566 - \frac{(23.92)(1,843.21)}{20} = 10.17744$$

Therefore, the least squares estimates of the slope and intercept are

$$\hat{\beta}_1 = \frac{S_{xy}}{S_{xx}} = \frac{10.17744}{0.68088} = 14.94748$$

and

$$\hat{\beta}_0 = \bar{y} - \hat{\beta}_1 \bar{x} = 92.1605 - (14.94748)1.196 = 74.28331$$

The fitted simple linear regression model (with the coefficients reported to three decimal places) is

$$\hat{y} = 74.283 + 14.947x$$

This model is plotted in Fig. 11-4, along with the sample data.

Computer software programs are widely used in regression modeling. These programs typically carry more decimal places in the calculations. Table 11-2 shows a portion of the output from Minitab for this problem. The estimates $\hat{\beta}_0$ and $\hat{\beta}_1$ are highlighted. In subsequent sections we will provide explanations for the information provided in this computer output.

Using the regression model of Example 11-1, we would predict oxygen purity of $\hat{y} = 89.23\%$ when the hydrocarbon level is $x = 1.00\%$. The purity 89.23% may be interpreted as

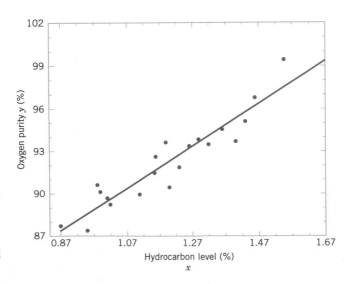

Figure 11-4   Scatter plot of oxygen purity y versus hydrocarbon level x and regression model $\hat{y} = 74.20 + 14.97x$.

Table 11-2   Minitab Output for the Oxygen Purity Data in Example 11-1

Regression Analysis

The regression equation is

Purity = 74.3 + 14.9 HC Level

Predictor	Coef	SE Coef	T	P
Constant	74.283 ←$\hat{\beta}_0$	1.593	46.62	0.000
HC Level	14.947 ←$\hat{\beta}_1$	1.317	11.35	0.000

S = 1.087   R-Sq = 87.7%   R-Sq (adj) = 87.1%

Analysis of Variance

Source	DF	SS	MS	F	P
Regression	1	152.13	152.13	128.86	0.000
Residual Error	18	21.25 ←SS_E	1.18 ←$\hat{\sigma}^2$		
Total	19	173.38			

Predicted Values for New Observations

New Obs	Fit	SE Fit	95.0% CI	95.0% PI
1	89.231	0.354	(88.486, 89.975)	(86.830, 91.632)

Values of Predictors for New Observations

New Obs	HC Level
1	1.00

an estimate of the true population mean purity when $x = 1.00\%$, or as an estimate of a new observation when $x = 1.00\%$. These estimates are, of course, subject to error; that is, it is unlikely that a future observation on purity would be exactly 89.23% when the hydrocarbon level is 1.00%. In subsequent sections we will see how to use confidence intervals and prediction intervals to describe the error in estimation from a regression model.

Estimating σ^2

There is actually another unknown parameter in our regression model, σ^2 (the variance of the error term ϵ). The residuals $e_i = y_i - \hat{y}_i$ are used to obtain an estimate of σ^2. The sum of squares of the residuals, often called the **error sum of squares,** is

$$SS_E = \sum_{i=1}^{n} e_i^2 = \sum_{i=1}^{n} (y_i - \hat{y}_i)^2 \tag{11-12}$$

We can show that the expected value of the error sum of squares is $E(SS_E) = (n-2)\sigma^2$. Therefore an **unbiased estimator** of σ^2 is

$$\hat{\sigma}^2 = \frac{SS_E}{n-2} \tag{11-13}$$

Computing SS_E using Equation 11-12 would be fairly tedious. A more convenient computing formula can be obtained by substituting $\hat{y}_i = \hat{\beta}_0 + \hat{\beta}_1 x_i$ into Equation 11-12 and simplifying.

The resulting computing formula is

$$SS_E = SS_T - \hat{\beta}_1 S_{xy} \qquad (11\text{-}14)$$

where $SS_T = \sum_{i=1}^{n}(\hat{y}_i - \bar{y})^2 = \sum_{i=1}^{n} y_i^2 - n\bar{y}^2$ is the **total sum of squares of the response variable** y. The error sum of squares and the estimate of σ^2 for the oxygen purity data, $\hat{\sigma}^2 = 1.18$, are highlighted in the Minitab output in Table 11-2.

EXERCISES FOR SECTION 11-2

11-1. An article in *Concrete Research* ("Near Surface Characteristics of Concrete: Intrinsic Permeability," Vol. 41, 1989), presented data on compressive strength x and intrinsic permeability y of various concrete mixes and cures. Summary quantities are $n = 14$, $\sum y_i = 572$, $\sum y_i^2 = 23{,}530$, $\sum x_i = 43$, $\sum x_i^2 = 157.42$, and $\sum x_i y_i = 1697.80$. Assume that the two variables are related according to the simple linear regression model.
(a) Calculate the least squares estimates of the slope and intercept.
(b) Use the equation of the fitted line to predict what permeability would be observed when the compressive strength is $x = 4.3$.
(c) Give a point estimate of the mean permeability when compressive strength is $x = 3.7$.
(d) Suppose that the observed value of permeability at $x = 3.7$ is $y = 46.1$. Calculate the value of the corresponding residual.

11-2. Regression methods were used to analyze the data from a study investigating the relationship between roadway surface temperature (x) and pavement deflection (y). Summary quantities were $n = 20$, $\sum y_i = 12.75$, $\sum y_i^2 = 8.86$, $\sum x_i = 1478$, $\sum x_i^2 = 143{,}215.8$, and $\sum x_i y_i = 1083.67$.

(a) Calculate the least squares estimates of the slope and intercept. Graph the regression line.
(b) Use the equation of the fitted line to predict what pavement deflection would be observed when the surface temperature is 85°F.
(c) What is the mean pavement deflection when the surface temperature is 90°F?
(d) What change in mean pavement deflection would be expected for a 1°F change in surface temperature?

11-3. Consider the regression model developed in Exercise 11-2.
(a) Suppose that temperature is measured in °C rather than °F. Write the new regression model that results.
(b) What change in expected pavement deflection is associated with a 1°C change in surface temperature?

11-4. Montgomery, Peck, and Vining (2001) present data concerning the performance of the 28 National Football League teams in 1976. It is suspected that the number of games won (y) is related to the number of yards gained rushing by an opponent (x). The data are shown in the following table.

Teams	Games Won (y)	Yards Rushing by Opponent (x)	Teams	Games Won (y)	Yards Rushing by Opponent (x)
Washington	10	2205	Detroit	6	1901
Minnesota	11	2096	Green Bay	5	2288
New England	11	1847	Houston	5	2072
Oakland	13	1903	Kansas City	5	2861
Pittsburgh	10	1457	Miami	6	2411
Baltimore	11	1848	New Orleans	4	2289
Los Angeles	10	1564	New York Giants	3	2203
Dallas	11	1821	New York Jets	3	2592
Atlanta	4	2577	Philadelphia	4	2053
Buffalo	2	2476	St. Louis	10	1979
Chicago	7	1984	San Diego	6	2048
Cincinnati	10	1917	San Francisco	8	1786
Cleveland	9	1761	Seattle	2	2876
Denver	9	1709	Tampa Bay	0	2560

(a) Calculate the least squares estimates of the slope and intercept. What is the estimate of σ^2? Graph the regression model.

(b) Find an estimate of the mean number of games won if the opponents can be limited to 1800 yards rushing.

(c) What change in the expected number of games won is associated with a decrease of 100 yards rushing by an opponent?

(d) To increase by 1 the mean number of games won, how much decrease in rushing yards must be generated by the defense?

(e) Given that $x = 1917$ yards (Cincinnati), find the fitted value of y and the corresponding residual.

11-5. An article in *Technometrics* by S. C. Narula and J. F. Wellington ("Prediction, Linear Regression, and a Minimum Sum of Relative Errors," Vol. 19, 1977) presents data on the selling price and annual taxes for 24 houses. The data are shown in the following table.

(a) Assuming that a simple linear regression model is appropriate, obtain the least squares fit relating selling price to taxes paid. What is the estimate of σ^2?

Sale Price/1000	Taxes (Local, School), County)/1000	Sale Price/1000	Taxes (Local, School), County)/1000
25.9	4.9176	30.0	5.0500
29.5	5.0208	36.9	8.2464
27.9	4.5429	41.9	6.6969
25.9	4.5573	40.5	7.7841
29.9	5.0597	43.9	9.0384
29.9	3.8910	37.5	5.9894
30.9	5.8980	37.9	7.5422
28.9	5.6039	44.5	8.7951
35.9	5.8282	37.9	6.0831
31.5	5.3003	38.9	8.3607
31.0	6.2712	36.9	8.1400
30.9	5.9592	45.8	9.1416

(b) Find the mean selling price given that the taxes paid are $x = 7.50$.

(c) Calculate the fitted value of y corresponding to $x = 5.8980$. Find the corresponding residual.

(d) Calculate the fitted $\hat{y}_i$ for each value of x_i used to fit the model. Then construct a graph of $\hat{y}_i$ versus the corresponding observed value y_i and comment on what this plot would look like if the relationship between y and x was a deterministic (no random error) straight line. Does the plot actually obtained indicate that taxes paid is an effective regressor variable in predicting selling price?

11-6. The number of pounds of steam used per month by a chemical plant is thought to be related to the average ambient temperature (in° F) for that month. The past year's usage and temperature are shown in the following table:

Month	Temp.	Usage/1000	Month	Temp.	Usage/1000
Jan.	21	185.79	July	68	621.55
Feb.	24	214.47	Aug.	74	675.06
Mar.	32	288.03	Sept.	62	562.03
Apr.	47	424.84	Oct.	50	452.93
May	50	454.58	Nov.	41	369.95
June	59	539.03	Dec.	30	273.98

(a) Assuming that a simple linear regression model is appropriate, fit the regression model relating steam usage (y) to the average temperature (x). What is the estimate of σ^2?

(b) What is the estimate of expected steam usage when the average temperature is 55°F?

(c) What change in mean steam usage is expected when the monthly average temperature changes by 1°F?

(d) Suppose the monthly average temperature is 47°F. Calculate the fitted value of y and the corresponding residual.

11-7. The data shown in the following table are highway gasoline mileage performance and engine displacement for a sample of 20 cars.

Make	Model	MPG (highway)	Engine Displacement (in³)	Make	Model	MPG (highway)	Engine Displacement (in³)
Acura	Legend	30	97	Ford	Taurus	27	153
BMW	735i	19	209	Ford	Tempo	33	90
Buick	Regal	29	173	Honda	Accord	30	119
Chevrolet	Cavalier	32	121	Mazda	RX-7	23	80
Chevrolet	Celebrity	30	151	Mercedes	260E	24	159
Chrysler	Conquest	24	156	Mercury	Tracer	29	97
Dodge	Aries	30	135	Nissan	Maxima	26	181
Dodge	Dynasty	28	181	Oldsmobile	Cutlass	29	173
Ford	Escort	31	114	Plymouth	Laser	37	122
Ford	Mustang	25	302	Pontiac	Grand Prix	29	173

(a) Fit a simple linear model relating highway miles per gallon (y) to engine displacement (x) using least squares.

(b) Find an estimate of the mean highway gasoline mileage performance for a car with 150 cubic inches engine displacement.

(c) Obtain the fitted value of y and the corresponding residual for a car, the Ford Escort, with engine displacement of 114 cubic inches.

11-8. An article in the *Tappi Journal* (March, 1986) presented data on green liquor Na_2S concentration (in grams per liter) and paper machine production (in tons per day). The data (read from a graph) are shown as follows:

y	40	42	49	46	44	48
x	825	830	890	895	890	910

y	46	43	53	52	54	57	58
x	915	960	990	1010	1012	1030	1050

(a) Fit a simple linear regression model with y = green liquor Na_2S concentration and x = production. Find an estimate of σ^2. Draw a scatter diagram of the data and the resulting least squares fitted model.

(b) Find the fitted value of y corresponding to $x = 910$ and the associated residual.

(c) Find the mean green liquor Na_2S concentration when the production rate is 950 tons per day.

11-9. An article in the *Journal of Sound and Vibration* (Vol. 151, 1991, pp. 383–394) described a study investigating the relationship between noise exposure and hypertension. The following data are representative of those reported in the article.

y	1	0	1	2	5	1	4	6	2	3
x	60	63	65	70	70	70	80	90	80	80

y	5	4	6	8	4	5	7	9	7	6
x	85	89	90	90	90	90	94	100	100	100

(a) Draw a scatter diagram of y (blood pressure rise in millimeters of mercury) versus x (sound pressure level in decibels). Does a simple linear regression model seem reasonable in this situation?

(b) Fit the simple linear regression model using least squares. Find an estimate of σ^2.

(c) Find the predicted mean rise in blood pressure level associated with a sound pressure level of 85 decibals.

11-10. An article in *Wear* (Vol. 152, 1992, pp. 171–181) presents data on the fretting wear of mild steel and oil viscosity. Representative data follow, with x = oil viscosity and y = wear volume (10^{-4} cubic millimeters).

y	240	181	193	155	172
x	1.6	9.4	15.5	20.0	22.0

y	110	113	75	94
x	35.5	43.0	40.5	33.0

(a) Construct a scatter plot of the data. Does a simple linear regression model appear to be plausible?

(b) Fit the simple linear regression model using least squares. Find an estimate of σ^2.

(c) Predict fretting wear when viscosity $x = 30$.

(d) Obtain the fitted value of y when $x = 22.0$ and calculate the corresponding residual.

11-11. An article in the *Journal of Environmental Engineering* (Vol. 115, No. 3, 1989, pp. 608–619) reported the results of a study on the occurrence of sodium and chloride in surface streams in central Rhode Island. The following data are chloride concentration y (in milligrams per liter) and roadway area in the watershed x (in percentage).

y	4.4	6.6	9.7	10.6	10.8	10.9
x	0.19	0.15	0.57	0.70	0.67	0.63

y	11.8	12.1	14.3	14.7	15.0	17.3
x	0.47	0.70	0.60	0.78	0.81	0.78

y	19.2	23.1	27.4	27.7	31.8	39.5
x	0.69	1.30	1.05	1.06	1.74	1.62

(a) Draw a scatter diagram of the data. Does a simple linear regression model seem appropriate here?

(b) Fit the simple linear regression model using the method of least squares. Find an estimate of σ^2.

(c) Estimate the mean chloride concentration for a watershed that has 1% roadway area.

(d) Find the fitted value corresponding to $x = 0.47$ and the associated residual.

11-12. A rocket motor is manufactured by bonding together two types of propellants, an igniter and a sustainer. The shear strength of the bond y is thought to be a linear function of the age of the propellant x when the motor is cast. Twenty observations are shown in the table on the next page.

(a) Draw a scatter diagram of the data. Does the straight-line regression model seem to be plausible?

(b) Find the least squares estimates of the slope and intercept in the simple linear regression model. Find an estimate of σ^2.

(c) Estimate the mean shear strength of a motor made from propellant that is 20 weeks old.

(d) Obtain the fitted values $\hat{y}_i$ that correspond to each observed value y_i. Plot $\hat{y}_i$ versus y_i, and comment on what this plot would look like if the linear relationship between

Observation Number	Strength y (psi)	Age x (weeks)	Observation Number	Strength y (psi)	Age x (weeks)
1	2158.70	15.50	11	2165.20	13.00
2	1678.15	23.75	12	2399.55	3.75
3	2316.00	8.00	13	1779.80	25.00
4	2061.30	17.00	14	2336.75	9.75
5	2207.50	5.00	15	1765.30	22.00
6	1708.30	19.00	16	2053.50	18.00
7	1784.70	24.00	17	2414.40	6.00
8	2575.00	2.50	18	2200.50	12.50
9	2357.90	7.50	19	2654.20	2.00
10	2277.70	11.00	20	1753.70	21.50

shear strength and age were perfectly deterministic (no error). Does this plot indicate that age is a reasonable choice of regressor variable in this model?

11-13. Show that in a simple linear regression model the point $(\bar{x}, \bar{y})$ lies exactly on the least squares regression line.

11-14. Consider the simple linear regression model $Y = \beta_0 + \beta_1 x + \epsilon$. Suppose that the analyst wants to use $z = x - \bar{x}$ as the regressor variable.

(a) Using the data in Exercise 11-12, construct one scatter plot of the (x_i, y_i) points and then another of the $(z_i = x_i - \bar{x}, y_i)$ points. Use the two plots to intuitively explain how the two models, $Y = \beta_0 + \beta_1 x + \epsilon$ and $Y = \beta_0^* + \beta_1^* z + \epsilon$, are related.

(b) Find the least squares estimates of β_0^* and β_1^* in the model $Y = \beta_0^* + \beta_1^* z + \epsilon$. How do they relate to the least squares estimates $\hat{\beta}_0$ and $\hat{\beta}_1$?

11-15. Suppose we wish to fit the model $y_i^* = \beta_0^* + \beta_1^*(x_i - \bar{x}) + \epsilon_i$, where $y_i^* = y_i - \bar{y}$ $(i = 1, 2, \ldots, n)$. Find the least squares estimates of β_0^* and β_1^*. How do they relate to $\hat{\beta}_0$ and $\hat{\beta}_1$?

11-16. Suppose we wish to fit a regression model for which the true regression line passes through the point $(0, 0)$. The appropriate model is $Y = \beta x + \epsilon$. Assume that we have n pairs of data $(x_1, y_1), (x_2, y_2), \ldots, (x_n, y_n)$. Find the least squares estimate of β.

11-17. Using the results of Exercise 11-16, fit the model $Y = \beta x + \epsilon$ to the chloride concentration-roadway area data in Exercise 11-11. Plot the fitted model on a scatter diagram of the data and comment on the appropriateness of the model.

11-3 PROPERTIES OF THE LEAST SQUARES ESTIMATORS

The statistical properties of the least squares estimators $\hat{\beta}_0$ and $\hat{\beta}_1$ may be easily described. Recall that we have assumed that the error term ϵ in the model $Y = \beta_0 + \beta_1 x + \epsilon$ is a random variable with mean zero and variance σ^2. Since the values of x are fixed, Y is a random variable with mean $\mu_{Y|x} = \beta_0 + \beta_1 x$ and variance σ^2. Therefore, the values of $\hat{\beta}_0$ and $\hat{\beta}_1$ depend on the observed y's; thus, the least squares estimators of the regression coefficients may be viewed as random variables. We will investigate the bias and variance properties of the least squares estimators $\hat{\beta}_0$ and $\hat{\beta}_1$.

Consider first $\hat{\beta}_1$. Because $\hat{\beta}_1$ is a linear combination of the observations Y_i, we can use properties of expectation to show that expected value of $\hat{\beta}_1$ is

$$E(\hat{\beta}_1) = \beta_1 \qquad (11\text{-}15)$$

Thus, $\hat{\beta}_1$ is an **unbiased estimator** of the true slope β_1.

Now consider the variance of $\hat{\beta}_1$. Since we have assumed that $V(\epsilon_i) = \sigma^2$, it follows that $V(Y_i) = \sigma^2$, and it can be shown that

$$V(\hat{\beta}_1) = \frac{\sigma^2}{S_{xx}} \tag{11-16}$$

For the intercept, we can show that

$$E(\hat{\beta}_0) = \beta_0 \quad \text{and} \quad V(\hat{\beta}_0) = \sigma^2 \left[\frac{1}{n} + \frac{\bar{x}^2}{S_{xx}} \right] \tag{11-17}$$

Thus, $\hat{\beta}_0$ is an unbiased estimator of the intercept β_0. The covariance of the random variables $\hat{\beta}_0$ and $\hat{\beta}_1$ is not zero. It can be shown (see Exercise 11-69) that $\text{cov}(\hat{\beta}_0, \hat{\beta}_1) = -\sigma^2 \bar{x}/S_{xx}$.

The estimate of σ^2 could be used in Equations 11-16 and 11-17 to provide estimates of the variance of the slope and the intercept. We call the square roots of the resulting variance estimators the **estimated standard errors** of the slope and intercept, respectively.

Definition

> In simple linear regression the **estimated standard error of the slope** and the **estimated standard error of the intercept** are
>
> $$se(\hat{\beta}_1) = \sqrt{\frac{\hat{\sigma}^2}{S_{xx}}} \quad \text{and} \quad se(\hat{\beta}_0) = \sqrt{\hat{\sigma}^2 \left[\frac{1}{n} + \frac{\bar{x}^2}{S_{xx}} \right]}$$
>
> respectively, where $\hat{\sigma}^2$ is computed from Equation 11-13.

The Minitab computer output in Table 11-2 reports the estimated standard errors of the slope and intercept under the column heading "*SE* coeff."

11-4 SOME COMMENTS ON USES OF REGRESSION (CD ONLY)

11-5 HYPOTHESIS TESTS IN SIMPLE LINEAR REGRESSION

An important part of assessing the adequacy of a linear regression model is testing statistical hypotheses about the model parameters and constructing certain confidence intervals. Hypothesis testing in simple linear regression is discussed in this section, and Section 11-6 presents methods for constructing confidence intervals. To test hypotheses about the slope and intercept of the regression model, we must make the additional assumption that the error component in the model, ϵ, is normally distributed. Thus, the complete assumptions are that the errors are normally and independently distributed with mean zero and variance σ^2, abbreviated NID$(0, \sigma^2)$.

11-5.1 Use of *t*-Tests

Suppose we wish to test the hypothesis that the slope equals a constant, say, $\beta_{1,0}$. The appropriate hypotheses are

$$H_0: \beta_1 = \beta_{1,0}$$
$$H_1: \beta_1 \neq \beta_{1,0} \tag{11-18}$$

where we have assumed a two-sided alternative. Since the errors ϵ_i are NID($0, \sigma^2$), it follows directly that the observations Y_i are NID($\beta_0 + \beta_1 x_i, \sigma^2$). Now $\hat{\beta}_1$ is a linear combination of independent normal random variables, and consequently, $\hat{\beta}_1$ is $N(\beta_1, \sigma^2/S_{xx})$, using the bias and variance properties of the slope discussed in Section 11-3. In addition, $(n-2)\hat{\sigma}^2/\sigma^2$ has a chi-square distribution with $n - 2$ degrees of freedom, and $\hat{\beta}_1$ is independent of $\hat{\sigma}^2$. As a result of those properties, the statistic

$$T_0 = \frac{\hat{\beta}_1 - \beta_{1,0}}{\sqrt{\hat{\sigma}^2/S_{xx}}} \tag{11-19}$$

follows the *t* distribution with $n - 2$ degrees of freedom under $H_0: \beta_1 = \beta_{1,0}$. We would reject $H_0: \beta_1 = \beta_{1,0}$ if

$$|t_0| > t_{\alpha/2,n-2} \tag{11-20}$$

where t_0 is computed from Equation 11-19. The denominator of Equation 11-19 is the standard error of the slope, so we could write the test statistic as

$$T_0 = \frac{\hat{\beta}_1 - \beta_{1,0}}{se(\hat{\beta}_1)}$$

A similar procedure can be used to test hypotheses about the intercept. To test

$$H_0: \beta_0 = \beta_{0,0}$$
$$H_1: \beta_0 \neq \beta_{0,0} \tag{11-21}$$

we would use the statistic

$$T_0 = \frac{\hat{\beta}_0 - \beta_{0,0}}{\sqrt{\hat{\sigma}^2\left[\dfrac{1}{n} + \dfrac{\bar{x}^2}{S_{xx}}\right]}} = \frac{\hat{\beta}_0 - \beta_{0,0}}{se(\hat{\beta}_0)} \tag{11-22}$$

and reject the null hypothesis if the computed value of this test statistic, t_0, is such that $|t_0| > t_{\alpha/2,n-2}$. Note that the denominator of the test statistic in Equation 11-22 is just the standard error of the intercept.

A very important special case of the hypotheses of Equation 11-18 is

$$H_0: \beta_1 = 0$$
$$H_1: \beta_1 \neq 0 \tag{11-23}$$

These hypotheses relate to the **significance of regression.** Failure to reject $H_0: \beta_1 = 0$ is equivalent to concluding that there is no linear relationship between x and Y. This situation is illustrated in Fig. 11-5. Note that this may imply either that x is of little value in explaining the variation in Y and that the best estimator of Y for any x is $\hat{y} = \bar{Y}$ (Fig. 11-5*a*) or that the true relationship between x and Y is not linear (Fig. 11-5*b*). Alternatively, if $H_0: \beta_1 = 0$ is rejected, this implies that x is of value in explaining the variability in Y (see Fig. 11-6). Rejecting $H_0: \beta_1 = 0$ could mean either that the straight-line model is adequate (Fig. 11-6*a*) or that,

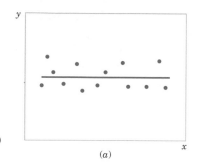

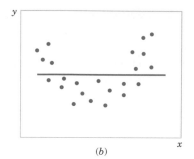

Figure 11-5   The
hypothesis H_0: $\beta_1 = 0$
is not rejected.

(a)   (b)

although there is a linear effect of x, better results could be obtained with the addition of higher order polynomial terms in x (Fig. 11-6b).

EXAMPLE 11-2   We will test for significance of regression using the model for the oxygen purity data from Example 11-1. The hypotheses are

$$H_0: \beta_1 = 0$$
$$H_1: \beta_1 \neq 0$$

and we will use $\alpha = 0.01$. From Example 11-1 and Table 11-2 we have

$$\hat{\beta}_1 = 14.97 \quad n = 20, \quad S_{xx} = 0.68088, \quad \hat{\sigma}^2 = 1.18$$

so the t-statistic in Equation 10-20 becomes

$$t_0 = \frac{\hat{\beta}_1}{\sqrt{\hat{\sigma}^2/S_{xx}}} = \frac{\hat{\beta}_1}{se(\hat{\beta}_1)} = \frac{14.947}{\sqrt{1.18/0.68088}} = 11.35$$

Since the reference value of t is $t_{0.005,18} = 2.88$, the value of the test statistic is very far into the critical region, implying that H_0: $\beta_1 = 0$ should be rejected. The P-value for this test is $P \simeq 1.23 \times 10^{-9}$. This was obtained manually with a calculator.

Table 11-2 presents the Minitab output for this problem. Notice that the t-statistic value for the slope is computed as 11.35 and that the reported P-value is $P = 0.000$. Minitab also reports the t-statistic for testing the hypothesis H_0: $\beta_0 = 0$. This statistic is computed from Equation 11-22, with $\beta_{0,0} = 0$, as $t_0 = 46.62$. Clearly, then, the hypothesis that the intercept is zero is rejected.

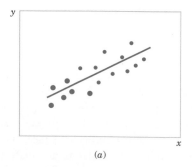

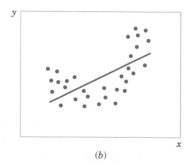

Figure 11-6   The
hypothesis H_0: $\beta_1 = 0$
is rejected.

(a)   (b)

11-5.2 Analysis of Variance Approach to Test Significance of Regression

A method called the **analysis of variance** can be used to test for significance of regression. The procedure partitions the total variability in the response variable into meaningful components as the basis for the test. The **analysis of variance identity** is as follows:

$$\sum_{i=1}^{n} (y_i - \bar{y})^2 = \sum_{i=1}^{n} (\hat{y}_i - \bar{y})^2 + \sum_{i=1}^{n} (y_i - \hat{y}_i)^2 \qquad (11\text{-}24)$$

The two components on the right-hand-side of Equation 11-24 measure, respectively, the amount of variability in y_i accounted for by the regression line and the residual variation left unexplained by the regression line. We usually call $SS_E = \sum_{i=1}^{n} (y_i - \hat{y}_i)^2$ the **error sum of squares** and $SS_R = \sum_{i=1}^{n} (\hat{y}_i - \bar{y})^2$ the **regression sum of squares.** Symbolically, Equation 11-24 may be written as

$$SS_T = SS_R + SS_E \qquad (11\text{-}25)$$

where $SS_T = \sum_{i=1}^{n} (y_i - \bar{y})^2$ is the **total corrected sum of squares** of y. In Section 11-2 we noted that $SS_E = SS_T - \hat{\beta}_1 S_{xy}$ (see Equation 11-14), so since $SS_T = \hat{\beta}_1 S_{xy} + SS_E$, we note that the regression sum of squares in Equation 10-26 is $SS_R = \hat{\beta}_1 S_{xy}$. The total sum of squares SS_T has $n - 1$ degrees of freedom, and SS_R and SS_E have 1 and $n - 2$ degrees of freedom, respectively.

We may show that $E[SS_E/(n - 2)] = \sigma^2$, $E(SS_R) = \sigma^2 + \beta_1^2 S_{xx}$ and that SS_E/σ^2 and SS_R/σ^2 are independent chi-square random variables with $n - 2$ and 1 degrees of freedom, respectively. Thus, if the null hypothesis H_0: $\beta_1 = 0$ is true, the statistic

$$F_0 = \frac{SS_R/1}{SS_E/(n - 2)} = \frac{MS_R}{MS_E} \qquad (11\text{-}26)$$

follows the $F_{1,n-2}$ distribution, and we would reject H_0 if $f_0 > f_{\alpha,1,n-2}$. The quantities $MS_R = SS_R/1$ and $MS_E = SS_E/(n - 2)$ are called **mean squares.** In general, a mean square is always computed by dividing a sum of squares by its number of degrees of freedom. The test procedure is usually arranged in an **analysis of variance table,** such as Table 11-3.

Table 11-3 Analysis of Variance for Testing Significance of Regression

Source of Variation	Sum of Squares	Degrees of Freedom	Mean Square	F_0
Regression	$SS_R = \hat{\beta}_1 S_{xy}$	1	MS_R	MS_R/MS_E
Error	$SS_E = SS_T - \hat{\beta}_1 S_{xy}$	$n - 2$	MS_E	
Total	SS_T	$n - 1$		

Note that $MS_E = \hat{\sigma}^2$.

EXAMPLE 11-3 | We will use the analysis of variance approach to test for significance of regression using the oxygen purity data model from Example 11-1. Recall that $SS_T = 173.38$, $\hat{\beta}_1 = 14.947$, $S_{xy} = 10.17744$, and $n = 20$. The regression sum of squares is

$$SS_R = \hat{\beta}_1 S_{xy} = (14.947)10.17744 = 152.13$$

and the error sum of squares is

$$SS_E = SS_T - SS_R = 173.38 - 152.13 = 21.25$$

The analysis of variance for testing $H_0: \beta_1 = 0$ is summarized in the Minitab output in Table 11-2. The test statistic is $f_0 = MS_R/MS_E = 152.13/1.18 = 128.86$, for which we find that the P-value is $P \simeq 1.23 \times 10^{-9}$, so we conclude that β_1 is not zero.

There are frequently minor differences in terminology among computer packages. For example, sometimes the regression sum of squares is called the "model" sum of squares, and the error sum of squares is called the "residual" sum of squares.

Note that the analysis of variance procedure for testing for significance of regression is equivalent to the t-test in Section 11-5.1. That is, either procedure will lead to the same conclusions. This is easy to demonstrate by starting with the t-test statistic in Equation 11-19 with $\beta_{1,0} = 0$, say

$$T_0 = \frac{\hat{\beta}_1}{\sqrt{\hat{\sigma}^2/S_{xx}}} \tag{11-27}$$

Squaring both sides of Equation 11-27 and using the fact that $\hat{\sigma}^2 = MS_E$ results in

$$T_0^2 = \frac{\hat{\beta}_1^2 S_{xx}}{MS_E} = \frac{\hat{\beta}_1 S_{xY}}{MS_E} = \frac{MS_R}{MS_E} \tag{11-28}$$

Note that T_0^2 in Equation 11-28 is identical to F_0 in Equation 11-26 It is true, in general, that the square of a t random variable with v degrees of freedom is an F random variable, with one and v degrees of freedom in the numerator and denominator, respectively. Thus, the test using T_0 is equivalent to the test based on F_0. Note, however, that the t-test is somewhat more flexible in that it would allow testing against a one-sided alternative hypothesis, while the F-test is restricted to a two-sided alternative.

EXERCISES FOR SECTION 11-5

11-18.   Consider the data from Exercise 11-1 on $x =$ compressive strength and $y =$ intrinsic permeability of concrete.
(a) Test for significance of regression using $\alpha = 0.05$. Find the P-value for this test. Can you conclude that the model specifies a useful linear relationship between these two variables?
(b) Estimate σ^2 and the standard deviation of $\hat{\beta}_1$.
(c) What is the standard error of the intercept in this model?

11-19.   Consider the data from Exercise 11-2 on $x =$ roadway surface temperature and $y =$ pavement deflection.

(a) Test for significance of regression using $\alpha = 0.05$. Find the P-value for this test. What conclusions can you draw?
(b) Estimate the standard errors of the slope and intercept.

11-20.   Consider the National Football League data in Exercise 11-4.
(a) Test for significance of regression using $\alpha = 0.01$. Find the P-value for this test. What conclusions can you draw?
(b) Estimate the standard errors of the slope and intercept.
(c) Test (using $\alpha = 0.01$) $H_0: \beta_1 = -0.01$ versus $H_1: \beta_1 \neq -0.01$. Would you agree with the statement that this is a test

of the claim that if you can decrease the opponent's rushing yardage by 100 yards the team will win one more game?

11-21. Consider the data from Exercise 11-5 on y = sales price and x = taxes paid.
(a) Test $H_0: \beta_1 = 0$ using the t-test; use $\alpha = 0.05$.
(b) Test $H_0: \beta_1 = 0$ using the analysis of variance with $\alpha = 0.05$. Discuss the relationship of this test to the test from part (a).
(c) Estimate the standard errors of the slope and intercept.
(d) Test the hypothesis that $\beta_0 = 0$.

11-22. Consider the data from Exercise 11-6 on y = steam usage and x = average temperature.
(a) Test for significance of regression using $\alpha = 0.01$. What is the P-value for this test? State the conclusions that result from this test.
(b) Estimate the standard errors of the slope and intercept.
(c) Test the hypothesis $H_0: \beta_1 = 10$ versus $H_1: \beta_1 \neq 10$ using $\alpha = 0.01$. Find the P-value for this test.
(d) Test $H_0: \beta_0 = 0$ versus $H_1: \beta_0 \neq 0$ using $\alpha = 0.01$. Find the P-value for this test and draw conclusions.

11-23. Exercise 11-7 gave 20 observations on y = highway gasoline mileage and x = engine displacement.
(a) Test for significance of regression using $\alpha = 0.01$. Find the P-value for this test. What conclusions can you reach?
(b) Estimate the standard errors of the slope and intercept.
(c) Test $H_0: \beta_1 = -0.05$ versus $H_1: \beta_1 < -0.05$ using $\alpha = 0.01$ and draw conclusions. What is the P-value for this test?
(d) Test the hypothesis $H_0: \beta_0 = 0$ versus $H_1: \beta_0 \neq 0$ using $\alpha = 0.01$. What is the P-value for this test?

11-24. Exercise 11-8 gave 13 observations on y = green liquor Na_2S concentration and x = production in a paper mill.
(a) Test for significance of regression using $\alpha = 0.05$. Find the P-value for this test.
(b) Estimate the standard errors of the slope and intercept.
(c) Test $H_0: \beta_0 = 0$ versus $H_1: \beta_0 \neq 0$ using $\alpha = 0.05$. What is the P-value for this test?

11-25. Exercise 11-9 presented data on y = blood pressure rise and x = sound pressure level.
(a) Test for significance of regression using $\alpha = 0.05$. What is the P-value for this test?
(b) Estimate the standard errors of the slope and intercept.
(c) Test $H_0: \beta_0 = 0$ versus $H_1: \beta_0 \neq 0$ using $\alpha = 0.05$. Find the P-value for this test.

11-26. Exercise 11-11 presented data on y = chloride concentration in surface streams and x = roadway area.
(a) Test the hypothesis $H_0: \beta_1 = 0$ versus $H_1: \beta_1 \neq 0$ using the analysis of variance procedure with $\alpha = 0.01$.
(b) Find the P-value for the test in part (a).
(c) Estimate the standard errors of $\hat{\beta}_1$ and $\hat{\beta}_0$.
(d) Test $H_0: \beta_0 = 0$ versus $H_1: \beta_0 \neq 0$ using $\alpha = 0.01$. What conclusions can you draw? Does it seem that the model might be a better fit to the data if the intercept were removed?

11-27. Refer to Exercise 11-12, which gives 20 observations on y = shear strength of a propellant and x = propellant age.
(a) Test for significance of regression with $\alpha = 0.01$. Find the P-value for this test.
(b) Estimate the standard errors of $\hat{\beta}_0$ and $\hat{\beta}_1$.
(c) Test $H_0: \beta_1 = -30$ versus $H_1: \beta_1 \neq -30$ using $\alpha = 0.01$. What is the P-value for this test?
(d) Test $H_0: \beta_0 = 0$ versus $H_1: \beta_0 \neq 0$ using $\alpha = 0.01$. What is the P-value for this test?
(e) Test $H_0: \beta_0 = 2500$ versus $H_1: \beta_0 > 2500$ using $\alpha = 0.01$. What is the P-value for this test?

11-28. Suppose that each value of x_i is multiplied by a positive constant a, and each value of y_i is multiplied by another positive constant b. Show that the t-statistic for testing $H_0: \beta_1 = 0$ versus $H_1: \beta_1 \neq 0$ is unchanged in value.

11-29. Consider the no-intercept model $Y = \beta x + \epsilon$ with the ϵ's NID$(0, \sigma^2)$. The estimate of σ^2 is $s^2 = \sum_{i=1}^{n} (y_i - \hat{\beta} x_i)^2 / (n - 1)$ and $V(\hat{\beta}) = \sigma^2 / \sum_{i=1}^{n} x_i^2$.
(a) Devise a test statistic for $H_0: \beta = 0$ versus $H_1: \beta \neq 0$.
(b) Apply the test in (a) to the model from Exercise 11-17.

11-30. The type II error probability for the t-test for $H_0: \beta_1 = \beta_{1,0}$ can be computed in a similar manner to the t-tests of Chapter 9. If the true value of β_1 is β_1', the value $d = |\beta_{1,0} - \beta_1'|/(\sigma \sqrt{(n-1)/S_{xx}}$ is calculated and used as the horizontal scale factor on the operating characteristic curves for the t-test, (Appendix Charts VIe through VIh) and the type II error probability is read from the vertical scale using the curve for $n - 2$ degrees of freedom. Apply this procedure to the football data of Exercise 11-4, using $\sigma = 2.4$ and $\beta_1' = -0.005$, where the hypotheses are $H_0: \beta_1 = -0.01$ versus $H_1: \beta_1 \neq -0.01$.

11-6 CONFIDENCE INTERVALS

11-6.1 Confidence Intervals on the Slope and Intercept

In addition to point estimates of the slope and intercept, it is possible to obtain **confidence interval** estimates of these parameters. The width of these confidence intervals is a measure of the overall quality of the regression line. If the error terms, ϵ_i, in the regression model are normally and independently distributed,

$$(\hat{\beta}_1 - \beta_1)/\sqrt{\hat{\sigma}^2/S_{xx}} \quad \text{and} \quad (\hat{\beta}_0 - \beta_0)/\sqrt{\hat{\sigma}^2\left[\frac{1}{n} + \frac{\bar{x}^2}{S_{xx}}\right]}$$

are both distributed as t random variables with $n - 2$ degrees of freedom. This leads to the following definition of $100(1 - \alpha)\%$ confidence intervals on the slope and intercept.

Definition

> Under the assumption that the observations are normally and independently distributed, a $100(1 - \alpha)\%$ **confidence interval on the slope** β_1 in simple linear regression is
>
> $$\hat{\beta}_1 - t_{\alpha/2, n-2}\sqrt{\frac{\hat{\sigma}^2}{S_{xx}}} \leq \beta_1 \leq \hat{\beta}_1 + t_{\alpha/2, n-2}\sqrt{\frac{\hat{\sigma}^2}{S_{xx}}} \tag{11-29}$$
>
> Similarly, a $100(1 - \alpha)\%$ **confidence interval on the intercept** β_0 is
>
> $$\hat{\beta}_0 - t_{\alpha/2, n-2}\sqrt{\hat{\sigma}^2\left[\frac{1}{n} + \frac{\bar{x}^2}{S_{xx}}\right]}$$
> $$\leq \beta_0 \leq \hat{\beta}_0 + t_{\alpha/2, n-2}\sqrt{\hat{\sigma}^2\left[\frac{1}{n} + \frac{\bar{x}^2}{S_{xx}}\right]} \tag{11-30}$$

EXAMPLE 11-4

We will find a 95% confidence interval on the slope of the regression line using the data in Example 11-1. Recall that $\hat{\beta}_1 = 14.947$, $S_{xx} = 0.68088$, and $\hat{\sigma}^2 = 1.18$ (see Table 11-2). Then, from Equation 10-31 we find

$$\hat{\beta}_1 - t_{0.025, 18}\sqrt{\frac{\hat{\sigma}^2}{S_{xx}}} \leq \beta_1 \leq \hat{\beta}_1 + t_{0.025, 18}\sqrt{\frac{\hat{\sigma}^2}{S_{xx}}}$$

or

$$14.947 - 2.101\sqrt{\frac{1.18}{0.68088}} \leq \beta_1 \leq 14.947 + 2.101\sqrt{\frac{1.18}{0.68088}}$$

This simplifies to

$$12.197 \leq \beta_1 \leq 17.697$$

11-6.2 Confidence Interval on the Mean Response

A confidence interval may be constructed on the mean response at a specified value of x, say, x_0. This is a confidence interval about $E(Y|x_0) = \mu_{Y|x_0}$ and is often called a confidence interval about the regression line. Since $E(Y|x_0) = \mu_{Y|x_0} = \beta_0 + \beta_1 x_0$, we may obtain a point estimate of the mean of Y at $x = x_0(\mu_{Y|x_0})$ from the fitted model as

$$\hat{\mu}_{Y|x_0} = \hat{\beta}_0 + \hat{\beta}_1 x_0$$

Now $\hat{\mu}_{Y|x_0}$ is an unbiased point estimator of $\mu_{Y|x_0}$, since $\hat{\beta}_0$ and $\hat{\beta}_1$ are unbiased estimators of β_0 and β_1. The variance of $\hat{\mu}_{Y|x_0}$ is

$$V(\hat{\mu}_{Y|x_0}) = \sigma^2 \left[\frac{1}{n} + \frac{(x_0 - \bar{x})^2}{S_{xx}} \right]$$

This last result follows from the fact that cov $(\bar{Y}, \hat{\beta}_1) = 0$ (Refer to Exercise 11-71). Also, $\hat{\mu}_{Y|x_0}$ is normally distributed, because $\hat{\beta}_1$ and $\hat{\beta}_0$ are normally distributed, and if we $\hat{\sigma}^2$ use as an estimate of σ^2, it is easy to show that

$$\frac{\hat{\mu}_{Y|x_0} - \mu_{Y|x_0}}{\sqrt{\hat{\sigma}^2 \left[\frac{1}{n} + \frac{(x_0 - \bar{x})^2}{S_{xx}} \right]}}$$

has a t distribution with $n - 2$ degrees of freedom. This leads to the following confidence interval definition.

Definition

> A $100(1 - \alpha)\%$ **confidence interval about the mean response** at the value of $x = x_0$, say $\mu_{Y|x_0}$, is given by
>
> $$\hat{\mu}_{Y|x_0} - t_{\alpha/2, n-2} \sqrt{\hat{\sigma}^2 \left[\frac{1}{n} + \frac{(x_0 - \bar{x})^2}{S_{xx}} \right]}$$
>
> $$\leq \mu_{Y|x_0} \leq \hat{\mu}_{Y|x_0} + t_{\alpha/2, n-2} \sqrt{\hat{\sigma}^2 \left[\frac{1}{n} + \frac{(x_0 - \bar{x})^2}{S_{xx}} \right]} \qquad (11\text{-}31)$$
>
> where $\hat{\mu}_{Y|x_0} = \hat{\beta}_0 + \hat{\beta}_1 x_0$ is computed from the fitted regression model.

Note that the width of the confidence interval for $\mu_{Y|x_0}$ is a function of the value specified for x_0. The interval width is a minimum for $x_0 = \bar{x}$ and widens as $|x_0 - \bar{x}|$ increases.

EXAMPLE 11-5   We will construct a 95% confidence interval about the mean response for the data in Example 11-1. The fitted model is $\hat{\mu}_{Y|x_0} = 74.283 + 14.947x_0$, and the 95% confidence interval on $\mu_{Y|x_0}$ is found from Equation 11-31 as

$$\hat{\mu}_{Y|x_0} \pm 2.101 \sqrt{1.18 \left[\frac{1}{20} + \frac{(x_0 - 1.1960)^2}{0.68088} \right]}$$

Suppose that we are interested in predicting mean oxygen purity when $x_0 = 1.00\%$. Then

$$\hat{\mu}_{Y|x_{1.00}} = 74.283 + 14.947(1.00) = 89.23$$

and the 95% confidence interval is

$$\left\{ 89.23 \pm 2.101 \sqrt{1.18 \left[\frac{1}{20} + \frac{(1.00 - 1.1960)^2}{0.68088} \right]} \right\}$$

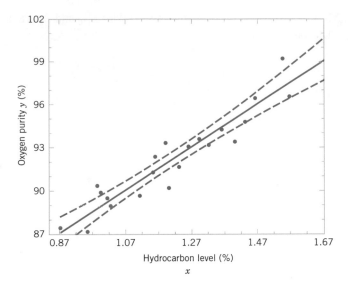

Figure 11-7 Scatter diagram of oxygen purity data from Example 11-1 with fitted regression line and 95 percent confidence limits on $\mu_{Y|x_0}$.

or

$$89.23 \pm 0.75$$

Therefore, the 95% confidence interval on $\mu_{Y|1.00}$ is

$$88.48 \leq \mu_{Y|1.00} \leq 89.98$$

Minitab will also perform these calculations. Refer to Table 11-2. The predicted value of y at $x = 1.00$ is shown along with the 95% CI on the mean of y at this level of x.

By repeating these calculations for several different values for x_0 we can obtain confidence limits for each corresponding value of $\mu_{Y|x_0}$. Figure 11-7 displays the scatter diagram with the fitted model and the corresponding 95% confidence limits plotted as the upper and lower lines. The 95% confidence level applies only to the interval obtained at one value of x and not to the entire set of x-levels. Notice that the width of the confidence interval on $\mu_{Y|x_0}$ increases as $|x_0 - \bar{x}|$ increases.

11-7 PREDICTION OF NEW OBSERVATIONS

An important application of a regression model is predicting new or future observations Y corresponding to a specified level of the regressor variable x. If x_0 is the value of the regressor variable of interest,

$$\hat{Y}_0 = \hat{\beta}_0 + \hat{\beta}_1 x_0 \tag{11-32}$$

is the point estimator of the new or future value of the response Y_0.

Now consider obtaining an interval estimate for this future observation Y_0. This new observation is independent of the observations used to develop the regression model. Therefore, the confidence interval for $\mu_{Y|x_0}$ in Equation 11-31 is inappropriate, since it is based only on the data used to fit the regression model. The confidence interval about $\mu_{Y|x_0}$ refers to the true mean response at $x = x_0$ (that is, a population parameter), not to future observations.

Let Y_0 be the future observation at $x = x_0$, and let $\hat{Y}_0$ given by Equation 11-32 be the estimator of Y_0. Note that the error in prediction

$$e_{\hat{p}} = Y_0 - \hat{Y}_0$$

is a normally distributed random variable with mean zero and variance

$$V(e_{\hat{p}}) = V(Y_0 - \hat{Y}_0) = \sigma^2 \left[1 + \frac{1}{n} + \frac{(x_0 - \bar{x})^2}{S_{xx}} \right]$$

because Y_0 is independent of $\hat{Y}_0$. If we use $\hat{\sigma}^2$ to estimate σ^2, we can show that

$$\frac{Y_0 - \hat{Y}_0}{\sqrt{\hat{\sigma}^2 \left[1 + \frac{1}{n} + \frac{(x_0 - \bar{x})^2}{S_{xx}} \right]}}$$

has a t distribution with $n - 2$ degrees of freedom. From this we can develop the following **prediction interval** definition.

Definition

> A $100(1 - \alpha)$ % **prediction interval on a future observation** Y_0 at the value x_0 is given by
>
> $$\hat{y}_0 - t_{\alpha/2,n-2} \sqrt{\hat{\sigma}^2 \left[1 + \frac{1}{n} + \frac{(x_0 - \bar{x})^2}{S_{xx}} \right]}$$
>
> $$\leq Y_0 \leq \hat{y}_0 + t_{\alpha/2,n-2} \sqrt{\hat{\sigma}^2 \left[1 + \frac{1}{n} + \frac{(x_0 - \bar{x})^2}{S_{xx}} \right]} \qquad (11\text{-}33)$$
>
> The value $\hat{y}_0$ is computed from the regression model $\hat{y}_0 = \hat{\beta}_0 + \hat{\beta}_1 x_0$.

Notice that the prediction interval is of minimum width at $x_0 = \bar{x}$ and widens as $|x_0 - \bar{x}|$ increases. By comparing Equation 11-33 with Equation 11-31, we observe that the prediction interval at the point x_0 is always wider than the confidence interval at x_0. This results because the prediction interval depends on both the error from the fitted model and the error associated with future observations.

EXAMPLE 11-6

To illustrate the construction of a prediction interval, suppose we use the data in Example 11-1 and find a 95% prediction interval on the next observation of oxygen purity at $x_0 = 1.00\%$. Using Equation 11-33 and recalling from Example 11-5 that $\hat{y}_0 = 89.23$, we find that the prediction interval is

$$89.23 - 2.101 \sqrt{1.18 \left[1 + \frac{1}{20} + \frac{(1.00 - 1.1960)^2}{0.68088} \right]}$$

$$\leq Y_0 \leq 89.23 + 2.101 \sqrt{1.18 \left[1 + \frac{1}{20} + \frac{(1.00 - 1.1960)^2}{0.68088} \right]}$$

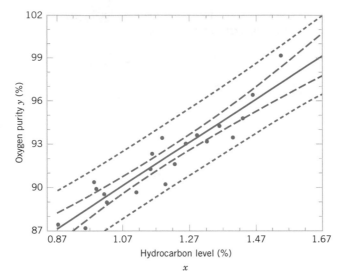

Figure 11-8   Scatter diagram of oxygen purity data from Example 11-1 with fitted regression line, 95% prediction limits (outer lines) and 95% confidence limits on $\mu_{Y|x_0}$.

which simplifies to

$$86.83 \leq y_0 \leq 91.63$$

Minitab will also calculate prediction intervals. Refer to the output in Table 11-2. The 95% PI on the future observation at $x_0 = 1.00$ is shown in the display.

By repeating the foregoing calculations at different levels of x_0, we may obtain the 95% prediction intervals shown graphically as the lower and upper lines about the fitted regression model in Fig. 11-8. Notice that this graph also shows the 95% confidence limits on $\mu_{Y|x_0}$ calculated in Example 11-5. It illustrates that the prediction limits are always wider than the confidence limits.

EXERCISES FOR SECTIONS 11-6 AND 11-7

11-31. Refer to the data in Exercise 11-1 on y = intrinsic permeability of concrete and x = compressive strength. Find a 95% confidence interval on each of the following:
(a) Slope   (b) Intercept
(c) Mean permeability when $x = 2.5$
(d) Find a 95% prediction interval on permeability when $x = 2.5$. Explain why this interval is wider than the interval in part (c).

11-32. Exercise 11-2 presented data on roadway surface temperature x and pavement deflection y. Find a 99% confidence interval on each of the following:
(a) Slope   (b) Intercept
(c) Mean deflection when temperature $x = 85°F$
(d) Find a 99% prediction interval on pavement deflection when the temperature is $90°F$.

11-33. Exercise 11-4 presented data on the number of games won by NFL teams in 1976. Find a 95% confidence interval on each of the following:

(a) Slope   (b) Intercept
(c) Mean number of games won when opponents rushing yardage is limited to $x = 1800$
(d) Find a 95% prediction interval on the number of games won when opponents rushing yards is 1800.

11-34. Refer to the data on y = house selling price and x = taxes paid in Exercise 11-5. Find a 95% confidence interval on each of the following:
(a) β_1 (b) β_0
(c) Mean selling price when the taxes paid are $x = 7.50$
(d) Compute the 95% prediction interval for selling price when the taxes paid are $x = 7.50$.

11-35. Exercise 11-6 presented data on y = steam usage and x = monthly average temperature.
(a) Find a 99% confidence interval for β_1.
(b) Find a 99% confidence interval for β_0.
(c) Find a 95% confidence interval on mean steam usage when the average temperature is $55°F$.

(d) Find a 95% prediction interval on steam usage when temperature is 55°F. Explain why this interval is wider than the interval in part (c).

11-36. Exercise 11-7 presented gasoline mileage performance for 20 cars, along with information about the engine displacement. Find a 95% confidence interval on each of the following:

(a) Slope   (b) Intercept
(c) Mean highway gasoline mileage when the engine displacement is $x = 150$ in^3
(d) Construct a 95% prediction interval on highway gasoline mileage when the engine displacement is $x = 150$ in^3.

11-37. Consider the data in Exercise 11-8 on $y = $ green liquor Na$_2$S concentration and $x = $ production in a paper mill. Find a 99% confidence interval on each of the following:

(a) β_1 (b) β_0
(c) Mean Na$_2$S concentration when production $x = 910$ tons/day
(d) Find a 99% prediction interval on Na$_2$S concentration when $x = 910$ tons/day.

11-38. Exercise 11-9 presented data on $y = $ blood pressure rise and $x = $ sound pressure level. Find a 95% confidence interval on each of the following:

(a) β_1 (b) β_0

(c) Mean blood pressure rise when the sound pressure level is 85 decibals
(d) Find a 95% prediction interval on blood pressure rise when the sound pressure level is 85 decibals.

11-39. Refer to the data in Exercise 11-10 on $y = $ wear volume of mild steel and $x = $ oil viscosity. Find a 95% confidence interval on each of the following:

(a) Intercept   (b) Slope
(c) Mean wear when oil viscosity $x = 30$

11-40. Exercise 11-11 presented data on chloride concentration y and roadway area x on watersheds in central Rhode Island. Find a 99% confidence interval on each of the following:

(a) β_1 (b) β_0
(c) Mean chloride concentration when roadway area $x = 1.0\%$
(d) Find a 99% prediction interval on chloride concentration when roadway area $x = 1.0\%$.

11-41. Refer to the data in Exercise 11-12 on rocket motor shear strength y and propellant age x. Find a 95% confidence interval on each of the following:

(a) Slope β_1 (b) Intercept β_0
(c) Mean shear strength when age $x = 20$ weeks
(d) Find a 95% prediction interval on shear strength when age $x = 20$ weeks.

11-8   ADEQUACY OF THE REGRESSION MODEL

Fitting a regression model requires several **assumptions.** Estimation of the model parameters requires the assumption that the errors are uncorrelated random variables with mean zero and constant variance. Tests of hypotheses and interval estimation require that the errors be normally distributed. In addition, we assume that the order of the model is correct; that is, if we fit a simple linear regression model, we are assuming that the phenomenon actually behaves in a linear or first-order manner.

The analyst should always consider the validity of these assumptions to be doubtful and conduct analyses to examine the adequacy of the model that has been tentatively entertained. In this section we discuss methods useful in this respect.

11-8.1   Residual Analysis

The **residuals** from a regression model are $e_i = y_i - \hat{y}_i$, $i = 1, 2, \ldots, n$, where y_i is an actual observation and $\hat{y}_i$ is the corresponding fitted value from the regression model. Analysis of the residuals is frequently helpful in checking the assumption that the errors are approximately normally distributed with constant variance, and in determining whether additional terms in the model would be useful.

As an approximate check of normality, the experimenter can construct a frequency histogram of the residuals or a **normal probability plot of residuals.** Many computer programs will produce a normal probability plot of residuals, and since the sample sizes in regression are often too small for a histogram to be meaningful, the normal probability plotting method

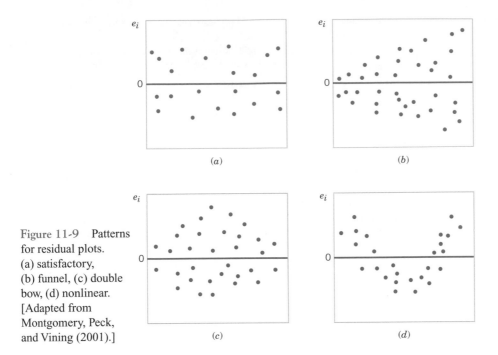

Figure 11-9   Patterns for residual plots. (a) satisfactory, (b) funnel, (c) double bow, (d) nonlinear. [Adapted from Montgomery, Peck, and Vining (2001).]

is preferred. It requires judgment to assess the abnormality of such plots. (Refer to the discussion of the "fat pencil" method in Section 6-7).

We may also **standardize** the residuals by computing $d_i = e_i/\sqrt{\hat{\sigma}^2}$, $i = 1, 2 \ldots, n$. If the errors are normally distributed, approximately 95% of the standardized residuals should fall in the interval $(-2, +2)$. Residuals that are far outside this interval may indicate the presence of an **outlier,** that is, an observation that is not typical of the rest of the data. Various rules have been proposed for discarding outliers. However, outliers sometimes provide important information about unusual circumstances of interest to experimenters and should not be automatically discarded. For further discussion of outliers, see Montgomery, Peck and Vining (2001).

It is frequently helpful to plot the residuals (1) in time sequence (if known), (2), against the $\hat{y}_i$, and (3) against the independent variable x. These graphs will usually look like one of the four general patterns shown in Fig. 11-9. Pattern (a) in Fig. 11-9 represents the ideal situation, while patterns (b), (c), and (d) represent anomalies. If the residuals appear as in (b), the variance of the observations may be increasing with time or with the magnitude of y_i or x_i. Data transformation on the response y is often used to eliminate this problem. Widely used variance-stabilizing transformations include the use of $\sqrt{y}$, ln y, or $1/y$ as the response. See Montgomery, Peck, and Vining (2001) for more details regarding methods for selecting an appropriate transformation. If a plot of the residuals against time has the appearance of (b), the variance of the observations is increasing with time. Plots of residuals against $\hat{y}_i$ and x_i that look like (c) also indicate inequality of variance. Residual plots that look like (d) indicate model inadequacy; that is, higher order terms should be added to the model, a transformation on the x-variable or the y-variable (or both) should be considered, or other regressors should be considered.

EXAMPLE 11-7   The regression model for the oxygen purity data in Example 11-1 is $\hat{y} = 74.283 + 14.947x$. Table 11-4 presents the observed and predicted values of y at each value of x from this data set, along with the corresponding residual. These values were computed using Minitab and show

Table 11-4 Oxygen Purity Data from Example 11-1, Predicted Values, and Residuals

	Hydrocarbon Level, x	Oxygen Purity, y	Predicted Value, $\hat{y}$	Residual $e = y - \hat{y}$		Hydrocarbon Level, x	Oxygen Purity, y	Predicted Value, $\hat{y}$	Residual $e = y - \hat{y}$
1	0.99	90.01	89.069009	0.940991	11	1.19	93.54	92.063189	1.476811
2	1.02	89.05	89.518136	−0.468136	12	1.15	92.52	91.614062	0.905938
3	1.15	91.43	91.464353	−0.034353	13	0.98	90.56	88.919300	1.640700
4	1.29	93.74	93.560279	0.179721	14	1.01	89.54	89.368427	0.171573
5	1.46	96.73	96.105332	0.624668	15	1.11	89.85	90.865517	−1.015517
6	1.36	94.45	94.608242	−0.158242	16	1.20	90.39	92.212898	−1.822898
7	0.87	87.59	87.272501	0.317499	17	1.26	93.25	93.111152	0.138848
8	1.23	91.77	92.662025	−0.892025	18	1.32	93.41	94.009406	−0.599406
9	1.55	99.42	97.452713	1.967287	19	1.43	94.98	95.656205	−0.676205
10	1.40	93.65	95.207078	−1.557078	20	0.95	87.33	88.470173	−1.140173

the number of decimal places typical of computer output. A normal probability plot of the residuals is shown in Fig. 11-10. Since the residuals fall approximately along a straight line in the figure, we conclude that there is no severe departure from normality. The residuals are also plotted against the predicted value $\hat{y}_i$ in Fig. 11-11 and against the hydrocarbon levels x_i in Fig. 11-12. These plots do not indicate any serious model inadequacies.

11-8.2 Coefficient of Determination (R^2)

The quantity

$$R^2 = \frac{SS_R}{SS_T} = 1 - \frac{SS_E}{SS_T} \tag{11-34}$$

is called the **coefficient of determination** and is often used to judge the adequacy of a regression model. Subsequently, we will see that in the case where X and Y are jointly distributed random variables, R^2 is the square of the correlation coefficient between X and Y. From

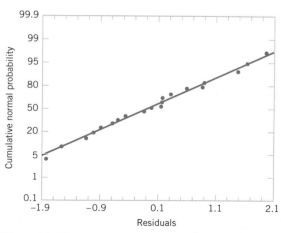

Figure 11-10   Normal probability plot of residuals, Example 11-7.

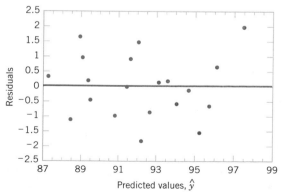

Figure 11-11   Plot of residuals versus predicted oxygen purity $\hat{y}$, Example 11-7.

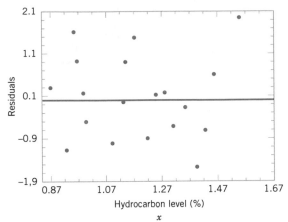

Figure 11-12   Plot of residuals versus hydrocarbon level x, Example 11-8.

the analysis of variance identity in Equations 11-24 and 11-25, $0 \leq R^2 \leq 1$. We often refer loosely to R^2 as the amount of variability in the data explained or accounted for by the regression model. For the oxygen purity regression model, we have $R^2 = SS_R/SS_T = 152.13/173.38 = 0.877$; that is, the model accounts for 87.7% of the variability in the data.

The statistic R^2 should be used with caution, because it is always possible to make R^2 unity by simply adding enough terms to the model. For example, we can obtain a "perfect" fit to n data points with a polynomial of degree $n - 1$. In addition, R^2 will always increase if we add a variable to the model, but this does not necessarily imply that the new model is superior to the old one. Unless the error sum of squares in the new model is reduced by an amount equal to the original error mean square, the new model will have a larger error mean square than the old one, because of the loss of one error degree of freedom. Thus, the new model will actually be worse than the old one.

There are several misconceptions about R^2. In general, R^2 does not measure the magnitude of the slope of the regression line. A large value of R^2 does not imply a steep slope. Furthermore, R^2 does not measure the appropriateness of the model, since it can be artificially inflated by adding higher order polynomial terms in x to the model. Even if y and x are related in a nonlinear fashion, R^2 will often be large. For example, R^2 for the regression equation in Fig. 11-6(b) will be relatively large, even though the linear approximation is poor. Finally, even though R^2 is large, this does not necessarily imply that the regression model will provide accurate predictions of future observations.

11-8.3   Lack-of-Fit Test (CD Only)

EXERCISES FOR SECTION 11-8

11-42.   Refer to the NFL team performance data in Exercise 11-4.
(a) Calculate R^2 for this model and provide a practical interpretation of this quantity.
(b) Prepare a normal probability plot of the residuals from the least squares model. Does the normality assumption seem to be satisfied?
(c) Plot the residuals versus $\hat{y}$ and against x. Interpret these graphs.

11-43.   Refer to the data in Exercise 11-5 on house selling price y and taxes paid x.

(a) Find the residuals for the least squares model.
(b) Prepare a normal probability plot of the residuals and interpret this display.
(c) Plot the residuals versus $\hat{y}$ and versus x. Does the assumption of constant variance seem to be satisfied?
(d) What proportion of total variability is explained by the regression model?

11-44.   Exercise 11-6 presents data on $y =$ steam usage and $x =$ average monthly temperature.
(a) What proportion of total variability is accounted for by the simple linear regression model?

(b) Prepare a normal probability plot of the residuals and interpret this graph.

(c) Plot residuals versus $\hat{y}$ and x. Do the regression assumptions appear to be satisfied?

11-45. Refer to the gasoline mileage data in Exercise 11-7.

(a) What proportion of total variability in highway gasoline mileage performance is accounted for by engine displacement?

(b) Plot the residuals versus $\hat{y}$ and x, and comment on the graphs.

(c) Prepare a normal probability plot of the residuals. Does the normality assumption appear to be satisfied?

11-46. Consider the data in Exercise 11-8 on $y =$ green liquor Na_2S concentration and $x =$ paper machine production. Suppose that a 14th sample point is added to the original data, where $y_{14} = 59$ and $x_{14} = 855$.

(a) Prepare a scatter diagram of y versus x. Fit the simple linear regression model to all 14 observations.

(b) Test for significance of regression with $\alpha = 0.05$.

(c) Estimate σ^2 for this model.

(d) Compare the estimate of σ^2 obtained in part (c) above with the estimate of σ^2 obtained from the original 13 points. Which estimate is larger and why?

(e) Compute the residuals for this model. Does the value of e_{14} appear unusual?

(f) Prepare and interpret a normal probability plot of the residuals.

(g) Plot the residuals versus $\hat{y}$ and versus x. Comment on these graphs.

11-47. Refer to Exercise 11-9, which presented data on blood pressure rise y and sound pressure level x.

(a) What proportion of total variability in blood pressure rise is accounted for by sound pressure level?

(b) Prepare a normal probability plot of the residuals from this least squares model. Interpret this plot.

(c) Plot residuals versus $\hat{y}$ and versus x. Comment on these plots.

11-48. Exercise 11-10 presents data on wear volume y and oil viscosity x.

(a) Calculate R^2 for this model. Provide an interpretation of this quantity.

(b) Plot the residuals from this model versus $\hat{y}$ and versus x. Interpret these plots.

(c) Prepare a normal probability plot of the residuals. Does the normality assumption appear to be satisfied?

11-49. Refer to Exercise 11-11, which presented data on chloride concentration y and roadway area x.

(a) What proportion of the total variability in chloride concentration is accounted for by the regression model?

(b) Plot the residuals versus $\hat{y}$ and versus x. Interpret these plots.

(c) Prepare a normal probability plot of the residuals. Does the normality assumption appear to be satisfied?

11-50. Consider the rocket propellant data in Exercise 11-12.

(a) Calculate R^2 for this model. Provide an interpretation of this quantity.

(b) Plot the residuals on a normal probability scale. Do any points seem unusual on this plot?

(c) Delete the two points identified in part (b) from the sample and fit the simple linear regression model to the remaining 18 points. Calculate the value of R^2 for the new model. Is it larger or smaller than the value of R^2 computed in part (a)? Why?

(d) Did the value of $\hat{\sigma}^2$ change dramatically when the two points identified above were deleted and the model fit to the remaining points? Why?

11-51. Show that an equivalent way to define the test for significance of regression in simple linear regression is to base the test on R^2 as follows: to test $H_0: \beta_1 = 0$ versus $H_1: \beta_1 \neq 0$, calculate

$$F_0 = \frac{R^2(n-2)}{1 - R^2}$$

and to reject $H_0: \beta_1 = 0$ if the computed value $f_0 > f_{\alpha,1,n-2}$.

11-52. Suppose that a simple linear regression model has been fit to $n = 25$ observations and $R^2 = 0.90$.

(a) Test for significance of regression at $\alpha = 0.05$. Use the results of Exercise 11-51.

(b) What is the smallest value of R^2 that would lead to the conclusion of a significant regression if $\alpha = 0.05$?

11-53. Consider the rocket propellant data in Exercise 11-12. Calculate the standardized residuals for these data. Does this provide any helpful information about the magnitude of the residuals?

11-54. **Studentized Residuals.** Show that the variance of the ith residual is

$$V(e_i) = \sigma^2 \left[1 - \left(\frac{1}{n} + \frac{(x_i - \bar{x})^2}{S_{xx}} \right) \right]$$

Hint:

$$\text{cov}(Y_i, \hat{Y}_i) = \sigma^2 \left[\frac{1}{n} + \frac{(x_i - \bar{x})^2}{S_{xx}} \right].$$

The ith studentized residual is defined as

$$r_i = \frac{e_i}{\sqrt{\hat{\sigma}^2 \left[1 - \left(\frac{1}{n} + \frac{(x_i - \bar{x})^2}{S_{xx}} \right) \right]}}$$

(a) Explain why r_i has unit standard deviation.

(b) Do the **standardized residuals** have unit standard deviation?

(c) Discuss the behavior of the studentized residual when the sample value x_i is very close to the middle of the range of x.

(d) Discuss the behavior of the studentized residual when the sample value x_i is very near one end of the range of x.

11-9   TRANSFORMATIONS TO A STRAIGHT LINE

We occasionally find that the straight-line regression model $Y = \beta_0 + \beta_1 x + \epsilon$ is inappropriate because the true regression function is nonlinear. Sometimes nonlinearity is visually determined from the scatter diagram, and sometimes, because of prior experience or underlying theory, we know in advance that the model is nonlinear. Occasionally, a scatter diagram will exhibit an apparent nonlinear relationship between Y and x. In some of these situations, a nonlinear function can be expressed as a straight line by using a suitable transformation. Such nonlinear models are called **intrinsically linear.**

As an example of a nonlinear model that is intrinsically linear, consider the exponential function

$$Y = \beta_0 e^{\beta_1 x} \epsilon$$

This function is intrinsically linear, since it can be transformed to a straight line by a logarithmic transformation

$$\ln Y = \ln \beta_0 + \beta_1 x + \ln \epsilon$$

This transformation requires that the transformed error terms $\ln \epsilon$ are normally and independently distributed with mean 0 and variance σ^2.

Another intrinsically linear function is

$$Y = \beta_0 + \beta_1 \left(\frac{1}{x} \right) + \epsilon$$

By using the reciprocal transformation $z = 1/x$, the model is linearized to

$$Y = \beta_0 + \beta_1 z + \epsilon$$

Sometimes several transformations can be employed jointly to linearize a function. For example, consider the function

$$Y = \frac{1}{\exp(\beta_0 + \beta_1 x + \epsilon)}$$

letting $Y^* = 1/Y$, we have the linearized form

$$\ln Y^* = \beta_0 + \beta_1 x + \epsilon$$

For examples of fitting these models, refer to Montgomery, Peck, and Vining (2001) or Myers (1990).

11-10   MORE ABOUT TRANSFORMATIONS (CD ONLY)

11-11   CORRELATION

Our development of regression analysis has assumed that x is a mathematical variable, measured with negligible error, and that Y is a random variable. Many applications of regression analysis involve situations in which both X and Y are random variables. In these situations, it is usually assumed that the observations (X_i, Y_i), $i = 1, 2, \ldots, n$ are jointly distributed random variables obtained from the distribution $f(x, y)$.

For example, suppose we wish to develop a regression model relating the shear strength of spot welds to the weld diameter. In this example, weld diameter cannot be controlled. We would randomly select n spot welds and observe a diameter (X_i) and a shear strength (Y_i) for each. Therefore (X_i, Y_i) are jointly distributed random variables.

We assume that the joint distribution of X_i and Y_i is the bivariate normal distribution presented in Chapter 5, and μ_Y and σ_Y^2 are the mean and variance of Y, μ_X and σ_X^2 are the mean and variance of X, and ρ is the **correlation coefficient** between Y and X. Recall that the correlation coefficient is defined as

$$\rho = \frac{\sigma_{XY}}{\sigma_X \sigma_Y} \tag{11-35}$$

where σ_{XY} is the covariance between Y and X.

The conditional distribution of Y for a given value of $X = x$ is

$$f_{Y|x}(y) = \frac{1}{\sqrt{2\pi}\sigma_{Y|x}} \exp\left[-\frac{1}{2}\left(\frac{y - \beta_0 - \beta_1 x}{\sigma_{Y|x}}\right)^2 \right] \tag{11-36}$$

where

$$\beta_0 = \mu_Y - \mu_X \rho \frac{\sigma_Y}{\sigma_X} \tag{11-37}$$

$$\beta_1 = \frac{\sigma_Y}{\sigma_X}\rho \tag{11-38}$$

and the variance of the conditional distribution of Y given $X = x$ is

$$\sigma_{Y|x}^2 = \sigma_Y^2(1 - \rho^2) \tag{11-39}$$

That is, the conditional distribution of Y given $X = x$ is normal with mean

$$E(Y|x) = \beta_0 + \beta_1 x \tag{11-40}$$

and variance $\sigma_{Y|x}^2$. Thus, the mean of the conditional distribution of Y given $X = x$ is a simple linear regression model. Furthermore, there is a relationship between the correlation coefficient ρ and the slope β_1. From Equation 11-38 we see that if $\rho = 0$, then $\beta_1 = 0$, which implies that there is no regression of Y on X. That is, knowledge of X does not assist us in predicting Y.

The method of maximum likelihood may be used to estimate the parameters β_0 and β_1. It can be shown that the maximum likelihood estimators of those parameters are

$$\hat{\beta}_0 = \overline{Y} - \hat{\beta}_1 \overline{X} \tag{11-41}$$

and

$$\hat{\beta}_1 = \frac{\sum\limits_{i=1}^{n} Y_i(X_i - \overline{X})}{\sum\limits_{i=1}^{n} (X_i - \overline{X})^2} = \frac{S_{XY}}{S_{XX}} \tag{11-42}$$

We note that the estimators of the intercept and slope in Equations 11-41 and 11-42 are identical to those given by the method of least squares in the case where X was assumed to be a mathematical variable. That is, the regression model with Y and X jointly normally distributed is equivalent to the model with X considered as a mathematical variable. This follows because the random variables Y given $X = x$ are independently and normally distributed with mean $\beta_0 + \beta_1 x$ and constant variance $\sigma_{Y|x}^2$. These results will also hold for any joint distribution of Y and X such that the conditional distribution of Y given X is normal.

It is possible to draw inferences about the correlation coefficient ρ in this model. The estimator of ρ is the **sample correlation coefficient**

$$R = \frac{\sum_{i=1}^{n} Y_i (X_i - \overline{X})}{\left[\sum_{i=1}^{n} (X_i - \overline{X})^2 \sum_{i=1}^{n} (Y_i - \overline{Y})^2 \right]^{1/2}} = \frac{S_{XY}}{(S_{XX} SS_T)^{1/2}} \tag{11-43}$$

Note that

$$\hat{\beta}_1 = \left(\frac{SS_T}{S_{XX}} \right)^{1/2} R \tag{11-44}$$

so the slope $\hat{\beta}_1$ is just the sample correlation coefficient R multiplied by a scale factor that is the square root of the "spread" of the Y values divided by the "spread" of the X values. Thus, $\hat{\beta}_1$ and R are closely related, although they provide somewhat different information. The sample correlation coefficient R measures the linear association between Y and X, while $\hat{\beta}_1$ measures the predicted change in the mean of Y for a unit change in X. In the case of a mathematical variable x, R has no meaning because the magnitude of R depends on the choice of spacing of x. We may also write, from Equation 11-44,

$$R^2 = \hat{\beta}_1^2 \frac{S_{XX}}{S_{YY}} = \frac{\hat{\beta}_1 S_{XY}}{SS_T} = \frac{SS_R}{SS_T}$$

which is just the coefficient of determination. That is, the coefficient of determination R^2 is just the square of the correlation coefficient between Y and X.

It is often useful to test the hypotheses

$$H_0: \rho = 0$$
$$H_1: \rho \neq 0 \tag{11-45}$$

The appropriate test statistic for these hypotheses is

$$T_0 = \frac{R \sqrt{n - 2}}{\sqrt{1 - R^2}} \tag{11-46}$$

which has the t distribution with $n - 2$ degrees of freedom if $H_0: \rho = 0$ is true. Therefore, we would reject the null hypothesis if $|t_0| > t_{\alpha/2, n-2}$. This test is equivalent to the test of the hypothesis H_0: $\beta_1 = 0$ given in Section 11-6.1. This equivalence follows directly from Equation 11-38.

The test procedure for the hypothesis is

$$H_0: \rho = \rho_0$$
$$H_1: \rho \neq \rho_0 \tag{11-47}$$

where $\rho_0 \neq 0$ is somewhat more complicated. For moderately large samples (say, $n \geq 25$) the statistic

$$Z = \text{arctanh } R = \frac{1}{2} \ln \frac{1 + R}{1 - R} \tag{11-48}$$

is approximately normally distributed with mean and variance

$$\mu_Z = \text{arctanh } \rho = \frac{1}{2} \ln \frac{1 + \rho}{1 - \rho} \quad \text{and} \quad \sigma_Z^2 = \frac{1}{n - 3}$$

respectively. Therefore, to test the hypothesis H_0: $\rho = \rho_0$, we may use the test statistic

$$Z_0 = (\text{arctanh } R - \text{arctanh } \rho_0)(n - 3)^{1/2} \tag{11-49}$$

and reject H_0: $\rho = \rho_0$ if the value of the test statistic in Equation 11-49 is such that $|z_0| > z_{\alpha/2}$.

It is also possible to construct an approximate $100(1 - \alpha)\%$ confidence interval for ρ, using the transformation in Equation 11-48. The approximate $100(1 - \alpha)\%$ confidence interval is

$$\tanh\left(\text{arctanh } r - \frac{z_{\alpha/2}}{\sqrt{n - 3}}\right) \leq \rho \leq \tanh\left(\text{arctanh } r + \frac{z_{\alpha/2}}{\sqrt{n - 3}}\right) \tag{11-50}$$

where $\tanh u = (e^u - e^{-u})/(e^u + e^{-u})$.

EXAMPLE 11-8

In Chapter 1 (Section 1-3) an application of regression analysis is described in which an engineer at a semiconductor assembly plant is investigating the relationship between pull strength of a wire bond and two factors: wire length and die height. In this example, we will consider only one of the factors, the wire length. A random sample of 25 units is selected and tested, and the wire bond pull strength and wire length are observed for each unit. The data are shown in Table 1-2. We assume that pull strength and wire length are jointly normally distributed.

Figure 11-13 shows a scatter diagram of wire bond strength versus wire length. We have used the Minitab option of displaying box plots of each individual variable on the scatter diagram. There is evidence of a linear relationship between the two variables.

The Minitab output for fitting a simple linear regression model to the data is shown on the following page.

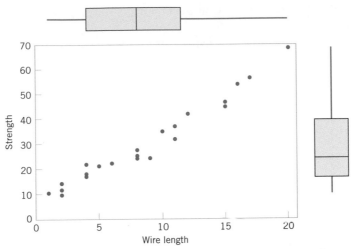

Figure 11-13   Scatter plot of wire bond strength versus wire length, Example 11-8.

Regression Analysis: Strength versus Length

The regression equation is
Strength = 5.11 + 2.90 Length

Predictor	Coef	SE Coef	T	P
Constant	5.115	1.146	4.46	0.000
Length	2.9027	0.1170	24.80	0.000

S = 3.093   R-Sq = 96.4%   R-Sq(adj) = 96.2%
PRESS = 272.144   R-Sq(pred) = 95.54%

Analysis of Variance

Source	DF	SS	MS	F	P
Regression	1	5885.9	5885.9	615.08	0.000
Residual Error	23	220.1	9.6		
Total	24	6105.9			

Now $S_{xx} = 698.56$ and $S_{xy} = 2027.7132$, and the sample correlation coefficient is

$$r = \frac{S_{xy}}{[S_{xx}SS_T]^{1/2}} = \frac{2027.7132}{[(698.560)(6105.9)]^{1/2}} = 0.9818$$

Note that $r^2 = (0.9818)^2 = 0.9640$ (which is reported in the Minitab output), or that approximately 96.40% of the variability in pull strength is explained by the linear relationship to wire length.

Now suppose that we wish to test the hypothesis

$$H_0: \rho = 0$$
$$H_1: \rho \neq 0$$

with $\alpha = 0.05$. We can compute the t-statistic of Equation 11-46 as

$$t_0 = \frac{r\sqrt{n-2}}{\sqrt{1-r^2}} = \frac{0.9818\sqrt{23}}{\sqrt{1-0.9640}} = 24.8$$

This statistic is also reported in the Minitab output as a test of $H_0: \beta_1 = 0$. Because $t_{0.025,23} = 2.069$, we reject H_0 and conclude that the correlation coefficient $\rho \neq 0$.

Finally, we may construct an approximate 95% confidence interval on ρ from Equation 11-50. Since arctanh r = arctanh 0.9818 = 2.3452, Equation 11-50 becomes

$$\tanh\left(2.3452 - \frac{1.96}{\sqrt{22}}\right) \leq \rho \leq \tanh\left(2.3452 + \frac{1.96}{\sqrt{22}}\right)$$

which reduces to

$$0.9585 \leq \rho \leq 0.9921$$

EXERCISES FOR SECTION 11–11

11-55. The final test and exam averages for 20 randomly selected students taking a course in engineering statistics and a course in operations research follow. Assume that the final averages are jointly normally distributed.
(a) Find the regression line relating the statistics final average to the OR final average.
(b) Test for significance of regression using $\alpha = 0.05$.

Statistics	86	75	69	75	90
OR	80	81	75	81	92
Statistics	94	83	86	71	65
OR	95	80	81	76	72
Statistics	84	71	62	90	83
OR	85	72	65	93	81
Statistics	75	71	76	84	97
OR	70	73	72	80	98

(c) Estimate the correlation coefficient.
(d) Test the hypothesis that $\rho = 0$, using $\alpha = 0.05$.
(e) Test the hypothesis that $\rho = 0.5$, using $\alpha = 0.05$.
(f) Construct a 95% confidence interval for the correlation coefficient.

11-56. The weight and systolic blood pressure of 26 randomly selected males in the age group 25 to 30 are shown in the following table. Assume that weight and blood pressure are jointly normally distributed.
(a) Find a regression line relating systolic blood pressure to weight.
(b) Test for significance of regression using $\alpha = 0.05$.

Subject	Weight	Systolic BP	Subject	Weight	Systolic BP
1	165	130	14	172	153
2	167	133	15	159	128
3	180	150	16	168	132
4	155	128	17	174	149
5	212	151	18	183	158
6	175	146	19	215	150
7	190	150	20	195	163
8	210	140	21	180	156
9	200	148	22	143	124
10	149	125	23	240	170
11	158	133	24	235	165
12	169	135	25	192	160
13	170	150	26	187	159

(c) Estimate the correlation coefficient.
(d) Test the hypothesis that $\rho = 0$, using $\alpha = 0.05$.
(e) Test the hypothesis that $\rho = 0.6$, using $\alpha = 0.05$.
(f) Construct a 95% confidence interval for the correlation coefficient.

11-57. Consider the NFL data introduced in Exercise 11-4.
(a) Estimate the correlation coefficient between the number of games won and the yards rushing by the opponents.
(b) Test the hypothesis $H_0: \rho = 0$ versus $H_1: \rho \neq 0$ using $\alpha = 0.05$. What is the P-value for this test?
(c) Construct a 95% confidence interval for ρ.
(d) Test the hypothesis $H_0: \rho = -0.7$ versus $H_1: \rho \neq -0.7$ using $\alpha = 0.05$. Find the P-value for this test.

11-58. Show that the t-statistic in Equation 11-46 for testing H_0: $\rho = 0$ is identical to the t-statistic for testing H_0: $\beta_1 = 0$.

11-59. A random sample of 50 observations was made on the diameter of spot welds and the corresponding weld shear strength.
(a) Given that $r = 0.62$, test the hypothesis that $\rho = 0$, using $\alpha = 0.01$. What is the P-value for this test?
(b) Find a 99% confidence interval for ρ.
(c) Based on the confidence interval in part (b), can you conclude that $\rho = 0.5$ at the 0.01 level of significance?

11-60. Suppose that a random sample of 10,000 (X, Y) pairs yielded a sample correlation coefficient of $r = 0.02$.
(a) What is the conclusion that you would reach if you tested H_0: $\rho = 0$ using $\alpha = 0.05$? What is the P-value for this test?
(b) Comment on the practical significance versus the statistical significance of your answer.

11-61. The following data gave $X = $ the water content of snow on April 1 and $Y = $ the yield from April to July (in inches) on the Snake River watershed in Wyoming for 1919 to 1935. (The data were taken from an article in *Research Notes,* Vol. 61, 1950, Pacific Northwest Forest Range Experiment Station, Oregon)

x	y	x	y
23.1	10.5	37.9	22.8
32.8	16.7	30.5	14.1
31.8	18.2	25.1	12.9
32.0	17.0	12.4	8.8
30.4	16.3	35.1	17.4
24.0	10.5	31.5	14.9
39.5	23.1	21.1	10.5
24.2	12.4	27.6	16.1
52.5	24.9		

(a) Estimate the correlation between Y and X.
(b) Test the hypothesis that $\rho = 0$, using $\alpha = 0.05$.
(c) Fit a simple linear regression model and test for significance of regression using $\alpha = 0.05$. What conclusions can you draw? How is the test for significance of regression related to the test on ρ in part (b)?
(d) Test the hypothesis H_0: $\beta_0 = 0$ versus H_1: $\beta_0 \neq 0$ and draw conclusions. Use $\alpha = 0.05$.
(e) Analyze the residuals and comment on model adequacy.

11-62. A random sample of $n = 25$ observations was made on the time to failure of an electronic component and the temperature in the application environment in which the component was used.
(a) Given that $r = 0.83$, test the hypothesis that $\rho = 0$, using $\alpha = 0.05$. What is the P-value for this test?
(b) Find a 95% confidence interval on ρ.

(c) Test the hypothesis H_0: $\rho = 0.8$ versus H_1: $\rho \neq 0.8$, using $\alpha = 0.05$. Find the P-value for this test.

Supplemental Exercises

11-63. Show that, for the simple linear regression model, the following statements are true:

(a) $\sum_{i=1}^{n} (y_i - \hat{y}_i) = 0$ (b) $\sum_{i=1}^{n} (y_i - \hat{y}_i)x_i = 0$

(c) $\dfrac{1}{n} \sum_{i=1}^{n} \hat{y}_i = \bar{y}$

11-64. An article in the *IEEE Transactions on Instrumentation and Measurement* ("Direct, Fast, and Accurate Measurement of V_T and K of MOS Transistor Using V_T-Sift Circuit," Vol. 40, 1991, pp. 951–955) described the use of a simple linear regression model to express drain current y (in milliamperes) as a function of ground-to-source voltage x (in volts). The data are as follows:

y	x	y	x
0.734	1.1	1.50	1.6
0.886	1.2	1.66	1.7
1.04	1.3	1.81	1.8
1.19	1.4	1.97	1.9
1.35	1.5	2.12	2.0

(a) Draw a scatter diagram of these data. Does a straight-line relationship seem plausible?
(b) Fit a simple linear regression model to these data.
(c) Test for significance of regression using $\alpha = 0.05$. What is the P-value for this test?
(d) Find a 95% confidence interval estimate on the slope.
(e) Test the hypothesis H_0: $\beta_0 = 0$ versus H_1: $\beta_0 \neq 0$ using $\alpha = 0.05$. What conclusions can you draw?

11-65. The strength of paper used in the manufacture of cardboard boxes (y) is related to the percentage of hardwood concentration in the original pulp (x). Under controlled conditions, a pilot plant manufactures 16 samples, each from a different batch of pulp, and measures the tensile strength. The data are shown in the table that follows:

(a) Fit a simple linear regression model to the data.
(b) Test for significance of regression using $\alpha = 0.05$.
(c) Construct a 90% confidence interval on the slope β_1.
(d) Construct a 90% confidence interval on the intercept β_0.
(e) Construct a 95% confidence interval on the mean strength at $x = 2.5$.
(f) Analyze the residuals and comment on model adequacy.

y	101.4	117.4	117.1	106.2
x	1.0	1.5	1.5	1.5
y	131.9	146.9	146.8	133.9
x	2.0	2.0	2.2	2.4
y	111.0	123.0	125.1	145.2
x	2.5	2.5	2.8	2.8
y	134.3	144.5	143.7	146.9
x	3.0	3.0	3.2	3.3

11-66. The vapor pressure of water at various temperatures follows:

Observation Number, i	Temperature (K)	Vapor pressure (mm Hg)
1	273	4.6
2	283	9.2
3	293	17.5
4	303	31.8
5	313	55.3
6	323	92.5
7	333	149.4
8	343	233.7
9	353	355.1
10	363	525.8
11	373	760.0

Customer	x	y	Customer	x	y
1	679	0.79	26	1434	0.31
2	292	0.44	27	837	4.20
3	1012	0.56	28	1748	4.88
4	493	0.79	29	1381	3.48
5	582	2.70	30	1428	7.58
6	1156	3.64	31	1255	2.63
7	997	4.73	32	1777	4.99
8	2189	9.50	33	370	0.59
9	1097	5.34	34	2316	8.19
10	2078	6.85	35	1130	4.79
11	1818	5.84	36	463	0.51
12	1700	5.21	37	770	1.74
13	747	3.25	38	724	4.10
14	2030	4.43	39	808	3.94
15	1643	3.16	40	790	0.96
16	414	0.50	41	783	3.29
17	354	0.17	42	406	0.44
18	1276	1.88	43	1242	3.24
19	745	0.77	44	658	2.14
20	795	3.70	45	1746	5.71
21	540	0.56	46	895	4.12
22	874	1.56	47	1114	1.90
23	1543	5.28	48	413	0.51
24	1029	0.64	49	1787	8.33
25	710	4.00	50	3560	14.94

(a) Draw a scatter diagram of these data. What type of relationship seems appropriate in relating y to x?

(b) Fit a simple linear regression model to these data.

(c) Test for significance of regression using $\alpha = 0.05$. What conclusions can you draw?

(d) Plot the residuals from the simple linear regression model versus $\hat{y}_i$. What do you conclude about model adequacy?

(e) The Clausis-Clapeyron relation states that $\ln(P_v) \propto -\frac{1}{T}$, where P_v is the vapor pressure of water. Repeat parts (a)–(d). using an appropriate transformation.

11-67. An electric utility is interested in developing a model relating peak hour demand (y in kilowatts) to total monthly energy usage during the month (x, in kilowatt hours). Data for 50 residential customers are shown in the following table.

(a) Draw a scatter diagram of y versus x.

(b) Fit the simple linear regression model.

(c) Test for significance of regression using $\alpha = 0.05$.

(d) Plot the residuals versus $\hat{y}_i$ and comment on the underlying regression assumptions. Specifically, does it seem that the equality of variance assumption is satisfied?

(e) Find a simple linear regression model using $\sqrt{y}$ as the response. Does this transformation on y stabilize the inequality of variance problem noted in part (d) above?

11-68. Consider the following data. Suppose that the relationship between Y and x is hypothesized to be $Y = (\beta_0 + \beta_1 x + \epsilon)^{-1}$. Fit an appropriate model to the data. Does the assumed model form seem reasonable?

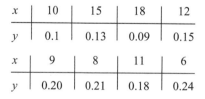

x	10	15	18	12
y	0.1	0.13	0.09	0.15

x	9	8	11	6
y	0.20	0.21	0.18	0.24

11-69. Consider the weight and blood pressure data in Exercise 11-56. Fit a no-intercept model to the data, and compare it to the model obtained in Exercise 11-56. Which model is superior?

11-70. The following data, adapted from Montgomery, Peck, and Vining (2001), present the number of certified mental defectives per 10,000 of estimated population in the United Kingdom (y) and the number of radio receiver licenses issued (x) by the BBC (in millions) for the years 1924 through 1937. Fit a regression model relating y and x. Comment on the model. Specifically, does the existence of a strong correlation imply a cause-and-effect relationship?

Year	y	x	Year	y	x
1924	8	1.350	1931	16	4.620
1925	8	1.960	1932	18	5.497
1926	9	2.270	1933	19	6.260
1927	10	2.483	1934	20	7.012
1928	11	2.730	1935	21	7.618
1929	11	3.091	1936	22	8.131
1930	12	3.674	1937	23	8.593

11-71. An article in *Air and Waste* ("Update on Ozone Trends in California's South Coast Air Basin," Vol. 43, 1993) studied the ozone levels on the South Coast air basin of California for the years 1976–1991. The author believes that the number of days that the ozone level exceeds 0.20 parts per million depends on the seasonal meteorological index (the seasonal average 850 millibar temperature). The data follow:

Year	Days	Index	Year	Days	Index
1976	91	16.7	1984	81	18.0
1977	105	17.1	1985	65	17.2
1978	106	18.2	1986	61	16.9
1979	108	18.1	1987	48	17.1
1980	88	17.2	1988	61	18.2
1981	91	18.2	1989	43	17.3
1982	58	16.0	1990	33	17.5
1983	82	17.2	1991	36	16.6

(a) Construct a scatter diagram of the data.
(b) Fit a simple linear regression model to the data. Test for significance of regression.
(c) Find a 95% CI on the slope β_1.
(d) Analyze the residuals and comment on model adequacy.

11-72. An article in the *Journal of Applied Polymer Science* (Vol. 56, pp. 471–476, 1995) studied the effect of the mole ratio of sebacic acid on the intrinsic viscosity of copolyesters. The data follow:

Mole ratio x	1.0	0.9	0.8	0.7	0.6	0.5	0.4	0.3
Viscosity y	0.45	0.20	0.34	0.58	0.70	0.57	0.55	0.44

(a) Construct a scatter diagram of the data.
(b) Fit a simple linear repression module.
(c) Test for significance of regression. Calculate R^2 for the model.
(d) Analyze the residuals and comment on model adequacy.

11-73. Suppose that we have n pairs of observations (x_i, y_i) such that the sample correlation coefficient r is unity (approximately). Now let $z_i = y_i^2$ and consider the sample correlation coefficient for the n-pairs of data (x_i, z_i). Will this sample correlation coefficient be approximately unity? Explain why or why not.

11-74. The grams of solids removed from a material (y) is thought to be related to the drying time. Ten observations obtained from an experimental study follow:

y	4.3	1.5	1.8	4.9	4.2	4.8	5.8	6.2	7.0	7.9
x	2.5	3.0	3.5	4.0	4.5	5.0	5.5	6.0	6.5	7.0

(a) Construct a scatter diagram for these data.
(b) Fit a simple linear regression model.
(c) Test for significance of regression.
(d) Based on these data, what is your estimate of the mean grams of solids removed at 4.25 hours? Find a 95% confidence interval on the mean.
(e) Analyze the residuals and comment on model adequacy.

11-75. Two different methods can be used for measuring the temperature of the solution in a Hall cell used in aluminum smelting, a thermocouple implanted in the cell and an indirect measurement produced from an IR device. The indirect method is preferable because the thermocouples are eventually destroyed by the solution. Consider the following 10 measurements:

Thermocouple	921	935	916	920	940
IR	918	934	924	921	945

Thermocouple	936	925	940	933	927
IR	930	919	943	932	935

(a) Construct a scatter diagram for these data, letting $x =$ thermocouple measurement and $y =$ IR measurement.
(b) Fit a simple linear regression model.
(c) Test for significance a regression and calculate R^2. What conclusions can you draw?
(d) Is there evidence to support a claim that both devices produce equivalent temperature measurements? Formulate and test an appropriate hypothesis to support this claim.
(e) Analyze the residuals and comment on model adequacy.

MIND-EXPANDING EXERCISES

11-76. Consider the simple linear regression model $Y = \beta_0 + \beta_1 x + \epsilon$, with $E(\epsilon) = 0$, $V(\epsilon) = \sigma^2$, and the errors ϵ uncorrelated.
(a) Show that $\text{cov}(\hat{\beta}_0, \hat{\beta}_1) = -\bar{x}\sigma^2/S_{xx}$.
(b) Show that $\text{cov}(\bar{Y}, \hat{\beta}_1) = 0$.

11-77. Consider the simple linear regression model $Y = \beta_0 + \beta_1 x + \epsilon$, with $E(\epsilon) = 0$, $V(\epsilon) = \sigma^2$, and the errors ϵ uncorrelated.
(a) Show that $E(\hat{\sigma}^2) = E(MS_E) = \sigma^2$.
(b) Show that $E(MS_R) = \sigma^2 + \beta_1^2 S_{xx}$.

11-78. Suppose that we have assumed the straight-line regression model

$$Y = \beta_0 + \beta_1 x_1 + \epsilon$$

but the response is affected by a second variable x_2 such that the true regression function is

$$E(Y) = \beta_0 + \beta_1 x_1 + \beta_2 x_2$$

Is the estimator of the slope in the simple linear regression model unbiased?

11-79. Suppose that we are fitting a line and we wish to make the variance of the regression coefficient $\hat{\beta}_1$ as small as possible. Where should the observations x_i, $i = 1, 2, \ldots, n$, be taken so as to minimize $V(\hat{\beta}_1)$? Discuss the practical implications of this allocation of the x_i.

11-80. Weighted Least Squares. Suppose that we are fitting the line $Y = \beta_0 + \beta_1 x + \epsilon$, but the variance of Y depends on the level of x; that is,

$$V(Y_i \mid x_i) = \sigma_i^2 = \frac{\sigma^2}{w_i} \qquad i = 1, 2, \ldots, n$$

where the w_i are constants, often called *weights*. Show that for an objective function in whole each squared residual is multiplied by the reciprocal of the variance of the corresponding observation, the resulting **weighted least squares normal equations** are

$$\hat{\beta}_0 \sum_{i=1}^{n} w_i + \hat{\beta}_1 \sum_{i=1}^{n} w_i x_i = \sum_{i=1}^{n} w_i y_i$$

$$\hat{\beta}_0 \sum_{i=1}^{n} w_i x_i + \hat{\beta}_1 \sum_{i=1}^{n} w_i x_i^2 = \sum_{i=1}^{n} w_i x_i y_i$$

Find the solution to these normal equations. The solutions are weighted least squares estimators of β_0 and β_1.

11-81. Consider a situation where both Y and X are random variables. Let s_x and s_y be the sample standard deviations of the observed x's and y's, respectively. Show that an alternative expression for the fitted simple linear regression model $\hat{y} = \hat{\beta}_0 + \hat{\beta}_1 x$ is

$$\hat{y} = \bar{y} + r \frac{s_y}{s_x}(x - \bar{x})$$

11-82. Suppose that we are interested in fitting a simple linear regression model $Y = \beta_0 + \beta_1 x + \epsilon$, where the intercept, β_0, is known.
(a) Find the least squares estimator of β_1.
(b) What is the variance of the estimator of the slope in part (a)?
(c) Find an expression for a $100(1 - \alpha)\%$ confidence interval for the slope β_1. Is this interval longer than the corresponding interval for the case where both the intercept and slope are unknown? Justify your answer.

IMPORTANT TERMS AND CONCEPTS

In the E-book, click on any term or concept below to go to that subject.

Analysis of variance test in regression

Confidence interval on mean response

Correlation coefficient

Empirical models

Confidence intervals on model parameters

Least squares estimation of regression model parameters

Model adequacy checking

Prediction interval on a future observation

Residual plots

Residuals

Scatter diagram

Significance of regression

Statistical tests on model parameters

Transformations

CD MATERIAL

Lack of fit test

Logistic regression

12 Multiple Linear Regression

CHAPTER OUTLINE

LEARNING OBJECTIVES

After careful study of this chapter, you should be able to do the following:

1. Use multiple regression techniques to build empirical models to engineering and scientific data

2. Understand how the method of least squares extends to fitting multiple regression models

3. Assess regression model adequacy
4. Test hypotheses and construct confidence intervals on the regression coefficients
5. Use the regression model to estimate the mean response and to make predictions and to construct confidence intervals and prediction intervals
6. Build regression models with polynomial terms
7. Use indicator variables to model categorical regressors
8. Use stepwise regression and other model building techniques to select the appropriate set of variables for a regression model

CD MATERIAL

9. Understand how ridge regression provides an effective way to estimate model parameters where there is multicollinearity.
10. Understand the basic concepts of fitting a nonlinear regression model.

Answers for many odd numbered exercises are at the end of the book. Answers to exercises whose numbers are surrounded by a box can be accessed in the e-Text by clicking on the box. Complete worked solutions to certain exercises are also available in the e-Text. These are indicated in the Answers to Selected Exercises section by a box around the exercise number. Exercises are also available for some of the text sections that appear on CD only. These exercises may be found within the e-Text immediately following the section they accompany.

12-1 MULTIPLE LINEAR REGRESSION MODEL

12-1.1 Introduction

Many applications of regression analysis involve situations in which there are more than one regressor variable. A regression model that contains more than one regressor variable is called a **multiple regression model.**

As an example, suppose that the effective life of a cutting tool depends on the cutting speed and the tool angle. A multiple regression model that might describe this relationship is

$$Y = \beta_0 + \beta_1 x_1 + \beta_2 x_2 + \epsilon \tag{12-1}$$

where Y represents the tool life, x_1 represents the cutting speed, x_2 represents the tool angle, and ϵ is a random error term. This is a **multiple linear regression model** with two regressors. The term *linear* is used because Equation 12-1 is a linear function of the unknown parameters β_0, β_1, and β_2.

The regression model in Equation 12-1 describes a plane in the three-dimensional space of Y, x_1, and x_2. Figure 12-1(a) shows this plane for the regression model

$$E(Y) = 50 + 10x_1 + 7x_2$$

where we have assumed that the expected value of the error term is zero; that is $E(\epsilon) = 0$. The parameter β_0 is the **intercept** of the plane. We sometimes call β_1 and β_2 **partial regression coefficients,** because β_1 measures the expected change in Y per unit change in x_1 when x_2 is held constant, and β_2 measures the expected change in Y per unit change in x_2 when x_1 is held constant. Figure 12-1(b) shows a **contour plot** of the regression model—that is, lines of constant $E(Y)$ as a function of x_1 and x_2. Notice that the contour lines in this plot are straight lines.

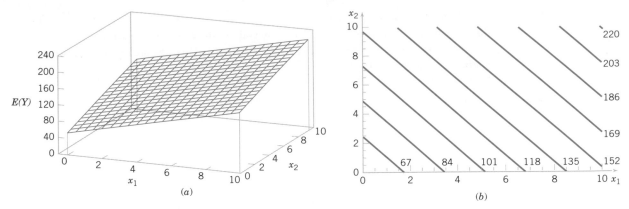

Figure 12-1   (a) The regression plane for the model $E(Y) = 50 + 10x_1 + 7x_2$. (b) The contour plot.

In general, the **dependent variable** or **response** Y may be related to k **independent** or **regressor variables.** The model

$$Y = \beta_0 + \beta_1 x_1 + \beta_2 x_2 + \cdots + \beta_k x_k + \epsilon \qquad (12\text{-}2)$$

is called a multiple linear regression model with k regressor variables. The parameters β_j, $j = 0$, $1, \ldots, k$, are called the regression coefficients. This model describes a hyperplane in the k-dimensional space of the regressor variables $\{x_j\}$. The parameter β_j represents the expected change in response Y per unit change in x_j when all the remaining regressors x_i $(i \neq j)$ are held constant.

Multiple linear regression models are often used as approximating functions. That is, the true functional relationship between Y and $x_1, x_2, \ldots, x_k$ is unknown, but over certain ranges of the independent variables the linear regression model is an adequate approximation.

Models that are more complex in structure than Equation 12-2 may often still be analyzed by multiple linear regression techniques. For example, consider the cubic polynomial model in one regressor variable.

$$Y = \beta_0 + \beta_1 x + \beta_2 x^2 + \beta_3 x^3 + \epsilon \qquad (12\text{-}3)$$

If we let $x_1 = x, x_2 = x^2, x_3 = x^3$, Equation 12-3 can be written as

$$Y = \beta_0 + \beta_1 x_1 + \beta_2 x_2 + \beta_3 x_3 + \epsilon \qquad (12\text{-}4)$$

which is a multiple linear regression model with three regressor variables.

Models that include **interaction** effects may also be analyzed by multiple linear regression methods. An interaction between two variables can be represented by a cross-product term in the model, such as

$$Y = \beta_0 + \beta_1 x_1 + \beta_2 x_2 + \beta_{12} x_1 x_2 + \epsilon \qquad (12\text{-}5)$$

If we let $x_3 = x_1 x_2$ and $\beta_3 = \beta_{12}$, Equation 12-5 can be written as

$$Y = \beta_0 + \beta_1 x_1 + \beta_2 x_2 + \beta_3 x_3 + \epsilon$$

which is a linear regression model.

Figure 12-2(a) and (b) shows the three-dimensional plot of the regression model

$$Y = 50 + 10x_1 + 7x_2 + 5x_1 x_2$$

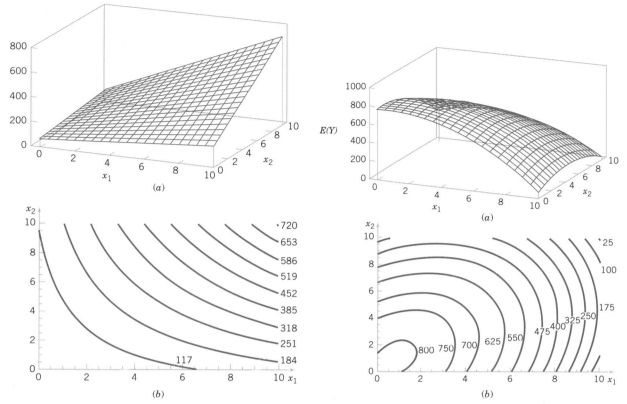

Figure 12-2   (a) Three-dimensional plot of the regression model $E(Y) = 50 + 10x_1 + 7x_2 + 5x_1x_2$. (b) The contour plot.

Figure 12-3   (a) Three-dimensional plot of the regression model $E(Y) = 800 + 10x_1 + 7x_2 - 8.5x_1^2 - 5x_2^2 + 4x_1x_2$. (b) The contour plot.

and the corresponding two-dimensional contour plot. Notice that, although this model is a linear regression model, the shape of the surface that is generated by the model is not linear. In general, **any regression model that is linear in parameters** (the β's) **is a linear regression model, regardless of the shape of the surface that it generates.**

Figure 12-2 provides a nice graphical interpretation of an interaction. Generally, interaction implies that the effect produced by changing one variable (x_1, say) depends on the level of the other variable (x_2). For example, Fig. 12-2 shows that changing x_1 from 2 to 8 produces a much smaller change in $E(Y)$ when $x_2 = 2$ than when $x_2 = 10$. Interaction effects occur frequently in the study and analysis of real-world systems, and regression methods are one of the techniques that we can use to describe them.

As a final example, consider the second-order model with interaction

$$Y = \beta_0 + \beta_1x_1 + \beta_2x_2 + \beta_{11}x_1^2 + \beta_{22}x_2^2 + \beta_{12}x_1x_2 + \epsilon \tag{12-6}$$

If we let $x_3 = x_1^2, x_4 = x_2^2, x_5 = x_1x_2, \beta_3 = \beta_{11}, \beta_4 = \beta_{22}$, and $\beta_5 = \beta_{12}$, Equation 12-6 can be written as a multiple linear regression model as follows:

$$Y = \beta_0 + \beta_1x_1 + \beta_2x_2 + \beta_3x_3 + \beta_4x_4 + \beta_5x_5 + \epsilon$$

Figure 12-3(a) and (b) show the three-dimensional plot and the corresponding contour plot for

$$E(Y) = 800 + 10x_1 + 7x_2 - 8.5x_1^2 - 5x_2^2 + 4x_1x_2$$

These plots indicate that the expected change in Y when x_1 is changed by one unit (say) is a function of *both* x_1 and x_2. The quadratic and interaction terms in this model produce a mound-shaped function. Depending on the values of the regression coefficients, the second-order model with interaction is capable of assuming a wide variety of shapes; thus, it is a very flexible regression model.

12-1.2   Least Squares Estimation of the Parameters

The **method of least squares** may be used to estimate the regression coefficients in the multiple regression model, Equation 12-2. Suppose that $n > k$ observations are available, and let x_{ij} denote the ith observation or level of variable x_j. The observations are

$$(x_{i1}, x_{i2}, \dots, x_{ik}, y_i), \qquad i = 1, 2, \dots, n \quad \text{and} \quad n > k$$

It is customary to present the data for multiple regression in a table such as Table 12-1.
    Each observation $(x_{i1}, x_{i2}, \dots, x_{ik}, y_i)$, satisfies the model in Equation 12-2, or

$$y_i = \beta_0 + \beta_1 x_{i1} + \beta_2 x_{i2} + \cdots + \beta_k x_{ik} + \epsilon_i$$
$$= \beta_0 + \sum_{j=1}^{k} \beta_j x_{ij} + \epsilon_i \qquad i = 1, 2, \dots, n \tag{12-7}$$

The least squares function is

$$L = \sum_{i=1}^{n} \epsilon_i^2 = \sum_{i=1}^{n} \left(y_i - \beta_0 - \sum_{j=1}^{k} \beta_j x_{ij} \right)^2 \tag{12-8}$$

We want to minimize L with respect to $\beta_0, \beta_1, \dots, \beta_k$. The **least squares estimates** of $\beta_0, \beta_1, \dots, \beta_k$ must satisfy

$$\left. \frac{\partial L}{\partial \beta_0} \right|_{\hat{\beta}_0, \hat{\beta}_1, \dots, \hat{\beta}_k} = -2 \sum_{i=1}^{n} \left(y_i - \hat{\beta}_0 - \sum_{j=1}^{k} \hat{\beta}_j x_{ij} \right) = 0 \tag{12-9a}$$

and

$$\left. \frac{\partial L}{\partial \beta_j} \right|_{\hat{\beta}_0, \hat{\beta}_1, \dots, \hat{\beta}_k} = -2 \sum_{i=1}^{n} \left(y_i - \hat{\beta}_0 - \sum_{j=1}^{k} \hat{\beta}_j x_{ij} \right) x_{ij} = 0 \quad j = 1, 2, \dots, k \tag{12-9b}$$

Simplifying Equation 12-9, we obtain the **least squares normal equations**

$$n\hat{\beta}_0 + \hat{\beta}_1 \sum_{i=1}^{n} x_{i1} + \hat{\beta}_2 \sum_{i=1}^{n} x_{i2} + \cdots + \hat{\beta}_k \sum_{i=1}^{n} x_{ik} = \sum_{i=1}^{n} y_i$$

$$\hat{\beta}_0 \sum_{i=1}^{n} x_{i1} + \hat{\beta}_1 \sum_{i=1}^{n} x_{i1}^2 + \hat{\beta}_2 \sum_{i=1}^{n} x_{i1} x_{i2} + \cdots + \hat{\beta}_k \sum_{i=1}^{n} x_{i1} x_{ik} = \sum_{i=1}^{n} x_{i1} y_i$$

$$\vdots \qquad \vdots \qquad \vdots \qquad \vdots \qquad \vdots$$

$$\hat{\beta}_0 \sum_{i=1}^{n} x_{ik} + \hat{\beta}_1 \sum_{i=1}^{n} x_{ik} x_{i1} + \hat{\beta}_2 \sum_{i=1}^{n} x_{ik} x_{i2} + \cdots + \hat{\beta}_k \sum_{i=1}^{n} x_{ik}^2 = \sum_{i=1}^{n} x_{ik} y_i \tag{12-10}$$

Note that there are $p = k + 1$ normal Equations, one for each of the unknown regression coefficients. The solution to the normal Equations will be the **least squares estimators** of the

Table 12-1 Data for Multiple Linear Regression

y	x_1	x_2	$\cdots$	x_k
y_1	x_{11}	x_{12}	$\cdots$	x_{1k}
y_2	x_{21}	x_{22}	$\cdots$	x_{2k}
$\vdots$	$\vdots$	$\vdots$		$\vdots$
y_n	x_{n1}	x_{n2}	$\cdots$	x_{nk}

regression coefficients, $\hat{\beta}_0, \hat{\beta}_1, \ldots, \hat{\beta}_k$. The normal equations can be solved by any method appropriate for solving a system of linear equations.

EXAMPLE 12-1

In Chapter 1, we used data on pull strength of a wire bond in a semiconductor manufacturing process, wire length, and die height to illustrate building an empirical model. We will use the same data, repeated for convenience in Table 12-2, and show the details of estimating the model parameters. A three-dimensional scatter plot of the data is presented in Fig. 1-13. Figure 12-4 shows a matrix of two-dimensional scatter plots of the data. These displays can be helpful in visualizing the relationships among variables in a multivariable data set.

Specifically, we will fit the multiple linear regression model

$$Y = \beta_0 + \beta_1 x_1 + \beta_2 x_2 + \epsilon$$

where Y = pull strength, x_1 = wire length, and x_2 = die height. From the data in Table 12-2 we calculate

$$n = 25, \quad \sum_{i=1}^{25} y_i = 725.82$$

$$\sum_{i=1}^{25} x_{i1} = 206, \quad \sum_{i=1}^{25} x_{i2} = 8{,}294$$

$$\sum_{i=1}^{25} x_{i1}^2 = 2{,}396, \quad \sum_{i=1}^{25} x_{i2}^2 = 3{,}531{,}848$$

$$\sum_{i=1}^{25} x_{i1} x_{i2} = 77{,}177, \quad \sum_{i=1}^{25} x_{i1} y_i = 8{,}008.37, \quad \sum_{i=1}^{25} x_{i2} y_i = 274{,}811.31$$

Table 12-2 Wire Bond Data for Example 11-1

Observation Number	Pull Strength y	Wire Length x_1	Die Height x_2	Observation Number	Pull Strength y	Wire Length x_1	Die Height x_2
1	9.95	2	50	14	11.66	2	360
2	24.45	8	110	15	21.65	4	205
3	31.75	11	120	16	17.89	4	400
4	35.00	10	550	17	69.00	20	600
5	25.02	8	295	18	10.30	1	585
6	16.86	4	200	19	34.93	10	540
7	14.38	2	375	20	46.59	15	250
8	9.60	2	52	21	44.88	15	290
9	24.35	9	100	22	54.12	16	510
10	27.50	8	300	23	56.63	17	590
11	17.08	4	412	24	22.13	6	100
12	37.00	11	400	25	21.15	5	400
13	41.95	12	500				

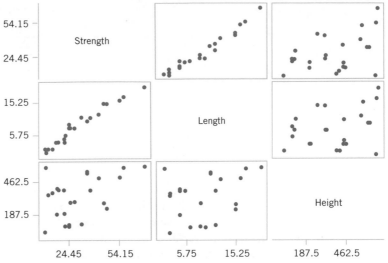

Figure 12-4   Matrix of scatter plots (from Minitab) for the wire bond pull strength data in Table 12-2.

For the model $Y = \beta_0 + \beta_1 x_1 + \beta_2 x_2 + \epsilon$, the normal equations 12-10 are

$$n\hat{\beta}_0 + \hat{\beta}_1 \sum_{i=1}^{n} x_{i1} + \hat{\beta}_2 \sum_{i=1}^{n} x_{i2} = \sum_{i=1}^{n} y_i$$

$$\hat{\beta}_0 \sum_{i=1}^{n} x_{i1} + \hat{\beta}_1 \sum_{i=1}^{n} x_{i1}^2 + \hat{\beta}_2 \sum_{i=1}^{n} x_{i1}x_{i2} = \sum_{i=1}^{n} x_{i1}y_i$$

$$\hat{\beta}_0 \sum_{i=1}^{n} x_{i2} + \hat{\beta}_1 \sum_{i=1}^{n} x_{i1}x_{i2} + \hat{\beta}_2 \sum_{i=1}^{n} x_{i2}^2 = \sum_{i=1}^{n} x_{i2}y_i$$

Inserting the computed summations into the normal equations, we obtain

$$25\hat{\beta}_0 + 206\hat{\beta}_1 + 8294\hat{\beta}_2 = 725.82$$

$$206\hat{\beta}_0 + 2396\hat{\beta}_1 + 77{,}177\hat{\beta}_2 = 8{,}008.37$$

$$8294\hat{\beta}_0 + 77{,}177\hat{\beta}_1 + 3{,}531{,}848\hat{\beta}_2 = 274{,}811.31$$

The solution to this set of equations is

$$\hat{\beta}_0 = 2.26379, \quad \hat{\beta}_1 = 2.74427, \quad \hat{\beta}_2 = 0.01253$$

Therefore, the fitted regression equation is

$$\hat{y} = 2.26379 + 2.74427x_1 + 0.01253x_2$$

This equation can be used to predict pull strength for pairs of values of the regressor variables wire length (x_1) and die height (x_2). This is essentially the same regression model given in Equation 1-7, Section 1-3. Figure 1-14 shows a three-dimentional plot of the plane of predicted values $\hat{y}$ generated from this equation.

12-1.3   Matrix Approach to Multiple Linear Regression

In fitting a multiple regression model, it is much more convenient to express the mathematical operations using **matrix notation.** Suppose that there are k regressor variables and n observations, $(x_{i1}, x_{i2}, \ldots, x_{ik}, y_i)$, $i = 1, 2, \ldots, n$ and that the model relating the regressors to the response is

$$y_i = \beta_0 + \beta_1 x_{i1} + \beta_2 x_{i2} + \cdots + \beta_k x_{ik} + \epsilon_i \qquad i = 1, 2, \ldots, n$$

This model is a system of n equations that can be expressed in matrix notation as

$$\mathbf{y} = \mathbf{X}\boldsymbol{\beta} + \boldsymbol{\epsilon} \tag{12-11}$$

where

$$\mathbf{y} = \begin{bmatrix} y_1 \\ y_2 \\ \vdots \\ y_n \end{bmatrix} \quad \mathbf{X} = \begin{bmatrix} 1 & x_{11} & x_{12} & \cdots & x_{1k} \\ 1 & x_{21} & x_{22} & \cdots & x_{2k} \\ \vdots & \vdots & \vdots & & \vdots \\ 1 & x_{n1} & x_{n2} & \cdots & x_{nk} \end{bmatrix} \quad \boldsymbol{\beta} = \begin{bmatrix} \beta_0 \\ \beta_1 \\ \vdots \\ \beta_k \end{bmatrix} \text{ and } \boldsymbol{\epsilon} = \begin{bmatrix} \epsilon_1 \\ \epsilon_2 \\ \vdots \\ \epsilon_n \end{bmatrix}$$

In general, $\mathbf{y}$ is an $(n \times 1)$ vector of the observations, $\mathbf{X}$ is an $(n \times p)$ matrix of the levels of the independent variables, $\boldsymbol{\beta}$ is a $(p \times 1)$ vector of the regression coefficients, and $\boldsymbol{\epsilon}$ is a $(n \times 1)$ vector of random errors.

We wish to find the vector of least squares estimators, $\hat{\boldsymbol{\beta}}$, that minimizes

$$L = \sum_{i=1}^{n} \epsilon_i^2 = \boldsymbol{\epsilon}'\boldsymbol{\epsilon} = (\mathbf{y} - \mathbf{X}\boldsymbol{\beta})'(\mathbf{y} - \mathbf{X}\boldsymbol{\beta})$$

The least squares estimator $\hat{\boldsymbol{\beta}}$ is the solution for $\boldsymbol{\beta}$ in the equations

$$\frac{\partial L}{\partial \boldsymbol{\beta}} = \mathbf{0}$$

We will not give the details of taking the derivatives above; however, the resulting equations that must be solved are

$$\mathbf{X}'\mathbf{X}\hat{\boldsymbol{\beta}} = \mathbf{X}'\mathbf{y} \tag{12-12}$$

Equations 12-12 are the least squares normal equations in matrix form. They are identical to the scalar form of the normal equations given earlier in Equations 12-10. To solve the normal equations, multiply both sides of Equations 12-12 by the inverse of $\mathbf{X}'\mathbf{X}$. Therefore, the least squares estimate of $\boldsymbol{\beta}$ is

$$\hat{\boldsymbol{\beta}} = (\mathbf{X}'\mathbf{X})^{-1}\mathbf{X}'\mathbf{y} \tag{12-13}$$

Note that there are $p = k + 1$ normal equations in $p = k + 1$ unknowns (the values of $\hat{\beta}_0, \hat{\beta}_1, \ldots, \hat{\beta}_k$). Furthermore, the matrix $\mathbf{X'X}$ is always nonsingular, as was assumed above, so the methods described in textbooks on determinants and matrices for inverting these matrices can be used to find $(\mathbf{X'X})^{-1}$. In practice, multiple regression calculations are almost always performed using a computer.

It is easy to see that the matrix form of the normal equations is identical to the scalar form. Writing out Equation 12-12 in detail, we obtain

$$
\begin{bmatrix}
n & \sum_{i=1}^{n} x_{i1} & \sum_{i=1}^{n} x_{i2} & \cdots & \sum_{i=1}^{n} x_{ik} \\
\sum_{i=1}^{n} x_{i1} & \sum_{i=1}^{n} x_{i1}^2 & \sum_{i=1}^{n} x_{i1}x_{i2} & \cdots & \sum_{i=1}^{n} x_{i1}x_{ik} \\
\vdots & \vdots & \vdots & & \vdots \\
\sum_{i=1}^{n} x_{ik} & \sum_{i=1}^{n} x_{ik}x_{i1} & \sum_{i=1}^{n} x_{ik}x_{i2} & \cdots & \sum_{i=1}^{n} x_{ik}^2
\end{bmatrix}
\begin{bmatrix}
\hat{\beta}_0 \\
\hat{\beta}_1 \\
\vdots \\
\hat{\beta}_k
\end{bmatrix}
=
\begin{bmatrix}
\sum_{i=1}^{n} y_i \\
\sum_{i=1}^{n} x_{i1}y_i \\
\vdots \\
\sum_{i=1}^{n} x_{ik}y_i
\end{bmatrix}
$$

If the indicated matrix multiplication is performed, the scalar form of the normal equations (that is, Equation 12-10) will result. In this form it is easy to see that $\mathbf{X'X}$ is a $(p \times p)$ symmetric matrix and $\mathbf{X'y}$ is a $(p \times 1)$ column vector. Note the special structure of the $\mathbf{X'X}$ matrix. The diagonal elements of $\mathbf{X'X}$ are the sums of squares of the elements in the columns of $\mathbf{X}$, and the off-diagonal elements are the sums of cross-products of the elements in the columns of $\mathbf{X}$. Furthermore, note that the elements of $\mathbf{X'y}$ are the sums of cross-products of the columns of $\mathbf{X}$ and the observations $\{y_i\}$.

The fitted regression model is

$$
\hat{y}_i = \hat{\beta}_0 + \sum_{j=1}^{k} \hat{\beta}_j x_{ij} \qquad i = 1, 2, \ldots, n \tag{12-14}
$$

In matrix notation, the fitted model is

$$
\hat{\mathbf{y}} = \mathbf{X}\hat{\boldsymbol{\beta}}
$$

The difference between the observation y_i and the fitted value $\hat{y}_i$ is a **residual,** say, $e_i = y_i - \hat{y}_i$. The $(n \times 1)$ vector of residuals is denoted by

$$
\mathbf{e} = \mathbf{y} - \hat{\mathbf{y}} \tag{12-15}
$$

EXAMPLE 12-2 | In Example 12-1, we illustrated fitting the multiple regression model

$$
y = \beta_0 + \beta_1 x_1 + \beta_2 x_2 + \epsilon
$$

where y is the observed pull strength for a wire bond, x_1 is the wire length, and x_2 is the die height. The 25 observations are in Table 12-2. We will now use the matrix approach to fit the regression model above to these data. The $\mathbf{X}$ matrix and $\mathbf{y}$ vector for this model are

$$\mathbf{X} = \begin{bmatrix} 1 & 2 & 50 \\ 1 & 8 & 110 \\ 1 & 11 & 120 \\ 1 & 10 & 550 \\ 1 & 8 & 295 \\ 1 & 4 & 200 \\ 1 & 2 & 375 \\ 1 & 2 & 52 \\ 1 & 9 & 100 \\ 1 & 8 & 300 \\ 1 & 4 & 412 \\ 1 & 11 & 400 \\ 1 & 12 & 500 \\ 1 & 2 & 360 \\ 1 & 4 & 205 \\ 1 & 4 & 400 \\ 1 & 20 & 600 \\ 1 & 1 & 585 \\ 1 & 10 & 540 \\ 1 & 15 & 250 \\ 1 & 15 & 290 \\ 1 & 16 & 510 \\ 1 & 17 & 590 \\ 1 & 6 & 100 \\ 1 & 5 & 400 \end{bmatrix} \quad \mathbf{y} = \begin{bmatrix} 9.95 \\ 24.45 \\ 31.75 \\ 35.00 \\ 25.02 \\ 16.86 \\ 14.38 \\ 9.60 \\ 24.35 \\ 27.50 \\ 17.08 \\ 37.00 \\ 41.95 \\ 11.66 \\ 21.65 \\ 17.89 \\ 69.00 \\ 10.30 \\ 34.93 \\ 46.59 \\ 44.88 \\ 54.12 \\ 56.63 \\ 22.13 \\ 21.15 \end{bmatrix}$$

The $\mathbf{X'X}$ matrix is

$$\mathbf{X'X} = \begin{bmatrix} 1 & 1 & \cdots & 1 \\ 2 & 8 & \cdots & 5 \\ 50 & 110 & \cdots & 400 \end{bmatrix} \begin{bmatrix} 1 & 2 & 50 \\ 1 & 8 & 110 \\ \vdots & \vdots & \vdots \\ 1 & 5 & 400 \end{bmatrix} = \begin{bmatrix} 25 & 206 & 8{,}294 \\ 206 & 2{,}396 & 77{,}177 \\ 8{,}294 & 77{,}177 & 3{,}531{,}848 \end{bmatrix}$$

and the $\mathbf{X'y}$ vector is

$$\mathbf{X'y} = \begin{bmatrix} 1 & 1 & \cdots & 1 \\ 2 & 8 & \cdots & 5 \\ 50 & 110 & \cdots & 400 \end{bmatrix} \begin{bmatrix} 9.95 \\ 24.45 \\ \vdots \\ 21.15 \end{bmatrix} = \begin{bmatrix} 725.82 \\ 8{,}008.37 \\ 274{,}811.31 \end{bmatrix}$$

The least squares estimates are found from Equation 12-13 as

$$\hat{\boldsymbol{\beta}} = (\mathbf{X'X})^{-1}\mathbf{X'y}$$

or

$$\begin{bmatrix} \hat{\beta}_0 \\ \hat{\beta}_1 \\ \hat{\beta}_2 \end{bmatrix} = \begin{bmatrix} 25 & 206 & 8{,}294 \\ 206 & 2{,}396 & 77{,}177 \\ 8{,}294 & 77{,}177 & 3{,}531{,}848 \end{bmatrix}^{-1} \begin{bmatrix} 725.82 \\ 8{,}008.37 \\ 274{,}811.31 \end{bmatrix}$$

$$= \begin{bmatrix} 0.214653 & -0.007491 & -0.000340 \\ -0.007491 & 0.001671 & -0.000019 \\ -0.000340 & -0.000019 & +0.0000015 \end{bmatrix} \begin{bmatrix} 725.82 \\ 8{,}008.47 \\ 274{,}811.31 \end{bmatrix} = \begin{bmatrix} 2.26379143 \\ 2.74426964 \\ 0.01252781 \end{bmatrix}$$

Therefore, the fitted regression model with the regression coefficients rounded to five decimal places is

$$\hat{y} = 2.26379 + 2.74427x_1 + 0.01253x_2$$

This is identical to the results obtained in Example 12-1.

This regression model can be used to predict values of pull strength for various values of wire length (x_1) and die height (x_2). We can also obtain the **fitted values** $\hat{y}_i$ by substituting each observation (x_{i1}, x_{i2}), $i = 1, 2, \ldots, n$, into the equation. For example, the first observation has $x_{11} = 2$ and $x_{12} = 50$, and the fitted value is

$$\hat{y}_1 = 2.26379 + 2.74427x_{11} + 0.01253x_{12}$$
$$= 2.26379 + 2.74427(2) + 0.01253(50)$$
$$= 8.38$$

The corresponding observed value is $y_1 = 9.95$. The *residual* corresponding to the first observation is

$$e_1 = y_1 - \hat{y}_1$$
$$= 9.95 - 8.38$$
$$= 1.57$$

Table 12-3 displays all 25 fitted values $\hat{y}_i$ and the corresponding residuals. The fitted values and residuals are calculated to the same accuracy as the original data.

Table 12-3 Observations, Fitted Values, and Residuals for Example 12-2

Observation Number	y_i	$\hat{y}_i$	$e_i = y_i - \hat{y}_i$	Observation Number	y_i	$\hat{y}_i$	$e_i = y_i - \hat{y}_i$
1	9.95	8.38	1.57	14	11.66	12.26	-0.60
2	24.45	25.60	-1.15	15	21.65	15.81	5.84
3	31.75	33.95	-2.20	16	17.89	18.25	-0.36
4	35.00	36.60	-1.60	17	69.00	64.67	4.33
5	25.02	27.91	-2.89	18	10.30	12.34	-2.04
6	16.86	15.75	1.11	19	34.93	36.47	-1.54
7	14.38	12.45	1.93	20	46.59	46.56	-0.03
8	9.60	8.40	1.20	21	44.88	47.06	-2.18
9	24.35	28.21	-3.86	22	54.12	52.56	1.56
10	27.50	27.98	-0.48	23	56.63	56.31	0.32
11	17.08	18.40	-1.32	24	22.13	19.98	2.15
12	37.00	37.46	-0.46	25	21.15	21.00	0.15
13	41.95	41.46	0.49				

Computers are almost always used in fitting multiple regression models. Table 12-4 presents some annotated output from Minitab for the least squares regression model for wire bond pull strength data. The upper part of the table contains the numerical estimates of the regression coefficients. The computer also calculates several other quantities that reflect important information about the regression model. In subsequent sections, we will define and explain the quantities in this output.

Estimating σ^2

Just as in simple linear regression, it is important to estimate σ^2, the variance of the error term ϵ, in a multiple regression model. Recall that in simple linear regression the estimate of σ^2 was obtained by dividing the sum of the squared residuals by $n - 2$. Now there are two parameters in the simple linear regression model, so in multiple linear regression with p parameters a logical estimator for σ^2 is

$$\hat{\sigma}^2 = \frac{\sum_{i=1}^{n} e_i^2}{n - p} = \frac{SS_E}{n - p} \qquad (12\text{-}16)$$

This is an **unbiased estimator** of σ^2. Just as in simple linear regression, the estimate of σ^2 is usually obtained from the **analysis of variance** for the regression model. The numerator of Equation 12-16 is called the **error** or **residual sum of squares**, and the denominator $n - p$ is called the **error** or **residual degrees of freedom**. Table 12-4 shows that the estimate of σ^2 for the wire bond pull strength regression model is $\hat{\sigma}^2 = 115.2/22 = 5.2364$. The Minitab output rounds the estimate to $\hat{\sigma}^2 = 5.2$.

12-1.4 Properties of the Least Squares Estimators

The statistical properties of the least squares estimators $\hat{\beta}_0, \hat{\beta}_1, \ldots, \hat{\beta}_k$ may be easily found, under certain assumptions on the error terms $\epsilon_1, \epsilon_2, \ldots, \epsilon_n$, in the regression model. Paralleling the assumptions made in Chapter 11, we assume that the errors ϵ_i are statistically independent with mean zero and variance σ^2. Under these assumptions, the least squares estimators $\hat{\beta}_0, \hat{\beta}_1, \ldots, \hat{\beta}_k$ are **unbiased estimators** of the regression coefficients $\beta_0, \beta_1, \ldots, \beta_k$. This property may be shown as follows:

$$\begin{aligned}
E(\hat{\beta}) &= E[(\mathbf{X}'\mathbf{X})^{-1}\mathbf{X}'\mathbf{Y}] \\
&= E[(\mathbf{X}'\mathbf{X})^{-1}\mathbf{X}'(\mathbf{X}\beta + \epsilon)] \\
&= E[(\mathbf{X}'\mathbf{X})^{-1}\mathbf{X}'\mathbf{X}\beta + (\mathbf{X}'\mathbf{X})^{-1}\mathbf{X}'\epsilon] \\
&= \beta
\end{aligned}$$

since $E(\epsilon) = 0$ and $(\mathbf{X}'\mathbf{X})^{-1}\mathbf{X}'\mathbf{X} = \mathbf{I}$, the identity matrix. Thus, $\hat{\beta}$ is an unbiased estimator of β.

The variances of the $\hat{\beta}$'s are expressed in terms of the elements of the inverse of the $\mathbf{X}'\mathbf{X}$ matrix. The inverse of $\mathbf{X}'\mathbf{X}$ times the constant σ^2 represents the **covariance matrix** of the regression coefficients $\hat{\beta}$. The diagonal elements of $\sigma^2 (\mathbf{X}'\mathbf{X})^{-1}$ are the variances of $\hat{\beta}_0$, $\hat{\beta}_1, \ldots, \hat{\beta}_k$, and the off-diagonal elements of this matrix are the covariances. For example, if we have $k = 2$ regressors, such as in the pull-strength problem,

$$\mathbf{C} = (\mathbf{X}'\mathbf{X})^{-1} = \begin{bmatrix} C_{00} & C_{01} & C_{02} \\ C_{10} & C_{11} & C_{12} \\ C_{20} & C_{21} & C_{22} \end{bmatrix}$$

Table 12-4   Minitab Multiple Regression Output for the Wire Bond Pull Strength Data

Regression Analysis: Strength versus Length, Height

The regression equation is
Strength = 2.26 + 2.74 Length + 0.0125 Height

Predictor		Coef	SE Coef	T	P	VIF
Constant	$\hat{\beta}_0 \rightarrow$ 2.264		1.060	2.14	0.044	
Length	$\hat{\beta}_1 \rightarrow$ 2.74427		0.09352	29.34	0.000	1.2
Height	$\hat{\beta}_2 \rightarrow$ 0.012528		0.002798	4.48	0.000	1.2

S = 2.288   R-Sq = 98.1%   R-Sq (adj) = 97.9%
PRESS = 156.163   R-Sq (pred) = 97.44%

Analysis of Variance

Source	DF	SS	MS	F	P
Regression	2	5990.8	2995.4	572.17	0.000
Residual Error	22	115.2	5.2 $\leftarrow \hat{\sigma}^2$		
Total	24	6105.9			

Source	DF	Seq SS
Length	1	5885.9
Height	1	104.9

Predicted Values for New Observations

New Obs	Fit	SE Fit	95.0% CI	95.0% PI
1	27.663	0.482	(26.663, 28.663)	(22.814, 32.512)

Values of Predictors for New Observations

New Obs	Length	Height
1	8.00	275

which is symmetric ($C_{10} = C_{01}$, $C_{20} = C_{02}$, and $C_{21} = C_{12}$) because $(\mathbf{X'X})^{-1}$ is symmetric, and we have

$$V(\hat{\beta}_j) = \sigma^2 C_{jj}, \qquad j = 0, 1, 2$$
$$\text{cov}(\hat{\beta}_i, \hat{\beta}_j) = \sigma^2 C_{ij}, \qquad i \neq j$$

In general, the covariance matrix of $\hat{\boldsymbol{\beta}}$ is a $(p \times p)$ symmetric matrix whose jjth element is the variance of $\hat{\beta}_j$ and whose i,jth element is the covariance between $\hat{\beta}_i$ and $\hat{\beta}_j$, that is,

$$\text{cov}(\hat{\boldsymbol{\beta}}) = \sigma^2 (\mathbf{X'X})^{-1} = \sigma^2 \mathbf{C}$$

The estimates of the variances of these regression coefficients are obtained by replacing σ^2 with an estimate. When σ^2 is replaced by it's estimate $\hat{\sigma}^2$, the square root of the estimated variance of the jth regression coefficient is called the **estimated standard error** of $\hat{\beta}_j$ or $se(\hat{\beta}_j) = \sqrt{\hat{\sigma}^2 C_{jj}}$. These standard errors are a useful measure of the **precision of estimation** for the regression coefficients; small standard errors imply good precision.

Multiple regression computer programs usually display these standard errors. For example, the Minitab output in Table 12-4 reports $se(\hat{\beta}_0) = 1.060$, $se(\hat{\beta}_1) = 0.09352$, and

$se(\hat{\beta}_1) = 0.002798$. The slope estimate is about twice the magnitude of its standard error, and $\hat{\beta}_1$ and $\hat{\beta}_2$ are considerably larger than $se(\hat{\beta}_1)$ and $se(\hat{\beta}_2)$. This implies reasonable precision of estimation, although the parameters β_1 and β_2 are much more precisely estimated than the intercept (this is not unusual in multiple regression).

EXERCISES FOR SECTION 12-1

12-1. A study was performed to investigate the shear strength of soil (y) as it related to depth in feet (x_1) and moisture content (x_2). Ten observations were collected, and the following summary quantities obtained: $n = 10$, $\sum x_{i1} = 223$, $\sum x_{i2} = 553$, $\sum y_i = 1{,}916$, $\sum x_{i1}^2 = 5{,}200.9$, $\sum x_{i2}^2 = 31{,}729$, $\sum x_{i1} x_{i2} = 12{,}352$, $\sum x_{i1} y_i = 43{,}550.8$, $\sum x_{i2} y_i = 104{,}736.8$, and $\sum y_i^2 = 371{,}595.6$.
(a) Set up the least squares normal equations for the model $Y = \beta_0 + \beta_1 x_1 + \beta_2 x_2 + \epsilon$.
(b) Estimate the parameters in the model in part (a).
(c) What is the predicted strength when $x_1 = 18$ feet and $x_2 = 43\%$?

12-2. A regression model is to be developed for predicting the ability of soil to absorb chemical contaminants. Ten observations have been taken on a soil absorption index (y) and two regressors: $x_1 = $ amount of extractable iron ore and $x_2 = $ amount of bauxite. We wish to fit the model $Y = \beta_0 + \beta_1 x_1 + \beta_2 x_2 + \epsilon$. Some necessary quantities are:

$$(\mathbf{X'X})^{-1} = \begin{bmatrix} 1.17991 & -7.30982 \text{ E-3} & 7.3006 \text{ E-4} \\ -7.30982 \text{ E-3} & 7.9799 \text{ E-5} & -1.23713 \text{ E-4} \\ 7.3006 \text{ E-4} & -1.23713 \text{ E-4} & 4.6576 \text{ E-4} \end{bmatrix}, \quad \mathbf{X'y} = \begin{bmatrix} 220 \\ 36{,}768 \\ 9{,}965 \end{bmatrix}$$

(a) Estimate the regression coefficients in the model specified above.
(b) What is the predicted value of the absorption index y when $x_1 = 200$ and $x_2 = 50$?

12-3. A chemical engineer is investigating how the amount of conversion of a product from a raw material (y) depends on reaction temperature (x_1) and the reaction time (x_2). He has developed the following regression models:
1. $\hat{y} = 100 + 2x_1 + 4x_2$
2. $\hat{y} = 95 + 1.5x_1 + 3x_2 + 2x_1 x_2$

Both models have been built over the range $0.5 \le x_2 \le 10$.
(a) What is the predicted value of conversion when $x_2 = 2$? Repeat this calculation for $x_2 = 8$. Draw a graph of the predicted values for both conversion models. Comment on the effect of the interaction term in model 2.
(b) Find the expected change in the mean conversion for a unit change in temperature x_1 for model 1 when $x_2 = 5$. Does this quantity depend on the specific value of reaction time selected? Why?
(c) Find the expected change in the mean conversion for a unit change in temperature x_1 for model 2 when $x_2 = 5$. Repeat this calculation for $x_2 = 2$ and $x_2 = 8$. Does the result depend on the value selected for x_2? Why?

12-4. The data in Table 12-5 are the 1976 team performance statistics for the teams in the National Football League (*Source: The Sporting News*).
(a) Fit a multiple regression model relating the number of games won to the teams' passing yardage (x_2), the percent rushing plays (x_7), and the opponents' yards rushing (x_8).
(b) Estimate σ^2.
(c) What are the standard errors of the regression coefficients?
(d) Use the model to predict the number of games won when $x_2 = 2000$ yards, $x_7 = 60\%$, and $x_8 = 1800$.

12-5. Table 12-6 presents gasoline mileage performance for 25 automobiles (*Source: Motor Trend, 1975*).
(a) Fit a multiple regression model relating gasoline mileage to engine displacement (x_1) and number of carburetor barrels (x_6).
(b) Estimate σ^2.
(c) Use the model developed in part (a) to predict mileage performance for a car with displacement $x_1 = 300$ and $x_6 = 2$.

12-6. The electric power consumed each month by a chemical plant is thought to be related to the average ambient temperature (x_1), the number of days in the month (x_2), the average product purity (x_3), and the tons of product produced (x_4). The past year's historical data are available and are presented in the following table:

y	x_1	x_2	x_3	x_4
240	25	24	91	100
236	31	21	90	95
270	45	24	88	110
274	60	25	87	88
301	65	25	91	94
316	72	26	94	99
300	80	25	87	97
296	84	25	86	96
267	75	24	88	110
276	60	25	91	105
288	50	25	90	100
261	38	23	89	98

(a) Fit a multiple linear regression model to these data.
(b) Estimate σ^2.

Table 12-5　National Football League 1976 Team Performance

Team	y	x_1	x_2	x_3	x_4	x_5	x_6	x_7	x_8	x_9
Washington	10	2113	1985	38.9	64.7	+4	868	59.7	2205	1917
Minnesota	11	2003	2855	38.8	61.3	+3	615	55.0	2096	1575
New England	11	2957	1737	40.1	60.0	+14	914	65.6	1847	2175
Oakland	13	2285	2905	41.6	45.3	−4	957	61.4	1903	2476
Pittsburgh	10	2971	1666	39.2	53.8	+15	836	66.1	1457	1866
Baltimore	11	2309	2927	39.7	74.1	+8	786	61.0	1848	2339
Los Angeles	10	2528	2341	38.1	65.4	+12	754	66.1	1564	2092
Dallas	11	2147	2737	37.0	78.3	−1	797	58.9	2476	2254
Atlanta	4	1689	1414	42.1	47.6	−3	714	57.0	2577	2001
Buffalo	2	2566	1838	42.3	54.2	−1	797	58.9	2476	2254
Chicago	7	2363	1480	37.3	48.0	+19	984	68.5	1984	2217
Cincinnati	10	2109	2191	39.5	51.9	+6	819	59.2	1901	1686
Cleveland	9	2295	2229	37.4	53.6	−5	1037	58.8	1761	2032
Denver	9	1932	2204	35.1	71.4	+3	986	58.6	1709	2025
Detroit	6	2213	2140	38.8	58.3	+6	819	59.2	1901	1686
Green Bay	5	1722	1730	36.6	52.6	−19	791	54.4	2288	1835
Houston	5	1498	2072	35.3	59.3	−5	776	49.6	2072	1914
Kansas City	5	1873	2929	41.1	55.3	+10	789	54.3	2861	2496
Miami	6	2118	2268	38.6	69.6	+6	582	58.7	2411	2670
New Orleans	4	1775	1983	39.3	78.3	+7	901	51.7	2289	2202
New York Giants	3	1904	1792	39.7	38.1	−9	734	61.9	2203	1988
New York Jets	3	1929	1606	39.7	68.8	−21	627	52.7	2592	2324
Philadelphia	4	2080	1492	35.5	68.8	−8	722	57.8	2053	2550
St. Louis	10	2301	2835	35.3	74.1	+2	683	59.7	1979	2110
San Diego	6	2040	2416	38.7	50.0	0	576	54.9	2048	2628
San Francisco	8	2447	1638	39.9	57.1	−8	848	65.3	1786	1776
Seattle	2	1416	2649	37.4	56.3	−22	684	43.8	2876	2524
Tampa Bay	0	1503	1503	39.3	47.0	−9	875	53.5	2560	2241

y: Games won (per 14 game season)
x_1: Rushing yards (season)
x_2: Passing yards (season)
x_3: Punting yards (yds/punt)
x_4: Field goal percentage (Field goals made/Field goals attempted—season)
x_5: Turnover differential (turnovers acquired—turnovers lost)
x_6: Penalty yards (season)
x_7: Percent rushing (rushing plays/total plays)

Table 12-6 Gasoline Mileage Performance for 25 Automobiles

Automobile	y	x_1	x_2	x_3	x_4	x_5	x_6	x_7	x_8	x_9	x_{10}	x_{11}
Apollo	18.90	350	165	260	8.0:1	2.56:1	4	3	200.3	69.9	3910	A
Nova	20.00	250	105	185	8.25:1	2.73:1	1	3	196.7	72.2	3510	A
Monarch	18.25	351	143	255	8.0:1	3.00:1	2	3	199.9	74.0	3890	A
Duster	20.07	225	95	170	8.4:1	2.76:1	1	3	194.1	71.8	3365	M
Jenson Conv.	11.2	440	215	330	8.2:1	2.88:1	4	3	184.5	69	4215	A
Skyhawk	22.12	231	110	175	8.0:1	2.56:1	2	3	179.3	65.4	3020	A
Scirocco	34.70	89.7	70	81	8.2:1	3.90:1	2	4	155.7	64	1905	M
Corolla SR-5	30.40	96.9	75	83	9.0:1	4.30:1	2	5	165.2	65	2320	M
Camaro	16.50	350	155	250	8.5:1	3.08:1	4	3	195.4	74.4	3885	A
Datsun B210	36.50	85.3	80	83	8.5:1	3.89:1	2	4	160.6	62.2	2009	M
Capri II	21.50	171	109	146	8.2:1	3.22:1	2	4	170.4	66.9	2655	M
Pacer	19.70	258	110	195	8.0:1	3.08:1	1	3	171.5	77	3375	A
Granada	17.80	302	129	220	8.0:1	3.0:1	2	3	199.9	74	3890	A
Eldorado	14.39	500	190	360	8.5:1	2.73:1	4	3	224.1	79.8	5290	A
Imperial	14.89	440	215	330	8.2:1	2.71:1	4	3	231.0	79.7	5185	A
Nova LN	17.80	350	155	250	8.5:1	3.08:1	4	3	196.7	72.2	3910	A
Starfire	23.54	231	110	175	8.0:1	2.56:1	2	3	179.3	65.4	3050	A
Cordoba	21.47	360	180	290	8.4:1	2.45:1	2	3	214.2	76.3	4250	A
Trans Am	16.59	400	185	NA	7.6:1	3.08:1	4	3	196	73	3850	A
Corolla E-5	31.90	96.9	75	83	9.0:1	4.30:1	2	5	165.2	61.8	2275	M
Mark IV	13.27	460	223	366	8.0:1	3.00:1	4	3	228	79.8	5430	A
Celica GT	23.90	133.6	96	120	8.4:1	3.91:1	2	5	171.5	63.4	2535	M
Charger SE	19.73	318	140	255	8.5:1	2.71:1	2	3	215.3	76.3	4370	A
Cougar	13.90	351	148	243	8.0:1	3.25:1	2	3	215.5	78.5	4540	A
Corvette	16.50	350	165	255	8.5:1	2.73:1	4	3	185.2	69	3660	A

y: Miles/gallon
x_1: Displacement (cubic inches)
x_2: Horsepower (foot-pounds)
x_3: Torque (foot-pounds)
x_4: Compression ratio
x_5: Rear axle ratio
x_6: Carburetor (barrels)
x_7: No. of transmission speeds
x_8: Overall length (inches)
x_9: Width (inches)
x_{10}: Weight (pounds)
x_{11}: Type of transmission (A—automatic, M—manual)

(c) Compute the standard errors of the regression coefficients.

(d) Predict power consumption for a month in which $x_1 = 75°F$, $x_2 = 24$ days, $x_3 = 90\%$, and $x_4 = 98$ tons.

12-7. A study was performed on wear of a bearing y and its relationship to $x_1 =$ oil viscosity and $x_2 =$ load. The following data were obtained.

y	x_1	x_2
293	1.6	851
230	15.5	816
172	22.0	1058
91	43.0	1201
113	33.0	1357
125	40.0	1115

(a) Fit a multiple linear regression model to these data.

(b) Estimate σ^2 and the standard errors of the regression coefficients.

(c) Use the model to predict wear when $x_1 = 25$ and $x_2 = 1000$.

(d) Fit a multiple linear regression model with an interaction term to these data.

(e) Estimate σ^2 and $se(\hat{\beta}_j)$ for this new model. How did these quantities change? Does this tell you anything about the value of adding the interaction term to the model?

(f) Use the model in (d) to predict when $x_1 = 25$ and $x_2 = 1000$. Compare this prediction with the predicted value from part (b) above.

12-8. The pull strength of a wire bond is an important characteristic. The following table gives information on pull strength (y), die height (x_1), post height (x_2), loop height (x_3), wire length (x_4), bond width on the die (x_5), and bond width on the post (x_6).

y	x_1	x_2	x_3	x_4	x_5	x_6
8.0	5.2	19.6	29.6	94.9	2.1	2.3
8.3	5.2	19.8	32.4	89.7	2.1	1.8
8.5	5.8	19.6	31.0	96.2	2.0	2.0
8.8	6.4	19.4	32.4	95.6	2.2	2.1
9.0	5.8	18.6	28.6	86.5	2.0	1.8
9.3	5.2	18.8	30.6	84.5	2.1	2.1
9.3	5.6	20.4	32.4	88.8	2.2	1.9
9.5	6.0	19.0	32.6	85.7	2.1	1.9
9.8	5.2	20.8	32.2	93.6	2.3	2.1
10.0	5.8	19.9	31.8	86.0	2.1	1.8
10.3	6.4	18.0	32.6	87.1	2.0	1.6
10.5	6.0	20.6	33.4	93.1	2.1	2.1
10.8	6.2	20.2	31.8	83.4	2.2	2.1
11.0	6.2	20.2	32.4	94.5	2.1	1.9
11.3	6.2	19.2	31.4	83.4	1.9	1.8
11.5	5.6	17.0	33.2	85.2	2.1	2.1
11.8	6.0	19.8	35.4	84.1	2.0	1.8
12.3	5.8	18.8	34.0	86.9	2.1	1.8
12.5	5.6	18.6	34.2	83.0	1.9	2.0

(a) Fit a multiple linear regression model using x_2, x_3, x_4, and x_5 as the regressors.

(b) Estimate σ^2.

(c) Find the $se(\hat{\beta}_j)$. How precisely are the regression coefficients estimated, in your opinion?

(d) Use the model from part (a) to predict pull strength when $x_2 = 20$, $x_3 = 30$, $x_4 = 90$, and $x_5 = 2.0$.

12-9. An engineer at a semiconductor company wants to model the relationship between the device HFE (y) and three parameters: Emitter-RS (x_1), Base-RS (x_2), and Emitter-to-Base RS (x_3). The data are shown in the following table.

x_1 Emitter-RS	x_2 Base-RS	x_3 E-B-RS	y HFE-1M-5V
14.620	226.00	7.000	128.40
15.630	220.00	3.375	52.62
14.620	217.40	6.375	113.90
15.000	220.00	6.000	98.01
14.500	226.50	7.625	139.90
15.250	224.10	6.000	102.60
16.120	220.50	3.375	48.14
15.130	223.50	6.125	109.60
15.500	217.60	5.000	82.68
15.130	228.50	6.625	112.60
15.500	230.20	5.750	97.52
16.120	226.50	3.750	59.06
15.130	226.60	6.125	111.80
15.630	225.60	5.375	89.09
15.380	229.70	5.875	101.00
14.380	234.00	8.875	171.90
15.500	230.00	4.000	66.80
14.250	224.30	8.000	157.10
14.500	240.50	10.870	208.40
14.620	223.70	7.375	133.40

(a) Fit a multiple linear regression model to the data.

(b) Estimate σ^2.

(c) Find the standard errors $se(\hat{\beta}_j)$.

(d) Predict HFE when $x_1 = 14.5$, $x_2 = 220$, and $x_3 = 5.0$.

12-10. Heat treating is often used to carburize metal parts, such as gears. The thickness of the carburized layer is considered a crucial feature of the gear and contributes to the overall reliability of the part. Because of the critical nature of this feature, two different lab tests are performed on each furnace load. One test is run on a sample pin that accompanies each load. The other test is a destructive test, where an actual part is cross-sectioned. This test involves running a carbon analysis on the surface of both the gear pitch (top of the gear tooth) and the gear root (between the gear teeth). Table 12-7 shows the results of the pitch carbon analysis test for 32 parts.

The regressors are furnace temperature (TEMP), carbon concentration and duration of the carburizing cycle

Table 12-7

TEMP	SOAKTIME	SOAKPCT	DIFFTIME	DIFFPCT	PITCH
1650	0.58	1.10	0.25	0.90	0.013
1650	0.66	1.10	0.33	0.90	0.016
1650	0.66	1.10	0.33	0.90	0.015
1650	0.66	1.10	0.33	0.95	0.016
1600	0.66	1.15	0.33	1.00	0.015
1600	0.66	1.15	0.33	1.00	0.016
1650	1.00	1.10	0.50	0.80	0.014
1650	1.17	1.10	0.58	0.80	0.021
1650	1.17	1.10	0.58	0.80	0.018
1650	1.17	1.10	0.58	0.80	0.019
1650	1.17	1.10	0.58	0.90	0.021
1650	1.17	1.10	0.58	0.90	0.019
1650	1.17	1.15	0.58	0.90	0.021
1650	1.20	1.15	1.10	0.80	0.025
1650	2.00	1.15	1.00	0.80	0.025
1650	2.00	1.10	1.10	0.80	0.026
1650	2.20	1.10	1.10	0.80	0.024
1650	2.20	1.10	1.10	0.80	0.025
1650	2.20	1.15	1.10	0.80	0.024
1650	2.20	1.10	1.10	0.90	0.025
1650	2.20	1.10	1.10	0.90	0.027
1650	2.20	1.10	1.50	0.90	0.026
1650	3.00	1.15	1.50	0.80	0.029
1650	3.00	1.10	1.50	0.70	0.030
1650	3.00	1.10	1.50	0.75	0.028
1650	3.00	1.15	1.66	0.85	0.032
1650	3.33	1.10	1.50	0.80	0.033
1700	4.00	1.10	1.50	0.70	0.039
1650	4.00	1.10	1.50	0.70	0.040
1650	4.00	1.15	1.50	0.85	0.035
1700	12.50	1.00	1.50	0.70	0.056
1700	18.50	1.00	1.50	0.70	0.068

(SOAKPCT, SOAKTIME), and carbon concentration and duration of the diffuse cycle (DIFFPCT, DIFFTIME).

(a) Fit a linear regression model relating the results of the pitch carbon analysis test (PITCH) to the five regressor variables.
(b) Estimate σ^2.
(c) Find the standard errors $se(\hat{\beta}_j)$.
(d) Use the model in part (a) to predict PITCH when TEMP = 1650, SOAKTIME = 1.00, SOAKPCT = 1.10, DIFFTIME = 1.00, and DIFFPCT = 0.80.

12-11. Statistics for 21 National Hockey League teams were obtained from the *Hockey Encyclopedia* and are shown in Table 12-8.

The variables and definitions are as follows:

Wins	Number of games won in a season.
Pts	Points awarded in a season. Two points for winning a game, one point for losing in overtime, zero points for losing in regular time.
GF	Goals for. Total goals scored during the season.
GA	Goals against. Goals scored against the team during the season.
PPG	Power play goals. Points scored while on power play.
PPcT	Power play percentage. The number of power play goals divided by the number of power play opportunities.

Table 12-8

Team	Wins	Pts	GF	GA	PPG	PPcT	SHG	PPGA	PKPcT	SHGA
Chicago	47	104	338	268	86	27.2	4	71	76.6	6
Minnesota	40	96	321	290	91	26.4	17	67	80.7	20
Toronto	28	68	23	330	79	22.3	13	83	75	9
St. Louis	25	65	285	316	67	21.2	9	63	81.3	12
Detroit	21	57	263	344	37	19.3	7	80	72.6	9
Edmonton	47	106	424	315	86	29.3	22	89	77.5	6
Calgary	32	78	321	317	90	27	7	59	77.1	6
Vancouver	30	75	303	309	65	23.8	5	56	80.8	13
Winnipeg	33	74	311	333	78	23.6	10	67	72.8	7
Los Angeles	27	66	308	365	81	23.8	10	94	68.2	14
Philadelphia	49	106	326	240	60	21.6	15	61	82	7
NY Islanders	42	96	302	226	69	25.8	10	55	83.4	3
Washington	39	94	306	283	75	20.9	3	53	81.6	11
NY Rangers	35	80	306	287	71	22.4	12	75	76	8
New Jersey	17	48	230	338	66	21.9	74	78	73.5	10
Pittsburgh	18	45	257	394	81	22.6	3	110	72.2	15
Boston	50	110	327	228	67	22.2	8	53	80.7	6
Montreal	42	98	350	286	64	22.2	8	68	73.8	8
Buffalo	38	89	318	285	67	21.5	12	48	82.5	9
Quebec	34	80	343	336	61	20.7	6	92	73.6	6
Hartford	19	45	261	403	51	19.3	6	70	76.1	9

SHG   Short-handed goals scored during the season.
PPGA   Power play goals against.
PKPcT   Penalty killing percentage. Measures a team's ability to prevent goals while its opponent is on a power play. Opponent power play goals divided by opponent's opportunities.
SHGA   Short-handed goals against. Fit a multiple linear regression model relating wins to the other variables. Estimate σ^2 and find the standard errors of the regression coefficients.

12-12.   Consider the linear regression model

$$Y_i = \beta_0' + \beta_1(x_{i1} - \bar{x}_1) + \beta_2(x_{i2} - \bar{x}_2) + \epsilon_i$$

where $\bar{x}_1 = \sum x_{i1}/n$ and $\bar{x}_2 = \sum x_{i2}/n$.
(a) Write out the least squares normal equations for this model.
(b) Verify that the least squares estimate of the intercept in this model is $\hat{\beta}_0' = \sum y_i/n = \bar{y}$.
(c) Suppose that we use $y_i - \bar{y}$ as the response variable in the model above. What effect will this have on the least squares estimate of the intercept?

12-2 HYPOTHESIS TESTS IN MULTIPLE LINEAR REGRESSION

In multiple linear regression problems, certain tests of hypotheses about the model parameters are useful in measuring model adequacy. In this section, we describe several important hypothesis-testing procedures. As in the simple linear regression case, hypothesis testing requires that the error terms ϵ_i in the regression model are normally and independently distributed with mean zero and variance σ^2.

12-2.1 Test for Significance of Regression

The test for significance of regression is a test to determine whether a linear relationship exists between the response variable y and a subset of the regressor variables $x_1, x_2, \ldots, x_k$. The appropriate hypotheses are

$$H_0: \beta_1 = \beta_2 = \cdots = \beta_k = 0$$

$$H_1: \beta_j \neq 0 \quad \text{for at least one } j \tag{12-17}$$

Rejection of $H_0: \beta_1 = \beta_2 = \cdots = \beta_k = 0$ implies that at least one of the regressor variables $x_1, x_2, \ldots, x_k$ contributes significantly to the model.

The test for significance of regression is a generalization of the procedure used in simple linear regression. The total sum of squares SS_T is partitioned into a sum of squares due to regression and a sum of squares due to error, say,

$$SS_T = SS_R + SS_E$$

Now if $H_0: \beta_1 = \beta_2 = \cdots = \beta_k = 0$ is true, SS_R/σ^2 is a chi-square random variable with k degrees of freedom. Note that the number of degrees of freedom for this chi-square random variable is equal to the number of regressor variables in the model. We can also show the SS_E/σ^2 is a chi-square random variable with $n - p$ degrees of freedom, and that SS_E and SS_R are independent. The test statistic for $H_0: \beta_1 = \beta_2 = \cdots = \beta_k = 0$ is

$$F_0 = \frac{SS_R/k}{SS_E/(n - p)} = \frac{MS_R}{MS_E} \tag{12-18}$$

We should reject H_0 if the computed value of the test statistic in Equation 12-18, f_0, is greater than $f_{\alpha,k,n-p}$. The procedure is usually summarized in an analysis of variance table such as Table 12-9.

We can find a computing formula for SS_E as follows:

$$SS_E = \sum_{i=1}^{n} (y_i - \hat{y}_i)^2 = \sum_{i=1}^{n} e_i^2 = \mathbf{e'e}$$

Substituting $\mathbf{e} = \mathbf{y} - \hat{\mathbf{y}} = \mathbf{y} - \mathbf{X}\hat{\boldsymbol{\beta}}$ into the above, we obtain

$$SS_E = \mathbf{y'y} - \hat{\boldsymbol{\beta}}'\mathbf{X'y} \tag{12-19}$$

Table 12-9   Analysis of Variance for Testing Significance of Regression in Multiple Regression

Source of Variation	Sum of Squares	Degrees of Freedom	Mean Square	F_0
Regression	SS_R	k	MS_R	MS_R/MS_E
Error or residual	SS_E	$n - p$	MS_E	
Total	SS_T	$n - 1$		

A computational formula for SS_R may be found easily. Now since $SS_T = \sum_{i=1}^{n} y_i^2 - (\sum_{i=1}^{n} y_i)^2/n = \mathbf{y'y} - (\sum_{i=1}^{n} y_i)^2/n$, we may rewrite Equation 12-19 as

$$SS_E = \mathbf{y'y} - \frac{\left(\sum_{i=1}^{n} y_i\right)^2}{n} - \left[\hat{\boldsymbol{\beta}}'\mathbf{X'y} - \frac{\left(\sum_{i=1}^{n} y_i\right)^2}{n}\right]$$

or

$$SS_E = SS_T - SS_R$$

Therefore, the regression sum of squares is

$$SS_R = \hat{\boldsymbol{\beta}}'\mathbf{X'y} - \frac{\left(\sum_{i=1}^{n} y_i\right)^2}{n} \tag{12-20}$$

EXAMPLE 12-3   We will test for significance of regression (with $\alpha = 0.05$) using the wire bond pull strength data from Example 12-1. The total sum of squares is

$$SS_T = \mathbf{y'y} - \frac{\left(\sum_{i=1}^{n} y_i\right)^2}{n}$$

$$= 27,177.9510 - \frac{(725.82)^2}{25} = 6105.9447$$

The regression sum of squares is computed from Equation 12-20 as follows:

$$SS_R = \hat{\boldsymbol{\beta}}'\mathbf{X'y} - \frac{\left(\sum_{i=1}^{n} y_i\right)^2}{n}$$

$$= 27,062.7775 - \frac{(725.82)^2}{25} = 5990.7712$$

and by subtraction

$$SS_E = SS_T - SS_R$$
$$= \mathbf{y'y} - \hat{\boldsymbol{\beta}}'\mathbf{X'y} = 115.1735$$

The analysis of variance is shown in Table 12-10. To test H_0: $\beta_1 = \beta_2 = 0$, we calculate the statistic

$$f_0 = \frac{MS_R}{MS_E} = \frac{2995.3856}{5.2352} = 572.17$$

Since $f_0 > f_{0.05,2,22} = 3.44$ (or since the P-value is considerably smaller than $\alpha = 0.05$), we reject the null hypothesis and conclude that pull strength is linearly related to either wire length or die height, or both. However, we note that this does not necessarily imply that the

Table 12-10   Test for Significance of Regression for Example 12-3

Source of Variation	Sum of Squares	Degrees of Freedom	Mean Square	f_0	P-value
Regression	5990.7712	2	2995.3856	572.17	1.08E-19
Error or residual	115.1735	22	5.2352		
Total	6105.9447	24			

relationship found is an appropriate model for predicting pull strength as a function of wire length and die height. Further tests of model adequacy are required before we can be comfortable using this model in practice.

Most multiple regression computer programs provide the test for significance of regression in their output display. The middle portion of Table 12-4 is the Minitab output for this example. Compare Tables 12-4 and 12-10 and note their equivalence apart from rounding. The P-value is rounded to zero in the computer output.

R^2 and Adjusted R^2

We may also use the **coefficient of multiple determination** R^2 as a global statistic to assess the fit of the model. Computationally,

$$R^2 = \frac{SS_R}{SS_T} = 1 - \frac{SS_E}{SS_T} \tag{12-21}$$

For the wire bond pull strength data, we find that $R^2 = SS_R/SS_T = 5990.7712/6105.9447 = 0.9811$. Thus the model accounts for about 98% of the variability in the pull strength response (refer to the Minitab output in Table 12-4). The R^2 statistic is somewhat problematic as a measure of the quality of the fit for a multiple regression model because it always increases when a variable is added to a model.

To illustrate, consider the model fit to wire bond pull strength data in Example 11-8. This was a simple linear regression model with $x_1 =$ wire length as the regressor. The value of R^2 for this model is $R^2 = 0.9640$. Therefore, adding $x_y =$ die height to the model increases R^2 by $0.9811 - 0.9640 = 0.0171$, a very small amount. Since R^2 always increases when a regressor is added, it can be difficult to judge whether the increase is telling us anything useful about the new regressor. It is particularly hard to interpret a small increase, such as observed in the pull strength data.

Many regression users prefer to use an **adjusted** R^2 statistic:

$$R_{adj}^2 = 1 - \frac{SS_E/(n-p)}{SS_T/(n-1)} \tag{12-22}$$

Because $SS_E/(n-p)$ is the error or residual mean square and $SS_T/(n-1)$ is a constant, R_{adj}^2 will only increase when a variable is added to the model if the new variable reduces the error mean square. Note that for the multiple regression model for the pull strength data $R_{adj}^2 = 0.979$ (see the Minitab output in Table 12-4), whereas in Example 11-8 the adjusted R^2 for the one-variable model is $R_{adj}^2 = 0.962$. Therefore, we would conclude that adding $x_2 =$ die height to the model does result in a meaningful reduction in unexplained variability in the response.

The **adjusted** R^2 statistic essentially penalizes the analyst for adding terms to the model. It is an easy way to guard against **overfitting,** that is, including regressors that are not really useful. Consequently, it is very useful in comparing and evaluating competing regression models. We will use R^2_{adj} for this when we discuss **variable selection** in regression in Section 12-6.3.

12-2.2 Tests on Individual Regression Coefficients and Subsets of Coefficients

We are frequently interested in testing hypotheses on the individual regression coefficients. Such tests would be useful in determining the potential value of each of the regressor variables in the regression model. For example, the model might be more effective with the inclusion of additional variables or perhaps with the deletion of one or more of the regressors presently in the model.

Adding a variable to a regression model always causes the sum of squares for regression to increase and the error sum of squares to decrease (this is why R^2 always increases when a variable is added). We must decide whether the increase in the regression sum of squares is large enough to justify using the additional variable in the model. Furthermore, adding an unimportant variable to the model can actually increase the error mean square, indicating that adding such a variable has actually made the model a poorer fit to the data (this is why R^2_{adj} is a better measure of global model fit then the ordinary R^2).

The hypotheses for testing the significance of any individual regression coefficient, say β_j, are

$$H_0\colon \beta_j = 0$$

$$H_1\colon \beta_j \neq 0 \tag{12-23}$$

If $H_0\colon \beta_j = 0$ is not rejected, this indicates that the regressor x_j can be deleted from the model. The test statistic for this hypothesis is

$$T_0 = \frac{\hat{\beta}_j}{\sqrt{\hat{\sigma}^2 C_{jj}}} = \frac{\hat{\beta}_j}{se(\hat{\beta}_j)} \tag{12-24}$$

where C_{jj} is the diagonal element of $(\mathbf{X}'\mathbf{X})^{-1}$ corresponding to $\hat{\beta}_j$. Notice that the denominator of Equation 12-24 is the standard error of the regression coefficient $\hat{\beta}_j$. The null hypothesis $H_0\colon \beta_j = 0$ is rejected if $|t_0| > t_{\alpha/2,n-p}$. This is called a **partial** or **marginal test** because the regression coefficient $\hat{\beta}_j$ depends on all the other regressor variables $x_i(i \neq j)$ that are in the model. More will be said about this in the following example.

EXAMPLE 12-4

Consider the wire bond pull strength data, and suppose that we want to test the hypothesis that the regression coefficient for x_2 (die height) is zero. The hypotheses are

$$H_0\colon \beta_2 = 0$$

$$H_1\colon \beta_2 \neq 0$$

The main diagonal element of the $(\mathbf{X}'\mathbf{X})^{-1}$ matrix corresponding to $\hat{\beta}_2$ is $C_{22} = 0.0000015$, so the t-statistic in Equation 12-24 is

$$t_0 = \frac{\hat{\beta}_2}{\sqrt{\hat{\sigma}^2 C_{22}}} = \frac{0.01253}{\sqrt{(5.2352)(0.0000015)}} = 4.4767$$

Note that we have used the estimate of σ^2 reported to four decimal places in Table 12-10. Since $t_{0.025,22} = 2.074$, we reject H_0: $\beta_2 = 0$ and conclude that the variable x_2 (die height) contributes significantly to the model. We could also have used a P-value to draw conclusions. The P-value for $t_0 = 4.4767$ is $P = 0.0002$, so with $\alpha = 0.05$ we would reject the null hypothesis. Note that this test measures the marginal or partial contribution of x_2 given that x_1 is in the model. That is, the t-test measures the contribution of adding the variable $x_2 =$ die height to a model that already contains $x_1 =$ wire length. Table 12-4 shows the value of the t-test computed by Minitab. The Minitab t-test statistic is reported to two decimal places. Note that the computer produces a t-test for each regression coefficient in the model. These t-tests indicate that both regressors contribute to the model.

There is another way to test the contribution of an individual regressor variable to the model. This approach determines the increase in the regression sum of squares obtained by adding a variable x_j (say) to the model, given that other variables $x_i (i \neq j)$ are already included in the regression equation.

The procedure used to do this is called the **general regression significance test**, or the **extra sum of squares method.** This procedure can also be used to investigate the contribution of a *subset* of the regressor variables to the model. Consider the regression model with k regressor variables

$$\mathbf{y} = \mathbf{X}\boldsymbol{\beta} + \boldsymbol{\epsilon} \tag{12-25}$$

where $\mathbf{y}$ is $(n \times 1)$, $\mathbf{X}$ is $(n \times p)$, $\boldsymbol{\beta}$ is $(p \times 1)$, $\boldsymbol{\epsilon}$ is $(n \times 1)$, and $p = k + 1$. We would like to determine if the subset of regressor variables $x_1, x_2, \ldots, x_r (r < k)$ as a whole contributes significantly to the regression model. Let the vector of regression coefficients be partitioned as follows:

$$\boldsymbol{\beta} = \begin{bmatrix} \boldsymbol{\beta}_1 \\ \boldsymbol{\beta}_2 \end{bmatrix} \tag{12-26}$$

where $\boldsymbol{\beta}_1$ is $(r \times 1)$ and $\boldsymbol{\beta}_2$ is $[(p - r) \times 1]$. We wish to test the hypotheses

$$H_0: \boldsymbol{\beta}_1 = \mathbf{0}$$
$$H_1: \boldsymbol{\beta}_1 \neq \mathbf{0} \tag{12-27}$$

where $\mathbf{0}$ denotes a vector of zeroes. The model may be written as

$$\mathbf{y} = \mathbf{X}\boldsymbol{\beta} + \boldsymbol{\epsilon} = \mathbf{X}_1\boldsymbol{\beta}_1 + \mathbf{X}_2\boldsymbol{\beta}_2 + \boldsymbol{\epsilon} \tag{12-28}$$

where $\mathbf{X}_1$ represents the columns of $\mathbf{X}$ associated with $\boldsymbol{\beta}_1$ and $\mathbf{X}_2$ represents the columns of $\mathbf{X}$ associated with $\boldsymbol{\beta}_2$.

For the **full model** (including both $\boldsymbol{\beta}_1$ and $\boldsymbol{\beta}_2$), we know that $\hat{\boldsymbol{\beta}} = (\mathbf{X}'\mathbf{X})^{-1}\mathbf{X}'\mathbf{y}$. In addition, the regression sum of squares for all variables including the intercept is

$$SS_R(\boldsymbol{\beta}) = \hat{\boldsymbol{\beta}}'\mathbf{X}'\mathbf{y} \quad (p = k + 1 \text{ degrees of freedom})$$

and

$$MS_E = \frac{\mathbf{y}'\mathbf{y} - \hat{\boldsymbol{\beta}}\mathbf{X}'\mathbf{y}}{n - p}$$

$SS_R(\boldsymbol{\beta})$ is called the regression sum of squares due to $\boldsymbol{\beta}$. To find the contribution of the terms in $\boldsymbol{\beta}_1$ to the regression, fit the model assuming the null hypothesis $H_0: \boldsymbol{\beta}_1 = \mathbf{0}$ to be true. The **reduced model** is found from Equation 12-28 as

$$\mathbf{y} = \mathbf{X}_2\boldsymbol{\beta}_2 + \boldsymbol{\epsilon} \tag{12-29}$$

The least squares estimate of $\boldsymbol{\beta}_2$ is $\hat{\boldsymbol{\beta}}_2 = (\mathbf{X}_2'\mathbf{X}_2)^{-1}\mathbf{X}_2'\mathbf{y}$, and

$$SS_R(\boldsymbol{\beta}_2) = \hat{\boldsymbol{\beta}}_2'\mathbf{X}_2'\mathbf{y} \quad (p - r \text{ degrees of freedom}) \tag{12-30}$$

The regression sum of squares due to $\boldsymbol{\beta}_1$ given that $\boldsymbol{\beta}_2$ is already in the model is

$$SS_R(\boldsymbol{\beta}_1|\boldsymbol{\beta}_2) = SS_R(\boldsymbol{\beta}) - SS_R(\boldsymbol{\beta}_2) \tag{12-31}$$

This sum of squares has r degrees of freedom. It is sometimes called the extra sum of squares due to $\boldsymbol{\beta}_1$. Note that $SS_R(\boldsymbol{\beta}_1|\boldsymbol{\beta}_2)$ is the increase in the regression sum of squares due to including the variables $x_1, x_2, \ldots, x_r$ in the model. Now $SS_R(\boldsymbol{\beta}_1|\boldsymbol{\beta}_2)$ is independent of MS_E, and the null hypothesis $\boldsymbol{\beta}_1 = \mathbf{0}$ may be tested by the statistic

$$F_0 = \frac{SS_R(\boldsymbol{\beta}_1|\boldsymbol{\beta}_2)/r}{MS_E} \tag{12-32}$$

If the computed value of the test statistic $f_0 > f_{\alpha,r,n-p}$, we reject H_0, concluding that at least one of the parameters in $\boldsymbol{\beta}_1$ is not zero and, consequently, at least one of the variables $x_1, x_2, \ldots, x_r$ in $\mathbf{X}_1$ contributes significantly to the regression model. Some authors call the test in Equation 12-32 a **partial F-test.**

The partial F-test is very useful. We can use it to measure the contribution of each individual regressor x_j as if it were the last variable added to the model by computing

$$SS_R(\beta_j|\beta_0, \beta_1, \ldots, \beta_{j-1}, \beta_{j+1}, \ldots, \beta_k), \quad j = 1, 2, \ldots, k$$

This is the increase in the regression sum of squares due to adding x_j to a model that already includes $x_1, \ldots, x_{j-1}, x_{j+1}, \ldots, x_k$. The partial F-test is a more general procedure in that we can measure the effect of sets of variables. In Section 12-6.3 we show how the partial F-test plays a major role in *model building*—that is, in searching for the best set of regressor variables to use in the model.

EXAMPLE 12-5 Consider the wire bond pull strength data in Example 12-1. We will investigate the contribution of the variable x_2 (die height) to the model using the partial F-test approach. That is, we wish to test

$$H_0: \beta_2 = 0$$
$$H_1: \beta_2 \neq 0$$

To test this hypothesis, we need the extra sum of squares due to β_2, or

$$SS_R(\beta_2|\beta_1,\beta_0) = SS_R(\beta_1,\beta_2,\beta_0) - SS_R(\beta_1,\beta_0)$$
$$= SS_R(\beta_1,\beta_2|\beta_0) - SS_R(\beta_1|\beta_0)$$

In Example 12-3 we have calculated

$$SS_R(\beta_1,\beta_2|\beta_0) = \hat{\boldsymbol{\beta}}'\mathbf{X}'\mathbf{y} - \frac{\left(\sum\limits_{i=1}^{n} y_i\right)}{n} = 5990.7712 \quad \text{(two degrees of freedom)}$$

and from Example 11-8, where we fit the model $Y = \beta_0 + \beta_1 x_1 + \epsilon$, we can calculate

$$SS_R(\beta_1|\beta_0) = \hat{\beta}_1 S_{xy} = (2.9027)(2027.7132)$$
$$= 5885.8521 \quad \text{(one degree of freedom)}$$

Therefore,

$$SS_R(\beta_2|\beta_1,\beta_0) = 5990.7712 - 5885.8521$$
$$= 104.9191 \quad \text{(one degree of freedom)}$$

This is the increase in the regression sum of squares due to adding x_2 to a model already containing x_1. To test $H_0: \beta_2 = 0$, calculate the test statistic

$$f_0 = \frac{SS_R(\beta_2|\beta_1,\beta_0)/1}{MS_E} = \frac{104.9191/1}{5.2352} = 20.04$$

Note that the MS_E from the full model, using both x_1 and x_2, is used in the denominator of the test statistic. Since $f_{0.05,1,22} = 4.30$, we reject $H_0: \beta_2 = 0$ and conclude that the regressor die height (x_2) contributes significantly to the model.

Table 12-4 shows the Minitab regression output for the wire bond pull strength data. Just below the analysis of variance summary in this table the quantity labeled "Seq SS" shows the sum of squares obtained by fitting x_1 alone (5885.9) and the sum of squares obtained by fitting x_2 after x_1. Notationally, these are referred to above as $SS_R(\beta_1|\beta_0)$ and $SS_R(\beta_2|\beta_1,\beta_0)$.

Since the partial F-test in the above example involves a single variable, it is equivalent to the t-test. To see this, recall from Example 12-5 that the t-test on $H_0: \beta_2 = 0$ resulted in the test statistic $t_0 = 4.4767$. Furthermore, the square of a t-random variable with ν degrees of freedom is an F-random variable with one and ν degrees of freedom, and we note that $t_0^2 = (4.4767)^2 = 20.04 = f_0$.

12-2.3 More About the Extra Sum of Squares Method (CD Only)

EXERCISES FOR SECTION 12-2

12-13. Consider the regression model fit to the soil shear strength data in Exercise 12-1.
(a) Test for significance of regression using $\alpha = 0.05$. What is the P-value for this test?
(b) Construct the t-test on each regression coefficient. What are your conclusions, using $\alpha = 0.05$?

12-14. Consider the absorption index data in Exercise 12-2. The total sum of squares for y is $SS_T = 742.00$.
(a) Test for significance of regression using $\alpha = 0.01$. What is the P-value for this test?
(b) Test the hypothesis H_0: $\beta_1 = 0$ versus H_1: $\beta_1 \neq 0$ using $\alpha = 0.01$. What is the P-value for this test?
What conclusion can you draw about the usefulness of x_1 as a regressor in this model?

12-15. Consider the NFL data in Exercise 12-4.
(a) Test for significance of regression using $\alpha = 0.05$. What is the P-value for this test?
(b) Conduct the t-test for each regression coefficient β_2, β_7, and β_8. Using $\alpha = 0.05$, what conclusions can you draw about the variables in this model?

12-16. Reconsider the NFL data in Exercise 12-4.
(a) Find the amount by which the regressor x_8 (opponents' yards rushing) increases the regression sum of squares.
(b) Use the results from part (a) above and Exercise 12-14 to conduct an F-test for H_0: $\beta_8 = 0$ versus H_1: $\beta_8 \neq 0$ using $\alpha = 0.05$. What is the P-value for this test? What conclusions can you draw?

12-17. Consider the gasoline mileage data in Exercise 12-5.
(a) Test for significance of regression using $\alpha = 0.05$. What conclusions can you draw?
(b) Find the t-test statistic for both regressors. Using $\alpha = 0.05$, what conclusions can you draw? Do both regressors contribute to the model?

12-18. A regression model $Y = \beta_0 + \beta_1 x_1 + \beta_2 x_2 + \beta_3 x_3 + \epsilon$ has been fit to a sample of $n = 25$ observations. The calculated t-ratios $\hat{\beta}_j / se(\hat{\beta}_j), j = 1, 2, 3$ are as follows: for β_1, $t_0 = 4.82$, for β_2, $t_0 = 8.21$ and for β_3, $t_0 = 0.98$.
(a) Find P-values for each of the t-statistics.
(b) Using $\alpha = 0.05$, what conclusions can you draw about the regressor x_3? Does it seem likely that this regressor contributes significantly to the model?

12-19. Consider the electric power consumption data in Exercise 12-6.
(a) Test for significance of regression using $\alpha = 0.05$. What is the P-value for this test?
(b) Use the t-test to assess the contribution of each regressor to the model. Using $\alpha = 0.05$, what conclusions can you draw?

12-20. Consider the bearing wear data in Exercise 12-7 with no interaction.

(a) Test for significance of regression using $\alpha = 0.05$. What is the P-value for this test? What are your conclusions?
(b) Compute the t-statistics for each regression coefficient. Using $\alpha = 0.05$, what conclusions can you draw?
(c) Use the extra sum of squares method to investigate the usefulness of adding $x_2 = $ load to a model that already contains $x_1 = $ oil viscosity. Use $\alpha = 0.05$.

12-21. Reconsider the bearing wear data from Exercises 12-7 and 12-20.
(a) Refit the model with an interaction term. Test for significance of regression using $\alpha = 0.05$.
(b) Use the extra sum of squares method to determine whether the interaction term contributes significantly to the model. Use $\alpha = 0.05$.
(c) Estimate σ^2 for the interaction model. Compare this to the estimate of σ^2 from the model in Exercise 12-20.

12-22. Consider the wire bond pull strength data in Exercise 12-8.
(a) Test for significance of regression using $\alpha = 0.05$. Find the P-value for this test. What conclusions can you draw?
(b) Calculate the t-test statistic for each regression coefficient. Using $\alpha = 0.05$, what conclusions can you draw? Do all variables contribute to the model?

12-23. Reconsider the semiconductor data in Exercise 12-9.
(a) Test for significance of regression using $\alpha = 0.05$. What conclusions can you draw?
(b) Calcuate the t-test statistic for each regression coefficient. Using $\alpha = 0.05$, what conclusions can you draw?

12-24. Exercise 12-10 presents data on heat treating gears.
(a) Test the regression model for significance of regression. Using $\alpha = 0.05$, find the P-value for the test and draw conclusions.
(b) Evaluate the contribution of each regressor to the model using the t-test with $\alpha = 0.05$.
(c) Fit a new model to the response PITCH using new regressors $x_1 = $ SOAKTIME $\times$ SOAKPCT and $x_2 = $ DIFFTIME $\times$ DIFFPCT.
(d) Test the model in part (c) for significance of regression using $\alpha = 0.05$. Also calculate the t-test for each regressor and draw conclusions.
(e) Estimate σ^2 for the model from part (c) and compare this to the estimate of σ^2 for the model in part (a). Which estimate is smaller? Does this offer any insight regarding which model might be preferable?

12-25. Data on National Hockey League team performance was presented in Exercise 12-11.
(a) Test the model from this exercise for significance of regression using $\alpha = 0.05$. What conclusions can you draw?

(b) Use the *t*-test to evaluate the contribution of each regressor to the model. Does it seem that all regressors are necessary? Use $\alpha = 0.05$.

(c) Fit a regression model relating the number of games won to the number of points scored and the number of power play goals. Does this seem to be a logical choice of regressors, considering your answer to part (b)? Test this new model for significance of regression and evaluate the contribution of each regressor to the model using the *t*-test. Use $\alpha = 0.05$.

12-3 CONFIDENCE INTERVALS IN MULTIPLE LINEAR REGRESSION

12-3.1 Confidence Intervals on Individual Regression Coefficients

In multiple regression models, it is often useful to construct confidence interval estimates for the regression coefficients $\{\beta_j\}$. The development of a procedure for obtaining these confidence intervals requires that the errors $\{\epsilon_i\}$ are normally and independently distributed with mean zero and variance σ^2. This is the same assumption required in hypothesis testing. Therefore, the observations $\{Y_i\}$ are normally and independently distributed with mean $\beta_0 + \sum_{j=1}^{k} \beta_j x_{ij}$ and variance σ^2. Since the least squares estimator $\hat{\boldsymbol{\beta}}$ is a linear combination of the observations, it follows that $\hat{\boldsymbol{\beta}}$ is normally distributed with mean vector $\boldsymbol{\beta}$ and covariance matrix $\sigma^2(\mathbf{X'X})^{-1}$. Then each of the statistics

$$T = \frac{\hat{\beta}_j - \beta_j}{\sqrt{\hat{\sigma}^2 C_{jj}}} \qquad j = 0, 1, \ldots, k \tag{12-33}$$

has a *t* distribution with $n - p$ degrees of freedom, where C_{jj} is the *jj*th element of the $(\mathbf{X'X})^{-1}$ matrix, and $\hat{\sigma}^2$ is the estimate of the error variance, obtained from Equation 12-16. This leads to the following $100(1 - \alpha)\%$ confidence interval for the regression coefficient $\beta_j, j = 0, 1, \ldots, k$.

Definition

> A $100(1 - \alpha)\%$ **confidence interval on the regression coefficient** $\beta_j, j = 0, 1, \ldots,$ k in the multiple linear regression model is given by
>
> $$\hat{\beta}_j - t_{\alpha/2, n-p}\sqrt{\hat{\sigma}^2 C_{jj}} \leq \beta_j \leq \hat{\beta}_j + t_{\alpha/2, n-p}\sqrt{\hat{\sigma}^2 C_{jj}} \tag{12-34}$$

Because $\sqrt{\hat{\sigma}^2 C_{jj}}$ is the standard error of the regression coefficient $\hat{\beta}_j$, we would also write the CI formula as $\hat{\beta}_j - t_{\alpha/2, n-p}\, se(\hat{\beta}_j) \leq \beta_j \leq \hat{\beta}_j + t_{\alpha/2, n-p}\, se(\hat{\beta}_j)$.

EXAMPLE 12-6

We will construct a 95% confidence interval on the parameter β_1 in the wire bond pull strength problem. The point estimate of β_1 is $\hat{\beta}_1 = 2.74427$ and the diagonal element of $(\mathbf{X'X})^{-1}$ corresponding to β_1 is $C_{11} = 0.001671$. The estimate of σ^2 is $\hat{\sigma}^2 = 5.2352$, and $t_{0.025, 22} = 2.074$. Therefore, the 95% CI on β_1 is computed from Equation 12-34 as

$$2.74427 - (2.074)\sqrt{(5.2352)(.001671)} \leq \beta_1 \leq 2.74427 + (2.074)\sqrt{(5.2352)(.001671)}$$

which reduces to

$$2.55029 \leq \beta_1 \leq 2.93825$$

12-3.2   Confidence Interval on the Mean Response

We may also obtain a confidence interval on the mean response at a particular point, say, $x_{01}, x_{02}, \ldots, x_{0k}$. To estimate the mean response at this point, define the vector

$$\mathbf{x}_0 = \begin{bmatrix} 1 \\ x_{01} \\ x_{02} \\ \vdots \\ x_{0k} \end{bmatrix}$$

The mean response at this point is $E(Y \mid \mathbf{x}_0) = \mu_{Y \mid \mathbf{x}_0} = \mathbf{x}_0' \boldsymbol{\beta}$, which is estimated by

$$\hat{\mu}_{Y \mid \mathbf{x}_0} = \mathbf{x}_0' \hat{\boldsymbol{\beta}} \tag{12-35}$$

This estimator is unbiased, since $E(\mathbf{x}_0' \hat{\boldsymbol{\beta}}) = \mathbf{x}_0' \boldsymbol{\beta} = E(Y \mid \mathbf{x}_0) = \mu_{Y \mid \mathbf{x}_0}$ and the variance of $\hat{\mu}_{Y \mid \mathbf{x}_0}$ is

$$V(\hat{\mu}_{Y \mid \mathbf{x}_0}) = \sigma^2 \mathbf{x}_0' (\mathbf{X}'\mathbf{X})^{-1} \mathbf{x}_0 \tag{12-36}$$

A $100(1 - \alpha)\%$ CI on $\mu_{Y \mid \mathbf{x}_0}$ can be constructed from the statistic

$$\frac{\hat{\mu}_{Y \mid \mathbf{x}_0} - \mu_{Y \mid \mathbf{x}_0}}{\sqrt{\hat{\sigma}^2 \mathbf{x}_0' (\mathbf{X}'\mathbf{X})^{-1} \mathbf{x}_0}} \tag{12-37}$$

Definition

For the multiple linear regression model, a $100(1 - \alpha)\%$ **confidence interval on the mean response** at the point $x_{01}, x_{02}, \ldots, x_{0k}$ is

$$\hat{\mu}_{Y \mid \mathbf{x}_0} - t_{\alpha/2, n-p} \sqrt{\hat{\sigma}^2 \mathbf{x}_0' (\mathbf{X}'\mathbf{X})^{-1} \mathbf{x}_0}$$

$$\leq \mu_{Y \mid \mathbf{x}_0} \leq \hat{\mu}_{Y \mid \mathbf{x}_0} + t_{\alpha/2, n-p} \sqrt{\hat{\sigma}^2 \mathbf{x}_0' (\mathbf{X}'\mathbf{X})^{-1} \mathbf{x}_0} \tag{12-38}$$

Equation 12-38 is a CI about the regression plane (or hyperplane). It is the multiple regression generalization of Equation 11-31.

EXAMPLE 12-7   The engineer in Example 12-1 would like to construct a 95% CI on the mean pull strength for a wire bond with wire length $x_1 = 8$ and die height $x_2 = 275$. Therefore,

$$\mathbf{x}_0 = \begin{bmatrix} 1 \\ 8 \\ 275 \end{bmatrix}$$

The estimated mean response at this point is found from Equation 12-35 as

$$\hat{\mu}_{Y|\mathbf{x}_0} = \mathbf{x}_0'\hat{\boldsymbol{\beta}} = \begin{bmatrix} 1 & 8 & 275 \end{bmatrix} \begin{bmatrix} 2.26379 \\ 2.74427 \\ 0.01253 \end{bmatrix} = 27.66$$

The variance of $\hat{\mu}_{Y|\mathbf{x}_0}$ is estimated by

$$\hat{\sigma}^2 \mathbf{x}_0'(\mathbf{X}'\mathbf{X})^{-1}\mathbf{x}_0 = 5.2352 \begin{bmatrix} 1 & 8 & 275 \end{bmatrix} \times \begin{bmatrix} .214653 & -.007491 & -.000340 \\ -.007491 & .001671 & -.000019 \\ -.000340 & -.000019 & .0000015 \end{bmatrix} \begin{bmatrix} 1 \\ 8 \\ 275 \end{bmatrix}$$

$$= 5.2352 (0.04444) = 0.23266$$

Therefore, a 95% CI on the mean pull strength at this point is found from Equation 12-38 as

$$27.66 - 2.074 \sqrt{0.23266} \le \mu_{Y|\mathbf{x}_0} \le 27.66 + 2.074 \sqrt{0.23266}$$

which reduces to

$$26.66 \le \mu_{Y|\mathbf{x}_0} \le 28.66$$

Some computer software packages will provide estimates of the mean for a point of interest $\mathbf{x}_0$ and the associated CI. Table 12-4 shows the Minitab output for Example 12-7. Both the estimate of the mean and the 95% CI are provided.

12-4 PREDICTION OF NEW OBSERVATIONS

A regression model can be used to predict new or **future observations** on the response variable Y corresponding to particular values of the independent variables, say, $x_{01}, x_{02}, \ldots, x_{0k}$. If $\mathbf{x}_0' = [1, x_{01}, x_{02}, \ldots, x_{0k}]$, a point estimate of the future observation Y_0 at the point x_{01}, $x_{02}, \ldots, x_{0k}$ is

$$\hat{y}_0 = \mathbf{x}_0'\hat{\boldsymbol{\beta}} \tag{12-39}$$

A $100(1 - \alpha)\%$ **prediction interval for this future observation** is

$$\hat{y}_0 - t_{\alpha/2,n-p} \sqrt{\hat{\sigma}^2(1 + \mathbf{x}_0'(\mathbf{X}'\mathbf{X})^{-1}\mathbf{x}_0)}$$
$$\le Y_0 \le \hat{y}_0 + t_{\alpha/2,n-p} \sqrt{\hat{\sigma}^2(1 + \mathbf{x}_0'(\mathbf{X}'\mathbf{X})^{-1}\mathbf{x}_0)} \tag{12-40}$$

This prediction interval is a generalization of the prediction interval given in Equation 11-33 for a future observation in simple linear regression. If you compare the prediction interval Equation 12-40 with the expression for the confidence interval on the mean, Equation 12-38, you will observe that the prediction interval is always wider than the confidence interval. The confidence interval expresses the error in estimating the mean of a distribution, while the prediction interval expresses the error in predicting a future observation from the distribution at

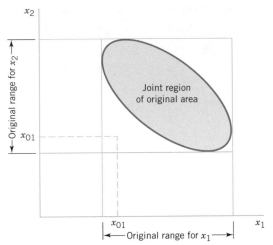

Figure 12-5   An example of extrapolation in multiple regression.

the point $\mathbf{x}_0$. This must include the error in estimating the mean at that point, as well as the inherent variability in the random variable Y at the same value $\mathbf{x} = \mathbf{x}_0$.

In predicting new observations and in estimating the mean response at a given point $x_{01}, x_{02}, \ldots, x_{0k}$, we must be careful about **extrapolating** beyond the region containing the original observations. It is very possible that a model that fits well in the region of the original data will no longer fit well outside of that region. In multiple regression it is often easy to inadvertently extrapolate, since the levels of the variables $(x_{i1}, x_{i2}, \ldots, x_{ik})$, $i = 1, 2, \ldots, n$, jointly define the region containing the data. As an example, consider Fig. 12-5, which illustrates the region containing the observations for a two-variable regression model. Note that the point (x_{01}, x_{02}) lies within the ranges of both regressor variables x_1 and x_2, but it is outside the region that is actually spanned by the original observations. Thus, either predicting the value of a new observation or estimating the mean response at this point is an extrapolation of the original regression model.

EXAMPLE 12-8

Suppose that the engineer in Example 12-1 wishes to construct a 95% prediction interval on the wire bond pull strength when the wire length is $x_1 = 8$ and the die height is $x_2 = 275$. Note that $\mathbf{x}_0' = [1 \quad 8 \quad 275]$, and the point estimate of the pull strength is $\hat{y}_0 = \mathbf{x}_0'\hat{\beta} = 27.66$. Also, in Example 12-7 we calculated $\mathbf{x}_0'(\mathbf{X}'\mathbf{X})^{-1}\mathbf{x}_0 = 0.04444$. Therefore, from Equation 12-40 we have

$$27.66 - 2.074\sqrt{5.2352(1 + 0.04444)} \leq Y_0 \leq 27.66 + 2.074\sqrt{5.2352(1 + 0.04444)}$$

and the 95% prediction interval is

$$22.81 \leq Y_0 \leq 32.51$$

Notice that the prediction interval is wider than the confidence interval on the mean response at the same point, calculated in Example 12-7. The Minitab output in Table 12-4 also displays this prediction interval.

EXERCISES FOR SECTIONS 12-3 AND 12-4

12-26. Consider the soil absorption data in Exercise 12-2.
(a) Find a 95% confidence interval on the regression coefficient β_1.
(b) Find a 95% confidence interval on mean soil absorption index when $x_1 = 200$ and $x_2 = 50$.
(c) Find a 95% prediction interval on the soil absorption index when $x_1 = 200$ and $x_2 = 50$.

12-27. Consider the NFL data in Exercise 12-4.
(a) Find a 95% confidence interval on β_8.
(b) What is the estimated standard error of $\hat{\mu}_{Y|\mathbf{x}_0}$ when $x_2 = 2000$ yards, $x_7 = 60\%$, and $x_8 = 1800$ yards?
(c) Find a 95% confidence interval on the mean number of games won when $x_2 = 2000$, $x_7 = 60$, and $x_8 = 1800$.

12-28. Consider the gasoline mileage data in Exercise 12-5.
(a) Find 99% confidence intervals on β_1 and β_6.
(b) Find a 99% confidence interval on the mean of Y when $x_1 = 300$ and $x_6 = 4$.
(c) Fit a new regression model to these data using x_1, x_2, x_6, and x_{10} as the regressors. Find 99% confidence intervals on the regression coefficients in this new model.
(d) Compare the lengths of the confidence intervals on β_1 and β_6 from part (c) with those found in part (a). Which intervals are longer? Does this offer any insight about adding the variables x_2 and x_{10} to the model?

12-29. Consider the electric power consumption data in Exercise 12-6.
(a) Find 95% confidence intervals on β_1, β_2, β_3, and β_4.
(b) Find a 95% confidence interval on the mean of Y when $x_1 = 75$, $x_2 = 24$, $x_3 = 90$, and $x_4 = 98$.
(c) Find a 95% prediction interval on the power consumption when $x_1 = 75$, $x_2 = 24$, $x_3 = 90$, and $x_4 = 98$.

12-30. Consider the bearing wear data in Exercise 12-7.
(a) Find 99% confidence intervals on β_1 and β_2.
(b) Recompute the confidence intervals in part (a) after the interaction term x_1x_2 is added to the model. Compare the lengths of these confidence intervals with those computed in part (a). Do the lengths of these intervals provide any information about the contribution of the interaction term in the model?

12-31. Consider the wire bond pull strength data in Exercise 12-8.
(a) Find 95% confidence interval on the regression coefficients.

(b) Find a 95% confidence interval on mean pull strength when $x_2 = 20$, $x_3 = 30$, $x_4 = 90$ and $x_5 = 2.0$.
(c) Find a 95% prediction interval on pull strength when $x_2 = 20$, $x_3 = 30$, $x_4 = 90$, and $x_5 = 2.0$.

12-32. Consider the semiconductor data in Exercise 12-9.
(a) Find 99% confidence intervals on the regression coefficients.
(b) Find a 99% prediction interval on HFE when $x_1 = 14.5$, $x_2 = 220$, and $x_3 = 5.0$.
(c) Find a 99% confidence interval on mean HFE when $x_1 = 14.5$, $x_2 = 220$, and $x_3 = 5.0$.

12-33. Consider the heat treating data from Exercise 12-10.
(a) Find 95% confidence intervals on the regression coefficients.
(b) Find a 95% confidence interval on mean PITCH when TEMP = 1650, SOAKTIME = 1.00, SOAKPCT = 1.10, DIFFTIME = 1.00, and DIFFPCT = 0.80.

12-34. Reconsider the heat treating data in Exercises 12-10 and 12-24, where we fit a model to PITCH using regressors $x_1 = $ SOAKTIME $\times$ SOAKPCT and $x_2 = $ DIFFTIME $\times$ DIFFPCT.
(a) Using the model with regressors x_1 and x_2, find a 95% confidence interval on mean PITCH when SOAKTIME = 1.00, SOAKPCT = 1.10, DIFFTIME = 1.00, and DIFFPCT = 0.80.
(b) Compare the length of this confidence interval with the length of the confidence interval on mean PITCH at the same point from Exercise 12-33 part (b), where an additive model in SOAKTIME, SOAKPCT, DIFFTIME, and DIFFPCT was used. Which confidence interval is shorter? Does this tell you anything about which model is preferable?

12-35. Consider the NHL data in Exercise 12-11.
(a) Find a 95% confidence interval on the regression coefficient for the variable "Pts."
(b) Fit a simple linear regression model relating the response variable "wins" to the regressor "Pts."
(c) Find a 95% confidence interval on the slope for the simple linear regression model from part (b).
(d) Compare the lengths of the two confidence intervals computed in parts (a) and (c). Which interval is shorter? Does this tell you anything about which model is preferable?

12-5 MODEL ADEQUACY CHECKING

12-5.1 Residual Analysis

The **residuals** from the multiple regression model, defined by $e_i = y_i - \hat{y}_i$, play an important role in judging model adequacy just as they do in simple linear regression. As noted in Section 11-7.1, several residual plots are often useful; these are illustrated in Example 12-9. It is also

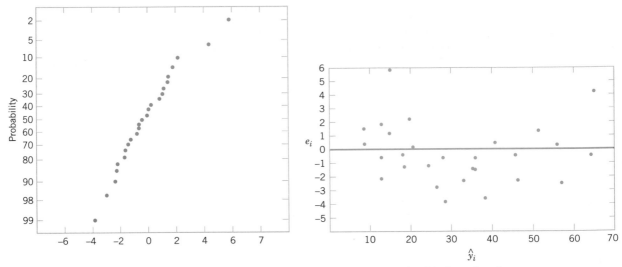

Figure 12-6   Normal probability plot of residuals.   Figure 12-7   Plot of residuals against $\hat{y}$.

helpful to plot the residuals against variables not presently in the model that are possible candidates for inclusion. Patterns in these plots may indicate that the model may be improved by adding the candidate variable.

EXAMPLE 12-9   The residuals for the model from Example 12-1 are shown in Table 12-3. A normal probability plot of these residuals is shown in Fig. 12-6. No severe deviations from normality are obviously apparent, although the two largest residuals ($e_{15} = 5.88$ and $e_{17} = 4.33$) do not fall extremely close to a straight line drawn through the remaining residuals.

The **standardized residuals**

$$d_i = \frac{e_i}{\sqrt{MS_E}} = \frac{e_i}{\sqrt{\hat{\sigma}^2}} \qquad (12\text{-}41)$$

are often more useful than the ordinary residuals when assessing residual magnitude. The standardized residuals corresponding to e_{15} and e_{17} are $d_{15} = 5.88/\sqrt{5.2352} = 2.57$ and $d_{17} = 4.33/\sqrt{4.2352} = 1.89$, and they do not seem unusually large. Inspection of the data does not reveal any error in collecting observations 15 and 17, nor does it produce any other reason to discard or modify these two points.

The residuals are plotted against $\hat{y}$ in Fig. 12-7, and against x_1 and x_2 in Figs. 12-8 and 12-9, respectively.[*] The two largest residuals, e_{15} and e_{17}, are apparent. Figure 12-8 gives some indication that the model underpredicts the pull strength for assemblies with short wire length ($x_1 \le 6$) and long wire length ($x_1 \ge 15$) and overpredicts the strength for assemblies with intermediate wire length ($7 \le x_1 \le 14$). The same impression is obtained from Fig. 12-7.

---

[*]There are other methods, described in Montgomery, Peck, and Vining (2001) and Myers (1990), that plot a modified version of the residual, called a **partial residual,** against each regressor. These partial residual plots are useful in displaying the relationship between the response y and each individual regressor.

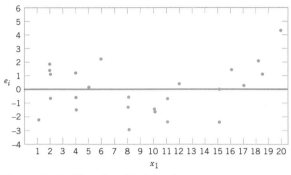

Figure 12-8   Plot of residuals against x_1.

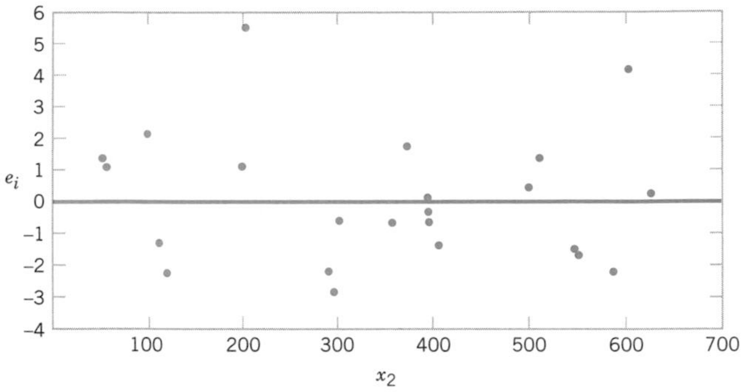

Figure 12-9   Plot of residuals against x_2.

Either the relationship between strength and wire length is not linear (requiring that a term involving x_1^2, say, be added to the model), or other regressor variables not presently in the model affected the response.

In Example 12-9 we used the standardized residuals $d_i = e_i / \sqrt{\hat{\sigma}^2}$ as a measure of residual magnitude. Some analysts prefer to plot standardized residuals instead of ordinary residuals, because the standardized residuals are scaled so that their standard deviation is approximately unity. Consequently, large residuals (that may indicate possible outliers or unusual observations) will be more obvious from inspection of the residual plots.

Many regression computer programs compute other types of scaled residuals. One of the most popular is the **studentized residual**

$$r_i = \frac{e_i}{\sqrt{\hat{\sigma}^2(1 - h_{ii})}} \qquad i = 1, 2, \ldots, n \qquad (12\text{-}42)$$

where h_{ii} is the ith diagonal element of the matrix

$$\mathbf{H} = \mathbf{X}(\mathbf{X}'\mathbf{X})^{-1}\mathbf{X}'$$

The **H** matrix is sometimes called the **"hat" matrix,** since

$$\hat{\mathbf{y}} = \mathbf{X}\hat{\boldsymbol{\beta}} = \mathbf{X}(\mathbf{X}'\mathbf{X})^{-1}\mathbf{X}'\mathbf{y} = \mathbf{H}\mathbf{y}$$

Thus **H** transforms the observed values of **y** into a vector of fitted values $\hat{\mathbf{y}}$.

Since each row of the matrix **X** corresponds to a vector, say $\mathbf{x}_i' = [1, x_{i1}, x_{i2}, \ldots, x_{ik}]$, another way to write the diagonal elements of the hat matrix is

$$h_{ii} = \mathbf{x}_i'(\mathbf{X}'\mathbf{X})^{-1}\mathbf{x}_i \qquad (12\text{-}43)$$

Note that apart from σ^2, h_{ii} is the variance of the fitted value $\hat{y}_i$. The quantities h_{ii} were used in the computation of the confidence interval on the mean response in Section 12-3.2.

Under the usual assumptions that the model errors are independently distributed with mean zero and variance σ^2, we can show that the variance of the ith residual e_i is

$$V(e_i) = \sigma^2(1 - h_{ii}), \qquad i = 1, 2, \ldots, n$$

Furthermore, the h_{ii} elements must fall in the interval $0 < h_{ii} \leq 1$. This implies that the standardized residuals understate the true residual magnitude; thus, the studentized residuals would be a better statistic to examine in evaluating potential **outliers.**

To illustrate, consider the two observations identified in Example 12-9 as having residuals that might be unusually large, observations 15 and 17. The standardized residuals are

$$d_{15} = \frac{e_{15}}{\sqrt{\hat{\sigma}^2}} = \frac{5.88}{\sqrt{5.2352}} = 2.57 \quad \text{and} \quad d_{17} = \frac{e_{17}}{\sqrt{MS_E}} = \frac{4.33}{\sqrt{5.2352}} = 1.89$$

Now $h_{15,15} = 0.0737$ and $h_{17,17} = 0.2593$, so the studentized residuals are

$$r_{15} = \frac{e_{15}}{\sqrt{\hat{\sigma}^2(1 - h_{15,15})}} = \frac{5.88}{\sqrt{5.2352(1 - 0.0737)}} = 2.67$$

and

$$r_{17} = \frac{e_{17}}{\sqrt{\hat{\sigma}^2(1 - h_{17,17})}} = \frac{4.33}{\sqrt{5.2352(1 - 0.2593)}} = 2.20$$

Notice that the studentized residuals are larger than the corresponding standardized residuals. However, the studentized residuals are still not so large as to cause us serious concern about possible outliers.

12-5.2   Influential Observations

When using multiple regression, we occasionally find that some subset of the observations is unusually influential. Sometimes these influential observations are relatively far away from the vicinity where the rest of the data were collected. A hypothetical situation for two variables is depicted in Fig. 12-10, where one observation in x-space is remote from the rest of the data. The disposition of points in the x-space is important in determining the properties of the model. For example, point (x_{i1}, x_{i2}) in Fig. 12-10 may be very influential in determining R^2, the estimates of the regression coefficients, and the magnitude of the error mean square.

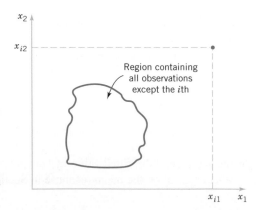

Figure 12-10   A point that is remote in x-space.

We would like to examine the influential points to determine whether they control many model properties. If these influential points are "bad" points, or erroneous in any way, they should be eliminated. On the other hand, there may be nothing wrong with these points, but at least we would like to determine whether or not they produce results consistent with the rest of the data. In any event, even if an influential point is a valid one, if it controls important model properties, we would like to know this, since it could have an impact on the use of the model.

Montgomery, Peck, and Vining (2001) and Myers (1990) describe several methods for detecting influential observations. An excellent diagnostic is the **distance measure** developed by Dennis R. Cook. This is a measure of the squared distance between the usual least squares estimate of $\boldsymbol{\beta}$ based on all n observations and the estimate obtained when the ith point is removed, say, $\hat{\boldsymbol{\beta}}_{(i)}$. The **Cook distance measure** is

$$D_i = \frac{(\hat{\boldsymbol{\beta}}_{(i)} - \hat{\boldsymbol{\beta}})' \mathbf{X}'\mathbf{X}(\hat{\boldsymbol{\beta}}_{(i)} - \hat{\boldsymbol{\beta}})}{p\hat{\sigma}^2} \qquad i = 1, 2, \ldots, n$$

Clearly, if the ith point is influential, its removal will result in $\hat{\boldsymbol{\beta}}_{(i)}$ changing considerably from the value $\hat{\boldsymbol{\beta}}$. Thus, a large value of D_i implies that the ith point is influential. The statistic D_i is actually computed using

$$D_i = \frac{r_i^2}{p} \frac{h_{ii}}{(1 - h_{ii})} \qquad i = 1, 2, \ldots, n \qquad (12\text{-}44)$$

From Equation 12-44 we see that D_i consists of the squared studentized residual, which reflects how well the model fits the ith observation y_i [recall that $r_i = e_i/\sqrt{\hat{\sigma}^2(1 - h_{ii})}$] and a component that measures how far that point is from the rest of the data [$h_{ii}/(1 - h_{ii})$ is a measure of the distance of the ith point from the centroid of the remaining $n - 1$ points]. A value of $D_i > 1$ would indicate that the point is influential. Either component of D_i (or both) may contribute to a large value.

EXAMPLE 12-10 Table 12-11 lists the values of the hat matrix diagonals h_{ii} and Cook's distance measure D_i for the wire bond pull strength data in Example 12-1. To illustrate the calculations, consider the first observation:

$$\begin{aligned}
D_1 &= \frac{r_1^2}{p} \cdot \frac{h_{11}}{(1 - h_{11})} \\
&= \frac{[e_1/\sqrt{MS_E(1 - h_{11})}]^2}{p} \cdot \frac{h_{11}}{(1 - h_{11})} \\
&= \frac{[1.57/\sqrt{5.2352(1 - 0.1573)}]^2}{3} \cdot \frac{0.1573}{(1 - 0.1573)} \\
&= 0.035
\end{aligned}$$

The Cook distance measure D_i does not identify any potentially influential observations in the data, for no value of D_i exceeds unity.

Table 12-11   Influence Diagnostics for the Wire Bond Pull Strength Data 2

Observations i	h_{ii}	Cook's Distance Measure D_i	Observations i	h_{ii}	Cook's Distance Measure D_i
1	0.1573	0.035	14	0.1129	0.003
2	0.1116	0.012	15	0.0737	0.187
3	0.1419	0.060	16	0.0879	0.001
4	0.1019	0.021	17	0.2593	0.565
5	0.0418	0.024	18	0.2929	0.155
6	0.0749	0.007	19	0.0962	0.018
7	0.1181	0.036	20	0.1473	0.000
8	0.1561	0.020	21	0.1296	0.052
9	0.1280	0.160	22	0.1358	0.028
10	0.0413	0.001	23	0.1824	0.002
11	0.0925	0.013	24	0.1091	0.040
12	0.0526	0.001	25	0.0729	0.000
13	0.0820	0.001			

EXERCISES FOR SECTION 12-5

12-36.   Consider the regression model for the NFL data in Exercise 12-4.
(a) What proportion of total variability is explained by this model?
(b) Construct a normal probability plot of the residuals. What conclusion can you draw from this plot?
(c) Plot the residuals versus $\hat{y}$ and versus each regressor, and comment on model adequacy.
(d) Are there any influential points in these data?

12-37.   Consider the gasoline mileage data in Exercise 12-5.
(a) What proportion of total variability is explained by this model?
(b) Construct a normal probability plot of the residuals and comment on the normality assumption.
(c) Plot residuals versus $\hat{y}$ and versus each regressor. Discuss these residual plots.
(d) Calculate Cook's distance for the observations in this data set. Are any observations influential?

12-38.   Consider the electric power consumption data in Exercise 12-6.
(a) Calculate R^2 for this model. Interpret this quantity.
(b) Plot the residuals versus $\hat{y}$. Interpret this plot.
(c) Construct a normal probability plot of the residuals and comment on the normality assumption.

12-39.   Consider the wear data in Exercise 12-7.
(a) Find the value of R^2 when the model uses the regressors x_1 and x_2.
(b) What happens to the value of R^2 when an interaction term x_1x_2 is added to the model? Does this necessarily imply that adding the interaction term is a good idea?

12-40.   For the regression model for the wire bond pull strength data in Exercise 12-8.
(a) Plot the residuals versus $\hat{y}$ and versus the regressors used in the model. What information do these plots provide?
(b) Construct a normal probability plot of the residuals. Are there reasons to doubt the normality assumption for this model?
(c) Are there any indications of influential observations in the data?

12-41.   Consider the semiconductor HFE data in Exercise 12-9.
(a) Plot the residuals from this model versus $\hat{y}$. Comment on the information in this plot.
(b) What is the value of R^2 for this model?
(c) Refit the model using log HFE as the response variable.
(d) Plot the residuals versus predicted log HFE for the model in part (c). Does this give any information about which model is preferable?
(e) Plot the residuals from the model in part (d) versus the regressor x_3. Comment on this plot.
(f) Refit the model to log HFE using x_1, x_2, and $1/x_3$, as the regressors. Comment on the effect of this change in the model.

12-42.   Consider the regression model for the heat treating data in Exercise 12-10.
(a) Calculate the percent of variability explained by this model.
(b) Construct a normal probability plot for the residuals. Comment on the normality assumption.
(c) Plot the residuals versus $\hat{y}$ and interpret the display.
(d) Calculate Cook's distance for each observation and provide an interpretation of this statistic.

12-43. In Exercise 12-24 we fit a model to the response PITCH in the heat treating data of Exercise 12-10 using new regressors x_1 = SOAKTIME × SOAKPCT and x_2 = DIFFTIME × DIFFPCT.

(a) Calculate the R^2 for this model and compare it to the value of R^2 from the original model in Exercise 12-10. Does this provide some information about which model is preferable?

(b) Plot the residuals from this model versus $\hat{y}$ and on a normal probability scale. Comment on model adequacy.

(c) Find the values of Cook's distance measure. Are any observations unusually influential?

12-44. Consider the regression model for the NHL data from Exercise 12-11.

(a) Fit a model using "pts" as the only regressor.

(b) How much variability is explained by this model?

(c) Plot the residuals versus $\hat{y}$ and comment on model adequacy.

(d) Plot the residuals versus "PPG," the points scored while in power play. Does this indicate that the model would be better if this variable were included?

12-45. The diagonal elements of the hat matrix are often used to denote **leverage**—that is, a point that is unusual in its location in the x-space and that may be influential. Generally, the ith point is called a **leverage point** if its hat diagonal h_{ii} exceeds $2p/n$, which is twice the average size of all the hat diagonals. Recall that $p = k + 1$.

(a) Table 12-11 contains the hat diagonal for the wire bond pull strength data used in Example 12-1. Find the average size of these elements.

(b) Based on the criterion above, are there any observations that are leverage points in the data set?

12-6 ASPECTS OF MULTIPLE REGRESSION MODELING

In this section we briefly discuss several other aspects of building multiple regression models. For more extensive presentations of these topics and additional examples refer to Montgomery, Peck, and Vining (2001) and Myers (1990).

12-6.1 Polynomial Regression Models

The linear model $\mathbf{y} = \mathbf{X\beta} + \boldsymbol{\epsilon}$ is a general model that can be used to fit any relationship that is **linear in the unknown parameters $\boldsymbol{\beta}$.** This includes the important class of **polynomial regression models.** For example, the second-degree polynomial in one variable

$$Y = \beta_0 + \beta_1 x + \beta_{11} x^2 + \epsilon \qquad (12\text{-}45)$$

and the second-degree polynomial in two variables

$$Y = \beta_0 + \beta_1 x_1 + \beta_2 x_2 + \beta_{11} x_1^2 + \beta_{22} x_2^2 + \beta_{12} x_1 x_2 + \epsilon \qquad (12\text{-}46)$$

are linear regression models.

Polynomial regression models are widely used when the response is curvilinear, because the general principles of multiple regression can be applied. The following example illustrates some of the types of analyses that can be performed.

EXAMPLE 12-11 | Sidewall panels for the interior of an airplane are formed in a 1500-ton press. The unit manufacturing cost varies with the production lot size. The data shown below give the average cost per unit (in hundreds of dollars) for this product (y) and the production lot size (x). The scatter diagram, shown in Fig. 12-11, indicates that a second-order polynomial may be appropriate.

y	1.81	1.70	1.65	1.55	1.48	1.40	1.30	1.26	1.24	1.21	1.20	1.18
x	20	25	30	35	40	50	60	65	70	75	80	90

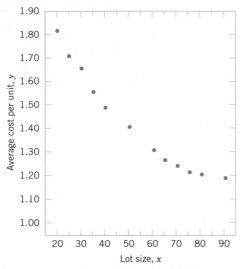

Figure 12-11   Data for Example 12-11.

We will fit the model

$$Y = \beta_0 + \beta_1 x + \beta_{11} x^2 + \epsilon$$

The **y** vector, **X** matrix, and $\boldsymbol{\beta}$ vector are as follows:

$$
\mathbf{y} = \begin{bmatrix} 1.81 \\ 1.70 \\ 1.65 \\ 1.55 \\ 1.48 \\ 1.40 \\ 1.30 \\ 1.26 \\ 1.24 \\ 1.21 \\ 1.20 \\ 1.18 \end{bmatrix}
\qquad
\mathbf{X} = \begin{bmatrix} 1 & 20 & 400 \\ 1 & 25 & 625 \\ 1 & 30 & 900 \\ 1 & 35 & 1225 \\ 1 & 40 & 1600 \\ 1 & 50 & 2500 \\ 1 & 60 & 3600 \\ 1 & 65 & 4225 \\ 1 & 70 & 4900 \\ 1 & 75 & 5625 \\ 1 & 80 & 6400 \\ 1 & 90 & 8100 \end{bmatrix}
\qquad
\boldsymbol{\beta} = \begin{bmatrix} \beta_0 \\ \beta_1 \\ \beta_{11} \end{bmatrix}
$$

Solving the normal equations $\mathbf{X'X\hat{\boldsymbol{\beta}}} = \mathbf{X'y}$ gives the fitted model

$$\hat{y} = 2.19826629 - 0.02252236x + 0.00012507x^2$$

The test for significance of regression is shown in Table 12-12. Since $f_0 = 2171.07$ is significant at 1%, we conclude that at least one of the parameters β_1 and β_{11} is not zero. Furthermore, the standard tests for model adequacy do not reveal any unusual behavior, and we would conclude that this is a reasonable model for the sidewall panel cost data.

Table 12-12   Test for Significance of Regression for the Second-Order Model in Example 12-11

Source of Variation	Sum of Squares	Degrees of Freedom	Mean Square	f_0	P-value
Regression	0.5254	2	0.262700	2171.07	5.18E-15
Error	0.0011	9	0.000121		
Total	0.5265	11			

In fitting polynomials, we generally like to use the **lowest-degree model** consistent with the data. In this example, it would seem logical to investigate the possibility of dropping the quadratic term from the model. That is, we would like to test

$$H_0: \beta_{11} = 0$$
$$H_1: \beta_{11} \neq 0$$

The general regression significance test can be used to test this hypothesis. We need to determine the "extra sum of squares" due to β_{11}, or

$$SS_R(\beta_{11}|\beta_1,\beta_0) = SS_R(\beta_1,\beta_{11}|\beta_0) - SS_R(\beta_1|\beta_0)$$

The sum of squares $SS_R(\beta_1,\beta_{11}|\beta_0) = 0.5254$ from Table 12-12. To find $SS_R(\beta_1|\beta_0)$, we fit a simple linear regression model to the original data, yielding

$$\hat{y} = 1.90036320 - 0.00910056x$$

It can be easily verified that the regression sum of squares for this model is

$$SS_R(\beta_1|\beta_0) = 0.4942$$

Therefore, the extra sum of the squares due to β_{11}, given that β_1 and β_0 are in the model, is

$$SS_R(\beta_{11}|\beta_1,\beta_0) = SS_R(\beta_1,\beta_{11}|\beta_0) - SS_R(\beta_1|\beta_0)$$
$$= 0.5254 - 0.4942$$
$$= 0.0312$$

The analysis of variance, with the test of $H_0: \beta_{11} = 0$ incorporated into the procedure, is displayed in Table 12-13. Note that the quadratic term contributes significantly to the model.

Table 12-13   Analysis of Variance for Example 12-11, Showing the Test for $H_0: \beta_{11} = 0$

Source of Variation	Sum of Squares	Degrees of Freedom	Mean Square	f_0	P-value
Regression	$SS_R(\beta_1,\beta_{11}|\beta_0) = 0.5254$	2	0.262700	2171.07	5.18E-15
Linear	$SS_R(\beta_1|\beta_0) = 0.4942$	1	0.494200	4084.30	1.17E-15
Quadratic	$SS_R(\beta_{11}|\beta_0,\beta_1) = 0.0312$	1	0.031200	258.18	5.51E-9
Error	0.0011	9	0.00121		
Total	0.5265	11			

12-6.2   Categorical Regressors and Indicator Variables

The regression models presented in previous sections have been based on **quantitative** variables, that is, variables that are measured on a numerical scale. For example, variables such as temperature, pressure, distance, and voltage are quantitative variables. Occasionally, we need to incorporate **categorical,** or **qualitative,** variables in a regression model. For example, suppose that one of the variables in a regression model is the operator who is associated with each observation y_i. Assume that only two operators are involved. We may wish to assign different levels to the two operators to account for the possibility that each operator may have a different effect on the response.

The usual method of accounting for the different levels of a qualitative variable is to use **indicator variables.** For example, to introduce the effect of two different operators into a regression model, we could define an indicator variable as follows:

$$x = \begin{cases} 0 \text{ if the observation is from operator 1} \\ 1 \text{ if the observation is from operator 2} \end{cases}$$

In general, a qualitative variable with r-levels can be modeled by $r - 1$ indicator variables, which are assigned the value of either zero or one. Thus, if there are *three* operators, the different levels will be accounted for by the *two* indicator variables defined as follows:

x_1	x_2	
0	0	if the observation is from operator 1
1	0	if the observation is from operator 2
0	1	if the observation is from operator 3

Indicator variables are also referred to as **dummy** variables. The following example [from Montgomery, Peck, and Vining (2001)] illustrates some of the uses of indicator variables; for other applications, see Montgomery, Peck, and Vining (2001).

EXAMPLE 12-12   A mechanical engineer is investigating the surface finish of metal parts produced on a lathe and its relationship to the speed (in revolutions per minute) of the lathe. The data are shown in Table 12-14. Note that the data have been collected using two different types of cutting tools. Since the type of cutting tool likely affects the surface finish, we will fit the model

$$Y = \beta_0 + \beta_1 x_1 + \beta_2 x_2 + \epsilon$$

where Y is the surface finish, x_1 is the lathe speed in revolutions per minute, and x_2 is an indicator variable denoting the type of cutting tool used; that is,

$$x_2 = \begin{cases} 0, \text{ for tool type 302} \\ 1, \text{ for tool type 416} \end{cases}$$

The parameters in this model may be easily interpreted. If $x_2 = 0$, the model becomes

$$Y = \beta_0 + \beta_1 x_1 + \epsilon$$

which is a straight-line model with slope β_1 and intercept β_0. However, if $x_2 = 1$, the model becomes

$$Y = \beta_0 + \beta_1 x_1 + \beta_2(1) + \epsilon = (\beta_0 + \beta_2) + \beta_1 x_1 + \epsilon$$

Table 12-14 Surface Finish Data for Example 12-13

Observation Number, i	Surface Finish y_i	RPM	Type of Cutting Tool	Observation Number, i	Surface Finish y_i	RPM	Type of Cutting Tool
1	45.44	225	302	11	33.50	224	416
2	42.03	200	302	12	31.23	212	416
3	50.10	250	302	13	37.52	248	416
4	48.75	245	302	14	37.13	260	416
5	47.92	235	302	15	34.70	243	416
6	47.79	237	302	16	33.92	238	416
7	52.26	265	302	17	32.13	224	416
8	50.52	259	302	18	35.47	251	416
9	45.58	221	302	19	33.49	232	416
10	44.78	218	302	20	32.29	216	416

which is a straight-line model with slope β_1 and intercept $\beta_0 + \beta_2$. Thus, the model $Y = \beta_0 + \beta_1 x + \beta_2 x_2 + \epsilon$ implies that surface finish is linearly related to lathe speed and that the slope β_1 does not depend on the type of cutting tool used. However, the type of cutting tool does affect the intercept, and β_2 indicates the change in the intercept associated with a change in tool type from 302 to 416.

The **X** matrix and **y** vector for this problem are as follows:

$$
\mathbf{X} = \begin{bmatrix}
1 & 225 & 0 \\
1 & 200 & 0 \\
1 & 250 & 0 \\
1 & 245 & 0 \\
1 & 235 & 0 \\
1 & 237 & 0 \\
1 & 265 & 0 \\
1 & 259 & 0 \\
1 & 221 & 0 \\
1 & 218 & 0 \\
1 & 224 & 1 \\
1 & 212 & 1 \\
1 & 248 & 1 \\
1 & 260 & 1 \\
1 & 243 & 1 \\
1 & 238 & 1 \\
1 & 224 & 1 \\
1 & 251 & 1 \\
1 & 232 & 1 \\
1 & 216 & 1
\end{bmatrix}
\qquad
\mathbf{y} = \begin{bmatrix}
45.44 \\
42.03 \\
50.10 \\
48.75 \\
47.92 \\
47.79 \\
52.26 \\
50.52 \\
45.58 \\
44.78 \\
33.50 \\
31.23 \\
37.52 \\
37.13 \\
34.70 \\
33.92 \\
32.13 \\
35.47 \\
33.49 \\
32.29
\end{bmatrix}
$$

The fitted model is

$$\hat{y} = 14.27620 + 0.14115x_1 - 13.28020x_2$$

Table 12-15   Analysis of Variance for Example 12-12

Source of Variation	Sum of Squares	Degrees of Freedom	Mean Square	f_0	P-value
Regression	1012.0595	2	506.0297	1103.69	1.02E-18
$SS_R(\beta_1|\beta_0)$	130.6091	1	130.6091	284.87	4.70E-12
$SS_R(\beta_2|\beta_1,\beta_0)$	881.4504	1	881.4504	1922.52	6.24E-19
Error	7.7943	17	0.4508		
Total	1019.8538	19			

The analysis of variance for this model is shown in Table 12-15. Note that the hypothesis $H_0: \beta_1 = \beta_2 = 0$ (significance of regression) would be rejected at any reasonable level of significance because the P-value is very small. This table also contains the sums of squares

$$SS_R = SS_R(\beta_1,\beta_2|\beta_0)$$
$$= SS_R(\beta_1|\beta_0) + SS_R(\beta_2|\beta_1,\beta_0)$$

so a test of the hypothesis $H_0: \beta_2 = 0$ can be made. Since this hypothesis is also rejected, we conclude that tool type has an effect on surface finish.

It is also possible to use indicator variables to investigate whether tool type affects both the slope and intercept. Let the model be

$$Y = \beta_0 + \beta_1 x_1 + \beta_2 x_2 + \beta_3 x_1 x_2 + \epsilon$$

where x_2 is the indicator variable. Now if tool type 302 is used, $x_2 = 0$, and the model is

$$Y = \beta_0 + \beta_1 x_1 + \epsilon$$

If tool type 416 is used, $x_2 = 1$, and the model becomes

$$Y = \beta_0 + \beta_1 x_1 + \beta_2 + \beta_3 x_1 + \epsilon$$
$$= (\beta_0 + \beta_2) + (\beta_1 + \beta_3)x_1 + \epsilon$$

Note that β_2 is the change in the intercept and that β_3 is the change in slope produced by a change in tool type.

Another method of analyzing these data is to fit separate regression models to the data for each tool type. However, the indicator variable approach has several advantages. First, only one regression model must be fit. Second, by pooling the data on both tool types, more degrees of freedom for error are obtained. Third, tests of both hypotheses on the parameters β_2 and β_3 are just special cases of the extra sum of squares method.

12-6.3   Selection of Variables and Model Building

An important problem in many applications of regression analysis involves selecting the set of regressor variables to be used in the model. Sometimes previous experience or underlying theoretical considerations can help the analyst specify the set of regressor variables to use in a particular situation. Usually, however, the problem consists of selecting an appropriate set of regressors from a set that quite likely includes all the important variables, but we are sure that not all these candidate regressors are necessary to adequately model the response Y.

In such a situation, we are interested in **variable selection;** that is, screening the candidate variables to obtain a regression model that contains the "best" subset of regressor variables. We would like the final model to contain enough regressor variables so that in the intended use of the model (prediction, for example) it will perform satisfactorily. On the other hand, to keep model maintenance costs to a minimum and to make the model easy to use, we would like the model to use as few regressor variables as possible. The compromise between these conflicting objectives is often called finding the "best" regression equation. However, in most problems, no single regression model is "best" in terms of the various evaluation criteria that have been proposed. A great deal of judgment and experience with the system being modeled is usually necessary to select an appropriate set of regressor variables for a regression equation.

No single algorithm will always produce a good solution to the variable selection problem. Most of the currently available procedures are search techniques, and to perform satisfactorily, they require interaction with judgment by the analyst. We now briefly discuss some of the more popular variable selection techniques. We assume that there are K candidate regressors, $x_1, x_2,$ $\ldots, x_K$, and a single response variable y. All models will include an intercept term β_0, so the model with *all* variables included would have $K + 1$ terms. Furthermore, the functional form of each candidate variable (for example, $x_1 = 1/x, x_2 = \ln x$, etc.) is assumed to be correct.

All Possible Regressions

This approach requires that the analyst fit all the regression equations involving one candidate variable, all regression equations involving two candidate variables, and so on. Then these equations are evaluated according to some suitable criteria to select the "best" regression model. If there are K candidate regressors, there are 2^K total equations to be examined. For example, if $K = 4$, there are $2^4 = 16$ possible regression equations; while if $K = 10$, there are $2^{10} = 1024$ possible regression equations. Hence, the number of equations to be examined increases rapidly as the number of candidate variables increases. However, there are some very efficient computing algorithms for all possible regressions available and they are widely implemented in statistical software, so it is a very practical procedure unless the number of candidate regressors is fairly large.

Several criteria may be used for evaluating and comparing the different regression models obtained. A commonly used criterion is based on the value of R^2 or the value of the adjusted R^2, R^2_{adj}. Basically, the analyst continues to increase the number of variables in the model until the increase in R^2 or the adjusted R^2_{adj} is small. Often, we will find that the R^2_{adj} will stabilize and actually begin to decrease as the number of variables in the model increases. Usually, the model that maximizes R^2_{adj} is considered to be a good candidate for the best regression equation. Because we can write $R^2_{adj} = 1 - \{MS_E/[SS_E/(n-1)]\}$ and $SS_E/(n-1)$ is a constant, the model that maximizes the R^2_{adj} value also minimizes the mean square error, so this is a very attractive criterion.

Another criterion used to evaluate regression models is the C_p statistic, which is a measure of the total mean square error for the regression model. We define the total standardized mean square error for the regression model as

$$\Gamma_p = \frac{1}{\sigma^2} \sum_{i=1}^{n} E[\hat{Y}_i - E(Y_i)]^2$$
$$= \frac{1}{\sigma^2} \left\{ \sum_{i=1}^{n} [E(Y_i) - E(\hat{Y}_i)]^2 + \sum_{i=1}^{n} V(\hat{Y}_i) \right\}$$
$$= \frac{1}{\sigma^2} [(\text{bias})^2 + \text{variance}]$$

We use the mean square error from the *full* $K + 1$ term model as an estimate of σ^2; that is, $\hat{\sigma}^2 = MS_E(K + 1)$. Then an estimator of Γ_p is [see Montgomery, Peck, and Vining (2001) or Myers (1990) for the details]:

$$C_p = \frac{SS_E(p)}{\hat{\sigma}^2} - n + 2p \tag{12-47}$$

If the p-term model has negligible bias, it can be shown that

$$E(C_p \mid \text{zero bias}) = p$$

Therefore, the values of C_p for each regression model under consideration should be evaluated relative to p. The regression equations that have negligible bias will have values of C_p that are close to p, while those with significant bias will have values of C_p that are significantly greater than p. We then choose as the "best" regression equation either a model with *minimum* C_p or a model with a slightly larger C_p, that does not contain as much bias (i.e., $C_p \cong p$).

The PRESS statistic can also be used to evaluate competing regression models. PRESS is an acronym for Prediction Error Sum of Squares, and it is defined as the sum of the squares of the differences between each observation y_i and the corresponding predicted value based on a model fit to the *remaining* $n - 1$ points, say $\hat{y}_{(i)}$. So PRESS provides a measure of how well the model is likely to perform when predicting **new data**, or data that was not used to fit the regression model. The computing formula for PRESS is

$$\text{PRESS} = \sum_{i=1}^{n} (y_i - \hat{y}_{(i)})^2 = \sum_{i=1}^{n} \left(\frac{e_i}{1 - h_{ii}} \right)^2$$

where $e_i = y_i - \hat{y}_i$ is the usual residual. Thus PRESS is easy to calculate from the standard least squares regression results. Models that have small values of PRESS are preferred.

EXAMPLE 12-13 Table 12-16 presents data on taste-testing 38 brands of pinot noir wine (the data were first reported in an article by Kwan, Kowalski, and Skogenboe in an article in the *Journal of Agricultural and Food Chemistry*, Vol. 27, 1979, and it also appears as one of the default data sets in Minitab). The response variable is $y =$ quality, and we wish to find the "best" regression equation that relates quality to the other five parameters.

Figure 12-12 is the matrix of scatter plots for the wine quality data, as constructed by Minitab. We notice that there are some indications of possible linear relationships between quality and the regressors, but there is no obvious visual impression of which regressors would be appropriate. Table 12-16 lists the all possible regressions output from Minitab. In this analysis, we asked Minitab to present the best three equations for each subset size. Note that Minitab reports the values of R^2, R^2_{adj}, C_p, and $S = \sqrt{MS_E}$ for each model. From Table 12-17 we see that the three-variable equation with $x_2 =$ aroma, $x_4 =$ flavor, and $x_5 =$ oakiness produces the minimum C_p equation, whereas the four-variable model, which adds

Table 12-16   Wine Quality Data

	x_1 Clarity	x_2 Aroma	x_3 Body	x_4 Flavor	x_5 Oakiness	y Quality
1	1.0	3.3	2.8	3.1	4.1	9.8
2	1.0	4.4	4.9	3.5	3.9	12.6
3	1.0	3.9	5.3	4.8	4.7	11.9
4	1.0	3.9	2.6	3.1	3.6	11.1
5	1.0	5.6	5.1	5.5	5.1	13.3
6	1.0	4.6	4.7	5.0	4.1	12.8
7	1.0	4.8	4.8	4.8	3.3	12.8
8	1.0	5.3	4.5	4.3	5.2	12.0
9	1.0	4.3	4.3	3.9	2.9	13.6
10	1.0	4.3	3.9	4.7	3.9	13.9
11	1.0	5.1	4.3	4.5	3.6	14.4
12	0.5	3.3	5.4	4.3	3.6	12.3
13	0.8	5.9	5.7	7.0	4.1	16.1
14	0.7	7.7	6.6	6.7	3.7	16.1
15	1.0	7.1	4.4	5.8	4.1	15.5
16	0.9	5.5	5.6	5.6	4.4	15.5
17	1.0	6.3	5.4	4.8	4.6	13.8
18	1.0	5.0	5.5	5.5	4.1	13.8
19	1.0	4.6	4.1	4.3	3.1	11.3
20	0.9	3.4	5.0	3.4	3.4	7.9
21	0.9	6.4	5.4	6.6	4.8	15.1
22	1.0	5.5	5.3	5.3	3.8	13.5
23	0.7	4.7	4.1	5.0	3.7	10.8
24	0.7	4.1	4.0	4.1	4.0	9.5
25	1.0	6.0	5.4	5.7	4.7	12.7
26	1.0	4.3	4.6	4.7	4.9	11.6
27	1.0	3.9	4.0	5.1	5.1	11.7
28	1.0	5.1	4.9	5.0	5.1	11.9
29	1.0	3.9	4.4	5.0	4.4	10.8
30	1.0	4.5	3.7	2.9	3.9	8.5
31	1.0	5.2	4.3	5.0	6.0	10.7
32	0.8	4.2	3.8	3.0	4.7	9.1
33	1.0	3.3	3.5	4.3	4.5	12.1
34	1.0	6.8	5.0	6.0	5.2	14.9
35	0.8	5.0	5.7	5.5	4.8	13.5
36	0.8	3.5	4.7	4.2	3.3	12.2
37	0.8	4.3	5.5	3.5	5.8	10.3
38	0.8	5.2	4.8	5.7	3.5	13.2

x_1 = clarity to the previous three regressors, results in maximum R_{adj}^2 (or minimum MS_E). The three-variable model is

$$\hat{y} = 6.47 + 0.580x_2 + 1.20x_4 - 0.602x_5$$

and the four-variable model is

$$\hat{y} = 4.99 + 1.79x_1 + 0.530x_2 + 1.26x_4 - 0.659x_5$$

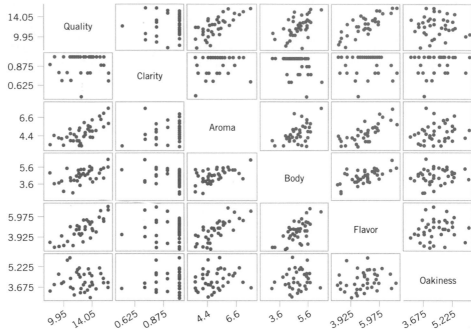

Figure 12-12
A Matrix of Scatter Plots from Minitab for the Wine Quality Data.

Table 12-17 Minitab All Possible Regressions Output for the Wine Quality Data

Best Subsets Regression: Quality versus Clarity, Aroma, . . .

Response is Quality

Vars	R-Sq	R-Sq (adj)	C–p	S	Clarity	Aroma	Body	Flavor	Oakiness
1	62.4	61.4	9.0	1.2712				X	
1	50.0	48.6	23.2	1.4658		X			
1	30.1	28.2	46.0	1.7335			X		
2	66.1	64.2	6.8	1.2242				X	X
2	65.9	63.9	7.1	1.2288		X		X	
2	63.3	61.2	10.0	1.2733	X			X	
3	70.4	67.8	3.9	1.1613		X		X	X
3	68.0	65.2	6.6	1.2068	X			X	X
3	66.5	63.5	8.4	1.2357			X	X	X
4	71.5	68.0	4.7	1.1568	X	X		X	X
4	70.5	66.9	5.8	1.1769		X	X	X	X
4	69.3	65.6	7.1	1.1996	X		X	X	X
5	72.1	67.7	6.0	1.1625	X	X	X	X	X

These models should now be evaluated further using residuals plots and the other techniques discussed earlier in the chapter, to see if either model is satisfactory with respect to the underlying assumptions and to determine if one of them is preferable. It turns out that the residual plots do not reveal any major problems with either model. The value of PRESS for the three-variable model is 56.0525 and for the four-variable model it is 60.3927. Since PRESS is smaller in the model with three regressors, and since it is the model with the smallest number of predictors, it would likely be the preferred choice.

Stepwise Regression

Stepwise regression is probably the most widely used variable selection technique. The procedure iteratively constructs a sequence of regression models by adding or removing variables at each step. The criterion for adding or removing a variable at any step is usually expressed in terms of a partial F-test. Let f_{in} be the value of the F-random variable for adding a variable to the model, and let f_{out} be the value of the F-random variable for removing a variable from the model. We must have $f_{in} \geq f_{out}$, and usually $f_{in} = f_{out}$.

Stepwise regression begins by forming a one-variable model using the regressor variable that has the highest correlation with the response variable Y. This will also be the regressor producing the largest F-statistic. For example, suppose that at this step, x_1 is selected. At the second step, the remaining $K - 1$ candidate variables are examined, and the variable for which the partial F-statistic

$$F_j = \frac{SS_R(\beta_j | \beta_1, \beta_0)}{MS_E(x_j, x_1)} \qquad (12\text{-}48)$$

is a maximum is added to the equation, provided that $f_j > f_{in}$. In equation 12-48, $MS_E (x_j, x_1)$ denotes the mean square for error for the model containing both x_1 and x_j. Suppose that this procedure indicates that x_2 should be added to the model. Now the stepwise regression algorithm determines whether the variable x_1 added at the first step should be removed. This is done by calculating the F-statistic

$$F_1 = \frac{SS_R(\beta_1 | \beta_2, \beta_0)}{MS_E(x_1, x_2)} \qquad (12\text{-}49)$$

If the calculated value $f_1 < f_{out}$, the variable x_1 is removed; otherwise it is retained, and we would attempt to add a regressor to the model containing both x_1 and x_2.

In general, at each step the set of remaining candidate regressors is examined, and the regressor with the largest partial F-statistic is entered, provided that the observed value of f exceeds f_{in}. Then the partial F-statistic for each regressor in the model is calculated, and the regressor with the smallest observed value of F is deleted if the observed $f < f_{out}$. The procedure continues until no other regressors can be added to or removed from the model.

Stepwise regression is almost always performed using a computer program. The analyst exercises control over the procedure by the choice of f_{in} and f_{out}. Some stepwise regression computer programs require that numerical values be specified for f_{in} and f_{out}. Since the number of degrees of freedom on MS_E depends on the number of variables in the model, which changes from step to step, a fixed value of f_{in} and f_{out} causes the type I and type II error rates to vary. Some computer programs allow the analyst to specify the type I error levels for f_{in} and f_{out}. However, the "advertised" significance level is not the true level, because the variable selected is the one that maximizes (or minimizes) the partial F-statistic at that stage. Sometimes it is useful to experiment with different values of f_{in} and f_{out} (or different advertised

Table 12-18   Minitab Stepwise Regression Output for the Wine Quality Data

Stepwise Regression: Quality versus Clarity, Aroma, . . .

Alpha-to-Enter: 0.15 Alpha-to-Remove: 0.15

Response is Quality on 5 predictors, with N = 38

Step	1	2	3
Constant	4.941	6.912	6.467
Flavor	1.57	1.64	1.20
T-Value	7.73	8.25	4.36
P-Value	0.000	0.000	0.000
Oakiness		−0.54	−0.60
T-Value		−1.95	−2.28
P-Value		0.059	0.029
Aroma			0.58
T-Value			2.21
P-Value			0.034
S	1.27	1.22	1.16
R-Sq	62.42	66.11	70.38
R-Sq(adj)	61.37	64.17	67.76
C–p	9.0	6.8	3.9

type I error rates) in several different runs to see if this substantially affects the choice of the final model.

EXAMPLE 12-14   Table 12-18 gives the Minitab stepwise regression output for the wine quality data. Minitab uses fixed values of α for entering and removing variables. The default level is $\alpha = 0.15$ for both decisions. The output in Table 12-18 uses the default value. Notice that the variables were entered in the order Flavor (step 1), Oakiness (step 2), and Aroma (step 3) and that no variables were removed. No other variable could be entered, so the algorithm terminated. This is the three-variable model found by all possible regressions that results in a minimum value of C_p.

Forward Selection

The **forward selection** procedure is a variation of stepwise regression and is based on the principle that regressors should be added to the model one at a time until there are no remaining candidate regressors that produce a significant increase in the regression sum of squares. That is, variables are added one at a time as long as their partial F-value exceeds f_{in}. Forward selection is a simplification of stepwise regression that omits the partial F-test for deleting variables from the model that have been added at previous steps. This is a potential weakness of forward selection; that is, the procedure does not explore the effect that adding a regressor at the current step has on regressor variables added at earlier steps. Notice that if we were to apply forward selection to the wine quality data, we would obtain exactly the same results as we did with stepwise regression in Example 12-14, since stepwise regression terminated without deleting a variable.

Table 12-19   Minitab Backward Elimination Output for the Wine Quality Data

Stepwise Regression: Quality versus Clarity, Aroma, ...

Backward elimination. Alpha-to-Remove: 0.1

Response is Quality on 5 predictors, with N = 38

Step	1	2	3
Constant	3.997	4.986	6.467
Clarity	2.3	1.8	
T-Value	1.35	1.12	
P-Value	0.187	0.269	
Aroma	0.48	0.53	0.58
T-Value	1.77	2.00	2.21
P-Value	0.086	0.054	0.034
Body	0.27		
T-Value	0.82		
P-Value	0.418		
Flavor	1.17	1.26	1.20
T-Value	3.84	4.52	4.36
P-Value	0.001	0.000	0.000
Oakiness	−0.68	−0.66	−0.60
T-Value	−2.52	−2.46	−2.28
P-Value	0.017	0.019	0.029
S	1.16	1.16	1.16
R-Sq	72.06	71.47	70.38
R-Sq(adj)	67.69	68.01	67.76
C−p	6.0	4.7	3.9

Backward Elimination

The **backward elimination** algorithm begins with all K candidate regressors in the model. Then the regressor with the smallest partial F-statistic is deleted if this F-statistic is insignificant, that is, if $f < f_{out}$. Next, the model with $K − 1$ regressors is fit, and the next regressor for potential elimination is found. The algorithm terminates when no further regressor can be deleted.

Table 12-19 shows the Minitab output for backward elimination applied to the wine quality data. The α value for removing a variable is $\alpha = 0.10$. Notice that this procedure removes Body at step 1 and then Clarity at step 2, terminating with the three-variable model found previously.

Some Comments on Final Model Selection

We have illustrated several different approaches to the selection of variables in multiple linear regression. The final model obtained from any model-building procedure should be subjected to the usual adequacy checks, such as residual analysis, lack-of-fit testing, and examination of the effects of influential points. The analyst may also consider augmenting the original set of candidate variables with cross-products, polynomial terms, or other transformations of the original variables that might improve the model. A major criticism of variable selection methods such as stepwise regression is that the analyst may conclude there is one "best" regression equation. Generally, this is not the case, because several equally good regression models can often be used. One way to avoid this problem is to use several different model-building techniques and see if different models result. For example, we have found the same model for the wine quality data using stepwise regression, forward selection, and backward elimination. The same model was also one of the two best found from all possible regressions. The results from variable selection methods frequently do not agree, so this is a good indication that the three-variable model is the best regression equation.

If the number of candidate regressors is not too large, the all-possible regressions method is recommended. We usually recommend using the minimum MS_E and C_p evaluation criteria in conjunction with this procedure. The all-possible regressions approach can find the "best" regression equation with respect to these criteria, while stepwise-type methods offer no such assurance. Furthermore, the all-possible regressions procedure is not distorted by dependencies among the regressors, as stepwise-type methods are.

12-6.4   Multicollinearity

In multiple regression problems, we expect to find dependencies between the response variable Y and the regressors x_j. In most regression problems, however, we find that there are also dependencies among the regressor variables x_j. In situations where these dependencies are strong, we say that **multicollinearity** exists. Multicollinearity can have serious effects on the estimates of the regression coefficients and on the general applicability of the estimated model.

The effects of multicollinearity may be easily demonstrated. The diagonal elements of the matrix $\mathbf{C} = (\mathbf{X'X})^{-1}$ can be written as

$$C_{jj} = \frac{1}{(1 - R_j^2)} \qquad j = 1, 2, \ldots, k$$

where R_j^2 is the coefficient of multiple determination resulting from regressing x_j on the other $k - 1$ regressor variables. Clearly, the stronger the linear dependency of x_j on the remaining regressor variables, and hence the stronger the multicollinearity, the larger the value of R_j^2 will be. Recall that $V(\hat{\beta}_j) = \sigma^2 C_{jj}$. Therefore, we say that the variance of $\hat{\beta}_j$ is "inflated" by the quantity $(1 - R_j^2)^{-1}$. Consequently, we define the **variance inflation factor** for β_j as

$$VIF(\beta_j) = \frac{1}{(1 - R_j^2)} \qquad j = 1, 2, \ldots, k \qquad (12\text{-}50)$$

These factors are an important measure of the extent to which multicollinearity is present.

Although the estimates of the regression coefficients are very imprecise when multicollinearity is present, the fitted model equation may still be useful. For example, suppose we wish to predict new observations on the response. If these predictions are interpolations in the original region of the x-space where the multicollinearity is in effect, satisfactory predictions will often be obtained, because while individual β_j may be poorly estimated, the function $\sum_{j=1}^{k} \beta_j x_{ij}$ may be estimated quite well. On the other hand, if the prediction of new observations requires extrapolation beyond the original region of the x-space where the data were collected, generally we would expect to obtain poor results. Extrapolation usually requires good estimates of the individual model parameters.

Multicollinearity arises for several reasons. It will occur when the analyst collects data such that a linear constraint holds approximately among the columns of the **X** matrix. For example, if four regressor variables are the components of a mixture, such a constraint will always exist because the sum of the components is always constant. Usually, these constraints do not hold exactly, and the analyst might not know that they exist.

The presence of multicollinearity can be detected in several ways. Two of the more easily understood of these will be discussed briefly.

1. The **variance inflation factors,** defined in equation 12-50, are very useful measures of multicollinearity. The larger the variance inflation factor, the more severe the multicollinearity. Some authors have suggested that if any variance inflation factor exceeds 10, multicollinearity is a problem. Other authors consider this value too liberal and suggest that the variance inflation factors should not exceed 4 or 5. Minitab will calculate the variance inflation factors. Table 12-4 presents the Minitab multiple regression output for the wire bond pull strength data. Since both VIF_1 and VIF_2 are small, there is no problem with multicollinearity.

2. If the F-test for significance of regression is significant, but tests on the individual regression coefficients are not significant, multicollinearity may be present.

Several remedial measures have been proposed for solving the problem of multicollinearity. Augmenting the data with new observations specifically designed to break up the approximate linear dependencies that currently exist is often suggested. However, this is sometimes impossible because of economic reasons or because of the physical constraints that relate the x_j. Another possibility is to delete certain variables from the model, but this approach has the disadvantage of discarding the information contained in the deleted variables.

Since multicollinearity primarily affects the stability of the regression coefficients, it would seem that estimating these parameters by some method that is less sensitive to multicollinearity than ordinary least squares would be helpful. Several methods have been suggested. One alternative to ordinary least squares, **ridge regression,** can be useful in combating multicollinearity. For more details on ridge regression, see Section 12-6.5 on the CD material or the more extensive presentations in Montgomery, Peck, and Vining (2001) and Myers (1990).

12-6.5 Ridge Regression (CD Only)

12-6.6 Nonlinear Regression (CD Only)

EXERCISES FOR SECTION 12-6

12-46. An article entitled "A Method for Improving the Accuracy of Polynomial Regression Analysis" in the *Journal of Quality Technology* (1971, pp. 149–155) reported the following data on y = ultimate shear strength of a rubber compound (psi) and x = cure temperature (°F).

y	770	800	840	810
x	280	284	292	295
y	735	640	590	560
x	298	305	308	315

(a) Fit a second-order polynomial to these data.
(b) Test for significance of regression using $\alpha = 0.05$.
(c) Test the hypothesis that $\beta_{11} = 0$ using $\alpha = 0.05$.

(d) Compute the residuals from part (a) and use them to evaluate model adequacy.

12-47. Consider the following data, which result from an experiment to determine the effect of x = test time in hours at a particular temperature on y = change in oil viscosity:

y	−1.42	−1.39	−1.55	−1.89	−2.43
x	.25	.50	.75	1.00	1.25
y	−3.15	−4.05	−5.15	−6.43	−7.89
x	1.50	1.75	2.00	2.25	2.50

(a) Fit a second-order polynomial to the data.
(b) Test for significance of regression using $\alpha = 0.05$.
(c) Test the hypothesis that $\beta_{11} = 0$ using $\alpha = 0.05$.

(d) Compute the residuals from part (a) and use them to evaluate model adequacy.

12-48. When fitting polynomial regression models, we often subtract $\bar{x}$ from each x value to produce a "centered" regressor $x' = x - \bar{x}$. This reduces the effects of dependencies among the model terms and often leads to more accurate estimates of the regression coefficients. Using the data from Exercise 12-46, fit the model $Y = \beta_0^* + \beta_1^* x' + \beta_{11}^* (x')^2 + \epsilon$. Use the results to estimate the coefficients in the uncentered model $Y = \beta_0 + \beta_1 x + \beta_{11} x^2 + \epsilon$.

12-49. Suppose that we use a standardized variable $x' = (x - \bar{x})/s_x$, where s_x is the standard deviation of x, in constructing a polynomial regression model. Using the data in Exercise 12-46 and the standardized variable approach, fit the model $Y = \beta_0^* + \beta_1^* x' + \beta_{11}^* (x')^2 + \epsilon$.

(a) What value of y do you predict when $x = 285°F$?

(b) Estimate the regression coefficients in the unstandardized model $Y = \beta_0 + \beta_1 x + \beta_{11} x^2 + \epsilon$.

(c) What can you say about the relationship between SS_E and R^2 for the standardized and unstandardized models?

(d) Suppose that $y' = (y - \bar{y})/s_y$ is used in the model along with x'. Fit the model and comment on the relationship between SS_E and R^2 in the standardized model and the unstandardized model.

12-50. The following data were collected during an experiment to determine the change in thrust efficiency (y, in percent) as the divergence angle of a rocket nozzle (x) changes:

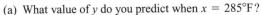

y	24.60	24.71	23.90	39.50	39.60	57.12
x	4.0	4.0	4.0	5.0	5.0	6.0

y	67.11	67.24	67.15	77.87	80.11	84.67
x	6.5	6.5	6.75	7.0	7.1	7.3

(a) Fit a second-order model to the data.

(b) Test for significance of regression and lack of fit using $\alpha = 0.05$.

(c) Test the hypothesis that $\beta_{11} = 0$, using $\alpha = 0.05$.

(d) Plot the residuals and comment on model adequacy.

(e) Fit a cubic model, and test for the significance of the cubic term using $\alpha = 0.05$.

12-51. An article in the *Journal of Pharmaceuticals Sciences* (Vol. 80, 1991, pp. 971–977) presents data on the observed mole fraction solubility of a solute at a constant temperature and the dispersion, dipolar, and hydrogen bonding Hansen partial solubility parameters. The data are as shown in the following table, where y is the negative logarithm of the mole fraction solubility, x_1 is the dispersion partial solubility, x_2 is the dipolar partial solubility, and x_3 is the hydrogen bonding partial solubility.

(a) Fit the model $Y = \beta_0 + \beta_1 x_1 + \beta_2 x_2 + \beta_3 x_3 +$
$\beta_{12} x_1 x_2 + \beta_{13} x_1 x_3 + \beta_{23} x_2 x_3 + \beta_{11} x_1^2 + \beta_{22} x_2^2 + \beta_{33} x_3^2 + \epsilon$.

(b) Test for significance of regression using $\alpha = 0.05$.

(c) Plot the residuals and comment on model adequacy.

Observation Number	y	x_1	x_2	x_3
1	0.22200	7.3	0.0	0.0
2	0.39500	8.7	0.0	0.3
3	0.42200	8.8	0.7	1.0
4	0.43700	8.1	4.0	0.2
5	0.42800	9.0	0.5	1.0
6	0.46700	8.7	1.5	2.8
7	0.44400	9.3	2.1	1.0
8	0.37800	7.6	5.1	3.4
9	0.49400	10.0	0.0	0.3
10	0.45600	8.4	3.7	4.1
11	0.45200	9.3	3.6	2.0
12	0.11200	7.7	2.8	7.1
13	0.43200	9.8	4.2	2.0
14	0.10100	7.3	2.5	6.8
15	0.23200	8.5	2.0	6.6
16	0.30600	9.5	2.5	5.0
17	0.09230	7.4	2.8	7.8
18	0.11600	7.8	2.8	7.7
19	0.07640	7.7	3.0	8.0
20	0.43900	10.3	1.7	4.2
21	0.09440	7.8	3.3	8.5
22	0.11700	7.1	3.9	6.6
23	0.07260	7.7	4.3	9.5
24	0.04120	7.4	6.0	10.9
25	0.25100	7.3	2.0	5.2
26	0.00002	7.6	7.8	20.7

(d) Use the extra sum of squares method to test the contribution of the second-order terms using $\alpha = 0.05$.

12-52. Consider the gasoline mileage data in Exercise 12-5.

(a) Discuss how you would model the information about the type of transmission in the car.

(b) Fit a regression model to the gasoline mileage using engine displacement, horsepower, and the type of transmission in the car as the regressors.

(c) Is there evidence that the type of transmission affects gasoline mileage performance?

12-53. Consider the tool life data in Example 12-12. Test the hypothesis that two different regression models (with different slopes and intercepts) are required to adequately model the data. Use indicator variables in answering this question.

12-54. Use the National Football League Team Performance data in Exercise 12-4 to build regression models using the following techniques:

(a) All possible regressions. Find the equations that minimize MS_E and that minimize C_p.

(b) Stepwise regression.

(c) Forward selection.

(d) Backward elimination.

(e) Comment on the various models obtained. Which model seems "best," and why?

12-55. Use the gasoline mileage data in Exercise 12-5 to build regression models using the following techniques:

(a) All possible regressions. Find the minimum C_p and minimum MS_E equations.

(b) Stepwise regression.

(c) Forward selection.

(d) Backward elimination.

(e) Comment on the various models obtained.

12-56. Consider the electric power data in Exercise 12-6. Build regression models for the data using the following techniques:

(a) All possible regressions.

(b) Stepwise regression.

(c) Forward selection.

(d) Backward elimination.

(e) Comment on the models obtained. Which model would you prefer?

12-57. Consider the wire bond pull strength data in Exercise 12-8. Build regression models for the data using the following methods:

(a) Stepwise regression.

(b) Forward selection.

(c) Backward elimination.

(d) Comment on the models obtained. Which model would you prefer?

12-58. Consider the NHL data in Exercise 12-11. Build regression models for these data using the following methods:

(a) Stepwise regression.

(b) Forward selection.

(c) Backward elimination.

(d) Which model would you prefer?

12-59. Consider the data in Exercise 12-51. Use all the terms in the full quadratic model as the candidate regressors.

(a) Use forward selection to identify a model.

(b) Use backward elimination to identify a model.

(c) Compare the two models obtained in parts (a) and (b). Which model would you prefer and why?

12-60. Find the minimum C_p equation and the equation that maximizes the adjusted R^2 statistic for the wire bond pull strength data in Exercise 12-8. Does the same equation satisfy both criteria?

12-61. For the NHL data in Exercise 12-11.

(a) Find the equation that minimizes C_p.

(b) Find the equation that minimizes MS_E.

(c) Find the equation that maximizes the adjusted R^2. Is this the same equation you found in part (b)?

12-62. We have used a sample of 30 observations to fit a regression model. The full model has nine regressors, the variance estimate is $\hat{\sigma}^2 = MS_E = 100$, and $R^2 = 0.92$.

(a) Calculate the F-statistic for testing significance of regression. Using $\alpha = 0.05$, what would you conclude?

(b) Suppose that we fit another model using only four of the original regressors and that the error sum of squares for this new model is 2200. Find the estimate of σ^2 for this new reduced model. Would you conclude that the reduced model is superior to the old one? Why?

(c) Find the value of C_p for the reduced model in part (b). Would you conclude that the reduced model is better than the old model?

12-63. A sample of 25 observations is used to fit a regression model in seven variables. The estimate of σ^2 for this full model is $MS_E = 10$.

(a) A forward selection algorithm has put three of the original seven regressors in the model. The error sum of squares for the three-variable model is $SS_E = 300$. Based on C_p, would you conclude that the three-variable model has any remaining bias?

(b) After looking at the forward selection model in part (a), suppose you could add one more regressor to the model. This regressor will reduce the error sum of squares to 275. Will the addition of this variable improve the model? Why?

Supplemental Exercises

12-64. The data shown in Table 12-20 on page 465 represent the thrust of a jet-turbine engine (y) and six candidate regressors: x_1 = primary speed of rotation, x_2 = secondary speed of rotation, x_3 = fuel flow rate, x_4 = pressure, x_5 = exhaust temperature, and x_6 = ambient temperature at time of test.

(a) Fit a multiple linear regression model using x_3 = fuel flow rate, x_4 = pressure, and x_5 = exhaust temperature as the regressors.

(b) Test for significance of regression using $\alpha = 0.01$. Find the P-value for this test. What are your conclusions?

(c) Find the t-test statistic for each regressor. Using $\alpha = 0.01$, explain carefully the conclusion you can draw from these statistics.

(d) Find R^2 and the adjusted statistic for this model.

(e) Construct a normal probability plot of the residuals and interpret this graph.

(f) Plot the residuals versus $\hat{y}$. Are there any indications of inequality of variance or nonlinearity?

(g) Plot the residuals versus x_3. Is there any indication of nonlinearity?

(h) Predict the thrust for an engine for which $x_3 = 1670$, $x_4 = 170$, and $x_5 = 1589$.

12-65. Consider the engine thrust data in Exercise 12-64. Refit the model using $y^* = \ln y$ as the response variable and $x_3^* = \ln x_3$ as the regressor (along with x_4 and x_5).

(a) Test for significance of regression using $\alpha = 0.01$. Find the P-value for this test and state your conclusions.

Observation Number	x_1	x_2	x_3	x_4	x_5	y
1	3	3	3	3	0	0.787
2	8	30	8	8	0	0.293
3	3	6	6	6	0	1.710
4	4	4	4	12	0	0.203
5	8	7	6	5	0	0.806
6	10	20	5	5	0	4.713
7	8	6	3	3	25	0.607
8	6	24	4	4	25	9.107
9	4	10	12	4	25	9.210
10	16	12	8	4	25	1.365
11	3	10	8	8	25	4.554
12	8	3	3	3	25	0.293
13	3	6	3	3	50	2.252
14	3	8	8	3	50	9.167
15	4	8	4	8	50	0.694
16	5	2	2	2	50	0.379
17	2	2	2	3	50	0.485
18	10	15	3	3	50	3.345
19	15	6	2	3	50	0.208
20	15	6	2	3	75	0.201
21	10	4	3	3	75	0.329
22	3	8	2	2	75	4.966
23	6	6	6	4	75	1.362
24	2	3	8	6	75	1.515
25	3	3	8	8	75	0.751

(b) Use the t-statistic to test $H_0: \beta_j = 0$ versus $H_1: \beta_j \neq 0$ for each variable in the model. If $\alpha = 0.01$, what conclusions can you draw?

(c) Plot the residuals versus $\hat{y}^*$ and versus x_3^*. Comment on these plots. How do they compare with their counterparts obtained in Exercise 12-64 parts (f) and (g)?

12-66. Transient points of an electronic inverter are influenced by many factors. The table at the top of page 464 gives data on the transient point (y, volts) of PMOS-NMOS inverters and five regressors: x_1 = width of the NMOS device, x_2 = length of the NMOS device, x_3 = width of the PMOS device, x_4 = length of the PMOS device, and x_5 = temperature (°C).

(a) Fit the multiple linear regression model to these data. Test for significance of regression using $\alpha = 0.01$. Find the P-value for this test and use it to draw your conclusions.

(b) Test the contribution of each variable to the model using the t-test with $\alpha = 0.05$. What are your conclusions?

(c) Delete x_5 from the model. Test the new model for significance of regression. Also test the relative contribution of each regressor to the new model with the t-test. Using $\alpha = 0.05$, what are your conclusions?

(d) Notice that the MS_E for the model in part (c) is smaller than the MS_E for the full model in part (a). Explain why this has occurred.

(e) Calculate the studentized residuals. Do any of these seem unusually large?

(f) Suppose that you learn that the second observation was incorrectly recorded. Delete this observation and refit the model using $x_1, x_2, x_3,$ and x_4 as the regressors. Notice that the R^2 for this model is considerably higher than the R^2 for either of the models fitted previously. Explain why the R^2 for this model has increased.

(g) Test the model from part (f) for significance of regression using $\alpha = 0.05$. Also investigate the contribution of each regressor to the model using the t-test with $\alpha = 0.05$. What conclusions can you draw?

(h) Plot the residuals from the model in part (f) versus $\hat{y}$ and versus each of the regressors $x_1, x_2, x_3,$ and x_4. Comment on the plots.

12-67. Consider the inverter data in Exercise 12-66. Delete observation 2 from the original data. Define new variables as follows: $y^* = \ln y$, $x_1^* = 1/\sqrt{x_1}$, $x_2^* = \sqrt{x_2}$, $x_3^* = 1/\sqrt{x_3}$, and $x_4^* = \sqrt{x_4}$.

(a) Fit a regression model using these transformed regressors (do not use x_5).

(b) Test the model for significance of regression using $\alpha = 0.05$. Use the t-test to investigate the contribution of each variable to the model ($\alpha = 0.05$). What are your conclusions?

(c) Plot the residuals versus $\hat{y}^*$ and versus each of the transformed regressors. Comment on the plots.

12-68. Following are data on y = green liquor (g/l) and x = paper machine speed (feet per minute) from a Kraft paper machine. (The data were read from a graph in an article in the *Tappi Journal*, March 1986.)

y	16.0	15.8	15.6	15.5	14.8
x	1700	1720	1730	1740	1750
y	14.0	13.5	13.0	12.0	11.0
x	1760	1770	1780	1790	1795

(a) Fit the model $Y = \beta_0 + \beta_1 x + \beta_2 x^2 + \epsilon$ using least squares.

(b) Test for significance of regression using $\alpha = 0.05$. What are your conclusions?

(c) Test the contribution of the quadratic term to the model, over the contribution of the linear term, using an F-statistic. If $\alpha = 0.05$, what conclusion can you draw?

(d) Plot the residuals from the model in part (a) versus $\hat{y}$. Does the plot reveal any inadequacies?

(e) Construct a normal probability plot of the residuals. Comment on the normality assumption.

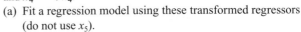

Table   12-20

Observation Number	y	x_1	x_2	x_3	x_4	x_5	x_6
1	4540	2140	20640	30250	205	1732	99
2	4315	2016	20280	30010	195	1697	100
3	4095	1905	19860	29780	184	1662	97
4	3650	1675	18980	29330	164	1598	97
5	3200	1474	18100	28960	144	1541	97
6	4833	2239	20740	30083	216	1709	87
7	4617	2120	20305	29831	206	1669	87
8	4340	1990	19961	29604	196	1640	87
9	3820	1702	18916	29088	171	1572	85
10	3368	1487	18012	28675	149	1522	85
11	4445	2107	20520	30120	195	1740	101
12	4188	1973	20130	29920	190	1711	100
13	3981	1864	19780	29720	180	1682	100
14	3622	1674	19020	29370	161	1630	100
15	3125	1440	18030	28940	139	1572	101
16	4560	2165	20680	30160	208	1704	98
17	4340	2048	20340	29960	199	1679	96
18	4115	1916	19860	29710	187	1642	94
19	3630	1658	18950	29250	164	1576	94
20	3210	1489	18700	28890	145	1528	94
21	4330	2062	20500	30190	193	1748	101
22	4119	1929	20050	29960	183	1713	100
23	3891	1815	19680	29770	173	1684	100
24	3467	1595	18890	29360	153	1624	99
25	3045	1400	17870	28960	134	1569	100
26	4411	2047	20540	30160	193	1746	99
27	4203	1935	20160	29940	184	1714	99
28	3968	1807	19750	29760	173	1679	99
29	3531	1591	18890	29350	153	1621	99
30	3074	1388	17870	28910	133	1561	99
31	4350	2071	20460	30180	198	1729	102
32	4128	1944	20010	29940	186	1692	101
33	3940	1831	19640	29750	178	1667	101
34	3480	1612	18710	29360	156	1609	101
35	3064	1410	17780	28900	136	1552	101
36	4402	2066	20520	30170	197	1758	100
37	4180	1954	20150	29950	188	1729	99
38	3973	1835	19750	29740	178	1690	99
39	3530	1616	18850	29320	156	1616	99
40	3080	1407	17910	28910	137	1569	100

12-69. Consider the jet engine thrust data in Exercise 12-64.

(a) Use all possible regressions to select the best regression equation, where the model with the minimum value of MS_E is to be selected as "best."

(b) Repeat part (a) using the C_P criterion to identify the best equation.

(c) Use stepwise regression to select a subset regression model.

(d) Compare the models obtained in parts (a), (b), and (c) above.

12-70. Consider the electronic inverter data in Exercise 12-66 and 12-67. Define the response and regressors variables as in Exercise 12-67, and delete the second observation in the sample.

(a) Use all possible regressions to find the equation that minimizes C_p.

(b) Use all possible regressions to find the equation that minimizes MS_E.

(c) Use stepwise regression to select a subset regression model.

(d) Compare the models you have obtained.

12-71. Consider the three-variable regression model for the jet engine thrust data in Exercise 12-65. Calculate the variance inflation factors for this model. Would you conclude that multicollinearity is a problem in this model?

12-72. A multiple regression model was used to relate $y =$ viscosity of a chemical product to $x_1 =$ temperature and $x_2 =$ reaction time. The data set consisted of $n = 15$ observations.

(a) The estimated regression coefficients were $\hat{\beta}_0 = 300.00$, $\hat{\beta}_1 = 0.85$, and $\hat{\beta}_2 = 10.40$. Calculate an estimate of mean viscosity when $x_1 = 100°F$ and $x_2 = 2$ hours.

(b) The sums of squares were $SS_T = 1230.50$ and $SS_E = 120.30$. Test for significance of regression using $\alpha = 0.05$. What conclusion can you draw?

(c) What proportion of total variability in viscosity is accounted for by the variables in this model?

(d) Suppose that another regressor, $x_3 =$ stirring rate, is added to the model. The new value of the error sum of squares is $SS_E = 117.20$. Has adding the new variable resulted in a smaller value of MS_E? Discuss the significance of this result.

(e) Calculate an F-statistic to assess the contribution of x_3 to the model. Using $\alpha = 0.05$, what conclusions do you reach?

12-73. An article in the *Journal of the American Ceramics Society* (Vol. 75, 1992, pp. 112–116) describes a process for immobilizing chemical or nuclear wastes in soil by dissolving the contaminated soil into a glass block. The authors mix CaO and Na_2O with soil and model viscosity and electrical conductivity. The electrical conductivity model involves six regressors, and the sample consists of $n = 14$ observations.

(a) For the six-regressor model, suppose that $SS_T = 0.50$ and $R^2 = 0.94$. Find SS_E and SS_R, and use this information to test for significance of regression with $\alpha = 0.05$. What are your conclusions?

(b) Suppose that one of the original regressors is deleted from the model, resulting in $R^2 = 0.92$. What can you conclude about the contribution of the variable that was removed? Answer this question by calculating an F-statistic.

(c) Does deletion of the regressor variable in part (b) result in a smaller value of MS_E for the five-variable model, in comparison to the original six-variable model? Comment on the significance of your answer.

MIND-EXPANDING EXERCISES

12-74. Consider a multiple regression model with k regressors. Show that the test statistic for significance of regression can be written as

$$F_0 = \frac{R^2/k}{(1 - R^2)/(n - k - 1)}$$

Suppose that $n = 20$, $k = 4$, and $R^2 = 0.90$. If $\alpha = 0.05$, what conclusion would you draw about the relationship between y and the four regressors?

12-75. A regression model is used to relate a response y to $k = 4$ regressors with $n = 20$. What is the smallest value of R^2 that will result in a significant regression if $\alpha = 0.05$? Use the results of the previous exercise. Are you surprised by how small the value of R^2 is?

12-76. Show that we can express the residuals from a multiple regression model as $e = (I - H)y$, where $H = X(X'X)^{-1}X'$.

12-77. Show that the variance of the ith residual e_i in a multiple regression model is $\sigma(1 - h_{ii})$ and that the covariance between e_i and e_j is $-\sigma^2 h_{ij}$, where the h's are the elements of $H = X(X X)^{-1}X'$.

12-78. Consider the multiple linear regression model $y = X\beta + \epsilon$. If $\hat{\beta}$ denotes the least squares estimator of β, show that $\hat{\beta} = \beta + R\epsilon$, where $R = (X'X)^{-1}X'$.

12-79. Constrained Least Squares. Suppose we wish to find the least squares estimator of β in the model $y = X\beta + \epsilon$ subject to a set of equality constraints, say, $T\beta = c$.

(a) Show that the estimator is

$$\hat{\beta}_c = \hat{\beta} + (X'X)^{-1}$$
$$\times T'[T(X'X)^{-1}T']^{-1}(c - T\hat{\beta})$$

where $\hat{\beta} = (X'X)^{-1}X'y$.

(b) Discuss situations where this model might be appropriate.

12-80. Piecewise Linear Regression (I). Suppose that y is piecewise linearly related to x. That is, different linear relationships are appropriate over the intervals $-\infty < x \leq x^*$ and $x^* < x < \infty$. Show how indicator variables can be used to fit such a piecewise linear regression model, assuming that the point x^* is known.

12-81. Piecewise Linear Regression (II). Consider the piecewise linear regression model described in Exercise 12-79. Suppose that at the point x^* a discontinuity occurs in the regression function. Show how indicator variables can be used to incorporate the discontinuity into the model.

12-82. Piecewise Linear Regression (III). Consider the piecewise linear regression model described in Exercise 12-79. Suppose that the point x^* is not known with certainty and must be estimated. Suggest an approach that could be used to fit the piecewise linear regression model.

IMPORTANT TERMS AND CONCEPTS

In the E-book, click on any term or concept below to go to that subject.

All possible regressions
Analysis of variance test in multiple regression
Categorical variables as regressors

Confidence interval on the mean response
Extra sum of squares method
Inference (test and intervals) on individual model parameters
Influential observations
Model parameters and their interpretation

in multiple regression
Outliers
Polynomial terms in a regression model
Prediction interval on a future observation
Residual analysis and model adequacy checking

Significance of regression
Stepwise regression and related methods

CD MATERIAL

Ridge regression
Nonlinear regression models

13 Design and Analysis of Single-Factor Experiments: The Analysis of Variance

LEARNING OBJECTIVES

After careful study of this chapter, you should be able to do the following:

1. Design and conduct engineering experiments involving a single factor with an arbitrary number of levels

2. Understand how the analysis of variance is used to analyze the data from these experiments

3. Assess model adequacy with residual plots

4. Use multiple comparison procedures to identify specific differences between means

5. Make decisions about sample size in single-factor experiments

6. Understand the difference between fixed and random factors

7. Estimate variance components in an experiment involving random factors

8. Understand the blocking principle and how it is used to isolate the effect of nuisance factors

9. Design and conduct experiments involving the randomized complete block design

CD MATERIAL

10. Use operating characteristic curves to make sample size decisions in single-factor random effects experiment

11. Use Tukey's test, orthogonal contrasts and graphical methods to identify specific differences between means.

Answers for most odd numbered exercises are at the end of the book. Answers to exercises whose numbers are surrounded by a box can be accessed in the e-Text by clicking on the box. Complete worked solutions to certain exercises are also available in the e-Text. These are indicated in the Answers to Selected Exercises section by a box around the exercise number. Exercises are also available for some of the text sections that appear on CD only. These exercises may be found within the e-Text immediately following the section they accompany.

13-1 DESIGNING ENGINEERING EXPERIMENTS

Experiments are a natural part of the engineering and scientific decision-making process. Suppose, for example, that a civil engineer is investigating the effects of different curing methods on the mean compressive strength of concrete. The experiment would consist of making up several test specimens of concrete using each of the proposed curing methods and then testing the compressive strength of each specimen. The data from this experiment could be used to determine which curing method should be used to provide maximum mean compressive strength.

If there are only two curing methods of interest, this experiment could be designed and analyzed using the statistical hypothesis methods for two samples introduced in Chapter 10. That is, the experimenter has a single **factor** of interest—curing methods—and there are only two **levels** of the factor. If the experimenter is interested in determining which curing method produces the maximum compressive strength, the number of specimens to test can be determined from the operating characteristic curves in Appendix Chart VI, and the t-test can be used to decide if the two means differ.

Many single-factor experiments require that more than two levels of the factor be considered. For example, the civil engineer may want to investigate five different curing methods. In this chapter we show how the **analysis of variance** (frequently abbreviated ANOVA) can be used for comparing means when there are more than two levels of a single factor. We will also discuss **randomization** of the experimental runs and the important role this concept plays in the overall experimentation strategy. In the next chapter, we will show how to design and analyze experiments with several factors.

Statistically based experimental design techniques are particularly useful in the engineering world for improving the performance of a manufacturing process. They also have extensive application in the development of new processes. Most processes can be described in terms of several **controllable variables,** such as temperature, pressure, and feed rate. By using designed experiments, engineers can determine which subset of the process variables has the greatest influence on process performance. The results of such an experiment can lead to

1. Improved process yield

2. Reduced variability in the process and closer conformance to nominal or target requirements

3. Reduced design and development time

4. Reduced cost of operation

Experimental design methods are also useful in **engineering design** activities, where new products are developed and existing ones are improved. Some typical applications of statistically designed experiments in engineering design include

1. Evaluation and comparison of basic design configurations
2. Evaluation of different materials
3. Selection of design parameters so that the product will work well under a wide variety of field conditions (or so that the design will be robust)
4. Determination of key product design parameters that affect product performance

The use of experimental design in the engineering design process can result in products that are easier to manufacture, products that have better field performance and reliability than their competitors, and products that can be designed, developed, and produced in less time.

Designed experiments are usually employed **sequentially.** That is, the first experiment with a complex system (perhaps a manufacturing process) that has many controllable variables is often a screening experiment designed to determine which variables are most important. Subsequent experiments are used to refine this information and determine which adjustments to these critical variables are required to improve the process. Finally, the objective of the experimenter is optimization, that is, to determine which levels of the critical variables result in the best process performance.

Every experiment involves a sequence of activities:

1. **Conjecture**—the original hypothesis that motivates the experiment.
2. **Experiment**—the test performed to investigate the conjecture.
3. **Analysis**—the statistical analysis of the data from the experiment.
4. **Conclusion**—what has been learned about the original conjecture from the experiment. Often the experiment will lead to a revised conjecture, and a new experiment, and so forth.

The statistical methods introduced in this chapter and Chapter 14 are essential to good experimentation. **All experiments are designed experiments;** unfortunately, some of them are poorly designed, and as a result, valuable resources are used ineffectively. Statistically designed experiments permit efficiency and economy in the experimental process, and the use of statistical methods in examining the data results in **scientific objectivity** when drawing conclusions.

13-2 THE COMPLETELY RANDOMIZED SINGLE-FACTOR EXPERIMENT

13-2.1 An Example

A manufacturer of paper used for making grocery bags is interested in improving the tensile strength of the product. Product engineering thinks that tensile strength is a function of the hardwood concentration in the pulp and that the range of hardwood concentrations of practical interest is between 5 and 20%. A team of engineers responsible for the study decides to investigate four levels of hardwood concentration: 5%, 10%, 15%, and 20%. They decide to make up six test specimens at each concentration level, using a pilot plant. All 24 specimens are tested on a laboratory tensile tester, in random order. The data from this experiment are shown in Table 13-1.

Table 13-1   Tensile Strength of Paper (psi)

Hardwood Concentration (%)	Observations						Totals	Averages
	1	2	3	4	5	6		
5	7, 7⁶	8	15	11	9	10	60	10.00
10	12	17	13	18	19	15	94	15.67
15	14	18	19	17	16	18	102	17.00
20	19	25	22	23	18	20	127	21.17
							383	15.96

This is an example of a completely randomized single-factor experiment with four levels of the factor. The levels of the factor are sometimes called **treatments,** and each treatment has six observations or **replicates.** The role of **randomization** in this experiment is extremely important. By randomizing the order of the 24 runs, the effect of any nuisance variable that may influence the observed tensile strength is approximately balanced out. For example, suppose that there is a warm-up effect on the tensile testing machine; that is, the longer the machine is on, the greater the observed tensile strength. If all 24 runs are made in order of increasing hardwood concentration (that is, all six 5% concentration specimens are tested first, followed by all six 10% concentration specimens, etc.), any observed differences in tensile strength could also be due to the warm-up effect.

It is important to graphically analyze the data from a designed experiment. Figure 13-1(a) presents box plots of tensile strength at the four hardwood concentration levels. This figure indicates that changing the hardwood concentration has an effect on tensile strength; specifically, higher hardwood concentrations produce higher observed tensile strength. Furthermore, the distribution of tensile strength at a particular hardwood level is reasonably symmetric, and the variability in tensile strength does not change dramatically as the hardwood concentration changes.

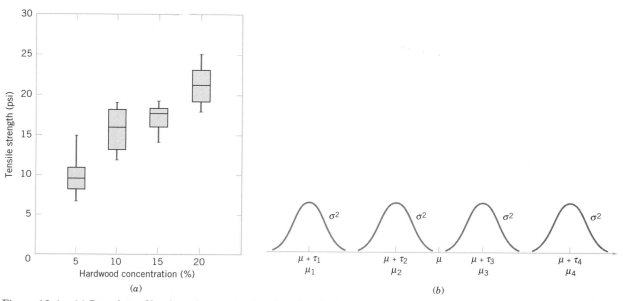

Figure 13-1   (a) Box plots of hardwood concentration data. (b) Display of the model in Equation 13-1 for the completely randomized single-factor experiment.

Graphical interpretation of the data is always useful. Box plots show the variability of the observations *within* a treatment (factor level) and the variability *between* treatments. We now discuss how the data from a single-factor randomized experiment can be analyzed statistically.

13-2.2 The Analysis of Variance

Suppose we have a different levels of a single factor that we wish to compare. Sometimes, each factor level is called a **treatment,** a very general term that can be traced to the early applications of experimental design methodology in the agricultural sciences. The **response** for each of the a treatments is a random variable. The observed data would appear as shown in Table 13-2. An entry in Table 13-2, say y_{ij}, represents the jth observation taken under treatment i. We initially consider the case in which there are an equal number of observations, n, on each treatment.

We may describe the observations in Table 13-2 by the **linear statistical model**

$$Y_{ij} = \mu + \tau_i + \epsilon_{ij} \begin{cases} i = 1, 2, \ldots, a \\ j = 1, 2, \ldots, n \end{cases} \qquad (13\text{-}1)$$

where Y_{ij} is a random variable denoting the (ij)th observation, μ is a parameter common to all treatments called the **overall mean,** τ_i is a parameter associated with the ith treatment called the ith **treatment effect,** and ϵ_{ij} is a random error component. Notice that the model could have been written as

$$Y_{ij} = \mu_i + \epsilon_{ij} \begin{cases} i = 1, 2, \ldots, a \\ j = 1, 2, \ldots, n \end{cases}$$

where $\mu_i = \mu + \tau_i$ is the mean of the ith treatment. In this form of the model, we see that each treatment defines a population that has mean μ_i, consisting of the overall mean μ plus an effect τ_i that is due to that particular treatment. We will assume that the errors ϵ_{ij} are normally and independently distributed with mean zero and variance σ^2. Therefore, each treatment can be thought of as a normal population with mean μ_i and variance σ^2. See Fig. 13-1(b).

Equation 13-1 is the underlying model for a single-factor experiment. Furthermore, since we require that the observations are taken in random order and that the environment (often called the experimental units) in which the treatments are used is as uniform as possible, this experimental design is called a **completely randomized design.**

Table 13-2   Typical Data for a Single-Factor Experiment

Treatment	Observations				Totals	Averages
1	y_{11}	y_{12}	$\cdots$	y_{1n}	$y_{1\cdot}$	$\bar{y}_{1\cdot}$
2	y_{21}	y_{22}	$\cdots$	y_{2n}	$y_{2\cdot}$	$\bar{y}_{2\cdot}$
$\vdots$	$\vdots$	$\vdots$	$\vdots$	$\vdots$	$\vdots$	$\vdots$
a	y_{a1}	y_{a2}	$\cdots$	y_{an}	$y_{a\cdot}$	$\bar{y}_{a\cdot}$
					$y_{\cdot\cdot}$	$\bar{y}_{\cdot\cdot}$

The *a* factor levels in the experiment could have been chosen in two different ways. First, the experimenter could have specifically chosen the *a* treatments. In this situation, we wish to test hypotheses about the treatment means, and conclusions cannot be extended to similar treatments that were not considered. In addition, we may wish to estimate the treatment effects. This is called the **fixed-effects model.** Alternatively, the *a* treatments could be a random sample from a larger population of treatments. In this situation, we would like to be able to extend the conclusions (which are based on the sample of treatments) to all treatments in the population, whether or not they were explicitly considered in the experiment. Here the treatment effects τ_i are random variables, and knowledge about the particular ones investigated is relatively unimportant. Instead, we test hypotheses about the variability of the τ_i and try to estimate this variability. This is called the **random effects,** or **components of variance,** model.

In this section we develop the **analysis of variance** for the fixed-effects model. The analysis of variance is not new to us; it was used previously in the presentation of regression analysis. However, in this section we show how it can be used to test for equality of treatment effects. In the fixed-effects model, the treatment effects τ_i are usually defined as deviations from the overall mean μ, so that

$$\sum_{i=1}^{a} \tau_i = 0 \tag{13-2}$$

Let $y_{i\cdot}$ represent the total of the observations under the *i*th treatment and $\bar{y}_{i\cdot}$ represent the average of the observations under the *i*th treatment. Similarly, let $y_{\cdot\cdot}$ represent the grand total of all observations and $\bar{y}_{\cdot\cdot}$ represent the grand mean of all observations. Expressed mathematically,

$$y_{i\cdot} = \sum_{j=1}^{n} y_{ij} \qquad \bar{y}_{i\cdot} = y_{i\cdot}/n \qquad i = 1, 2, \ldots, a$$

$$y_{\cdot\cdot} = \sum_{i=1}^{a} \sum_{j=1}^{n} y_{ij} \qquad \bar{y}_{\cdot\cdot} = y_{\cdot\cdot}/N \tag{13-3}$$

where $N = an$ is the total number of observations. Thus, the "dot" subscript notation implies summation over the subscript that it replaces.

We are interested in testing the equality of the *a* treatment means $\mu_1, \mu_2, \ldots, \mu_a$. Using Equation 13-2, we find that this is equivalent to testing the hypotheses

$$H_0: \tau_1 = \tau_2 = \cdots = \tau_a = 0$$
$$H_1: \tau_i \neq 0 \quad \text{for at least one } i \tag{13-4}$$

Thus, if the null hypothesis is true, each observation consists of the overall mean μ plus a realization of the random error component ϵ_{ij}. This is equivalent to saying that all N observations are taken from a normal distribution with mean μ and variance σ^2. Therefore, if the null hypothesis is true, changing the levels of the factor has no effect on the mean response.

The ANOVA partitions the total variability in the sample data into two component parts. Then, the test of the hypothesis in Equation 13-4 is based on a comparison of two independent estimates of the population variance. The total variability in the data is described by the **total sum of squares**

$$SS_T = \sum_{i=1}^{a} \sum_{j=1}^{n} (y_{ij} - \bar{y}_{\cdot\cdot})^2$$

The partition of the total sum of squares is given in the following definition.

Definition

> **The sum of squares identity is**
>
> $$\sum_{i=1}^{a} \sum_{j=1}^{n} (y_{ij} - \bar{y}_{..})^2 = n \sum_{i=1}^{a} (\bar{y}_{i.} - \bar{y}_{..})^2 + \sum_{i=1}^{a} \sum_{j=1}^{n} (y_{ij} - \bar{y}_{i.})^2 \qquad (13\text{-}5)$$
>
> or symbolically
>
> $$SS_T = SS_{\text{Treatments}} + SS_E \qquad (13\text{-}6)$$

The identity in Equation 13-5 (which is developed in Section 13-4.4 on the CD) shows that the total variability in the data, measured by the total corrected sum of squares SS_T, can be partitioned into a sum of squares of differences between treatment means and the grand mean denoted $SS_{\text{Treatments}}$ and a sum of squares of differences of observations within a treatment from the treatment mean denoted SS_E. Differences between observed treatment means and the grand mean measure the differences between treatments, while differences of observations within a treatment from the treatment mean can be due only to random error.

We can gain considerable insight into how the analysis of variance works by examining the expected values of $SS_{\text{Treatments}}$ and SS_E. This will lead us to an appropriate statistic for testing the hypothesis of no differences among treatment means (or all $\tau_i = 0$).

> The expected value of the treatment sum of squares is
>
> $$E(SS_{\text{Treatments}}) = (a - 1)\sigma^2 + n \sum_{i=1}^{a} \tau_i^2$$
>
> and the expected value of the error sum of squares is
>
> $$E(SS_E) = a(n - 1)\sigma^2$$

There is also a partition of the number of degrees of freedom that corresponds to the sum of squares identity in Equation 13-5. That is, there are $an = N$ observations; thus, SS_T has $an - 1$ degrees of freedom. There are a levels of the factor, so $SS_{\text{Treatments}}$ has $a - 1$ degrees of freedom. Finally, within any treatment there are n replicates providing $n - 1$ degrees of freedom with which to estimate the experimental error. Since there are a treatments, we have $a(n - 1)$ degrees of freedom for error. Therefore, the degrees of freedom partition is

$$an - 1 = a - 1 + a(n - 1)$$

The ratio

$$MS_{\text{Treatments}} = SS_{\text{Treatments}}/(a - 1)$$

is called the **mean square for treatments.** Now if the null hypothesis $H_0: \tau_1 = \tau_2 = \cdots = \tau_a = 0$ is true, $MS_{\text{Treatments}}$ is an unbiased estimator of σ^2 because $\sum_{i=1}^{a} \tau_i = 0$. However, if H_1 is true, $MS_{\text{Treatments}}$ estimates σ^2 plus a positive term that incorporates variation due to the systematic difference in treatment means.

Note that the **error mean square**

$$MS_E = SS_E/[a(n-1)]$$

is an unbiased estimator of σ^2 regardless of whether or not H_0 is true. We can also show that $MS_{\text{Treatments}}$ and MS_E are independent. Consequently, we can show that if the null hypothesis H_0 is true, the ratio

$$F_0 = \frac{SS_{\text{Treatments}}/(a-1)}{SS_E/[a(n-1)]} = \frac{MS_{\text{Treatments}}}{MS_E} \qquad (13\text{-}7)$$

has an F-distribution with $a-1$ and $a(n-1)$ degrees of freedom. Furthermore, from the expected mean squares, we know that MS_E is an unbiased estimator of σ^2. Also, under the null hypothesis, $MS_{\text{Treatments}}$ is an unbiased estimator of σ^2. However, if the null hypothesis is false, the expected value of $MS_{\text{Treatments}}$ is greater than σ^2. Therefore, under the alternative hypothesis, the expected value of the numerator of the test statistic (Equation 13-7) is greater than the expected value of the denominator. Consequently, we should reject H_0 if the statistic is large. This implies an upper-tail, one-tail critical region. Therefore, we would reject H_0 if $f_0 > f_{\alpha, a-1, a(n-1)}$ where f_0 is the computed value of F_0 from Equation 13-7.

Efficient computational formulas for the sums of squares may be obtained by expanding and simplifying the definitions of $SS_{\text{Treatments}}$ and SS_T. This yields the following results.

Definition

The sums of squares computing formulas for the ANOVA with equal sample sizes in each treatment are

$$SS_T = \sum_{i=1}^{a} \sum_{j=1}^{n} y_{ij}^2 - \frac{y_{..}^2}{N} \qquad (13\text{-}8)$$

and

$$SS_{\text{Treatments}} = \sum_{i=1}^{a} \frac{y_i^2}{n} - \frac{y_{..}^2}{N} \qquad (13\text{-}9)$$

The error sum of squares is obtained by subtraction as

$$SS_E = SS_T - SS_{\text{Treatments}} \qquad (13\text{-}10)$$

The computations for this test procedure are usually summarized in tabular form as shown in Table 13-3. This is called an **analysis of variance** (or **ANOVA**) **table.**

Table 13-3 The Analysis of Variance for a Single-Factor Experiment, Fixed-Effects Model

Source of Variation	Sum of Squares	Degrees of Freedom	Mean Square	F_0
Treatments	$SS_{\text{Treatments}}$	$a-1$	$MS_{\text{Treatments}}$	$\dfrac{MS_{\text{Treatments}}}{MS_E}$
Error	SS_E	$a(n-1)$	MS_E	
Total	SS_T	$an-1$		

EXAMPLE 13-1   Consider the paper tensile strength experiment described in Section 13-2.1. We can use the analysis of variance to test the hypothesis that different hardwood concentrations do not affect the mean tensile strength of the paper.

The hypotheses are

$$H_0: \tau_1 = \tau_2 = \tau_3 = \tau_4 = 0$$

$$H_1: \tau_i \neq 0 \text{ for at least one } i \quad .$$

We will use $\alpha = 0.01$. The sums of squares for the analysis of variance are computed from Equations 13-8, 13-9, and 13-10 as follows:

$$SS_T = \sum_{i=1}^{4} \sum_{j=1}^{6} y_{ij}^2 - \frac{y_{..}^2}{N}$$

$$= (7)^2 + (8)^2 + \cdots + (20)^2 - \frac{(383)^2}{24} = 512.96$$

$$SS_{\text{Treatments}} = \sum_{i=1}^{4} \frac{y_{i.}^2}{n} - \frac{y_{..}^2}{N}$$

$$= \frac{(60)^2 + (94)^2 + (102)^2 + (127)^2}{6} - \frac{(383)^2}{24} = 382.79$$

$$SS_E = SS_T - SS_{\text{Treatments}}$$
$$= 512.96 - 382.79 = 130.17$$

The ANOVA is summarized in Table 13-4. Since $f_{0.01,3,20} = 4.94$, we reject H_0 and conclude that hardwood concentration in the pulp significantly affects the mean strength of the paper. We can also find a P-value for this test statistic as follows:

$$P = P(F_{3,20} > 19.60) \simeq 3.59 \times 10^{-6}$$

Since $P \simeq 3.59 \times 10^{-6}$ is considerably smaller than $\alpha = 0.01$, we have strong evidence to conclude that H_0 is not true.

Minitab Output

Many software packages have the capability to analyze data from designed experiments using the analysis of variance. Table 13-5 presents the output from the Minitab one-way analysis of variance routine for the paper tensile strength experiment in Example 13-1. The results agree closely with the manual calculations reported previously in Table 13-4.

Table 13-4   ANOVA for the Tensile Strength Data

Source of Variation	Sum of Squares	Degrees of Freedom	Mean Square	f_0	P-value
Hardwood concentration	382.79	3	127.60	19.60	3.59 E-6
Error	130.17	20	6.51		
Total	512.96	23			

Table 13-5   Minitab Analysis of Variance Output for Example 13-1

One-Way ANOVA: Strength versus CONC

Analysis of Variance for Strength

Source	DF	SS	MS	F	P
Conc	3	382.79	127.60	19.61	0.000
Error	20	130.17	6.51		
Total	23	512.96			

Individual 95% CIs For Mean
Based on Pooled StDev

Level	N	Mean	StDev	—- + ——- + —— + —— + —
5	6	10.000	2.828	(—*—)
10	6	15.667	2.805	(—*—)
15	6	17.000	1.789	(—*—)
20	6	21.167	2.639	(—*—)

—- + ——— + ——— + ——- + -
10.0   15.0   20.0   25.0

Pooled StDev = 2.551

Fisher's pairwise comparisons

Family error rate = 0.192
Individual error rate = 0.0500

Critical value = 2.086

Intervals for (column level mean) − (row level mean)

	5	10	15
10	−8.739		
	−2.594		
15	−10.072	−4.406	
	−3.928	1.739	
20	−14.239	−8.572	−7.239
	−8.094	−2.428	−1.094

(handwritten) $\alpha = 0.01$

$f_{0.01, 3, 20} = 4.94$

The Minitab output also presents 95% **confidence intervals** on each individual treatment mean. The mean of the ith treatment is defined as

$$\mu_i = \mu + \tau_i \quad i = 1, 2, \ldots, a$$

A point estimator of μ_i is $\hat{\mu}_i = \overline{Y}_{i\cdot}$. Now, if we assume that the errors are normally distributed, each treatment average is normally distributed with mean μ_i and variance σ^2/n. Thus, if σ^2 were known, we could use the normal distribution to construct a CI. Using MS_E as an estimator of σ^2 (The square root of MS_E is the "Pooled StDev" referred to in the Minitab output), we would base the CI on the t-distribution, since

$$T = \frac{\overline{Y}_{i\cdot} - \mu_i}{\sqrt{MS_E/n}}$$

has a t-distribution with $a(n-1)$ degrees of freedom. This leads to the following definition of the confidence interval.

Definition

A $100(1 - \alpha)$ percent confidence interval on the mean of the ith treatment μ_i is

$$\bar{y}_{i\cdot} - t_{\alpha/2,a(n-1)} \sqrt{\frac{MS_E}{n}} \leq \mu_i \leq \bar{y}_{i\cdot} + t_{\alpha/2,a(n-1)} \sqrt{\frac{MS_E}{n}} \qquad (13\text{-}11)$$

Equation 13-11 is used to calculate the 95% CIs shown graphically in the Minitab output of Table 13-5. For example, at 20% hardwood the point estimate of the mean is $\bar{y}_{4\cdot} = 21.167$, $MS_E = 6.51$, and $t_{0.025,20} = 2.086$, so the 95% CI is

$$\left[\bar{y}_{4\cdot} \pm t_{0.025,20} \sqrt{MS_E/n}\right]$$
$$\left[21.167 \pm (2.086) \sqrt{6.51/6}\right]$$

or

$$19.00 \text{ psi} \leq \mu_4 \leq 23.34 \text{ psi}$$

It can also be interesting to find confidence intervals on the difference in two treatment means, say, $\mu_i - \mu_j$. The point estimator of $\mu_i - \mu_j$ is $\bar{Y}_{i\cdot} - \bar{Y}_{j\cdot}$, and the variance of this estimator is

$$V(\bar{Y}_{i\cdot} - \bar{Y}_{j\cdot}) = \frac{\sigma^2}{n} + \frac{\sigma^2}{n} = \frac{2\sigma^2}{n}$$

Now if we use MS_E to estimate σ^2,

$$T = \frac{\bar{Y}_{i\cdot} - \bar{Y}_{j\cdot} - (\mu_i - \mu_j)}{\sqrt{2MS_E/n}}$$

has a t-distribution with $a(n - 1)$ degrees of freedom. Therefore, a CI on $\mu_i - \mu_j$ may be based on the t-distribution.

Definition

A $100(1 - \alpha)$ percent confidence interval on the difference in two treatment means $\mu_i - \mu_j$ is

$$\bar{y}_{i\cdot} - \bar{y}_{j\cdot} - t_{\alpha/2,a(n-1)} \sqrt{\frac{2MS_E}{n}} \leq \mu_i - \mu_j \leq \bar{y}_{i\cdot} - \bar{y}_{j\cdot} + t_{\alpha/2,a(n-1)} \sqrt{\frac{2MS_E}{n}}$$
$$(13\text{-}12)$$

A 95% CI on the difference in means $\mu_3 - \mu_2$ is computed from Equation 13-12 as follows:

$$\left[\bar{y}_{3\cdot} - \bar{y}_{2\cdot} \pm t_{0.025,20} \sqrt{2MS_E/n}\right]$$
$$\left[17.00 - 15.67 \pm (2.086) \sqrt{2(6.51)/6}\right]$$

or

$$-1.74 \leq \mu_3 - \mu_2 \leq 4.40$$

Since the CI includes zero, we would conclude that there is no difference in mean tensile strength at these two particular hardwood levels.

The bottom portion of the computer output in Table 13-5 provides additional information concerning which specific means are different. We will discuss this in more detail in Section 13-2.3.

An Unbalanced Experiment

In some single-factor experiments, the number of observations taken under each treatment may be different. We then say that the design is **unbalanced.** In this situation, slight modifications must be made in the sums of squares formulas. Let n_i observations be taken under treatment i ($i = 1, 2, \ldots, a$), and let the total number of observations $N = \sum_{i=1}^{a} n_i$. The computational formulas for SS_T and $SS_{\text{Treatments}}$ are as shown in the following definition.

Definition

> The sums of squares computing formulas for the ANOVA with unequal sample sizes n_i in each treatment are
>
> $$SS_T = \sum_{i=1}^{a} \sum_{j=1}^{n_i} y_{ij}^2 - \frac{y_{..}^2}{N} \qquad (13\text{-}13)$$
>
> $$SS_{\text{Treatments}} = \sum_{i=1}^{a} \frac{y_{i.}^2}{n_i} - \frac{y_{..}^2}{N} \qquad (13\text{-}14)$$
>
> and
>
> $$SS_E = SS_T - SS_{\text{Treatments}} \qquad (13\text{-}15)$$

Choosing a balanced design has two important advantages. First, the ANOVA is relatively insensitive to small departures from the assumption of equality of variances if the sample sizes are equal. This is not the case for unequal sample sizes. Second, the power of the test is maximized if the samples are of equal size.

13-2.3 Multiple Comparisons Following the ANOVA

When the null hypothesis $H_0: \tau_1 = \tau_2 = \cdots = \tau_a = 0$ is rejected in the ANOVA, we know that some of the treatment or factor level means are different. However, the ANOVA doesn't identify which means are different. Methods for investigating this issue are called **multiple comparisons methods.** Many of these procedures are available. Here we describe a very simple one, **Fisher's least significant difference (LSD) method.** In Section 13-2.4 on the CD, we describe three other procedures. Montgomery (2001) presents these and other methods and provides a comparative discussion.

The Fisher LSD method compares all pairs of means with the null hypotheses $H_0: \mu_i = \mu_j$ (for all $i \neq j$) using the t-statistic

$$t_0 = \frac{\bar{y}_{i.} - \bar{y}_{j.}}{\sqrt{\dfrac{2MS_E}{n}}}$$

Assuming a two-sided alternative hypothesis, the pair of means μ_i and μ_j would be declared significantly different if

$$|\bar{y}_{i.} - \bar{y}_{j.}| > \text{LSD}$$

where LSD, the **least significant difference**, is

$$
\text{LSD} = t_{\alpha/2, a(n-1)} \sqrt{\frac{2MS_E}{n}} \tag{13-16}
$$

If the sample sizes are different in each treatment, the LSD is defined as

$$
\text{LSD} = t_{\alpha/2, N-a} \sqrt{MS_E \left(\frac{1}{n_i} + \frac{1}{n_j} \right)}
$$

EXAMPLE 13-2   We will apply the Fisher LSD method to the hardwood concentration experiment. There are $a = 4$ means, $n = 6$, $MS_E = 6.51$, and $t_{0.025, 20} = 2.086$. The treatment means are

$$
\bar{y}_{1.} = 10.00 \text{ psi}
$$
$$
\bar{y}_{2.} = 15.67 \text{ psi}
$$
$$
\bar{y}_{3.} = 17.00 \text{ psi}
$$
$$
\bar{y}_{4.} = 21.17 \text{ psi}
$$

The value of LSD is LSD $= t_{0.025, 20} \sqrt{2MS_E/n} = 2.086\sqrt{2(6.51)/6} = 3.07$. Therefore, any pair of treatment averages that differs by more than 3.07 implies that the corresponding pair of treatment means are different.

The comparisons among the observed treatment averages are as follows:

$$
\begin{aligned}
4 \text{ vs. } 1 &= 21.17 - 10.00 = 11.17 > 3.07 \\
4 \text{ vs. } 2 &= 21.17 - 15.67 = 5.50 > 3.07 \\
4 \text{ vs. } 3 &= 21.17 - 17.00 = 4.17 > 3.07 \\
3 \text{ vs. } 1 &= 17.00 - 10.00 = 7.00 > 3.07 \\
3 \text{ vs. } 2 &= 17.00 - 15.67 = 1.33 < 3.07 \\
2 \text{ vs. } 1 &= 15.67 - 10.00 = 5.67 > 3.07
\end{aligned}
$$

From this analysis, we see that there are significant differences between all pairs of means except 2 and 3. This implies that 10 and 15% hardwood concentration produce approximately the same tensile strength and that all other concentration levels tested produce different tensile strengths. It is often helpful to draw a graph of the treatment means, such as in Fig. 13-2, with the means that are *not* different underlined. This graph clearly reveals the results of the experiment and shows that 20% hardwood produces the maximum tensile strength.

The Minitab output in Table 13-5 shows the Fisher LSD method under the heading "Fisher's pairwise comparisons." The critical value reported is actually the value of $t_{0.025, 20} =$

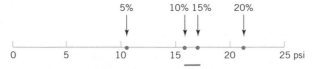

Figure 13-2   Results of Fisher's LSD method in Example 13-2.

2.086. Minitab implements Fisher's LSD method by computing **confidence intervals** on all pairs of treatment means using Equation 13-12. The lower and upper 95% confidence limits are shown at the bottom of the table. Notice that the only pair of means for which the confidence interval includes zero is for μ_{10} and μ_{15}. This implies that μ_{10} and μ_{15} are not significantly different, the same result found in Example 13-2.

Table 13-5 also provides a "family error rate," equal to 0.192 in this example. When all possible pairs of means are tested, the probability of at least one type I error can be much greater than for a single test. We can interpret the family error rate as follows. The probability is $1 - 0.192 = 0.808$ that there are no type I errors in the six comparisons. The family error rate in Table 13-5 is based on the distribution of the range of the sample means. See Montgomery (2001) for details. Alternatively, Minitab permits you to specify a family error rate and will then calculate an individual error rate for each comparison.

13-2.4   More About Multiple Comparisons (CD Only)

13-2.5   Residual Analysis and Model Checking

The analysis of variance assumes that the observations are normally and independently distributed with the same variance for each treatment or factor level. These assumptions should be checked by examining the residuals. A **residual** is the difference between an observation y_{ij} and its estimated (or fitted) value from the statistical model being studied, denoted as $\hat{y}_{ij}$. For the completely randomized design $\hat{y}_{ij} = \bar{y}_{i\cdot}$ and each residual is $e_{ij} = y_{ij} - \bar{y}_{i\cdot}$, that is, the difference between an observation and the corresponding observed treatment mean. The residuals for the paper tensile strength experiment are shown in Table 13-6. Using $\bar{y}_{i\cdot}$ to calculate each residual essentially removes the effect of hardwood concentration from the data; consequently, the residuals contain information about unexplained variability.

The normality assumption can be checked by constructing a **normal probability plot** of the residuals. To check the assumption of equal variances at each factor level, plot the residuals against the factor levels and compare the spread in the residuals. It is also useful to plot the residuals against $\bar{y}_{i\cdot}$ (sometimes called the fitted value); the variability in the residuals should not depend in any way on the value of $\bar{y}_{i\cdot}$. Most statistics software packages will construct these plots on request. When a pattern appears in these plots, it usually suggests the need for a transformation, that is, analyzing the data in a different metric. For example, if the variability in the residuals increases with $\bar{y}_{i\cdot}$, a transformation such as $\log y$ or $\sqrt{y}$ should be considered. In some problems, the dependency of residual scatter on the observed mean $\bar{y}_{i\cdot}$ is very important information. It may be desirable to select the factor level that results in maximum response; however, this level may also cause more variation in response from run to run.

The independence assumption can be checked by plotting the residuals against the time or run order in which the experiment was performed. A pattern in this plot, such as sequences of positive and negative residuals, may indicate that the observations are not independent.

Table 13-6   Residuals for the Tensile Strength Experiment

Hardwood Concentration (%)	Residuals					
5	−3.00	−2.00	5.00	1.00	−1.00	0.00
10	−3.67	1.33	−2.67	2.33	3.33	−0.67
15	−3.00	1.00	2.00	0.00	−1.00	1.00
20	−2.17	3.83	0.83	1.83	−3.17	−1.17

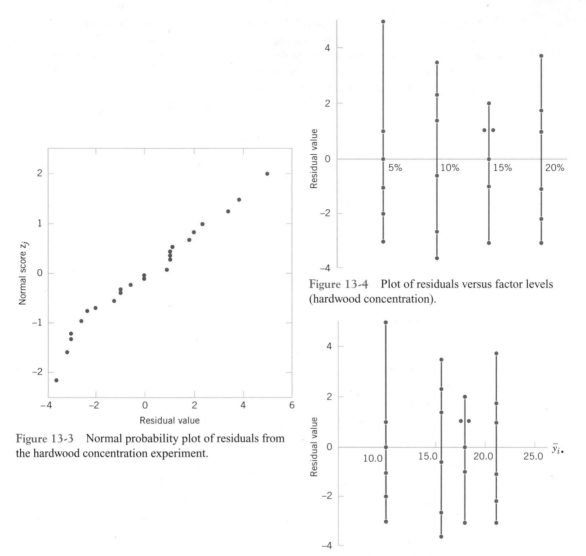

Figure 13-3   Normal probability plot of residuals from the hardwood concentration experiment.

Figure 13-4   Plot of residuals versus factor levels (hardwood concentration).

Figure 13-5   Plot of residuals versus $\bar{y}_{i.}$

This suggests that time or run order is important or that variables that change over time are important and have not been included in the experimental design.

A normal probability plot of the residuals from the paper tensile strength experiment is shown in Fig. 13-3. Figures 13-4 and 13-5 present the residuals plotted against the factor levels and the fitted value $\bar{y}_{i.}$ respectively. These plots do not reveal any model inadequacy or unusual problem with the assumptions.

13-2.6   Determining Sample Size

In any experimental design problem, the choice of the sample size or number of replicates to use is important. **Operating characteristic curves** can be used to provide guidance in making this selection. Recall that the operating characteristic curve is a plot of the probability of a type II error (β) for various sample sizes against a measure of the difference in means that it is important to detect. Thus, if the experimenter knows the magnitude of the difference in means that is of potential importance, the operating characteristic curves can be used to determine how many replicates are required to achieve adequate sensitivity.

The power of the ANOVA test is

$$
\begin{aligned}
1 - \beta &= P\{\text{Reject } H_0 \mid H_0 \text{ is false}\} \\
&= P\{F_0 > f_{\alpha, a-1, a(n-1)} \mid H_0 \text{ is false}\}
\end{aligned}
\tag{13-17}
$$

To evaluate this probability statement, we need to know the distribution of the test statistic F_0 if the null hypothesis is false. It can be shown that, if H_0 is false, the statistic $F_0 = MS_{\text{Treatments}}/MS_E$ is distributed as a **noncentral F random variable,** with $a - 1$ and $a(n - 1)$ degrees of freedom and a noncentrality parameter δ. If $\delta = 0$, the noncentral F-distribution becomes the usual or *central* F-distribution.

Operating characteristic curves are used to evaluate β defined in Equation 13-17. These curves plot β against a parameter Φ, where

$$
\Phi^2 = \frac{n \sum_{i=1}^{a} \tau_i^2}{a\sigma^2}
\tag{13-18}
$$

The parameter Φ^2 is (apart from n) the noncentrality parameter δ. Curves are available for $\alpha = 0.05$ and $\alpha = 0.01$ and for several values of the number of degrees of freedom for numerator (denoted v_1) and denominator (denoted v_2). Figure 13-6 gives representative O.C. curves, one for $a = 4$ ($v_1 = 3$) and one for $a = 5$ ($v_1 = 4$) treatments. Notice that for each value of a there are curves for $\alpha = 0.05$ and $\alpha = 0.01$. O.C. curves for other values of a are in Section 13-2.7 on the CD.

In using the operating curves, we must define the difference in means that we wish to detect in terms of $\sum_{i=1}^{a} \tau_i^2$. Also, the error variance σ^2 is usually unknown. In such cases, we must choose ratios of $\sum_{i=1}^{a} \tau_i^2/\sigma^2$ that we wish to detect. Alternatively, if an estimate of σ^2 is available, one may replace σ^2 with this estimate. For example, if we were interested in the sensitivity of an experiment that has already been performed, we might use MS_E as the estimate of σ^2.

EXAMPLE 13-3

Suppose that five means are being compared in a completely randomized experiment with $\alpha = 0.01$. The experimenter would like to know how many replicates to run if it is important to reject H_0 with probability at least 0.90 if $\sum_{i=1}^{5} \tau_i^2/\sigma^2 = 5.0$. The parameter Φ^2 is, in this case,

$$
\Phi^2 = \frac{n \sum_{i=1}^{a} \tau_i^2}{a\sigma^2} = \frac{n}{5}(5) = n
$$

and for the operating characteristic curve with $v_1 = a - 1 = 5 - 1 = 4$, and $v_2 = a(n - 1) = 5(n - 1)$ error degrees of freedom refer to the lower curve in Figure 13-6. As a first guess, try $n = 4$ replicates. This yields $\Phi^2 = 4$, $\Phi = 2$, and $v_2 = 5(3) = 15$ error degrees of freedom. Consequently, from Figure 13-6, we find that $\beta \simeq 0.38$. Therefore, the power of the test is approximately $1 - \beta = 1 - 0.38 = 0.62$, which is less than the required 0.90, and so we

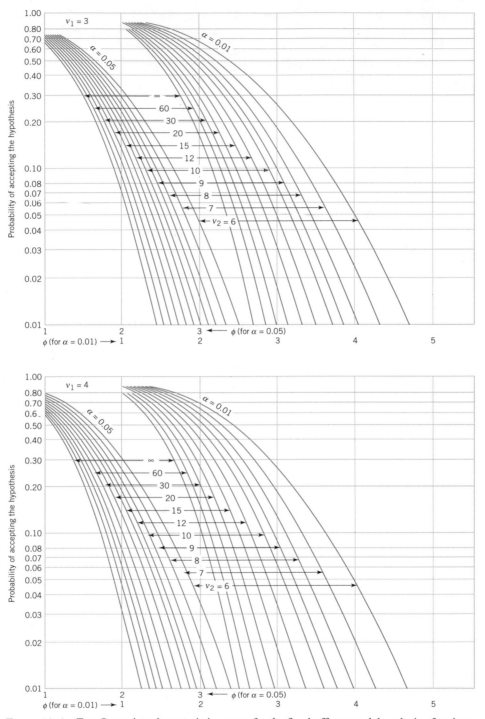

Figure 13-6 Two Operating characteristic curves for the fixed-effects model analysis of variance.

conclude that $n = 4$ replicates is not sufficient. Proceeding in a similar manner, we can construct the following table:

n	Φ^2	Φ	$a(n-1)$	β	Power $= (1-\beta)$
4	4	2.00	15	0.38	0.62
5	5	2.24	20	0.18	0.82
6	6	2.45	25	0.06	0.94

Thus, at least $n = 6$ replicates must be run in order to obtain a test with the required power.

13-2.7 Technical Details about the Analysis of Variance (CD Only)

EXERCISES FOR SECTION 13-2

13-1. In *Design and Analysis of Experiments,* 5th edition (John Wiley & Sons, 2001) D. C. Montgomery describes an experiment in which the tensile strength of a synthetic fiber is of interest to the manufacturer. It is suspected that strength is related to the percentage of cotton in the fiber. Five levels of cotton percentage are used, and five replicates are run in random order, resulting in the data below.

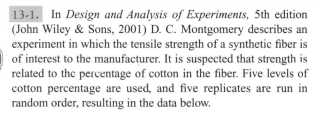

Cotton	Observations				
Percentage	1	2	3	4	5
15	7	7	15	11	9
20	12	17	12	18	18
25	14	18	18	19	19
30	19	25	22	19	23
35	7	10	11	15	11

(a) Does cotton percentage affect breaking strength? Draw comparative box plots and perform an analysis of variance. Use $\alpha = 0.05$.

(b) Plot average tensile strength against cotton percentage and interpret the results.

(c) Analyze the residuals and comment on model adequacy.

13-2. In "Orthogonal Design for Process Optimization and Its Application to Plasma Etching" (*Solid State Technology,* May 1987), G. Z. Yin and D. W. Jillie describe an experiment to determine the effect of C_2F_6 flow rate on the uniformity of the etch on a silicon wafer used in integrated circuit manufacturing. Three flow rates are used in the experiment, and the resulting uniformity (in percent) for six replicates is shown below.

C_2F_6 Flow	Observations					
(SCCM)	1	2	3	4	5	6
125	2.7	4.6	2.6	3.0	3.2	3.8
160	4.9	4.6	5.0	4.2	3.6	4.2
200	4.6	3.4	2.9	3.5	4.1	5.1

(a) Does C_2F_6 flow rate affect etch uniformity? Construct box plots to compare the factor levels and perform the analysis of variance. Use $\alpha = 0.05$.

(b) Do the residuals indicate any problems with the underlying assumptions?

13-3. The compressive strength of concrete is being studied, and four different mixing techniques are being investigated. The following data have been collected.

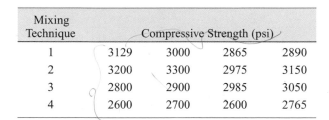

Mixing Technique	Compressive Strength (psi)			
1	3129	3000	2865	2890
2	3200	3300	2975	3150
3	2800	2900	2985	3050
4	2600	2700	2600	2765

(a) Test the hypothesis that mixing techniques affect the strength of the concrete. Use $\alpha = 0.05$.

(b) Find the P-value for the F-statistic computed in part (a).

(c) Analyze the residuals from this experiment.

13-4. An experiment was run to determine whether four specific firing temperatures affect the density of a certain type of brick. The experiment led to the following data.

Temperature (°F)	Density						
100	21.8	21.9	21.7	21.6	21.7	21.5	21.8
125	21.7	21.4	21.5	21.5	—	—	—
150	21.9	21.8	21.8	21.6	21.5	—	—
175	21.9	21.7	21.8	21.7	21.6	21.8	

(a) Does the firing temperature affect the density of the bricks? Use $\alpha = 0.05$.

(b) Find the P-value for the F-statistic computed in part (a).

(c) Analyze the residuals from the experiment.

13-5. An electronics engineer is interested in the effect on tube conductivity of five different types of coating for cathode ray tubes in a telecommunications system display device. The following conductivity data are obtained.

Coating Type	Conductivity			
1	143	141	150	146
2	152	149	137	143
3	134	133	132	127
4	129	127	132	129
5	147	148	144	142

(a) Is there any difference in conductivity due to coating type? Use $\alpha = 0.01$.
(b) Analyze the residuals from this experiment.
(c) Construct a 95% interval estimate of the coating type 1 mean. Construct a 99% interval estimate of the mean difference between coating types 1 and 4.

13-6. The response time in milliseconds was determined for three different types of circuits in an electronic calculator. The results are recorded here.

Circuit Type	Response				
1	19	22	20	18	25
2	20	21	33	27	40
3	16	15	18	26	17

(a) Using $\alpha = 0.01$, test the hypothesis that the three circuit types have the same response time.
(b) Analyze the residuals from this experiment.
(c) Find a 95% confidence interval on the response time for circuit three.

13-7. An article in the *ACI Materials Journal* (Vol. 84, 1987, pp. 213–216) describes several experiments investigating the rodding of concrete to remove entrapped air. A 3-inch × 6-inch cylinder was used, and the number of times this rod was used is the design variable. The resulting compressive strength of the concrete specimen is the response. The data are shown in the following table.

Rodding Level	Compressive Strength		
10	1530	1530	1440
15	1610	1650	1500
20	1560	1730	1530
25	1500	1490	1510

(a) Is there any difference in compressive strength due to the rodding level?

(b) Find the *P*-value for the *F*-statistic in part (a).
(c) Analyze the residuals from this experiment. What conclusions can you draw about the underlying model assumptions?

13-8. An article in *Environment International* (Vol. 18, No. 4, 1992) describes an experiment in which the amount of radon released in showers was investigated. Radon-enriched water was used in the experiment, and six different orifice diameters were tested in shower heads. The data from the experiment are shown in the following table.

Orifice Diameter	Radon Released (%)			
0.37	80	83	83	85
0.51	75	75	79	79
0.71	74	73	76	77
1.02	67	72	74	74
1.40	62	62	67	69
1.99	60	61	64	66

(a) Does the size of the orifice affect the mean percentage of radon released? Use $\alpha = 0.05$.
(b) Find the *P*-value for the *F*-statistic in part (a).
(c) Analyze the residuals from this experiment.
(d) Find a 95% confidence interval on the mean percent of radon released when the orifice diameter is 1.40.

13-9. A paper in the *Journal of the Association of Asphalt Paving Technologists* (Vol. 59, 1990) describes an experiment to determine the effect of air voids on percentage retained strength of asphalt. For purposes of the experiment, air voids are controlled at three levels; low (2–4%), medium (4–6%), and high (6–8%). The data are shown in the following table.

Air Voids	Retained Strength (%)						
Low	106	90	103	90	79	88	92	95
Medium	80	69	94	91	70	83	87	83
High	78	80	62	69	76	85	69	85

(a) Do the different levels of air voids significantly affect mean retained strength? Use $\alpha = 0.01$.
(b) Find the *P*-value for the *F*-statistic in part (a).
(c) Analyze the residuals from this experiment.
(d) Find a 95% confidence interval on mean retained strength where there is a high level of air voids.
(e) Find a 95% confidence interval on the difference in mean retained strength at the low and high levels of air voids.

13-10. An article in the *Materials Research Bulletin* (Vol. 26, No. 11, 1991) investigated four different methods of preparing the superconducting compound $PbMo_6S_8$. The authors contend that the presence of oxygen during the preparation process affects the material's superconducting transition temperature T_c. Preparation methods 1 and 2 use techniques that are designed to eliminate the presence of oxygen, while methods 3 and 4 allow oxygen to be present. Five observations on T_c (in °K) were made for each method, and the results are as follows:

Preparation Method	Transition Temperature T_c(°K)				
1	14.8	14.8	14.7	14.8	14.9
2	14.6	15.0	14.9	14.8	14.7
3	12.7	11.6	12.4	12.7	12.1
4	14.2	14.4	14.4	12.2	11.7

(a) Is there evidence to support the claim that the presence of oxygen during preparation affects the mean transition temperature? Use $\alpha = 0.05$.
(b) What is the P-value for the F-test in part (a)?
(c) Analyze the residuals from this experiment.
(d) Find a 95% confidence interval on mean T_c when method 1 is used to prepare the material.

13-11. Use Fisher's LSD method with $\alpha = 0.05$ to analyze the means of the five different levels of cotton content in Exercise 13-1.

13-12. Use Fisher's LSD method with $\alpha = 0.05$ test to analyze the means of the three flow rates in Exercise 13-2.

13-13. Use Fisher's LSD method with $\alpha = 0.05$ to analyze the mean compressive strength of the four mixing techniques in Exercise 13-3.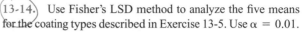

13-14. Use Fisher's LSD method to analyze the five means for the coating types described in Exercise 13-5. Use $\alpha = 0.01$.

13-15. Use Fisher's LSD method to analyze the mean response times for the three circuits described in Exercise 13-6. Use $\alpha = 0.01$.

13-16. Use Fisher's LSD method to analyze the mean amounts of radon released in the experiment described in Exercise 13-8. Use $\alpha = 0.05$.

13-17. Apply Fisher's LSD method to the air void experiment described in Exercise 13-9. Using $\alpha = 0.05$, which treatment means are different?

13-18. Apply Fisher's LSD method to the superconducting material experiment described in Exercise 13-10. Which preparation methods differ, if $\alpha = 0.05$?

13-19. Suppose that four normal populations have common variance $\sigma^2 = 25$ and means $\mu_1 = 50$, $\mu_2 = 60$, $\mu_3 = 50$, and $\mu_4 = 60$. How many observations should be taken on each population so that the probability of rejecting the hypothesis of equality of means is at least 0.90? Use $\alpha = 0.05$.

13-20. Suppose that five normal populations have common variance $\sigma^2 = 100$ and means $\mu_1 = 175$, $\mu_2 = 190$, $\mu_3 = 160$, $\mu_4 = 200$, and $\mu_5 = 215$. How many observations per population must be taken so that the probability of rejecting the hypothesis of equality of means is at least 0.95? Use $\alpha = 0.01$.

13-3   THE RANDOM-EFFECTS MODEL

13-3.1   Fixed versus Random Factors

In many situations, the factor of interest has a large number of possible levels. The analyst is interested in drawing conclusions about the entire population of factor levels. If the experimenter randomly selects a of these levels from the population of factor levels, we say that the factor is a **random factor**. Because the levels of the factor actually used in the experiment were chosen randomly, the conclusions reached will be valid for the entire population of factor levels. We will assume that the population of factor levels is either of infinite size or is large enough to be considered infinite. Notice that this is a very different situation than we encountered in the fixed effects case, where the conclusions apply only for the factor levels used in the experiment.

13-3.2   ANOVA and Variance Components

The linear statistical model is

$$Y_{ij} = \mu + \tau_i + \epsilon_{ij} \begin{cases} i = 1, 2, \ldots, a \\ j = 1, 2, \ldots, n \end{cases} \qquad (13\text{-}19)$$

where the treatment effects τ_i and the errors ϵ_{ij} are independent random variables. Note that the model is identical in structure to the fixed-effects case, but the parameters have a different interpretation. If the variance of the treatment effects τ_i is σ_τ^2, by independence the variance of the response is

$$V(Y_{ij}) = \sigma_\tau^2 + \sigma^2$$

The variances σ_τ^2 and σ^2 are called **variance components,** and the model, Equation 13-19, is called the **components of variance model** or the **random-effects model.** To test hypotheses in this model, we assume that the errors ϵ_{ij} are normally and independently distributed with mean 0 and variance σ^2 and that the treatment effects τ_i are normally and independently distributed with mean zero and variance σ_τ^2.*

For the random-effects model, testing the hypothesis that the individual treatment effects are zero is meaningless. It is more appropriate to test hypotheses about σ_τ^2. Specifically,

$$H_0: \sigma_\tau^2 = 0$$
$$H_1: \sigma_\tau^2 > 0$$

If $\sigma_\tau^2 = 0$, all treatments are identical; but if $\sigma_\tau^2 > 0$, there is variability between treatments.

The ANOVA decomposition of total variability is still valid; that is,

$$SS_T = SS_{\text{Treatments}} + SS_E \tag{13-20}$$

However, the expected values of the mean squares for treatments and error are somewhat different than in the fixed-effect case.

In the random-effects model for a single-factor, completely randomized experiment, the expected mean square for treatments is

$$E(MS_{\text{Treatments}}) = E\left(\frac{SS_{\text{Treatments}}}{a-1}\right)$$

$$= \sigma^2 + n\sigma_\tau^2 \tag{13-21}$$

and the expected mean square for error is

$$E(MS_E) = E\left[\frac{SS_E}{a(n-1)}\right]$$

$$= \sigma^2 \tag{13-22}$$

From examining the expected mean squares, it is clear that both MS_E and $MS_{\text{Treatments}}$ estimate σ^2 when $H_0: \sigma_\tau^2 = 0$ is true. Furthermore, MS_E and $MS_{\text{Treatments}}$ are independent. Consequently, the ratio

$$F_0 = \frac{MS_{\text{Treatments}}}{MS_E} \tag{13-23}$$

---

*The assumption that the $\{\tau_i\}$ are independent random variables implies that the usual assumption of $\sum_{i=1}^{a} \tau_i = 0$ from the fixed-effects model does not apply to the random-effects model.

is an F random variable with $a - 1$ and $a(n - 1)$ degrees of freedom when H_0 is true. The null hypothesis would be rejected at the α-level of significance if the computed value of the test statistic $f_0 > f_{\alpha, a-1, a(n-1)}$.

The computational procedure and construction of the ANOVA table for the random-effects model are identical to the fixed-effects case. The conclusions, however, are quite different because they apply to the entire population of treatments.

Usually, we also want to estimate the variance components (σ^2 and σ_τ^2) in the model. The procedure that we will use to estimate σ^2 and σ_τ^2 is called the **analysis of variance method** because it uses the information in the analysis of variance table. It does not require the normality assumption on the observations. The procedure consists of equating the expected mean squares to their observed values in the ANOVA table and solving for the variance components. When equating observed and expected mean squares in the one-way classification random-effects model, we obtain

$$MS_{\text{Treatments}} = \sigma^2 + n\sigma_\tau^2 \quad \text{and} \quad MS_E = \sigma^2$$

Therefore, the estimators of the variance components are

and

$$\hat{\sigma}^2 = MS_E \qquad (13\text{-}24)$$

$$\hat{\sigma}_\tau^2 = \frac{MS_{\text{Treatments}} - MS_E}{n} \qquad (13\text{-}25)$$

Sometimes the analysis of variance method produces a negative estimate of a variance component. Since variance components are by definition nonnegative, a negative estimate of a variance component is disturbing. One course of action is to accept the estimate and use it as evidence that the true value of the variance component is zero, assuming that sampling variation led to the negative estimate. While this approach has intuitive appeal, it will disturb the statistical properties of other estimates. Another alternative is to reestimate the negative variance component with a method that always yields nonnegative estimates. Still another possibility is to consider the negative estimate as evidence that the assumed linear model is incorrect, requiring that a study of the model and its assumptions be made to find a more appropriate model.

EXAMPLE 13-4

In *Design and Analysis of Experiments*, 5th edition (John Wiley, 2001), D. C. Montgomery describes a single-factor experiment involving the random-effects model in which a textile manufacturing company weaves a fabric on a large number of looms. The company is interested in loom-to-loom variability in tensile strength. To investigate this variability, a manufacturing engineer selects four looms at random and makes four strength determinations on fabric samples chosen at random from each loom. The data are shown in Table 13-7 and the ANOVA is summarized in Table 13-8.

From the analysis of variance, we conclude that the looms in the plant differ significantly in their ability to produce fabric of uniform strength. The variance components are estimated by $\hat{\sigma}^2 = 1.90$ and

$$\hat{\sigma}_\tau^2 = \frac{29.73 - 1.90}{4} = 6.96$$

Table 13-7   Strength Data for Example 13-4

Loom	\multicolumn{4}{c}{Observations}				Total	Average
	1	2	3	4	Total	Average
1	98	97	99	96	390	97.5
2	91	90	93	92	366	91.5
3	96	95	97	95	383	95.8
4	95	96	99	98	388	97.0
					1527	95.45

Table 13-8   Analysis of Variance for the Strength Data

Source of Variation	Sum of Squares	Degrees of Freedom	Mean Square	f_0	P-value
Looms	89.19	3	29.73	15.68	1.88 E-4
Error	22.75	12	1.90		
Total	111.94	15			

Therefore, the variance of strength in the manufacturing process is estimated by

$$\widehat{V(Y_{ij})} = \hat{\sigma}_\tau^2 + \hat{\sigma}^2 = 6.96 + 1.90 = 8.86$$

Most of this variability is attributable to differences between looms.

This example illustrates an important application of the analysis of variance—the isolation of different sources of variability in a manufacturing process. Problems of excessive variability in critical functional parameters or properties frequently arise in quality-improvement programs. For example, in the previous fabric strength example, the process mean is estimated by $\bar{y} = 95.45$ psi, and the process standard deviation is estimated by $\hat{\sigma}_y = \sqrt{\hat{V}(Y_{ij})} = \sqrt{8.86} = 2.98$ psi. If strength is approximately normally distributed, the distribution of strength in the outgoing product would look like the normal distribution shown in Fig. 13-7(a). If the lower specification limit (LSL) on strength is at 90 psi, a substantial proportion of the process output is **fallout**—that is, scrap or defective material that must be sold as second quality, and so on. This fallout is directly related to the excess variability resulting from **differences between looms.** Variability in loom performance could be caused by faulty setup, poor maintenance, inadequate supervision, poorly trained operators, and so forth. The engineer or manager responsible for quality improvement must identify and remove these sources of variability from the process. If this can be done, strength variability will be greatly reduced, perhaps as low as $\hat{\sigma}_Y = \sqrt{\hat{\sigma}^2} = \sqrt{1.90} = 1.38$ psi, as shown in Fig. 13-7(b). In this improved process, reducing the variability in strength has greatly reduced the fallout, resulting in lower cost, higher quality, a more satisfied customer, and enhanced competitive position for the company.

13-3.3   Determining Sample Size in the Random Model (CD Only)

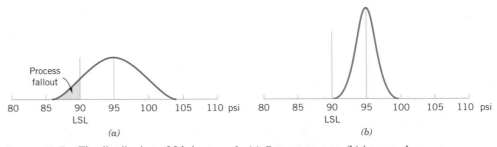

Figure 13-7   The distribution of fabric strength. (a) Current process, (b) improved process.

EXERCISES FOR SECTION 13-3

13-21. A textile mill has a large number of looms. Each loom is supposed to provide the same output of cloth per minute. To investigate this assumption, five looms are chosen at random, and their output is measured at different times. The following data are obtained:

Loom	Output (lb/min)				
1	4.0	4.1	4.2	4.0	4.1
2	3.9	3.8	3.9	4.0	4.0
3	4.1	4.2	4.1	4.0	3.9
4	3.6	3.8	4.0	3.9	3.7
5	3.8	3.6	3.9	3.8	4.0

(a) Are the looms similar in output? Use $\alpha = 0.05$.
(b) Estimate the variability between looms.
(c) Estimate the experimental error variance.
(d) Analyze the residuals from this experiment and check for model adequacy.

13-22. An article in the *Journal of the Electrochemical Society* (Vol. 139, No. 2, 1992, pp. 524–532) describes an experiment to investigate the low-pressure vapor deposition of polysilicon. The experiment was carried out in a large-capacity reactor at Sematech in Austin, Texas. The reactor has several wafer positions, and four of these positions are selected at random. The response variable is film thickness uniformity. Three replicates of the experiment were run, and the data are as follows:

Water Position	Uniformity		
1	2.76	5.67	4.49
2	1.43	1.70	2.19
3	2.34	1.97	1.47
4	0.94	1.36	1.65

(a) Is there a difference in the wafer positions? Use $\alpha = 0.05$.
(b) Estimate the variability due to wafer positions.
(c) Estimate the random error component.
(d) Analyze the residuals from this experiment and comment on model adequacy.

13-23. An article in the *Journal of Quality Technology* (Vol. 13, No. 2, 1981, pp. 111–114) describes an experiment that investigates the effects of four bleaching chemicals on pulp brightness. These four chemicals were selected at random from a large population of potential bleaching agents. The data are as follows:

Chemical	Pulp Brightness				
1	77.199	74.466	92.746	76.208	82.876
2	80.522	79.306	81.914	80.346	73.385
3	79.417	78.017	91.596	80.802	80.626
4	78.001	78.358	77.544	77.364	77.386

(a) Is there a difference in the chemical types? Use $\alpha = 0.05$.
(b) Estimate the variability due to chemical types.
(c) Estimate the variability due to random error.
(d) Analyze the residuals from this experiment and comment on model adequacy.

13-24. Consider the vapor-deposition experiment described in Exercise 13-22.
(a) Estimate the total variability in the uniformity response.
(b) How much of the total variability in the uniformity response is due to the difference between positions in the reactor?
(c) To what level could the variability in the uniformity response be reduced, if the position-to-position variability in the reactor could be eliminated? Do you believe this is a significant reduction?

13-4 RANDOMIZED COMPLETE BLOCK DESIGN

13-4.1 Design and Statistical Analysis

In many experimental design problems, it is necessary to design the experiment so that the variability arising from a nuisance factor can be controlled. For example, consider the situation of Example 10-9, where two different methods were used to predict the shear strength of steel plate girders. Because each girder has different strength (potentially), and this variability in strength was not of direct interest, we designed the experiment by using the two test methods on each girder and then comparing the average difference in strength readings on each girder to zero using the paired *t*-test. The paired *t*-test is a procedure for comparing two treatment means when all experimental runs cannot be made under

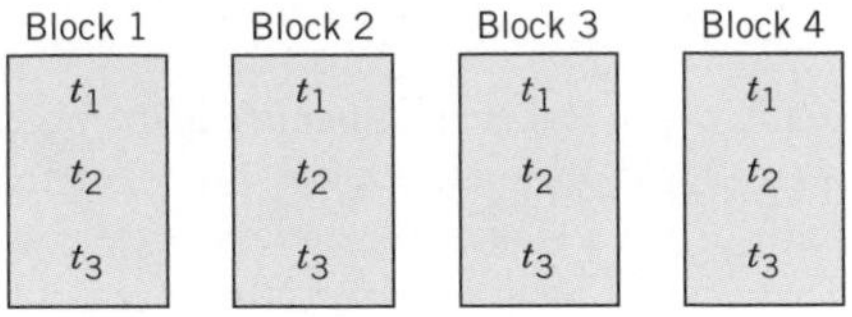

Block 1	Block 2	Block 3	Block 4
t_1	t_1	t_1	t_1
t_2	t_2	t_2	t_2
t_3	t_3	t_3	t_3

Figure 13-8 A randomized complete block design.

Table 13-9 A Randomized Complete Block Design

Treatments (Method)	Block (Girder)			
	1	2	3	4
1	y_{11}	y_{12}	y_{13}	y_{14}
2	y_{21}	y_{22}	y_{23}	y_{24}
3	y_{31}	y_{32}	y_{33}	y_{34}

homogeneous conditions. Alternatively, we can view the paired t-test as a method for reducing the background noise in the experiment by blocking out a **nuisance factor** effect. The block is the nuisance factor, and in this case, the nuisance factor is the actual **experimental unit**—the steel girder specimens used in the experiment.

The randomized block design is an extension of the paired t-test to situations where the factor of interest has more than two levels; that is, more than two treatments must be compared. For example, suppose that three methods could be used to evaluate the strength readings on steel plate girders. We may think of these as three treatments, say t_1, t_2, and t_3. If we use four girders as the experimental units, a **randomized complete block design** would appear as shown in Fig. 13-8. The design is called a randomized complete block design because each block is large enough to hold all the treatments and because the actual assignment of each of the three treatments within each block is done randomly. Once the experiment has been conducted, the data are recorded in a table, such as is shown in Table 13-9. The observations in this table, say y_{ij}, represent the response obtained when method i is used on girder j.

The general procedure for a randomized complete block design consists of selecting b blocks and running a complete replicate of the experiment in each block. The data that result from running a randomized complete block design for investigating a single factor with a levels and b blocks are shown in Table 13-10. There will be a observations (one per factor level) in each block, and the order in which these observations are run is randomly assigned within the block.

We will now describe the statistical analysis for a randomized complete block design. Suppose that a single factor with a levels is of interest and that the experiment is run in b blocks. The observations may be represented by the **linear statistical model**

$$Y_{ij} = \mu + \tau_i + \beta_j + \epsilon_{ij} \begin{cases} i = 1, 2, \ldots, a \\ j = 1, 2, \ldots, b \end{cases} \qquad (13\text{-}26)$$

Table 13-10 A Randomized Complete Block Design with a Treatments and b Blocks

Treatments	Blocks				Totals	Averages
	1	2	...	b		
1	y_{11}	y_{12}	...	y_{1b}	$y_{1\cdot}$	$\bar{y}_{1\cdot}$
2	y_{21}	y_{22}	...	y_{2b}	$y_{2\cdot}$	$\bar{y}_{2\cdot}$
⋮	⋮	⋮		⋮	⋮	⋮
a	y_{a1}	y_{a2}	...	y_{ab}	$y_{a\cdot}$	$\bar{y}_{a\cdot}$
Totals	$y_{\cdot 1}$	$y_{\cdot 2}$	...	$y_{\cdot b}$	$y_{\cdot\cdot}$	
Averages	$\bar{y}_{\cdot 1}$	$\bar{y}_{\cdot 2}$	...	$\bar{y}_{\cdot b}$		$\bar{y}_{\cdot\cdot}$

where μ is an overall mean, τ_i is the effect of the ith treatment, β_j is the effect of the jth block, and ϵ_{ij} is the random error term, which is assumed to be normally and independently distributed with mean zero and variance σ^2. Treatments and blocks will initially be considered as fixed factors. Furthermore, the treatment and block effects are defined as deviations from the overall mean, so $\sum_{i=1}^{a} \tau_i = 0$ and $\sum_{j=1}^{b} \beta_j = 0$. We also assume that treatments and blocks do not interact. That is, the effect of treatment i is the same regardless of which block (or blocks) it is tested in. We are interested in testing the equality of the treatment effects. That is

$$H_0\colon \tau_1 = \tau_2 = \cdots = \tau_a = 0$$
$$H_1\colon \tau_i \neq 0 \text{ at least one } i$$

Testing the hypothesis that all the treatment effects τ_i are equal to zero is equivalent to testing the hypothesis that the treatment means are equal. To see this, note that the mean of the ith treatment is μ_i, defined as

$$\mu_i = E\left(\frac{\sum_{j=1}^{b} Y_{ij}}{b}\right) = \frac{1}{b}\sum_{j=1}^{b} E(Y_{ij}) = \frac{1}{b}\sum_{j=1}^{b} E(\mu + \tau_i + \beta_j + \epsilon_{ij})$$

$$= \frac{1}{b}\sum_{j=1}^{b} E(\mu + \tau_i + \beta_j) = \mu + \tau_i + \frac{1}{b}\sum_{j=1}^{b} \beta_j$$

and since $\sum_{j=1}^{b} \beta_j = 0$, we have the mean of the ith treatment defined as

$$\mu_i = \mu + \tau_i, \quad i = 1, 2, \ldots, a$$

Therefore, testing the hypothesis that the a treatment means are equal is equivalent to testing that all the treatment effects τ_i are equal to zero.

The analysis of variance can be extended to the randomized complete block design. The procedure uses a sum of squares identity that partitions the total sum of squares into three components.

The **sum of squares identity for the randomized complete block design** is

$$\sum_{i=1}^{a}\sum_{j=1}^{b} (y_{ij} - \bar{y}..)^2 = b\sum_{i=1}^{a} (\bar{y}_{i\cdot} - \bar{y}..)^2 + a\sum_{j=1}^{b} (\bar{y}_{\cdot j} - \bar{y}..)^2$$
$$+ \sum_{i=1}^{a}\sum_{j=1}^{b} (y_{ij} - \bar{y}_{\cdot j} - \bar{y}_{i\cdot} + \bar{y}..)^2 \qquad (13\text{-}27)$$

or symbolically

$$SS_T = SS_{\text{Treatments}} + SS_{\text{Blocks}} + SS_E$$

Furthermore, the degrees of freedom corresponding to these sums of squares are

$$ab - 1 = (a - 1) + (b - 1) + (a - 1)(b - 1)$$

For the randomized block design, the relevant mean squares are

$$MS_{\text{Treatments}} = \frac{SS_{\text{Treatments}}}{a - 1}$$

$$MS_{\text{Blocks}} = \frac{SS_{\text{Blocks}}}{b - 1}$$

$$MS_E = \frac{SS_E}{(a - 1)(b - 1)}$$

The expected values of these mean squares can be shown to be as follows:

$$E(MS_{\text{Treatments}}) = \sigma^2 + \frac{b \sum_{i=1}^{a} \tau_i^2}{a - 1}$$

$$E(MS_{\text{Blocks}}) = \sigma^2 + \frac{a \sum_{j=1}^{b} \beta_j^2}{b - 1}$$

$$E(MS_E) = \sigma^2$$

Therefore, if the null hypothesis H_0 is true so that all treatment effects $\tau_i = 0$, $MS_{\text{Treatments}}$ is an unbiased estimator of σ^2, while if H_0 is false, $MS_{\text{Treatments}}$ overestimates σ^2. The mean square for error is always an unbiased estimate of σ^2. To test the null hypothesis that the treatment effects are all zero, we use the ratio

$$F_0 = \frac{MS_{\text{Treatments}}}{MS_E} \tag{13-28}$$

which has an F-distribution with $a - 1$ and $(a - 1)(b - 1)$ degrees of freedom if the null hypothesis is true. We would reject the null hypothesis at the α-level of significance if the computed value of the test statistic in Equation 13-28 is $f_0 > f_{\alpha, a-1, (a-1)(b-1)}$.

In practice, we compute SS_T, $SS_{\text{Treatments}}$ and SS_{Blocks} and then obtain the error sum of squares SS_E by subtraction. The appropriate computing formulas are as follows.

Definition

The computing formulas for the sums of squares in the analysis of variance for a randomized complete block design are

$$SS_T = \sum_{i=1}^{a} \sum_{j=1}^{b} y_{ij}^2 - \frac{y_{..}^2}{ab} \tag{13-29}$$

$$SS_{\text{Treatments}} = \frac{1}{b} \sum_{i=1}^{a} y_{i.}^2 - \frac{y_{..}^2}{ab} \tag{13-30}$$

$$SS_{\text{Blocks}} = \frac{1}{a} \sum_{j=1}^{b} y_{.j}^2 - \frac{y_{..}^2}{ab} \tag{13-31}$$

and

$$SS_E = SS_T - SS_{\text{Treatments}} - SS_{\text{Blocks}} \tag{13-32}$$

Table 13-11   ANOVA for a Randomized Complete Block Design

Source of Variation	Sum of Squares	Degrees of Freedom	Mean Square	F_0
Treatments	$SS_{\text{Treatments}}$	$a - 1$	$\dfrac{SS_{\text{Treatments}}}{a - 1}$	$\dfrac{MS_{\text{Treatments}}}{MS_E}$
Blocks	SS_{Blocks}	$b - 1$	$\dfrac{SS_{\text{Blocks}}}{b - 1}$	
Error	SS_E (by subtraction)	$(a - 1)(b - 1)$	$\dfrac{SS_E}{(a - 1)(b - 1)}$	
Total	SS_T	$ab - 1$		

The computations are usually arranged in an ANOVA table, such as is shown in Table 13-11. Generally, a computer software package will be used to perform the analysis of variance for the randomized complete block design.

EXAMPLE 13-5   An experiment was performed to determine the effect of four different chemicals on the strength of a fabric. These chemicals are used as part of the permanent press finishing process. Five fabric samples were selected, and a randomized complete block design was run by testing each chemical type once in random order on each fabric sample. The data are shown in Table 13-12. We will test for differences in means using an ANOVA with $\alpha = 0.01$.

The sums of squares for the analysis of variance are computed as follows:

$$SS_T = \sum_{i=1}^{4} \sum_{j=1}^{5} y_{ij}^2 - \frac{y_{..}^2}{ab}$$

$$= (1.3)^2 + (1.6)^2 + \cdots + (3.4)^2 - \frac{(39.2)^2}{20} = 25.69$$

$$SS_{\text{Treatments}} = \sum_{i=1}^{4} \frac{y_{i.}^2}{b} - \frac{y_{..}^2}{ab}$$

$$= \frac{(5.7)^2 + (8.8)^2 + (6.9)^2 + (17.8)^2}{5} - \frac{(39.2)^2}{20} = 18.04$$

Table 13-12   Fabric Strength Data—Randomized Complete Block Design

Chemical Type	Fabric Sample					Treatment Totals $y_{i.}$	Treatment Averages $\bar{y}_{i.}$
	1	2	3	4	5		
1	1.3	1.6	0.5	1.2	1.1	5.7	1.14
2	2.2	2.4	0.4	2.0	1.8	8.8	1.76
3	1.8	1.7	0.6	1.5	1.3	6.9	1.38
4	3.9	4.4	2.0	4.1	3.4	17.8	3.56
Block totals $y_{.j}$	9.2	10.1	3.5	8.8	7.6	39.2($y_{..}$)	
Block averages $\bar{y}_{.j}$	2.30	2.53	0.88	2.20	1.90		1.96($\bar{y}_{..}$)

Table 13-13   Analysis of Variance for the Randomized Complete Block Experiment

Source of Variation	Sum of Squares	Degrees of Freedom	Mean Square	f_0	P-value
Chemical types (treatments)	18.04	3	6.01	75.13	4.79 E-8
Fabric samples (blocks)	6.69	4	1.67		
Error	0.96	12	0.08		
Total	25.69	19			

$$SS_{\text{Blocks}} = \sum_{j=1}^{5} \frac{y_{\cdot j}^2}{a} - \frac{y_{\cdot\cdot}^2}{ab}$$

$$= \frac{(9.2)^2 + (10.1)^2 + (3.5)^2 + (8.8)^2 + (7.6)^2}{4} - \frac{(39.2)^2}{20} = 6.69$$

$$SS_E = SS_T - SS_{\text{Blocks}} - SS_{\text{Treatments}}$$
$$= 25.69 - 6.69 - 18.04 = 0.96$$

The ANOVA is summarized in Table 13-13. Since $f_0 = 75.13 > f_{0.01,3,12} = 5.95$ (the P-value is 4.79×10^{-8}), we conclude that there is a significant difference in the chemical types so far as their effect on strength is concerned.

When Is Blocking Necessary?

Suppose an experiment is conducted as a randomized block design, and blocking was not really necessary. There are ab observations and $(a - 1)(b - 1)$ degrees of freedom for error. If the experiment had been run as a completely randomized single-factor design with b replicates, we would have had $a(b - 1)$ degrees of freedom for error. Therefore, blocking has cost $a(b - 1) - (a - 1)(b - 1) = b - 1$ degrees of freedom for error. Thus, since the loss in error degrees of freedom is usually small, if there is a reasonable chance that block effects may be important, the experimenter should use the randomized block design.

For example, consider the experiment described in Example 13-5 as a single-factor experiment with no blocking. We would then have 16 degrees of freedom for error. In the randomized block design, there are 12 degrees of freedom for error. Therefore, blocking has cost only 4 degrees of freedom, which is a very small loss considering the possible gain in information that would be achieved if block effects are really important. The block effect in Example 13-5 is large, and if we had not blocked, SS_{Blocks} would have been included in the error sum of squares for the completely randomized analysis. This would have resulted in a much larger MS_E, making it more difficult to detect treatment differences. As a general rule, when in doubt as to the importance of block effects, the experimenter should block and gamble that the block effect does exist. If the experimenter is wrong, the slight loss in the degrees of freedom for error will have a negligible effect, unless the number of degrees of freedom is very small.

Computer Solution

Table 13-14 presents the computer output from Minitab for the randomized complete block design in Example 13-5. We used the analysis of variance menu for balanced designs to solve this problem. The results agree closely with the hand calculations from Table 13-13. Notice that Minitab computes an F-statistic for the blocks (the fabric samples). The validity of this ratio as a test statistic for the null hypothesis of no block effects is doubtful because the blocks represent a **restriction on randomization;** that is, we have only randomized within the blocks. If the blocks are not chosen at random, or if they are not run in random order, the

Table 13-14   Minitab Analysis of Variance for the Randomized Complete Block Design in Example 13-5

Analysis of Variance (Balanced Designs)

Factor	Type	Levels	Values				
Chemical	fixed	4	1	2	3	4	
Fabric S	fixed	5	1	2	3	4	5

Analysis of Variance for strength

Source	DF	SS	MS	F	P
Chemical	3	18.0440	6.0147	75.89	0.000
Fabric S	4	6.6930	1.6733	21.11	0.000
Error	12	0.9510	0.0792		
Total	19	25.6880			

F-test with denominator: Error
Denominator MS = 0.079250 with 12 degrees of freedom

Numerator	DF	MS	F	P
Chemical	3	6.015	75.89	0.000
Fabric S	4	1.673	21.11	0.000

F-ratio for blocks may not provide reliable information about block effects. For more discussion see Montgomery (2001, Chapter 4).

13-4.2 Multiple Comparisons

When the ANOVA indicates that a difference exists between the treatment means, we may need to perform some follow-up tests to isolate the specific differences. Any multiple comparison method, such as Fisher's LSD method, could be used for this purpose.

We will illustrate Fisher's LSD method. The four chemical type averages from Example 13-5 are:

$$\bar{y}_{1.} = 1.14 \qquad \bar{y}_{2.} = 1.76 \qquad \bar{y}_{3.} = 1.38 \qquad \bar{y}_{4.} = 3.56$$

Each treatment average uses $b = 5$ observations (one from each block). We will use $\alpha = 0.05$, so $t_{0.025,12} = 2.179$. Therefore the value of the LSD is

$$\text{LSD} = t_{0.025,12}\sqrt{\frac{2MS_E}{b}} = 2.179\sqrt{\frac{2(0.08)}{5}} = 0.39$$

Any pair of treatment averages that differ by 0.39 or more indicates that this pair of treatment means is significantly different. The comparisons are shown below:

$$4 \text{ vs. } 1 = \bar{y}_{4.} - \bar{y}_{1.} = 3.56 - 1.14 = 2.42 > 0.39$$
$$4 \text{ vs. } 3 = \bar{y}_{4.} - \bar{y}_{3.} = 3.56 - 1.38 = 2.18 > 0.39$$
$$4 \text{ vs. } 2 = \bar{y}_{4.} - \bar{y}_{2.} = 3.56 - 1.76 = 1.80 > 0.39$$
$$2 \text{ vs. } 1 = \bar{y}_{2.} - \bar{y}_{1.} = 1.76 - 1.14 = 0.62 > 0.39$$
$$2 \text{ vs. } 3 = \bar{y}_{2.} - \bar{y}_{3.} = 1.76 - 1.38 = 0.38 < 0.39$$
$$3 \text{ vs. } 1 = \bar{y}_{3.} - \bar{y}_{1.} = 1.38 - 1.14 = 0.24 < 0.39$$

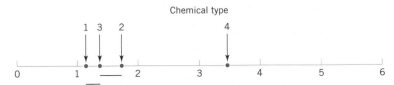

Figure 13-9 Results of Fisher's LSD method.

Figure 13-9 presents the results graphically. The underlined pairs of means are not different. The LSD procedure indicates that chemical type 4 results in significantly different strengths than the other three types do. Chemical types 2 and 3 do not differ, and types 1 and 3 do not differ. There may be a small difference in strength between types 1 and 2.

13-4.3 Residual Analysis and Model Checking

In any designed experiment, it is always important to examine the residuals and to check for violation of basic assumptions that could invalidate the results. As usual, the residuals for the randomized complete block design are just the difference between the observed and estimated (or fitted) values from the statistical model, say,

$$e_{ij} = y_{ij} - \hat{y}_{ij} \tag{13-33}$$

and the fitted values are

$$\hat{y}_{ij} = \bar{y}_{i.} + \bar{y}_{.j} - \bar{y}_{..}$$

The fitted value represents the estimate of the mean response when the ith treatment is run in the jth block. The residuals from the chemical type experiment are shown in Table 13-15.

Figures 13-10, 13-11, 13-12, and 13-13 present the important residual plots for the experiment. These residual plots are usually constructed by computer software packages. There is some indication that fabric sample (block) 3 has greater variability in strength when treated with the four chemicals than the other samples. Chemical type 4, which provides the greatest strength, also has somewhat more variability in strength. Followup experiments may be necessary to confirm these findings, if they are potentially important.

13-4.4 Randomized Complete Block Design with Random Factors (CD Only)

Table 13-15 Residuals from the Randomized Complete Block Design

Chemical Type	Fabric Sample				
	1	2	3	4	5
1	−0.18	−0.10	0.44	−0.18	0.02
2	0.10	0.08	−0.28	0.00	0.10
3	0.08	−0.24	0.30	−0.12	−0.02
4	0.00	0.28	−0.48	0.30	−0.10

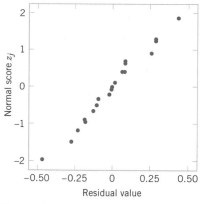

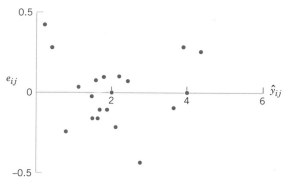

Figure 13-10 Normal probability plot of residuals from the randomized complete block design.

Figure 13-11 Residuals by treatment.

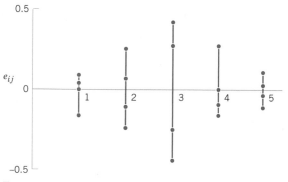

Figure 13-12 Residuals by block.

Figure 13-13 Residuals versus $\hat{y}_{ij}$.

EXERCISES FOR SECTION 13-4

13-25. In "The Effect of Nozzle Design on the Stability and Performance of Turbulent Water Jets" (*Fire Safety Journal*, Vol. 4, August 1981), C. Theobald describes an experiment in which a shape measurement was determined for several different nozzle types at different levels of jet efflux velocity. Interest in this experiment focuses primarily on nozzle type, and velocity is a nuisance factor. The data are as follows:

Nozzle Type	Jet Efflux Velocity (m/s)					
	11.73	14.37	16.59	20.43	23.46	28.74
1	0.78	0.80	0.81	0.75	0.77	0.78
2	0.85	0.85	0.92	0.86	0.81	0.83
3	0.93	0.92	0.95	0.89	0.89	0.83
4	1.14	0.97	0.98	0.88	0.86	0.83
5	0.97	0.86	0.78	0.76	0.76	0.75

(a) Does nozzle type affect shape measurement? Compare the nozzles with box plots and the analysis of variance.

(b) Use Fisher's LSD method to determine specific differences between the nozzles. Does a graph of the average (or standard deviation) of the shape measurements versus nozzle type assist with the conclusions?

(c) Analyze the residuals from this experiment.

13-26. In *Design and Analysis of Experiments,* 5th edition (John Wiley & Sons, 2001), D. C. Montgomery describes an experiment that determined the effect of four different types of tips in a hardness tester on the observed hardness of a metal alloy. Four specimens of the alloy were obtained, and each tip was tested once on each specimen, producing the following data:

(a) Is there any difference in hardness measurements between the tips?

Type of	Specimen			
Tip	1	2	3	4
1	9.3	9.4	9.6	10.0
2	9.4	9.3	9.8	9.9
3	9.2	9.4	9.5	9.7
4	9.7	9.6	10.0	10.2

(b) Use Fisher's LSD method to investigate specific differences between the tips.

(c) Analyze the residuals from this experiment.

13-27. An article in the *American Industrial Hygiene Association Journal* (Vol. 37, 1976, pp. 418–422) describes a field test for detecting the presence of arsenic in urine samples. The test has been proposed for use among forestry workers because of the increasing use of organic arsenics in that industry. The experiment compared the test as performed by both a trainee and an experienced trainer to an analysis at a remote laboratory. Four subjects were selected for testing and are considered as blocks. The response variable is arsenic content (in ppm) in the subject's urine. The data are as follows:

	Subject			
Test	1	2	3	4
Trainee	0.05	0.05	0.04	0.15
Trainer	0.05	0.05	0.04	0.17
Lab	0.04	0.04	0.03	0.10

(a) Is there any difference in the arsenic test procedure?

(b) Analyze the residuals from this experiment.

13-28. An article in the *Food Technology Journal* (Vol. 10, 1956, pp. 39–42) describes a study on the protopectin content of tomatoes during storage. Four storage times were selected, and samples from nine lots of tomatoes were analyzed. The protopectin content (expressed as hydrochloric acid soluble fraction mg/kg) is in the following table.

(a) The researchers in this study hypothesized that mean protopectin content would be different at different storage times. Can you confirm this hypothesis with a statistical test using $\alpha = 0.05$?

(b) Find the P-value for the test in part (a).

(c) Which specific storage times are different? Would you agree with the statement that protopectin content decreases as storage time increases?

(d) Analyze the residuals from this experiment.

13-29. An experiment was conducted to investigate leaking current in a SOS MOSFETS device. The purpose of the experiment was to investigate how leakage current varies as the channel length changes. Four channel lengths were selected. For each channel length, five different widths were also used, and width is to be considered a nuisance factor. The data are as follows:

Channel	Width				
Length	1	2	3	4	5
1	0.7	0.8	0.8	0.9	1.0
2	0.8	0.8	0.9	0.9	1.0
3	0.9	1.0	1.7	2.0	4.0
4	1.0	1.5	2.0	3.0	20.0

(a) Test the hypothesis that mean leakage voltage does not depend on the channel length, using $\alpha = 0.05$.

(b) Analyze the residuals from this experiment. Comment on the residual plots.

13-30. Consider the leakage voltage experiment described in Exercise 13-29. The observed leakage voltage for channel length 4 and width 5 was erroneously recorded. The correct observation is 4.0. Analyze the corrected data from this experiment. Is there evidence to conclude that mean leakage voltage increases with channel length?

Supplemental Exercises

13-31. An article in the *IEEE Transactions on Components, Hybrids, and Manufacturing Technology* (Vol. 15, No. 2, 1992, pp. 146–153) describes an experiment in which the contact resistance of a brake-only relay was studied

Storage	Lot								
Time	1	2	3	4	5	6	7	8	9
0 days	1694.0	989.0	917.3	346.1	1260.0	965.6	1123.0	1106.0	1116.0
7 days	1802.0	1074.0	278.8	1375.0	544.0	672.2	818.0	406.8	461.6
14 days	1568.0	646.2	1820.0	1150.0	983.7	395.3	422.3	420.0	409.5
21 days	415.5	845.4	377.6	279.4	447.8	272.1	394.1	356.4	351.2

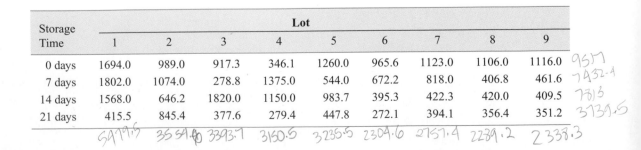

for three different materials (all were silver-based alloys). The data are as follows.

Alloy	Contact Resistance				
1	95	97	99	98	99
	99	99	94	95	98
2	104	102	102	105	99
	102	111	103	100	103
3	119	130	132	136	141
	172	145	150	144	135

(a) Does the type of alloy affect mean contact resistance? Use $\alpha = 0.01$.
(b) Use Fisher's LSD method to determine which means differ.
(c) Find a 99% confidence interval on the mean contact resistance for alloy 3.
(d) Analyze the residuals for this experiment.

13-32. An article in *Lubrication Engineering* (December 1990) describes the results of an experiment designed to investigate the effects of carbon material properties on the progression of blisters on carbon face seals. The carbon face seals are used extensively in equipment such as air turbine starters. Five different carbon materials were tested, and the surface roughness was measured. The data are as follows:

Carbon Material Type	Surface Roughness					
EC10	0.50	0.55	0.55	0.36		
EC10A	0.31	0.07	0.25	0.18	0.56	0.20
EC4	0.20	0.28	0.12			
EC1	0.10	0.16				

(a) Does carbon material type have an effect on mean surface roughness? Use $\alpha = 0.05$.
(b) Find the residuals for this experiment. Does a normal probability plot of the residuals indicate any problem with the normality assumption?
(c) Plot the residuals versus $\hat{y}_{ij}$. Comment on the plot.
(d) Find a 95% confidence interval on the difference between the mean surface roughness between the EC10 and the EC1 carbon grades.

13-33. Apply the Fisher LSD method to the experiment in Exercise 13-32. Summarize your conclusions regarding the effect of material type on surface roughness.

13-34. An article in the *Journal of Quality Technology* (Vol. 14, No. 2, 1982, pp. 80–89) describes an experiment in

which three different methods of preparing fish are evaluated on the basis of sensory criteria and a quality score is assigned. Assume that these methods have been randomly selected from a large population of preparation methods. The data are in the following table:

Method	Score			
1	24.4	23.2	25.0	19.7
	22.2	24.4	23.8	18.0
2	22.1	19.5	17.3	19.7
	22.3	23.2	21.4	22.6
3	23.3	22.8	22.4	23.7
	20.4	23.5	20.8	24.1

(a) Is there any difference in preparation methods? Use $\alpha = 0.05$.
(b) Calculate the *P*-value for the *F*-statistic in part (a).
(c) Analyze the residuals from this experiment and comment on model adequacy.
(d) Estimate the components of variance.

13-35. An article in the *Journal of Agricultural Engineering Research* (Vol. 52, 1992, pp. 53–76) describes an experiment to investigate the effect of drying temperature of wheat grain on the baking quality of bread. Three temperature levels were used, and the response variable measured was the volume of the loaf of bread produced. The data are as follows:

Temperature (°C)	Volume (CC)				
70.0	1245	1235	1285	1245	1235
75.0	1235	1240	1200	1220	1210
80.0	1225	1200	1170	1155	1095

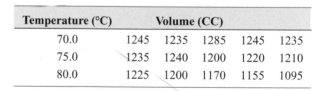

(a) Does drying temperature affect mean bread volume? Use $\alpha = 0.01$.
(b) Find the *P*-value for this test.
(c) Use the Fisher's LSD method to determine which means are different.
(d) Analyze the residuals from this experiment and comment on model adequacy.

13-36. An article in *Agricultural Engineering* (December 1964, pp. 672–673) describes an experiment in which the daily weight gain of swine is evaluated at different levels of housing temperature. The mean weight of each group of swine at the start of the experiment is considered to be a nuisance factor. The data from this experiment are as follows:

a = 3
b = 6

Mean	Housing Air Temperatures (degrees F)					
Weight						
(lbs)	50	60	70	80	90	100
100	−1.37	1.58	2.00	1.97	1.40	0.39 *8.71*
150	1.47	1.75	2.16	1.82	1.14	−0.19 *8.15*
200	1.19	1.91	2.22	1.67	0.88	−0.77 *7.1*
	4.03	*5.24*	*6.38*	*5.46*	*3.42*	*−0.57*

23.96

(a) Does housing air temperature affect mean weight gain? Use α = 0.05.

(b) Use Fisher's LSD method to determine which temperature levels are different.

(c) Analyze the residuals from this experiment and comment on model adequacy.

13-37. An article in *Communications of the ACM* (Vol. 30, No. 5, 1987) studied different algorithms for estimating software development costs. Six algorithms were applied to eight software development projects and the percent error in estimating the development cost was observed. The data are in the table at the bottom of the page.

(a) Do the algorithms differ in their mean cost estimation accuracy? Use α = 0.05.

(b) Analyze the residuals from this experiment.

(c) Which algorithm would you recommend for use in practice?

13-38.   Consider an ANOVA situation with $a = 4$ means $\mu_1 = 1$, $\mu_2 = 5$, $\mu_3 = 8$, and $\mu_4 = 4$. Suppose that $\sigma^2 = 4$, $n = 4$, and $\alpha = 0.05$.

(a) Find the power of the ANOVA F-test.

(b) How large would the sample size have to be if we want the power of the F-test for detecting this difference in means to be at least 0.90?

13-39.   Consider an ANOVA situation with $a = 5$ treatments. Let $\sigma^2 = 9$ and $\alpha = 0.05$, and suppose that $n = 4$.

(a) Find the power of the ANOVA F-test when $\mu_1 = \mu_2 = \mu_3 = 1$, $\mu_4 = 3$, and $\mu_5 = 2$.

(b) What sample size is required if we want the power of the F-test in this situation to be at least 0.90?

| | | Project | | | | | | | |
Algorithm		1	2	3	4	5	6	7	8
1(SLIM)		1244	21	82	2221	905	839	527	122
2(COCOMO-A)		281	129	396	1306	336	910	473	199
3(COCOMO-R)		220	84	458	543	300	794	488	142
4(COCOMO-C)		225	83	425	552	291	826	509	153
5(FUNCTION POINTS)		19	11	−34	121	15	103	87	−17
6(ESTIMALS)		−20	35	−53	170	104	199	142	41

MIND-EXPANDING EXERCISES

13-40. Show that in the fixed-effects model analysis of variance $E(MS_E) = \sigma^2$. How would your development change if the random-effects model had been specified?

13-41. Consider testing the equality of the means of two normal populations where the variances are unknown but are assumed equal. The appropriate test procedure is the two-sample t-test. Show that the two-sample t-test is equivalent to the single-factor analysis of variance F-test.

13-42. Consider the ANOVA with $a = 2$ treatments. Show that the MS_E in this analysis is equal to the pooled variance estimate used in the two-sample t-test.

13-43. Show that the variance of the linear combination

$$\sum_{i=1}^{a} c_i Y_{i\cdot} \text{ is } \sigma^2 \sum_{i=1}^{a} n_i c_i^2.$$

13-44. In a fixed-effects model, suppose that there are n observations for each of four treatments. Let Q_1^2, Q_2^2, and Q_3^2 be single-degree-of-freedom sums of squares for the orthogonal contrasts. Prove that $SS_{\text{Treatments}} = Q_1^2 + Q_2^2 + Q_3^2$.

13-45. Consider the single-factor completely randomized design with a treatments and n replicates. Show that if the difference between any two treatment means is as large as D, the minimum value that the OC curve parameter Φ^2 can take on is

$$\Phi^2 = \frac{nD^2}{2a\sigma^2}$$

13-46. Consider the single-factor completely randomized design. Show that a $100(1 - \alpha)$ percent confidence interval for σ^2 is

$$\frac{(N - a)MS_E}{\chi^2_{\alpha/2, N-a}} \le \sigma^2 \le \frac{(N - a)MS_E}{\chi^2_{1-\alpha/2, N-a}}$$

where N is the total number of observations in the experimental design.

13-47. Consider the random-effect model for the single-factor completely randomized design. Show that

a $100(1 - \alpha)$% confidence interval on the ratio of variance components σ_τ^2/σ^2 is given by

$$L \le \frac{\sigma_\tau^2}{\sigma^2} \le U$$

where

$$L = \frac{1}{n} \left[\frac{MS_{\text{Treatments}}}{MS_E} \times \left(\frac{1}{f_{\alpha/2, a-1, N-a}} \right) - 1 \right]$$

and

$$U = \frac{1}{n} \left[\frac{MS_{\text{Treatments}}}{MS_E} \times \left(\frac{1}{f_{1-\alpha/2, a-1, N-a}} \right) - 1 \right]$$

13-48. Consider a random-effects model for the single-factor completely randomized design. Show that a $100(1 - \alpha)$% confidence interval on the ratio $\sigma_\tau^2/(\sigma^2 + \sigma_\tau^2)$ is

$$\frac{L}{1 + L} \le \frac{\sigma_\tau^2}{\sigma^2 + \sigma_\tau^2} \le \frac{U}{1 + U}$$

where L and U are as defined in Exercise 13-47.

13-49. Continuation of Exercise 13-48. Use the results of Exercise 13-48 to find a $100(1 - \alpha)$% confidence interval for $\sigma^2/(\sigma^2 + \sigma_\tau^2)$.

13-50. Consider the fixed-effect model of the completely randomized single-factor design. The model parameters are restricted by the constraint $\sum_{i=1}^{a} \tau_i = 0$. (Actually, other restrictions could be used, but this one is simple and results in intuitively pleasing estimates for the model parameters.) For the case of unequal sample size $n_1, n_2, \ldots, n_a$, the restriction is $\sum_{i=1}^{a} n_i \tau_i = 0$. Use this to show that

$$E(MS_{\text{Treatments}}) = \sigma^2 + \frac{\sum_{i=1}^{a} n_i \tau_i^2}{a - 1}$$

Does this suggest that the null hypothesis in this model is $H_0: n_1\tau_1 = n_2\tau_2 = \cdots = n_a\tau_a = 0$?

13-51. Sample Size Determination. In the single-factor completely randomized design, the accuracy of a

MIND-EXPANDING EXERCISES

$100(1 - \alpha)\%$ confidence interval on the difference in any two treatment means is $t_{\alpha/2, a(n-1)} \sqrt{2MS_E/n}$.

(a) Show that if A is the desired accuracy of the interval, the sample size required is

$$n = \frac{2F_{\alpha/2, 1, a(n-1)} MS_E}{A^2}$$

(b) Suppose that in comparing $a = 5$ means we have a preliminary estimate of σ^2 of 4. If we want the 95% confidence interval on the difference in means to have an accuracy of 2, how many replicates should we use?

IMPORTANT TERMS AND CONCEPTS

In the E-book, click on any term or concept below to go to that subject.

Analysis of variance
Blocking
Complete randomized experiment
Expected mean squares

Fisher's least significant difference method
Fixed factor
Multiple comparisons
Nuisance factors
Random factor
Randomization

Randomized complete block design
Residual analysis and model adequacy checking
Sample size and replication in an experiment
Variance component

CD MATERIAL

Graphical comparison of means
Orthogonal contrasts
Tukey's test

14 Design of Experiments with Several Factors

CHAPTER OUTLINE

LEARNING OBJECTIVES

After careful study of this chapter, you should be able to do the following:

1. Design and conduct engineering experiments involving several factors using the factorial design approach

2. Know how to analyze and interpret main effects and interactions

3. Understand how the ANOVA is used to analyze the data from these experiments

4. Assess model adequacy with residual plots

5. Know how to use the two-level series of factorial designs

6. Understand how two-level factorial designs can be run in blocks
7. Design and conduct two-level fractional factorial designs

CD MATERIAL

8. Incorporate random factors in factorial experiments.
9. Test for curvature in two-level factorial designs by using center points.
10. Use response surface methodology for process optimization experiments.

Answers for most odd numbered exercises are at the end of the book. Answers to exercises whose numbers are surrounded by a box can be accessed in the e-text by clicking on the box. Complete worked solutions to certain exercises are also available in the e-Text. These are indicated in the Answers to Selected Exercises section by a box around the exercise number. Exercises are also available for some of the text sections that appear on CD only. These exercises may be found in the Mind-Expanding Exercises at the end of the chapter.

14-1   INTRODUCTION

An **experiment** is just a **test** or series of tests. Experiments are performed in all engineering and scientific disciplines and are an important part of the way we learn about how systems and processes work. The validity of the conclusions that are drawn from an experiment depends to a large extent on how the experiment was conducted. Therefore, the **design** of the experiment plays a major role in the eventual solution of the problem that initially motivated the experiment.

In this chapter we focus on experiments that include two or more factors that the experimenter thinks may be important. The **factorial experimental design** will be introduced as a powerful technique for this type of problem. Generally, in a factorial experimental design, experimental trials (or runs) are performed at all combinations of factor levels. For example, if a chemical engineer is interested in investigating the effects of reaction time and reaction temperature on the yield of a process, and if two levels of time (1 and 1.5 hours) and two levels of temperature (125 and 150°F) are considered important, a factorial experiment would consist of making experimental runs at each of the four possible combinations of these levels of reaction time and reaction temperature.

Most of the statistical concepts introduced in Chapter 13 for single-factor experiments can be extended to the factorial experiments of this chapter. The **analysis of variance** (**ANOVA**), in particular, will continue to be used as one of the primary tools for statistical data analysis. We will also introduce several graphical methods that are useful in analyzing the data from designed experiments.

14-2   SOME APPLICATIONS OF DESIGNED EXPERIMENTS (CD ONLY)

14-3   FACTORIAL EXPERIMENTS

When several factors are of interest in an experiment, a **factorial experimental design** should be used. As noted previously, in these experiments factors are varied together.

Definition

> By a **factorial experiment** we mean that in each complete trial or replicate of the experiment all possible combinations of the levels of the factors are investigated.

Thus, if there are two factors A and B with a levels of factor A and b levels of factor B, each replicate contains all ab treatment combinations.

The effect of a factor is defined as the change in response produced by a change in the level of the factor. It is called a **main effect** because it refers to the primary factors in the study. For example, consider the data in Table 14-1. This is a factorial experiment with two factors, A and B, each at two levels (A_{low}, A_{high}, and B_{low}, B_{high}). The main effect of factor A is the difference between the average response at the high level of A and the average response at the low level of A, or

$$A = \frac{30 + 40}{2} - \frac{10 + 20}{2} = 20$$

That is, changing factor A from the low level to the high level causes an average response increase of 20 units. Similarly, the main effect of B is

$$B = \frac{20 + 40}{2} - \frac{10 + 30}{2} = 10$$

In some experiments, the difference in response between the levels of one factor is not the same at all levels of the other factors. When this occurs, there is an **interaction** between the factors. For example, consider the data in Table 14-2. At the low level of factor B, the A effect is

$$A = 30 - 10 = 20$$

and at the high level of factor B, the A effect is

$$A = 0 - 20 = -20$$

Since the effect of A depends on the level chosen for factor B, there is interaction between A and B.

When an interaction is large, the corresponding main effects have very little practical meaning. For example, by using the data in Table 14-2, we find the main effect of A as

$$A = \frac{30 + 0}{2} - \frac{10 + 20}{2} = 0$$

and we would be tempted to conclude that there is no factor A effect. However, when we examined the effects of A at *different levels of factor B*, we saw that this was not the case. The effect of factor A depends on the levels of factor B. Thus, knowledge of the AB interaction is more useful than knowledge of the main effect. A significant interaction can mask the significance of main effects. Consequently, when interaction is present, the main effects of the factors involved in the interaction may not have much meaning.

Table 14-1 A Factorial Experiment with Two Factors

Factor A	Factor B	
	B_{low}	B_{high}
A_{low}	10	20
A_{high}	30	40

Table 14-2 A Factorial Experiment with Interaction

Factor A	Factor B	
	B_{low}	B_{high}
A_{low}	10	20
A_{high}	30	0

It is easy to estimate the interaction effect in factorial experiments such as those illustrated in Tables 14-1 and 14-2. In this type of experiment, when both factors have two levels, the AB interaction effect is the difference in the diagonal averages. This represents one-half the difference between the A effects at the two levels of B. For example, in Table 14-1, we find the AB interaction effect to be

$$AB = \frac{20 + 30}{2} - \frac{10 + 40}{2} = 0$$

Thus, there is no interaction between A and B. In Table 14-2, the AB interaction effect is

$$AB = \frac{20 + 30}{2} - \frac{10 + 0}{2} = 20$$

As we noted before, the interaction effect in these data is very large.

The concept of interaction can be illustrated graphically in several ways. Figure 14-1 plots the data in Table 14-1 against the levels of A for both levels of B. Note that the B_{low} and B_{high} lines are approximately parallel, indicating that factors A and B do not interact significantly. Figure 14-2 presents a similar plot for the data in Table 14-2. In this graph, the B_{low} and B_{high} lines are not parallel, indicating the interaction between factors A and B. Such graphical displays are called **two-factor interaction plots.** They are often useful in presenting the results of experiments, and many computer software programs used for analyzing data from designed experiments will construct these graphs automatically.

Figures 14-3 and 14-4 present another graphical illustration of the data from Tables 14-1 and 14-2. In Fig. 14-3 we have shown a **three-dimensional surface plot** of the data from Table 14-1. These data contain no interaction, and the surface plot is a plane lying above the A-B space. The slope of the plane in the A and B directions is proportional to the main effects of factors A and B, respectively. Figure 14-4 is a surface plot of the data from Table 14-2. Notice that the effect of the interaction in these data is to "twist" the plane, so that there is curvature in the response function. **Factorial experiments are the only way to discover interactions between variables.**

An alternative to the factorial design that is (unfortunately) used in practice is to change the factors *one at a time* rather than to vary them simultaneously. To illustrate this one-factor-at-a-time procedure, suppose that an engineer is interested in finding the values of temperature and pressure that maximize yield in a chemical process. Suppose that we fix temperature at 155°F (the current operating level) and perform five runs at different levels of time, say,

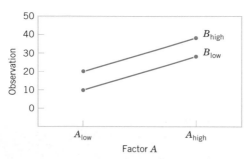

Figure 14-1   Factorial experiment, no interaction.

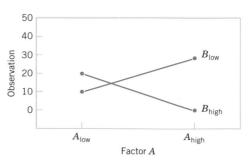

Figure 14-2   Factorial experiment, with interaction.

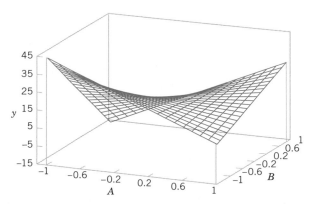

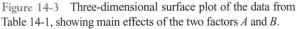

Figure 14-3   Three-dimensional surface plot of the data from Table 14-1, showing main effects of the two factors A and B.

Figure 14-4   Three-dimensional surface plot of the data from Table 14-2 showing the effect of the A and B interaction.

0.5, 1.0, 1.5, 2.0, and 2.5 hours. The results of this series of runs are shown in Fig. 14-5. This figure indicates that maximum yield is achieved at about 1.7 hours of reaction time. To optimize temperature, the engineer then fixes time at 1.7 hours (the apparent optimum) and performs five runs at different temperatures, say, 140, 150, 160, 170, and 180°F. The results of this set of runs are plotted in Fig. 14-6. Maximum yield occurs at about 155°F. Therefore, we would conclude that running the process at 155°F and 1.7 hours is the best set of operating conditions, resulting in yields of around 75%.

Figure 14-7 displays the contour plot of actual process yield as a function of temperature and time with the one-factor-at-a-time experiments superimposed on the contours. Clearly, this one-factor-at-a-time approach has failed dramatically here, as the true optimum is at least 20 yield points higher and occurs at much lower reaction times and higher temperatures. The failure to discover the importance of the shorter reaction times is particularly important because this could have significant impact on production volume or capacity, production planning, manufacturing cost, and total productivity.

The one-factor-at-a-time approach has failed here because it cannot detect the interaction between temperature and time. Factorial experiments are the only way to detect interactions. Furthermore, the one-factor-at-a-time method is inefficient. It will require more

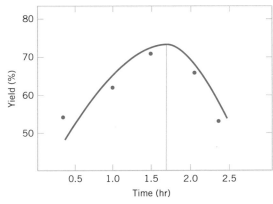

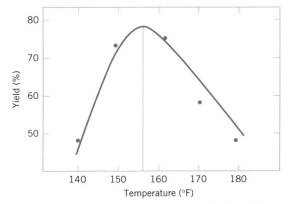

Figure 14-5   Yield versus reaction time with temperature constant at 155 °F.

Figure 14-6   Yield versus temperature with reaction time constant at 1.7 hours.

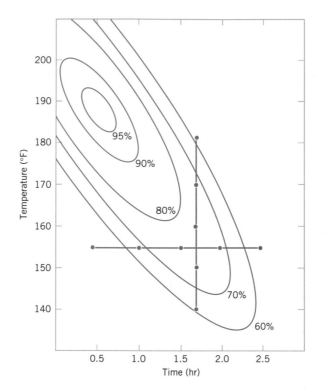

Figure 14-7
Optimization
experiment using the
one-factor-at-a-
time method.

experimentation than a factorial, and as we have just seen, there is no assurance that it will produce the correct results.

14-4 TWO-FACTOR FACTORIAL EXPERIMENTS

The simplest type of factorial experiment involves only two factors, say, A and B. There are a levels of factor A and b levels of factor B. This two-factor factorial is shown in Table 14-3. The experiment has n **replicates,** and each replicate contains all ab treatment combinations.

Table 14-3   Data Arrangement for a Two-Factor Factorial Design

		Factor B					
		1	2	$\cdots$	b	Totals	Averages
Factor A	1	$y_{111}, y_{112},$ $\ldots, y_{11n}$	$y_{121}, y_{122},$ $\ldots, y_{12n}$		$y_{1b1}, y_{1b2},$ $\ldots, y_{1bn}$	$y_{1\cdot\cdot}$	$\bar{y}_{1\cdot\cdot}$
	2	$y_{211}, y_{212},$ $\ldots, y_{21n}$	$y_{221}, y_{222},$ $\ldots, y_{22n}$		$y_{2b1}, y_{2b2},$ $\ldots, y_{2bn}$	$y_{2\cdot\cdot}$	$\bar{y}_{2\cdot\cdot}$
	$\vdots$						
	a	$y_{a11}, y_{a12},$ $\ldots, y_{a1n}$	$y_{a21}, y_{a22},$ $\ldots, y_{a2n}$		$y_{ab1}, y_{ab2},$ $\ldots, y_{abn}$	$y_{a\cdot\cdot}$	$\bar{y}_{a\cdot\cdot}$
Totals		$y_{\cdot1\cdot}$	$y_{\cdot2\cdot}$		$y_{\cdot b\cdot}$	$y_{\cdots}$	
Averages		$\bar{y}_{\cdot1\cdot}$	$\bar{y}_{\cdot2\cdot}$		$\bar{y}_{\cdot b\cdot}$		$\bar{y}_{\cdots}$

The observation in the ijth cell for the kth replicate is denoted by y_{ijk}. In performing the experiment, the abn observations would be run in **random order.** Thus, like the single-factor experiment studied in Chapter 13, the two-factor factorial is a *completely randomized design*.

The observations may be described by the linear statistical model

$$Y_{ijk} = \mu + \tau_i + \beta_j + (\tau\beta)_{ij} + \epsilon_{ijk} \begin{cases} i = 1, 2, \ldots, a \\ j = 1, 2, \ldots, b \\ k = 1, 2, \ldots, n \end{cases} \tag{14-1}$$

where μ is the overall mean effect, τ_i is the effect of the ith level of factor A, β_j is the effect of the jth level of factor B, $(\tau\beta)_{ij}$ is the effect of the interaction between A and B, and ϵ_{ijk} is a random error component having a normal distribution with mean zero and variance σ^2. We are interested in testing the hypotheses of no main effect for factor A, no main effect for B, and no AB interaction effect. As with the single-factor experiments of Chapter 13, the analysis of variance (ANOVA) will be used to test these hypotheses. Since there are two factors in the experiment, the test procedure is sometimes called the two-way analysis of variance.

14-4.1   Statistical Analysis of the Fixed-Effects Model

Suppose that A and B are **fixed factors.** That is, the a levels of factor A and the b levels of factor B are specifically chosen by the experimenter, and inferences are confined to these levels only. In this model, it is customary to define the effects τ_i, β_j, and $(\tau\beta)_{ij}$ as deviations from the mean, so that $\sum_{i=1}^{a}\tau_i = 0$, $\sum_{j=1}^{b}\beta_j = 0$, $\sum_{i=1}^{a}(\tau\beta)_{ij} = 0$, and $\sum_{j=1}^{b}(\tau\beta)_{ij} = 0$.

The **analysis of variance** can be used to test hypotheses about the main factor effects of A and B and the AB interaction. To present the ANOVA, we will need some symbols, some of which are illustrated in Table 14-3. Let $y_{i..}$ denote the total of the observations taken at the ith level of factor A; $y_{.j.}$ denote the total of the observations taken at the jth level of factor B; $y_{ij.}$ denote the total of the observations in the ijth cell of Table 14-3; and $y_{...}$ denote the grand total of all the observations. Define $\bar{y}_{i..}$, $\bar{y}_{.j.}$, $\bar{y}_{ij.}$, and $\bar{y}_{...}$ as the corresponding row, column, cell, and grand averages. That is,

$$y_{i..} = \sum_{j=1}^{b}\sum_{k=1}^{n} y_{ijk} \qquad \bar{y}_{i..} = \frac{y_{i..}}{bn} \qquad i = 1, 2, \ldots, a$$

$$y_{.j.} = \sum_{i=1}^{a}\sum_{k=1}^{n} y_{ijk} \qquad \bar{y}_{.j.} = \frac{y_{.j.}}{an} \qquad j = 1, 2, \ldots, b$$

$$y_{ij.} = \sum_{k=1}^{n} y_{ijk} \qquad \bar{y}_{ij.} = \frac{y_{ij.}}{n} \qquad \begin{matrix} i = 1, 2, \ldots, a \\ j = 1, 2, \ldots, b \end{matrix}$$

$$y_{...} = \sum_{i=1}^{a}\sum_{j=1}^{b}\sum_{k=1}^{n} y_{ijk} \qquad \bar{y}_{...} = \frac{y_{...}}{abn}$$

The hypotheses that we will test are as follows:

1. H_0: $\tau_1 = \tau_2 = \cdots = \tau_a = 0$ (no main effect of factor A)
H_1: at least one $\tau_i \neq 0$

2. H_0: $\beta_1 = \beta_2 = \cdots = \beta_b = 0$ (no main effect of factor B)   (14-2)
H_1: at least one $\beta_j \neq 0$

3. H_0: $(\tau\beta)_{11} = (\tau\beta)_{12} = \cdots = (\tau\beta)_{ab} = 0$ (no interaction)
H_1: at least one $(\tau\beta)_{ij} \neq 0$

As before, the ANOVA tests these hypotheses by decomposing the total variability in the data into component parts and then comparing the various elements in this decomposition. Total variability is measured by the total sum of squares of the observations

$$SS_T = \sum_{i=1}^{a} \sum_{j=1}^{b} \sum_{k=1}^{n} (y_{ijk} - \bar{y}...)^2$$

and the sum of squares decomposition is defined below.

The **sum of squares identity for a two-factor ANOVA** is

$$\sum_{i=1}^{a} \sum_{j=1}^{b} \sum_{k=1}^{n} (y_{ijk} - \bar{y}...)^2 = bn \sum_{i=1}^{a} (\bar{y}_{i..} - \bar{y}...)^2$$

$$+ an \sum_{j=1}^{b} (\bar{y}_{.j.} - \bar{y}...)^2$$

$$+ n \sum_{i=1}^{a} \sum_{j=1}^{b} (\bar{y}_{ij.} - \bar{y}_{i..} - \bar{y}_{.j.} + \bar{y}...)^2$$

$$+ \sum_{i=1}^{a} \sum_{j=1}^{b} \sum_{k=1}^{n} (y_{ijk} - \bar{y}_{ij.})^2 \qquad (14\text{-}3)$$

or symbolically,

$$SS_T = SS_A + SS_B + SS_{AB} + SS_E \qquad (14\text{-}4)$$

Equations 14-3 and 14-4 state that the total sum of squares SS_T is partitioned into a sum of squares for the row factor A (SS_A), a sum of squares for the column factor B (SS_B), a sum of squares for the interaction between A and B (SS_{AB}), and an error sum of squares (SS_E). There are $abn - 1$ total degrees of freedom. The main effects A and B have $a - 1$ and $b - 1$ degrees of freedom, while the interaction effect AB has $(a - 1)(b - 1)$ degrees of freedom. Within each of the ab cells in Table 14-3, there are $n - 1$ degrees of freedom between the n replicates, and observations in the same cell can differ only because of random error. Therefore, there are $ab(n - 1)$ degrees of freedom for error. Therefore, the degrees of freedom are partitioned according to

$$abn - 1 = (a - 1) + (b - 1) + (a - 1)(b - 1) + ab(n - 1)$$

If we divide each of the sum of squares on the right-hand side of Equation 14-4 by the corresponding number of degrees of freedom, we obtain the **mean squares** for A, B, the interaction, and error:

$$MS_A = \frac{SS_A}{a - 1} \qquad MS_B = \frac{SS_B}{b - 1} \qquad MS_{AB} = \frac{SS_{AB}}{(a - 1)(b - 1)} \qquad MS_E = \frac{SS_E}{ab(n - 1)}$$

Assuming that factors A and B are fixed factors, it is not difficult to show that the **expected values** of these mean squares are

$$E(MS_A) = E\left(\frac{SS_A}{a - 1}\right) = \sigma^2 + \frac{bn \sum_{i=1}^{a} \tau_i^2}{a - 1} \qquad E(MS_B) = E\left(\frac{SS_B}{b - 1}\right) = \sigma^2 + \frac{an \sum_{j=1}^{b} \beta_j^2}{b - 1}$$

$$E(MS_{AB}) = E\left(\frac{SS_{AB}}{(a - 1)(b - 1)}\right) = \sigma^2 + \frac{n \sum_{i=1}^{a} \sum_{j=1}^{b} (\tau\beta)_{ij}^2}{(a - 1)(b - 1)}$$

$$E(MS_E) = E\left(\frac{SS_E}{ab(n - 1)}\right) = \sigma^2$$

From examining these expected mean squares, it is clear that if the null hypotheses about main effects H_0: $\tau_i = 0$, H_0: $\beta_j = 0$, and the interaction hypothesis H_0: $(\tau\beta)_{ij} = 0$ are all true, all four mean squares are unbiased estimates of σ^2.

To test that the row factor effects are all equal to zero (H_0: $\tau_i = 0$), we would use the ratio

$$F_0 = \frac{MS_A}{MS_E}$$

which has an F-distribution with $a - 1$ and $ab(n - 1)$ degrees of freedom if H_0: $\tau_i = 0$ is true. This null hypothesis is rejected at the α level of significance if $f_0 > f_{\alpha, a-1, ab(n-1)}$. Similarly, to test the hypothesis that all the column factor effects are equal to zero (H_0: $\beta_j = 0$), we would use the ratio

$$F_0 = \frac{MS_B}{MS_E}$$

which has an F-distribution with $b - 1$ and $ab(n - 1)$ degrees of freedom if H_0: $\beta_j = 0$ is true. This null hypothesis is rejected at the α level of significance if $f_0 > f_{\alpha, b-1, ab(n-1)}$. Finally, to test the hypothesis H_0: $(\tau\beta)_{ij} = 0$, which is the hypothesis that all interaction effects are zero, we use the ratio

$$F_0 = \frac{MS_{AB}}{MS_E}$$

which has an F-distribution with $(a-1)(b-1)$ and $ab(n-1)$ degrees of freedom if the null hypothesis H_0: $(\tau\beta)_{ij} = 0$. This hypothesis is rejected at the α level of significance if $f_0 > f_{\alpha,(a-1)(b-1),ab(n-1)}$.

It is usually best to conduct the test for interaction first and then to evaluate the main effects. If interaction is not significant, interpretation of the tests on the main effects is straightforward. However, as noted in Section 14-4, when interaction is significant, the main effects of the factors involved in the interaction may not have much practical interpretative value. Knowledge of the interaction is usually more important than knowledge about the main effects.

Computational formulas for the sums of squares are easily obtained.

Definition

Computing formulas for the sums of squares in a two-factor analysis of variance.

$$SS_T = \sum_{i=1}^{a} \sum_{j=1}^{b} \sum_{k=1}^{n} y_{ijk}^2 - \frac{y_{...}^2}{abn} \tag{14-5}$$

$$SS_A = \sum_{i=1}^{a} \frac{y_{i..}^2}{bn} - \frac{y_{...}^2}{abn} \tag{14-6}$$

$$SS_B = \sum_{j=1}^{b} \frac{y_{.j.}^2}{an} - \frac{y_{...}^2}{abn} \tag{14-7}$$

$$SS_{AB} = \sum_{i=1}^{a} \sum_{j=1}^{b} \frac{y_{ij.}^2}{n} - \frac{y_{...}^2}{abn} - SS_A - SS_B \tag{14-8}$$

$$SS_E = SS_T - SS_{AB} - SS_A - SS_B \tag{14-9}$$

The computations are usually displayed in an ANOVA table, such as Table 14-4.

EXAMPLE 14-1   Aircraft primer paints are applied to aluminum surfaces by two methods: dipping and spraying. The purpose of the primer is to improve paint adhesion, and some parts can be primed using either application method. The process engineering group responsible for this operation is interested in learning whether three different primers differ in their adhesion properties.

Table 14-4   ANOVA Table for a Two-Factor Factorial, Fixed-Effects Model

Source of Variation	Sum of Squares	Degrees of Freedom	Mean Square	F_0
A treatments	SS_A	$a-1$	$MS_A = \dfrac{SS_A}{a-1}$	$\dfrac{MS_A}{MS_E}$
B treatments	SS_B	$b-1$	$MS_B = \dfrac{SS_B}{b-1}$	$\dfrac{MS_B}{MS_E}$
Interaction	SS_{AB}	$(a-1)(b-1)$	$MS_{AB} = \dfrac{SS_{AB}}{(a-1)(b-1)}$	$\dfrac{MS_{AB}}{MS_E}$
Error	SS_E	$ab(n-1)$		
Total	SS_T	$abn-1$	$MS_E = \dfrac{SS_E}{ab(n-1)}$	

Table 14-5   Adhesion Force Data for Example 14-1

Primer Type	Dipping		Spraying		$y_{i\cdot\cdot}$
1	4.0, 4.5, 4.3	(12.8)	5.4, 4.9, 5.6	(15.9)	28.7
2	5.6, 4.9, 5.4	(15.9)	5.8, 6.1, 6.3	(18.2)	34.1
3	3.8, 3.7, 4.0	(11.5)	5.5, 5.0, 5.0	(15.5)	27.0
$y_{\cdot j\cdot}$	40.2		49.6		$89.8 = y_{\cdots}$

A factorial experiment was performed to investigate the effect of paint primer type and application method on paint adhesion. For each combination of primer type and application method, three specimens were painted, then a finish paint was applied, and the adhesion force was measured. The data from the experiment are shown in Table 14-5. The circled numbers in the cells are the cell totals $y_{ij\cdot}$. The sums of squares required to perform the ANOVA are computed as follows:

$$SS_T = \sum_{i=1}^{a} \sum_{j=1}^{b} \sum_{k=1}^{n} y_{ijk}^2 - \frac{y_{\cdots}^2}{abn}$$

$$= (4.0)^2 + (4.5)^2 + \cdots + (5.0)^2 - \frac{(89.8)^2}{18} = 10.72$$

$$SS_{\text{types}} = \sum_{i=1}^{a} \frac{y_{i\cdot\cdot}^2}{bn} - \frac{y_{\cdots}^2}{abn}$$

$$= \frac{(28.7)^2 + (34.1)^2 + (27.0)^2}{6} - \frac{(89.8)^2}{18} = 4.58$$

$$SS_{\text{methods}} = \sum_{j=1}^{b} \frac{y_{\cdot j\cdot}^2}{an} - \frac{y_{\cdots}^2}{abn}$$

$$= \frac{(40.2)^2 + (49.6)^2}{9} - \frac{(89.8)^2}{18} = 4.91$$

$$SS_{\text{interaction}} = \sum_{i=1}^{a} \sum_{j=1}^{b} \frac{y_{ij\cdot}^2}{n} - \frac{y_{\cdots}^2}{abn} - SS_{\text{types}} - SS_{\text{methods}}$$

$$= \frac{(12.8)^2 + (15.9)^2 + (11.5)^2 + (15.9)^2 + (18.2)^2 + (15.5)^2}{3}$$

$$- \frac{(89.8)^2}{18} - 4.58 - 4.91 = 0.24$$

and

$$SS_E = SS_T - SS_{\text{types}} - SS_{\text{methods}} - SS_{\text{interaction}}$$
$$= 10.72 - 4.58 - 4.91 - 0.24 = 0.99$$

The ANOVA is summarized in Table 14-6. The experimenter has decided to use $\alpha = 0.05$. Since $f_{0.05,2,12} = 3.89$ and $f_{0.05,1,12} = 4.75$, we conclude that the main effects of primer type and

Table 14-6   ANOVA for Example 14-1

Source of Variation	Sum of Squares	Degrees of Freedom	Mean Square	f_0	P-Value
Primer types	4.58	2	2.29	28.63	$2.7 \times$ E-5
Application methods	4.91	1	4.91	61.38	$5.0 \times$ E-7
Interaction	0.24	2	0.12	1.50	0.2621
Error	0.99	12	0.08		
Total	10.72	17			

application method affect adhesion force. Furthermore, since $1.5 < f_{0.05,2,12}$, there is no indication of interaction between these factors. The last column of Table 14-6 shows the P-value for each F-ratio. Notice that the P-values for the two test statistics for the main effects are considerably less than 0.05, while the P-value for the test statistic for the interaction is greater than 0.05.

A graph of the cell adhesion force averages $\{\bar{y}_{ij\cdot}\}$ versus levels of primer type for each application method is shown in Fig. 14-8. The no-interaction conclusion is obvious in this graph, because the two lines are nearly parallel. Furthermore, since a large response indicates greater adhesion force, we conclude that spraying is the best application method and that primer type 2 is most effective.

Tests on Individual Means

When both factors are fixed, comparisons between the individual means of either factor may be made using any multiple comparison technique such as Fisher's LSD method (described in Chapter 13). When there is no interaction, these comparisons may be made using either the row averages $\bar{y}_{i\cdot\cdot}$ or the column averages $\bar{y}_{\cdot j\cdot}$. However, when interaction is significant, comparisons between the means of one factor (say, A) may be obscured by the AB interaction. In this case, we could apply a procedure such as Fisher's LSD method to the means of factor A, with factor B set at a particular level.

Minitab Output

Table 14-7 shows some of the output from the Minitab analysis of variance procedure for the aircraft primer paint experiment in Example 14-1. The upper portion of the table gives factor name and level information, and the lower portion of the table presents the analysis of variance for the adhesion force response. The results are identical to the manual calculations displayed in Table 14-6 apart from rounding.

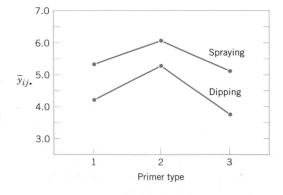

Figure 14-8 Graph of average adhesion force versus primer types for both application methods.

Table 14-7 Analysis of Variance From Minitab for Example 14-1

ANOVA (Balanced Designs)

Factor	Type	Levels	Values		
Primer	fixed	3	1	2	3
Method	fixed	2	Dip	Spray	

Analysis of Variance for Adhesion

Source	DF	SS	MS	F	P
Primer	2	4.5811	2.2906	27.86	0.000
Method	1	4.9089	4.9089	59.70	0.000
Primer *Method	2	0.2411	0.1206	1.47	0.269
Error	12	0.9867	0.0822		
Total	17	10.7178			

14-4.2 Model Adequacy Checking

Just as in the single-factor experiments discussed in Chapter 13, the **residuals** from a factorial experiment play an important role in assessing **model adequacy.** The residuals from a two-factor factorial are

$$e_{ijk} = y_{ijk} - \bar{y}_{ij.}$$

That is, the residuals are just the difference between the observations and the corresponding cell averages.

Table 14-8 presents the residuals for the aircraft primer paint data in Example 14-1. The normal probability plot of these residuals is shown in Fig. 14-9. This plot has tails that do not fall exactly along a straight line passing through the center of the plot, indicating some potential problems with the normality assumption, but the deviation from normality does not appear severe. Figures 14-10 and, 14-11 plot the residuals versus the levels of primer types and application methods, respectively. There is some indication that primer type 3 results in slightly lower variability in adhesion force than the other two primers. The graph of residuals versus fitted values in Fig. 14-12 does not reveal any unusual or diagnostic pattern.

14-4.3 One Observation per Cell

In some cases involving a two-factor factorial experiment, we may have only one replicate—that is, only one observation per cell. In this situation, there are exactly as many parameters in the analysis of variance model as observations, and the error degrees of freedom are zero. Thus, we cannot test hypotheses about the main effects and interactions unless some additional

Table 14-8 Residuals for the Aircraft Primer Experiment in Example 14-1

Primer Type	Application Method	
	Dipping	Spraying
1	−0.27, 0.23, 0.03	0.10, −0.40, 0.30
2	0.30, −0.40, 0.10	−0.27, 0.03, 0.23
3	−0.03, −0.13, 0.17	0.33, −0.17, −0.17

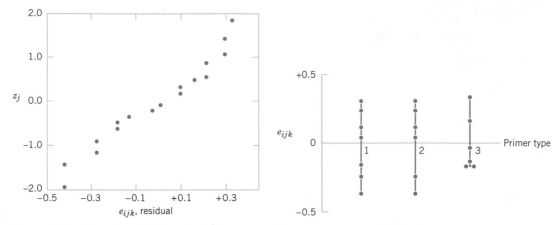

Figure 14-9   Normal probability plot of the residuals from Example 14-1.

Figure 14-10   Plot of residuals versus primer type.

assumptions are made. One possible assumption is to assume the interaction effect is negligible and use the interaction mean square as an error mean square. Thus, the analysis is equivalent to the analysis used in the randomized block design. This no-interaction assumption can be dangerous, and the experimenter should carefully examine the data and the residuals for indications as to whether or not interaction is present. For more details, see Montgomery (2001).

14-4.4   Factorial Experiments with Random Factors: Overview

In Section 13-3 we introduced the concept of a **random factor.** This is, of course, a situation in which the factor of interest has a large number of possible levels and the experimenter chooses a subset of these levels at random from this population. Conclusions are then drawn about the population of factor levels.

Random factors can occur in factorial experiments. If all the factors are random, the analysis of variance model is called a **random-effects model.** If some factors are fixed and other factors are random, the analysis of variance model is called a **mixed model.** The statistical analysis of random and mixed models is very similar to that of the standard fixed-effects models that are the primary focus of this chapter. The primary differences are in the types of hypotheses that are tested, the construction of test statistics for these hypotheses, and the estimation of model parameters. Some additional details on these topics are presented in Section 14-6 on the CD. For a more in-depth presentation, refer to Montgomery (2001) and Neter, Wasserman, Nachtsheim, and Kutner (1996).

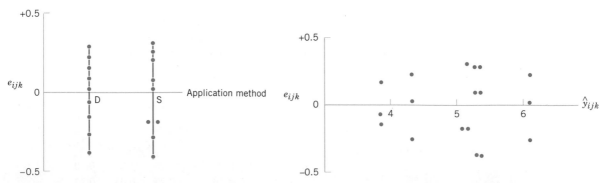

Figure 14-11   Plot of residuals versus application method.

Figure 14-12   Plot of residuals versus predicted values $\hat{y}_{ijk}$.

EXERCISES FOR SECTION 14-4

14-1. In his book (*Design and Analysis of Experiments,* 5th edition, 2001 John Wiley & Sons), D. C. Montgomery presents the results of an experiment involving a storage battery used in the launching mechanism of a shoulder-fired ground-to-air missile. Three material types can be used to make the battery plates. The objective is to design a battery that is relatively unaffected by the ambient temperature. The output response from the battery is effective life in hours. Three temperature levels are selected, and a factorial experiment with four replicates is run. The data are as follows:

Material	Temperature (°F)					
	Low		Medium		High	
1	130	155	34	40	20	70
	74	180	80	75	82	58
2	150	188	136	122	25	70
	159	126	106	115	58	45
3	138	110	174	120	96	104
	168	160	150	139	82	60

(a) Test the appropriate hypotheses and draw conclusions using the analysis of variance with $\alpha = 0.05$.
(b) Graphically analyze the interaction.
(c) Analyze the residuals from this experiment.

14-2. An engineer who suspects that the surface finish of metal parts is influenced by the type of paint used and the drying time. He selected three drying times—20, 25, and 30 minutes—and used two types of paint. Three parts are tested with each combination of paint type and drying time. The data are as follows:

Paint	Drying Time (min)		
	20	25	30
1	74	73	78
	64	61	85
	50	44	92
2	92	98	66
	86	73	45
	68	88	85

(a) State and test the appropriate hypotheses using the analysis of variance with $\alpha = 0.05$.
(b) Analyze the residuals from this experiment.

14-3. An article in *Industrial Quality Control* (1956, pp. 5–8) describes an experiment to investigate the effect of two factors (glass type and phosphor type) on the brightness of a television tube. The response variable measured is the current (in microamps) necessary to obtain a specified brightness level. The data are shown in the following table:

Glass Type	Phosphor Type		
	1	2	3
1	280	300	290
	290	310	285
	285	295	290
2	230	260	220
	235	240	225
	240	235	230

(a) State the hypotheses of interest in this experiment.
(b) Test the above hypotheses and draw conclusions using the analysis of variance with $\alpha = 0.05$.
(c) Analyze the residuals from this experiment.

14-4. An experiment was conducted to determine whether either firing temperature or furnace position affects the baked density of a carbon anode. The data are as follows:

Position	Temperature (°C)		
	800	825	850
1	570	1063	565
	565	1080	510
	583	1043	590
2	528	988	526
	547	1026	538
	521	1004	532

(a) State the hypotheses of interest.
(b) Test the above hypotheses using the analysis of variance with $\alpha = 0.05$. What are your conclusions?
(c) Analyze the residuals from this experiment.

14-5. Continuation of Exercise 14-4. Using Fisher's LSD method, investigate the differences between the mean baked anode density at the three different levels of temperature in Exercise 14-4. Use $\alpha = 0.05$.

14-6. Johnson and Leone (*Statistics and Experimental Design in Engineering and the Physical Sciences,* John Wiley, 1977) describe an experiment conducted to investigate warping of copper plates. The two factors studied were temperature and the copper content of the plates. The response variable is the amount of warping. The data are as follows:

Temperature	Copper Content (%)			
(°C)	40	60	80	100
50	17, 20	16, 21	24, 22	28, 27
75	12, 9	18, 13	17, 12	27, 31
100	16, 12	18, 21	25, 23	30, 23
125	21, 17	23, 21	23, 22	29, 31

(a) Is there any indication that either factor affects the amount of warping? Is there any interaction between the factors? Use $\alpha = 0.05$.

(b) Analyze the residuals from this experiment.

(c) Plot the average warping at each level of copper content and compare the levels using Fisher's LSD method. Describe the differences in the effects of the different levels of copper content on warping. If low warping is desirable, what level of copper content would you specify?

(d) Suppose that temperature cannot be easily controlled in the environment in which the copper plates are to be used. Does this change your answer for part (c)?

14-7. Consider a two-factor factorial experiment. Develop a formula for finding a $100(1 - \alpha)\%$ confidence interval on the difference between any two means for either a row or column factor. Apply this formula to find a 95% CI on the difference in mean warping at the levels of copper content 60 and 80% in Exercise 14-6.

14-8. An article in the *Journal of Testing and Evaluation* (Vol. 16, no. 6, 1988, pp. 508–515) investigated the effects of cyclic loading frequency and environment conditions on fatigue crack growth at a constant 22 MPa stress for a particular

material. The data from the experiment follow. The response variable is fatigue crack growth rate.

		Environment		
		Air	H$_2$O	Salt H$_2$O
	10	2.29	2.06	1.90
		2.47	2.05	1.93
		2.48	2.23	1.75
		2.12	2.03	2.06
Frequency	1	2.65	3.20	3.10
		2.68	3.18	3.24
		2.06	3.96	3.98
		2.38	3.64	3.24
	0.1	2.24	11.00	9.96
		2.71	11.00	10.01
		2.81	9.06	9.36
		2.08	11.30	10.40

(a) Is there indication that either factor affects crack growth rate? Is there any indication of interaction? Use $\alpha = 0.05$.

(b) Analyze the residuals from this experiment.

(c) Repeat the analysis in part (a) using $ln(y)$ as the response. Analyze the residuals from this new response variable and comment on the results.

14-9. An article in the *IEEE Transactions on Electron Devices* (November 1986, p. 1754) describes a study on the effects of two variables—polysilicon doping and anneal conditions (time and temperature)—on the base current of a bipolar transistor. The data from this experiment follows below Exercise 14-10.

(a) Is there any evidence to support the claim that either polysilicon doping level or anneal conditions affect base current? Do these variables interact? Use $\alpha = 0.05$.

(b) Graphically analyze the interaction.

(c) Analyze the residuals from this experiment.

14-10. Consider the experiment described in Exercise 14-9. Use Fisher's LSD method to isolate the effects of anneal conditions on base current, with $\alpha = 0.05$.

		Anneal (temperature/time)				
		900/60	900/180	950/60	1000/15	1000/30
Polysilicon doping	1×10^{20}	4.40	8.30	10.15	10.29	11.01
		4.60	8.90	10.20	10.30	10.58
	2×10^{20}	3.20	7.81	9.38	10.19	10.81
		3.50	7.75	10.02	10.10	10.60

14-5 GENERAL FACTORIAL EXPERIMENTS

Many experiments involve more than two factors. In this section we introduce the case where there are a levels of factor A, b levels of factor B, c levels of factor C, and so on, arranged in a factorial experiment. In general, there will be $abc \ldots n$ total observations, if there are n replicates of the complete experiment.

Table 14-9   Analysis of Variance Table for the Three-Factor Fixed Effects Model

Source of Variation	Sum of Squares	Degrees of Freedom	Mean Square	Expected Mean Squares	F_0
A	SS_A	$a-1$	MS_A	$\sigma^2 + \dfrac{bcn\sum\tau_i^2}{a-1}$	$\dfrac{MS_A}{MS_E}$
B	SS_B	$b-1$	MS_B	$\sigma^2 + \dfrac{acn\sum\beta_j^2}{b-1}$	$\dfrac{MS_B}{MS_E}$
C	SS_C	$c-1$	MS_C	$\sigma^2 + \dfrac{abn\sum\gamma_k^2}{c-1}$	$\dfrac{MS_C}{MS_E}$
AB	SS_{AB}	$(a-1)(b-1)$	MS_{AB}	$\sigma^2 + \dfrac{cn\sum\sum(\tau\beta)_{ij}^2}{(a-1)(b-1)}$	$\dfrac{MS_{AB}}{MS_E}$
AC	SS_{AC}	$(a-1)(c-1)$	MS_{AC}	$\sigma^2 + \dfrac{bn\sum\sum(\tau\gamma)_{ik}^2}{(a-1)(c-1)}$	$\dfrac{MS_{AC}}{MS_E}$
BC	SS_{BC}	$(b-1)(c-1)$	MS_{BC}	$\sigma^2 + \dfrac{an\sum\sum(\beta\gamma)_{jk}^2}{(b-1)(c-1)}$	$\dfrac{MS_{BC}}{MS_E}$
ABC	SS_{ABC}	$(a-1)(b-1)(c-1)$	MS_{ABC}	$\sigma^2 + \dfrac{n\sum\sum\sum(\tau\beta\gamma)_{ijk}^2}{(a-1)(b-1)(c-1)}$	$\dfrac{MS_{ABC}}{MS_E}$
Error	SS_E	$abc(n-1)$	MS_E	σ^2	
Total	SS_T	$abcn-1$			

For example, consider the **three-factor-factorial experiment,** with underlying model

$$Y_{ijkl} = \mu + \tau_i + \beta_j + \gamma_k + (\tau\beta)_{ij} + (\tau\gamma)_{ik} + (\beta\gamma)_{jk}$$

$$+ (\tau\beta\gamma)_{ijk} + \epsilon_{ijkl} \quad \begin{cases} i = 1,2,\ldots,a \\ j = 1,2,\ldots,b \\ k = 1,2,\ldots,c \\ l = 1,2,\ldots,n \end{cases} \quad (14\text{-}10)$$

Notice that the model contains three main effects, three two-factor interactions, a three-factor interaction, and an error term. Assuming that A, B, and C are fixed factors, the analysis of variance is shown in Table 14-9. Note that there must be at least two replicates ($n \geq 2$) to compute an error sum of squares. The F-test on main effects and interactions follows directly from the expected mean squares. These ratios follow F distributions under the respective null hypotheses.

EXAMPLE 14-2   A mechanical engineer is studying the surface roughness of a part produced in a metal-cutting operation. Three factors, feed rate (A), depth of cut (B), and tool angle (C), are of interest. All three factors have been assigned two levels, and two replicates of a factorial design are run. The coded data are shown in Table 14-10.

   The ANOVA is summarized in Table 14-11. Since manual ANOVA computions are tedious for three-factor experiments, we have used Minitab for the solution of this problem.

Table 14-10 Coded Surface Roughness Data for Example 14-2

Feed Rate (A)	Depth of Cut (B)				$y_{i\cdots}$
	0.025 inch		0.040 inch		
	Tool Angle (C)		Tool Angle (C)		
	15°	25°	15°	25°	
20 inches per minute	9	11	9	10	
	7	10	11	8	75
30 inches per minute	10	10	12	16	
	12	13	15	14	102

The F-ratios for all three main effects and the interactions are formed by dividing the mean square for the effect of interest by the error mean square. Since the experimenter has selected $\alpha = 0.05$, the critical value for each of these F-ratios is $f_{0.05,1,8} = 5.32$. Alternately, we could use the P-value approach. The P-values for all the test statistics are shown in the last column of Table 14-11. Inspection of these P-values is revealing. There is a strong main effect of feed rate, since the F-ratio is well into the critical region. However, there is some indication of an effect due to the depth of cut, since $P = 0.0710$ is not much greater than $\alpha = 0.05$. The next largest effect is the AB or feed rate × depth of cut interaction. Most likely, both feed rate and depth of cut are important process variables.

Obviously, factorial experiments with three or more factors can require many runs, particularly if some of the factors have several (more than two) levels. This point of view leads us to the class of factorial designs considered in Section 14-7 with all factors at two levels. These designs are easy to set up and analyze, and they may be used as the basis of many other useful experimental designs.

Table 14-11 Minitab ANOVA for Example 14-2

ANOVA (Balanced Designs)

Factor	Type	Levels	Values	
Feed	fixed	2	20	30
Depth	fixed	2	0.025	0.040
Angle	fixed	2	15	25

Analysis of Variance for Roughness

Source	DF	SS	MS	F	P
Feed	1	45.563	45.563	18.69	0.003
Depth	1	10.563	10.563	4.33	0.071
Angle	1	3.063	3.063	1.26	0.295
Feed*Depth	1	7.563	7.563	3.10	0.116
Feed*Angle	1	0.062	0.062	0.03	0.877
Depth*Angle	1	1.563	1.563	0.64	0.446
Feed*Depth*Angle	1	5.062	5.062	2.08	0.188
Error	8	19.500	2.437		
Total	15	92.938			

EXERCISES FOR SECTION 14-5

14-11. The percentage of hardwood concentration in raw pulp, the freeness, and the cooking time of the pulp are being investigated for their effects on the strength of paper. The data from a three-factor factorial experiment are shown in the following table.

(a) Analyze the data using the analysis of variance assuming that all factors are fixed. Use $\alpha = 0.05$.
(b) Find P-values for the F-ratios in part (a).
(c) The residuals are found by $e_{ijkl} = y_{ijkl} - \bar{y}_{ijk\cdot}$. Graphically analyze the residuals from this experiment.

Percentage of Hardwood Concentration	Cooking Time 1.5 hours Freeness			Cooking Time 2.0 hours Freeness		
	350	500	650	350	500	650
10	96.6	97.7	99.4	98.4	99.6	100.6
	96.0	96.0	99.8	98.6	100.4	100.9
15	98.5	96.0	98.4	97.5	98.7	99.6
	97.2	96.9	97.6	98.1	96.0	99.0
20	97.5	95.6	97.4	97.6	97.0	98.5
	96.6	96.2	98.1	98.4	97.8	99.8

14-12. The quality control department of a fabric finishing plant is studying the effects of several factors on dyeing for a blended cotton/synthetic cloth used to manufacture shirts. Three operators, three cycle times, and two temperatures were selected, and three small specimens of cloth were dyed under each set of conditions. The finished cloth was compared to a standard, and a numerical score was assigned. The results are shown in the following table.

(a) State and test the appropriate hypotheses using the analysis of variance with $\alpha = 0.05$.
(b) The residuals may be obtained from $e_{ijkl} = y_{ijkl} - \bar{y}_{ijk\cdot}$. Graphically analyze the residuals from this experiment.

Cycle Time	Temperature 300° Operator			Temperature 350° Operator		
	1	2	3	1	2	3
40	23	27	31	24	38	34
	24	28	32	23	36	36
	25	26	28	28	35	39
50	36	34	33	37	34	34
	35	38	34	39	38	36
	36	39	35	35	36	31
60	28	35	26	26	36	28
	24	35	27	29	37	26
	27	34	25	25	34	34

14-6 FACTORIAL EXPERIMENTS WITH RANDOM FACTORS (CD ONLY)

14-7 2^k FACTORIAL DESIGNS

Factorial designs are frequently used in experiments involving several factors where it is necessary to study the joint effect of the factors on a response. However, several special cases of the general factorial design are important because they are widely employed in research work and because they form the basis of other designs of considerable practical value.

The most important of these special cases is that of k factors, each at only two levels. These levels may be quantitative, such as two values of temperature, pressure, or time; or they may be qualitative, such as two machines, two operators, the "high" and "low" levels of a factor, or perhaps the presence and absence of a factor. A complete replicate of such a design requires $2 \times 2 \times \cdots \times 2 = 2^k$ observations and is called a **2^k factorial design.**

The 2^k design is particularly useful in the early stages of experimental work, when many factors are likely to be investigated. It provides the smallest number of runs for which k factors can be studied in a complete factorial design. Because there are only two levels for each factor, we must assume that the response is approximately linear over the range of the factor levels chosen.

14-7.1 2^2 Design

The simplest type of 2^k design is the 2^2—that is, two factors A and B, each at two levels. We usually think of these levels as the low and high levels of the factor. The 2^2 design is shown in Fig. 14-13. Note that the design can be represented geometrically as a square with the $2^2 = 4$ runs, or treatment combinations, forming the corners of the square. In the 2^2 design it is customary to denote the low and high levels of the factors A and B by the signs $-$ and $+$, respectively. This is sometimes called the **geometric notation** for the design.

A special notation is used to label the treatment combinations. In general, a treatment combination is represented by a series of lowercase letters. If a letter is present, the corresponding factor is run at the high level in that treatment combination; if it is absent, the factor is run at its low level. For example, treatment combination a indicates that factor A is at the high level and factor B is at the low level. The treatment combination with both factors at the low level is represented by (1). This notation is used throughout the 2^k design series. For example, the treatment combination in a 2^4 with A and C at the high level and B and D at the low level is denoted by ac.

The effects of interest in the 2^2 design are the main effects A and B and the two-factor interaction AB. Let the letters (1), a, b, and ab also represent the totals of all n observations taken at these design points. It is easy to estimate the effects of these factors. To estimate the main effect of A, we would average the observations on the right side of the square in Fig. 14-13 where A is at the high level, and subtract from this the average of the observations on the left side of the square, where A is at the low level, or

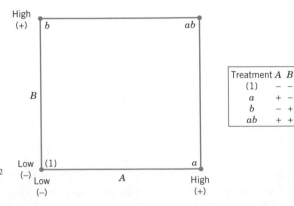

Figure 14-13   The 2^2 factorial design.

Treatment	A	B
(1)	$-$	$-$
a	$+$	$-$
b	$-$	$+$
ab	$+$	$+$

$$A = \bar{y}_{A+} - \bar{y}_{A-}$$
$$= \frac{a + ab}{2n} - \frac{b + (1)}{2n}$$
$$= \frac{1}{2n}[a + ab - b - (1)] \qquad (14\text{-}11)$$

Similarly, the main effect of B is found by averaging the observations on the top of the square, where B is at the high level, and subtracting the average of the observations on the bottom of the square, where B is at the low level:

$$B = \bar{y}_{B+} - \bar{y}_{B-}$$
$$= \frac{b + ab}{2n} - \frac{a + (1)}{2n}$$
$$= \frac{1}{2n}[b + ab - a - (1)] \qquad (14\text{-}12)$$

Finally, the AB interaction is estimated by taking the difference in the diagonal averages in Fig. 14-12, or

$$AB = \frac{ab + (1)}{2n} - \frac{a + b}{2n}$$
$$= \frac{1}{2n}[ab + (1) - a - b] \qquad (14\text{-}13)$$

The quantities in brackets in Equations 14-11, 14-12, and 14-13 are called **contrasts.** For example, the A contrast is

$$\text{Contrast}_A = a + ab - b - (1)$$

In these equations, the contrast coefficients are always either $+1$ or -1. A table of plus and minus signs, such as Table 14-12, can be used to determine the sign on each treatment

Table 14-12   Signs for Effects in the 2^2 Design

Treatment Combination	Factorial Effect			
	I	A	B	AB
(1)	+	−	−	+
a	+	+	−	−
b	+	−	+	−
ab	+	+	+	+

combination for a particular contrast. The column headings for Table 14-12 are the main effects A and B, the AB interaction, and I, which represents the total. The row headings are the treatment combinations. Note that the signs in the AB column are the product of signs from columns A and B. To generate a contrast from this table, multiply the signs in the appropriate column of Table 14-12 by the treatment combinations listed in the rows and add. For example, $\text{contrast}_{AB} = [(1)] + [-a] + [-b] + [ab] = ab + (1) - a - b$.

Contrasts are used in calculating both the effect estimates and the sums of squares for A, B, and the AB interaction. The sums of squares formulas are

$$SS_A = \frac{[a + ab - b - (1)]^2}{4n}$$

$$SS_B = \frac{[b + ab - a - (1)]^2}{4n} \tag{14-14}$$

$$SS_{AB} = \frac{[ab + (1) - a - b]^2}{4n}$$

The analysis of variance is completed by computing the total sum of squares SS_T (with $4n - 1$ degrees of freedom) as usual, and obtaining the error sum of squares SS_E [with $4(n - 1)$ degrees of freedom] by subtraction.

EXAMPLE 14-3

An article in the *AT&T Technical Journal* (Vol. 65, March/April 1986, pp. 39–50) describes the application of two-level factorial designs to integrated circuit manufacturing. A basic processing step in this industry is to grow an epitaxial layer on polished silicon wafers. The wafers are mounted on a susceptor and positioned inside a bell jar. Chemical vapors are introduced through nozzles near the top of the jar. The susceptor is rotated, and heat is applied. These conditions are maintained until the epitaxial layer is thick enough.

Table 14-13 presents the results of a 2^2 factorial design with $n = 4$ replicates using the factors A = deposition time and B = arsenic flow rate. The two levels of deposition time are $-$ =short and $+$ =long, and the two levels of arsenic flow rate are $-$ =55% and $+$ =59%. The response variable is epitaxial layer thickness (μm). We may find the estimates of the effects using Equations 14-11, 14-12, and 14-13 as follows:

$$A = \frac{1}{2n}[a + ab - b - (1)]$$

$$= \frac{1}{2(4)}[59.299 + 59.156 - 55.686 - 56.081] = 0.836$$

$$B = \frac{1}{2n}[b + ab - a - (1)]$$

$$= \frac{1}{2(4)}[55.686 + 59.156 - 59.299 - 56.081] = 0.067$$

$$AB = \frac{1}{2n}[ab + (1) - a - b]$$

$$AB = \frac{1}{2(4)}[59.156 + 56.081 - 59.299 - 55.686] = 0.032$$

Table 14-13 The 2^2 Design for the Epitaxial Process Experiment

Treatment Combination	Design Factors			Thickness (μm)				Thickness (μm)	
	A	B	AB					Total	Average
(1)	−	−	+	14.037	14.165	13.972	13.907	56.081	14.020
a	+	−	−	14.821	14.757	14.843	14.878	59.299	14.825
b	−	+	−	13.880	13.860	14.032	13.914	55.686	13.922
ab	+	+	+	14.888	14.921	14.415	14.932	59.156	14.789

The numerical estimates of the effects indicate that the effect of deposition time is large and has a positive direction (increasing deposition time increases thickness), since changing deposition time from low to high changes the mean epitaxial layer thickness by 0.836 μm. The effects of arsenic flow rate (B) and the AB interaction appear small.

The importance of these effects may be confirmed with the analysis of variance. The sums of squares for A, B, and AB are computed as follows:

$$SS_A = \frac{[a + ab - b - (1)]^2}{16} = \frac{[6.688]^2}{16} = 2.7956$$

$$SS_B = \frac{[b + ab - a - (1)]^2}{16} = \frac{[-0.538]^2}{16} = 0.0181$$

$$SS_{AB} = \frac{[ab + (1) - a - b]^2}{16} = \frac{[0.252]^2}{16} = 0.0040$$

$$SS_T = 14.037^2 + \cdots + 14.932^2 - \frac{(56.081 + \cdots + 59.156)^2}{16}$$

$$= 3.0672$$

The analysis of variance is summarized in Table 14-14 and confirms our conclusions obtained by examining the magnitude and direction of the effects. Deposition time is the only factor that significantly affects epitaxial layer thickness, and from the direction of the effect estimates we know that longer deposition times lead to thicker epitaxial layers.

Residual Analysis

It is easy to obtain the residuals from a 2^k design by fitting a **regression model** to the data. For the epitaxial process experiment, the regression model is

$$Y = \beta_0 + \beta_1 x_1 + \epsilon$$

Table 14-14 Analysis of Variance for the Epitaxial Process Experiment

Source of Variation	Sum of Squares	Degrees of Freedom	Mean Square	f_0	P-Value
A (deposition time)	2.7956	1	2.7956	134.40	7.07 E-8
B (arsenic flow)	0.0181	1	0.0181	0.87	0.38
AB	0.0040	1	0.0040	0.19	0.67
Error	0.2495	12	0.0208		
Total	3.0672	15			

since the only active variable is deposition time, which is represented by a coded variable x_1. The low and high levels of deposition time are assigned values $x_1 = -1$ and $x_1 = +1$, respectively. The least squares fitted model is

$$\hat{y} = 14.389 + \left(\frac{0.836}{2}\right)x_1$$

where the intercept $\hat{\beta}_0$ is the grand average of all 16 observations $(\bar{y})$ and the slope $\hat{\beta}_1$ is one-half the effect estimate for deposition time. (The regression coefficient is one-half the effect estimate because regression coefficients measure the effect of a unit change in x_1 on the mean of Y, and the effect estimate is based on a two-unit change from -1 to $+1$.)

This model can be used to obtain the predicted values at the four points that form the corners of the square in the design. For example, consider the point with low deposition time $(x_1 = -1)$ and low arsenic flow rate. The predicted value is

$$\hat{y} = 14.389 + \left(\frac{0.836}{2}\right)(-1) = 13.971 \ \mu m$$

and the residuals for the four runs at that design point are

$$e_1 = 14.037 - 13.971 = 0.066$$
$$e_2 = 14.165 - 13.971 = 0.194$$
$$e_3 = 13.972 - 13.971 = 0.001$$
$$e_4 = 13.907 - 13.971 = -0.064$$

The remaining predicted values and residuals at the other three design points are calculated in a similar manner.

A normal probability plot of these residuals is shown in Fig. 14-14. This plot indicates that one residual $e_{15} = -0.392$ is an **outlier.** Examining the four runs with high deposition time and high arsenic flow rate reveals that observation $y_{15} = 14.415$ is considerably smaller than the other three observations at that treatment combination. This adds some additional evidence to the tentative conclusion that observation 15 is an outlier. Another possibility is that some process variables affect the *variability* in epitaxial layer thickness. If we could discover which variables produce this effect, we could perhaps adjust these variables to levels that would minimize the variability in epitaxial layer thickness. This could have important implications in subsequent manufacturing stages. Figures 14-15 and 14-16 are plots of residuals versus deposition time and arsenic flow rate, respectively. Apart from that unusually large residual associated with y_{15}, there is no strong evidence that either deposition time or arsenic flow rate influences the variability in epitaxial layer thickness.

Figure 14-17 shows the standard deviation of epitaxial layer thickness at all four runs in the 2^2 design. These standard deviations were calculated using the data in Table 14-13. Notice that the standard deviation of the four observations with A and B at the high level is considerably larger than the standard deviations at any of the other three design points. Most of this difference is attributable to the unusually low thickness measurement associated with y_{15}. The standard deviation of the four observations with A and B at the low level is also somewhat larger than the standard deviations at the remaining two runs. This could indicate that other process variables not included in this experiment may affect the variability in epitaxial layer thickness. Another experiment to study this possibility, involving other process variables, could be designed and conducted. (The original paper in the *AT&T Technical Journal* shows that two additional factors, not considered in this example, affect process variability.)

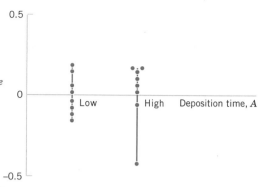

Figure 14-15   Plot of residuals versus deposition time.

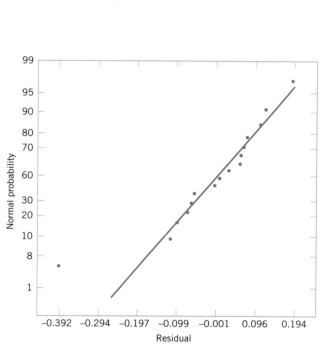

Figure 14-14   Normal probability plot of residuals for the epitaxial process experiment.

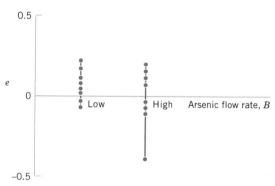

Figure 14-16   Plot of residuals versus arsenic flow rate.

14-7.2 2^k Design for $k \geq 3$ Factors

The methods presented in the previous section for factorial designs with $k = 2$ factors each at two levels can be easily extended to more than two factors. For example, consider $k = 3$ factors, each at two levels. This design is a 2^3 factorial design, and it has eight runs or treatment combinations. Geometrically, the design is a cube as shown in Fig. 14-18(a), with the eight runs forming the corners of the cube. Figure 14-18(b) lists the eight runs in a table, with each row representing one of the runs are the − and + settings indicating the low and high levels

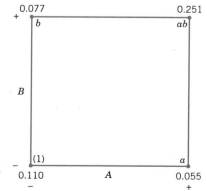

Figure 14-17   The standard deviation of epitaxial layer thickness at the four runs in the 2^2 design.

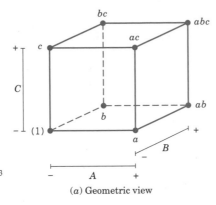

Run	A	B	C
1	–	–	–
2	+	–	–
3	–	+	–
4	+	+	–
5	–	–	+
6	+	–	+
7	–	+	+
8	+	+	+

Figure 14-18   The 2^3 design.

(a) Geometric view

(b) The 2^3 design matrix

for each of the three factors. This table is sometimes called the **design matrix.** This design allows three main effects to be estimated (A, B, and C) along with three two-factor interactions (AB, AC, and BC) and a three-factor interaction (ABC).

The main effects can easily be estimated. Remember that the lowercase letters (1), a, b, ab, c, ac, bc, and abc represent the total of all n replicates at each of the eight runs in the design. As seen in Fig. 14-19(a), the main effect of A can be estimated by averaging the four treatment combinations on the right-hand side of the cube, where A is at the high level, and by

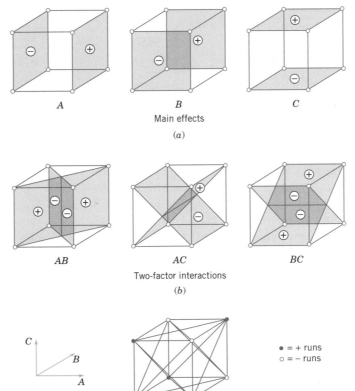

Figure 14-19
Geometric presentation of contrasts corresponding to the main effects and interaction in the 2^3 design. (a) Main effects. (b) Two-factor interactions. (c) Three-factor interaction.

subtracting from this quantity the average of the four treatment combinations on the left-hand side of the cube where A is at the low level. This gives

$$A = \bar{y}_{A+} - \bar{y}_{A-}$$
$$= \frac{a + ab + ac + abc}{4n} - \frac{(1) + b + c + bc}{4n}$$

This equation can be rearranged as

$$A = \frac{1}{4n} [a + ab + ac + abc - (1) - b - c - bc] \qquad (14\text{-}15)$$

In a similar manner, the effect of B is the difference in averages between the four treatment combinations in the back face of the cube (Fig. 14-19a), and the four in the front. This yields

$$B = \bar{y}_{B+} - \bar{y}_{B-}$$
$$= \frac{1}{4n} [b + ab + bc + abc - (1) - a - c - ac] \qquad (14\text{-}16)$$

The effect of C is the difference in average response between the four treatment combinations in the top face of the cube in Figure 14-19(a) and the four in the bottom, that is,

$$C = \bar{y}_{C+} - \bar{y}_{C-}$$
$$= \frac{1}{4n} [c + ac + bc + abc - (1) - a - b - ab] \qquad (14\text{-}17)$$

The two-factor interaction effects may be computed easily. A measure of the AB interaction is the difference between the average A effects at the two levels of B. By convention, one-half of this difference is called the AB interaction. Symbolically,

B	Average A Effect
High ($+$)	$\dfrac{[(abc - bc) + (ab - b)]}{2n}$
Low ($-$)	$\dfrac{\{(ac - c) + [a - (1)]\}}{2n}$
Difference	$\dfrac{[abc - bc + ab - b - ac + c - a + (1)]}{2n}$

Since the *AB* interaction is one-half of this difference,

$$AB = \frac{1}{4n} \left[abc - bc + ab - b - ac + c - a + (1) \right] \qquad (14\text{-}18)$$

We could write Equation 14-18 as follows:

$$AB = \frac{abc + ab + c + (1)}{4n} - \frac{bc + b + ac + a}{4n}$$

In this form, the *AB* interaction is easily seen to be the difference in averages between runs on two diagonal planes in the cube in Fig. 14-19(b). Using similar logic and referring to Fig. 14-19(b), we find that the *AC* and *BC* interactions are

$$AC = \frac{1}{4n} \left[(1) - a + b - ab - c + ac - bc + abc \right] \qquad (14\text{-}19)$$

$$BC = \frac{1}{4n} \left[(1) + a - b - ab - c - ac + bc + abc \right] \qquad (14\text{-}20)$$

The *ABC* interaction is defined as the average difference between the *AB* interaction for the two different levels of *C*. Thus,

$$ABC = \frac{1}{4n} \left\{ [abc - bc] - [ac - c] - [ab - b] + [a - (1)] \right\}$$

or

$$ABC = \frac{1}{4n} \left[abc - bc - ac + c - ab + b + a - (1) \right] \qquad (14\text{-}21)$$

As before, we can think of the *ABC* interaction as the difference in two averages. If the runs in the two averages are isolated, they define the vertices of the two tetrahedra that comprise the cube in Fig. 14-19(c).

In Equations 14-15 through 14-21, the quantities in brackets are **contrasts** in the treatment combinations. A table of plus and minus signs can be developed from the contrasts and is shown in Table 14-15. Signs for the main effects are determined directly from the test matrix in Figure 14-18(b). Once the signs for the main effect columns have been established, the signs for the remaining columns can be obtained by multiplying the appropriate

Table 14-15   Algebraic Signs for Calculating Effects in the 2^3 Design

Treatment Combination	Factorial Effect							
	I	A	B	AB	C	AC	BC	ABC
(1)	+	−	−	+	−	+	+	−
a	+	+	−	−	−	−	+	+
b	+	−	+	−	−	+	−	+
ab	+	+	+	+	−	−	−	−
c	+	−	−	+	+	−	−	+
ac	+	+	−	−	+	+	−	−
bc	+	−	+	−	+	−	+	−
abc	+	+	+	+	+	+	+	+

main effect row by row. For example, the signs in the AB column are the products of the A and B column signs in each row. The contrast for any effect can easily be obtained from this table.

Table 14-15 has several interesting properties:

1. Except for the identity column I, each column has an equal number of plus and minus signs.

2. The sum of products of signs in any two columns is zero; that is, the columns in the table are **orthogonal.**

3. Multiplying any column by column I leaves the column unchanged; that is, I is an **identity element.**

4. The product of any two columns yields a column in the table, for example $A \times B = AB$, and $AB \times ABC = A^2B^2C = C$, since any column multiplied by itself is the identity column.

The estimate of any main effect or interaction in a 2^k design is determined by multiplying the treatment combinations in the first column of the table by the signs in the corresponding main effect or interaction column, by adding the result to produce a contrast, and then by dividing the contrast by one-half the total number of runs in the experiment. For any 2^k design with n replicates, the effect estimates are computed from

$$\text{Effect} = \frac{\text{Contrast}}{n2^{k-1}} \qquad (14\text{-}22)$$

and the sum of squares for any effect is

$$SS = \frac{(\text{Contrast})^2}{n2^k} \qquad (14\text{-}23)$$

EXAMPLE 14-4

Consider the surface roughness experiment originally described in Example 14-2. This is a 2^3 factorial design in the factors feed rate (A), depth of cut (B), and tool angle (C), with $n = 2$ replicates. Table 14-16 presents the observed surface roughness data.

The main effects may be estimated using Equations 14-15 through 14-21. The effect of A, for example, is

$$A = \frac{1}{4n}\left[a + ab + ac + abc - (1) - b - c - bc\right]$$

$$= \frac{1}{4(2)}\left[22 + 27 + 23 + 40 - 16 - 20 - 21 - 18\right]$$

$$= \frac{1}{8}\left[27\right] = 3.375$$

and the sum of squares for A is found using Equation 14-23:

$$SS_A = \frac{(\text{Contrast}_A)^2}{n2^k} = \frac{(27)^2}{2(8)} = 45.5625$$

It is easy to verify that the other effects are

$$
\begin{aligned}
B &= 1.625 \\
C &= 0.875 \\
AB &= 1.375 \\
AC &= 0.125 \\
BC &= -0.625 \\
ABC &= 1.125
\end{aligned}
$$

Examining the magnitude of the effects clearly shows that feed rate (factor A) is dominant, followed by depth of cut (B) and the AB interaction, although the interaction effect is relatively small. The analysis of variance, summarized in Table 14-17, confirms our interpretation of the effect estimates.

Minitab will analyze 2^k factorial designs. The output from the Minitab DOE (Design of Experiments) module for this experiment is shown in Table 14-18. The upper portion of the table displays the effect estimates and regression coefficients for each factorial effect. However, a

Table 14-16   Surface Roughness Data for Example 14-4

Treatment Combinations	Design Factors			Surface Roughness	Totals
	A	B	C		
(1)	-1	-1	-1	9, 7	16
a	1	-1	-1	10, 12	22
b	-1	1	-1	9, 11	20
ab	1	1	-1	12, 15	27
c	-1	-1	1	11, 10	21
ac	1	-1	1	10, 13	23
bc	-1	1	1	10, 8	18
abc	1	1	1	16, 14	30

Table 14-17   Analysis of Variance for the Surface Finish Experiment

Source of Variation	Sum of Squares	Degrees of Freedom	Mean Square	f_0	P-Value
A	45.5625	1	45.5625	18.69	0.0025
B	10.5625	1	10.5625	4.33	0.0709
C	3.0625	1	3.0625	1.26	0.2948
AB	7.5625	1	7.5625	3.10	0.1162
AC	0.0625	1	0.0625	0.03	0.8784
BC	1.5625	1	1.5625	0.64	0.4548
ABC	5.0625	1	5.0625	2.08	0.1875
Error	19.5000	8	2.4375		
Total	92.9375	15			

t-statistic is reported for each effect instead of the F-statistic used in Table 14-17. Now the square of a t random variable with d degrees of freedom is an F random variable with 1 numerator and d denominator degrees of freedom. Thus the square of the t-statistic reported by Minitab will be equal (apart from rounding errors) to the F-statistic in Table 14-17. To illustrate, for the main effect of feed Minitab reports $t = 4.32$ (with eight degrees of freedom), and $t^2 = (4.32)^2 = 18.66$, which is approximately equal to the F-ratio for feed reported in Table 14-17 ($F = 18.69$). This F-ratio has one numerator and eight denominator degrees of freedom.

The lower panel of the Minitab output in Table 14-18 is an analysis of variance summary focusing on the types of terms in the model. A regression model approach is used in the presentation. You might find it helpful to review Section 12-2.2, particularly the material on the partial F-test. The row entitled "main effects" under source refers to the three main effects

Table 14-18   Minitab Analysis for Example 14-4

Estimated Effects and Coefficients for Roughness

Term	Effect	Coef	StDev Coef	T	P
Constant		11.0625	0.3903	28.34	0.000
Feed	3.3750	1.6875	0.3903	4.32	0.003
Depth	1.6250	0.8125	0.3903	2.08	0.071
Angle	0.8750	0.4375	0.3903	1.12	0.295
Feed*Depth	1.3750	0.6875	0.3903	1.76	0.116
Feed*Angle	0.1250	0.0625	0.3903	0.16	0.877
Depth*Angle	−0.6250	−0.3125	0.3903	−0.80	0.446
Feed*Depth*Angle	1.1250	0.5625	0.3903	1.44	0.188

Analysis of Variance for Roughness

Source	DF	Seq SS	Adj SS	Adj MS	F	P
Main Effects	3	59.188	59.188	19.729	8.09	0.008
2-Way Interactions	3	9.187	9.187	3.062	1.26	0.352
3-Way Interactions	1	5.062	5.062	5.062	2.08	0.188
Residual Error	8	19.500	19.500	2.437		
Pure Error	8	19.500	19.500	2.437		
Total	15	92.938				

feed, depth, and angle, each having a single degree of freedom, giving the total 3 in the column headed "DF." The column headed "Seq SS" (an abbreviation for sequential sum of squares) reports how much the model sum of squares increases when each group of terms is added to a model that contains the terms listed *above* the groups. The first number in the "Seq SS" column presents the model sum of squares for fitting a model having only the three main effects. The row labeled "2-Way Interactions" refers to AB, AC, and BC, and the sequential sum of squares reported here is the increase in the model sum of squares if the interaction terms are added to a model containing only the main effects. Similarly, the sequential sum of squares for the three-way interaction is the increase in the model sum of squares that results from adding the term ABC to a model containing all other effects. The column headed "Adj SS" (an abbreviation for adjusted sum of squares) reports how much the model sum of squares increases when each group of terms is added to a model that contains *all* the other terms. Now since any 2^k design with an equal number of replicates in each cell is an orthogonal design, the adjusted sum of squares will equal the sequential sum of squares. Therefore, the F-tests for each row in the Minitab analysis of variance table are testing the significance of each group of terms (main effects, two-factor interactions, and three-factor interactions) as if they were the last terms to be included in the model. Clearly, only the main effect terms are significant. The t-tests on the individual factor effects indicate that feed rate and depth of cut have large main effects, and there may be some mild interaction between these two factors. Therefore, the Minitab output is in agreement with the results given previously.

Residual Analysis

We may obtain the residuals from a 2^k design by using the method demonstrated earlier for the 2^2 design. As an example, consider the surface roughness experiment. The three largest effects are A, B, and the AB interaction. The regression model used to obtain the predicted values is

$$Y = \beta_0 + \beta_1 x_1 + \beta_2 x_2 + \beta_{12} x_1 x_2 + \epsilon$$

where x_1 represents factor A, x_2 represents factor B, and $x_1 x_2$ represents the AB interaction. The regression coefficients β_1, β_2, and β_{12} are estimated by one-half the corresponding effect estimates, and β_0 is the grand average. Thus

$$\hat{y} = 11.0625 + \left(\frac{3.375}{2}\right) x_1 + \left(\frac{1.625}{2}\right) x_2 + \left(\frac{1.375}{2}\right) x_1 x_2$$
$$= 11.0625 + 1.6875 x_1 + 0.8125 x_2 + 0.6875 x_1 x_2$$

Note that the regression coefficients are presented by Minitab in the upper panel of Table 14-18. The predicted values would be obtained by substituting the low and high levels of A and B into this equation. To illustrate this, at the treatment combination where A, B, and C are all at the low level, the predicted value is

$$\hat{y} = 11.065 + 1.6875(-1) + 0.8125(-1) + 0.6875(-1)(-1) = 9.25$$

Since the observed values at this run are 9 and 7, the residuals are $9 - 9.25 = -0.25$ and $7 - 9.25 = -2.25$. Residuals for the other 14 runs are obtained similarly.

A normal probability plot of the residuals is shown in Fig. 14-20. Since the residuals lie approximately along a straight line, we do not suspect any problem with normality in the data. There are no indications of severe outliers. It would also be helpful to plot the residuals versus the predicted values and against each of the factors A, B, and C.

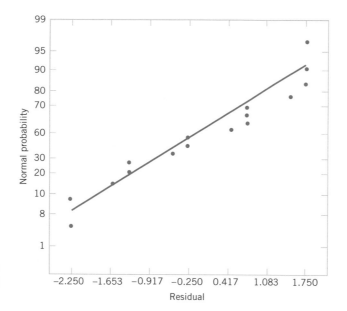

Figure 14-20
Normal probability
plot of residuals from
the surface roughness
experiment.

Projection of 2^k Designs

Any 2^k design will collapse or project into another 2^k design in fewer variables if one or more of the original factors are dropped. Sometimes this can provide additional insight into the remaining factors. For example, consider the surface roughness experiment. Since factor C and all its interactions are negligible, we could eliminate factor C from the design. The result is to collapse the cube in Fig. 14-18 into a square in the $A - B$ plane; therefore, each of the four runs in the new design has four replicates. In general, if we delete h factors so that $r = k - h$ factors remain, the original 2^k design with n replicates will project into a 2^r design with $n2^h$ replicates.

14-7.3   Single Replicate of the 2^k Design

As the number of factors in a factorial experiment grows, the number of effects that can be estimated also grows. For example, a 2^4 experiment has 4 main effects, 6 two-factor interactions, 4 three-factor interactions, and 1 four-factor interaction, while a 2^6 experiment has 6 main effects, 15 two-factor interactions, 20 three-factor interactions, 15 four-factor interactions, 6 five-factor interactions, and 1 six-factor interaction. In most situations the **sparsity of effects principle** applies; that is, the system is usually dominated by the main effects and low-order interactions. The three-factor and higher order interactions are usually negligible. Therefore, when the number of factors is moderately large, say, $k \geq 4$ or 5, a common practice is to run only a single replicate of the 2^k design and then pool or combine the higher order interactions as an estimate of error. Sometimes a single replicate of a 2^k design is called an **unreplicated** 2^k factorial design.

When analyzing data from unreplicated factorial designs, occasionally real high-order interactions occur. The use of an error mean square obtained by pooling high-order interactions is inappropriate in these cases. A simple method of analysis can be used to overcome this problem. Construct a plot of the estimates of the effects on a normal probability scale. The effects that are negligible are normally distributed, with mean zero and variance σ^2 and will tend to fall along a straight line on this plot, whereas significant effects will have nonzero means and will not lie along the straight line. We will illustrate this method in the next example.

EXAMPLE 14-5

An article in *Solid State Technology* ("Orthogonal Design for Process Optimization and Its Application in Plasma Etching," May 1987, pp. 127–132) describes the application of factorial designs in developing a nitride etch process on a single-wafer plasma etcher. The process uses C_2F_6 as the reactant gas. It is possible to vary the gas flow, the power applied to the cathode, the pressure in the reactor chamber, and the spacing between the anode and the cathode (gap). Several response variables would usually be of interest in this process, but in this example we will concentrate on etch rate for silicon nitride.

We will use a single replicate of a 2^4 design to investigate this process. Since it is unlikely that the three- and four-factor interactions are significant, we will tentatively plan to combine them as an estimate of error. The factor levels used in the design are shown below:

Level	Design Factor			
	Gap (cm)	Pressure (mTorr)	C_2F_6 Flow (SCCM)	Power (w)
Low (−)	0.80	450	125	275
High (+)	1.20	550	200	325

Table 14-19 presents the data from the 16 runs of the 2^4 design. Table 14-20 is the table of plus and minus signs for the 2^4 design. The signs in the columns of this table can be used to estimate the factor effects. For example, the estimate of factor A is

$$A = \frac{1}{8}[a + ab + ac + abc + ad + abd + acd + abcd - (1) - b$$

$$- c - bc - d - bd - cd - bcd]$$

$$= \frac{1}{8}[669 + 650 + 642 + 635 + 749 + 868 + 860 + 729$$

$$- 550 - 604 - 633 - 601 - 1037 - 1052 - 1075 - 1063]$$

$$= -101.625$$

Table 14-19   The 2^4 Design for the Plasma Etch Experiment

A (Gap)	B (Pressure)	C (C_2F_6 Flow)	D (Power)	Etch Rate (Å/min)
−1	−1	−1	−1	550
1	−1	−1	−1	669
−1	1	−1	−1	604
1	1	−1	−1	650
−1	−1	1	−1	633
1	−1	1	−1	642
−1	1	1	−1	601
1	1	1	−1	635
−1	−1	−1	1	1037
1	−1	−1	1	749
−1	1	−1	1	1052
1	1	−1	1	868
−1	−1	1	1	1075
1	−1	1	1	860
−1	1	1	1	1063
1	1	1	1	729

Table 14-20   Contrast Constants for the 2^4 Design

	A	B	AB	C	AC	BC	ABC	D	AD	BD	ABD	CD	ACD	BCD	ABCD
(1)	−	−	+	−	+	+	−	−	+	+	−	+	−	−	+
a	+	−	−	−	−	+	+	−	−	+	+	+	+	−	−
b	−	+	−	−	+	−	+	−	+	−	+	+	−	+	−
ab	+	+	+	−	−	−	−	−	−	−	−	+	+	+	+
c	−	−	+	+	−	−	+	−	+	+	−	−	+	+	−
ac	+	−	−	+	+	−	−	−	−	+	+	−	−	+	+
bc	−	+	−	+	−	+	−	−	+	−	+	−	+	−	+
abc	+	+	+	+	+	+	+	−	−	−	−	−	−	−	+
d	−	−	+	−	+	+	−	+	−	−	+	−	+	+	−
ad	+	−	−	−	−	+	+	+	+	−	−	−	−	+	+
bd	−	+	−	−	+	−	+	+	−	+	−	−	+	−	+
abd	+	+	+	−	−	−	−	+	+	+	+	−	−	−	−
cd	−	−	+	+	−	−	+	+	−	−	+	+	−	−	+
acd	+	−	−	+	+	−	−	+	+	−	−	+	+	−	−
bcd	−	+	−	+	−	+	−	+	−	+	−	+	−	+	−
abcd	+	+	+	+	+	+	+	+	+	+	+	+	+	+	+

Thus, the effect of increasing the gap between the anode and the cathode from 0.80 to 1.20 centimeters is to decrease the etch rate by 101.625 angstroms per minute.

It is easy to verify (using Minitab, for example) that the complete set of effect estimates is

$$
\begin{aligned}
A   &= -101.625 & AD   &= -153.625 \\
B   &= -1.625   & BD   &= -0.625 \\
AB &= -7.875   & ABD &= 4.125 \\
C   &= 7.375   & CD   &= -2.125 \\
AC &= -24.875 & ACD &= 5.625 \\
BC &= -43.875 & BCD &= -25.375 \\
ABC &= -15.625 & ABCD &= -40.125 \\
D   &= 306.125
\end{aligned}
$$

The normal probability plot of these effects from the plasma etch experiment is shown in Fig. 14-21. Clearly, the main effects of A and D and the AD interaction are significant, because they fall far from the line passing through the other points. The analysis of variance summarized in Table 14-21 confirms these findings. Notice that in the analysis of variance we have pooled the three- and four-factor interactions to form the error mean square. If the normal probability plot had indicated that any of these interactions were important, they would not have been included in the error term.

Since $A = -101.625$, the effect of increasing the gap between the cathode and anode is to decrease the etch rate. However, $D = 306.125$; thus, applying higher power levels will increase the etch rate. Figure 14-22 is a plot of the AD interaction. This plot indicates that the effect of changing the gap width at low power settings is small, but that increasing the gap at high power settings dramatically reduces the etch rate. High etch rates are obtained at high power settings and narrow gap widths.

The residuals from the experiment can be obtained from the regression model

$$
\hat{y} = 776.0625 - \left(\frac{101.625}{2}\right)x_1 + \left(\frac{306.125}{2}\right)x_4 - \left(\frac{153.625}{2}\right)x_1 x_4
$$

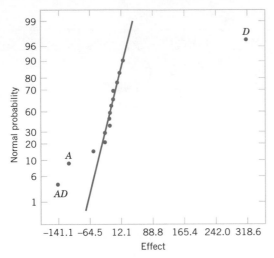

Figure 14-21   Normal probability plot of effects from the plasma etch experiment.

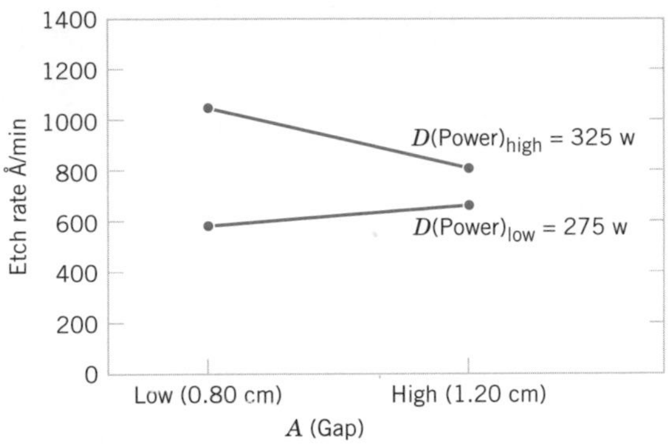

Figure 14-22   *AD* (Gap-Power) interaction from the plasma etch experiment.

For example, when both *A* and *D* are at the low level, the predicted value is

$$\hat{y} = 776.0625 - \left(\frac{101.625}{2}\right)(-1) + \left(\frac{306.125}{2}\right)(-1) - \left(\frac{153.625}{2}\right)(-1)(-1)$$
$$= 597$$

and the four residuals at this treatment combination are

$$e_1 = 550 - 597 = -47 \qquad e_2 = 604 - 597 = 7$$
$$e_3 = 633 - 597 = 36 \qquad e_4 = 601 - 597 = 4$$

The residuals at the other three treatment combinations (*A* high, *D* low), (*A* low, *D* high), and (*A* high, *D* high) are obtained similarly. A normal probability plot of the residuals is shown in Fig. 14-23. The plot is satisfactory.

Table 14-21   Analysis of Variance for the Plasma Etch Experiment

Source of Variation	Sum of Squares	Degrees of Freedom	Mean Square	f_0	P-Value
A	41,310.563	1	41,310.563	20.28	0.0064
B	10.563	1	10.563	<1	—
C	217.563	1	217.563	<1	—
D	374,850.063	1	374,850.063	183.99	0.0000
AB	248.063	1	248.063	<1	—
AC	2,475.063	1	2,475.063	1.21	0.3206
AD	94,402.563	1	94,402.563	46.34	0.0010
BC	7,700.063	1	7,700.063	3.78	0.1095
BD	1.563	1	1.563	<1	—
CD	18.063	1	18.063	<1	—
Error	10,186.813	5	2,037.363		
Total	531,420.938	15			

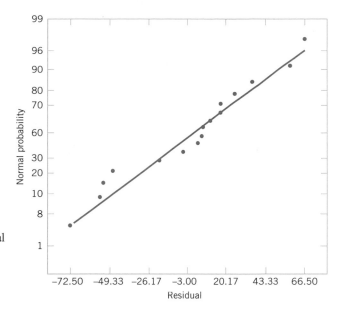

Figure 14-23 Normal probability plot of residuals from the plasma etch experiment.

14-7.4 Addition of Center Points to a 2^k Design (CD Only)

EXERCISES FOR SECTION 14-7

14-13. An engineer is interested in the effect of cutting speed (*A*), metal hardness (*B*), and cutting angle (*C*) on the life of a cutting tool. Two levels of each factor are chosen, and two replicates of a 2^3 factorial design are run. The tool life data (in hours) are shown in the following table:

Treatment Combination	Replicate	
	I	II
(1)	221	311
a	325	435
b	354	348
ab	552	472
c	440	453
ac	406	377
bc	605	500
abc	392	419

(a) Analyze the data from this experiment.
(b) Find an appropriate regression model that explains tool life in terms of the variables used in the experiment.
(c) Analyze the residuals from this experiment.

14-14. Four factors are thought to influence the taste of a soft-drink beverage: type of sweetener (*A*), ratio of syrup to water (*B*), carbonation level (*C*), and temperature (*D*). Each factor can be run at two levels, producing a 2^4 design. At each run in the design,

samples of the beverage are given to a test panel consisting of 20 people. Each tester assigns the beverage a point score from 1 to 10. Total score is the response variable, and the objective is to find a formulation that maximizes total score. Two replicates of this design are run, and the results are shown in the table. Analyze the data and draw conclusions. Use $\alpha = 0.05$ in the statistical tests.

Treatment Combination	Replicate	
	I	II
(1)	159	163
a	168	175
b	158	163
ab	166	168
c	175	178
ac	179	183
bc	173	168
abc	179	182
d	164	159
ad	187	189
bd	163	159
abd	185	191
cd	168	174
acd	197	199
bcd	170	174
abcd	194	198

14-15. Consider the experiment in Exercise 14-14. Determine an appropriate model and plot the residuals against the levels of factors *A*, *B*, *C*, and *D*. Also construct a normal probability plot of the residuals. Comment on these plots.

14-16. The data shown here represent a single replicate of a 2^5 design that is used in an experiment to study the compressive strength of concrete. The factors are mix (*A*), time (*B*), laboratory (*C*), temperature (*D*), and drying time (*E*).

(1)	=	700	*e*	=	800
a	=	900	*ae*	=	1200
b	=	3400	*be*	=	3500
ab	=	5500	*abe*	=	6200
c	=	600	*ce*	=	600
ac	=	1000	*ace*	=	1200
bc	=	3000	*bce*	=	3006
abc	=	5300	*abce*	=	5500
d	=	1000	*de*	=	1900
ad	=	1100	*ade*	=	1500
bd	=	3000	*bde*	=	4000
abd	=	6100	*abde*	=	6500
cd	=	800	*cde*	=	1500
acd	=	1100	*acde*	=	2000
bcd	=	3300	*bcde*	=	3400
abcd	=	6000	*abcde*	=	6800

(a) Estimate the factor effects.
(b) Which effects appear important? Use a normal probability plot.
(c) If it is desirable to maximize the strength, in which direction would you adjust the process variables?
(d) Analyze the residuals from this experiment.

14-17. An article in the *IEEE Transactions on Semiconductor Manufacturing* (Vol. 5, no. 3, 1992, pp. 214–222) describes an experiment to investigate the surface charge on a silicon wafer. The factors thought to influence induced surface charge are cleaning method (spin rinse dry or SRD and spin dry or SD) and the position on the wafer where the charge was measured. The surface charge ($\times 10^{11}$ q/cm^3) response data are as shown.

		Test Position	
		L	**R**
Cleaning Method	**SD**	1.66	1.84
		1.90	1.84
		1.92	1.62
	SRD	−4.21	−7.58
		−1.35	−2.20
		−2.08	−5.36

(a) Estimate the factor effects.
(b) Which factors appear important? Use $\alpha = 0.05$.
(c) Analyze the residuals from this experiment.

14-18. An experiment described by M. G. Natrella in the National Bureau of Standards *Handbook of Experimental Statistics* (No. 91, 1963) involves flame testing fabrics after applying fire-retardant treatments. The four factors considered are type of fabric (*A*), type of fire-retardant treatment (*B*), laundering condition (*C*—the low level is no laundering, the high level is after one laundering), and method of conducting the flame test (*D*). All factors are run at two levels, and the response variable is the inches of fabric burned on a standard size test sample. The data are:

(1)	= 42		*d*	= 40	
a	= 31		*ad*	= 30	
b	= 45		*bd*	= 50	
ab	= 29		*abd*	= 25	
c	= 39		*cd*	= 40	
ac	= 28		*acd*	= 25	
bc	= 46		*bcd*	= 50	
abc	= 32		*abcd*	= 23	

(a) Estimate the effects and prepare a normal plot of the effects.
(b) Construct an analysis of variance table based on the model tentatively identified in part (a).
(c) Construct a normal probability plot of the residuals and comment on the results.

14-19. An experiment was run in a semiconductor fabrication plant in an effort to increase yield. Five factors, each at two levels, were studied. The factors (and levels) were *A* = aperture setting (small, large), *B* = exposure time (20% below nominal, 20% above nominal), *C* = development time (30 and 45 seconds), *D* = mask dimension (small, large), and *E* = etch time (14.5 and 15.5 minutes). The following unreplicated 2^5 design was run:

(1)	=	7	*e*	=	8
a	=	9	*ae*	=	12
b	=	34	*be*	=	35
ab	=	55	*abe*	=	52
c	=	16	*ce*	=	15
ac	=	20	*ace*	=	22
bc	=	40	*bce*	=	45
abc	=	60	*abce*	=	65
d	=	8	*de*	=	6
ad	=	10	*ade*	=	10
bd	=	32	*bde*	=	30
abd	=	50	*abde*	=	53
cd	=	18	*cde*	=	15
acd	=	21	*acde*	=	20
bcd	=	44	*bcde*	=	41
abcd	=	61	*abcde*	=	63

(a) Construct a normal probability plot of the effect estimates. Which effects appear to be large?

(b) Conduct an analysis of variance to confirm your findings for part (a).

(c) Construct a normal probability plot of the residuals. Is the plot satisfactory?

(d) Plot the residuals versus the predicted yields and versus each of the five factors. Comment on the plots.

(e) Interpret any significant interactions.

(f) What are your recommendations regarding process operating conditions?

(g) Project the 2^5 design in this problem into a 2^r for $r < 5$ design in the important factors. Sketch the design and show the average and range of yields at each run. Does this sketch aid in data interpretation?

14-20. Consider the data from Exercise 14-13. I suppose that the data from the second replicate was not available. Analyze the data from replicate I only and comment on your findings.

14-21. An experiment has run a single replicate of a 2^4 design and calculated the following factor effects:

$A = 80.25$	$AB = 53.25$	$ABC = -2.95$
$B = -65.50$	$AC = 11.00$	$ABD = -8.00$
$C = -9.25$	$AD = 9.75$	$ACD = 10.25$
$D = -20.50$	$BC = 18.36$	$BCD = -7.95$
	$BD = 15.10$	$ABCD = -6.25$
	$CD = -1.25$	

(a) Construct a normal probability plot of the effects.

(b) Identify a tentative model, based on the plot of effects in part (a).

(c) Estimate the regression coefficients in this model, assuming that $\bar{y} = 400$.

14-22. A 2^4 factorial design was run in a chemical process. The design factors are A = time, B = concentration, C = pressure, and D = temperature. The response variable is

yield. The data follows:

Run	A	B	C	D	Yield (pounds)	Factor Levels	−	+
1	−	−	−	−	12	A (hours)	25	3
2	+	−	−	−	18	B (%)	14	18
3	−	+	−	−	13	C (psi)	60	80
4	+	+	−	−	16	D (°C)	200	250
5	−	−	+	−	17			
6	+	−	+	−	15			
7	−	+	+	−	20			
8	+	+	+	−	15			
9	−	−	−	+	10			
10	+	−	−	+	25			
11	−	+	−	+	13			
12	+	+	−	+	24			
13	−	−	+	+	19			
14	+	−	+	+	21			
15	−	+	+	+	17			
16	+	+	+	+	23			

(a) Estimate the factor effects. Based on a normal probability plot of the effect estimates, identify a model for the data from this experiment.

(b) Conduct an ANOVA based on the model identified in part (a). What are your conclusions?

(c) Analyze the residuals and comment on model adequacy.

(d) Find a regression model to predict yield in terms of the actual factor levels.

(e) Can this design be projected into a 2^3 design with two replicates? If so, sketch the design and show the average and range of the two yield values at each cube corner. Discuss the practical value of this plot.

14-8 BLOCKING AND CONFOUNDING IN THE 2^k DESIGN

It is often impossible to run all the observations in a 2^k factorial design under homogeneous conditions. Blocking is the design technique that is appropriate for this general situation. However, in many situations the block size is smaller than the number of runs in the complete replicate. In these cases, **confounding** is a useful procedure for running the 2^k design in 2^p blocks where the number of runs in a block is less than the number of treatment combinations in one complete replicate. The technique causes certain interaction effects to be indistinguishable from blocks or **confounded with blocks.** We will illustrate confounding in the 2^k factorial design in 2^p blocks, where $p < k$.

Consider a 2^2 design. Suppose that each of the $2^2 = 4$ treatment combinations requires four hours of laboratory analysis. Thus, two days are required to perform the experiment. If days are considered as blocks, we must assign two of the four treatment combinations to each day.

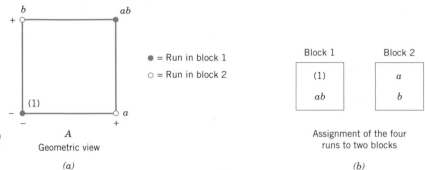

Figure 14-24   A 2^2 design in two blocks. (a) Geometric view. (b) Assignment of the four runs to two blocks.

This design is shown in Fig. 14-24. Notice that block 1 contains the treatment combinations (1) and ab and that block 2 contains a and b. The contrasts for estimating the main effects of factors A and B are

$$\text{Contrast}_A = ab + a - b - (1)$$

$$\text{Contrast}_B = ab + b - a - (1)$$

Note that these contrasts are unaffected by blocking since in each contrast there is one plus and one minus treatment combination from each block. That is, any difference between block 1 and block 2 that increases the readings in one block by an additive constant cancels out. The contrast for the AB interaction is

$$\text{Contrast}_{AB} = ab + (1) - a - b$$

Since the two treatment combinations with the plus signs, ab and (1), are in block 1 and the two with the minus signs, a and b, are in block 2, the block effect and the AB interaction are identical. That is, the AB interaction is confounded with blocks.

The reason for this is apparent from the table of plus and minus signs for the 2^2 design shown in Table 14-12. From the table we see that all treatment combinations that have a plus on AB are assigned to block 1, whereas all treatment combinations that have a minus sign on AB are assigned to block 2.

This scheme can be used to confound any 2^k design in two blocks. As a second example, consider a 2^3 design, run in two blocks. From the table of plus and minus signs, shown in Table 14-15, we assign the treatment combinations that are minus in the ABC column to block 1 and those that are plus in the ABC column to block 2. The resulting design is shown in Fig. 14-25.

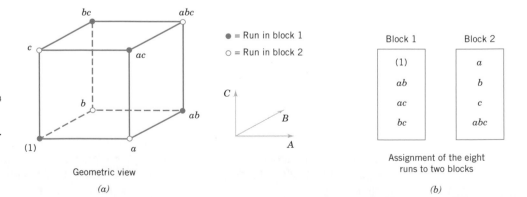

Figure 14-25   The 2^3 design in two blocks with ABC confounded. (a) Geometric View. (b) Assignment of the eight runs to two blocks.

There is a more general method of constructing the blocks. The method employs a **defining contrast,** say

$$L = \alpha_1 x_1 + \alpha_2 x_2 + \cdots + \alpha_k x_k \qquad (14\text{-}24)$$

where x_i is the level of the ith factor appearing in a treatment combination and α_i is the exponent appearing on the ith factor in the effect that is to be confounded with blocks. For the 2^k system, we have either $\alpha_i = 0$ or 1, and either $x_i = 0$ (low level) or $x_i = 1$ (high level). Treatment combinations that produce the same value of L (modulus 2) will be placed in the same block. Since the only possible values of L (mod 2) are 0 and 1, this will assign the 2^k treatment combinations to exactly two blocks.

As an example, consider the 2^3 design with ABC confounded with blocks. Here x_1 corresponds to A, x_2 to B, x_3 to C, and $\alpha_1 = \alpha_2 = \alpha_3 = 1$. Thus, the defining contrast that would be used to confound ABC with blocks is

$$L = x_1 + x_2 + x_3$$

To assign the treatment combinations to the two blocks, we substitute the treatment combinations into the defining contrast as follows:

$$(1)\colon L = 1(0) + 1(0) + 1(0) = 0 = 0 \ (\text{mod } 2)$$
$$a\colon \quad L = 1(1) + 1(0) + 1(0) = 1 = 1 \ (\text{mod } 2)$$
$$b\colon \quad L = 1(0) + 1(1) + 1(0) = 1 = 1 \ (\text{mod } 2)$$
$$ab\colon L = 1(1) + 1(1) + 1(0) = 2 = 0 \ (\text{mod } 2)$$
$$c\colon \quad L = 1(0) + 1(0) + 1(1) = 1 = 1 \ (\text{mod } 2)$$
$$ac\colon L = 1(1) + 1(0) + 1(1) = 2 = 0 \ (\text{mod } 2)$$
$$bc\colon L = 1(0) + 1(1) + 1(1) = 2 = 0 \ (\text{mod } 2)$$
$$abc\colon L = 1(1) + 1(1) + 1(1) = 3 = 1 \ (\text{mod } 2)$$

Thus (1), ab, ac, and bc are run in block 1, and a, b, c, and abc are run in block 2. This same design is shown in Fig. 14-25.

A shortcut method is useful in constructing these designs. The block containing the treatment combination (1) is called the **principal block.** Any element [except (1)] in the principal block may be generated by multiplying two other elements in the principal block modulus 2 on the exponents. For example, consider the principal block of the 2^3 design with ABC confounded, shown in Fig. 14-25. Note that

$$ab \cdot ac = a^2 bc = bc$$
$$ab \cdot bc = ab^2 c = ac$$
$$ac \cdot bc = abc^2 = ab$$

Treatment combinations in the other block (or blocks) may be generated by multiplying one element in the new block by each element in the principal block modulus 2 on the exponents. For the 2^3 with ABC confounded, since the principal block is (1), ab, ac, and bc, we know that the treatment combination b is in the other block. Thus, elements of this second block are

$$b \cdot (1) \quad\quad = b$$

$$b \cdot ab = ab^2 = a$$

$$b \cdot ac \quad\quad = abc$$

$$b \cdot bc = b^2c = c$$

EXAMPLE 14-6   An experiment is performed to investigate the effect of four factors on the terminal miss distance of a shoulder-fired ground-to-air-missile. The four factors are target type (A), seeker type (B), target altitude (C), and target range (D). Each factor may be conveniently run at two levels, and the optical tracking system will allow terminal miss distance to be measured to the nearest foot. Two different operators or gunners are used in the flight test and, since there may be differences between operators, the test engineers decided to conduct the 2^4 design in two blocks with $ABCD$ confounded. Thus, the defining contrast is

$$L = x_1 + x_2 + x_3 + x_4$$

The experimental design and the resulting data are shown in Fig. 14-26. The effect estimates obtained from Minitab are shown in Table 14-22. A normal probability plot of the effects in Fig. 14-27 reveals that A (target type), D (target range), AD, and AC have large effects. A confirming analysis of variance, pooling the three-factor interactions as error, is shown in Table 14-23. Since the AC and AD interactions are significant, it is logical to conclude that A (target type), C (target altitude), and D (target range) all have important effects on the miss distance and that there are interactions between target type and altitude and target type and range. Notice that the $ABCD$ effect is treated as blocks in this analysis.

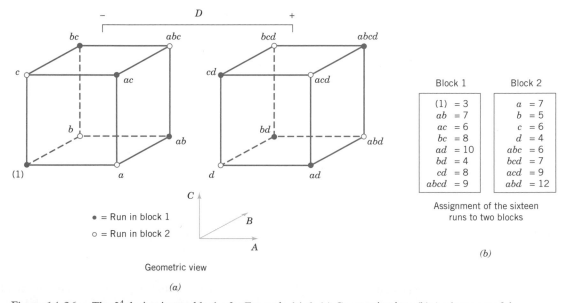

Figure 14-26   The 2^4 design in two blocks for Example 14-6. (a) Geometric view. (b) Assignment of the 16 runs to two blocks.

Table 14-22   Minitab Effect Estimates for Example 14-6

Estimated Effects and Coefficients for Distance		
Term	Effect	Coef
Constant		6.938
Block		0.063
A	2.625	1.312
B	0.625	0.313
C	0.875	0.438
D	1.875	0.938
AB	−0.125	−0.063
AC	−2.375	−1.187
AD	1.625	0.813
BC	−0.375	−0.188
BD	−0.375	−0.187
CD	−0.125	−0.062
ABC	−0.125	−0.063
ABD	0.875	0.438
ACD	−0.375	−0.187
BCD	−0.375	−0.187

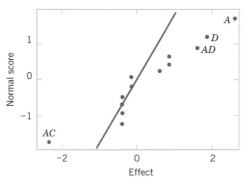

Figure 14-27   Normal probability plot of the effects from Minitab, Example 14-6.

It is possible to confound the 2^k design in four blocks of 2^{k-2} observations each. To construct the design, two effects are chosen to confound with blocks, and their defining contrasts are obtained. A third effect, the **generalized interaction** of the two effects initially chosen, is also confounded with blocks. The generalized interaction of two effects is found by multiplying their respective letters and reducing the exponents modulus 2.

For example, consider the 2^4 design in four blocks. If AC and BD are confounded with blocks, their generalized interaction is $(AC)(BD) = ABCD$. The design is constructed by using

Table 14-23   Analysis of Variance for Example 14-6

Source of Variation	Sum of Squares	Degrees of Freedom	Mean Square	f_0	P-Value
Blocks (*ABCD*)	0.0625	1	0.0625	0.06	—
A	27.5625	1	27.5625	25.94	0.0070
B	1.5625	1	1.5625	1.47	0.2920
C	3.0625	1	3.0625	2.88	0.1648
D	14.0625	1	14.0625	13.24	0.0220
AB	0.0625	1	0.0625	0.06	—
AC	22.5625	1	22.5625	21.24	0.0100
AD	10.5625	1	10.5625	9.94	0.0344
BC	0.5625	1	0.5625	0.53	—
BD	0.5625	1	0.5625	0.53	—
CD	0.0625	1	0.0625	0.06	—
Error (*ABC* + *ABD* + *ACD* + *BCD*)	4.2500	4	1.0625		
Total	84.9375	15			

the defining contrasts for AC and BD:

$$L_1 = x_1 + x_3$$
$$L_2 = x_2 + x_4$$

It is easy to verify that the four blocks are

Block 1 $L_1 = 0, L_2 = 0$	Block 2 $L_1 = 1, L_2 = 0$	Block 3 $L_1 = 0, L_2 = 1$	Block 4 $L_1 = 1, L_2 = 1$
(1)	a	b	ab
ac	c	abc	bc
bd	abd	d	ad
$abcd$	bcd	acd	cd

This general procedure can be extended to confounding the 2^k design in 2^p blocks, where $p < k$. Start by selecting p effects to be confounded, such that no effect chosen is a generalized interaction of the others. Then the blocks can be constructed from the p defining contrasts $L_1, L_2, \ldots, L_p$ that are associated with these effects. In addition to the p effects chosen to be confounded, exactly $2^p - p - 1$ additional effects are confounded with blocks; these are the generalized interactions of the original p effects chosen. Care should be taken so as not to confound effects of potential interest.

For more information on confounding in the 2^k factorial design, refer to Montgomery (2001, Chapter 7). This book contains guidelines for selecting factors to confound with blocks so that main effects and low-order interactions are not confounded. In particular, the book contains a table of suggested confounding schemes for designs with up to seven factors and a range of block sizes, some of which are as small as two runs.

EXERCISES FOR SECTION 14-8

14-23. Consider the data from the first replicate of Exercise 14-13. Suppose that these observations could not all be run under the same conditions. Set up a design to run these observations in two blocks of four observations each, with ABC confounded. Analyze the data.

14-24. Consider the data from the first replicate of Exercise 14-14. Construct a design with two blocks of eight observations each, with $ABCD$ confounded. Analyze the data.

14-25. Repeat Exercise 14-24 assuming that four blocks are required. Confound ABD and ABC (and consequently CD) with blocks.

14-26. Construct a 2^5 design in two blocks. Select the $ABCDE$ interaction to be confounded with blocks.

14-27. Construct a 2^5 design in four blocks. Select the appropriate effects to confound so that the highest possible interactions are confounded with blocks.

14-28. Consider the data from Exercise 14-18. Construct the design that would have been used to run this experiment in two blocks of eight runs each. Analyze the data and draw conclusions.

14-29. An article in *Industrial and Engineering Chemistry* ("Factorial Experiments in Pilot Plant Studies," 1951, pp. 1300–1306) reports on an experiment to investigate the effect of temperature (A), gas throughput (B), and concentration (C) on the strength of product solution in a recirculation unit. Two blocks were used with ABC confounded, and the experiment was replicated twice. The data are as follows:

Replicate 1	
Block 1	Block 2
(1) = 99	$a = 18$
$ab = 52$	$b = 51$
$ac = 42$	$c = 108$
$bc = 95$	$abc = 35$

Replicate 2	
Block 3	Block 4
(1) = 46	$a = 18$
$ab = 47$	$b = 62$
$ac = 22$	$c = 104$
$bc = 67$	$abc = 36$

(a) Analyze the data from this experiment.

(b) Analyze the residuals and comment on model adequacy.

(c) Comment on the efficiency of this design. Note that we have replicated the experiment twice, yet we have no information on the *ABC* interaction.

(d) Suggest a better design, specifically, one that would provide some information on *all* interactions.

14-30. Consider the 2^6 factorial design. Set up a design to be run in four blocks of 16 runs each. Show that a design that confounds three of the four-factor interactions with blocks is the best possible blocking arrangement.

14-9 FRACTIONAL REPLICATION OF THE 2^k DESIGN

As the number of factors in a 2^k factorial design increases, the number of runs required increases rapidly. For example, a 2^5 requires 32 runs. In this design, only 5 degrees of freedom correspond to main effects, and 10 degrees of freedom correspond to two-factor interactions. Sixteen of the 31 degrees of freedom are used to estimate high-order interactions—that is, three-factor and higher order interactions. Often there is little interest in these high-order interactions, particularly when we first begin to study a process or system. If we can assume that certain high-order interactions are negligible, a **fractional factorial design** involving fewer than the complete set of 2^k runs can be used to obtain information on the main effects and low-order interactions. In this section, we will introduce fractional replications of the 2^k design.

A major use of fractional factorials is in **screening experiments.** These are experiments in which many factors are considered with the purpose of identifying those factors (if any) that have large effects. Screening experiments are usually performed in the early stages of a project when it is likely that many of the factors initially considered have little or no effect on the response. The factors that are identified as important are then investigated more thoroughly in subsequent experiments.

14-9.1 One-Half Fraction of the 2^k Design

A one-half fraction of the 2^k design contains 2^{k-1} runs and is often called a 2^{k-1} fractional factorial design. As an example, consider the 2^{3-1} design—that is, a one-half fraction of the 2^3. This design has only four runs, in contrast to the full factorial that would require eight runs. The table of plus and minus signs for the 2^3 design is shown in Table 14-24. Suppose we select the four treatment combinations a, b, c, and abc, as our one-half fraction. These treatment combinations are shown in the top half of Table 14-24 and in Fig. 14-28(a).

Notice that the 2^{3-1} design is formed by selecting only those treatment combinations that yield a plus on the *ABC* effect. Thus, *ABC* is called the **generator** of this particular fraction.

Table 14-24 Plus and Minus Signs for the 2^3 Factorial Design

Treatment Combination	Factorial Effect							
	I	A	B	C	AB	AC	BC	ABC
a	+	+	−	−	−	−	+	+
b	+	−	+	−	−	+	−	+
c	+	−	−	+	+	−	−	+
abc	+	+	+	+	+	+	+	+
ab	+	+	+	−	+	−	−	−
ac	+	+	−	+	−	+	−	−
bc	+	−	+	+	−	−	+	−
(1)	+	−	−	−	+	+	+	−

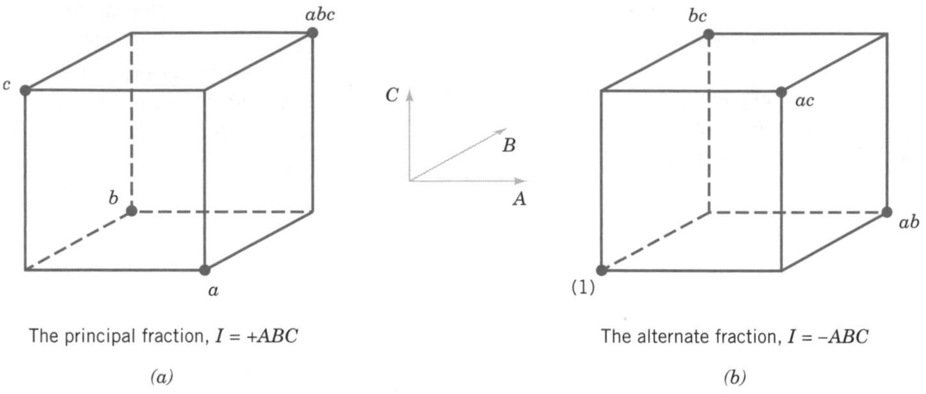

Figure 14-28 The one-half fractions of the 2^3 design. (a) The principal fraction, $I = +ABC$. (b) The alternate fraction, $I = -ABC$.

The principal fraction, $I = +ABC$

(a)

The alternate fraction, $I = -ABC$

(b)

Furthermore, the identity element I is also plus for the four runs, so we call

$$I = ABC$$

the defining relation for the design.

The treatment combinations in the 2^{3-1} design yields three degrees of freedom associated with the main effects. From the upper half of Table 14-24, we obtain the estimates of the main effects as linear combinations of the observations, say,

$$A = \tfrac{1}{2}[a - b - c + abc]$$
$$B = \tfrac{1}{2}[-a + b - c + abc]$$
$$C = \tfrac{1}{2}[-a - b + c + abc]$$

It is also easy to verify that the estimates of the two-factor interactions should be the following linear combinations of the observations:

$$BC = \tfrac{1}{2}[a - b - c + abc]$$
$$AC = \tfrac{1}{2}[-a + b - c + abc]$$
$$AB = \tfrac{1}{2}[-a - b + c + abc]$$

Thus, the linear combination of observations in column A, ℓ_A, estimates both the main effect of A and the BC interaction. That is, the linear combination ℓ_A estimates the sum of these two effects $A + BC$. Similarly, ℓ_B estimates $B + AC$, and ℓ_C estimates $C + AB$. Two or more effects that have this property are called **aliases.** In our 2^{3-1} design, A and BC are aliases, B and AC are aliases, and C and AB are aliases. Aliasing is the direct result of fractional replication. In many practical situations, it will be possible to select the fraction so that the main effects and low-order interactions that are of interest will be aliased only with high-order interactions (which are probably negligible).

The alias structure for this design is found by using the defining relation $I = ABC$. Multiplying any effect by the defining relation yields the aliases for that effect. In our example, the alias of A is

$$A = A \cdot ABC = A^2BC = BC$$

since $A \cdot I = A$ and $A^2 = I$. The aliases of B and C are

$$B = B \cdot ABC = AB^2C = AC$$

and

$$C = C \cdot ABC = ABC^2 = AB$$

Now suppose that we had chosen the other one-half fraction, that is, the treatment combinations in Table 14-24 associated with minus on ABC. These four runs are shown in the lower half of Table 14-24 and in Fig. 14-28(b). The defining relation for this design is $I = -ABC$. The aliases are $A = -BC$, $B = -AC$, and $C = -AB$. Thus, estimates of A, B, and C that result from this fraction really estimate $A - BC$, $B - AC$, and $C - AB$. In practice, it usually does not matter which one-half fraction we select. The fraction with the plus sign in the defining relation is usually called the **principal fraction,** and the other fraction is usually called the **alternate fraction.**

Note that if we had chosen AB as the generator for the fractional factorial,

$$A = A \cdot AB = B$$

and the two main effects of A and B would be aliased. This typically loses important information.

Sometimes we use **sequences** of fractional factorial designs to estimate effects. For example, suppose we had run the principal fraction of the 2^{3-1} design with generator ABC. From this design we have the following effect estimates:

$$\ell_A = A + BC$$

$$\ell_B = B + AC$$

$$\ell_C = C + AB$$

Suppose that we are willing to assume at this point that the two-factor interactions are negligible. If they are, the 2^{3-1} design has produced estimates of the three main effects A, B, and C. However, if after running the principal fraction we are uncertain about the interactions, it is possible to estimate them by running the *alternate* fraction. The alternate fraction produces the following effect estimates:

$$\ell_A' = A - BC$$

$$\ell_B' = B - AC$$

$$\ell_C' = C - AB$$

We may now obtain de-aliased estimates of the main effects and two-factor interactions by adding and subtracting the linear combinations of effects estimated in the two individual fractions. For example, suppose we want to de-alias A from the two-factor interaction BC. Since $\ell_A = A + BC$ and $\ell_A' = A - BC$, we can combine these effect estimates as follows:

$$\frac{1}{2}(\ell_A + \ell_A') = \frac{1}{2}(A + BC + A - BC) = A$$

and

$$\frac{1}{2}(\ell_A - \ell_A') = \frac{1}{2}(A + BC - A + BC) = BC$$

For all three pairs of effect estimates, we would obtain the following results:

Effect, i	from $\frac{1}{2}(l_i + l_i')$	from $\frac{1}{2}(l_i - l_i')$
$i = A$	$\frac{1}{2}(A + BC + A - BC) = A$	$\frac{1}{2}[A + BC - (A - BC)] = BC$
$i = B$	$\frac{1}{2}(B + AC + B - AC) = B$	$\frac{1}{2}[B + AC - (B - AC)] = AC$
$i = C$	$\frac{1}{2}(C + AB + C - AB) = C$	$\frac{1}{2}[C + AB - (C - AB)] = AB$

Thus, by combining a sequence of two fractional factorial designs, we can isolate both the main effects and the two-factor interactions. This property makes the fractional factorial design highly useful in experimental problems since we can run sequences of small, efficient experiments, combine information across *several* experiments, and take advantage of learning about the process we are experimenting with as we go along. This is an illustration of the concept of sequential experimentation.

A 2^{k-1} design may be constructed by writing down the treatment combinations for a full factorial with $k - 1$ factors, called the **basic design,** and then adding the kth factor by identifying its plus and minus levels with the plus and minus signs of the highest order interaction. Therefore, a 2^{3-1} fractional factorial is constructed by writing down the basic design as a full 2^2 factorial and then equating factor C with the $\pm AB$ interaction. Thus, to construct the principal fraction, we would use $C = +AB$ as follows:

Basic Design		Fractional Design		
Full 2^2		$2^{3-1}, I = +ABC$		
A	B	A	B	$C = AB$
$-$	$-$	$-$	$-$	$+$
$+$	$-$	$+$	$-$	$-$
$-$	$+$	$-$	$+$	$-$
$+$	$+$	$+$	$+$	$+$

To obtain the alternate fraction we would equate the last column to $C = -AB$.

EXAMPLE 14-7   To illustrate the use of a one-half fraction, consider the plasma etch experiment described in Example 14-5. Suppose that we decide to use a 2^{4-1} design with $I = ABCD$ to investigate the four factors gap (A), pressure (B), C_2F_6 flow rate (C), and power setting (D). This design would be constructed by writing down as the basic design a 2^3 in the factors A, B, and C and then setting the levels of the fourth factor $D = ABC$. The design and the resulting etch rates are shown in Table 14-25. The design is shown graphically in Fig. 14-29.

Table 14-25   The 2^{4-1} Design with Defining Relation $I = ABCD$

A	B	C	$D = ABC$	Treatment Combination	Etch Rate
$-$	$-$	$-$	$-$	(1)	550
$+$	$-$	$-$	$+$	ad	749
$-$	$+$	$-$	$+$	bd	1052
$+$	$+$	$-$	$-$	ab	650
$-$	$-$	$+$	$+$	cd	1075
$+$	$-$	$+$	$-$	ac	642
$-$	$+$	$+$	$-$	bc	601
$+$	$+$	$+$	$+$	$abcd$	729

In this design, the main effects are aliased with the three-factor interactions; note that the alias of A is

$$A \cdot I = A \cdot ABCD \qquad \text{or} \qquad A = A^2BCD = BCD$$

and similarly $B = ACD$, $C = ABD$, and $D = ABC$.

The two-factor interactions are aliased with each other. For example, the alias of AB is CD:

$$AB \cdot I = AB \cdot ABCD \qquad \text{or} \qquad AB = A^2B^2CD = CD$$

The other aliases are $AC = BD$ and $AD = BC$.

The estimates of the main effects and their aliases are found using the four columns of signs in Table 14-25. For example, from column A we obtain the estimated effect

$$\ell_A = A + BCD = \tfrac{1}{4}(-550 + 749 - 1052 + 650 - 1075 + 642 - 601 + 729)$$

$$= -127.00$$

The other columns produce

$$\ell_B = B + ACD = 4.00 \quad \ell_C = C + ABD = 11.50 \quad \text{and} \quad \ell_D = D + ABC = 290.50$$

Clearly, ℓ_A and ℓ_D are large, and if we believe that the three-factor interactions are negligible, the main effects A (gap) and D (power setting) significantly affect etch rate.

The interactions are estimated by forming the AB, AC, and AD columns and adding them to the table. For example, the signs in the AB column are $+$, $-$, $-$, $+$, $+$, $-$, $-$, $+$, and this column produces the estimate

$$\ell_{AB} = AB + CD = \tfrac{1}{4}(550 - 749 - 1052 + 650 + 1075 - 642 - 601 + 729) = -10$$

From the AC and AD columns we find

$$\ell_{AC} = AC + BD = -25.50 \qquad \text{and} \qquad \ell_{AD} = AD + BC = -197.50$$

The ℓ_{AD} estimate is large; the most straightforward interpretation of the results is that since A and D are large, this is the AD interaction. Thus, the results obtained from the 2^{4-1} design agree with the full factorial results in Example 14-5.

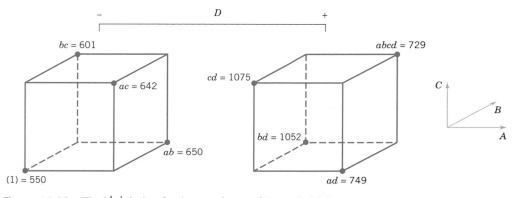

Figure 14-29   The 2^{4-1} design for the experiment of Example 14-7.

Table 14-26   Effect Estimates from Minitab,
Example 14-7

Fractional Factorial Fit		
Estimated Effects and Coefficients for Etch Rt		
Term	Effect	Coef
Constant		756.00
Gap	−127.00	−63.50
Pressure	4.00	2.00
F.	11.50	5.75
Power	290.50	145.25
Gap*Pressure	−10.00	−5.00
Gap*F.	−25.50	−12.75
Gap*Power	−197.50	−98.75

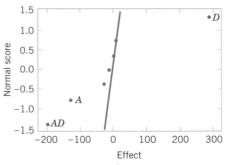

Figure 14-30   Normal probability plot of the effects from Minitab, Example 14-7.

Computer Solution

Fractional factorial designs are usually analyzed with a software package. Table 14-26 shows the effect estimates obtained from Minitab for Example 14-7. They are in agreement with the hand calculation reported earlier.

Normal Probability Plot of Effects

The normal probability plot is very useful in assessing the significance of effects from a fractional factorial design, particularly when many effects are to be estimated. We strongly recommend examining this plot. Figure 14-30 presents the normal probability plot of the effects from Example 14-7. This plot was obtained from Minitab. Notice that the A, D, and AD interaction effects stand out clearly in this graph.

Residual Analysis

The residuals can be obtained from a fractional factorial by the regression model method shown previously. Note that the Minitab output for Example 14-7 in Table 14-26 shows the regression coefficients. The residuals should be graphically analyzed as we have discussed before, both to assess the validity of the underlying model assumptions and to gain additional insight into the experimental situation.

Projection of the 2^{k-1} Design

If one or more factors from a one-half fraction of a 2^k can be dropped, the design will project into a full factorial design. For example, Fig. 14-31 presents a 2^{3-1} design. Notice that this design will project into a full factorial in any two of the three original factors. Thus, if we think that at most two of the three factors are important, the 2^{3-1} design is an excellent design for identifying the significant factors. This **projection property** is highly useful in factor screening, because it allows negligible factors to be eliminated, resulting in a stronger experiment in the active factors that remain.

In the 2^{4-1} design used in the plasma etch experiment in Example 14-7, we found that two of the four factors (B and C) could be dropped. If we eliminate these two factors, the remaining columns in Table 14-25 form a 2^2 design in the factors A and D, with two replicates. This design is shown in Fig. 14-32. The main effects of A and D and the strong two-factor AD interaction are clearly evident from this graph.

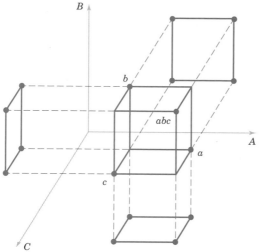

Figure 14-31   Projection of a 2^{3-1} design into three 2^2 designs.

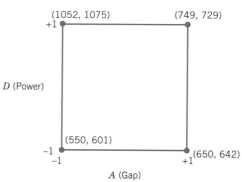

Figure 14-32   The 2^2 design obtained by dropping factors B and C from the plasma etch experiment in Example 14-7.

Design Resolution

The concept of design resolution is a useful way to catalog fractional factorial designs according to the alias patterns they produce. Designs of resolution III, IV, and V are particularly important. The definitions of these terms and an example of each follow.

1. **Resolution III Designs.** These are designs in which no main effects are aliased with any other main effect, but main effects are aliased with two-factor interactions and some two-factor interactions may be aliased with each other. The 2^{3-1} design with $I = ABC$ is a resolution III design. We usually employ a Roman numeral subscript to indicate design resolution; thus, this one-half fraction is a 2^{3-1}_{III} design.

2. **Resolution IV Designs.** These are designs in which no main effect is aliased with any other main effect or two-factor interactions, but two-factor interactions are aliased with each other. The 2^{4-1} design with $I = ABCD$ used in Example 14-7 is a resolution IV design (2^{4-1}_{IV}).

3. **Resolution V Designs.** These are designs in which no main effect or two-factor interaction is aliased with any other main effect or two-factor interaction, but two-factor interactions are aliased with three-factor interactions. The 2^{5-1} design with $I = ABCDE$ is a resolution V design (2^{5-1}_{V}).

Resolution III and IV designs are particularly useful in factor screening experiments. A resolution IV design provides good information about main effects and will provide some information about all two-factor interactions.

14-9.2   Smaller Fractions: The 2^{k-p} Fractional Factorial

Although the 2^{k-1} design is valuable in reducing the number of runs required for an experiment, we frequently find that smaller fractions will provide almost as much useful information at even greater economy. In general, a 2^k design may be run in a $1/2^p$ fraction called a 2^{k-p} fractional factorial design. Thus, a $1/4$ fraction is called a 2^{k-2} design, a $1/8$ fraction is called a 2^{k-3} design, a $1/16$ fraction a 2^{k-4} design, and so on.

To illustrate the 1/4 fraction, consider an experiment with six factors and suppose that the engineer is primarily interested in main effects but would also like to get some information about the two-factor interactions. A 2^{6-1} design would require 32 runs and would have 31 degrees of freedom for estimating effects. Since there are only six main effects and 15 two-factor interactions, the one-half fraction is inefficient—it requires too many runs. Suppose we consider a 1/4 fraction, or a 2^{6-2} design. This design contains 16 runs and, with 15 degrees of freedom, will allow all six main effects to be estimated, with some capability for examining the two-factor interactions.

To generate this design, we would write down a 2^4 design in the factors A, B, C, and D as the basic design and then add two columns for E and F. To find the new columns we could select the two **design generators** $I = ABCE$ and $I = BCDF$. Thus, column E would be found from $E = ABC$, and column F would be $F = BCD$. That is, columns $ABCE$ and $BCDF$ are equal to the identity column. However, we know that the product of any two columns in the table of plus and minus signs for a 2^k design is just another column in the table; therefore, the product of $ABCE$ and $BCDF$ or $ABCE(BCDF) = AB^2C^2DEF = ADEF$ is also an identity column. Consequently, the **complete defining relation** for the 2^{6-2} design is

$$I = ABCE = BCDF = ADEF$$

We refer to each term in a defining relation (such as $ABCE$ above) as a **word.** To find the alias of any effect, simply multiply the effect by each word in the foregoing defining relation. For example, the alias of A is

$$A = BCE = ABCDF = DEF$$

The complete alias relationships for this design are shown in Table 14-27. In general, the resolution of a 2^{k-p} design is equal to the number of letters in the shortest word in the complete defining relation. Therefore, this is a resolution IV design; main effects are aliased with three-factor and higher interactions, and two-factor interactions are aliased with each other. This design would provide good information on the main effects and would give some idea about the strength of the two-factor interactions. The construction and analysis of the design are illustrated in Example 14-8.

EXAMPLE 14-8   Parts manufactured in an injection-molding process are showing excessive shrinkage, which is causing problems in assembly operations upstream from the injection-molding area. In an effort to reduce the shrinkage, a quality-improvement team has decided to use a designed experiment to study the injection-molding process. The team investigates six factors—mold temperature (A), screw speed (B), holding time (C), cycle time (D), gate size (E), and holding

Table 14-27   Alias Structure for the 2_{IV}^{6-2} Design with $I = ABCE = BCDF = ADEF$

$A = BCE = DEF = ABCDF$	$AB = CE = ACDF = BDEF$
$B = ACE = CDF = ABDEF$	$AC = BE = ABDF = CDEF$
$C = ABE = BDF = ACDEF$	$AD = EF = BCDE = ABCF$
$D = BCF = AEF = ABCDE$	$AE = BC = DF = ABCDEF$
$E = ABC = ADF = BCDEF$	$AF = DE = BCEF = ABCD$
$F = BCD = ADE = ABCEF$	$BD = CF = ACDE = ABEF$
$ABD = CDE = ACF = BEF$	$BF = CD = ACEF = ABDE$
$ACD = BDE = ABF = CEF$	

Table 14-28   A 2_{IV}^{6-2} Design for the Injection-Molding Experiment in Example 14-8

Run	A	B	C	D	$E = ABC$	$F = BCD$	Observed Shrinkage ($\times 10$)
1	−	−	−	−	−	−	6
2	+	−	−	−	+	−	10
3	−	+	−	−	+	+	32
4	+	+	−	−	−	+	60
5	−	−	+	−	+	+	4
6	+	−	+	−	−	+	15
7	−	+	+	−	−	−	26
8	+	+	+	−	+	−	60
9	−	−	−	+	−	+	8
10	+	−	−	+	+	+	12
11	−	+	−	+	+	−	34
12	+	+	−	+	−	−	60
13	−	−	+	+	+	−	16
14	+	−	+	+	−	−	5
15	−	+	+	+	−	+	37
16	+	+	+	+	+	+	52

pressure (F)—each at two levels, with the objective of learning how each factor affects shrinkage and obtaining preliminary information about how the factors interact.

The team decides to use a 16-run two-level fractional factorial design for these six factors. The design is constructed by writing down a 2^4 as the basic design in the factors A, B, C, and D and then setting $E = ABC$ and $F = BCD$ as discussed above. Table 14-28 shows the design, along with the observed shrinkage ($\times 10$) for the test part produced at each of the 16 runs in the design.

A normal probability plot of the effect estimates from this experiment is shown in Fig. 14-33. The only large effects are A (mold temperature), B (screw speed), and the AB interaction. In light of the alias relationship in Table 14-27, it seems reasonable to tentatively adopt these conclusions. The plot of the AB interaction in Fig. 14-34 shows that the process is

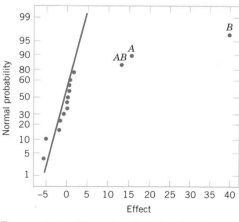

Figure 14-33   Normal probability plot of effects for Example 14-8.

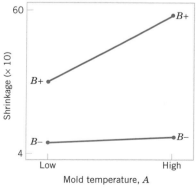

Figure 14-34   Plot of AB (mold temperature–screw speed) interaction for Example 14-8.

insensitive to temperature if the screw speed is at the low level but sensitive to temperature if the screw speed is at the high level. With the screw speed at a low level, the process should produce an average shrinkage of around 10% regardless of the temperature level chosen.

Based on this initial analysis, the team decides to set both the mold temperature and the screw speed at the low level. This set of conditions should reduce the mean shrinkage of parts to around 10%. However, the variability in shrinkage from part to part is still a potential problem. In effect, the mean shrinkage can be adequately reduced by the above modifications; however, the part-to-part variability in shrinkage over a production run could still cause problems in assembly. One way to address this issue is to see if any of the process factors affect the variability in parts shrinkage.

Figure 14-35 presents the normal probability plot of the residuals. This plot appears satisfactory. The plots of residuals versus each factor were then constructed. One of these plots, that for residuals versus factor C (holding time), is shown in Fig. 14-36. The plot reveals much less scatter in the residuals at the low holding time than at the high holding time. These residuals were obtained in the usual way from a model for predicted shrinkage

$$\hat{y} = \hat{\beta}_0 + \hat{\beta}_1 x_1 + \hat{\beta}_2 x_2 + \hat{\beta}_{12} x_1 x_2$$
$$= 27.3125 + 6.9375 x_1 + 17.8125 x_2 + 5.9375 x_1 x_2$$

where x_1, x_2, and $x_1 x_2$ are coded variables that correspond to the factors A and B and the AB interaction. The residuals are then

$$e = y - \hat{y}$$

The regression model used to produce the residuals essentially removes the location effects of A, B, and AB from the data; the residuals therefore contain information about unexplained variability. Figure 14-36 indicates that there is a pattern in the variability and

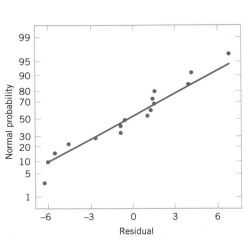

Figure 14-35   Normal probability plot of residuals for Example 14-8.

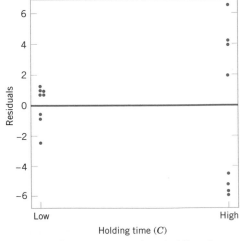

Figure 14-36   Residuals versus holding time (C) for Example 14-8.

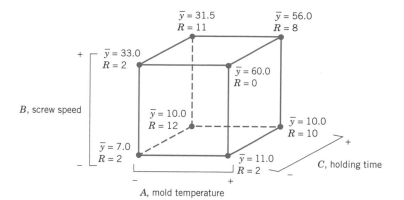

Figure 14-37 Average shrinkage and range of shrinkage in factors A, B, and C for Example 14-8.

that the variability in the shrinkage of parts may be smaller when the holding time is at the low level.

Figure 14-37 shows the data from this experiment projected onto a cube in the factors A, B, and C. The average observed shrinkage and the range of observed shrinkage are shown at each corner of the cube. From inspection of this figure, we see that running the process with the screw speed (B) at the low level is the key to reducing average parts shrinkage. If B is low, virtually any combination of temperature (A) and holding time (C) will result in low values of average parts shrinkage. However, from examining the ranges of the shrinkage values at each corner of the cube, it is immediately clear that setting the holding time (C) at the low level is the most appropriate choice if we wish to keep the part-to-part variability in shrinkage low during a production run.

The concepts used in constructing the 2^{6-2} fractional factorial design in Example 14-8 can be extended to the construction of any 2^{k-p} fractional factorial design. In general, a 2^k fractional factorial design containing 2^{k-p} runs is called a $1/2^p$ fraction of the 2^k design or, more simply, a 2^{k-p} fractional factorial design. These designs require the selection of p independent generators. The defining relation for the design consists of the p generators initially chosen and their $2^p - p - 1$ generalized interactions.

The alias structure may be found by multiplying each effect column by the defining relation. Care should be exercised in choosing the generators so that effects of potential interest are not aliased with each other. Each effect has $2^p - 1$ aliases. For moderately large values of k, we usually assume higher order interactions (say, third- or fourth-order or higher) to be negligible, and this greatly simplifies the alias structure.

It is important to select the p generators for the 2^{k-p} fractional factorial design in such a way that we obtain the best possible alias relationships. A reasonable criterion is to select the generators so that the resulting 2^{k-p} design has the highest possible design resolution. Montgomery (2001) presents a table of recommended generators for 2^{k-p} fractional factorial designs for $k \leq 15$ factors and up to as many as $n \leq 128$ runs. A portion of his table is reproduced here as Table 14-29. In this table, the generators are shown with either $+$ or $-$ choices; selection of all generators as $+$ will give a principal fraction, while if any generators are $-$ choices, the design will be one of the alternate fractions for the same family. The suggested generators in this table will result in a design of the highest possible resolution. Montgomery (2001) also gives a table of alias relationships for these designs.

Table 14-29 Selected 2^{k-p} Fractional Factorial Designs

Number of Factors k	Fraction	Number of Runs	Design Generators	Number of Factors k	Fraction	Number of Runs	Design Generators
3	2_{III}^{3-1}	4	$C = \pm AB$	10			$H = \pm ABCG$
4	2_{IV}^{4-1}	8	$D = \pm ABC$				$J = \pm ACDE$
5	2_{V}^{5-1}	16	$E = \pm ABCD$		2_{V}^{10-3}	128	$K = \pm ACDF$
	2_{III}^{5-2}	8	$D = \pm AB$				$G = \pm BCDF$
			$E = \pm AC$				$H = \pm ACDF$
6	2_{VI}^{6-1}	32	$F = \pm ABCDE$				$J = \pm ABDE$
	2_{IV}^{6-2}	16	$E = \pm ABC$		2_{IV}^{10-4}	64	$K = \pm ABCE$
			$F = \pm BCD$				$F = \pm ABCD$
	2_{III}^{6-3}	8	$D = \pm AB$				$G = \pm ABCE$
			$E = \pm AC$				$H = \pm ABDE$
			$F = \pm BC$				$J = \pm ACDE$
7	2_{VII}^{7-1}	64	$G = \pm ABCDEF$		2_{IV}^{10-5}	32	$K = \pm BCDE$
	2_{IV}^{7-2}	32	$E = \pm ABC$				$E = \pm ABC$
			$G = \pm ABDE$				$F = \pm BCD$
	2_{IV}^{7-3}	16	$E = \pm ABC$				$G = \pm ACD$
			$F = \pm BCD$				$H = \pm ABD$
			$G = \pm ACD$				$J = \pm ABCD$
	2_{III}^{7-4}	8	$D = \pm AB$		2_{III}^{10-6}	16	$K = \pm AB$
			$E = \pm AC$	11			$G = \pm CDE$
			$F = \pm BC$				$H = \pm ABCD$
			$G = \pm ABC$				$J = \pm ABF$
8	2_{V}^{8-2}	64	$G = \pm ABCD$				$K = \pm BDEF$
			$H = \pm ABEF$		2_{IV}^{11-5}	64	$L = \pm ADEF$
	2_{IV}^{8-3}	32	$F = \pm ABC$				$F = \pm ABC$
			$G = \pm ABD$				$G = \pm BCD$
			$H = \pm BCDE$				$H = \pm CDE$
	2_{IV}^{8-4}	16	$E = \pm BCD$				$J = \pm ACD$
			$F = \pm ACD$				$K = \pm ADE$
			$G = \pm ABC$		2_{IV}^{11-6}	32	$L = \pm BDE$
			$H = \pm ABD$				$E = \pm ABC$
9	2_{VI}^{9-2}	128	$H = \pm ACDFG$				$F = \pm BCD$
			$J = \pm BCEFG$				$G = \pm ACD$
	2_{IV}^{9-3}	64	$G = \pm ABCD$				$H = \pm ABD$
			$H = \pm ACEF$				$J = \pm ABCD$
			$J = \pm CDEF$				$K = \pm AB$
	2_{IV}^{9-4}	32	$F = \pm BCDE$		2_{III}^{11-7}	16	$L = \pm AC$
			$G = \pm ACDE$				
			$H = \pm ABDE$				
			$J = \pm ABCE$				
	2_{III}^{9-5}	16	$E = \pm ABC$				
			$F = \pm BCD$				
			$G = \pm ACD$				
			$H = \pm ABD$				
			$J = \pm ABCD$				

Source: Montgomery (2001)

EXAMPLE 14-9

To illustrate the use of Table 14-29, suppose that we have seven factors and that we are interested in estimating the seven main effects and obtaining some insight regarding the two-factor interactions. We are willing to assume that three-factor and higher interactions are negligible. This information suggests that a resolution IV design would be appropriate.

Table 14-29 shows that two resolution IV fractions are available: the 2_{IV}^{7-2} with 32 runs and the 2_{IV}^{7-3} with 16 runs. The aliases involving main effects and two- and three-factor interactions for the 16-run design are presented in Table 14-30. Notice that all seven main effects are aliased with three-factor interactions. All the two-factor interactions are aliased in groups of three. Therefore, this design will satisfy our objectives; that is, it will allow the estimation of the main effects, and it will give some insight regarding two-factor interactions. It is not necessary to run the 2_{IV}^{7-2} design, which would require 32 runs. The construction of the 2_{IV}^{7-3} design is shown in Table 14-31. Notice that it was constructed by starting with the 16-run 2^4 design in A, B, C, and D as the basic design and then adding the three columns $E = ABC$, $F = BCD$, and $G = ACD$ as suggested in Table 14-29. Thus, the generators for this design are $I = ABCE$, $I = BCDF$, and $I = ACDG$. The complete defining relation is $I = ABCE = BCDF = ADEF = ACDG = BDEG = CEFG = ABFG$. This defining relation was used to produce the aliases in Table 14-30. For example, the alias relationship of A is

$$A = BCE = ABCDF = DEF = CDG = ABDEG = ACEFG = BFG$$

which, if we ignore interactions higher than three factors, agrees with Table 14-30.

For seven factors, we can reduce the number of runs even further. The 2^{7-4} design is an eight-run experiment accommodating seven variables. This is a 1/16th fraction and is obtained by first writing down a 2^3 design as the basic design in the factors A, B, and C, and then forming the four new columns from $I = ABD$, $I = ACE$, $I = BCF$, and $I = ABCG$, as suggested in Table 14-29. The design is shown in Table 14-32.

The complete defining relation is found by multiplying the generators together two, three, and finally four at a time, producing

$$I = ABD = ACE = BCF = ABCG = BCDE = ACDF = CDG = ABEF$$
$$= BEG = AFG = DEF = ADEG = CEFG = BDFG = ABCDEFG$$

The alias of any main effect is found by multiplying that effect through each term in the

Table 14-30   Generators, Defining Relation, and Aliases for the 2_{IV}^{7-3} Fractional Factorial Design

Generators and Defining Relation		
$E = ABC$, $F = BCD$, $G = ACD$		
$I = ABCE = BCDF = ADEF = ACDG = BDEG = ABFG = CEFG$		
Aliases		
$A = BCE = DEF = CDG = BFG$		$AB = CE = FG$
$B = ACE = CDF = DEG = AFG$		$AC = BE = DG$
$C = ABE = BDF = ADG = EFG$		$AD = EF = CG$
$D = BCF = AEF = ACG = BEG$		$AE = BC = DF$
$E = ABC = ADF = BDG = CFG$		$AF = DE = BG$
$F = BCD = ADE = ABG = CEG$		$AG = CD = BF$
$G = ACD = BDE = ABF = CEF$		$BD = CF = EG$
$ABD = CDE = ACF = BEF = BCG = AEG = DFG$		

Table 14-31   A 2_{IV}^{7-3} Fractional Factorial Design

Run	Basic Design				$E = ABC$	$F = BCD$	$G = ACD$
	A	B	C	D			
1	−	−	−	−	−	−	−
2	+	−	−	−	+	−	+
3	−	+	−	−	+	+	−
4	+	+	−	−	−	+	+
5	−	−	+	−	+	+	+
6	+	−	+	−	−	+	−
7	−	+	+	−	−	−	+
8	+	+	+	−	+	−	−
9	−	−	−	+	−	+	−
10	+	−	−	+	+	+	−
11	−	+	−	+	+	−	+
12	+	+	−	+	−	−	−
13	−	−	+	+	+	−	−
14	+	−	+	+	−	−	+
15	−	+	+	+	−	+	−
16	+	+	+	+	+	+	+

defining relation. For example, the alias of A is

$$A = BD = CE = ABCF = BCG = ABCDE = CDF = ACDG$$
$$= BEF = ABEG = FG = ADEF = DEG = ACEFG = ABDFG = BCDEFG$$

This design is of resolution III, since the main effect is aliased with two-factor interactions. If we assume that all three-factor and higher interactions are negligible, the aliases of the seven main effects are

$$\ell_A = A + BD + CE + FG$$

$$\ell_B = B + AD + CF + EG$$

$$\ell_C = C + AE + BF + DG$$

$$\ell_D = D + AB + CG + EF$$

$$\ell_E = E + AC + BG + DF$$

$$\ell_F = F + BC + AG + DE$$

$$\ell_G = G + CD + BE + AF$$

Table 14-32   A 2_{III}^{7-4} Fractional Factorial Design

A	B	C	$D = AB$	$E = AC$	$F = BC$	$G = ABC$
−	−	−	+	+	+	−
+	−	−	−	−	+	+
−	+	−	−	+	−	+
+	+	−	+	−	−	−
−	−	+	+	−	−	+
+	−	+	−	+	−	−
−	+	+	−	−	+	−
+	+	+	+	+	+	+

This 2_{III}^{7-4} design is called a **saturated fractional factorial,** because all the available degrees of freedom are used to estimate main effects. It is possible to combine sequences of these resolution III fractional factorials to separate the main effects from the two-factor interactions. The procedure is illustrated in Montgomery (2001) and in Box, Hunter, and Hunter (1978).

EXERCISES FOR SECTION 14-9

14-31.   R. D. Snee ("Experimenting with a Large Number of Variables," in *Experiments in Industry: Design, Analysis and Interpretation of Results,* by R. D. Snee, L. D. Hare, and J. B. Trout, eds., ASQC, 1985) describes an experiment in which a 2^{5-1} design with $I = ABCDE$ was used to investigate the effects of five factors on the color of a chemical product. The factors are A = solvent/reactant, B = catalyst/reactant, C = temperature, D = reactant purity, and E = reactant pH. The results obtained are as follows:

$e = -0.63$	$d = 6.79$
$a = 2.51$	$ade = 6.47$
$b = -2.68$	$bde = 3.45$
$abe = 1.66$	$abd = 5.68$
$c = 2.06$	$cde = 5.22$
$ace = 1.22$	$acd = 4.38$
$bce = -2.09$	$bcd = 4.30$
$abc = 1.93$	$abcde = 4.05$

(a) Prepare a normal probability plot of the effects. Which factors are active?

(b) Calculate the residuals. Construct a normal probability plot of the residuals and plot the residuals versus the fitted values. Comment on the plots.

(c) If any factors are negligible, collapse the 2^{5-1} design into a full factorial in the active factors. Comment on the resulting design, and interpret the results.

14-32.   Montgomery (2001) describes a 2^{4-1} fractional factorial design used to study four factors in a chemical process. The factors are A = temperature, B = pressure, C = concentration, and D = stirring rate, and the response is filtration rate. The design and the data are as follows:

(a) Write down the alias relationships.

(b) Estimate the factor effects. Which factor effects appear large?

(c) Project this design into a full factorial in the three apparently important factors and provide a practical interpretation of the results.

14-33.   An article in *Industrial and Engineering Chemistry* ("More on Planning Experiments to Increase Research Efficiency," 1970, pp. 60–65) uses a 2^{5-2} design to investigate the effect on process yield of A = condensation temperature, B = amount of material 1, C = solvent volume, D = condensation time, and E = amount of material 2. The results obtained are as follows:

$e = 23.2$	$cd = 23.8$
$ab = 15.5$	$ace = 23.4$
$ad = 16.9$	$bde = 16.8$
$bc = 16.2$	$abcde = 18.1$

(a) Verify that the design generators used were $I = ACE$ and $I = BDE$.

(b) Write down the complete defining relation and the aliases from the design.

(c) Estimate the main effects.

(d) Prepare an analysis of variance table. Verify that the AB and AD interactions are available to use as error.

(e) Plot the residuals versus the fitted values. Also construct a normal probability plot of the residuals. Comment on the results.

Run	A	B	C	$D = ABC$	Treatment Combination	Filtration Rate
1	−	−	−	−	(1)	45
2	+	−	−	+	ad	100
3	−	+	−	+	bd	45
4	+	+	−	−	ab	65
5	−	−	+	+	cd	75
6	+	−	+	−	ac	60
7	−	+	+	−	bc	80
8	+	+	+	+	$abcd$	96

14-34. Consider the 2^{6-2} design in Table 14-28. Suppose that after analyzing the original data, we find that factors C and E can be dropped. What type of 2^k design is left in the remaining variables?

14-35. Consider the 2^{6-2} design in Table 14-28. Suppose that after the original data analysis, we find that factors D and F can be dropped. What type of 2^k design is left in the remaining variables? Compare the results with Exercise 14-34. Can you explain why the answers are different?

14-36. Suppose that in Exercise 14-22 it was possible to run only a $\frac{1}{2}$ fraction of the 2^4 design. Construct the design and use only the data from the eight runs you have generated to perform the analysis.

14-37. Suppose that in Exercise 14-16 only a ¼ fraction of the 2^5 design could be run. Construct the design and analyze the data that are obtained by selecting only the response for the eight runs in your design.

14-38. Construct the 2^{8-4}_{IV} design recommended in Table 14-29. What are the aliases of the main effects and two-factor interactions?

14-39. Construct a 2^{6-3}_{III} fractional factorial design. Write down the aliases, assuming that only main effects and two-factor interactions are of interest.

14-40. Consider the problem in Exercise 14-19. Suppose that only half of the 32 runs could be made.
(a) Choose the half that you think should be run.
(b) Write out the alias relationships for your design.
(c) Estimate the factor effects.
(d) Plot the effect estimates on normal probability paper and interpret the results.
(e) Set up an analysis of variance for the factors identified as potentially interesting from the normal probability plot in part (d).
(f) Analyze the residuals from the model.
(g) Provide a practical interpretation of the results.

14-10 RESPONSE SURFACE METHODS AND DESIGNS (CD ONLY)

Supplemental Exercises

14-41. An article in *Process Engineering* (No. 71, 1992, pp. 46–47) presents a two-factor factorial experiment used to investigate the effect of pH and catalyst concentration on product viscosity (cSt). The data are as follows:

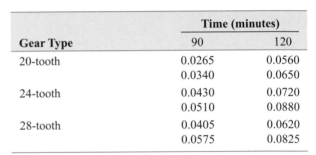

	Catalyst Concentration	
	2.5	2.7
pH 5.6	192, 199, 189, 198	178, 186, 179, 188
5.9	185, 193, 185, 192	197, 196, 204, 204

(a) Test for main effects and interactions using $\alpha = 0.05$. What are your conclusions?
(b) Graph the interaction and discuss the information provided by this plot.
(c) Analyze the residuals from this experiment.

14-42. Heat treating of metal parts is a widely used manufacturing process. An article in the *Journal of Metals* (Vol. 41, 1989) describes an experiment to investigate flatness distortion from heat treating for three types of gears and two heat-treating times. The data are as follows:

	Time (minutes)	
Gear Type	90	120
20-tooth	0.0265	0.0560
	0.0340	0.0650
24-tooth	0.0430	0.0720
	0.0510	0.0880
28-tooth	0.0405	0.0620
	0.0575	0.0825

(a) Is there any evidence that flatness distortion is different for the different gear types? Is there any indication that heat treating time affects the flatness distortion? Do these factors interact? Use $\alpha = 0.05$.
(b) Construct graphs of the factor effects that aid in drawing conclusions from this experiment.
(c) Analyze the residuals from this experiment. Comment on the validity of the underlying assumptions.

14-43. An article in the *Textile Research Institute Journal* (Vol. 54, 1984, pp. 171–179) reported the results of an experiment that studied the effects of treating fabric with selected inorganic salts on the flammability of the material. Two application levels of each salt were used, and a vertical burn test was used on each sample. (This finds the temperature at which each sample ignites.) The burn test data follow.

	Salt					
Level	Untreated	MgCl$_2$	NaCl	CaCO$_3$	CaCl$_2$	Na$_2$CO$_3$
1	812	752	739	733	725	751
	827	728	731	728	727	761
	876	764	726	720	719	755
2	945	794	741	786	756	910
	881	760	744	771	781	854
	919	757	727	779	814	848

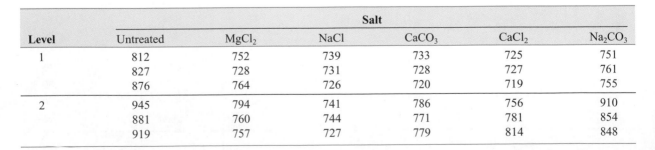

(a) Test for differences between salts, application levels, and interactions. Use $\alpha = 0.01$.

(b) Draw a graph of the interaction between salt and application level. What conclusions can you draw from this graph?

(c) Analyze the residuals from this experiment.

14-44. An article in the *IEEE Transactions on Components, Hybrids, and Manufacturing Technology* (Vol. 15, 1992) describes an experiment for investigating a method for aligning optical chips onto circuit boards. The method involves placing solder bumps onto the bottom of the chip. The experiment used three solder bump sizes and three alignment methods. The response variable is alignment accuracy (in micrometers). The data are as follows:

Solder Bump Size	Alignment Method		
(diameter in μm)	1	2	3
75	4.60	1.55	1.05
	4.53	1.45	1.00
130	2.33	1.72	0.82
	2.44	1.76	0.95
260	4.95	2.73	2.36
	4.55	2.60	2.46

(a) Is there any indication that either solder bump size or alignment method affects the alignment accuracy? Is there any evidence of interaction between these factors? Use $\alpha = 0.05$.

(b) What recommendations would you make about this process?

(c) Analyze the residuals from this experiment. Comment on model adequacy.

14-45. An article in *Solid State Technology* (Vol. 29, 1984, pp. 281–284) describes the use of factorial experiments in photolithography, an important step in the process of manufacturing integrated circuits. The variables in this experiment (all at two levels) are prebake temperature (A), prebake time (B), and exposure energy (C), and the response variable is delta line width, the difference between the line on the mask and the printed line on the device. The data are as follows: $(1) = -2.30$, $a = -9.87$, $b = -18.20$, $ab = -30.20$, $c = -23.80$, $ac = -4.30$, $bc = -3.80$, and $abc = -14.70$.

(a) Estimate the factor effects.

(b) Use a normal probability plot of the effect estimates to identity factors that may be important.

(c) What model would you recommend for predicting the delta line width response, based on the results of this experiment?

(d) Analyze the residuals from this experiment, and comment on model adequacy.

14-46. An article in the *Journal of Coatings Technology* (Vol. 60, 1988, pp. 27–32) describes a 2^4 factorial design used for studying a silver automobile basecoat. The response variable is distinctness of image (DOI). The variables used in the experiment are

A = Percentage of polyester by weight of polyester/melamine (low value = 50%, high value = 70%)

B = Percentage of cellulose acetate butyrate carboxylate (low value = 15%, high value = 30%)

C = Percentage of aluminum stearate (low value = 1%, high value = 3%)

D = Percentage of acid catalyst (low value = 0.25%, high value = 0.50%)

The responses are $(1) = 63.8$, $a = 77.6$, $b = 68.8$, $ab = 76.5$, $c = 72.5$, $ac = 77.2$, $bc = 77.7$, $abc = 84.5$, $d = 60.6$, $ad = 64.9$, $bd = 72.7$, $abd = 73.3$, $cd = 68.0$, $acd = 76.3$, $bcd = 76.0$, and $abcd = 75.9$.

(a) Estimate the factor effects.

(b) From a normal probability plot of the effects, identify a tentative model for the data from this experiment.

(c) Using the apparently negligible factors as an estimate of error, test for significance of the factors identified in part (b). Use $\alpha = 0.05$.

(d) What model would you use to describe the process, based on this experiment? Interpret the model.

(e) Analyze the residuals from the model in part (d) and comment on your findings.

14-47. An article in the *Journal of Manufacturing Systems* (Vol. 10, 1991, pp. 32–40) describes an experiment to investigate the effect of four factors P = waterjet pressure, F = abrasive flow rate, G = abrasive grain size, and V = jet traverse speed on the surface roughness of a waterjet cutter. A 2^4 design follows.

	Factors				Surface
Run	V (in/min)	F (lb/min)	P (kpsi)	G (Mesh No.)	Roughness (μm)
1	6	2.0	38	80	104
2	2	2.0	38	80	98
3	6	2.0	30	80	103
4	2	2.0	30	80	96
5	6	1.0	38	80	137
6	2	1.0	38	80	112
7	6	1.0	30	80	143
8	2	1.0	30	80	129
9	6	2.0	38	170	88
10	2	2.0	38	170	70
11	6	2.0	30	170	110
12	2	2.0	30	170	110
13	6	1.0	38	170	102
14	2	1.0	38	170	76
15	6	1.0	30	170	98
16	2	1.0	30	170	68

(a) Estimate the factor effects.
(b) Form a tentative model by examining a normal probability plot of the effects.
(c) Is the model in part (b) a reasonable description of the process? Is lack of fit significant? Use $\alpha = 0.05$.
(d) Interpret the results of this experiment.
(e) Analyze the residuals from this experiment.

14-48. Construct a 2_{IV}^{4-1} design for the problem in Exercise 14-46. Select the data for the eight runs that would have been required for this design. Analyze these runs and compare your conclusions to those obtained in Exercise 14-46 for the full factorial.

14-49. Construct a 2_{IV}^{4-1} design for the problem in Exercise 14-47. Select the data for the eight runs that would have been required for this design. Analyze these data and compare your conclusions to those obtained in Exercise 14-47 for the full factorial.

14-50. Construct a 2_{IV}^{8-4} design in 16 runs. What are the alias relationships in this design?

14-51. Construct a 2_{III}^{5-2} design in eight runs. What are the alias relationships in this design?

14-52. In a process development study on yield, four factors were studied, each at two levels: time (A), concentration (B), pressure (C), and temperature (D). A single replicate of a 2^4 design was run, and the data are shown in the table below.

(a) Plot the effect estimates on a normal probability scale. Which factors appear to have large effects?
(b) Conduct an analysis of variance using the normal probability plot in part (a) for guidance in forming an error term. What are your conclusions?
(c) Analyze the residuals from this experiment. Does your analysis indicate any potential problems?
(d) Can this design be collapsed into a 2^3 design with two replicates? If so, sketch the design with the average and range of yield shown at each point in the cube. Interpret the results.

14-53. An article in the *Journal of Quality Technology* (Vol. 17, 1985, pp. 198–206) describes the use of a replicated fractional factorial to investigate the effect of five factors on the free height of leaf springs used in an automotive application. The factors are A = furnace temperature, B = heating time, C = transfer time, D = hold down time, and E = quench oil temperature. The data are shown in the following table.

(a) What is the generator for this fraction? Write out the alias structure.
(b) Analyze the data. What factors influence mean free height?
(c) Calculate the range of free height for each run. Is there any indication that any of these factors affect variability in free height?
(d) Analyze the residuals from this experiment and comment on your findings.

Run Number	Actual Run Order	A	B	C	D	Yield (lbs)	Factor Levels Low ($-$)		Factor Levels High ($+$)
1	5	$-$	$-$	$-$	$-$	12	A (h) 2.5		3
2	9	$+$	$-$	$-$	$-$	18	B (%) 14		18
3	8	$-$	$+$	$-$	$-$	13	C (psi) 60		80
4	13	$+$	$+$	$-$	$-$	16	D (°C) 225		250
5	3	$-$	$-$	$+$	$-$	17			
6	7	$+$	$-$	$+$	$-$	15			
7	14	$-$	$+$	$+$	$-$	20			
8	1	$+$	$+$	$+$	$-$	15			
9	6	$-$	$-$	$-$	$+$	10			
10	11	$+$	$-$	$-$	$+$	25			
11	2	$-$	$+$	$-$	$+$	13			
12	15	$+$	$+$	$-$	$+$	24			
13	4	$-$	$-$	$+$	$+$	19			
14	16	$+$	$-$	$+$	$+$	21			
15	10	$-$	$+$	$+$	$+$	17			
16	12	$+$	$+$	$+$	$+$	23			

A	B	C	D	E		Free Height	
−	−	−	−	−	7.78	7.78	7.81
+	−	−	+	−	8.15	8.18	7.88
−	+	−	+	−	7.50	7.56	7.50
+	+	−	−	−	7.59	7.56	7.75
−	−	+	+	−	7.54	8.00	7.88
+	−	+	−	−	7.69	8.09	8.06
−	+	+	−	−	7.56	7.52	7.44
+	+	+	+	−	7.56	7.81	7.69
−	−	−	−	+	7.50	7.56	7.50
+	−	−	+	+	7.88	7.88	7.44
−	+	−	+	+	7.50	7.56	7.50
+	+	−	−	+	7.63	7.75	7.56
−	−	+	+	+	7.32	7.44	7.44
+	−	+	−	+	7.56	7.69	7.62
−	+	+	−	+	7.18	7.18	7.25
+	+	+	+	+	7.81	7.50	7.59

14-54. An article in *Rubber Chemistry and Technology* (Vol. 47, 1974, pp. 825–836) describes an experiment that studies the Mooney viscosity of rubber to several variables, including silica filler (parts per hundred) and oil filler (parts per hundred). Data typical of that reported in this experiment follow, where

$$x_1 = \frac{\text{silica} - 60}{15}, \qquad x_2 = \frac{\text{oil} - 21}{15}$$

Coded levels		
x_1	x_2	y
−1	−1	13.71
1	−1	14.15
−1	1	12.87
1	1	13.53
−1	−1	13.90
1	−1	14.88
−1	1	12.25
−1	1	13.35

(a) What type of experimental design has been used?
(b) Analyze the data and draw appropriate conclusions.

MIND-EXPANDING EXERCISES

14-55. Consider an unreplicated 2^k factorial, and suppose that one of the treatment combinations is missing. One logical approach to this problem is to estimate the missing value with a number that makes the highest order interaction estimate zero. Apply this technique to the data in Example 14-5, assuming that ab is missing. Compare the results of the analysis of these data with the results in Example 14-5.

14-56. What blocking scheme would you recommend if it were necessary to run a 2^4 design in four blocks of four runs each?

14-57. Consider a 2^2 design in two blocks with AB confounded with blocks. Prove algebraically that $SS_{AB} = SS_{\text{Blocks}}$.

14-58. Consider a 2^3 design. Suppose that the largest number of runs that can be made in one block is four, but we can afford to perform a total of 32 observations.

(a) Suggest a blocking scheme that will provide some information on all interactions.

(b) Show an outline (source of variability, degrees of freedom only) for the analysis of variance for this design.

14-59. Construct a 2^{5-1} design. Suppose that it is necessary to run this design in two blocks of eight runs each. Show how this can be done by confounding a two-factor interaction (and its aliased three-factor interaction) with blocks.

14-60. Construct a 2_{IV}^{7-2} design. Show how this design may be confounded in four blocks of eight runs each. Are any two-factor interactions confounded with blocks?

14-61. Construct a 2_{IV}^{7-3} design. Show how this design can be confounded in two blocks of eight runs each without losing information on any of the two-factor interactions.

14-62. Set up a 2_{III}^{7-4} design using $D = AB$, $E = AC$, $F = BC$, and $G = ABC$ as the design generators. Ignore all interaction above the two factors.

(a) Verify that each main effect is aliased with three two-factor interactions.

(b) Suppose that a second 2_{III}^{7-4} design with generators $D = -AB$, $E = -AC$, $F = -BC$, and $G = ABC$ is run. What are the aliases of the main effects in this design?

(c) What factors may be estimated if the two sets of factor effect estimates above are combined?

To work Exercises 14-63 through 14-67 you will need to read Section 14.6 on the CD.

14-63. Consider the experiment described in Example 14-1. Suppose that both factors were random.
(a) Analyze the data and draw appropriate conclusions.
(b) Estimate the variance components.

14-64. For the breaking strength data in Table S14-1, suppose that the operators were chosen at random, but machines were a fixed factor. Does this influence the analysis or your conclusions?

14-65. A company employs two time-study engineers. Their supervisor wishes to determine whether the standards set by them are influenced by an interaction between engineers and operators. She selects three operators at random and conducts an experiment in which the engineers set standard times for the same job. She obtains the data shown here:

Engineer	Operator		
	1	2	3
1	2.59	2.38	2.40
	2.78	2.49	2.72
2	2.15	2.85	2.66
	2.86	2.72	2.87

(a) State the appropriate hypotheses.
(b) Use the analysis of variance to test these hypotheses with $\alpha = 0.05$.
(c) Graphically analyze the residuals from this experiment.
(d) Estimate the appropriate variance components.

14-66. Consider the experiment on baked anode density described in Exercise 14-4. Suppose that positions on the furnace were chosen at random and temperature is a fixed factor.
(a) State the appropriate hypotheses.
(b) Use the analysis of variance to test these hypotheses with $\alpha = 0.05$.
(c) Estimate the variance components.

14-67. Consider the experiment described in Exercise 14-65. How does the analysis (and conclusions) change if both factors are random? Use $\alpha = 0.05$.

To work Exercises 14-68 and 14-69 you will need to read Section 14-7.4 on the CD.

MIND-EXPANDING EXERCISES

14-68. Consider the experiment in Exercise 14-19. Suppose that a center point had been run (replicated five times) and the responses were 45, 40, 41, 47, and 43.

(a) Estimate the experimental error using the center points. Compare this to the estimate obtained originally in Exercise 14-19 by pooling apparently nonsignificant effects.

(b) Test for lack-of-fit, using $\alpha = 0.05$.

14-69. Consider the data from Exercise 14-13, replicate 1 only. Suppose that a center point with four replicates is added to these eight factorial runs and the responses are 425, 400, 437, and 418.

(a) Estimate the facter effects.

(b) Test for lack of fit using $\alpha = 0.05$

(c) Test for main effects and interactions using $\alpha = 0.05$.

(d) Analyze residuals and draw conclusions.

To work problem 14-70 through 14-74 you will need to read Section 14-10 on the CD.

14-70. An article in *Rubber Age* (Vol. 89, 1961, pp. 453–458) describes an experiment on the manufacture of a product in which two factors were varied. The factors are reaction time (hr) and temperature (°C). These factors are coded as $x_1 = (\text{time} - 12)/8$ and $x_2 = (\text{temperature} - 250)/30$. The following data were observed where y is the yield (in percent):

Run Number	x_1	x_2	y
1	−1	0	83.8
2	1	0	81.7
3	0	0	82.4
4	0	0	82.9
5	0	−1	84.7
6	0	1	75.9
7	0	0	81.2
8	−1.414	−0.414	81.3
9	−1.414	1.414	83.1
10	1.414	−1.414	85.3
11	1.414	1.414	72.7
12	0	0	82.0

(a) Plot the points at which the experimental runs were made.

(b) Fit a second-order model to the data. Is the second-order model adequate?

(c) Plot the yield response surface. What recommendations would you make about the operating conditions for this process?

14-71. Consider the first-order model

$$\hat{y} = 50 + 1.5x_1 - 0.8x_2$$

where $-1 \le x_i \le 1$. Find the direction of steepest ascent.

14-72. A manufacturer of cutting tools has developed two empirical equations for tool life (y_1) and tool cost (y_2). Both models are functions of tool hardness (x_1) and manufacturing time (x_2). The equations are

$$\hat{y}_1 = 10 + 5x_1 + 2x_2$$
$$\hat{y}_2 = 23 + 3x_1 + 4x_2$$

and both equations are valid over the range $-1.5 \le x_i \le 1.5$. Suppose that tool life must exceed 12 hours and cost must be below $27.50.

(a) Is there a feasible set of operating conditions?

(b) Where would you run this process?

14-73. An article in *Tappi* (Vol. 43, 1960, pp. 38–44) describes an experiment that investigated the ash value of paper pulp (a measure of inorganic impurities). Two variables, temperature T in degrees Celsius and time t in hours, were studied, and some of the results are shown in the following table. The coded predictor variables shown are

$$x_1 = \frac{(T - 775)}{115}, \qquad x_2 = \frac{(t - 3)}{1.5}$$

and the response y is (dry ash value in %) $\times 10^3$.

x_1	x_2	y	x_1	x_2	y
−1	−1	211	0	−1.5	168
1	−1	92	0	1.5	179
−1	1	216	0	0	122
1	1	99	0	0	175
−1.5	0	222	0	0	157
1.5	0	48	0	0	146

(a) What type of design has been used in this study? Is the design rotatable?

MIND-EXPANDING EXERCISES

(b) Fit a quadratic model to the data. Is this model satisfactory?

(c) If it is important to minimize the ash value, where would you run the process?

14-74.   In their book *Empirical Model Building and Response Surfaces* (John Wiley, 1987), G. E. P. Box and N. R. Draper describe an experiment with three factors. The data shown in the following table are a variation of the original experiment on page 247 of their book. Suppose that these data were collected in a semiconductor manufacturing process.

(a) The response y_1 is the average of three readings on resistivity for a single wafer. Fit a quadratic model to this response.

(b) The response y_2 is the standard deviation of the three resistivity measurements. Fit a linear model to this response.

(c) Where would you recommend that we set x_1, x_2, and x_3 if the objective is to hold mean resistivity at 500 and minimize the standard deviation?

x_1	x_2	x_3	y_1	y_2	x_1	x_2	x_3	y_1	y_2
−1	−1	−1	24.00	12.49	1	0	0	501.67	92.50
0	−1	−1	120.33	8.39	−1	1	0	264.00	63.50
1	−1	−1	213.67	42.83	0	1	0	427.00	88.61
−1	0	−1	86.00	3.46	1	1	0	730.67	21.08
0	0	−1	136.63	80.41	−1	−1	1	220.67	133.82
1	0	−1	340.67	16.17	0	−1	1	239.67	23.46
−1	1	−1	112.33	27.57	1	−1	1	422.00	18.52
0	1	−1	256.33	4.62	−1	0	1	199.00	29.44
1	1	−1	271.67	23.63	0	0	1	485.33	44.67
−1	−1	0	81.00	0.00	1	0	1	673.67	158.21
0	−1	0	101.67	17.67	−1	1	1	176.67	55.51
1	−1	0	357.00	32.91	0	1	1	501.00	138.94
−1	0	0	171.33	15.01	1	1	1	1010.00	142.45
0	0	0	372.00	0.00					

IMPORTANT TERMS AND CONCEPTS

In the E-book, click on any term or concept below to go to that subject.

Analysis of variance (ANOVA)

Blocking and nuisance factors

Confounding

Factorial Experiment

Fractional factorial design

Interaction

Main effect

Normal probability plot of factor effects

Orthogonal design

Regression model

Residual analysis

Two-level factorial design

CD MATERIAL

Center points in a factorial

Central composite design

Mixed model

Random model

Response surface

Steepest ascent

Variance components

15

Nonparametric Statistics

CHAPTER OUTLINE

LEARNING OBJECTIVES

After careful study of this chapter, you should be able to do the following:

1. Determine situations where nonparametric procedures are better alternatives to the t-test and ANOVA

2. Use one- and two-sample nonparametric tests

3. Use nonparametric alternatives to the single-factor ANOVA

4. Understand how nonparametric tests compare to the t-test in terms of relative efficiency

---

Answers for most odd numbered exercises are at the end of the book. Answers to exercises whose numbers are surrounded by a box can be accessed in the e-text by clicking on the box. Complete worked solutions to certain exercises are also available in the e-Text. These are indicated in the Answers to Selected Exercises section by a box around the exercise number. Exercises are also available for some of the text sections that appear on CD only. These exercises may be found within the e-Text immediately following the section they accompany.

15-1 INTRODUCTION

Most of the hypothesis-testing and confidence interval procedures discussed in previous chapters are based on the assumption that we are working with random samples from normal populations. Traditionally, we have called these procedures **parametric methods** because they are based on a particular parametric family of distributions—in this case, the normal. Alternately, sometimes we say that these procedures are not *distribution-free* because they depend on the assumption of normality. Fortunately, most of these procedures are relatively insensitive to slight departures from normality. In general, the *t*- and *F*-tests and the *t*-confidence intervals will have actual levels of significance or confidence levels that differ from the nominal or advertised levels chosen by the experimenter, although the difference between the actual and advertised levels is usually fairly small when the underlying population is not too different from the normal.

In this chapter we describe procedures called **nonparametric** and **distribution-free methods,** and we usually make no assumptions about the distribution of the underlying population other than that it is continuous. These procedures have actual level of significance α or confidence level $100(1 - α)\%$ for many different types of distributions. These procedures have considerable appeal. One of their advantages is that the data need not be quantitative but can be categorical (such as yes or no, defective or nondefective) or rank data. Another advantage is that nonparametric procedures are usually very quick and easy to perform.

The procedures described in this chapter are competitors of the parametric *t*- and *F*-procedures described earlier. Consequently, it is important to compare the performance of both parametric and nonparametric methods under the assumptions of both normal and non-normal populations. In general, nonparametric procedures do not utilize all the information provided by the sample. As a result, a nonparametric procedure will be less efficient than the corresponding parametric procedure when the underlying population is normal. This loss of efficiency is reflected by a requirement of a larger sample size for the nonparametric procedure than would be required by the parametric procedure in order to achieve the same power. On the other hand, this loss of efficiency is usually not large, and often the difference in sample size is very small. When the underlying distributions are not close to normal, nonparametric methods have much to offer. They often provide considerable improvement over the normal-theory parametric methods.

Generally, if both parametric and nonparametric methods are applicable to a particular problem, we should use the more efficient parametric procedure. However, the assumptions for the parametric method may be difficult or impossible to justify. For example, the data may be in the form of **ranks.** These situations frequently occur in practice. For instance, a panel of judges may be used to evaluate 10 different formulations of a soft-drink beverage for overall quality, with the "best" formulation assigned rank 1, the "next-best" formulation assigned rank 2, and so forth. It is unlikely that rank data satisfy the normality assumption. Many nonparametric methods involve the analysis of ranks and consequently are ideally suited to this type of problem.

15-2 SIGN TEST

15-2.1 Description of the Test

The **sign test** is used to test hypotheses about the **median** $\tilde{\mu}$ of a continuous distribution. The median of a distribution is a value of the random variable X such that the probability is 0.5 that an observed value of X is less than or equal to the median, and the probability is 0.5 that an observed value of X is greater than or equal to the median. That is, $P(X \leq \tilde{\mu}) = P(X \geq \tilde{\mu}) = 0.5$.

Since the normal distribution is symmetric, the mean of a normal distribution equals the median. Therefore, the sign test can be used to test hypotheses about the mean of a normal distribution. This is the same problem for which we used the t-test in Chapter 9. We will discuss the relative merits of the two procedures in Section 15-2.4. Note that, although the t-test was designed for samples from a normal distribution, the sign test is appropriate for samples from any continuous distribution. Thus, the sign test is a nonparametric procedure.

Suppose that the hypotheses are

$$H_0: \tilde{\mu} = \tilde{\mu}_0$$
$$H_1: \tilde{\mu} < \tilde{\mu}_0 \tag{15-1}$$

The test procedure is easy to describe. Suppose that $X_1, X_2, \ldots, X_n$ is a random sample from the population of interest. Form the differences

$$X_i - \tilde{\mu}_0, \qquad i = 1, 2, \ldots, n \tag{15-2}$$

Now if the null hypothesis $H_0: \tilde{\mu} = \tilde{\mu}_0$ is true, any difference $X_i - \tilde{\mu}_0$ is equally likely to be positive or negative. An appropriate test statistic is the number of these differences that are positive, say R^+. Therefore, to test the null hypothesis we are really testing that the number of plus signs is a value of a binomial random variable that has the parameter $p = 1/2$. A P-value for the observed number of plus signs r^+ can be calculated directly from the binomial distribution. For instance, in testing the hypotheses in Equation 15-1, we will reject H_0 in favor of H_1 only if the proportion of plus signs is sufficiently less than $1/2$ (or equivalently, whenever the observed number of plus signs r^+ is too small). Thus, if the computed P-value

$$P = P\left(R^+ \le r^+ \text{ when } p = \frac{1}{2}\right)$$

is less than or equal to some preselected significance level α, we will reject H_0 and conclude H_1 is true.

To test the other one-sided hypothesis

$$H_0: \tilde{\mu} = \tilde{\mu}_0$$
$$H_1: \tilde{\mu} > \tilde{\mu}_0 \tag{15-3}$$

we will reject H_0 in favor of H_1 only if the observed number of plus signs, say r^+, is large or, equivalently, whenever the observed fraction of plus signs is significantly greater than $1/2$. Thus, if the computed P-value

$$P = P\left(R^+ \ge r^+ \text{ when } p = \frac{1}{2}\right)$$

is less than α, we will reject H_0 and conclude that H_1 is true.

The two-sided alternative may also be tested. If the hypotheses are

$$H_0: \tilde{\mu} = \tilde{\mu}_0$$
$$H_1: \tilde{\mu} \ne \tilde{\mu}_0 \tag{15-4}$$

we should reject H_0: $\tilde{\mu} = \tilde{\mu}_0$ if the proportion of plus signs is significantly different (either less than or greater than) from $1/2$. This is equivalent to the observed number of plus signs r^+ being either sufficiently large or sufficiently small. Thus, if $r^+ < n/2$ the P-value is

$$P = 2P\left(R^+ \le r^+ \text{ when } p = \frac{1}{2}\right)$$

and if $r^+ > n/2$ the P-value is

$$P = 2P\left(R^+ \ge r^+ \text{ when } p = \frac{1}{2}\right)$$

If the P-value is less than some preselected level α, we will reject H_0 and conclude that H_1 is true.

EXAMPLE 15-1

Montgomery, Peck, and Vining (2001) report on a study in which a rocket motor is formed by binding an igniter propellant and a sustainer propellant together inside a metal housing. The shear strength of the bond between the two propellant types is an important characteristic. The results of testing 20 randomly selected motors are shown in Table 15-1. We would like to test the hypothesis that the median shear strength is 2000 psi, using $\alpha = 0.05$.

This problem can be solved using the eight-step hypothesis-testing procedure introduced in Chapter 9:

1. The parameter of interest is the median of the distribution of propellant shear strength.
2. H_0: $\tilde{\mu} = 2000$ psi
3. H_1: $\tilde{\mu} \neq 2000$ psi
4. $\alpha = 0.05$

Table 15-1 Propellant Shear Strength Data

Observation i	Shear Strength x_i	Differences $x_i - 2000$	Sign
1	2158.70	+158.70	+
2	1678.15	−321.85	−
3	2316.00	+316.00	+
4	2061.30	+61.30	+
5	2207.50	+207.50	+
6	1708.30	−291.70	−
7	1784.70	−215.30	−
8	2575.10	+575.10	+
9	2357.90	+357.90	+
10	2256.70	+256.70	+
11	2165.20	+165.20	+
12	2399.55	+399.55	+
13	1779.80	−220.20	−
14	2336.75	+336.75	+
15	1765.30	−234.70	−
16	2053.50	+53.50	+
17	2414.40	+414.40	+
18	2200.50	+200.50	+
19	2654.20	+654.20	+
20	1753.70	−246.30	−

5. The test statistic is the observed number of plus differences in Table 15-1, or $r^+ = 14$.

6. We will reject H_0 if the P-value corresponding to $r^+ = 14$ is less than or equal to $\alpha = 0.05$.

7. Computations: Since $r^+ = 14$ is greater than $n/2 = 20/2 = 10$, we calculate the P-value from

$$P = 2P\left(R^+ \geq 14 \text{ when } p = \frac{1}{2}\right)$$

$$= 2 \sum_{r=14}^{20} \binom{20}{r}(0.5)^r(0.5)^{20-r}$$

$$= 0.1153$$

8. Conclusions: Since $P = 0.1153$ is not less than $\alpha = 0.05$, we cannot reject the null hypothesis that the median shear strength is 2000 psi. Another way to say this is that the observed number of plus signs $r^+ = 14$ was not large or small enough to indicate that median shear strength is different from 2000 psi at the $\alpha = 0.05$ level of significance.

It is also possible to construct a table of critical values for the sign test. This table is shown as Appendix Table VII. The use of this table for the two-sided alternative hypothesis in Equation 15-4 is simple. As before, let R^+ denote the number of the differences $(X_i - \tilde{\mu}_0)$ that are positive and let R^- denote the number of these differences that are negative. Let $R = \min(R^+, R^-)$. Appendix Table VII presents critical values r_α^* for the sign test that ensure that P (type I error) = P (reject H_0 when H_0 is true) = α for $\alpha = 0.01$, $\alpha = 0.05$ and $\alpha = 0.10$. If the observed value of the test statistic $r \leq r_\alpha^*$, the null hypothesis $H_0: \tilde{\mu} = \tilde{\mu}_0$ should be rejected.

To illustrate how this table is used, refer to the data in Table 15-1 that was used in Example 15-1. Now $r^+ = 14$ and $r^- = 6$; therefore, $r = \min(14, 6) = 6$. From Appendix Table VII with $n = 20$ and $\alpha = 0.05$, we find that $r_{0.05}^* = 5$. Since $r = 6$ is not less than or equal to the critical value $r_{0.05}^* = 5$, we cannot reject the null hypothesis that the median shear strength is 2000 psi.

We can also use Appendix Table VII for the sign test when a one-sided alternative hypothesis is appropriate. If the alternative is $H_1: \tilde{\mu} > \tilde{\mu}_0$, reject $H_0: \tilde{\mu} = \tilde{\mu}_0$ if $r^- \leq r_\alpha^*$; if the alternative is $H_1: \tilde{\mu} > \tilde{\mu}_0$, reject $H_0: \tilde{\mu} = \tilde{\mu}_0$ if $r^+ \leq r_\alpha^*$. The level of significance of a one-sided test is one-half the value for a two-sided test. Appendix Table VII shows the one-sided significance levels in the column headings immediately below the two-sided levels.

Finally, note that when a test statistic has a discrete distribution such as R does in the sign test, it may be impossible to choose a critical value r_α^* that has a level of significance exactly equal to α. The approach used in Appendix Table VII is to choose r_α^* to yield an α that is as close to the advertised significance level α as possible.

Ties in the Sign Test

Since the underlying population is assumed to be continuous, there is a zero probability that we will find a "tie"—that is, a value of X_i exactly equal to $\tilde{\mu}_0$. However, this may sometimes happen in practice because of the way the data are collected. When ties occur, they should be set aside and the sign test applied to the remaining data.

The Normal Approximation

When $p = 0.5$, the binomial distribution is well approximated by a normal distribution when n is at least 10. Thus, since the mean of the binomial is np and the variance is $np(1 - p)$, the distribution of R^+ is approximately normal with mean $0.5n$ and variance $0.25n$ whenever n is moderately large. Therefore, in these cases the null hypothesis H_0: $\tilde{\mu} = \tilde{\mu}_0$ can be tested using the statistic

$$Z_0 = \frac{R^+ - 0.5n}{0.5\sqrt{n}} \qquad (15\text{-}5)$$

The two-sided alternative would be rejected if the observed value of the test statistic $|z_0| > z_{\alpha/2}$, and the critical regions of the one-sided alternative would be chosen to reflect the sense of the alternative. (If the alternative is H_1: $\tilde{\mu} > \tilde{\mu}_0$, reject H_0 if $z_0 > z_\alpha$, for example.)

EXAMPLE 15-2   We will illustrate the normal approximation procedure by applying it to the problem in Example 15-1. Recall that the data for this example are in Table 15-1. The eight-step procedure follows:

1.   The parameter of interest is the median of the distribution of propellant shear strength.
2. H_0: $\tilde{\mu} = 2000$ psi
3. H_1: $\tilde{\mu} \neq 2000$ psi
4. $\alpha = 0.05$
5.   The test statistic is

$$z_0 = \frac{r^+ - 0.5n}{0.5\sqrt{n}}$$

6.   Since $\alpha = 0.05$, we will reject H_0 in favor of H_1 if $|z_0| > z_{0.025} = 1.96$.
7.   Computations: Since $r^+ = 14$, the test statistic is

$$z_0 = \frac{14 - 0.5(20)}{0.5\sqrt{20}} = 1.789$$

8.   Conclusions: Since $z_0 = 1.789$ is not greater than $z_{0.025} = 1.96$, we cannot reject the null hypothesis. Thus, our conclusions are identical to those in Example 15-1.

15-2.2   Sign Test for Paired Samples

The sign test can also be applied to paired observations drawn from continuous populations. Let (X_{1j}, X_{2j}), $j = 1, 2, \ldots, n$ be a collection of paired observations from two continuous populations, and let

$$D_j = X_{1j} - X_{2j} \qquad j = 1, 2, \ldots, n$$

be the paired differences. We wish to test the hypothesis that the two populations have a common median, that is, that $\tilde{\mu}_1 = \tilde{\mu}_2$. This is equivalent to testing that the median of the differences $\tilde{\mu}_D = 0$. This can be done by applying the sign test to the n observed differences d_j, as illustrated in the following example.

EXAMPLE 15-3   An automotive engineer is investigating two different types of metering devices for an electronic fuel injection system to determine whether they differ in their fuel mileage performance. The system is installed on 12 different cars, and a test is run with each metering device on each car. The observed fuel mileage performance data, corresponding differences, and their signs are shown in Table 15-2. We will use the sign test to determine whether the median fuel mileage performance is the same for both devices using $\alpha = 0.05$. The eight-step-procedure follows:

1. The parameters of interest are the median fuel mileage performance for the two metering devices.

2. $H_0: \tilde{\mu}_1 = \tilde{\mu}_2$, or, equivalently, $H_0: \tilde{\mu}_D = 0$

3. $H_1: \tilde{\mu}_1 \neq \tilde{\mu}_2$, or, equivalently, $H_1: \tilde{\mu}_D \neq 0$

4. $\alpha = 0.05$

5. We will use Appendix Table VII to conduct the test, so the test statistic is $r = \min(r^+, r^-)$.

6. Since $\alpha = 0.05$ and $n = 12$, Appendix Table VII gives the critical values as $r^*_{0.05} = 2$. We will reject H_0 in favor of H_1 if $r \leq 2$.

7. Computations: Table 15-2 shows the differences and their signs, and we note that $r^+ = 8$, $r^- = 4$, and so $r = \min(8, 4) = 4$.

8. Conclusions: Since $r = 4$ is not less than or equal to the critical value $r^*_{0.05} = 2$, we cannot reject the null hypothesis that the two devices provide the same median fuel mileage performance.

Table 15-2   Performance of Flow Metering Devices

	Metering Device			
Car	1	2	Difference, d_j	Sign
1	17.6	16.8	0.8	+
2	19.4	20.0	−0.6	−
3	19.5	18.2	1.3	+
4	17.1	16.4	0.7	+
5	15.3	16.0	−0.7	−
6	15.9	15.4	0.5	+
7	16.3	16.5	−0.2	−
8	18.4	18.0	0.4	+
9	17.3	16.4	0.9	+
10	19.1	20.1	−1.0	−
11	17.8	16.7	1.1	+
12	18.2	17.9	0.3	+

15-2.3 Type II Error for the Sign Test

The sign test will control the probability of type I error at an advertised level α for testing the null hypothesis H_0: $\tilde{\mu} = \tilde{\mu}$ for any continuous distribution. As with any hypothesis-testing procedure, it is important to investigate the probability of a type II error, β. The test should be able to effectively detect departures from the null hypothesis, and a good measure of this effectiveness is the value of β for departures that are important. A small value of β implies an effective test procedure.

In determining β, it is important to realize not only that a particular value of $\tilde{\mu}$, say $\tilde{\mu}_0 + \Delta$, must be used but also that the **form** of the underlying distribution will affect the calculations. To illustrate, suppose that the underlying distribution is normal with $\sigma = 1$ and we are testing the hypothesis H_0: $\tilde{\mu} = 2$ versus H_1: $\tilde{\mu} > 2$. (Since $\tilde{\mu} = \mu$ in the normal distribution, this is equivalent to testing that the mean equals 2.) Suppose that it is important to detect a departure from $\tilde{\mu} = 2$ to $\tilde{\mu} = 3$. The situation is illustrated graphically in Fig. 15-1(a). When the alternative hypothesis is true (H_1: $\tilde{\mu} = 3$), the probability that the random variable X is less than or equal to the value 2 is

$$p = P(X \le 2) = P(Z \le -1) = \Phi(-1) = 0.1587$$

Suppose we have taken a random sample of size 12. At the $\alpha = 0.05$ level, Appendix Table VII indicates that we would reject H_0: $\tilde{\mu} = 2$ if $r^- \le r^*_{0.05} = 2$. Therefore, β is the probability that we do not reject H_0: $\tilde{\mu} = 2$ when in fact $\tilde{\mu} = 3$, or

$$\beta = 1 - \sum_{x=0}^{2} \binom{12}{x}(0.1587)^x(0.8413)^{12-x} = 0.2944$$

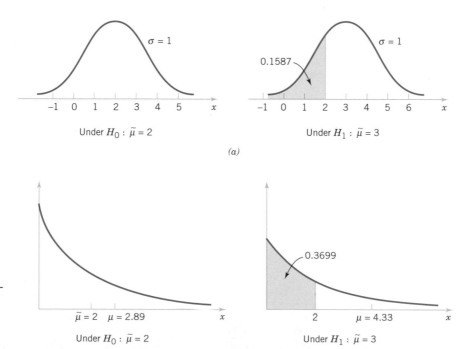

Figure 15-1   Calculation of β for the sign test. (a) Normal distributions. (b) Exponential distributions.

If the distribution of X had been exponential rather than normal, the situation would be as shown in Fig. 15-1(b), and the probability that the random variable X is less than or equal to the value $x = 2$ when $\tilde{\mu} = 3$ (note that when the median of an exponential distribution is 3, the mean is 4.33) is

$$p = P(X \le 2) = \int_0^2 \frac{1}{4.33} e^{-\frac{1}{4.33}x} \, dx = 0.3699$$

In this case,

$$\beta = 1 - \sum_{x=0}^{2} \binom{12}{x}(0.3699)^x(0.6301)^{12-x} = 0.8794$$

Thus, β for the sign test depends not only on the alternative value of $\tilde{\mu}$ but also on the area to the right of the value specified in the null hypothesis under the population probability distribution. This area is highly dependent on the shape of that particular probability distribution.

15-2.4 Comparison to the t-Test

If the underlying population is normal, either the sign test or the t-test could be used to test $H_0\colon \tilde{\mu} = \tilde{\mu}_0$. The t-test is known to have the smallest value of β possible among all tests that have significance level α for the one-sided alternative and for tests with symmetric critical regions for the two-sided alternative, so it is superior to the sign test in the normal distribution case. When the population distribution is symmetric and nonnormal (but with finite mean $\mu = \tilde{\mu}$), the t-test will have a smaller β (or a higher power) than the sign test, unless the distribution has very heavy tails compared with the normal. Thus, the sign test is usually considered a test procedure for the median rather than as a serious competitor for the t-test. The **Wilcoxon signed-rank test** discussed in the next section is preferable to the sign test and compares well with the t-test for symmetric distributions.

EXERCISES FOR SECTION 15-2

15-1. Ten samples were taken from a plating bath used in an electronics manufacturing process, and the bath pH was determined. The sample pH values are 7.91, 7.85, 6.82, 8.01, 7.46, 6.95, 7.05, 7.35, 7.25, 7.42. Manufacturing engineering believes that pH has a median value of 7.0. Do the sample data indicate that this statement is correct? Use the sign test with $\alpha = 0.05$ to investigate this hypothesis. Find the P-value for this test.

15-2. The titanium content in an aircraft-grade alloy is an important determinant of strength. A sample of 20 test coupons reveals the following titanium content (in percent):

8.32, 8.05, 8.93, 8.65, 8.25, 8.46, 8.52, 8.35, 8.36, 8.41, 8.42, 8.30, 8.71, 8.75, 8.60, 8.83, 8.50, 8.38, 8.29, 8.46

The median titanium content should be 8.5%. Use the sign test with $\alpha = 0.05$ to investigate this hypothesis. Find the P-value for this test.

15-3. The impurity level (in ppm) is routinely measured in an intermediate chemical product. The following data were observed in a recent test:

2.4, 2.5, 1.7, 1.6, 1.9, 2.6, 1.3, 1.9, 2.0, 2.5, 2.6, 2.3, 2.0, 1.8, 1.3, 1.7, 2.0, 1.9, 2.3, 1.9, 2.4, 1.6

Can you claim that the median impurity level is less than 2.5 ppm? State and test the appropriate hypothesis using the sign test with $\alpha = 0.05$. What is the P-value for this test?

15-4. Consider the data in Exercise 15-1. Use the normal approximation for the sign test to test H_0: $\tilde{\mu} = 7.0$ versus H_1: $\tilde{\mu} \neq 7.0$. What is the P-value for this test?

15-5. Consider the compressive strength data in Exercise 8-26.

(a) Use the sign test to investigate the claim that the median strength is at least 2250 psi. Use $\alpha = 0.05$.

(b) Use the normal approximation to test the same hypothesis that you formulated in part (a). What is the P-value for this test?

15-6. Consider the margarine fat content data in Exercise 8-25. Use the sign test to test H_0: $\tilde{\mu} = 17.0$ versus H_1: $\tilde{\mu} \neq 17.0$, with $\alpha = 0.05$. Find the P-value for the test statistic and use this quantity to make your decision.

15-7. Consider the data in Exercise 15-2. Use the normal approximation for the sign test to test H_0: $\tilde{\mu} = 8.5$ versus H_1: $\tilde{\mu} \neq 8.5$, with $\alpha = 0.05$. What is the P-value for this test?

15-8. Consider the data in Exercise 15-3. Use the normal approximation for the sign test to test H_0: $\tilde{\mu} = 2.5$ versus H_1: $\tilde{\mu} < 2.5$. What is the P-value for this test?

15-9. Two different types of tips can be used in a Rockwell hardness tester. Eight coupons from test ingots of a nickel-based alloy are selected, and each coupon is tested twice, once with each tip. The Rockwell C-scale hardness readings are shown in the following table. Use the sign test with $\alpha = 0.05$ to determine whether or not the two tips produce equivalent hardness readings.

Coupon	Tip 1	Tip 2
1	63	60
2	52	51
3	58	56
4	60	59
5	55	58
6	57	54
7	53	52
8	59	61

15-10. Two different formulations of primer paint can be used on aluminum panels. The drying time of these two formulations is an important consideration in the manufacturing process. Twenty panels are selected; half of each panel is painted with primer 1, and the other half is painted with primer 2. The drying times are observed and reported in the following table. Is there evidence that the median drying times of the two formulations are different? Use the sign test with $\alpha = 0.01$.

Panel	Drying Times (in hr)	
	Formulation 1	Formulation 2
1	1.6	1.8
2	1.3	1.5
3	1.5	1.5
4	1.6	1.7
5	1.7	1.6
6	1.9	2.0
7	1.8	2.1
8	1.6	1.7
9	1.4	1.6
10	1.8	1.9
11	1.9	2.0
12	1.8	1.9
13	1.7	1.5
14	1.5	1.7
15	1.6	1.6
16	1.4	1.2
17	1.3	1.6
18	1.6	1.8
19	1.5	1.6
20	1.8	2.0

15-11. Use the normal approximation to the sign test for the data in Exercise 15-10. What conclusions can you draw?

15-12. The diameter of a ball bearing was measured by 12 inspectors, each using two different kinds of calipers. The results were as follows:

Inspector	Caliper 1	Caliper 2
1	0.265	0.264
2	0.265	0.265
3	0.266	0.264
4	0.267	0.266
5	0.267	0.267
6	0.265	0.268
7	0.267	0.264
8	0.267	0.265
9	0.265	0.265
10	0.268	0.267
11	0.268	0.268
12	0.265	0.269

Is there a significant difference between the medians of the population of measurements represented by the two samples? Use $\alpha = 0.05$.

15-13. Consider the blood cholesterol data in Exercise 10-39. Use the sign test to determine whether there is any difference between the medians of the two groups of measurements, with $\alpha = 0.05$. What practical conclusion would you draw from this study?

15-14. Use the normal approximation for the sign test for the data in Exercise 15-12. With $\alpha = 0.05$, what conclusions can you draw?

15-15. Use the normal approximation to the sign test for the data in Exercise 15-13. With $\alpha = 0.05$, what conclusions can you draw?

15-16. The distribution time between arrivals in a telecommunication system is exponential, and the system manager wishes to test the hypothesis that H_0: $\tilde{\mu} = 3.5$ minutes versus H_1: $\tilde{\mu} > 3.5$ minutes.
(a) What is the value of the mean of the exponential distribution under H_0: $\tilde{\mu} = 3.5$?
(b) Suppose that we have taken a sample of $n = 10$ observations and we observe $r^- = 3$. Would the sign test reject H_0 at $\alpha = 0.05$?

(c) What is the type II error probability of this test if $\tilde{\mu} = 4.5$?

15-17. Suppose that we take a sample of $n = 10$ measurements from a normal distribution with $\sigma = 1$. We wish to test H_0: $\mu = 0$ against H_1: $\mu > 0$. The normal test statistic is $Z_0 = \bar{X}/(\sigma/\sqrt{n})$, and we decide to use a critical region of 1.96 (that is, reject H_0 if $z_0 \geq 1.96$).
(a) What is α for this test?
(b) What is β for this test, if $\mu = 1$?
(c) If a sign test is used, specify the critical region that gives an α value consistent with α for the normal test.
(d) What is the β value for the sign test, if $\mu = 1$? Compare this with the result obtained in part (b).

15-18. Consider the test statistic for the sign test in Exercise 15-9. Find the P-value for this statistic.

15-19. Consider the test statistic for the sign test in Exercise 15-10. Find the P-value for this statistic. Compare it to the P-value for the normal approximation test statistic computed in Exercise 15-11.

15-3 WILCOXON SIGNED-RANK TEST

The sign test makes use only of the plus and minus signs of the differences between the observations and the median $\tilde{\mu}_0$ (or the plus and minus signs of the differences between the observations in the paired case). It does not take into account the size or magnitude of these differences. Frank Wilcoxon devised a test procedure that uses both direction (sign) and magnitude. This procedure, now called the **Wilcoxon signed-rank test,** is discussed and illustrated in this section.

The Wilcoxon signed-rank test applies to the case of **symmetric continuous distributions.** Under these assumptions, the mean equals the median, and we can use this procedure to test the null hypothesis that $\mu = \mu_0$. We now show how to do this.

15-3.1 Description of the Test

We are interested in testing H_0: $\mu = \mu_0$ against the usual alternatives. Assume that X_1, $X_2, \ldots, X_n$ is a random sample from a continuous and symmetric distribution with mean (and median) μ. Compute the differences $X_i - \mu_0$, $i = 1, 2, \ldots, n$. Rank the absolute differences $|X_i - \mu_0|$, $i = 1, 2, \ldots, n$ in ascending order, and then give the ranks the signs of their corresponding differences. Let W^+ be the sum of the positive ranks and W^- be the absolute value of the sum of the negative ranks, and let $W = \min(W^+, W^-)$. Appendix Table VIII contains critical values of W, say w_α^*. If the alternative hypothesis is H_1: $\mu \neq \mu_0$, then if the observed value of the statistic $w \leq w_\alpha^*$, the null hypothesis H_0: $\mu = \mu_0$ is rejected. Appendix Table VIII provides significance levels of $\alpha = 0.10$, $\alpha = 0.05$, $\alpha = 0.02$, $\alpha = 0.01$ for the two-sided test.

For one-sided tests, if the alternative is H_1: $\mu > \mu_0$, reject H_0: $\mu = \mu_0$ if $w^- \leq w_\alpha^*$; and if the alternative is H_1: $\mu < \mu_0$, reject H_0: $\mu = \mu_0$ if $w^+ \leq w_\alpha^*$. The significance levels for one-sided tests provided in Appendix Table VIII are $\alpha = 0.05$, 0.025, 0.01, and 0.005.

EXAMPLE 15-4

We will illustrate the Wilcoxon signed-rank test by applying it to the propellant shear strength data from Table 15-1. Assume that the underlying distribution is a continuous symmetric distribution. The eight-step procedure is applied as follows:

1. The parameter of interest is the mean (or median) of the distribution of propellant shear strength.

2. H_0: μ = 2000 psi

3. H_1: $\mu \neq$ 2000 psi

4. α = 0.05

5. The test statistic is

$$w = \min(w^+, w^-)$$

6. We will reject H_0 if $w \leq w^*_{0.05}$ = 52 from Appendix Table VIII.

7. Computations: The signed ranks from Table 15-1 are shown in the following table:

Observation	Difference $x_i - 2000$	Signed Rank
16	+53.50	+1
4	+61.30	+2
1	+158.70	+3
11	+165.20	+4
18	+200.50	+5
5	+207.50	+6
7	−215.30	−7
13	−220.20	−8
15	−234.70	−9
20	−246.30	−10
10	+256.70	+11
6	−291.70	−12
3	+316.00	+13
2	−321.85	−14
14	+336.75	+15
9	+357.90	+16
12	+399.55	+17
17	+414.40	+18
8	+575.10	+19
19	+654.20	+20

The sum of the positive ranks is w^+ = (1 + 2 + 3 + 4 + 5 + 6 + 11 + 13 + 15 + 16 + 17 + 18 + 19 + 20) = 150, and the sum of the absolute values of the negative ranks is w^- = (7 + 8 + 9 + 10 + 12 + 14) = 60. Therefore,

$$w = \min(150, 60) = 60$$

8. Conclusions: Since w = 60 is not less than or equal to the critical value $w_{0.05}$ = 52, we cannot reject the null hypothesis that the mean (or median, since the population is assumed to be symmetric) shear strength is 2000 psi.

Ties in the Wilcoxon Signed-Rank Test

Because the underlying population is continuous, ties are theoretically impossible, although they will sometimes occur in practice. If several observations have the same absolute magnitude, they are assigned the average of the ranks that they would receive if they differed slightly from one another.

15-3.2   Large-Sample Approximation

If the sample size is moderately large, say $n > 20$, it can be shown that W^+ (or W^-) has approximately a normal distribution with mean

$$\mu_{W^+} = \frac{n(n + 1)}{4}$$

and variance

$$\sigma^2_{W^+} = \frac{n(n + 1)(2n + 1)}{24}$$

Therefore, a test of H_0: $\mu = \mu_0$ can be based on the statistic

$$Z_0 = \frac{W^+ - n(n + 1)/4}{\sqrt{n(n + 1)(2n + 1)/24}} \qquad (15\text{-}6)$$

An appropriate critical region for either the two-sided or one-sided alternative hypotheses can be chosen from a table of the standard normal distribution.

15-3.3   Paired Observations

The Wilcoxon signed-rank test can be applied to paired data. Let (X_{1j}, X_{2j}), $j = 1, 2, \ldots, n$ be a collection of paired observations from two continuous distributions that differ only with respect to their means. (It is not necessary that the distributions of X_1 and X_2 be symmetric.) This assures that the distribution of the differences $D_j = X_{1j} - X_{2j}$ is continuous and symmetric. Thus, the null hypothesis is H_0: $\mu_1 = \mu_2$, which is equivalent to H_0: $\mu_D = 0$. We initially consider the two-sided alternative H_1: $\mu_1 \neq \mu_2$ (or H_1: $\mu_D \neq 0$).

To use the Wilcoxon signed-rank test, the differences are first ranked in ascending order of their absolute values, and then the ranks are given the signs of the differences. Ties are assigned average ranks. Let W^+ be the sum of the positive ranks and W^- be the absolute value of the sum of the negative ranks, and $W = \min(W^+, W^-)$. If the observed value $w \leq w^*_\alpha$, the null hypothesis H_0: $\mu_1 = \mu_2$ (or H_0: $\mu_D = 0$) is rejected where w^*_α is chosen from Appendix Table VIII.

For one-sided tests, if the alternative is H_1: $\mu_1 > \mu_2$ (or H_1: $\mu_D > 0$), reject H_0 if $w^- \leq w^*_\alpha$; and if H_1: $\mu_1 < \mu_2$ (or H_1: $\mu_D < 0$), reject H_0 if $w^+ \leq w^*_\alpha$. Be sure to use the one-sided test significance levels shown in Appendix Table VIII.

EXAMPLE 15-5   We will apply the Wilcoxon signed-rank test to the fuel-metering device test data used previously in Example 15-3. The eight-step hypothesis-testing procedure can be applied as follows:

1. The parameters of interest are the mean fuel mileage performance for the two metering devices.

2. H_0: $\mu_1 = \mu_2$ or, equivalently, H_0: $\mu_D = 0$

3. H_1: $\mu_1 \neq \mu_2$ or, equivalently, H_1: $\mu_D \neq 0$

4. $\alpha = 0.05$

5. The test statistic is

$$w = \min(w^+, w^-)$$

where w^+ and w^- are the sums of the positive and negative ranks of the differences in Table 15-2.

6. Since $\alpha = 0.05$ and $n = 12$, Appendix Table VIII gives the critical value as $w^*_{0.05} = 13$. We will reject H_0: $\mu_D = 0$ if $w \leq 13$.

7. Computations: Using the data in Table 15-2, we compute the following signed ranks:

Car	Difference	Signed Rank
7	-0.2	-1
12	0.3	2
8	0.4	3
6	0.5	4
2	-0.6	-5
4	0.7	6.5
5	-0.7	-6.5
1	0.8	8
9	0.9	9
10	-1.0	-10
11	1.1	11
3	1.3	12

Note that $w^+ = 55.5$ and $w^- = 22.5$. Therefore,

$$w = \min(55.5, 22.5) = 22.5$$

8. Conclusions: Since $w = 22.5$ is not less than or equal to $w^*_{0.05} = 13$, we cannot reject the null hypothesis that the two metering devices produce the same mileage performance.

15-3.4 Comparison to the *t*-Test

When the underlying population is normal, either the *t*-test or the Wilcoxon signed-rank test can be used to test hypotheses about μ. As mentioned earlier, the *t*-test is the best test in such situations in the sense that it produces a minimum value of β for all tests with significance level α. However, since it is not always clear that the normal distribution is appropriate, and since in many situations it is inappropriate, it is of interest to compare the two procedures for both normal and nonnormal populations.

Unfortunately, such a comparison is not easy. The problem is that β for the Wilcoxon signed-rank test is very difficult to obtain, and β for the *t*-test is difficult to obtain for nonnormal distributions. Because type II error comparisons are difficult, other measures of comparison have been developed. One widely used measure is **asymptotic relative efficiency** (ARE).

The ARE of one test relative to another is the limiting ratio of the sample sizes necessary to obtain identical error probabilities for the two procedures. For example, if the ARE of one test relative to the competitor is 0.5, when sample sizes are large, the first test will require twice as large a sample as the second one to obtain similar error performance. While this does not tell us anything for small sample sizes, we can say the following:

1. For normal populations, the ARE of the Wilcoxon signed-rank test relative to the *t*-test is approximately 0.95.

2. For nonnormal populations, the ARE is at least 0.86, and in many cases it will exceed unity.

Although these are large-sample results, we generally conclude that the Wilcoxon signed-rank test will never be much worse than the *t*-test and that in many cases where the population is non-normal it may be superior. Thus, the Wilcoxon signed-rank test is a useful alternate to the *t*-test.

EXERCISES FOR SECTION 15-3

15-20. Consider the data in Exercise 15-1 and assume that the distribution of pH is symmetric and continuous. Use the Wilcoxon signed-rank test with $\alpha = 0.05$ to test the hypothesis $H_0: \mu = 7$ against $H_1: \mu \neq 7$.

15-21. Consider the data in Exercise 15-2. Suppose that the distribution of titanium content is symmetric and continuous. Use the Wilcoxon signed-rank test with $\alpha = 0.05$ to test the hypothesis $H_0: \mu = 8.5$ versus $H_1: \mu \neq 8.5$.

15-22. Consider the data in Exercise 15-2. Use the large-sample approximation for the Wilcoxon signed-rank test to test the hypothesis $H_0: \mu = 8.5$ versus $H_1: \mu \neq 8.5$. Use $\alpha = 0.05$. Assume that the distribution of titanium content is continuous and symmetric.

15-23. Consider the data in Exercise 15-3. Use the Wilcoxon signed-rank test to test the hypothesis $H_0: \mu = 2.5$ ppm versus $H_1: \mu < 2.5$ ppm with $\alpha = 0.05$. Assume that the distribution of impurity level is continuous and symmetric.

15-24. Consider the Rockwell hardness test data in Exercise 15-9. Assume that both distributions are continuous and use the Wilcoxon signed-rank test to test that the mean difference in hardness readings between the two tips is zero. Use $\alpha = 0.05$.

15-25. Consider the paint drying time data in Exercise 15-10. Assume that both populations are continuous, and use the Wilcoxon signed-rank test to test that the difference in mean drying times between the two formulations is zero. Use $\alpha = 0.01$.

15-26. Apply the Wilcoxon signed-rank test to the measurement data in Exercise 15-12. Use $\alpha = 0.05$ and assume that the two distributions of measurements are continuous.

15-27. Apply the Wilcoxon signed-rank test to the blood cholesterol data from Exercise 10-39. Use $\alpha = 0.05$ and assume that the two distributions are continuous.

15-4 WILCOXON RANK-SUM TEST

Suppose that we have two independent continuous populations X_1 and X_2 with means μ_1 and μ_2. Assume that the distributions of X_1 and X_2 have the same shape and spread and differ only (possibly) in their locations. The Wilcoxon rank-sum test can be used to test the hypothesis $H_0: \mu_1 = \mu_2$. This procedure is sometimes called the Mann-Whitney test, although the Mann-Whitney test statistic is usually expressed in a different form.

15-4.1 Description of the Test

Let $X_{11}, X_{12}, \ldots, X_{1n_1}$ and $X_{21}, X_{22}, \ldots, X_{2n_2}$ be two independent random samples of sizes $n_1 \leq n_2$ from the continuous populations X_1 and X_2 described earlier. We wish to test the hypotheses

$$H_0: \mu_1 = \mu_2$$
$$H_1: \mu_1 \neq \mu_2$$

The test procedure is as follows. Arrange all $n_1 + n_2$ observations in ascending order of magnitude and assign ranks to them. If two or more observations are tied (identical), use the mean of the ranks that would have been assigned if the observations differed.

Let W_1 be the sum of the ranks in the smaller sample (1), and define W_2 to be the sum of the ranks in the other sample. Then,

$$W_2 = \frac{(n_1 + n_2)(n_1 + n_2 + 1)}{2} - W_1 \qquad (15\text{-}7)$$

Now if the sample means do not differ, we will expect the sum of the ranks to be nearly equal for both samples after adjusting for the difference in sample size. Consequently, if the sums of the ranks differ greatly, we will conclude that the means are not equal.

Appendix Table IX contains the critical value of the rank sums for $\alpha = 0.05$ and $\alpha = 0.01$ assuming the two-sided alternative above. Refer to Appendix Table IX with the appropriate sample sizes n_1 and n_2, and the critical value w_α can be obtained. The null H_0: $\mu_1 = \mu_2$ is rejected in favor of H_1: $\mu_1 \neq \mu_2$ if either of the observed values w_1 or w_2 is less than or equal to the tabulated critical value w_α.

The procedure can also be used for one-sided alternatives. If the alternative is H_1: $\mu_1 < \mu_2$, reject H_0 if $w_1 \leq w_\alpha$; for H_1: $\mu_1 > \mu_2$, reject H_0 if $w_2 \leq w_\alpha$. For these one-sided tests, the tabulated critical values w_α correspond to levels of significance of $\alpha = 0.025$ and $\alpha = 0.005$.

EXAMPLE 15-6

The mean axial stress in tensile members used in an aircraft structure is being studied. Two alloys are being investigated. Alloy 1 is a traditional material, and alloy 2 is a new aluminum-lithium alloy that is much lighter than the standard material. Ten specimens of each alloy type are tested, and the axial stress is measured. The sample data are assembled in Table 15-3. Using $\alpha = 0.05$, we wish to test the hypothesis that the means of the two stress distributions are identical.

We will apply the eight-step hypothesis-testing procedure to this problem:

1. The parameters of interest are the means of the two distributions of axial stress.
2. H_0: $\mu_1 = \mu_2$
3. H_1: $\mu_1 \neq \mu_2$
4. $\alpha = 0.05$
5. We will use the Wilcoxon rank-sum test statistic in Equation 15-7,

$$w_2 = \frac{(n_1 + n_2)(n_1 + n_2 + 1)}{2} - w_1$$

6. Since $\alpha = 0.05$ and $n_1 = n_2 = 10$, Appendix Table IX gives the critical value as $w_{0.05} = 78$. If either w_1 or w_2 is less than or equal to $w_{0.05} = 78$, we will reject H_0: $\mu_1 = \mu_2$.

Table 15-3 Axial Stress for Two Aluminum-Lithium Alloys

Alloy 1		Alloy 2	
238 psi	3254 psi	3261 psi	3248 psi
3195	3229	3187	3215
3246	3225	3209	3226
3190	3217	3212	3240
3204	3241	3258	3234

7. Computations: The data from Table 15-3 are analyzed in ascending order and ranked as follows:

Alloy Number	Axial Stress	Rank
2	3187 psi	1
1	3190	2
1	3195	3
1	3204	4
2	3209	5
2	3212	6
2	3215	7
1	3217	8
1	3225	9
2	3226	10
1	3229	11
2	3234	12
1	3238	13
2	3240	14
1	3241	15
1	3246	16
2	3248	17
1	3254	18
2	3258	19
2	3261	20

The sum of the ranks for alloy 1 is

$$w_1 = 2 + 3 + 4 + 8 + 9 + 11 + 13 + 15 + 16 + 18 = 99$$

and for alloy 2

$$w_2 = \frac{(n_1 + n_2)(n_1 + n_2 + 1)}{2} - w_1 = \frac{(10 + 10)(10 + 10 + 1)}{2} - 99 = 111$$

8. Conclusions: Since neither w_1 nor w_2 is less than or equal to $w_{0.05} = 78$, we cannot reject the null hypothesis that both alloys exhibit the same mean axial stress.

15-4.2 Large-Sample Approximation

When both n_1 and n_2 are moderately large, say, greater than 8, the distribution of w_1 can be well approximated by the normal distribution with mean

$$\mu_{W_1} = \frac{n_1(n_1 + n_2 + 1)}{2}$$

and variance

$$\sigma_{W_1}^2 = \frac{n_1 n_2(n_1 + n_2 + 1)}{12}$$

Therefore, for n_1 and $n_2 > 8$, we could use

$$Z_0 = \frac{W_1 - \mu_{W_1}}{\sigma_{W_1}} \qquad (15\text{-}8)$$

as a statistic, and the appropriate critical region is $|z_0| > z_{\alpha/2}, z_0 > z_\alpha$, or $z_0 < -z_\alpha$, depending on whether the test is a two-tailed, upper-tail, or lower-tail test.

15-4.3 Comparison to the t-Test

In Section 15-3.4 we discussed the comparison of the t-test with the Wilcoxon signed-rank test. The results for the two-sample problem are identical to the one-sample case. That is, when the normality assumption is correct, the Wilcoxon rank-sum test is approximately 95% as efficient as the t-test in large samples. On the other hand, regardless of the form of the distributions, the Wilcoxon rank-sum test will always be at least 86% as efficient. The efficiency of the Wilcoxon test relative to the t-test is usually high if the underlying distribution has heavier tails than the normal, because the behavior of the t-test is very dependent on the sample mean, which is quite unstable in heavy-tailed distributions.

EXERCISES FOR SECTION 15-4

15-28.   An electrical engineer must design a circuit to deliver the maximum amount of current to a display tube to achieve sufficient image brightness. Within her allowable design constraints, she has developed two candidate circuits and tests prototypes of each. The resulting data (in microamperes) are as follows:

Circuit 1:	251, 255, 258, 257, 250, 251, 254, 250, 248
Circuit 2:	250, 253, 249, 256, 259, 252, 260, 251

Use the Wilcoxon rank-sum test to test $H_0: \mu_1 = \mu_2$ against the alternative $H_1: \mu_1 > \mu_2$. Use $\alpha = 0.025$.

15-29.   One of the authors travels regularly to Seattle, Washington. He uses either Delta or Alaska. Flight delays are sometimes unavoidable, but he would be willing to give most of his business to the airline with the best on-time arrival record. The number of minutes that his flight arrived late for the last six trips on each airline follows. Is there evidence that either airline has superior on-time arrival performance? Use $\alpha = 0.01$ and the Wilcoxon rank-sum test.

Delta:	13, 10, 1, −4, 0, 9 (minutes late)
Alaska:	15, 8, 3, −1, −2, 4 (minutes late)

15-30.   The manufacturer of a hot tub is interested in testing two different heating elements for his product. The element that produces the maximum heat gain after 15 minutes would be preferable. He obtains 10 samples of each heating unit and tests each one. The heat gain after 15 minutes (in °F) follows.

Is there any reason to suspect that one unit is superior to the other? Use $\alpha = 0.05$ and the Wilcoxon rank-sum test.

Unit 1:	25, 27, 29, 31, 30, 26, 24, 32, 33, 38
Unit 2:	31, 33, 32, 35, 34, 29, 38, 35, 37, 30

15-31.   Use the normal approximation for the Wilcoxon rank-sum test for the problem in Exercise 15-28. Assume that $\alpha = 0.05$. Find the approximate P-value for this test statistic.

15-32.   Use the normal approximation for the Wilcoxon rank-sum test for the heat gain experiment in Exercise 15-30. Assume that $\alpha = 0.05$. What is the approximate P-value for this test statistic?

15-33.   Consider the chemical etch rate data in Exercise 10-21. Use the Wilcoxon rank-sum test to investigate the claim that the mean etch rate is the same for both solutions. If $\alpha = 0.05$, what are your conclusions?

15-34.   Use the Wilcoxon rank-sum test for the pipe deflection temperature experiment described in Exercise 10-20. If $\alpha = 0.05$, what are your conclusions?

15-35.   Use the normal approximation for the Wilcoxon rank-sum test for the problem in Exercise 10-21. Assume that $\alpha = 0.05$. Find the approximate P-value for this test.

15-36.   Use the normal approximation for the Wilcoxon rank-sum test for the problem in Exercise 10-20. Assume that $\alpha = 0.05$. Find the approximate P-value for this test.

15-5 NONPARAMETRIC METHODS IN THE ANALYSIS OF VARIANCE

15-5.1 Kruskal-Wallis Test

The single-factor analysis of variance model developed in Chapter 13 for comparing a population means is

$$Y_{ij} = \mu + \tau_i + \epsilon_{ij} \begin{cases} i = 1, 2, \ldots, a \\ j = 1, 2, \ldots, n_i \end{cases} \tag{15-9}$$

In this model, the error terms ϵ_{ij} are assumed to be normally and independently distributed with mean zero and variance σ^2. The assumption of normality led directly to the F-test described in Chapter 13. The Kruskal-Wallis test is a nonparametric alternative to the F-test; it requires only that the ϵ_{ij} have the same continuous distribution for all factor levels $i = 1, 2, \ldots, a$.

Suppose that $N = \sum_{i=1}^{a} n_i$ is the total number of observations. Rank all N observations from smallest to largest, and assign the smallest observation rank 1, the next smallest rank 2, . . . , and the largest observation rank N. If the null hypothesis

$$H_0: \mu_1 = \mu_2 = \cdots = \mu_a$$

is true, the N observations come from the same distribution, and all possible assignments of the N ranks to the a samples are equally likely, we would expect the ranks 1, 2, . . . , N to be mixed throughout the a samples. If, however, the null hypothesis H_0 is false, some samples will consist of observations having predominantly small ranks, while other samples will consist of observations having predominantly large ranks. Let R_{ij} be the rank of observation Y_{ij}, and let $R_{i\cdot}$ and $\bar{R}_{i\cdot}$ denote the total and average of the n_i ranks in the ith treatment. When the null hypothesis is true,

$$E(R_{ij}) = \frac{N + 1}{2}$$

and

$$E(\bar{R}_{i\cdot}) = \frac{1}{n_i} \sum_{j=1}^{n_i} E(R_{ij}) = \frac{N + 1}{2}$$

The Kruskal-Wallis test statistic measures the degree to which the actual observed average ranks $\bar{R}_{i\cdot}$ differ from their expected value $(N + 1)/2$. If this difference is large, the null hypothesis H_0 is rejected. The test statistic is

$$H = \frac{12}{N(N + 1)} \sum_{i=1}^{a} n_i \left(\bar{R}_{i\cdot} - \frac{N + 1}{2} \right)^2 \tag{15-10}$$

An alternative computing formula that is occasionally more convenient is

$$H = \frac{12}{N(N+1)} \sum_{i=1}^{a} \frac{R_{i\cdot}^2}{n_i} - 3(N+1) \tag{15-11}$$

We would usually prefer Equation 15-11 to Equation 15-10 because it involves the rank totals rather than the averages.

The null hypothesis H_0 should be rejected if the sample data generate a large value for H. The null distribution for H has been obtained by using the fact that under H_0 each possible assignment of ranks to the a treatments is equally likely. Thus, we could enumerate all possible assignments and count the number of times each value of H occurs. This has led to tables of the critical values of H, although most tables are restricted to small sample sizes n_i. In practice, we usually employ the following large-sample approximation. Whenever H_0 is true and either

$$a = 3 \quad \text{and} \quad n_i \geq 6 \quad \text{for } i = 1, 2, 3$$
$$a > 3 \quad \text{and} \quad n_i \geq 5 \quad \text{for } i = 1, 2, \ldots, a$$

H has approximately a chi-square distribution with $a-1$ degrees of freedom. Since large values of H imply that H_0 is false, we will reject H_0 if the observed value

$$h \geq \chi_{\alpha, a-1}^2$$

The test has approximate significance level α.

Ties in the Kruskal-Wallis Test

When observations are tied, assign an average rank to each of the tied observations. When there are ties, we should replace the test statistic in Equation 15-11 by

$$H = \frac{1}{S^2} \left[\sum_{i=1}^{a} \frac{R_{i\cdot}^2}{n_i} - \frac{N(N+1)^2}{4} \right] \tag{15-12}$$

where n_i is the number of observations in the ith treatment, N is the total number of observations, and

$$S^2 = \frac{1}{N-1} \left[\sum_{i=1}^{a} \sum_{j=1}^{n_i} R_{ij}^2 - \frac{N(N+1)^2}{4} \right] \tag{15-13}$$

Note that S^2 is just the variance of the ranks. When the number of ties is moderate, there will be little difference between Equations 15-11 and 15-12 and the simpler form (Equation 15-11) may be used.

EXAMPLE 15-7   Montgomery (2001) presented data from an experiment in which five different levels of cotton content in a synthetic fiber were tested to determine whether cotton content has any effect on fiber tensile strength. The sample data and ranks from this experiment are shown in Table 15-4. We will apply the Kruskal-Wallis test to these data, using $\alpha = 0.01$.

Table 15-4   Data and Ranks for the Tensile Testing Experiment

Percentage of Cotton							$r_{i.}$
15	y_{1j}	7	7	15	11	9	
ranks	r_{1j}	2.0	2.0	12.5	7.0	4.0	27.5
20	y_{2j}	12	17	12	18	18	
ranks	r_{2j}	9.5	14.0	9.5	16.5	16.5	66.0
25	y_{3j}	14	18	18	19	19	
ranks	r_{3j}	11.0	16.5	16.5	20.5	20.5	85.0
30	y_{4j}	19	25	22	19	23	
ranks	r_{4j}	20.5	25.0	23.0	20.5	24.0	113.0
35	y_{5j}	7	10	11	15	11	
ranks	r_{5j}	2.0	5.0	7.0	12.5	7.0	33.5

Since there is a fairly large number of ties, we use Equation 15-12 as the test statistic. From Equation 15-13 we find

$$s^2 = \frac{1}{N-1}\left[\sum_{i=1}^{a}\sum_{j=1}^{n_i} r_{ij}^2 - \frac{N(N+1)^2}{4}\right]$$

$$= \frac{1}{24}\left[5510 - \frac{25(26)^2}{2}\right]$$

$$= 53.54$$

and the test statistic is

$$h = \frac{1}{s^2}\left[\sum_{i=1}^{a}\frac{r_{i.}^2}{n_i} - \frac{N(N+1)^2}{4}\right]$$

$$= \frac{1}{53.54}\left[5245.7 - \frac{25(26)2}{4}\right]$$

$$= 19.06$$

Since $h > \chi_{0.01,4}^2 = 13.28$, we would reject the null hypothesis and conclude that treatments differ. This same conclusion is given by the usual analysis of variance F-test.

15-5.2   Rank Transformation

The procedure used in the previous section whereby the observations are replaced by their ranks is called the **rank transformation.** It is a very powerful and widely useful technique. If we were to apply the ordinary F-test to the ranks rather than to the original data, we would obtain

$$F_0 = \frac{H/(a-1)}{(N-1-H)/(N-a)}$$

as the test statistic. Note that as the Kruskal-Wallis statistic H increases or decreases, F_0 also increases or decreases. Now, since the distribution of F_0 is approximated by the F-distribution, the Kruskal-Wallis test is approximately equivalent to applying the usual analysis of variance to the ranks.

The rank transformation has wide applicability in experimental design problems for which no nonparametric alternative to the analysis of variance exists. If the data are ranked and the ordinary F-test is applied, an approximate procedure results, but one that has good statistical properties. When we are concerned about the normality assumption or the effect of outliers or "wild" values, we recommend that the usual analysis of variance be performed on both the original data and the ranks. When both procedures give similar results, the analysis of variance assumptions are probably satisfied reasonably well, and the standard analysis is satisfactory. When the two procedures differ, the rank transformation should be preferred since it is less likely to be distorted by nonnormality and unusual observations. In such cases, the experimenter may want to investigate the use of transformations for nonnormality and examine the data and the experimental procedure to determine whether outliers are present and why they have occurred.

EXERCISES FOR SECTION 15-5

15-37. Montgomery (2001) presented the results of an experiment to compare four different mixing techniques on the tensile strength of portland cement. The results are shown in the following table. Is there any indication that mixing technique affects the strength? Use $\alpha = 0.05$.

Mixing Technique	Tensile Strength (lb/in.^{2})			
1	3129	3000	2865	2890
2	3200	3000	2975	3150
3	2800	2900	2985	3050
4	2600	2700	2600	2765

15-38. An article in the *Quality Control Handbook*, 3rd edition (McGraw-Hill, 1962) presents the results of an experiment performed to investigate the effect of three different conditioning methods on the breaking strength of cement briquettes. The data are shown in the following table. Using $\alpha = 0.05$, is there any indication that conditioning method affects breaking strength?

Conditioning Method	Breaking Strength (lb/in.^{2})				
1	553	550	568	541	537
2	553	599	579	545	540
3	492	530	528	510	571

15-39. In *Statistics for Research* (John Wiley & Sons, 1983), S. Dowdy and S. Wearden presented the results of an experiment to measure the performance of hand-held chain saws. The experimenters measured the kickback angle through which the saw is deflected when it begins to cut a 3-inch stock synthetic board. Shown in the following table are deflection angles for five saws chosen at random from each of four different manufacturers. Is there any evidence that the manufacturers' products differ with respect to kickback angle? Use $\alpha = 0.01$.

Manufacturer	Kickback Angle				
A	42	17	24	39	43
B	28	50	44	32	61
C	57	45	48	41	54
D	29	40	22	34	30

15-40. Consider the data in Exercise 13-2. Use the Kruskal-Wallis procedure with $\alpha = 0.05$ to test for differences between mean uniformity at the three different gas flow rates.

15-41. Find the approximate P-value for the test statistic computed in Exercise 15-37.

15-42. Find the approximate P-value for the test statistic computed in Exercise 15-40.

Supplemental Exercises

15-43. The surface finish of 10 metal parts produced in a grinding process is as follows: (in microinches): 10.32, 9.68, 9.92, 10.10, 10.20, 9.87, 10.14, 9.74, 9.80, 10.26. Do the data support the claim that the median value of surface finish is 10 microinches? Use the sign test with $\alpha = 0.05$. What is the P-value for this test?

15-44. Use the normal appoximation for the sign test for the problem in Exercise 15-43. Find the P-value for this test. What are your conclusions if $\alpha = 0.05$?

15-45. Fluoride emissions (in ppm) from a chemical plant are monitored routinely. The following are 15 observations based on air samples taken randomly during one month of production: 7, 3, 4, 2, 5, 6, 9, 8, 7, 3, 4, 4, 3, 2, 6. Can you claim that the median fluoride impurity level is less than 6 ppm? State and test the appropriate hypotheses using the sign test with $\alpha = 0.05$. What is the P-value for this test?

15-46. Use the normal approximation for the sign test for the problem in Exercise 15-45. What is the P-value for this test?

15-47. Consider the data in Exercise 10-42. Use the sign test with $\alpha = 0.05$ to determine whether there is a difference in median impurity readings between the two analytical tests.

15-48. Consider the data in Exercise 15-43. Use the Wilcoxon signed-rank test for this problem with $\alpha = 0.05$. What hypotheses are being tested in this problem?

15-49. Consider the data in Exercise 15-45. Use the Wilcoxon signed-rank test for this problem with $\alpha = 0.05$. What conclusions can you draw? Does the hypothesis you are testing now differ from the one tested originally in Exercise 15-45?

15-50. Use the Wilcoxon signed-rank test with $\alpha = 0.05$ for the diet-modification experiment described in Exercise 10-41. State carefully the conclusions that you can draw from this experiment.

15-51. Use the Wilcoxon rank-sum test with $\alpha = 0.01$ for the fuel-economy study described in Exercise 10-83. What conclusions can you draw about the difference in mean mileage performance for the two vehicles in this study?

15-52. Use the large-sample approximation for the Wilcoxon rank-sum test for the fuel-economy data in Exercise 10-83. What conclusions can you draw about the difference in means if $\alpha = 0.01$? Find the P-value for this test.

15-53. Use the Wilcoxon rank-sum test with $\alpha = 0.025$ for the fill-capability experiment described in Exercise 10-85. What conclusions can you draw about the capability of the two fillers?

15-54. Use the large-sample approximation for the Wilcoxon rank-sum test with $\alpha = 0.025$ for the fill-capability experiment described in Exercise 10-85. Find the P-value for this test. What conclusions can you draw?

15-55. Consider the contact resistance experiment in Exercise 13-31. Use the Kruskal-Wallis test to test for differences in mean contact resistance among the three alloys. If $\alpha = 0.01$, what are your conclusions? Find the P-value for this test.

15-56. Consider the experiment described in Exercise 13-28. Use the Kruskal-Wallis test for this experiment with $\alpha = 0.05$. What conclusions would you draw? Find the P-value for this test.

15-57. Consider the bread quality experiment in Exercise 13-35. Use the Kruskal-Wallis test with $\alpha = 0.01$ to analyze the data from this experiment. Find the P-value for this test. What conclusions can you draw?

MIND-EXPANDING EXERCISES

15-58. For the large-sample approximation to the Wilcoxon signed-rank test, derive the mean and standard deviation of the test statistic used in the procedure.

15-59. **Testing for Trends.** A turbocharger wheel is manufactured using an investment-casting process. The shaft fits into the wheel opening, and this wheel opening is a critical dimension. As wheel wax patterns are formed, the hard tool producing the wax patterns wears. This may cause growth in the wheel-opening dimension. Ten wheel-opening measurements, in time order of production, are 4.00 (millimeters), 4.02, 4.03, 4.01, 4.00, 4.03, 4.04, 4.02, 4.03, 4.03.

(a) Suppose that p is the probability that observation X_{i+5} exceeds observation X_i. If there is no upward or downward trend, X_{i+5} is no more or less likely to exceed X_i or lie below X_i. What is the value of p?

(b) Let V be the number of values of i for which $X_{i+5} > X_i$. If there is no upward or downward trend

in the measurements, what is the probability distribution of V?

(c) Use the data above and the results of parts (a) and (b) to test H_0: there is no trend, versus H_1: there is upward trend. Use $\alpha = 0.05$.

Note that this test is a modification of the sign test. It was developed by Cox and Stuart.

15-60. Consider the Wilcoxon signed-rank test, and suppose that $n = 5$. Assume that H_0: $\mu = \mu_0$ is true.

(a) How many different sequences of signed ranks are possible? Enumerate these sequences.

(b) How many different values of W^+ are there? Find the probability associated with each value of W^+.

(c) Suppose that we define the critical region of the test to be to reject H_0 if $w^+ > w_\alpha^*$ and $w_\alpha^* = 13$. What is the approximate α level of this test?

(d) Does this exercise show how the critical values for the Wilcoxon signed-rank test were developed? Explain.

IMPORTANT TERMS AND CONCEPTS

In the E-book, click on any term or concept below to go to that subject.

Asymptotic relative efficiency

Kruskal-Wallis test

Nonparametric or distribution free procedures

Normal approximation to nonparametric tests

Paired observations

Ranks

Rank transformation in ANOVA

Sign test

Tests on the mean of a symmetric continuous distribution

Tests on the median of a continuous distribution

Wilcoxon rank-sum test

Wilcoxon signed-rank test

16 Statistical Quality Control

CHAPTER OUTLINE

LEARNING OBJECTIVES

After careful study of this chapter, you should be able to do the following:

1. Understand the role of statistical tools in quality improvement

2. Understand the different types of variability, rational subgroups, and how a control chart is used to detect assignable causes

3. Understand the general form of a Shewhart control chart and how to apply zone rules (such as the Western Electric rules) and pattern analysis to detect assignable causes

4. Construct and interpret control charts for variables such as $\overline{X}$, R, S, and individuals charts

5. Construct and interpret control charts for attributes such as P and U charts

6. Calculate and interpret process capability ratios

7. Calculate the ARL performance for a Shewhart control chart
8. Construct and interpret a cumulative sum control chart
9. Use other statistical process control problem-solving tools

---

Answers for most odd numbered exercises are at the end of the book. Answers to exercises whose numbers are surrounded by a box can be accessed in the e-text by clicking on the box. Complete worked solutions to certain exercises are also available in the e-text. These are indicated in the Answers to Selected Exercises section by a box around the exercise number. Exercises are also available for some of the text sections that appear on CD only. These exercises may be found within the e-Text immediately following the section they accompany.

16-1   QUALITY IMPROVEMENT AND STATISTICS

The quality of products and services has become a major decision factor in most businesses today. Regardless of whether the consumer is an individual, a corporation, a military defense program, or a retail store, when the consumer is making purchase decisions, he or she is likely to consider quality of equal importance to cost and schedule. Consequently, **quality improvement** has become a major concern to many U.S. corporations. This chapter is about **statistical quality control,** a collection of tools that are essential in quality-improvement activities.

Quality means **fitness for use.** For example, you or I may purchase automobiles that we expect to be free of manufacturing defects and that should provide reliable and economical transportation, a retailer buys finished goods with the expectation that they are properly packaged and arranged for easy storage and display, or a manufacturer buys raw material and expects to process it with no rework or scrap. In other words, all consumers expect that the products and services they buy will meet their requirements. Those requirements define fitness for use.

Quality or fitness for use is determined through the interaction of **quality of design** and **quality of conformance.** By quality of design we mean the different grades or levels of performance, reliability, serviceability, and function that are the result of deliberate engineering and management decisions. By quality of conformance, we mean the systematic **reduction of variability** and **elimination of defects** until every unit produced is identical and defect-free.

Some confusion exists in our society about quality improvement; some people still think that it means gold-plating a product or spending more money to develop a product or process. This thinking is wrong. Quality improvement means the systematic **elimination of waste.** Examples of waste include scrap and rework in manufacturing, inspection and testing, errors on documents (such as engineering drawings, checks, purchase orders, and plans), customer complaint hotlines, warranty costs, and the time required to do things over again that could have been done right the first time. A successful quality-improvement effort can eliminate much of this waste and lead to lower costs, higher productivity, increased customer satisfaction, increased business reputation, higher market share, and ultimately higher profits for the company.

Statistical methods play a vital role in quality improvement. Some applications are outlined below:

1.   In product design and development, statistical methods, including designed experiments, can be used to compare different materials, components, or ingredients, and to help determine both system and component tolerances. This application can significantly lower development costs and reduce development time.

2. Statistical methods can be used to determine the capability of a manufacturing process. Statistical process control can be used to systematically improve a process by reducing variability.

3. Experimental design methods can be used to investigate improvements in the process. These improvements can lead to higher yields and lower manufacturing costs.

4. Life testing provides reliability and other performance data about the product. This can lead to new and improved designs and products that have longer useful lives and lower operating and maintenance costs.

Some of these applications have been illustrated in earlier chapters of this book. It is essential that engineers, scientists, and managers have an in-depth understanding of these statistical tools in any industry or business that wants to be a high-quality, low-cost producer. In this chapter we provide an introduction to the basic methods of statistical quality control that, along with experimental design, form the basis of a successful quality-improvement effort.

16-2 STATISTICAL QUALITY CONTROL

The field of statistical quality control can be broadly defined as those statistical and engineering methods that are used in measuring, monitoring, controlling, and improving quality. Statistical quality control is a field that dates back to the 1920s. Dr. Walter A. Shewhart of the Bell Telephone Laboratories was one of the early pioneers of the field. In 1924 he wrote a memorandum showing a modern control chart, one of the basic tools of statistical process control. Harold F. Dodge and Harry G. Romig, two other Bell System employees, provided much of the leadership in the development of statistically based sampling and inspection methods. The work of these three men forms much of the basis of the modern field of statistical quality control. World War II saw the widespread introduction of these methods to U.S. industry. Dr. W. Edwards Deming and Dr. Joseph M. Juran have been instrumental in spreading statistical quality-control methods since World War II.

The Japanese have been particularly successful in deploying statistical quality-control methods and have used statistical methods to gain significant advantage over their competitors. In the 1970s American industry suffered extensively from Japanese (and other foreign) competition; that has led, in turn, to renewed interest in statistical quality-control methods in the United States. Much of this interest focuses on *statistical process control* and *experimental design*. Many U.S. companies have begun extensive programs to implement these methods in their manufacturing, engineering, and other business organizations.

16-3 STATISTICAL PROCESS CONTROL

It is impractical to inspect quality into a product; the product must be built right the first time. The manufacturing process must therefore be stable or repeatable and capable of operating with little variability around the target or nominal dimension. Online statistical process control is a powerful tool for achieving process stability and improving capability through the reduction of variability.

It is customary to think of **statistical process control (SPC)** as a set of problem-solving tools that may be applied to any process. The major tools of SPC* are

1. Histogram
2. Pareto chart
3. Cause-and-effect diagram
4. Defect-concentration diagram
5. Control chart
6. Scatter diagram
7. Check sheet

Although these tools are an important part of SPC, they comprise only the technical aspect of the subject. An equally important element of SPC is attitude—a desire of all individuals in the organization for continuous improvement in quality and productivity through the systematic reduction of variability. The control chart is the most powerful of the SPC tools.

16-4   INTRODUCTION TO CONTROL CHARTS

16-4.1   Basic Principles

In any production process, regardless of how well-designed or carefully maintained it is, a certain amount of inherent or natural variability will always exist. This natural variability or "background noise" is the cumulative effect of many small, essentially unavoidable causes. When the background noise in a process is relatively small, we usually consider it an acceptable level of process performance. In the framework of statistical quality control, this natural variability is often called a "stable system of chance causes." A process that is operating with only **chance causes** of variation present is said to be in statistical control. In other words, the chance causes are an inherent part of the process.

Other kinds of variability may occasionally be present in the output of a process. This variability in key quality characteristics usually arises from three sources: improperly adjusted machines, operator errors, or defective raw materials. Such variability is generally large when compared to the background noise, and it usually represents an unacceptable level of process performance. We refer to these sources of variability that are not part of the chance cause pattern as **assignable causes.** A process that is operating in the presence of assignable causes is said to be out of control.[†]

Production processes will often operate in the in-control state, producing acceptable product for relatively long periods of time. Occasionally, however, assignable causes will occur, seemingly at random, resulting in a "shift" to an out-of-control state where a large proportion of the process output does not conform to requirements. A major objective of statistical process control is to quickly detect the occurrence of assignable causes or process shifts so that investigation of the process and corrective action may be undertaken before many

---

* Some prefer to include the experimental design methods discussed previously as part of the SPC toolkit. We did not do so, because we think of SPC as an online approach to quality improvement using techniques founded on passive observation of the process, while design of experiments is an active approach in which deliberate changes are made to the process variables. As such, designed experiments are often referred to as offline quality control.

[†] The terminology *chance* and *assignable* causes was developed by Dr. Walter A. Shewhart. Today, some writers use *common* cause instead of *chance* cause and *special* cause instead of *assignable* cause.

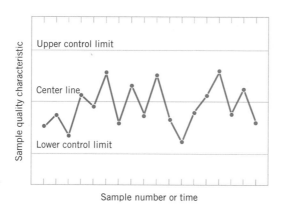

Figure 16-1   A typical control chart.

nonconforming units are manufactured. The control chart is an online process-monitoring technique widely used for this purpose.

Recall the following from Chapter 1. Figure 1-10 illustrates that adjustments to common causes of variation increase the variation of a process whereas Fig. 1-11 illustrates that actions should be taken in response to assignable causes of variation. Control charts may also be used to estimate the parameters of a production process and, through this information, to determine the capability of a process to meet specifications. The control chart can also provide information that is useful in improving the process. Finally, remember that the eventual goal of statistical process control is the *elimination of variability in the process*. Although it may not be possible to eliminate variability completely, the control chart helps reduce it as much as possible.

A typical control chart is shown in Fig. 16-1, which is a graphical display of a quality characteristic that has been measured or computed from a sample versus the sample number or time. Often, the samples are selected at periodic intervals such as every hour. The chart contains a center line (CL) that represents the average value of the quality characteristic corresponding to the in-control state. (That is, only chance causes are present.) Two other horizontal lines, called the upper control limit (UCL) and the lower control limit (LCL), are also shown on the chart. These control limits are chosen so that if the process is in control, nearly all of the sample points will fall between them. In general, as long as the points plot within the control limits, the process is assumed to be in control, and no action is necessary. However, a point that plots outside of the control limits is interpreted as evidence that the process is out of control, and investigation and corrective action are required to find and eliminate the assignable cause or causes responsible for this behavior. The sample points on the control chart are usually connected with straight-line segments so that it is easier to visualize how the sequence of points has evolved over time.

Even if all the points plot inside the control limits, if they behave in a systematic or non-random manner, this is an indication that the process is out of control. For example, if 18 of the last 20 points plotted above the center line but below the upper control limit and only two of these points plotted below the center line but above the lower control limit, we would be very suspicious that something was wrong. If the process is in control, all the plotted points should have an essentially random pattern. Methods designed to find sequences or nonrandom patterns can be applied to control charts as an aid in detecting out-of-control conditions. A particular nonrandom pattern usually appears on a control chart for a reason, and if that reason can be found and eliminated, process performance can be improved.

There is a close connection between control charts and hypothesis testing. Essentially, the control chart is a test of the hypothesis that the process is in a state of statistical control. A point plotting within the control limits is equivalent to failing to reject the hypothesis of statistical control, and a point plotting outside the control limits is equivalent to rejecting the hypothesis of statistical control.

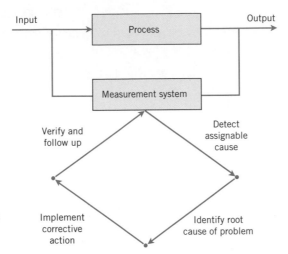

Figure 16-2 Process improvement using the control chart.

We give a general *model* for a control chart. Let W be a sample statistic that measures some quality characteristic of interest, and suppose that the mean of W is μ_W and the standard deviation of W is σ_W.* Then the center line, the upper control limit, and the lower control limit become

$$UCL = \mu_W + k\sigma_W$$
$$CL = \mu_W$$
$$LCL = \mu_W - k\sigma_W \qquad\qquad (16\text{-}1)$$

where k is the "distance" of the control limits from the center line, expressed in standard deviation units. A common choice is $k = 3$. This general theory of control charts was first proposed by Dr. Walter A. Shewhart, and control charts developed according to these principles are often called **Shewhart control charts.**

The control chart is a device for describing exactly what is meant by statistical control; as such, it may be used in a variety of ways. In many applications, it is used for online process monitoring. That is, sample data are collected and used to construct the control chart, and if the sample values of $\bar{x}$ (say) fall within the control limits and do not exhibit any systematic pattern, we say the process is in control at the level indicated by the chart. Note that we may be interested here in determining *both* whether the past data came from a process that was in control and whether future samples from this process indicate statistical control.

The most important use of a control chart is to *improve* the process. We have found that, generally

1. Most processes do not operate in a state of statistical control.

2. Consequently, the routine and attentive use of control charts will identify assignable causes. If these causes can be eliminated from the process, variability will be reduced and the process will be improved.

This process-improvement activity using the control chart is illustrated in Fig. 16-2. Notice that:

---

*Note that "sigma" refers to the standard deviation of the statistic plotted on the chart (i.e., σ_W), not the standard deviation of the quality characteristic.

3. The control chart will only *detect* assignable causes. Management, operator, and engineering *action* will usually be necessary to eliminate the assignable cause. An action plan for responding to control chart signals is vital.

In identifying and eliminating assignable causes, it is important to find the underlying **root cause** of the problem and to attack it. A cosmetic solution will not result in any real, long-term process improvement. Developing an effective system for corrective action is an essential component of an effective SPC implementation.

We may also use the control chart as an *estimating device*. That is, from a control chart that exhibits statistical control, we may estimate certain process parameters, such as the mean, standard deviation, and fraction nonconforming or fallout. These estimates may then be used to determine the *capability* of the process to produce acceptable products. Such **process capability studies** have considerable impact on many management decision problems that occur over the product cycle, including make-or-buy decisions, plant and process improvements that reduce process variability, and contractual agreements with customers or suppliers regarding product quality.

Control charts may be classified into two general types. Many quality characteristics can be measured and expressed as numbers on some continuous scale of measurement. In such cases, it is convenient to describe the quality characteristic with a measure of central tendency and a measure of variability. Control charts for central tendency and variability are collectively called **variables control charts.** The $\overline{X}$ chart is the most widely used chart for monitoring central tendency, whereas charts based on either the sample range or the sample standard deviation are used to control process variability. Many quality characteristics are not measured on a continuous scale or even a quantitative scale. In these cases, we may judge each unit of product as either conforming or nonconforming on the basis of whether or not it possesses certain attributes, or we may count the number of nonconformities (defects) appearing on a unit of product. Control charts for such quality characteristics are called **attributes control charts.**

Control charts have had a long history of use in industry. There are at least five reasons for their popularity:

1. Control charts are a proven technique for improving productivity. A successful control chart program will reduce scrap and rework, which are the primary productivity killers in *any* operation. If you reduce scrap and rework, productivity increases, cost decreases, and production capacity (measured in the number of *good* parts per hour) increases.

2. Control charts are effective in defect prevention. The control chart helps keep the process in control, which is consistent with the "do it right the first time" philosophy. It is never cheaper to sort out the "good" units from the "bad" later on than it is to build them correctly initially. If you do not have effective process control, you are paying someone to make a nonconforming product.

3. Control charts prevent unnecessary process adjustments. A control chart can distinguish between background noise and abnormal variation; no other device, including a human operator, is as effective in making this distinction. If process operators adjust the process based on periodic tests unrelated to a control chart program, they will often overreact to the background noise and make unneeded adjustments. These unnecessary adjustments can result in a deterioration of process performance. In other words, the control chart is consistent with the "if it isn't broken, don't fix it" philosophy.

4. Control charts provide diagnostic information. Frequently, the pattern of points on the control chart will contain information that is of diagnostic value to an experienced operator or engineer. This information allows the operator to implement a change in the process that will improve its performance.

5. **Control charts provide information about process capability.** The control chart provides information about the value of important process parameters and their stability over time. This allows an estimate of process capability to be made. This information is of tremendous use to product and process designers.

Control charts are among the most effective management control tools, and they are as important as cost controls and material controls. Modern computer technology has made it easy to implement control charts in any type of process, because data collection and analysis can be performed on a microcomputer or a local area network terminal in realtime, online at the work center.

16-4.2   Design of a Control Chart

To illustrate these ideas, we give a simplified example of a control chart. In manufacturing automobile engine piston rings, the inside diameter of the rings is a critical quality characteristic. The process mean inside ring diameter is 74 millimeters, and it is known that the standard deviation of ring diameter is 0.01 millimeters. A control chart for average ring diameter is shown in Fig. 16-3. Every hour a random sample of five rings is taken, the average ring diameter of the sample (say $\bar{x}$) is computed, and $\bar{x}$ is plotted on the chart. Because this control chart utilizes the sample mean $\overline{X}$ to monitor the process mean, it is usually called an $\overline{X}$ control chart. Note that all the points fall within the control limits, so the chart indicates that the process is in statistical control.

Consider how the control limits were determined. The process average is 74 millimeters, and the process standard deviation is $\sigma = 0.01$ millimeters. Now if samples of size $n = 5$ are taken, the standard deviation of the sample average $\overline{X}$ is

$$\sigma_{\overline{X}} = \frac{\sigma}{\sqrt{n}} = \frac{0.01}{\sqrt{5}} = 0.0045$$

Therefore, if the process is in control with a mean diameter of 74 millimeters, by using the central limit theorem to assume that $\overline{X}$ is approximately normally distributed, we would expect approximately $100(1 - \alpha)\%$ of the sample mean diameters $\overline{X}$ to fall between $74 + z_{\alpha/2}(0.0045)$ and $74 - z_{\alpha/2}(0.0045)$. As discussed above, we customarily choose the

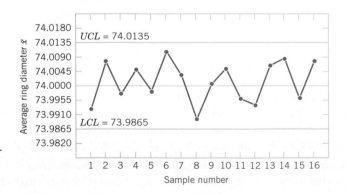

Figure 16-3 $\overline{X}$ control chart for piston ring diameter.

constant $z_{\alpha/2}$ to be 3, so the upper and lower control limits become

$$UCL = 74 + 3(0.0045) = 74.0135$$

and

$$LCL = 74 - 3(0.0045) = 73.9865$$

as shown on the control chart. These are the 3-sigma control limits referred to above. Note that the use of 3-sigma limits implies that $\alpha = 0.0027$; that is, the probability that the point plots outside the control limits when the process is in control is 0.0027. The width of the control limits is inversely related to the sample size n for a given multiple of sigma. Choosing the control limits is equivalent to setting up the critical region for testing the hypothesis

$$H_0: \mu = 74$$
$$H_1: \mu \neq 74$$

where $\sigma = 0.01$ is known. Essentially, the control chart tests this hypothesis repeatedly at different points in time.

In designing a control chart, we must specify both the sample size to use and the frequency of sampling. In general, larger samples will make it easier to detect small shifts in the process. When choosing the sample size, we must keep in mind the size of the shift that we are trying to detect. If we are interested in detecting a relatively large process shift, we use smaller sample sizes than those that would be employed if the shift of interest were relatively small.

We must also determine the frequency of sampling. The most desirable situation from the point of view of detecting shifts would be to take large samples very frequently; however, this is usually not economically feasible. The general problem is one of *allocating sampling effort*. That is, either we take small samples at short intervals or larger samples at longer intervals. Current industry practice tends to favor smaller, more frequent samples, particularly in high-volume manufacturing processes or where a great many types of assignable causes can occur. Furthermore, as automatic sensing and measurement technology develops, it is becoming possible to greatly increase frequencies. Ultimately, every unit can be tested as it is manufactured. This capability will not eliminate the need for control charts because the test system will not prevent defects. The increased data will increase the effectiveness of process control and improve quality.

16-4.3 Rational Subgroups

A fundamental idea in the use of control charts is to collect sample data according to what Shewhart called the **rational subgroup** concept. Generally, this means that subgroups or samples should be selected so that to the extent possible, the variability of the observations within a subgroup should include all the chance or natural variability and exclude the assignable variability. Then, the control limits will represent bounds for all the chance variability and not the assignable variability. Consequently, assignable causes will tend to generate points that are outside of the control limits, while chance variability will tend to generate points that are within the control limits.

When control charts are applied to production processes, the time order of production is a logical basis for rational subgrouping. Even though time order is preserved, it is still possible to form subgroups erroneously. If some of the observations in the subgroup are taken at the end of one 8-hour shift and the remaining observations are taken at the start of the next 8-hour shift, any differences between shifts might not be detected. Time order is frequently a good basis for forming subgroups because it allows us to detect assignable causes that occur over time.

Two general approaches to constructing rational subgroups are used. In the first approach, each subgroup consists of units that were produced at the same time (or as closely together as possible). This approach is used when the primary purpose of the control chart is to detect process shifts. It minimizes variability due to assignable causes *within* a sample, and it maximizes variability *between* samples if assignable causes are present. It also provides better estimates of the standard deviation of the process in the case of variables control charts. This approach to rational subgrouping essentially gives a "snapshot" of the process at each point in time where a sample is collected.

In the second approach, each sample consists of units of product that are representative of *all* units that have been produced since the last sample was taken. Essentially, each subgroup is a *random sample* of *all* process output over the sampling interval. This method of rational subgrouping is often used when the control chart is employed to make decisions about the acceptance of all units of product that have been produced since the last sample. In fact, if the process shifts to an out-of-control state and then back in control again *between* samples, it is sometimes argued that the first method of rational subgrouping defined above will be ineffective against these types of shifts, and so the second method must be used.

When the rational subgroup is a random sample of all units produced over the sampling interval, considerable care must be taken in interpreting the control charts. If the process mean drifts between several levels during the interval between samples, the range of observations within the sample may consequently be relatively large. It is the within-sample variability that determines the width of the control limits on an $\overline{X}$ chart, so this practice will result in wider limits on the $\overline{X}$ chart. This makes it harder to detect shifts in the mean. In fact, we can often make *any* process appear to be in statistical control just by stretching out the interval between observations in the sample. It is also possible for shifts in the process average to cause points on a control chart for the range or standard deviation to plot out of control, even though no shift in process variability has taken place.

There are other bases for forming rational subgroups. For example, suppose a process consists of several machines that pool their output into a common stream. If we sample from this common stream of output, it will be very difficult to detect whether or not some of the machines are out of control. A logical approach to rational subgrouping here is to apply control chart techniques to the output for each individual machine. Sometimes this concept needs to be applied to different heads on the same machine, different workstations, different operators, and so forth.

The rational subgroup concept is very important. The proper selection of samples requires careful consideration of the process, with the objective of obtaining as much useful information as possible from the control chart analysis.

16-4.4   Analysis of Patterns on Control Charts

A control chart may indicate an out-of-control condition either when one or more points fall beyond the control limits, or when the plotted points exhibit some nonrandom pattern of behavior. For example, consider the $\overline{X}$ chart shown in Fig. 16-4. Although all 25 points fall within the control limits, the points do not indicate statistical control because their pattern is very nonrandom in appearance. Specifically, we note that 19 of the 25 points plot below the center line, while only 6 of them plot above. If the points are truly random, we should expect a more even distribution of them above and below the center line. We also observe that following the fourth point, five points in a row increase in magnitude. This arrangement of points is called a **run.** Since the observations are increasing, we could call it a run up; similarly, a sequence of decreasing points

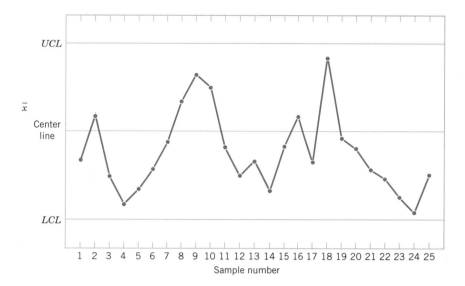

Figure 16-4   An $\overline{X}$ control chart.

is called a run down. This control chart has an unusually long run up (beginning with the fourth point) and an unusually long run down (beginning with the eighteenth point).

In general, we define a run as a sequence of observations of the same type. In addition to runs up and runs down, we could define the types of observations as those above and below the center line, respectively, so two points in a row above the center line would be a run of length 2.

A run of length 8 or more points has a very low probability of occurrence in a random sample of points. Consequently, any type of run of length 8 or more is often taken as a signal of an out-of-control condition. For example, eight consecutive points on one side of the center line will indicate that the process is out of control.

Although runs are an important measure of nonrandom behavior on a control chart, other types of patterns may also indicate an out-of-control condition. For example, consider the $\overline{X}$ chart in Fig. 16-5. Note that the plotted sample averages exhibit a cyclic behavior, yet they all fall within the control limits. Such a pattern may indicate a problem with the process, such as operator fatigue, raw material deliveries, and heat or stress buildup. The yield may be improved by eliminating or reducing the sources of variability causing this cyclic behavior (see Fig. 16-6). In Fig. 16-6, *LSL* and *USL* denote the lower and upper specification limits of the process. These limits represent bounds within which acceptable product must fall and they are often based on customer requirements.

The problem is one of **pattern recognition,** that is, recognizing systematic or nonrandom patterns on the control chart and identifying the reason for this behavior. The ability to interpret

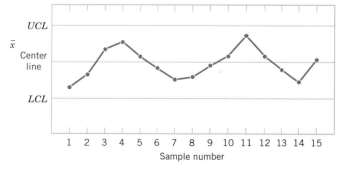

Figure 16-5   An $\overline{X}$ chart with a cyclic pattern.

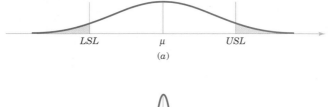

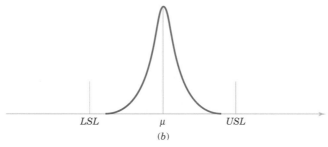

Figure 16-6   (a)
Variability with the
cyclic pattern.
(b) Variability with
the cyclic pattern elim-
inated.

a particular pattern in terms of assignable causes requires experience and knowledge of the process. That is, we must not only know the statistical principles of control charts, but we must also have a good understanding of the process.

The Western Electric Handbook (1956) suggests a set of decision rules for detecting non-random patterns on control charts. Specifically, the **Western Electric rules** would conclude that the process is out of control if either

1.   One point plots outside 3-sigma control limits.

2.   Two out of three consecutive points plot beyond a 2-sigma limit.

3.   Four out of five consecutive points plot at a distance of 1-sigma or beyond from the center line.

4.   Eight consecutive points plot on one side of the center line.

We have found these rules very effective in practice for enhancing the sensitivity of control charts. Rules 2 and 3 apply to one side of the center line at a time. That is, a point above the *upper* 2-sigma limit followed immediately by a point below the *lower* 2-sigma limit would not signal an out-of-control alarm.

Figure 16-7 shows an $\overline{X}$ control chart for the piston ring process with the 1-sigma, 2-sigma, and 3-sigma limits used in the Western Electric procedure. Notice that these inner

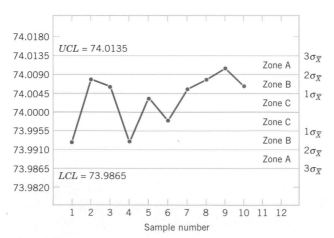

Figure 16-7   The
Western Electric zone
rules.

limits (sometimes called **warning limits**) partition the control chart into three zones A, B, and C on each side of the center line. Consequently, the Western Electric rules are sometimes called the **run rules** for control charts. Notice that the last four points fall in zone B or beyond. Thus, since four of five consecutive points exceed the 1-sigma limit, the Western Electric procedure will conclude that the pattern is nonrandom and the process is out of control.

16-5 $\overline{X}$ AND R OR S CONTROL CHARTS

When dealing with a quality characteristic that can be expressed as a measurement, it is customary to monitor both the mean value of the quality characteristic and its variability. Control over the average quality is exercised by the control chart for averages, usually called the $\overline{X}$ chart. Process variability can be controlled by either a range chart (R chart) or a standard deviation chart (S chart), depending on how the population standard deviation is estimated.

Suppose that the process mean and standard deviation μ and σ are known and that we can assume that the quality characteristic has a normal distribution. Consider the $\overline{X}$ chart. As discussed previously, we can use μ as the center line for the control chart, and we can place the upper and lower 3-sigma limits at

$$
\begin{aligned}
UCL &= \mu + 3\sigma/\sqrt{n} \\
LCL &= \mu - 3\sigma/\sqrt{n} \\
CL &= \mu
\end{aligned}
\tag{16-2}
$$

When the parameters μ and σ are unknown, we usually estimate them on the basis of preliminary samples, taken when the process is thought to be in control. We recommend the use of at least 20 to 25 preliminary samples. Suppose m preliminary samples are available, each of size n. Typically, n will be 4, 5, or 6; these relatively small sample sizes are widely used and often arise from the construction of rational subgroups. Let the sample mean for the ith sample be $\overline{X}_i$. Then we estimate the mean of the population, μ, by the **grand mean**

$$
\hat{\mu} = \overline{\overline{X}} = \frac{1}{m} \sum_{i=1}^{m} \overline{X}_i
\tag{16-3}
$$

Thus, we may take $\overline{\overline{X}}$ as the center line on the $\overline{X}$ control chart.

We may estimate σ from either the standard deviation or the range of the observations within each sample. The sample size is relatively small, so there is little loss in efficiency in estimating σ from the sample ranges.

The relationship between the range R of a sample from a normal population with known parameters and the standard deviation of that population is needed. Since R is a random variable, the quantity $W = R/\sigma$, called the relative range, is also a random variable. The parameters of the distribution of W have been determined for any sample size n. The mean of the distribution of W is called d_2, and a table of d_2 for various n is given in Appendix Table X.

The standard deviation of W is called d_3. Because $R = \sigma W$

$$\mu_R = d_2\sigma \qquad \sigma_R = d_3\sigma \tag{16-4}$$

Let R_i be the range of the ith sample, and let

$$\overline{R} = \frac{1}{m} \sum_{i=1}^{m} R_i \tag{16-5}$$

be the average range. Then $\overline{R}$ is an estimator of μ_R and from Equation 16-4 an unbiased estimator of σ is

$$\hat{\sigma} = \frac{\overline{R}}{d_2} \tag{16-6}$$

Therefore, we may use as our upper and lower control limits for the $\overline{X}$ chart

$$UCL = \overline{\overline{X}} + \frac{3}{d_2\sqrt{n}}\overline{R} \qquad LCL = \overline{\overline{X}} - \frac{3}{d_2\sqrt{n}}\overline{R} \tag{16-7}$$

Define the constant

$$A_2 = \frac{3}{d_2\sqrt{n}} \tag{16-8}$$

Now, once we have computed the sample values $\overline{\overline{x}}$ and $\overline{r}$, the $\overline{X}$ control chart may be defined as follows:

$\overline{X}$ **Control Chart**
(from $\overline{R}$)

The center line and upper and lower control limits for an $\overline{X}$ control chart are

$$UCL = \overline{\overline{x}} + A_2\overline{r} \qquad CL = \overline{\overline{x}} \qquad LCL = \overline{\overline{x}} - A_2\overline{r} \tag{16-9}$$

where the constant A_2 is tabulated for various sample sizes in Appendix Table X.

The parameters of the R chart may also be easily determined. The center line will obviously be $\overline{R}$. To determine the control limits, we need an estimate of σ_R, the standard deviation of R. Once again, assuming the process is in control, the distribution of the relative range, W, will be useful. We may estimate σ_R from Equation 16-4 as

$$\hat{\sigma}_R = d_3\hat{\sigma} = d_3\frac{\overline{R}}{d_2} \tag{16-10}$$

and we would use as the upper and lower control limits on the R chart

$$UCL = \overline{R} + \frac{3d_3}{d_2} \overline{R} = \left(1 + \frac{3d_3}{d_2}\right)\overline{R}$$

$$LCL = \overline{R} - \frac{3d_3}{d_2} \overline{R} = \left(1 - \frac{3d_3}{d_2}\right)\overline{R} \qquad (16\text{-}11)$$

Setting $D_3 = 1 - 3d_3/d_2$ and $D_4 = 1 + 3d_3/d_2$ leads to the following definition.

R Chart

> The center line and upper and lower control limits for an R chart are
>
> $$UCL = D_4\overline{r} \qquad CL = \overline{r} \qquad LCL = D_3\overline{r} \qquad (16\text{-}12)$$
>
> where $\overline{r}$ is the sample average range, and the constants D_3 and D_4 are tabulated for various sample sizes in Appendix Table X.

The LCL for an R chart can be a negative number. In that case, it is customary to set LCL to zero. Because the points plotted on an R chart are nonnegative, no points can fall below an LCL of zero.

When preliminary samples are used to construct limits for control charts, these limits are customarily treated as trial values. Therefore, the m sample means and ranges should be plotted on the appropriate charts, and any points that exceed the control limits should be investigated. If assignable causes for these points are discovered, they should be eliminated and new limits for the control charts determined. In this way, the process may be eventually brought into statistical control and its inherent capabilities assessed. Other changes in process centering and dispersion may then be contemplated. Also, we often study the R chart first because if the process variability is not constant over time the control limits calculated for the $\overline{X}$ chart can be misleading.

Rather than base control charts on ranges, a more modern approach is to calculate the standard deviation of each subgroup and plot these standard deviations to monitor the process standard deviation σ. This is called an S chart. When an S chart is used, it is common to use these standard deviations to develop control limits for the $\overline{X}$ chart. Typically, the sample size used for subgroups is small (fewer than 10) and in that case there is usually little difference in the $\overline{X}$ chart generated from ranges or standard deviations. However, because computer software is often used to implement control charts, S charts are quite common. Details to construct these charts follow.

In Section 7-2.2 on the CD, it was shown that S is a biased estimator of σ. That is, $E(S) = c_4\sigma$ where c_4 is a constant that is near, but not equal to, 1. Furthermore, a calculation similar to the one used for $E(S)$ can derive the standard deviation of the statistic S with the result $\sigma\sqrt{1 - c_4^2}$. Therefore, the center line and three-sigma control limits for S are

$$LCL = c_4\sigma - 3\sigma\sqrt{1 - c_4^2} \qquad CL = c_4\sigma$$

$$UCL = c_4\sigma + 3\sigma\sqrt{1 - c_4^2} \qquad (16\text{-}13)$$

Assume that there are m preliminary samples available, each of size n, and let S_i denote the standard deviation of the ith sample. Define

$$\bar{S} = \frac{1}{m} \sum_{i=1}^{m} S_i \qquad (16\text{-}14)$$

Because $E(\bar{S}) = c_4\sigma$, an unbiased estimator of σ is $\bar{S}/c_4$ That is,

$$\hat{\sigma} = \bar{S}/c_4 \qquad (16\text{-}15)$$

A control chart for standard deviations follows.

S Chart

$$UCL = \bar{s} + 3\frac{\bar{s}}{c_4}\sqrt{1 - c_4^2} \qquad CL = \bar{s} \qquad LCL = \bar{s} - 3\frac{\bar{s}}{c_4}\sqrt{1 - c_4^2} \quad (16\text{-}16)$$

The *LCL* for an *S* chart can be a negative number, in that case, it is customary to set *LCL* to zero. When an *S* chart is used, the estimate for σ in Equation 16-15 is commonly used to calculate the control limits for an $\overline{X}$ chart. This produces the following control limits for an $\overline{X}$ chart.

**$\overline{X}$ Control Chart
(from $\bar{S}$)**

$$UCL = \bar{\bar{x}} + 3\frac{\bar{s}}{c_4\sqrt{n}} \qquad CL = \bar{\bar{x}} \qquad LCL = \bar{s} - 3\frac{\bar{s}}{c_4\sqrt{n}} \qquad (16\text{-}17)$$

EXAMPLE 16-1

A component part for a jet aircraft engine is manufactured by an investment casting process. The vane opening on this casting is an important functional parameter of the part. We will illustrate the use of $\overline{X}$ and R control charts to assess the statistical stability of this process. Table 16-1 presents 20 samples of five parts each. The values given in the table have been coded by using the last three digits of the dimension; that is, 31.6 should be 0.50316 inch.

The quantities $\bar{\bar{x}} = 33.3$ and $\bar{r} = 5.8$ are shown at the foot of Table 16-1. The value of A_2 for samples of size 5 is $A_2 = 0.577$. Then the trial control limits for the $\overline{X}$ chart are

$$\bar{x} \pm A_2\bar{r} = 33.32 \pm (0.577)(5.8) = 33.32 \pm 3.35$$

or

$$UCL = 36.67 \qquad LCL = 29.97$$

Table 16-1 Vane-Opening Measurements

Sample Number	x_1	x_2	x_3	x_4	x_5	$\overline{x}$	r	s
1	33	29	31	32	33	31.6	4	1.67332
2	33	31	35	37	31	33.4	6	2.60768
3	35	37	33	34	36	35.0	4	1.58114
4	30	31	33	34	33	32.2	4	1.64317
5	33	34	35	33	34	33.8	2	0.83666
6	38	37	39	40	38	38.4	3	1.14018
7	30	31	32	34	31	31.6	4	1.51658
8	29	39	38	39	39	36.8	10	4.38178
9	28	33	35	36	43	35.0	15	5.43139
10	38	33	32	35	32	34.0	6	2.54951
11	28	30	28	32	31	29.8	4	1.78885
12	31	35	35	35	34	34.0	4	1.73205
13	27	32	34	35	37	33.0	10	3.80789
14	33	33	35	37	36	34.8	4	1.78885
15	35	37	32	35	39	35.6	7	2.60768
16	33	33	27	31	30	30.8	6	2.48998
17	35	34	34	30	32	33.0	5	2.00000
18	32	33	30	30	33	31.6	3	1.51658
19	25	27	34	27	28	28.2	9	3.42053
20	35	35	36	33	30	33.8	6	2.38747
						$\overline{\overline{x}} = 33.32$	$\overline{r} = 5.8$	$\overline{s} = 2.345$

For the R chart, the trial control limits are

$$UCL = D_4\overline{r} = (2.115)(5.8) = 12.27$$
$$LCL = D_3\overline{r} = (0)(5.8) = 0$$

The $\overline{X}$ and R control charts with these trial control limits are shown in Fig. 16-8. Notice that samples 6, 8, 11, and 19 are out of control on the $\overline{X}$ chart and that sample 9 is out of control on the R chart. (These points are labeled with a "1" because they violate the first Western Electric rule.)

For the S chart, the value of $c_4 = 0.94$. Therefore,

$$\frac{3\overline{s}}{c_4}\sqrt{1 - c_4^2} = \frac{3(2.345)}{0.94}\sqrt{1 - 0.94^2} = 2.553$$

and the trial control limits are

$$UCL = 2.345 + 2.553 = 4.898$$
$$LCL = 2.345 - 2.553 = -0.208$$

The LCL is set to zero. If $\overline{s}$ is used to determine the control limits for the $\overline{X}$ chart,

$$\overline{\overline{x}} \pm \frac{3\overline{s}}{c_4\sqrt{n}} = 33.32 \pm \frac{3(2.345)}{0.94} = 33.32 \pm 3.35$$

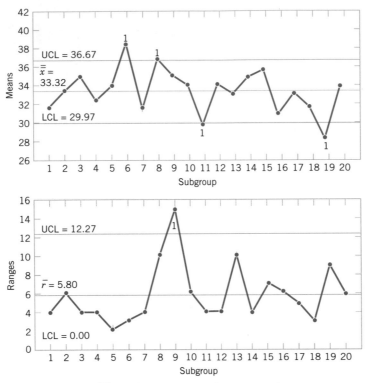

Figure 16-8   The $\overline{X}$ and R control charts for vane opening.

and this result is nearly the same as from $\overline{r}$. The S chart is shown in Fig. 16-9. Because the control limits for the $\overline{X}$ chart calculated from $\overline{s}$ are nearly the same as from $\overline{r}$, the chart is not shown.

Suppose that all of these assignable causes can be traced to a defective tool in the wax-molding area. We should discard these five samples and recompute the limits for the $\overline{X}$ and R charts. These new revised limits are, for the $\overline{X}$ chart,

$$UCL = \overline{\overline{x}} + A_2\overline{r} = 33.21 + (0.577)(5.0) = 36.10$$

$$LCL = \overline{\overline{x}} - A_2\overline{r} = 33.21 - (0.577)(5.0) = 30.33$$

and for the R chart,

$$UCL = D_4\overline{r} = (2.115)(5.0) = 10.57$$

$$LCL = D_3\overline{r} = (0)(5.0) = 0$$

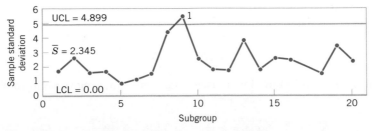

Figure 16-9.   The S control chart for vane opening.

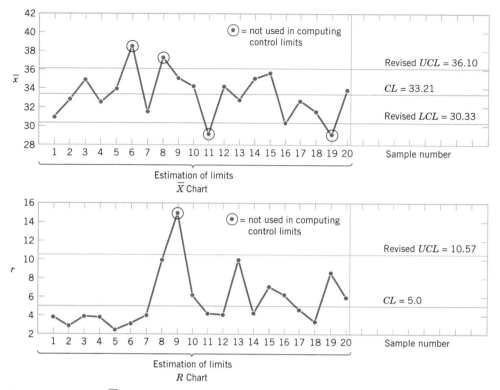

Figure 16-10   The $\overline{X}$ and R control charts for vane opening, revised limits.

The revised control charts are shown in Fig. 16-10. Notice that we have treated the first 20 preliminary samples as **estimation data** with which to establish control limits. These limits can now be used to judge the statistical control of future production. As each new sample becomes available, the values of $\bar{x}$ and r should be computed and plotted on the control charts. It may be desirable to revise the limits periodically, even if the process remains stable. The limits should always be revised when process improvements are made.

Computer Construction of $\overline{X}$ and R Control Charts

Many computer programs construct $\overline{X}$ and R control charts. Figures 16-8 and 16-10 show charts similar to those produced by Minitab for the vane-opening data. This program will allow the user to select any multiple of sigma as the width of the control limits and use the Western Electric rules to detect out-of-control points. The program will also prepare a summary report as in Table 16-2 and exclude subgroups from the calculation of the control limits.

Table 16-2   Summary Report from Minitab for the Vane-Opening Data

Test Results for Xbar Chart
TEST 1. One point more than 3.00 sigmas from center line.
Test Failed at points: 6 8 11 19

Test Results for R Chart
TEST 1. One point more than 3.00 sigmas from center line.
Test Failed at points: 9

EXERCISES FOR SECTION 16-5

16-1. An extrusion die is used to produce aluminum rods. The diameter of the rods is a critical quality characteristic. The following table shows $\bar{x}$ and r values for 20 samples of five rods each. Specifications on the rods are 0.5035 ± 0.0010 inch. The values given are the last three digits of the measurement; that is, 34.2 is read as 0.50342.

Sample	$\bar{x}$	r
1	34.2	3
2	31.6	4
3	31.8	4
4	33.4	5
5	35.0	4
6	32.1	2
7	32.6	7
8	33.8	9
9	34.8	10
10	38.6	4
11	35.4	8
12	34.0	6
13	36.0	4
14	37.2	7
15	35.2	3
16	33.4	10
17	35.0	4
18	34.4	7
19	33.9	8
20	34.0	4

(a) Using all the data, find trial control limits for $\bar{X}$ and R charts, construct the chart, and plot the data.
(b) Use the trial control limits from part (a) to identify out-of-control points. If necessary, revise your control limits, assuming that any samples that plot outside the control limits can be eliminated.

16-2. Twenty-five samples of size 5 are drawn from a process at one-hour intervals, and the following data are obtained:

$$\sum_{i=1}^{25} \bar{x}_i = 362.75 \qquad \sum_{i=1}^{25} r_i = 8.60 \qquad \sum_{i=1}^{25} s_i = 3.64$$

(a) Find trial control limits for $\bar{X}$ and R charts.
(b) Repeat part (a) for $\bar{X}$ and S charts.

16-3. The pull strength of a wire-bonded lead for an integrated circuit monitored. The following table provides data for 20 samples each of size three.
(a) Use all the data to determine trial control limits for $\bar{X}$ and R charts, construct the control limits, and plot the data.

(b) Use the control limits from part (a) to identify out-of-control points. If necessary, revise your control limits assuming that any samples that plot outside of the control limits can be eliminated.
(c) Repeat parts (a) and (b) for $\bar{X}$ and S charts.

Sample Number	x_1	x_2	x_3
1	15.4	15.6	15.3
2	15.4	17.1	15.2
3	16.1	16.1	13.5
4	13.5	12.5	10.2
5	18.3	16.1	17.0
6	19.2	17.2	19.4
7	14.1	12.4	11.7
8	15.6	13.3	13.6
9	13.9	14.9	15.5
10	18.7	21.2	20.1
11	15.3	13.1	13.7
12	16.6	18.0	18.0
13	17.0	15.2	18.1
14	16.3	16.5	17.7
15	8.4	7.7	8.4
16	11.1	13.8	11.9
17	16.5	17.1	18.5
18	18.0	14.1	15.9
19	17.8	17.3	12.0
20	11.5	10.8	11.2

16-4. Samples of size $n = 6$ are collected from a process every hour. After 20 samples have been collected, we calculate $\bar{\bar{x}} = 20.0$ and $\bar{r}/d_2 = 1.4$.
(a) Find trial control limits for $\bar{X}$ and R charts.
(b) If $\bar{s}/c_4 = 1.5$, determine trial control limits for $\bar{X}$ and S charts.

16-5. Control charts for $\bar{X}$ and R are to be set up for an important quality characteristic. The sample size is $n = 5$, and $\bar{x}$ and r are computed for each of 35 preliminary samples. The summary data are

$$\sum_{i=1}^{35} \bar{x}_i = 7805 \qquad \sum_{i=1}^{35} r_i = 1200$$

(a) Find trial control limits for $\bar{X}$ and R charts.
(b) Assuming that the process is in control, estimate the process mean and standard deviation.

16-6. Control charts are to be constructed for samples of size $n = 4$, and $\bar{x}$ and s are computed for each of 20 preliminary samples as follows:

$$\sum_{i=1}^{20} \bar{x}_i = 4460 \qquad \sum_{i=1}^{20} s_i = 271.6$$

(a) Determine trial control limits for $\overline{X}$ and S charts.

(b) Assuming the process is in control, estimate the process mean and standard deviation.

16-7. The thickness of a metal part is an important quality parameter. Data on thickness (in inches) are given in the following table, for 25 samples of five parts each.

Sample Number	x_1	x_2	x_3	x_4	x_5
1	0.0629	0.0636	0.0640	0.0635	0.0640
2	0.0630	0.0631	0.0622	0.0625	0.0627
3	0.0628	0.0631	0.0633	0.0633	0.0630
4	0.0634	0.0630	0.0631	0.0632	0.0633
5	0.0619	0.0628	0.0630	0.0619	0.0625
6	0.0613	0.0629	0.0634	0.0625	0.0628
7	0.0630	0.0639	0.0625	0.0629	0.0627
8	0.0628	0.0627	0.0622	0.0625	0.0627
9	0.0623	0.0626	0.0633	0.0630	0.0624
10	0.0631	0.0631	0.0633	0.0631	0.0630
11	0.0635	0.0630	0.0638	0.0635	0.0633
12	0.0623	0.0630	0.0630	0.0627	0.0629
13	0.0635	0.0631	0.0630	0.0630	0.0630
14	0.0645	0.0640	0.0631	0.0640	0.0642
15	0.0619	0.0644	0.0632	0.0622	0.0635
16	0.0631	0.0627	0.0630	0.0628	0.0629
17	0.0616	0.0623	0.0631	0.0620	0.0625
18	0.0630	0.0630	0.0626	0.0629	0.0628
19	0.0636	0.0631	0.0629	0.0635	0.0634
20	0.0640	0.0635	0.0629	0.0635	0.0634
21	0.0628	0.0625	0.0616	0.0620	0.0623
22	0.0615	0.0625	0.0619	0.0619	0.0622
23	0.0630	0.0632	0.0630	0.0631	0.0630
24	0.0635	0.0629	0.0635	0.0631	0.0633
25	0.0623	0.0629	0.0630	0.0626	0.0628

(a) Using all the data, find trial control limits for $\overline{X}$ and R charts, construct the chart, and plot the data. Is the process in statistical control?

(b) Repeat part (a) for $\overline{X}$ and S charts.

(c) Use the trial control limits from part (a) to identify out-of-control points. List the sample numbers of the out-of-control points.

16-8. The copper content of a plating bath is measured three times per day, and the results are reported in ppm. The $\overline{x}$ and r values for 25 days are shown in the following table:

(a) Using all the data, find trial control limits for $\overline{X}$ and R charts, construct the chart, and plot the data. Is the process in statistical control?

(b) If necessary, revise the control limits computed in part (a), assuming that any samples that plot outside the control limits can be eliminated.

Day	$\overline{x}$	r	Day	$\overline{x}$	r
1	5.45	1.21	14	7.01	1.45
2	5.39	0.95	15	5.83	1.37
3	6.85	1.43	16	6.35	1.04
4	6.74	1.29	17	6.05	0.83
5	5.83	1.35	18	7.11	1.35
6	7.22	0.88	19	7.32	1.09
7	6.39	0.92	20	5.90	1.22
8	6.50	1.13	21	5.50	0.98
9	7.15	1.25	22	6.32	1.21
10	5.92	1.05	23	6.55	0.76
11	6.45	0.98	24	5.90	1.20
12	5.38	1.36	25	5.95	1.19
13	6.03	0.83			

16-6 CONTROL CHARTS FOR INDIVIDUAL MEASUREMENTS

In many situations, the sample size used for process control is $n = 1$; that is, the sample consists of an individual unit. Some examples of these situations are as follows:

1. Automated inspection and measurement technology is used, and every unit manufactured is analyzed.

2. The production rate is very slow, and it is inconvenient to allow sample sizes of $n > 1$ to accumulate before being analyzed.

3. Repeat measurements on the process differ only because of laboratory or analysis error, as in many chemical processes.

4. In process plants, such as papermaking, measurements on some parameters such as coating thickness *across* the roll will differ very little and produce a standard deviation that is much too small if the objective is to control coating thickness *along* the roll.

In such situations, the **individuals control chart** is useful. The control chart for individuals uses the **moving range** of two successive observations to estimate the process variability. The moving range is defined as $MR_i = |X_i - X_{i-1}|$.

An estimate of σ is

$$\hat{\sigma} = \frac{\overline{MR}}{d_2} = \frac{\overline{MR}}{1.128} \tag{16-18}$$

because $d_2 = 1.128$ when two consecutive observations are used to calculate a moving range. It is also possible to establish a control chart on the moving range using D_3 and D_4 for $n = 2$. The parameters for these charts are defined as follows.

Individuals Control Chart

The center line and upper and lower control limits for a control chart for individuals are

$$UCL = \bar{x} + 3\frac{\overline{mr}}{d_2} = \bar{x} + 3\frac{\overline{mr}}{1.128}$$

$$CL = \bar{x} \tag{16-19}$$

$$LCL = \bar{x} - 3\frac{\overline{mr}}{d_2} = \bar{x} - 3\frac{\overline{mr}}{1.128}$$

and for a control chart for moving ranges

$$UCL = D_4\overline{mr} = 3.267\,\overline{mr}$$

$$CL = \overline{mr}$$

$$LCL = D_3\overline{mr} = 0$$

The procedure is illustrated in the following example.

EXAMPLE 16-2

Table 16-3 shows 20 observations on concentration for the output of a chemical process. The observations are taken at one-hour intervals. If several observations are taken at the same time, the observed concentration reading will differ only because of measurement error. Since the measurement error is small, only one observation is taken each hour.

To set up the control chart for individuals, note that the sample average of the 20 concentration readings is $\bar{x} = 99.1$ and that the average of the moving ranges of two observations shown in the last column of Table 16-3 is $\overline{mr} = 2.59$. To set up the moving-range chart, we note that $D_3 = 0$ and $D_4 = 3.267$ for $n = 2$. Therefore, the moving-range chart has center line $\overline{mr} = 2.59$, $LCL = 0$, and $UCL = D_4\overline{mr} = (3.267)(2.59) = 8.46$. The control chart is shown as the lower control chart in Fig. 16-11 on page 618. This control chart was constructed by Minitab. Because no points exceed the upper control limit, we may now set up the control chart for individual concentration measurements. If a moving range of $n = 2$ observations is used,

$d_2 = 1.128$. For the data in Table 16-3 we have

$$UCL = \bar{x} + 3\frac{\overline{mr}}{d_2} = 99.1 + 3\frac{2.59}{1.128} = 105.99$$

$$CL = \bar{x} = 99.1$$

$$LCL = \bar{x} - 3\frac{\overline{mr}}{d_2} = 99.1 - 3\frac{2.59}{1.128} = 92.21$$

The control chart for individual concentration measurements is shown as the upper control chart in Fig. 16-11. There is no indication of an out-of-control condition.

The chart for individuals can be interpreted much like an ordinary $\overline{X}$ control chart. A shift in the process average will result in either a point (or points) outside the control limits, or a pattern consisting of a run on one side of the center line.

Some care should be exercised in interpreting patterns on the moving-range chart. The moving ranges are correlated, and this correlation may often induce a pattern of runs or cycles on the chart. The individual measurements are assumed to be uncorrelated, however, and any apparent pattern on the individuals' control chart should be carefully investigated.

The control chart for individuals is very insensitive to small shifts in the process mean. For example, if the size of the shift in the mean is one standard deviation, the average number of points to detect this shift is 43.9. This result is shown later in the chapter. While the performance of the control chart for individuals is much better for large shifts, in many situations the shift of interest is not large and more rapid shift detection is desirable. In these cases, we recommend the *cumulative sum control chart* (discussed in Section 16-10) or an *exponentially weighted moving-average chart* (Montgomery, 2001).

Table 16-3   Chemical Process Concentration Measurements

Observation	Concentration x	Moving Range mr
1	102.0	
2	94.8	7.2
3	98.3	3.5
4	98.4	0.1
5	102.0	3.6
6	98.5	3.5
7	99.0	0.5
8	97.7	1.3
9	100.0	2.3
10	98.1	1.9
11	101.3	3.2
12	98.7	2.6
13	101.1	2.4
14	98.4	2.7
15	97.0	1.4
16	96.7	0.3
17	100.3	3.6
18	101.4	1.1
19	97.2	4.2
20	101.0	3.8
	$\bar{x} = 99.1$	$\overline{mr} = 2.59$

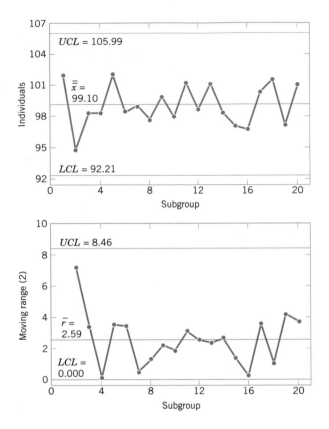

Figure 16-11
Control charts for
individuals and the
moving range (from
Minitab) for the
chemical process
concentration data.

Some individuals have suggested that limits narrower than 3-sigma be used on the chart for individuals to enhance its ability to detect small process shifts. This is a dangerous suggestion, for narrower limits will dramatically increase false alarms such that the charts may be ignored and become useless. If you are interested in detecting small shifts, use the cumulative sum or exponentially weighted moving-average control chart referred to on the previous page.

EXERCISES FOR SECTION 16-6

16-9. Twenty successive hardness measurements are made on a metal alloy, and the data are shown in the following table.

Observation	Hardness	Observation	Hardness
1	51	11	51
2	52	12	57
3	54	13	58
4	55	14	50
5	55	15	53
6	51	16	52
7	52	17	54
8	50	18	50
9	51	19	56
10	56	20	53

(a) Using all the data, compute trial control limits for individual observations and moving-range charts. Construct the chart and plot the data. Determine whether the process is in statistical control. If not, assume assignable causes can be found to eliminate these samples and revise the control limits.

(b) Estimate the process mean and standard deviation for the in-control process.

16-10. In a semiconductor manufacturing process CVD metal thickness was measured on 30 wafers obtained over approximately two weeks. Data are shown in the following table.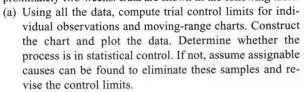

(a) Using all the data, compute trial control limits for individual observations and moving-range charts. Construct the chart and plot the data. Determine whether the process is in statistical control. If not, assume assignable causes can be found to eliminate these samples and revise the control limits.

(b) Estimate the process mean and standard deviation for the in-control process.

Wafer	x	Wafer	x
1	16.8	16	15.4
2	14.9	17	14.3
3	18.3	18	16.1
4	16.5	19	15.8
5	17.1	20	15.9
6	17.4	21	15.2
7	15.9	22	16.7
8	14.4	23	15.2
9	15.0	24	14.7
10	15.7	25	17.9
11	17.1	26	14.8
12	15.9	27	17.0
13	16.4	28	16.2
14	15.8	29	15.6
15	15.4	30	16.3

16-11. The diameter of holes is measured in consecutive order by an automatic sensor. The results of measuring 25 holes are in the following table.

Sample	Diameter	Sample	Diameter
1	9.94	14	9.99
2	9.93	15	10.12
3	10.09	16	9.81
4	9.98	17	9.73
5	10.11	18	10.14
6	9.99	19	9.96
7	10.11	20	10.06
8	9.84	21	10.11
9	9.82	22	9.95
10	10.38	23	9.92
11	9.99	24	10.09
12	10.41	25	9.85
13	10.36		

(a) Using all the data, compute trial control limits for individual observations and moving-range charts. Construct the control chart and plot the data. Determine whether the process is in statistical control. If not, assume assignable causes can be found to eliminate these samples and revise the control limits.

(b) Estimate the process mean and standard deviation for the in-control process.

16-12. The viscosity of a chemical intermediate is measured every hour. Twenty samples each of size $n = 1$, are in the following table.

Sample	Viscosity
1	495
2	491
3	501
4	501
5	512
6	540
7	492
8	504
9	542
10	508
11	493
12	507
13	503
14	475
15	497
16	499
17	468
18	486
19	511
20	487

(a) Using all the data, compute trial control limits for individual observations and moving-range charts. Determine whether the process is in statistical control. If not, assume assignable causes can be found to eliminate these samples and revise the control limits.

(b) Estimate the process mean and standard deviation for the in-control process.

16-7 PROCESS CAPABILITY

It is usually necessary to obtain some information about the **process capability,** that is, the performance of the process when it is operating in control. Two graphical tools, the **tolerance chart** (or tier chart) and the **histogram,** are helpful in assessing process capability.

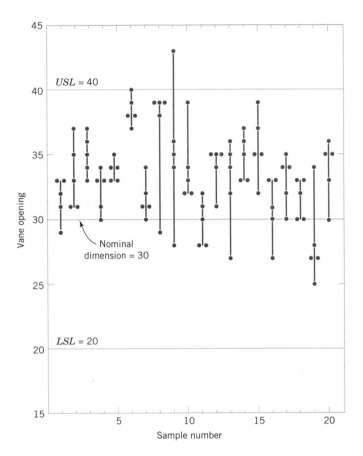

Figure 16-12
Tolerance diagram of
vane openings.

The tolerance chart for all 20 samples from the vane-manufacturing process is shown in Fig. 16-12. The specifications on vane opening are 0.5030 ± 0.0010 in. In terms of the coded data, the upper specification limit is $USL = 40$ and the lower specification limit is $LSL = 20$, and these limits are shown on the chart in Fig. 16-12. Each measurement is plotted on the tolerance chart. Measurements from the same subgroup are connected with lines. The tolerance chart is useful in revealing patterns over time in the individual measurements, or it may show that a particular value of $\bar{x}$ or r was produced by one or two unusual observations in the sample. For example, note the two unusual observations in sample 9 and the single unusual observation in sample 8. Note also that it is appropriate to plot the specification limits on the tolerance chart, since it is a chart of individual measurements. **It is never appropriate to plot specification limits on a control chart or to use the specifications in determining the control limits.** Specification limits and control limits are unrelated. Finally, note from Fig. 16-12 that the process is running off-center from the nominal dimension of 30 (or 0.5030 inch).

The histogram for the vane-opening measurements is shown in Fig. 16-13. The observations from samples 6, 8, 9, 11, and 19 (corresponding to out of-control points on either the $\bar{X}$ or R chart) have been deleted from this histogram. The general impression from examining this histogram is that the process is capable of meeting the specification but that it is running off-center.

Another way to express process capability is in terms of an index that is defined as follows.

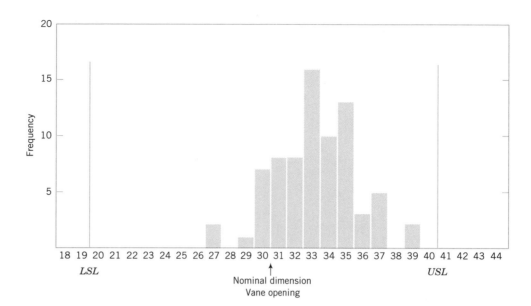

Figure 16-13
Histogram for vane
opening.

**Process Capability
Ratio**

The **process capability ratio** (*PCR*) is

$$PCR = \frac{USL - LSL}{6\sigma} \tag{16-20}$$

The numerator of *PCR* is the width of the specifications. The limits 3σ on either side of the process mean are sometimes called **natural tolerance limits,** for these represent limits that an in-control process should meet with most of the units produced. Consequently, 6σ is often referred to as the width of the process. For the vane opening, where our sample size is 5, we could estimate σ as

$$\hat{\sigma} = \frac{\bar{r}}{d_2} = \frac{5.0}{2.326} = 2.15$$

Therefore, the *PCR* is estimated to be

$$PCR = \frac{USL - LSL}{6\hat{\sigma}} = \frac{40 - 20}{6(2.15)} = 1.55$$

The *PCR* has a natural interpretation: $(1/PCR)100$ is just the percentage of the specifications' width used by the process. Thus, the vane-opening process uses approximately $(1/1.55)100 = 64.5\%$ of the specifications' width.

Figure 16-14(a) shows a process for which the *PCR* exceeds unity. Since the process natural tolerance limits lie inside the specifications, very few defective or nonconforming units will be produced. If $PCR = 1$, as shown in Fig. 16-14(b), more nonconforming units result. In fact, for a normally distributed process, if $PCR = 1$, the fraction

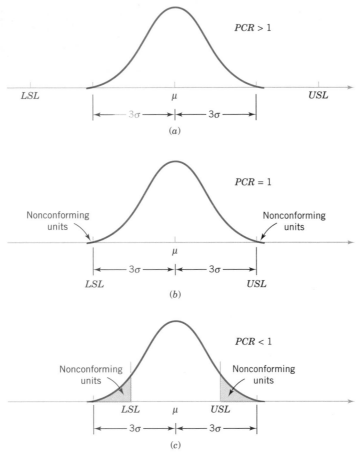

Figure 16-14
Process fallout and the process capability ratio (PCR).

nonconforming is 0.27%, or 2700 parts per million. Finally, when the *PCR* is less than unity, as in Fig. 16-14(c), the process is very yield-sensitive and a large number of nonconforming units will be produced.

The definition of the *PCR* given in Equation 16-19 implicitly assumes that the process is centered at the nominal dimension. If the process is running off-center, its **actual capability** will be less than indicated by the *PCR*. It is convenient to think of *PCR* as a measure of **potential capability**, that is, capability with a centered process. If the process is not centered, a measure of actual capability is often used. This ratio, called PCR_k, is defined below.

PCR_k

$$PCR_k = \min\left[\frac{USL - \mu}{3\sigma}, \frac{\mu - LSL}{3\sigma}\right] \tag{16-21}$$

In effect, PCR_k is a one-sided process capability ratio that is calculated relative to the specification limit nearest to the process mean. For the vane-opening process, we find that the estimate of the process capability ratio PCR_k is

$$\widehat{PCR}_k = \min\left[\frac{USL - \bar{\bar{x}}}{3\hat{\sigma}}, \frac{\bar{\bar{x}} - LSL}{3\hat{\sigma}}\right]$$

$$= \min\left[\frac{40 - 33.19}{3(2.15)} = 1.06, \frac{33.19 - 20}{3(2.15)} = 2.04\right] = 1.06$$

Note that if $PCR = PCR_k$, the process is centered at the nominal dimension. Since $\widehat{PCR}_k = 1.06$ for the vane-opening process and $\widehat{PCR} = 1.55$, the process is obviously running off-center, as was first noted in Figs. 16-14 and 16-17. This off-center operation was ultimately traced to an oversized wax tool. Changing the tooling resulted in a substantial improvement in the process (Montgomery, 2001).

The fractions of nonconforming output (or fallout) below the lower specification limit and above the upper specification limit are often of interest. Suppose that the output from a normally distributed process in statistical control is denoted as X. The fractions are determined from

$$P(X < LSL) = P(Z < (LSL - \mu)/\sigma) \qquad P(X > USL) = P(Z > (USL - \mu)/\sigma)$$

EXAMPLE 16-3

For an electronic manufacturing process a current has specifications of 100 ± 10 milliamperes. The process mean μ and standard deviation σ are 107.0 and 1.5, respectively. The process mean is nearer to the *USL*. Consequently,

$$PCR = (110 - 90)/(6 \cdot 1.5) = 2.22 \qquad \text{and} \qquad PCR_k = (110 - 107)/(3 \cdot 1.5) = 0.67$$

The small PCR_k indicates that the process is likely to produce currents outside of the specification limits. From the normal distribution in Appendix Table II

$$P(X < LSL) = P(Z < (90 - 107)/1.5) = P(Z < -11.33) = 0$$
$$P(X > USL) = P(Z > (110 - 107)/1.5) = P(Z > 2) = 0.023$$

For this example, the relatively large probability of exceeding the *USL* is a warning of potential problems with this criterion even if none of the measured observations in a preliminary sample exceed this limit. We emphasize that the fraction-nonconforming calculation assumes that the observations are normally distributed and the process is in control. Departures from normality can seriously affect the results. The calculation should be interpreted as an approximate guideline for process performance. To make matters worse, μ and σ need to be estimated from the data available and a small sample size can result in poor estimates that further degrade the calculation.

Montgomery (2001) provides guidelines on appropriate values of the *PCR* and a table relating fallout for a normally distributed process in statistical control to the value of *PCR*. Many U.S. companies use $PCR = 1.33$ as a minimum acceptable target and $PCR = 1.66$ as a minimum target for strength, safety, or critical characteristics. Some companies require that internal processes and those at suppliers achieve a $PCR_k = 2.0$. Figure 16-15 illustrates a process with $PCR = PCR_k = 2.0$. Assuming a normal distribution, the calculated fallout for this process is 0.0018 parts per million. A process with $PCR_k = 2.0$ is referred to as a **six-sigma process** because the distance from the process mean to the nearest specification is six standard deviations. The reason that such a large process capability is often required is that it

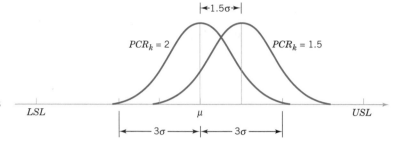

Figure 16-15   Mean of a six-sigma process shifts by 1.5 standard deviations.

is difficult to maintain a process mean at the center of the specifications for long periods of time. A common model that is used to justify the importance of a six-sigma process is illustrated by referring to Fig. 16-15. If the process mean shifts off-center by 1.5 standard deviations, the PCR_k decreases to $4.5\sigma/3\sigma = 1.5$. Assuming a normally distributed process, the fallout of the shifted process is **3.4 parts per million.** Consequently, the mean of a 6-sigma process can shift 1.5 standard deviations from the center of the specifications and still maintain a fallout of 3.4 parts per million.

In addition, some U.S. companies, particularly the automobile industry, have adopted the terminology $C_p = PCR$ and $C_{pk} = PCR_k$. Because C_p has another meaning in statistics (in multiple regression) we prefer the traditional notation PCR and PCR_k.

We repeat that process capability calculations are meaningful only for stable processes; that is, processes that are in control. A process capability ratio indicates whether or not the natural or chance variability in a process is acceptable relative to the specifications.

EXERCISES FOR SECTION 16-7

16-13.   A normally distributed process uses 66.7% of the specification band. It is centered at the nominal dimension, located halfway between the upper and lower specification limits.
(a) Estimate PCR and PCR_k. Interpret these ratios.
(b) What fallout level (fraction defective) is produced?

16-14.   Reconsider Exercise 16-1. Use the revised control limits and process estimates.
(a) Estimate PCR and PCR_k. Interpret these ratios.
(b) What percentage of defectives is being produced by this process?

16-15.   Reconsider Exercise 16-2, where the specification limits are 14.50 ± 0.50.
(a) What conclusions can you draw about the ability of the process to operate within these limits? Estimate the percentage of defective items that will be produced.
(b) Estimage PCR and PCR_k. Interpret these ratios.

16-16.   Reconsider Exercise 16-3. Using the process estimates, what is the fallout level if the coded specifications are 10 ± 5 mm? Estimate PCR and interpret this ratio.

16-17.   A normally distributed process uses 85% of the specification band. It is centered at the nominal dimension, located halfway between the upper and lower specification limits.

(a) Estimate PCR and PCR_k. Interpret these ratios.
(b) What fallout level (fraction defective) is produced?

16-18.   Reconsider Exercise 16-5. Suppose that the quality characteristic is normally distributed with specification at 220 ± 40. What is the fallout level? Estimate PCR and PCR_k and interpret these ratios.

16-19.   Reconsider Exercise 16-6. Suppose that the variable is normally distributed with specifications at 220 ± 50. What is the proportion out of specifications? Estimate and interpret PCR and PCR_k.

16-20.   Reconsider Exercise 16-4(a). Assuming that both charts exhibit statistical control and that the process specifications are at 20 ± 5, estimate PCR and PCR_k and interpret these ratios.

16-21.   Reconsider Exercise 16-8. Given that the specifications are at 6.0 ± 1.0, estimate PCR and PCR_k and interpret these ratios.

16-22.   Reconsider 16-7(b). What are the natural tolerance limits of this process?

16-23.   Reconsider 16-12. The viscosity specifications are at 500 ± 25. Calculate estimates of the process capability ratios PCR and PCR_k for this process and provide an interpretation.

16-8 ATTRIBUTE CONTROL CHARTS

16-8.1 *P* Chart (Control Chart for Proportions)

Often it is desirable to classify a product as either defective or nondefective on the basis of comparison with a standard. This classification is usually done to achieve economy and simplicity in the inspection operation. For example, the diameter of a ball bearing may be checked by determining whether it will pass through a gauge consisting of circular holes cut in a template. This kind of measurement would be much simpler than directly measuring the diameter with a device such as a micrometer. Control charts for attributes are used in these situations. Attribute control charts often require a considerably larger sample size than do their variable measurements counterparts. In this section, we will discuss the **fraction-defective control chart,** or *P* **chart.** Sometimes the *P* chart is called the **control chart for fraction nonconforming.**

Suppose *D* is the number of defective units in a random sample of size *n*. We assume that *D* is a binomial random variable with unknown parameter *p*. The fraction defective

$$\hat{P} = \frac{D}{n}$$

of each sample is plotted on the chart. Furthermore, the variance of the statistic $\hat{P}$ is

$$\sigma_{\hat{P}}^2 = \frac{p(1-p)}{n}$$

Therefore, a *P* chart for fraction defective could be constructed using *p* as the center line and control limits at

$$UCL = p + 3\sqrt{\frac{p(1-p)}{n}} \qquad LCL = p - 3\sqrt{\frac{p(1-p)}{n}} \qquad (16\text{-}22)$$

However, the true process fraction defective is almost always unknown and must be estimated using the data from preliminary samples.

Suppose that *m* preliminary samples each of size *n* are available, and let D_i be the number of defectives in the *i*th sample. The $\hat{P}_i = D_i/n$ is the sample fraction defective in the *i*th sample. The average fraction defective is

$$\overline{P} = \frac{1}{m}\sum_{i=1}^{m}\hat{P}_i = \frac{1}{mn}\sum_{i=1}^{m}D_i \qquad (16\text{-}23)$$

Now $\overline{P}$ may be used as an estimator of *p* in the center line and control limit calculations.

P Chart

The center line and upper and lower control limits for the *P* chart are

$$UCL = \overline{p} + 3\sqrt{\frac{\overline{p}(1-\overline{p})}{n}} \quad CL = \overline{p} \quad LCL = \overline{p} - 3\sqrt{\frac{\overline{p}(1-\overline{p})}{n}} \quad (16\text{-}24)$$

where $\overline{p}$ is the observed value of the average fraction defective.

Table 16-4   Number of Defectives in Samples of 100
Ceramic Substrates

Sample	No. of Defectives	Sample	No. of Defectives
1	44	11	36
2	48	12	52
3	32	13	35
4	50	14	41
5	29	15	42
6	31	16	30
7	46	17	46
8	52	18	38
9	44	19	26
10	48	20	30

These control limits are based on the normal approximation to the binomial distribution. When p is small, the normal approximation may not always be adequate. In such cases, we may use control limits obtained directly from a table of binomial probabilities. If $\bar{p}$ is small, the lower control limit obtained from the normal approximation may be a negative number. If this should occur, it is customary to consider zero as the lower control limit.

EXAMPLE 16-4   Suppose we wish to construct a fraction-defective control chart for a ceramic substrate production line. We have 20 preliminary samples, each of size 100; the number of defectives in each sample is shown in Table 16-4. Assume that the samples are numbered in the sequence of production. Note that $\bar{p} = (800/2000) = 0.40$; therefore, the trial parameters for the control chart are

$$UCL = 0.40 + 3\sqrt{\frac{(0.40)(0.60)}{100}} = 0.55 \qquad CL = 0.40$$

$$LCL = 0.40 - 3\sqrt{\frac{(0.40)(0.60)}{100}} = 0.25$$

The control chart is shown in Fig. 16-16. All samples are in control. If they were not, we would search for assignable causes of variation and revise the limits accordingly. This chart can be used for controlling future production.

Although this process exhibits statistical control, its defective rate ($\bar{p} = 0.40$) is very poor. We should take appropriate steps to investigate the process to determine why such a large number of defective units is being produced. Defective units should be analyzed to determine the specific types of defects present. Once the defect types are known, process changes should be investigated to determine their impact on defect levels. Designed experiments may be useful in this regard.

Computer software also produces an **NP chart.** This is just a control chart of $n\hat{P} = D$, the number of defectives in a sample. The points, center line, and control limits for this chart are just multiples (times n) of the corresponding elements of a P chart. The use of an NP chart avoids the fractions in a P chart.

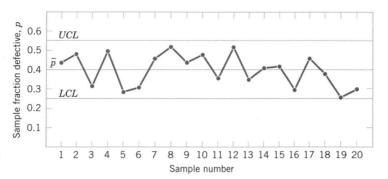

Figure 16-16   *P* chart
for a ceramic substrate.

16-8.2   *U* Chart (Control Chart for Defects per Unit)

It is sometimes necessary to monitor the number of defects in a unit of product rather than the fraction defective. Suppose that in the production of cloth it is necessary to control the number of defects per yard or that in assembling an aircraft wing the number of missing rivets must be controlled. In these situations we may use the control chart for defects per unit, or the *U* chart. Many defects-per-unit situations can be modeled by the Poisson distribution.

If each sample consists of n units and there are C total defects in the sample,

$$U = \frac{C}{n}$$

is the average number of defects per unit. A *U* chart may be constructed for such data.

If the number of defects in a unit is a Poisson random variable with parameter λ, the mean and variance of this distribution are both λ. Each point on the chart is U, the average number of defects per unit from a sample of n units. Therefore, the mean of U is λ and the variance of U is λ/n.

$$UCL = \lambda + 3\sqrt{\frac{\lambda}{n}}$$

$$LCL = \lambda - 3\sqrt{\frac{\lambda}{n}} \qquad (16\text{-}25)$$

If there are m preliminary samples, and the number of defects per unit in these samples are U_1, $U_2, \ldots, U_m$, the estimator of the average number of defects per unit is

$$\overline{U} = \frac{1}{m}\sum_{i=1}^{m} U_i \qquad (16\text{-}26)$$

The parameters of the *U* chart are defined as follows.

U Chart

> The center line and upper and lower control limits on the *U* chart are
>
> $$UCL = \overline{u} + 3\sqrt{\frac{\overline{u}}{n}} \qquad CL = \overline{u} \qquad LCL = \overline{u} - 3\sqrt{\frac{\overline{u}}{n}} \qquad (16\text{-}27)$$
>
> where $\overline{u}$ is the average number of defects per unit.

These control limits are based on the normal approximation to the Poisson distribution. When λ is small, the normal approximation may not always be adequate. In such cases, we may use control limits obtained directly from a table of Poisson probabilities. If $\bar{u}$ is small, the lower control limit obtained from the normal approximation may be a negative number. If this should occur, it is customary to consider zero as the lower control limit.

EXAMPLE 16-5

Printed circuit boards are assembled by a combination of manual assembly and automation. A flow solder machine is used to make the mechanical and electrical connections of the leaded components to the board. The boards are run through the flow solder process almost continuously, and every hour five boards are selected and inspected for process-control purposes. The number of defects in each sample of five boards is noted. Results for 20 samples are shown in Table 16-5.

The center line for the U chart is

$$\bar{u} = \frac{1}{20} \sum_{i=1}^{20} u_i = \frac{32}{20} = 1.6$$

and the upper and lower control limits are

$$UCL = \bar{u} + 3 \sqrt{\frac{\bar{u}}{n}} = 1.6 + 3 \sqrt{\frac{1.6}{5}} = 3.3$$

$$LCL = \bar{u} - 3 \sqrt{\frac{\bar{u}}{n}} = 1.6 - 3 \sqrt{\frac{1.6}{5}} < 0$$

The control chart is plotted in Fig. 16-17. Because LCL is negative, it is set to 0. From the control chart in Fig. 16-17, we see that the process is in control. However, eight defects per group of five circuit boards are too many (about $8/5 = 1.6$ defects/board), and the process needs improvement. An investigation needs to be made of the specific types of defects found on the printed circuit boards. This will usually suggest potential avenues for process improvement.

Computer software also produces a **C chart.** This is just a control chart of C, the total of defects in a sample. The points, center line, and control limits for this chart are just multiples

Table 16-5 Number of Defects in Samples of Five Printed Circuit Boards

Sample	Number of Defects	Defects per Unit u_i	Sample	Number of Defects	Defects per Unit u_i
1	6	1.2	11	9	1.8
2	4	0.8	12	15	3.0
3	8	1.6	13	8	1.6
4	10	2.0	14	10	2.0
5	9	1.8	15	8	1.6
6	12	2.4	16	2	0.4
7	16	3.2	17	7	1.4
8	2	0.4	18	1	0.2
9	3	0.6	19	7	1.4
10	10	2.0	20	13	2.6

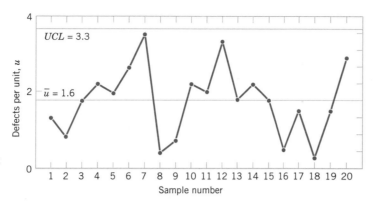

Figure 16-17   *U* chart of defects per unit on printed circuit boards.

(times *n*) of the corresponding elements of a *U* chart. The use of a *C* chart avoids the fractions that can occur in a *U* chart.

EXERCISES FOR SECTION 16-8

16-24. Suppose the following fraction defective has been found in successive samples of size 100 (read down):

0.09	0.03	0.12
0.10	0.05	0.14
0.13	0.13	0.06
0.08	0.10	0.05
0.14	0.14	0.14
0.09	0.07	0.11
0.10	0.06	0.09
0.15	0.09	0.13
0.13	0.08	0.12
0.06	0.11	0.09

(a) Using all the data, compute trial control limits for a fraction-defective control chart, construct the chart, and plot the data.
(b) Determine whether the process is in statistical control. If not, assume assignable causes can be found and out-of-control points eliminated. Revise the control limits.

16-25. The following represent the number of solder defects observed on 24 samples of five printed circuit boards: 7, 6, 8, 10, 24, 6, 5, 4, 8, 11, 15, 8, 4, 16, 11, 12, 8, 6, 5, 9, 7, 14, 8, 21.

(a) Using all the data, compute trial control limits for a *U* control chart, construct the chart, and plot the data.
(b) Can we conclude that the process is in control using a *U* chart? If not, assume assignable causes can be found, list points and revise the control limits.

16-26. The following represent the number of defects per 1000 feet in rubber-covered wire: 1, 1, 3, 7, 8, 10, 5, 13, 0, 19, 24, 6, 9, 11, 15, 8, 3, 6, 7, 4, 9, 20, 11, 7, 18, 10, 6, 4, 0, 9, 7, 3, 1, 8, 12. Do the data come from a controlled process?

16-27. Consider the data in Exercise 16-25. Set up a *C* chart for this process. Compare it to the *U* chart in Exercise 16-25. Comment on your findings.

16-28. The following are the numbers of defective solder joints found during successive samples of 500 solder joints:

Day	No. of Defectives	Day	No. of Defectives
1	106	12	37
2	116	13	25
3	164	14	88
4	89	15	101
5	99	16	64
6	40	17	51
7	112	18	74
8	36	19	71
9	69	20	43
10	74	21	80
11	42		

(a) Using all the data, compute trial control limits for a fraction-defective control chart, construct the chart, and plot the data.
(b) Determine whether the process is in statistical control. If not, assume assignable causes can be found and out-of-control points eliminated. Revise the control limits.

16-9 CONTROL CHART PERFORMANCE

Specifying the control limits is one of the critical decisions that must be made in designing a control chart. By moving the control limits further from the center line, we decrease the risk of a type I error—that is, the risk of a point falling beyond the control limits, indicating an out-of-control condition when no assignable cause is present. However, widening the control limits will also increase the risk of a type II error—that is, the risk of a point falling between the control limits when the process is really out of control. If we move the control limits closer to the center line, the opposite effect is obtained: The risk of type I error is increased, while the risk of type II error is decreased.

The control limits on a Shewhart control chart are customarily located a distance of plus or minus three standard deviations of the variable plotted on the chart from the center line. That is, the constant k in equation 16-1 should be set equal to 3. These limits are called **3-sigma control limits.**

A way to evaluate decisions regarding sample size and sampling frequency is through the **average run length (ARL)** of the control chart. Essentially, the ARL is the average number of points that must be plotted before a point indicates an out-of-control condition. For any Shewhart control chart, the ARL can be calculated from the mean of a geometric random variable (Montgomery 2001). Suppose that p is the probability that any point exceeds the control limits. Then

$$ARL = \frac{1}{p} \qquad (16\text{-}28)$$

Thus, for an $\overline{X}$ chart with 3-sigma limits, $p = 0.0027$ is the probability that a single point falls outside the limits when the process is in control, so

$$ARL = \frac{1}{p} = \frac{1}{0.0027} \cong 370$$

is the average run length of the $\overline{X}$ chart when the process is in control. That is, even if the process remains in control, an out-of-control signal will be generated every 370 points, on the average.

Consider the piston ring process discussed in Section 16-4.2, and suppose we are sampling every hour. Thus, we will have a **false alarm** about every 370 hours on the average. Suppose we are using a sample size of $n = 5$ and that when the process goes out of control the mean shifts to 74.0135 millimeters. Then, the probability that $\overline{X}$ falls between the control limits of Fig. 16-3 is equal to

$$P[73.9865 \leq \overline{X} \leq 74.0135 \text{ when } \mu = 74.0135]$$
$$= P\left[\frac{73.9865 - 74.0135}{0.0045} \leq Z \leq \frac{74.0135 - 74.0135}{0.0045}\right]$$
$$= P[-6 \leq Z \leq 0] = 0.5$$

Therefore, p in Equation 16-28 is 0.50, and the out-of-control ARL is

$$ARL = \frac{1}{p} = \frac{1}{0.5} = 2$$

Table 16-6 Average Run Length (ARL) for an $\overline{X}$ Chart with 3-Sigma Control Limits

Magnitude of Process Shift	ARL $n = 1$	ARL $n = 4$
0	370.4	370.4
0.5σ	155.2	43.9
1.0σ	43.9	6.3
1.5σ	15.0	2.0
2.0σ	6.3	1.2
3.0σ	2.0	1.0

That is, the control chart will require two samples to detect the process shift, on the average, so two hours will elapse between the shift and its detection (*again on the average*). Suppose this approach is unacceptable, because production of piston rings with a mean diameter of 74.0135 millimeters results in excessive scrap costs and delays final engine assembly. How can we reduce the time needed to detect the out-of-control condition? One method is to sample more frequently. For example, if we sample every half hour, only one hour will elapse (on the average) between the shift and its detection. The second possibility is to increase the sample size. For example, if we use $n = 10$, the control limits in Fig. 16-3 narrow to 73.9905 and 74.0095. The probability of $\overline{X}$ falling between the control limits when the process mean is 74.0135 millimeters is approximately 0.1, so $p = 0.9$, and the out-of-control ARL is

$$\text{ARL} = \frac{1}{p} = \frac{1}{0.9} = 1.11$$

Thus, the larger sample size would allow the shift to be detected about twice as quickly as the old one. If it became important to detect the shift in the first hour after it occurred, two control chart designs would work:

Design 1	Design 2
Sample size: $n = 5$	Sample size: $n = 10$
Sampling frequency: every half hour	Sampling frequency: every hour

Table 16-6 provides average run lengths for an $\overline{X}$ chart with 3-sigma control limits. The average run lengths are calculated for shifts in the process mean from 0 to 3.0σ and for sample sizes of $n = 1$ and $n = 4$ by using $1/p$, where p is the probability that a point plots outside of the control limits. Figure 16-18 illustrates a shift in the process mean of 2σ.

Figure 16-18 Process mean shift of 2σ.

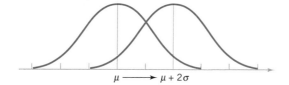

$\mu \longrightarrow \mu + 2\sigma$

EXERCISES FOR SECTION 16-9

16-29.   Consider the $\overline{X}$ control chart in Fig. 16-3. Suppose that the mean shifts to 74.010 millimeters.
(a) What is the probability that this shift will be detected on the next sample?
(b) What is the ARL after the shift?

16-30.   An $\overline{X}$ chart uses samples of size 4. The center line is at 100, and the upper and lower 3-sigma control limits are at 106 and 94, respectively.
(a) What is the process σ?
(b) Suppose the process mean shifts to 96. Find the probability that this shift will be detected on the next sample.
(c) Find the ARL to detect the shift in part (b).

16-31.   Consider the revised $\overline{X}$ control chart in Exercise 16-1 with $\hat{\sigma} = 2.466$, $UCL = 37.404$, $LCL = 30.780$, and $n = 5$. Suppose that the mean shifts to 36.
(a) What is the probability that this shift will be detected on the next sample?
(b) What is the ARL after the shift?

16-32.   Consider the $\overline{X}$ control chart in Exercise 16-2(a) with $\bar{r} = 0.344$, $UCL = 14.708$, $LCL = 14.312$, and $n = 5$. Suppose that the mean shifts to 14.6.
(a) What is the probability that this shift will be detected on the next sample?
(b) What is the ARL after the shift?

16-33.   Consider the $\overline{X}$ control chart in Exercise 16-3(a) $\bar{r} = 2.25$, $UCL = 17.40$, $LCL = 12.79$, and $n = 3$. Suppose that the mean shifts to 13.

(a) What is the probability that this shift will be detected on the next sample?
(b) What is the ARL after the shift?

16-34.   Consider the $\overline{X}$ control chart in Exercise 16-4(a) with $\hat{\sigma} = 1.40$, $UCL = 21.71$, $LCL = 18.29$, and $n = 6$. Suppose that the mean shifts to 17.
(a) What is the probability that this shift will be detected on the next sample?
(b) What is the ARL after the shift?

16-35.   Consider the $\overline{X}$ control chart in Exercise 16-5 with $\bar{r} = 34.286$, $UCL = 242.780$, $LCL = 203.220$, and $n = 5$. Suppose that the mean shifts to 210.
(a) What is the probability that this shift will be detected on the next sample?
(b) What is the ARL after the shift?

16-36.   Consider the revised $\overline{X}$ control chart in Exercise 16-7 with $\bar{r} = 0.000924$, $UCL = 0.0635$, $LCL = 0.0624$, and $n = 5$. Suppose that the mean shifts to 0.0625.
(a) What is the probability that this shift will be detected on the next sample?
(b) What is the ARL after the shift?

16-37.   Consider the revised $\overline{X}$ control chart in Exercise 16-8 with $\hat{\sigma} = 0.669$, $UCL = 7.443$, $LCL = 5.125$, and $n = 3$. Suppose that the mean shifts to 5.5.
(a) What is the probability that this shift will be detected on the next sample?
(b) What is the ARL after the shift?

16-10   CUMULATIVE SUM CONTROL CHART

In Sections 16-5 and 16-6 we have presented basic types of **Shewhart control charts.** A major disadvantage of any Shewhart control chart is that the chart is relatively insensitive to small shifts in the process, say, on the order of about 1.5σ or less. One reason for this relatively poor performance in detecting small process shifts is that the Shewhart chart makes use of only the information in the last plotted point, and it ignores the information in the sequence of points. This problem can be addressed, to some extent by adding criteria such as the **Western Electric rules** to a Shewhart chart, but the use of these rules reduces the simplicity and ease of interpretation of the chart. These rules would also cause the in-control average run length of a Shewhart chart to drop below 370. This increase in the false alarm rate can have serious practical consequences.

A very effective alternative to the Shewhart control chart is the **cumulative sum control chart** (or **CUSUM**). This chart has much better performance (in terms of ARL) for detecting small shifts than the Shewhart chart, but it does not cause the in-control ARL to drop significantly. This section will illustrate the use of the CUSUM for sample averages and individual measurements.

The CUSUM chart plots the cumulative sums of the deviations of the sample values from a target value. For example, suppose that samples of size $n \geq 1$ are collected, and $\overline{X}_j$ is the average of the jth sample. Then if μ_0 is the target for the process mean, the cumulative sum control chart is formed by plotting the quantity

$$S_i = \sum_{j=1}^{i} (\overline{X}_j - \mu_0) \qquad (16\text{-}29)$$

against the sample number i. Now, S_i is called the cumulative sum up to and including the ith sample. Because they combine information from *several* samples, cumulative sum charts are more effective than Shewhart charts for detecting small process shifts. Furthermore, they are particularly effective with samples of $n = 1$. This makes the cumulative sum control chart a good candidate for use in the chemical and process industries where rational subgroups are frequently of size 1, as well as in discrete parts manufacturing with automatic measurement of each part and online control using a microcomputer directly at the work center.

If the process remains in control at the target value μ_0, the cumulative sum defined in equation 16-29 should fluctuate around zero. However, if the mean shifts upward to some value $\mu_1 > \mu_0$, say, an upward or positive drift will develop in the cumulative sum S_i. Conversely, if the mean shifts downward to some $\mu_1 < \mu_0$, a downward or negative drift in S_i will develop. Therefore, if a trend develops in the plotted points either upward or downward, we should consider this as evidence that the process mean has shifted, and a search for the assignable cause should be performed.

This theory can easily be demonstrated by applying the CUSUM to the chemical process concentration data in Table 16-3. Since the concentration readings are individual measurements, we would take $\overline{X}_j = X_j$ in computing the CUSUM. Suppose that the target value for the concentration is $\mu_0 = 99$. Then the CUSUM is

$$
\begin{aligned}
S_i &= \sum_{j=1}^{i} (X_j - 99) \\
&= (X_i - 99) + \sum_{j=1}^{i-1} (X_j - 99) \\
&= (X_i - 99) + S_{i-1}
\end{aligned}
$$

Table 16-7 shows the computation of this CUSUM, where the starting value of the CUSUM, S_0, is taken to be zero. Figure 16-19 plots the CUSUM from the last column of Table 16-7. Notice that the CUSUM fluctuates around the value of 0.

The graph in Fig. 16-19 is not a control chart because it lacks control limits. There are two general approaches to devising control limits for CUSUMS. The older of these two methods is the V-mask procedure. A typical V mask is shown in Fig. 16-20(a). It is a V-shaped notch in a plane that can be placed at different locations on the CUSUM chart. The decision procedure consists of placing the V mask on the cumulative sum control chart with the point O on the last value of s_i and the line OP parallel to the horizontal axis. If all the previous cumulative sums, $s_1, s_2, \ldots, s_{i-1}$, lie within the two arms of the V mask, the process is in control. However, if any s_i lies outside the arms of the mask, the process is considered to be out of control. In actual use, the V mask would be applied to each new point on the CUSUM chart as soon as it was plotted. In the example shown in Fig. 16-20(b), an upward shift in the mean is indicated, since at least one of the points that have occurred earlier than sample 22 now lies below the lower arm of the mask, when the V mask is centered on the thirtieth observation. If the point lies above the upper arm, a downward shift in the mean is

Table 16-7   CUSUM Computations for the Chemical Process Concentration Data in Table 16-3

Observation, i	x_i	$x_i - 99$	$s_i = (x_i - 99) + s_{i-1}$
1	102.0	3.0	3.0
2	94.8	−4.2	−1.2
3	98.3	−0.7	−1.9
4	98.4	−0.6	−2.5
5	102.0	3.0	0.5
6	98.5	−0.5	0.0
7	99.0	0.0	0.0
8	97.7	−1.3	−1.3
9	100.0	1.0	−0.3
10	98.1	−0.9	−1.2
11	101.3	2.3	1.1
12	98.7	−0.3	0.8
13	101.1	2.1	2.9
14	98.4	−0.6	2.3
15	97.0	−2.0	0.3
16	96.7	−2.3	−2.0
17	100.3	1.3	−0.7
18	101.4	2.4	1.7
19	97.2	−1.8	−0.1
20	101.0	2.0	1.9

indicated. Thus, the V mask forms a visual frame of reference similar to the control limits on an ordinary Shewhart control chart. For the technical details of designing the V mask, see Montgomery (2001).

While some computer programs plot CUSUMS with the V-mask control scheme, we feel that the other approach to CUSUM control, the **tabular CUSUM,** is superior.

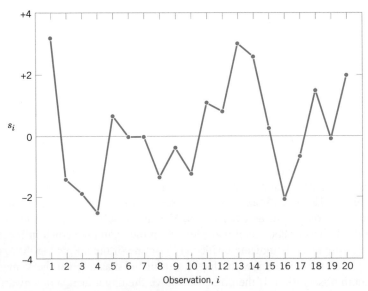

Figure 16-19   Plot of the cumulative sum for the concentration data, Table 16-7.

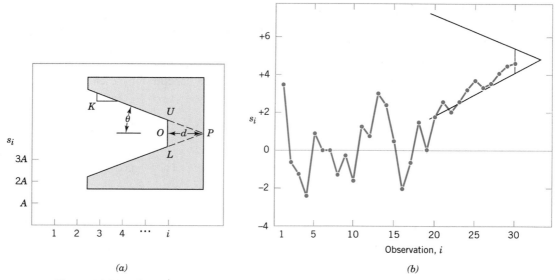

Figure 16-20   The cumulative sum control chart. (a) The V-mask and scaling. (b) The cumulative sum control chart in operation.

The tabular procedure is particularly attractive when the CUSUM is implemented on a computer.

Let $S_H(i)$ be an upper one-sided CUSUM for period i and $S_L(i)$ be a lower one-sided CUSUM for period i. These quantities are calculated from

CUSUM Control Chart

$$s_H(i) = \max[0, \bar{x}_i - (\mu_0 + K) + s_H(i-1)] \qquad (16\text{-}30)$$

and

$$s_L(i) = \max[0, (\mu_0 - K) - \bar{x}_i + s_L(i-1)] \qquad (16\text{-}31)$$

where the starting values $s_H(0) = s_L(0) = 0$.

In Equations 16-30 and 16-31 K is called the **reference value,** which is usually chosen about halfway between the target μ_0 and the value of the mean corresponding to the out-of-control state, $\mu_1 = \mu_0 + \Delta$. That is, K is about one-half the magnitude of the shift we are interested in, or

$$K = \frac{\Delta}{2}$$

Notice that $S_H(i)$ and $S_L(i)$ accumulate deviations from the target value that are greater than K, with both quantities reset to zero upon becoming negative. If either $S_H(i)$ or $S_L(i)$ exceeds a constant H, the process is out of control. This constant H is usually called the **decision interval.**

EXAMPLE 16-6 | **A Tabular CUSUM**

We will illustrate the tabular CUSUM by applying it to the chemical process concentration data in Table 16-7. The process target is $\mu_0 = 99$, and we will use $K = 1$ as the reference value and $H = 10$ as the decision interval. The reasons for these choices will be explained later.

Table 16-8 shows the tabular CUSUM scheme for the chemical process concentration data. To illustrate the calculations, note that

$$s_H(i) = \max[0, x_i - (\mu_0 + K) + s_H(i-1)] = \max[0, x_i - (99 + 1) + s_H(i-1)]$$
$$= \max[0, x_i - 100 + s_H(i-1)]$$
$$s_L(i) = \max[0, (\mu_0 - K) - x_i + s_L(i-1)] = \max[0, (99 - 1) - x_i + s_L(i-1)]$$
$$= \max[0, 98 - x_i + s_L(i-1)]$$

Therefore, for observation 1 the CUSUMS are

$$s_H(1) = \max[0, x_1 - 100 + s_H(0)] = \max[0, 102.0 - 100 + 0] = 2.0$$

and

$$s_L(1) = \max[0, 98 - x_1 + s_L(0)] = \max[0, 98 - 102.0 + 0] = 0$$

as shown in Table 16-8. The quantities n_H and n_L in Table 16-8 indicate the number of periods that the CUSUM $s_H(i)$ or $s_L(i)$ have been nonzero. Notice that the CUSUMS in this example never exceed the decision interval $H = 10$. We would therefore conclude that the process is in control.

When the tabular CUSUM indicates that the process is out of control, we should search for the assignable cause, take any corrective actions indicated, and restart the CUSUMS at

Table 16-8 The Tabular CUSUM for the Chemical Process Concentration Data

Observation i	x_i	Upper CUSUM $x_i - 100$	$s_H(i)$	n_H	Lower CUSUM $98 - x_i$	$s_L(i)$	n_L
1	102.0	2.0	2.0	1	−4.0	0.0	0
2	94.8	−5.2	0.0	0	3.2	3.2	1
3	98.3	−1.7	0.0	0	−0.3	2.9	2
4	98.4	−1.6	0.0	0	−0.4	2.5	3
5	102.0	2.0	2.0	1	−4.0	0.0	0
6	98.5	−1.5	0.5	2	−0.5	0.0	0
7	99.0	−1.0	0.0	0	−1.0	0.0	0
8	97.7	−2.3	0.0	0	0.3	0.3	1
9	100.0	0.0	0.0	0	−2.0	0.0	0
10	98.1	−1.9	0.0	0	−0.1	0.0	0
11	101.3	1.3	1.3	1	−3.3	0.0	0
12	98.7	−1.3	0.0	0	−0.7	0.0	0
13	101.1	1.1	1.1	1	−3.1	0.0	0
14	98.4	−1.6	0.0	0	−0.4	0.0	0
15	97.0	−3.0	0.0	0	1.0	1.0	1
16	96.7	−3.3	0.0	0	1.3	2.3	2
17	100.3	0.3	0.3	1	−2.3	0.0	0
18	101.4	1.4	1.7	2	−3.4	0.0	0
19	97.2	−2.8	0.0	0	0.8	0.8	1
20	101.0	1.0	1.0	0	−3.0	0.0	0

zero. It may be helpful to have an estimate of the new process mean following the shift. This can be computed from

$$\hat{\mu} = \begin{cases} \mu_0 + K + \dfrac{s_H(i)}{n_H}, & \text{if } s_H(i) > H \\[2ex] \mu_0 - K - \dfrac{s_L(i)}{n_L}, & \text{if } s_L(i) > H \end{cases} \tag{16-32}$$

It is also useful to present a graphical display of the tabular CUSUMS, which are sometimes called CUSUM status charts. They are constructed by plotting $s_H(i)$ and $s_L(i)$ versus the sample number. Figure 16-21 shows the CUSUM status chart for the data in Example 16-6. Each vertical bar represents the value of $s_H(i)$ and $s_L(i)$ in period i. With the decision interval plotted on the chart, the CUSUM status chart resembles a Shewhart control chart. We have also plotted the sample statistics x_i for each period on the CUSUM status chart as the solid dots. This frequently helps the user of the control chart to visualize the actual process performance that has led to a particular value of the CUSUM.

The tabular CUSUM is designed by choosing values for the reference value K and the decision interval H. We recommend that these parameters be selected to provide good average run-length values. There have been many analytical studies of CUSUM ARL performance. Based on these studies, we may give some general recommendations for selecting H and K. Define $H = h\sigma_{\overline{X}}$ and $K = k\sigma_{\overline{X}}$, where $\sigma_{\overline{X}}$ is the standard deviation of the sample variable used in forming the CUSUM (if $n = 1$, $\sigma_{\overline{X}} = \sigma_X$). Using $h = 4$ or $h = 5$ and $k = 1/2$ will generally provide a CUSUM that has good ARL properties against a shift of about $1\sigma_{\overline{X}}$ (or $1\sigma_X$) in the process mean. If much larger or smaller shifts are of interest, set $k = \delta/2$, where δ is the size of the shift in standard deviation units. Some practitioners prefer to use a standardized variable $y_i = (\overline{x}_i - \mu_0)/\sigma_{\overline{X}}$ as the basis of the CUSUM. In that case, Equations 16-30 and 16-31 become

$$s_H(i) = \max[0, y_i - K + s_H(i - 1)] \qquad \text{and} \qquad s_L(i) = \max[0, K - y_i + s_L(i - 1)]$$

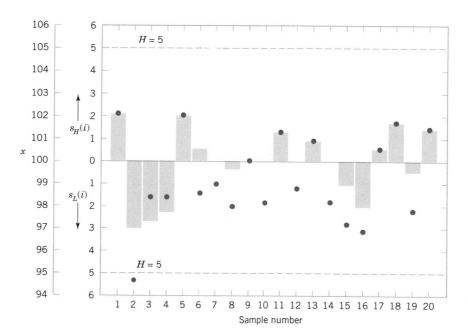

Figure 16-21   The CUSUM status chart for Example 15-6.

Table 16-9 Average Run Lengths for a CUSUM Control Chart With $K = 1/2$

Shift in Mean (multiple of $\sigma_{\overline{X}}$)	$h = 4$	$h = 5$
0	168	465
0.25	74.2	139
0.50	26.6	38.0
0.75	13.3	17.0
1.00	8.38	10.4
1.50	4.75	5.75
2.00	3.34	4.01
2.50	2.62	3.11
3.00	2.19	2.57
4.00	1.71	2.01

For this scheme, we would usually select $K = 1/2$ and $H = 4$ or $H = 5$.

To illustrate how well the recommendations of $h = 4$ or $h = 5$ with $k = 1/2$ work, consider these average run lengths in Table 16-9. Notice that a shift of $1\sigma_{\overline{X}}$ would be detected in either 8.38 samples (with $k = 1/2$ and $h = 4$) or 10.4 samples (with $k = 1/2$ and $h = 5$). By comparison, Table 16-1 shows that an $\overline{X}$ chart would require approximately 43.9 samples, on the average, to detect this shift.

These design rules were used for the CUSUM in Example 16-6. We assumed that the process standard deviation $\sigma = 2$. (This is a reasonable value; see Example 16-2.) Then with $k = 1/2$ and $h = 5$, we would use

$$K = k\sigma = \tfrac{1}{2}(2) = 1 \quad \text{and} \quad H = h\sigma = 5(2) = 10$$

in the tabular CUSUM procedure.

Finally, we should note that supplemental procedures such as the Western Electric rules cannot be safely applied to the CUSUM, because successive values of $S_H(i)$ and $S_L(i)$ are not independent. In fact, the CUSUM can be thought of as a weighted average, where the weights are stochastic or random. In effect, all the CUSUM values are highly correlated, thereby causing the Western Electric rules to give too many false alarms.

EXERCISES FOR SECTION 16-10

16-38. The purity of a chemical product is measured every two hours. The results of 20 consecutive measurements are as follows:

Sample	Purity	Sample	Purity
1	89.11	11	88.55
2	90.59	12	90.43
3	91.03	13	91.04
4	89.46	14	88.17
5	89.78	15	91.23
6	90.05	16	90.92
7	90.63	17	88.86
8	90.75	18	90.87
9	89.65	19	90.73
10	90.15	20	89.78

(a) Set up a CUSUM control chart for this process. Use $\sigma = 0.8$ in setting up the procedure, and assume that the desired process target is 90. Does the process appear to be in control?

(b) Suppose that the next five observations are 90.75, 90.00, 91.15, 90.95, and 90.86. Apply the CUSUM in part (a) to these new observations. Is there any evidence that the process has shifted out of control?

16-39. The diameter of holes is measured in consecutive order by an automatic sensor. The results of measuring 25 holes follow.

(a) Estimate the process standard deviation.

(b) Set up a CUSUM control procedure, assuming that the target diameter is 10.0 millimeters. Does the process

Sample	Diameter	Sample	Diameter
1	9.94	14	9.99
2	9.93	15	10.12
3	10.09	16	9.81
4	9.98	17	9.73
5	10.11	18	10.14
6	9.99	19	9.96
7	10.11	20	10.06
8	9.84	21	10.11
9	9.82	22	9.95
10	10.38	23	9.92
11	9.99	24	10.09
12	10.41	25	9.85
13	10.36		

appear to be operating in a state of statistical control at the desired target level?

 16-40. The concentration of a chemical product is measured by taking four samples from each batch of material. The average concentration of these measurements is shown for the last 20 batches in the following table:

Batch	Concentration	Batch	Concentration
1	104.5	11	95.4
2	99.9	12	94.5
3	106.7	13	104.5
4	105.2	14	99.7
5	94.8	15	97.7
6	94.6	16	97.0
7	104.4	17	95.8
8	99.4	18	97.4
9	100.3	19	99.0
10	100.3	20	102.6

(a) Suppose that the process standard deviation is $\sigma = 8$ and that the target value of concentration for this process is 100. Design a CUSUM scheme for the process. Does the process appear to be in control at the target?

(b) How many batches would you expect to be produced with off-target concentration before it would be detected by the CUSUM control chart if the concentration shifted to 104? Use Table 16-9.

16-41. Consider a standardized CUSUM with $H = 5$ and $K = 1/2$. Samples are taken every two hours from the process. The target value for the process is $\mu_0 = 50$ and $\sigma = 2$. Use Table 16-9.

(a) If the sample size is $n = 1$, how many samples would be required to detect a shift in the process mean to $\mu = 51$ on average?

(b) If the sample size is increased to $n = 4$, how does this affect the average run length to detect the shift to $\mu = 51$ that you determined in part (a)?

16-42. A process has a target of $\mu_0 = 100$ and a standard deviation of $\sigma = 4$. Samples of size $n = 1$ are taken every two hours. Use Table 16-9.

(a) Suppose the process mean shifts to $\mu = 102$. How many hours of production will occur before the process shift is detected by a standardized CUSUM with $H = 5$ and $K = 1/2$?

(b) It is important to detect the shift defined in part (a) more quickly. A proposal is made to reduce the sampling frequency to 0.5 hour. How will this affect the CUSUM control procedure? How much more quickly will the shift be detected?

(c) Suppose that the 0.5 hour sampling interval in part (b) is adopted. How often will false alarms occur with this new sampling interval? How often did they occur with the old interval of two hours?

(d) A proposal is made to increase the sample size to $n = 4$ and retain the two-hour sampling interval. How does this suggestion compare in terms of average detection time to the suggestion of decreasing the sampling interval to 0.5 hour?

16-11 OTHER SPC PROBLEM-SOLVING TOOLS

While the control chart is a very powerful tool for investigating the causes of variation in a process, it is most effective when used with other SPC problem-solving tools. In this section we illustrate some of these tools, using the printed circuit board defect data in Example 16-4.

Figure 16-17 shows a U chart for the number of defects in samples of five printed circuit boards. The chart exhibits statistical control, but the number of defects must be reduced. The average number of defects per board is $8/5 = 1.6$, and this level of defects would require extensive rework.

The first step in solving this problem is to construct a **Pareto diagram** of the individual defect types. The Pareto diagram, shown in Fig. 16-22, indicates that insufficient solder and solder balls are the most frequently occurring defects, accounting for $(109/160) \, 100 = 68\%$ of the

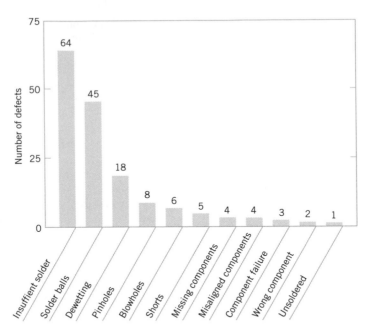

Figure 16-22   Pareto diagram for printed circuit board defects.

observed defects. Furthermore, the first five defect categories on the Pareto chart are all solder-related defects. This points to the flow solder process as a potential opportunity for improvement.

To improve the flow solder process, a team consisting of the flow solder operator, the shop supervisor, the manufacturing engineer responsible for the process, and a quality engineer meets to study potential causes of solder defects. They conduct a brainstorming session and produce the cause-and-effect diagram shown in Fig. 16-23. The cause-and-effect diagram is widely used to display the various potential causes of defects in products and their interrelationships. They are useful in summarizing knowledge about the process.

As a result of the brainstorming session, the team tentatively identifies the following variables as potentially influential in creating solder defects:

1.   Flux specific gravity

2.   Solder temperature

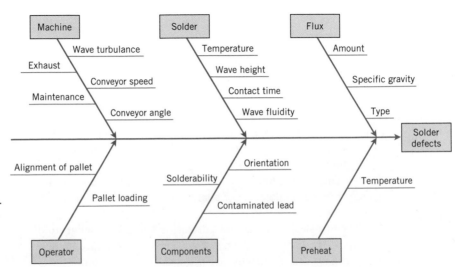

Figure 16-23   Cause-and-effect diagram for the printed circuit board flow solder process.

Front

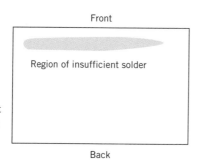

Region of insufficient solder

Back

Figure 16-24 Defect concentration diagram for a printed circuit board.

3. Conveyor speed
4. Conveyor angle
5. Solder wave height
6. Preheat temperature
7. Pallet loading method

A statistically **designed experiment** could be used to investigate the effect of these seven variables on solder defects.

In addition, the team constructed a **defect concentration diagram** for the product. A defect concentration diagram is just a sketch or drawing of the product, with the most frequently occurring defects shown on the part. This diagram is used to determine whether defects occur in the same location on the part. The defect concentration diagram for the printed circuit board is shown in Fig. 16-24. This diagram indicates that most of the insufficient solder defects are near the front edge of the board, where it makes initial contact with the solder wave. Further investigation showed that one of the pallets used to carry the boards across the wave was bent, causing the front edge of the board to make poor contact with the solder wave.

When the defective pallet was replaced, a designed experiment was used to investigate the seven variables discussed earlier. The results of this experiment indicated that several of these factors were influential and could be adjusted to reduce solder defects. After the results of the experiment were implemented, the percentage of solder joints requiring rework was reduced from 1% to under 100 parts per million (0.01%).

16-12 IMPLEMENTING SPC

The methods of statistical process control can provide significant payback to those companies that can successfully implement them. While SPC seems to be a collection of statistically based problem-solving tools, there is more to the successful use of SPC than simply learning and using these tools. Management involvement and commitment to the quality-improvement process is the most vital component of SPC's potential success. Management is a role model, and others in the organization will look to management for guidance and as an example. A team approach is also important, for it is usually difficult for one person alone to introduce process improvements. Many of the "magnificent seven" problem-solving tools are helpful in building an improvement team, including cause-and-effect diagrams, Pareto charts, and defect concentration diagrams. The basic SPC problem-solving tools must become widely known and widely used throughout the organization. Continuous training in SPC and quality improvement is necessary to achieve this widespread knowledge of the tools.

The objective of an SPC-based quality-improvement program is continuous improvement on a weekly, quarterly, and annual basis. SPC is not a one-time program to be applied when the business is in trouble and later abandoned. Quality improvement must become part of the culture of the organization.

The control chart is an important tool for process improvement. Processes do not naturally operate in an in-control state, and the use of control charts is an important step that must be taken early in an SPC program to eliminate assignable causes, reduce process variability, and stabilize process performance. To improve quality and productivity, we must begin to manage with facts and data, and not just rely on judgment. Control charts are an important part of this change in management approach.

In implementing a company-wide SPC program, we have found that the following elements are usually present in all successful efforts:

1. Management leadership
2. A team approach
3. Education of employees at all levels
4. Emphasis on continuous improvement
5. A mechanism for recognizing success

We cannot overemphasize the importance of management leadership and the team approach. Successful quality improvement is a "top-down" management-driven activity. It is also important to measure progress and success and to spread knowledge of this success throughout the organization. When successful improvements are communicated throughout the company, this can provide motivation and incentive to improve other processes and to make continuous improvement a normal part of the way of doing business.

The philosophy of W. Edwards **Deming** provides an important framework for implementing quality and productivity improvement. Deming's philosophy is summarized in his 14 points for management. The adherence to these management principles has been an important factor in Japan's industrial success and continues to be the catalyst in that nation's quality- and productivity-improvement efforts. This philosophy has also now spread rapidly in the West. **Deming's 14 points** are as follows.

1. **Create a constancy of purpose focused on the improvement of products and services.** Constantly try to improve product design and performance. Investment in research, development, and innovation will have a long-term payback to the organization.

2. **Adopt a new philosophy of rejecting poor workmanship, defective products, or bad service.** It costs as much to produce a defective unit as it does to produce a good one (and sometimes more). The cost of dealing with scrap, rework, and other losses created by defectives is an enormous drain on company resources.

3. **Do not rely on mass inspection to "control" quality.** All inspection can do is sort out defectives, and at this point it is too late because we have already paid to produce these defectives. Inspection occurs too late in the process, it is expensive, and it is often ineffective. Quality results from the prevention of defectives through process improvement, not inspection.

4. **Do not award business to suppliers on the basis of price alone, but also consider quality.** Price is a meaningful measure of a supplier's product only if it is considered in relation to a measure of quality. In other words, the total cost of the item must be considered, not just the purchase price. When quality is considered, the lowest bidder is frequently not the low-cost supplier. Preference should be given to suppliers who use modern methods of quality improvement in their business and who can demonstrate process control and capability.

5. **Focus on continuous improvement.** Constantly try to improve the production and service system. Involve the workforce in these activities and make use of statistical methods, particularly the SPC problem-solving tools discussed in the previous section.

6. **Practice modern training methods and invest in training for all employees.** Everyone should be trained in the technical aspects of their job, as well as in modern quality- and productivity-improvement methods. The training should encourage all employees to practice these methods every day.

7. **Practice modern supervision methods.** Supervision should not consist merely of passive surveillance of workers, but should be focused on helping the employees improve the system in which they work. The first goal of supervision should be to improve the work system and the product.

8. **Drive out fear.** Many workers are afraid to ask questions, report problems, or point out conditions that are barriers to quality and effective production. In many organizations the economic loss associated with fear is large; only management can eliminate fear.

9. **Break down the barriers between functional areas of the business.** Teamwork among different organizational units is essential for effective quality and productivity improvement to take place.

10. **Eliminate targets, slogans, and numerical goals for the workforce.** A target such as "zero defects" is useless without a plan as to how to achieve this objective. In fact, these slogans and "programs" are usually counterproductive. Work to improve the system and provide information on that.

11. **Eliminate numerical quotas and work standards.** These standards have historically been set without regard to quality. Work standards are often symptoms of management's inability to understand the work process and to provide an effective management system focused on improving this process.

12. **Remove the barriers that discourage employees from doing their jobs.** Management must listen to employee suggestions, comments, and complaints. The person who is doing the job is the one who knows the most about it, and usually has valuable ideas about how to make the process work more effectively. The workforce is an important participant in the business, and not just an opponent in collective bargaining.

13. **Institute an ongoing program of training and education for all employees.** Education in simple, powerful statistical techniques should be mandatory for all employees. Use of the basic SPC problem-solving tools, particularly the control chart, should become widespread in the business. As these charts become widespread, and as employees understand their uses, they will be more likely to look for the causes of poor quality and to identify process improvements. Education is a way of making everyone partners in the quality-improvement process.

14. **Create a structure in top management that will vigorously advocate the first 13 points.**

As we read Deming's 14 points, we notice two things. First, there is a strong emphasis on change. Second, the role of management in guiding this change process is of dominating importance. But what should be changed, and how should this change process be started? For example, if we want to improve the yield of a semiconductor manufacturing process, what should we do? It is in this area that statistical methods most frequently come into play. To improve the semiconductor process, we must determine which controllable factors in the process influence the number of defective units produced. To answer this question, we must collect data on the process and see how the system reacts to changes in the process variables. Statistical methods, including the SPC and experimental design techniques in this book, can contribute to this knowledge.

SUPPLEMENTAL EXERCISES

16-43. The diameter of fuse pins used in an aircraft engine application is an important quality characteristic. Twenty-five samples of three pins each are shown as follows:

Sample Number	Diameter		
1	64.030	64.002	64.019
2	63.995	63.992	64.001
3	63.988	64.024	64.021
4	64.002	63.996	63.993
5	63.992	64.007	64.015
6	64.009	63.994	63.997
7	63.995	64.006	63.994
8	63.985	64.003	63.993
9	64.008	63.995	64.009
10	63.998	74.000	63.990
11	63.994	63.998	63.994
12	64.004	64.000	64.007
13	63.983	64.002	63.998
14	64.006	63.967	63.994
15	64.012	64.014	63.998
16	64.000	63.984	64.005
17	63.994	64.012	63.986
18	64.006	64.010	64.018
19	63.984	64.002	64.003
20	64.000	64.010	64.013
21	63.988	64.001	64.009
22	64.004	63.999	63.990
23	64.010	63.989	63.990
24	64.015	64.008	63.993
25	63.982	63.984	63.995

(a) Set up $\overline{X}$ and R charts for this process. If necessary, revise limits so that no observations are out-of-control.
(b) Estimate the process mean and standard deviation.
(c) Suppose the process specifications are at 64 ± 0.02. Calculate an estimate of PCR. Does the process meet a minimum capability level of $PCR \geq 1.33$?
(d) Calculate an estimate of PCR_k. Use this ratio to draw conclusions about process capability.

(e) To make this process a six-sigma process, the variance σ^2 would have to be decreased such that $PCR_k = 2.0$. What should this new variance value be?
(f) Suppose the mean shifts to 64.01. What is the probability that this shift will be detected on the next sample? What is the ARL after the shift?

16-44. Rework Exercise 16-43 with $\overline{X}$ and S charts.

16-45. Plastic bottles for liquid laundry detergent are formed by blow molding. Twenty samples of $n = 100$ bottles are inspected in time order of production, and the fraction defective in each sample is reported. The data are as follows:

Sample	Fraction Defective
1	0.12
2	0.15
3	0.18
4	0.10
5	0.12
6	0.11
7	0.05
8	0.09
9	0.13
10	0.13
11	0.10
12	0.07
13	0.12
14	0.08
15	0.09
16	0.15
17	0.10
18	0.06
19	0.12
20	0.13

(a) Set up a P chart for this process. Is the process in statistical control?
(b) Suppose that instead of $n = 100$, $n = 200$. Use the data given to set up a P chart for this process. Revise the control limits if necessary.
(c) Compare your control limits for the P charts in parts (a) and (b). Explain why they differ. Also, explain why your assessment about statistical control differs for the two sizes of n.

16-46. Cover cases for a personal computer are manufactured by injection molding. Samples of five cases are taken from the process periodically, and the number of defects is noted. Twenty-five samples follow:

Sample	No. of Defects	Sample	No. of Defects
1	3	14	8
2	2	15	0
3	0	16	2
4	1	17	4
5	4	18	3
6	3	19	5
7	2	20	0
8	4	21	2
9	1	22	1
10	0	23	9
11	2	24	3
12	3	25	2
13	2		

(a) Using all the data, find trial control limits for this U chart for the process.
(b) Use the trial control limits from part (a) to identify out-of-control points. If necessary, revise your control limits.
(c) Suppose that instead of samples of 5 cases, the sample size was 10. Repeat parts (a) and (b). Explain how this change alters your answers to parts (a) and (b).

16-47. Consider the data in Exercise 16-46.

(a) Using all the data, find trial control limits for a C chart for this process.
(b) Use the trial control limits of part (a) to identify out-of-control points. If necessary, revise your control limits.
(c) Suppose that instead of samples of 5 cases, the sample was 10 cases. Repeat parts (a) and (b). Explain how this alters your answers to parts (a) and (b).

16-48. Suppose that a process is in control and an $\overline{X}$ chart is used with a sample size of 4 to monitor the process. Suddenly there is a mean shift of 1.5σ.
(a) If 3-sigma control limits are in use on the $\overline{X}$ chart, what is the probability that this shift will remain undetected for three consecutive samples?
(b) If 2-sigma control limits are in use on the $\overline{X}$ chart, what is the probability that this shift will remain undetected for three consecutive samples?
(c) Compare your answers to parts (a) and (b) and explain why they differ. Also, which limits you would recommend using and why?

16-49. Consider the control chart for individuals with 3-sigma limits.

(a) Suppose that a shift in the process mean of magnitude σ occurs. Verify that the ARL for detecting the shift is ARL = 43.9.
(b) Find the ARL for detecting a shift of magnitude 2σ in the process mean.
(c) Find the ARL for detecting a shift of magnitude 3σ in the process mean.
(d) Compare your answers to parts (a), (b), and (c) and explain why the ARL for detection is decreasing as the magnitude of the shift increases.

16-50. Consider a control chart for individuals, applied to a continuous 24-hour chemical process with observations taken every hour.
(a) If the chart has 3-sigma limits, verify that the in-control ARL is ARL = 370. How many false alarms would occur each 30-day month, on the average, with this chart?
(b) Suppose that the chart has 2-sigma limits. Does this reduce the ARL for detecting a shift in the mean of magnitude σ? (Recall that the ARL for detecting this shift with 3-sigma limits is 43.9.)
(c) Find the in-control ARL if 2-sigma limits are used on the chart. How many false alarms would occur each month with this chart? Is this in-control ARL performance satisfactory? Explain your answer.

16-51. The depth of a keyway is an important part quality characteristic. Samples of size $n = 5$ are taken every four hours from the process and 20 samples are summarized as follows:

Sample	$\overline{X}$	r
1	139.7	1.1
2	139.8	1.4
3	140.0	1.3
4	140.1	1.6
5	139.8	0.9
6	139.9	1.0
7	139.7	1.4
8	140.2	1.2
9	139.3	1.1
10	140.7	1.0
11	138.4	0.8
12	138.5	0.9
13	137.9	1.2
14	138.5	1.1
15	140.8	1.0
16	140.5	1.3
17	139.4	1.4
18	139.9	1.0
19	137.5	1.5
20	139.2	1.3

(a) Using all the data, find trial control limits for $\overline{X}$ and R charts. Is the process in control?

(b) Use the trial control limits from part (a) to identify out-of-control points. If necessary, revise your control limits. Then, estimate the process standard deviation.

(c) Suppose that the specifications are at 140 ± 2. Using the results from part (b), what statements can you make about process capability? Compute estimates of the appropriate process capability ratios.

(d) To make this process a "6-sigma process," the variance σ^2 would have to be decreased such that $PCR_k = 2.0$. What should this new variance value be?

(e) Suppose the mean shifts to 139.7. What is the probability that this shift will be detected on the next sample? What is the ARL after the shift?

16-52. A process is controlled by a P chart using samples of size 100. The center line on the chart is 0.05.

(a) What is the probability that the control chart detects a shift to 0.08 on the first sample following the shift?

(b) What is the probability that the control chart does not detect a shift to 0.08 on the first sample following the shift but does detect it on the second sample?

(c) Suppose that instead of a shift in the mean to 0.07, the mean shifts to 0.10. Repeat parts (a) and (b).

(d) Compare your answers for a shift to 0.07 and for a shift to 0.10. Explain why they differ. Also, explain why a shift to 0.10 is easier to detect.

16-53. Suppose the average number of defects in a unit is known to be 8. If the mean number of defects in a unit shifts to 16, what is the probability that it will be detected by the U chart on the first sample following the shift

(a) if the sample size is $n = 4$?

(b) if the sample size is $n = 10$?

Use a normal approximation for U.

16-54. Suppose the average number of defects in a unit is known to be 10. If the mean number of defects in a unit shifts to 14, what is the probability that it will be detected by the U chart on the first sample following the shift

(a) if the sample size is $n = 1$?

(b) if the sample size is $n = 4$?

Use a normal approximation for U.

16-55. Suppose that an $\overline{X}$ control chart with 2-sigma limits is used to control a process. Find the probability that a false out-of-control signal will be produced on the next sample. Compare this with the corresponding probability for the chart with 3-sigma limits and discuss. Comment on when you would prefer to use 2-sigma limits instead of 3-sigma limits.

16-56. Consider the $\overline{X}$ control chart with 2-sigma limits in Exercise 16-50.

(a) Find the probability of no signal on the first sample but a signal on the second.

(b) What is the probability that there will not be a signal in three samples?

16-57. Suppose a process has a $PCR = 2$, but the mean is exactly three standard deviations above the upper specification limit. What is the probability of making a product outside the specification limits?

16-58. Consider the hardness measurement data in Exercise 16-9. Set up a CUSUM scheme for this process using $\mu = 50$ and $\sigma = 2$, so that $K = 1$ and $H = 10$. Is the process in control?

16-59. Consider the data in Exercise 16-10. Set up a CUSUM scheme for this process assuming that $\mu = 16$ is the process target. Explain how you determined your estimate of σ and the CUSUM parameters K and H.

16-60. Reconsider the data in Exercise 16-12. Construct a CUSUM control chart for this process using $\mu_0 = 500$ as the process target. Explain how you determined your estimate of σ and the CUSUM parameters H and K.

MIND-EXPANDING EXERCISES

16-61. Suppose a process is in control, and 3-sigma control limits are in use on the $\overline{X}$ chart. Let the mean shift by 1.5σ. What is the probability that this shift will remain undetected for three consecutive samples? What would its probability be if 2-sigma control limits were used? The sample size is 4.

16-62. Consider an $\overline{X}$ control chart with k-sigma control limits. Develop a general expression for the probability that a point will plot outside the control limits when the process mean has shifted by δ units from the center line.

16-63. Suppose that an $\overline{X}$ chart is used to control a normally distributed process and that samples of size n are taken every n hours and plotted on the chart, which has k-sigma limits.

(a) Find a general expression for the expected number of samples and time that will be taken until a false action signal is generated.

(b) Suppose that the process mean shifts to an out-of-control state, say $\mu_1 = \mu_0 + \delta\sigma$. Find an expression for the expected number of samples that will be taken until a false action is generated.

MIND-EXPANDING EXERCISES

(c) Evaluate the in-control ARL for $k = 3$. How, does this change if $k = 2$? What do you think about the use of 2-sigma limits in practice?

(d) Evaluate the out-of-control ARL for a shift of 1 sigma, given that $n = 5$.

16-64. Suppose a P chart with center line at $\bar{p}$ with k-sigma control limits is used to control a process. There is a critical fraction defective p_c that must be detected with probability 0.50 on the first sample following the shift to this state. Derive a general formula for the sample size that should be used on this chart.

16-65. Suppose that a P chart with center line at $\bar{p}$ and k-sigma control limits is used to control a process. What is the smallest sample size that can be used on this control chart to ensure that the lower control limit is positive?

16-66. A process is controlled by a P chart using samples of size 100. The center line on the chart is 0.05. What is the probability that the control chart detects a shift to 0.08 on the first sample following the shift? What is the probability that the shift is detected by at least the third sample following the shift?

16-67. Consider a process where specifications on a quality characteristic are 100 ± 15. We know that the standard deviation of this normally distributed quality characteristic is 5. Where should we center the process to minimize the fraction defective produced? Now suppose the mean shifts to 105 and we are using a sample size of 4 on an $\bar{X}$ chart. What is the probability that such a shift will be detected on the first sample following the shift? What is the average number of samples until an out-of-control point occurs? Compare this result to the average number of observations until a defective occurs (assuming normality).

16-68. The NP Control Chart. An alternative to the control chart for fraction defective is a control chart based on the number of defectives, or the NP control chart. The chart has centerline at $n\bar{p}$, and the control limits are

$$UCL = n\bar{p} + 3\sqrt{n\bar{p}(1 - \bar{p})}$$

$$LCL = n\bar{p} - 3\sqrt{n\bar{p}(1 - \bar{p})}$$

and the number of defectives for each sample is plotted on the chart.

(a) Verify that the control limits given above are correct.

(b) Apply this control chart to the data in Example 16-4.

(c) Will this chart always provide results that are equivalent to the usual P chart?

16-69. The EWMA Control Chart. The exponentially weighted moving average (or EWMA) is defined as follows:

$$Z_t = \lambda \bar{X}_t + (1 - \lambda)\bar{Z}_{t-1}$$

where $0 < \lambda \le 1$, and the starting value of the EWMA at time $t = 0$ is $Z_0 = \mu_0$ (the process target). An EWMA control chart is constructed by plotting the Z_t values on a chart with center line at μ_0 and appropriate control limits.

(a) Verify that $E(Z_t) = \mu_0$

(b) Let $\sigma_{z_t}^2$ be $V(Z_t)$, and show that

$$\sigma_{z_t}^2 = \frac{\sigma^2}{n}\left(\frac{\lambda}{2 - \lambda}\right)\left[1 - (1 - \lambda)^{2t}\right]$$

(c) Use the results of part (b) to determine the control limits for the EWMA chart.

(d) As $\lambda \to 1$, the EWMA control chart should perform like a standard Shewhart $\bar{X}$ chart. Do you agree with this statement? Why?

(e) As $\lambda \to 0$, the EWMA control chart should perform like a CUSUM. Provide an argument as to why this is so.

(f) Apply this procedure to the data in Example 16-2.

16-70. Standardized Control Chart. Consider the P chart with the usual 3-sigma control limits. Suppose that we define a new variable:

$$Z_i = \frac{\hat{P}_i - \bar{P}}{\sqrt{\dfrac{\bar{P}(1 - \bar{P})}{n}}}$$

as the quantity to plot on a control chart. It is proposed that this new chart will have a center line at 0 with the upper and lower control limits at ± 3. Verify that this standardized control chart will be equivalent to the original p chart.

16-71. Unequal Sample Sizes. One application of the standardized control chart introduced in Exercise 16-70 is to allow unequal sample sizes on the control chart. Provide details concerning how this procedure would be implemented and illustrate using the following data:

Sample, i	1	2	3	4	5	6	7	8	9	10
n_i	20	25	20	25	50	30	25	25	25	20
p_i	0.2	0.16	0.25	0.08	0.3	0.1	0.12	0.16	0.12	0.15

IMPORTANT TERMS AND CONCEPTS

In the E-book, click on any term or concept below to go to that subject.

ARL
Assignable causes
Attributes control charts
Average run length
C chart
Cause-and-effect diagram
Center line
Chance causes
Control chart

Control limits
Cumulative sum control chart
Defect concentration diagram
Defects-per-unit chart
Deming's 14 points
False alarm
Fraction-defective control chart
Implementing SPC
Individuals control chart
Moving range

NP chart
P chart
Pareto diagram
PCR
PCR_k
Problem-solving tools
Process capability
Process capability ratio
Quality control
R chart
Rational subgroup
Run rule

S chart
Shewhart control chart
Six-sigma process
Specification limits
Statistical process control
Statistical quality control
U chart
V mask
Variables control charts
Warning limits
Western Electric rules
$\overline{X}$ chart

APPENDICES

Appendix A
Statistical
Tables and
Charts

Table I Summary of Common Probability Distributions

Name	Probability Distribution	Mean	Variance	Section in Book
Discrete				
Uniform	$\dfrac{1}{n}, a \leq b$	$\dfrac{(b+a)}{2}$	$\dfrac{(b-a+1)^2 - 1}{12}$	3-5
Binomial	$\binom{n}{x} p^x (1-p)^{n-x},$ $x = 0, 1, \ldots, n, 0 \leq p \leq 1$	np	$np(1-p)$	3-6
Geometric	$(1-p)^{x-1} p,$ $x = 1, 2, \ldots, 0 \leq p \leq 1$	$1/p$	$(1-p)/p^2$	3-7.1
Negative binomial	$\binom{x-1}{r-1}(1-p)^{x-r} p^r$ $x = r, r+1, r+2, \ldots, 0 \leq p \leq 1$	r/p	$r(1-p)/p^2$	3-7.2
Hypergeometric	$\dfrac{\binom{K}{x}\binom{N-K}{n-x}}{\binom{N}{n}}$ $x = \max(0, n-N+K), 1, \ldots$ $\min(K, n), K \leq N, n \leq N$	$np,$ where $p = \dfrac{K}{N}$	$np(1-p)\left(\dfrac{N-n}{N-1}\right)$	3-8
Poisson	$\dfrac{e^{-\lambda}\lambda^x}{x!}, x = 0, 1, 2, \ldots, 0 < \lambda$	λ	λ	3-9
Continuous				
Uniform	$\dfrac{1}{b-a}, a \leq x \leq b$	$\dfrac{(b+a)}{2}$	$\dfrac{(b-a)^2}{12}$	4-5
Normal	$\dfrac{1}{\sigma\sqrt{2\pi}} e^{-\frac{1}{2}\left(\frac{x-\mu}{\sigma}\right)^2}$ $-\infty < x < \infty, -\infty < \mu < \infty, 0 < \sigma$	μ	σ^2	4-6
Exponential	$\lambda e^{-\lambda x}, 0 \leq x, 0 < \lambda$	$1/\lambda$	$1/\lambda^2$	4-8
Erlang	$\dfrac{\lambda^r x^{r-1} e^{-\lambda x}}{(r-1)!}, 0 < x, r = 1, 2, \ldots$	r/λ	r/λ^2	4-9.1
Gamma	$\dfrac{\lambda x^{r-1} e^{-\lambda x}}{\Gamma(r)}, 0 < x, 0 < r, 0 < \lambda$	r/λ	r/λ^2	4-9.2
Weibull	$\dfrac{\beta}{\delta}\left(\dfrac{x}{\delta}\right)^{\beta-1} e^{-(x/\delta)^\beta},$ $0 < x, 0 < \beta, 0 < \delta$	$\delta\Gamma\left(1 + \dfrac{1}{\beta}\right)$	$\delta^2\Gamma\left(1 + \dfrac{2}{\beta}\right)$ $-\delta^2\left[\Gamma\left(1 + \dfrac{1}{\beta}\right)\right]^2$	4-10
Lognormal	$\dfrac{1}{x\omega\sqrt{2\pi}} \exp\left(\dfrac{-[\ln(x) - \theta]^2}{2\omega^2}\right)$	$e^{\theta + \omega^2/2}$	$e^{2\theta + \omega^2}(e^{\omega^2} - 1)$	4-11

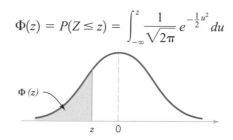

$$\Phi(z) = P(Z \le z) = \int_{-\infty}^{z} \frac{1}{\sqrt{2\pi}} e^{-\frac{1}{2}u^2}\, du$$

Table II   Cumulative Standard Normal Distribution

z	−0.09	−0.08	−0.07	−0.06	−0.05	−0.04	−0.03	−0.02	−0.01	−0.00
−3.9	0.000033	0.000034	0.000036	0.000037	0.000039	0.000041	0.000042	0.000044	0.000046	0.000048
−3.8	0.000050	0.000052	0.000054	0.000057	0.000059	0.000062	0.000064	0.000067	0.000069	0.000072
−3.7	0.000075	0.000078	0.000082	0.000085	0.000088	0.000092	0.000096	0.000100	0.000104	0.000108
−3.6	0.000112	0.000117	0.000121	0.000126	0.000131	0.000136	0.000142	0.000147	0.000153	0.000159
−3.5	0.000165	0.000172	0.000179	0.000185	0.000193	0.000200	0.000208	0.000216	0.000224	0.000233
−3.4	0.000242	0.000251	0.000260	0.000270	0.000280	0.000291	0.000302	0.000313	0.000325	0.000337
−3.3	0.000350	0.000362	0.000376	0.000390	0.000404	0.000419	0.000434	0.000450	0.000467	0.000483
−3.2	0.000501	0.000519	0.000538	0.000557	0.000577	0.000598	0.000619	0.000641	0.000664	0.000687
−3.1	0.000711	0.000736	0.000762	0.000789	0.000816	0.000845	0.000874	0.000904	0.000935	0.000968
−3.0	0.001001	0.001035	0.001070	0.001107	0.001144	0.001183	0.001223	0.001264	0.001306	0.001350
−2.9	0.001395	0.001441	0.001489	0.001538	0.001589	0.001641	0.001695	0.001750	0.001807	0.001866
−2.8	0.001926	0.001988	0.002052	0.002118	0.002186	0.002256	0.002327	0.002401	0.002477	0.002555
−2.7	0.002635	0.002718	0.002803	0.002890	0.002980	0.003072	0.003167	0.003264	0.003364	0.003467
−2.6	0.003573	0.003681	0.003793	0.003907	0.004025	0.004145	0.004269	0.004396	0.004527	0.004661
−2.5	0.004799	0.004940	0.005085	0.005234	0.005386	0.005543	0.005703	0.005868	0.006037	0.006210
−2.4	0.006387	0.006569	0.006756	0.006947	0.007143	0.007344	0.007549	0.007760	0.007976	0.008198
−2.3	0.008424	0.008656	0.008894	0.009137	0.009387	0.009642	0.009903	0.010170	0.010444	0.010724
−2.2	0.011011	0.011304	0.011604	0.011911	0.012224	0.012545	0.012874	0.013209	0.013553	0.013903
−2.1	0.014262	0.014629	0.015003	0.015386	0.015778	0.016177	0.016586	0.017003	0.017429	0.017864
−2.0	0.018309	0.018763	0.019226	0.019699	0.020182	0.020675	0.021178	0.021692	0.022216	0.022750
−1.9	0.023295	0.023852	0.024419	0.024998	0.025588	0.026190	0.026803	0.027429	0.028067	0.028717
−1.8	0.029379	0.030054	0.030742	0.031443	0.032157	0.032884	0.033625	0.034379	0.035148	0.035930
−1.7	0.036727	0.037538	0.038364	0.039204	0.040059	0.040929	0.041815	0.042716	0.043633	0.044565
−1.6	0.045514	0.046479	0.047460	0.048457	0.049471	0.050503	0.051551	0.052616	0.053699	0.054799
−1.5	0.055917	0.057053	0.058208	0.059380	0.060571	0.061780	0.063008	0.064256	0.065522	0.066807
−1.4	0.068112	0.069437	0.070781	0.072145	0.073529	0.074934	0.076359	0.077804	0.079270	0.080757
−1.3	0.082264	0.083793	0.085343	0.086915	0.088508	0.090123	0.091759	0.093418	0.095098	0.096801
−1.2	0.098525	0.100273	0.102042	0.103835	0.105650	0.107488	0.109349	0.111233	0.113140	0.115070
−1.1	0.117023	0.119000	0.121001	0.123024	0.125072	0.127143	0.129238	0.131357	0.133500	0.135666
−1.0	0.137857	0.140071	0.142310	0.144572	0.146859	0.149170	0.151505	0.153864	0.156248	0.158655
−0.9	0.161087	0.163543	0.166023	0.168528	0.171056	0.173609	0.176185	0.178786	0.181411	0.184060
−0.8	0.186733	0.189430	0.192150	0.194894	0.197662	0.200454	0.203269	0.206108	0.208970	0.211855
−0.7	0.214764	0.217695	0.220650	0.223627	0.226627	0.229650	0.232695	0.235762	0.238852	0.241964
−0.6	0.245097	0.248252	0.251429	0.254627	0.257846	0.261086	0.264347	0.267629	0.270931	0.274253
−0.5	0.277595	0.280957	0.284339	0.287740	0.291160	0.294599	0.298056	0.301532	0.305026	0.308538
−0.4	0.312067	0.315614	0.319178	0.322758	0.326355	0.329969	0.333598	0.337243	0.340903	0.344578
−0.3	0.348268	0.351973	0.355691	0.359424	0.363169	0.366928	0.370700	0.374484	0.378281	0.382089
−0.2	0.385908	0.389739	0.393580	0.397432	0.401294	0.405165	0.409046	0.412936	0.416834	0.420740
−0.1	0.424655	0.428576	0.432505	0.436441	0.440382	0.444330	0.448283	0.452242	0.456205	0.460172
0.0	0.464144	0.468119	0.472097	0.476078	0.480061	0.484047	0.488033	0.492022	0.496011	0.500000

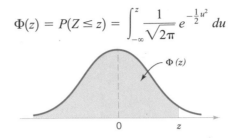

$$\Phi(z) = P(Z \le z) = \int_{-\infty}^{z} \frac{1}{\sqrt{2\pi}} e^{-\frac{1}{2}u^2} \, du$$

Table II   Cumulative Standard Normal Distribution (*continued*)

z	0.00	0.01	0.02	0.03	0.04	0.05	0.06	0.07	0.08	0.09
0.0	0.500000	0.503989	0.507978	0.511967	0.515953	0.519939	0.532922	0.527903	0.531881	0.535856
0.1	0.539828	0.543795	0.547758	0.551717	0.555760	0.559618	0.563559	0.567495	0.571424	0.575345
0.2	0.579260	0.583166	0.587064	0.590954	0.594835	0.598706	0.602568	0.606420	0.610261	0.614092
0.3	0.617911	0.621719	0.625516	0.629300	0.633072	0.636831	0.640576	0.644309	0.648027	0.651732
0.4	0.655422	0.659097	0.662757	0.666402	0.670031	0.673645	0.677242	0.680822	0.684386	0.687933
0.5	0.691462	0.694974	0.698468	0.701944	0.705401	0.708840	0.712260	0.715661	0.719043	0.722405
0.6	0.725747	0.729069	0.732371	0.735653	0.738914	0.742154	0.745373	0.748571	0.751748	0.754903
0.7	0.758036	0.761148	0.764238	0.767305	0.770350	0.773373	0.776373	0.779350	0.782305	0.785236
0.8	0.788145	0.791030	0.793892	0.796731	0.799546	0.802338	0.805106	0.807850	0.810570	0.813267
0.9	0.815940	0.818589	0.821214	0.823815	0.826391	0.828944	0.831472	0.833977	0.836457	0.838913
1.0	0.841345	0.843752	0.846136	0.848495	0.850830	0.853141	0.855428	0.857690	0.859929	0.862143
1.1	0.864334	0.866500	0.868643	0.870762	0.872857	0.874928	0.876976	0.878999	0.881000	0.882977
1.2	0.884930	0.886860	0.888767	0.890651	0.892512	0.894350	0.896165	0.897958	0.899727	0.901475
1.3	0.903199	0.904902	0.906582	0.908241	0.909877	0.911492	0.913085	0.914657	0.916207	0.917736
1.4	0.919243	0.920730	0.922196	0.923641	0.925066	0.926471	0.927855	0.929219	0.930563	0.931888
1.5	0.933193	0.934478	0.935744	0.936992	0.938220	0.939429	0.940620	0.941792	0.942947	0.944083
1.6	0.945201	0.946301	0.947384	0.948449	0.949497	0.950529	0.951543	0.952540	0.953521	0.954486
1.7	0.955435	0.956367	0.957284	0.958185	0.959071	0.959941	0.960796	0.961636	0.962462	0.963273
1.8	0.964070	0.964852	0.965621	0.966375	0.967116	0.967843	0.968557	0.969258	0.969946	0.970621
1.9	0.971283	0.971933	0.972571	0.973197	0.973810	0.974412	0.975002	0.975581	0.976148	0.976705
2.0	0.977250	0.977784	0.978308	0.978822	0.979325	0.979818	0.980301	0.980774	0.981237	0.981691
2.1	0.982136	0.982571	0.982997	0.983414	0.983823	0.984222	0.984614	0.984997	0.985371	0.985738
2.2	0.986097	0.986447	0.986791	0.987126	0.987455	0.987776	0.988089	0.988396	0.988696	0.988989
2.3	0.989276	0.989556	0.989830	0.990097	0.990358	0.990613	0.990863	0.991106	0.991344	0.991576
2.4	0.991802	0.992024	0.992240	0.992451	0.992656	0.992857	0.993053	0.993244	0.993431	0.993613
2.5	0.993790	0.993963	0.994132	0.994297	0.994457	0.994614	0.994766	0.994915	0.995060	0.995201
2.6	0.995339	0.995473	0.995604	0.995731	0.995855	0.995975	0.996093	0.996207	0.996319	0.996427
2.7	0.996533	0.996636	0.996736	0.996833	0.996928	0.997020	0.997110	0.997197	0.997282	0.997365
2.8	0.997445	0.997523	0.997599	0.997673	0.997744	0.997814	0.997882	0.997948	0.998012	0.998074
2.9	0.998134	0.998193	0.998250	0.998305	0.998359	0.998411	0.998462	0.998511	0.998559	0.998605
3.0	0.998650	0.998694	0.998736	0.998777	0.998817	0.998856	0.998893	0.998930	0.998965	0.998999
3.1	0.999032	0.999065	0.999096	0.999126	0.999155	0.999184	0.999211	0.999238	0.999264	0.999289
3.2	0.999313	0.999336	0.999359	0.999381	0.999402	0.999423	0.999443	0.999462	0.999481	0.999499
3.3	0.999517	0.999533	0.999550	0.999566	0.999581	0.999596	0.999610	0.999624	0.999638	0.999650
3.4	0.999663	0.999675	0.999687	0.999698	0.999709	0.999720	0.999730	0.999740	0.999749	0.999758
3.5	0.999767	0.999776	0.999784	0.999792	0.999800	0.999807	0.999815	0.999821	0.999828	0.999835
3.6	0.999841	0.999847	0.999853	0.999858	0.999864	0.999869	0.999874	0.999879	0.999883	0.999888
3.7	0.999892	0.999896	0.999900	0.999904	0.999908	0.999912	0.999915	0.999918	0.999922	0.999925
3.8	0.999928	0.999931	0.999933	0.999936	0.999938	0.999941	0.999943	0.999946	0.999948	0.999950
3.9	0.999952	0.999954	0.999956	0.999958	0.999959	0.999961	0.999963	0.999964	0.999966	0.999967

2(1 —

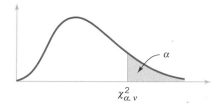

$\chi^2_{\alpha, \nu}$

Table III Percentage Points $\chi^2_{\alpha,\nu}$ of the Chi-Squared Distribution

ν \ α	.995	.990	.975	.950	.900	.500	.100	.050	.025	.010	.005
1	.00+	.00+	.00+	.00+	.02	.45	2.71	3.84	5.02	6.63	7.88
2	.01	.02	.05	.10	.21	1.39	4.61	5.99	7.38	9.21	10.60
3	.07	.11	.22	.35	.58	2.37	6.25	7.81	9.35	11.34	12.84
4	.21	.30	.48	.71	1.06	3.36	7.78	9.49	11.14	13.28	14.86
5	.41	.55	.83	1.15	1.61	4.35	9.24	11.07	12.83	15.09	16.75
6	.68	.87	1.24	1.64	2.20	5.35	10.65	12.59	14.45	16.81	18.55
7	.99	1.24	1.69	2.17	2.83	6.35	12.02	14.07	16.01	18.48	20.28
8	1.34	1.65	2.18	2.73	3.49	7.34	13.36	15.51	17.53	20.09	21.96
9	1.73	2.09	2.70	3.33	4.17	8.34	14.68	16.92	19.02	21.67	23.59
10	2.16	2.56	3.25	3.94	4.87	9.34	15.99	18.31	20.48	23.21	25.19
11	2.60	3.05	3.82	4.57	5.58	10.34	17.28	19.68	21.92	24.72	26.76
12	3.07	3.57	4.40	5.23	6.30	11.34	18.55	21.03	23.34	26.22	28.30
13	3.57	4.11	5.01	5.89	7.04	12.34	19.81	22.36	24.74	27.69	29.82
14	4.07	4.66	5.63	6.57	7.79	13.34	21.06	23.68	26.12	29.14	31.32
15	4.60	5.23	6.27	7.26	8.55	14.34	22.31	25.00	27.49	30.58	32.80
16	5.14	5.81	6.91	7.96	9.31	15.34	23.54	26.30	28.85	32.00	34.27
17	5.70	6.41	7.56	8.67	10.09	16.34	24.77	27.59	30.19	33.41	35.72
18	6.26	7.01	8.23	9.39	10.87	17.34	25.99	28.87	31.53	34.81	37.16
19	6.84	7.63	8.91	10.12	11.65	18.34	27.20	30.14	32.85	36.19	38.58
20	7.43	8.26	9.59	10.85	12.44	19.34	28.41	31.41	34.17	37.57	40.00
21	8.03	8.90	10.28	11.59	13.24	20.34	29.62	32.67	35.48	38.93	41.40
22	8.64	9.54	10.98	12.34	14.04	21.34	30.81	33.92	36.78	40.29	42.80
23	9.26	10.20	11.69	13.09	14.85	22.34	32.01	35.17	38.08	41.64	44.18
24	9.89	10.86	12.40	13.85	15.66	23.34	33.20	36.42	39.36	42.98	45.56
25	10.52	11.52	13.12	14.61	16.47	24.34	34.28	37.65	40.65	44.31	46.93
26	11.16	12.20	13.84	15.38	17.29	25.34	35.56	38.89	41.92	45.64	48.29
27	11.81	12.88	14.57	16.15	18.11	26.34	36.74	40.11	43.19	46.96	49.65
28	12.46	13.57	15.31	16.93	18.94	27.34	37.92	41.34	44.46	48.28	50.99
29	13.12	14.26	16.05	17.71	19.77	28.34	39.09	42.56	45.72	49.59	52.34
30	13.79	14.95	16.79	18.49	20.60	29.34	40.26	43.77	46.98	50.89	53.67
40	20.71	22.16	24.43	26.51	29.05	39.34	51.81	55.76	59.34	63.69	66.77
50	27.99	29.71	32.36	34.76	37.69	49.33	63.17	67.50	71.42	76.15	79.49
60	35.53	37.48	40.48	43.19	46.46	59.33	74.40	79.08	83.30	88.38	91.95
70	43.28	45.44	48.76	51.74	55.33	69.33	85.53	90.53	95.02	100.42	104.22
80	51.17	53.54	57.15	60.39	64.28	79.33	96.58	101.88	106.63	112.33	116.32
90	59.20	61.75	65.65	69.13	73.29	89.33	107.57	113.14	118.14	124.12	128.30
100	67.33	70.06	74.22	77.93	82.36	99.33	118.50	124.34	129.56	135.81	140.17

ν = degrees of freedom.

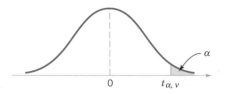

Table IV   Percentage Points $t_{\alpha,\nu}$ of the t-Distribution

ν \ α	.40	.25	.10	.05	.025	.01	.005	.0025	.001	.0005
1	.325	1.000	3.078	6.314	12.706	31.821	63.657	127.32	318.31	636.62
2	.289	.816	1.886	2.920	4.303	6.965	9.925	14.089	23.326	31.598
3	.277	.765	1.638	2.353	3.182	4.541	5.841	7.453	10.213	12.924
4	.271	.741	1.533	2.132	2.776	3.747	4.604	5.598	7.173	8.610
5	.267	.727	1.476	2.015	2.571	3.365	4.032	4.773	5.893	6.869
6	.265	.718	1.440	1.943	2.447	3.143	3.707	4.317	5.208	5.959
7	.263	.711	1.415	1.895	2.365	2.998	3.499	4.029	4.785	5.408
8	.262	.706	1.397	1.860	2.306	2.896	3.355	3.833	4.501	5.041
9	.261	.703	1.383	1.833	2.262	2.821	3.250	3.690	4.297	4.781
10	.260	.700	1.372	1.812	2.228	2.764	3.169	3.581	4.144	4.587
11	.260	.697	1.363	1.796	2.201	2.718	3.106	3.497	4.025	4.437
12	.259	.695	1.356	1.782	2.179	2.681	3.055	3.428	3.930	4.318
13	.259	.694	1.350	1.771	2.160	2.650	3.012	3.372	3.852	4.221
14	.258	.692	1.345	1.761	2.145	2.624	2.977	3.326	3.787	4.140
15	.258	.691	1.341	1.753	2.131	2.602	2.947	3.286	3.733	4.073
16	.258	.690	1.337	1.746	2.120	2.583	2.921	3.252	3.686	4.015
17	.257	.689	1.333	1.740	2.110	2.567	2.898	3.222	3.646	3.965
18	.257	.688	1.330	1.734	2.101	2.552	2.878	3.197	3.610	3.922
19	.257	.688	1.328	1.729	2.093	2.539	2.861	3.174	3.579	3.883
20	.257	.687	1.325	1.725	2.086	2.528	2.845	3.153	3.552	3.850
21	.257	.686	1.323	1.721	2.080	2.518	2.831	3.135	3.527	3.819
22	.256	.686	1.321	1.717	2.074	2.508	2.819	3.119	3.505	3.792
23	.256	.685	1.319	1.714	2.069	2.500	2.807	3.104	3.485	3.767
24	.256	.685	1.318	1.711	2.064	2.492	2.797	3.091	3.467	3.745
25	.256	.684	1.316	1.708	2.060	2.485	2.787	3.078	3.450	3.725
26	.256	.684	1.315	1.706	2.056	2.479	2.779	3.067	3.435	3.707
27	.256	.684	1.314	1.703	2.052	2.473	2.771	3.057	3.421	3.690
28	.256	.683	1.313	1.701	2.048	2.467	2.763	3.047	3.408	3.674
29	.256	.683	1.311	1.699	2.045	2.462	2.756	3.038	3.396	3.659
30	.256	.683	1.310	1.697	2.042	2.457	2.750	3.030	3.385	3.646
40	.255	.681	1.303	1.684	2.021	2.423	2.704	2.971	3.307	3.551
60	.254	.679	1.296	1.671	2.000	2.390	2.660	2.915	3.232	3.460
120	.254	.677	1.289	1.658	1.980	2.358	2.617	2.860	3.160	3.373
∞	.253	.674	1.282	1.645	1.960	2.326	2.576	2.807	3.090	3.291

ν = degrees of freedom.

Table V Percentage Points f_{α,ν_1,ν_2} of the F-Distribution

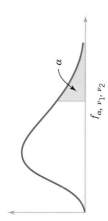

f_{α,ν_1,ν_2}

$f_{0.25,\nu_1,\nu_2}$

(handwritten annotation: ~1.55)

ν_2 \ ν_1	1	2	3	4	5	6	7	8	9	10	12	15	20	24	30	40	60	120	∞
1	5.83	7.50	8.20	8.58	8.82	8.98	9.10	9.19	9.26	9.32	9.41	9.49	9.58	9.63	9.67	9.71	9.76	9.80	9.85
2	2.57	3.00	3.15	3.23	3.28	3.31	3.34	3.35	3.37	3.38	3.39	3.41	3.43	3.43	3.44	3.45	3.46	3.47	3.48
3	2.02	2.28	2.36	2.39	2.41	2.42	2.43	2.44	2.44	2.44	2.45	2.46	2.46	2.46	2.47	2.47	2.47	2.47	2.47
4	1.81	2.00	2.05	2.06	2.07	2.08	2.08	2.08	2.08	2.08	2.08	2.08	2.08	2.08	2.08	2.08	2.08	2.08	2.08
5	1.69	1.85	1.88	1.89	1.89	1.89	1.89	1.89	1.89	1.89	1.89	1.89	1.88	1.88	1.88	1.88	1.87	1.87	1.87
6	1.62	1.76	1.78	1.79	1.79	1.78	1.78	1.78	1.77	1.77	1.77	1.76	1.76	1.75	1.75	1.75	1.74	1.74	1.74
7	1.57	1.70	1.72	1.72	1.71	1.71	1.70	1.70	1.70	1.69	1.68	1.68	1.67	1.67	1.66	1.66	1.65	1.65	1.65
8	1.54	1.66	1.67	1.66	1.66	1.65	1.64	1.64	1.63	1.63	1.62	1.62	1.61	1.60	1.60	1.59	1.59	1.58	1.58
9	1.51	1.62	1.63	1.63	1.62	1.61	1.60	1.60	1.59	1.59	1.58	1.57	1.56	1.56	1.55	1.54	1.54	1.53	1.53
10	1.49	1.60	1.60	1.59	1.59	1.58	1.57	1.56	1.56	1.55	1.54	1.53	1.52	1.52	1.51	1.51	1.50	1.49	1.48
11	1.47	1.58	1.58	1.57	1.56	1.55	1.54	1.53	1.53	1.52	1.51	1.50	1.49	1.49	1.48	1.47	1.47	1.46	1.45
12	1.46	1.56	1.56	1.55	1.54	1.53	1.52	1.51	1.51	1.50	1.49	1.48	1.47	1.46	1.45	1.45	1.44	1.43	1.42
13	1.45	1.55	1.55	1.53	1.52	1.51	1.50	1.49	1.49	1.48	1.47	1.46	1.45	1.44	1.43	1.42	1.42	1.41	1.40
14	1.44	1.53	1.53	1.52	1.51	1.50	1.49	1.48	1.47	1.46	1.45	1.44	1.43	1.42	1.41	1.41	1.40	1.39	1.38
15	1.43	1.52	1.52	1.51	1.49	1.48	1.47	1.46	1.46	1.45	1.44	1.43	1.41	1.41	1.40	1.39	1.38	1.37	1.36
16	1.42	1.51	1.51	1.50	1.48	1.47	1.46	1.45	1.44	1.44	1.43	1.41	1.40	1.39	1.38	1.37	1.36	1.35	1.34
17	1.42	1.51	1.50	1.49	1.47	1.46	1.45	1.44	1.43	1.43	1.41	1.40	1.39	1.38	1.37	1.36	1.35	1.34	1.33
18	1.41	1.50	1.49	1.48	1.46	1.45	1.44	1.43	1.42	1.42	1.40	1.39	1.38	1.37	1.36	1.35	1.34	1.33	1.32
19	1.41	1.49	1.49	1.47	1.46	1.44	1.43	1.42	1.41	1.41	1.40	1.38	1.37	1.36	1.35	1.34	1.33	1.32	1.30
20	1.40	1.49	1.48	1.47	1.45	1.44	1.43	1.42	1.41	1.40	1.39	1.37	1.36	1.35	1.34	1.33	1.32	1.31	1.29
21	1.40	1.48	1.48	1.46	1.44	1.43	1.42	1.41	1.40	1.39	1.38	1.37	1.35	1.34	1.33	1.32	1.31	1.30	1.28
22	1.40	1.48	1.47	1.45	1.44	1.42	1.41	1.40	1.39	1.39	1.37	1.36	1.34	1.33	1.32	1.31	1.30	1.29	1.28
23	1.39	1.47	1.47	1.45	1.43	1.42	1.41	1.40	1.39	1.38	1.37	1.35	1.34	1.33	1.32	1.31	1.30	1.28	
24	1.39	1.47	1.46	1.44	1.43	1.41	1.40	1.39	1.38	1.38	1.36	1.35	1.33	1.32	1.31	1.30	1.29	1.28	
25	1.39	1.47	1.46	1.44	1.42	1.41	1.40	1.39	1.38	1.37	1.36	1.34	1.33	1.32	1.31	1.29	1.28	1.27	
26	1.38	1.46	1.45	1.44	1.42	1.41	1.39	1.38	1.37	1.37	1.35	1.34	1.32	1.31	1.30	1.29	1.28	1.26	
27	1.38	1.46	1.45	1.43	1.42	1.40	1.39	1.38	1.37	1.36	1.35	1.33	1.32	1.31	1.30	1.28	1.27	1.26	
28	1.38	1.46	1.45	1.43	1.41	1.40	1.38	1.37	1.37	1.36	1.34	1.33	1.31	1.30	1.29	1.28	1.27	1.25	
29	1.38	1.45	1.45	1.43	1.41	1.40	1.38	1.37	1.36	1.35	1.34	1.32	1.31	1.30	1.29	1.27	1.26	1.25	
30	1.38	1.45	1.44	1.42	1.41	1.39	1.38	1.37	1.36	1.35	1.34	1.32	1.30	1.29	1.28	1.27	1.26	1.24	
40	1.36	1.44	1.42	1.40	1.39	1.37	1.36	1.35	1.34	1.33	1.31	1.30	1.28	1.26	1.25	1.24	1.22	1.21	
60	1.35	1.42	1.41	1.38	1.37	1.35	1.33	1.32	1.31	1.30	1.29	1.27	1.25	1.24	1.22	1.21	1.19	1.17	
120	1.34	1.40	1.39	1.37	1.35	1.33	1.31	1.30	1.29	1.28	1.26	1.24	1.22	1.21	1.19	1.18	1.16	1.13	
∞	1.32	1.39	1.37	1.35	1.33	1.31	1.29	1.28	1.27	1.25	1.24	1.22	1.19	1.18	1.16	1.14	1.12	1.08	

Degrees of freedom for the numerator (ν_1)

Degrees of freedom for the denominator (ν_2)

Table V Percentage Points of the F-Distribution (continued)

$f_{0.10,v_1,v_2}$

Degrees of freedom for the numerator (v_1)

v_2	1	2	3	4	5	6	7	8	9	10	12	15	20	24	30	40	60	120	∞
1	39.86	49.50	53.59	55.83	57.24	58.20	58.91	59.44	59.86	60.19	60.71	61.22	61.74	62.00	62.26	62.53	62.79	63.06	63.33
2	8.53	9.00	9.16	9.24	9.29	9.33	9.35	9.37	9.38	9.39	9.41	9.42	9.44	9.45	9.46	9.47	9.47	9.48	9.49
3	5.54	5.46	5.39	5.34	5.31	5.28	5.27	5.25	5.24	5.23	5.22	5.20	5.18	5.18	5.17	5.16	5.15	5.14	5.13
4	4.54	4.32	4.19	4.11	4.05	4.01	3.98	3.95	3.94	3.92	3.90	3.87	3.84	3.83	3.82	3.80	3.79	3.78	3.76
5	4.06	3.78	3.62	3.52	3.45	3.40	3.37	3.34	3.32	3.30	3.27	3.24	3.21	3.19	3.17	3.16	3.14	3.12	3.10
6	3.78	3.46	3.29	3.18	3.11	3.05	3.01	2.98	2.96	2.94	2.90	2.87	2.84	2.82	2.80	2.78	2.76	2.74	2.72
7	3.59	3.26	3.07	2.96	2.88	2.83	2.78	2.75	2.72	2.70	2.67	2.63	2.59	2.58	2.56	2.54	2.51	2.49	2.47
8	3.46	3.11	2.92	2.81	2.73	2.67	2.62	2.59	2.56	2.54	2.50	2.46	2.42	2.40	2.38	2.36	2.34	2.32	2.29
9	3.36	3.01	2.81	2.69	2.61	2.55	2.51	2.47	2.44	2.42	2.38	2.34	2.30	2.28	2.25	2.23	2.21	2.18	2.16
10	3.29	2.92	2.73	2.61	2.52	2.46	2.41	2.38	2.35	2.32	2.28	2.24	2.20	2.18	2.16	2.13	2.11	2.08	2.06
11	3.23	2.86	2.66	2.54	2.45	2.39	2.34	2.30	2.27	2.25	2.21	2.17	2.12	2.10	2.08	2.05	2.03	2.00	1.97
12	3.18	2.81	2.61	2.48	2.39	2.33	2.28	2.24	2.21	2.19	2.15	2.10	2.06	2.04	2.01	1.99	1.96	1.93	1.90
13	3.14	2.76	2.56	2.43	2.35	2.28	2.23	2.20	2.16	2.14	2.10	2.05	2.01	1.98	1.96	1.93	1.90	1.88	1.85
14	3.10	2.73	2.52	2.39	2.31	2.24	2.19	2.15	2.12	2.10	2.05	2.01	1.96	1.94	1.91	1.89	1.86	1.83	1.80
15	3.07	2.70	2.49	2.36	2.27	2.21	2.16	2.12	2.09	2.06	2.02	1.97	1.92	1.90	1.87	1.85	1.82	1.79	1.76
16	3.05	2.67	2.46	2.33	2.24	2.18	2.13	2.09	2.06	2.03	1.99	1.94	1.89	1.87	1.84	1.81	1.78	1.75	1.72
17	3.03	2.64	2.44	2.31	2.22	2.15	2.10	2.06	2.03	2.00	1.96	1.91	1.86	1.84	1.81	1.78	1.75	1.72	1.69
18	3.01	2.62	2.42	2.29	2.20	2.13	2.08	2.04	2.00	1.98	1.93	1.89	1.84	1.81	1.78	1.75	1.72	1.69	1.66
19	2.99	2.61	2.40	2.27	2.18	2.11	2.06	2.02	1.98	1.96	1.91	1.86	1.81	1.79	1.76	1.73	1.70	1.67	1.63
20	2.97	2.59	2.38	2.25	2.16	2.09	2.04	2.00	1.96	1.94	1.89	1.84	1.79	1.77	1.74	1.71	1.68	1.64	1.61
21	2.96	2.57	2.36	2.23	2.14	2.08	2.02	1.98	1.95	1.92	1.87	1.83	1.78	1.75	1.72	1.69	1.66	1.62	1.59
22	2.95	2.56	2.35	2.22	2.13	2.06	2.01	1.97	1.93	1.90	1.86	1.81	1.76	1.73	1.70	1.67	1.64	1.60	1.57
23	2.94	2.55	2.34	2.21	2.11	2.05	1.99	1.95	1.92	1.89	1.84	1.80	1.74	1.72	1.69	1.66	1.62	1.59	1.55
24	2.93	2.54	2.33	2.19	2.10	2.04	1.98	1.94	1.91	1.88	1.83	1.78	1.73	1.70	1.67	1.64	1.61	1.57	1.53
25	2.92	2.53	2.32	2.18	2.09	2.02	1.97	1.93	1.89	1.87	1.82	1.77	1.72	1.69	1.66	1.63	1.59	1.56	1.52
26	2.91	2.52	2.31	2.17	2.08	2.01	1.96	1.92	1.88	1.86	1.81	1.76	1.71	1.68	1.65	1.61	1.58	1.54	1.50
27	2.90	2.51	2.30	2.17	2.07	2.00	1.95	1.91	1.87	1.85	1.80	1.75	1.70	1.67	1.64	1.60	1.57	1.53	1.49
28	2.89	2.50	2.29	2.16	2.06	2.00	1.94	1.90	1.87	1.84	1.79	1.74	1.69	1.66	1.63	1.59	1.56	1.52	1.48
29	2.89	2.50	2.28	2.15	2.06	1.99	1.93	1.89	1.86	1.83	1.78	1.73	1.68	1.65	1.62	1.58	1.55	1.51	1.47
30	2.88	2.49	2.28	2.14	2.03	1.98	1.93	1.88	1.85	1.82	1.77	1.72	1.67	1.64	1.61	1.57	1.54	1.50	1.46
40	2.84	2.44	2.23	2.09	2.00	1.93	1.87	1.83	1.79	1.76	1.71	1.66	1.61	1.57	1.54	1.51	1.47	1.42	1.38
60	2.79	2.39	2.18	2.04	1.95	1.87	1.82	1.77	1.74	1.71	1.66	1.60	1.54	1.51	1.48	1.44	1.40	1.35	1.29
120	2.75	2.35	2.13	1.99	1.90	1.82	1.77	1.72	1.68	1.65	1.60	1.55	1.48	1.45	1.41	1.37	1.32	1.26	1.19
∞	2.71	2.30	2.08	1.94	1.85	1.77	1.72	1.67	1.63	1.60	1.55	1.49	1.42	1.38	1.34	1.30	1.24	1.17	1.00

Degrees of freedom for the denominator (v_2)

Table V   Percentage Points of the F-Distribution (continued)

$$f_{0.05,\nu_1,\nu_2}$$

| | Degrees of freedom for the numerator (ν_1) | | | | | | | | | | | | | | | | | | |
ν_2	1	2	3	4	5	6	7	8	9	10	12	15	20	24	30	40	60	120	∞
1	161.4	199.5	215.7	224.6	230.2	234.0	236.8	238.9	240.5	241.9	243.9	245.9	248.0	249.1	250.1	251.1	252.2	253.3	254.3
2	18.51	19.00	19.16	19.25	19.30	19.33	19.35	19.37	19.38	19.40	19.41	19.43	19.45	19.45	19.46	19.47	19.48	19.49	19.50
3	10.13	9.55	9.28	9.12	9.01	8.94	8.89	8.85	8.81	8.79	8.74	8.70	8.66	8.64	8.62	8.59	8.57	8.55	8.53
4	7.71	6.94	6.59	6.39	6.26	6.16	6.09	6.04	6.00	5.96	5.91	5.86	5.80	5.77	5.75	5.72	5.69	5.66	5.63
5	6.61	5.79	5.41	5.19	5.05	4.95	4.88	4.82	4.77	4.74	4.68	4.62	4.56	4.53	4.50	4.46	4.43	4.40	4.36
6	5.99	5.14	4.76	4.53	4.39	4.28	4.21	4.15	4.10	4.06	4.00	3.94	3.87	3.84	3.81	3.77	3.74	3.70	3.67
7	5.59	4.74	4.35	4.12	3.97	3.87	3.79	3.73	3.68	3.64	3.57	3.51	3.44	3.41	3.38	3.34	3.30	3.27	3.23
8	5.32	4.46	4.07	3.84	3.69	3.58	3.50	3.44	3.39	3.35	3.28	3.22	3.15	3.12	3.08	3.04	3.01	2.97	2.93
9	5.12	4.26	3.86	3.63	3.48	3.37	3.29	3.23	3.18	3.14	3.07	3.01	2.94	2.90	2.86	2.83	2.79	2.75	2.71
10	4.96	4.10	3.71	3.48	3.33	3.22	3.14	3.07	3.02	2.98	2.91	2.85	2.77	2.74	2.70	2.66	2.62	2.58	2.54
11	4.84	3.98	3.59	3.36	3.20	3.09	3.01	2.95	2.90	2.85	2.79	2.72	2.65	2.61	2.57	2.53	2.49	2.45	2.40
12	4.75	3.89	3.49	3.26	3.11	3.00	2.91	2.85	2.80	2.75	2.69	2.62	2.54	2.51	2.47	2.43	2.38	2.34	2.30
13	4.67	3.81	3.41	3.18	3.03	2.92	2.83	2.77	2.71	2.67	2.60	2.53	2.46	2.42	2.38	2.34	2.30	2.25	2.21
14	4.60	3.74	3.34	3.11	2.96	2.85	2.76	2.70	2.65	2.60	2.53	2.46	2.39	2.35	2.31	2.27	2.22	2.18	2.13
15	4.54	3.68	3.29	3.06	2.90	2.79	2.71	2.64	2.59	2.54	2.48	2.40	2.33	2.29	2.25	2.20	2.16	2.11	2.07
16	4.49	3.63	3.24	3.01	2.85	2.74	2.66	2.59	2.54	2.49	2.42	2.35	2.28	2.24	2.19	2.15	2.11	2.06	2.01
17	4.45	3.59	3.20	2.96	2.81	2.70	2.61	2.55	2.49	2.45	2.38	2.31	2.23	2.19	2.15	2.10	2.06	2.01	1.96
18	4.41	3.55	3.16	2.93	2.77	2.66	2.58	2.51	2.46	2.41	2.34	2.27	2.19	2.15	2.11	2.06	2.02	1.97	1.92
19	4.38	3.52	3.13	2.90	2.74	2.63	2.54	2.48	2.42	2.38	2.31	2.23	2.16	2.11	2.07	2.03	1.98	1.93	1.88
20	4.35	3.49	3.10	2.87	2.71	2.60	2.51	2.45	2.39	2.35	2.28	2.20	2.12	2.08	2.04	1.99	1.95	1.90	1.84
21	4.32	3.47	3.07	2.84	2.68	2.57	2.49	2.42	2.37	2.32	2.25	2.18	2.10	2.05	2.01	1.96	1.92	1.87	1.81
22	4.30	3.44	3.05	2.82	2.66	2.55	2.46	2.40	2.34	2.30	2.23	2.15	2.07	2.03	1.98	1.94	1.89	1.84	1.78
23	4.28	3.42	3.03	2.80	2.64	2.53	2.44	2.37	2.32	2.27	2.20	2.13	2.05	2.01	1.96	1.91	1.86	1.81	1.76
24	4.26	3.40	3.01	2.78	2.62	2.51	2.42	2.36	2.30	2.25	2.18	2.11	2.03	1.98	1.94	1.89	1.84	1.79	1.73
25	4.24	3.39	2.99	2.76	2.60	2.49	2.40	2.34	2.28	2.24	2.16	2.09	2.01	1.96	1.92	1.87	1.82	1.77	1.71
26	4.23	3.37	2.98	2.74	2.59	2.47	2.39	2.32	2.27	2.22	2.15	2.07	1.99	1.95	1.90	1.85	1.80	1.75	1.69
27	4.21	3.35	2.96	2.73	2.57	2.46	2.37	2.31	2.25	2.20	2.13	2.06	1.97	1.93	1.88	1.84	1.79	1.73	1.67
28	4.20	3.34	2.95	2.71	2.56	2.45	2.36	2.29	2.24	2.19	2.12	2.04	1.96	1.91	1.87	1.82	1.77	1.71	1.65
29	4.18	3.33	2.93	2.70	2.55	2.43	2.35	2.28	2.22	2.18	2.10	2.03	1.94	1.90	1.85	1.81	1.75	1.70	1.64
30	4.17	3.32	2.92	2.69	2.53	2.42	2.33	2.27	2.21	2.16	2.09	2.01	1.93	1.89	1.84	1.79	1.74	1.68	1.62
40	4.08	3.23	2.84	2.61	2.45	2.34	2.25	2.18	2.12	2.08	2.00	1.92	1.84	1.79	1.74	1.69	1.64	1.58	1.51
60	4.00	3.15	2.76	2.53	2.37	2.25	2.17	2.10	2.04	1.99	1.92	1.84	1.75	1.70	1.65	1.59	1.53	1.47	1.39
120	3.92	3.07	2.68	2.45	2.29	2.17	2.09	2.02	1.96	1.91	1.83	1.75	1.66	1.61	1.55	1.50	1.43	1.35	1.25
∞	3.84	3.00	2.60	2.37	2.21	2.10	2.01	1.94	1.88	1.83	1.75	1.67	1.57	1.52	1.46	1.39	1.32	1.22	1.00

Degrees of freedom for the denominator (ν_2)

Table V Percentage Points of the F-Distribution (continued)

$f_{0.025, \nu_1, \nu_2}$

$\nu_2 \backslash \nu_1$	1	2	3	4	5	6	7	8	9	10	12	15	20	24	30	40	60	120	∞
1	647.8	799.5	864.2	899.6	921.8	937.1	948.2	956.7	963.3	968.6	976.7	984.9	993.1	997.2	1001	1006	1010	1014	1018
2	38.51	39.00	39.17	39.25	39.30	39.33	39.36	39.37	39.39	39.40	39.41	39.43	39.45	39.46	39.46	39.47	39.48	39.49	39.50
3	17.44	16.04	15.44	15.10	14.88	14.73	14.62	14.54	14.47	14.42	14.34	14.25	14.17	14.12	14.08	14.04	13.99	13.95	13.90
4	12.22	10.65	9.98	9.60	9.36	9.20	9.07	8.98	8.90	8.84	8.75	8.66	8.56	8.51	8.46	8.41	8.36	8.31	8.26
5	10.01	8.43	7.76	7.39	7.15	6.98	6.85	6.76	6.68	6.62	6.52	6.43	6.33	6.28	6.23	6.18	6.12	6.07	6.02
6	8.81	7.26	6.60	6.23	5.99	5.82	5.70	5.60	5.52	5.46	5.37	5.27	5.17	5.12	5.07	5.01	4.96	4.90	4.85
7	8.07	6.54	5.89	5.52	5.29	5.12	4.99	4.90	4.82	4.76	4.67	4.57	4.47	4.42	4.36	4.31	4.25	4.20	4.14
8	7.57	6.06	5.42	5.05	4.82	4.65	4.53	4.43	4.36	4.30	4.20	4.10	4.00	3.95	3.89	3.84	3.78	3.73	3.67
9	7.21	5.71	5.08	4.72	4.48	4.32	4.20	4.10	4.03	3.96	3.87	3.77	3.67	3.61	3.56	3.51	3.45	3.39	3.33
10	6.94	5.46	4.83	4.47	4.24	4.07	3.95	3.85	3.78	3.72	3.62	3.52	3.42	3.37	3.31	3.26	3.20	3.14	3.08
11	6.72	5.26	4.63	4.28	4.04	3.88	3.76	3.66	3.59	3.53	3.43	3.33	3.23	3.17	3.12	3.06	3.00	2.94	2.88
12	6.55	5.10	4.47	4.12	3.89	3.73	3.61	3.51	3.44	3.37	3.28	3.18	3.07	3.02	2.96	2.91	2.85	2.79	2.72
13	6.41	4.97	4.35	4.00	3.77	3.60	3.48	3.39	3.31	3.25	3.15	3.05	2.95	2.89	2.84	2.78	2.72	2.66	2.60
14	6.30	4.86	4.24	3.89	3.66	3.50	3.38	3.29	3.21	3.15	3.05	2.95	2.84	2.79	2.73	2.67	2.61	2.55	2.49
15	6.20	4.77	4.15	3.80	3.58	3.41	3.29	3.20	3.12	3.06	2.96	2.86	2.76	2.70	2.64	2.59	2.52	2.46	2.40
16	6.12	4.69	4.08	3.73	3.50	3.34	3.22	3.12	3.05	2.99	2.89	2.79	2.68	2.63	2.57	2.51	2.45	2.38	2.32
17	6.04	4.62	4.01	3.66	3.44	3.28	3.16	3.06	2.98	2.92	2.82	2.72	2.62	2.56	2.50	2.44	2.38	2.32	2.25
18	5.98	4.56	3.95	3.61	3.38	3.22	3.10	3.01	2.93	2.87	2.77	2.67	2.56	2.50	2.44	2.38	2.32	2.26	2.19
19	5.92	4.51	3.90	3.56	3.33	3.17	3.05	2.96	2.88	2.82	2.72	2.62	2.51	2.45	2.39	2.33	2.27	2.20	2.13
20	5.87	4.46	3.86	3.51	3.29	3.13	3.01	2.91	2.84	2.77	2.68	2.57	2.46	2.41	2.35	2.29	2.22	2.16	2.09
21	5.83	4.42	3.82	3.48	3.25	3.09	2.97	2.87	2.80	2.73	2.64	2.53	2.42	2.37	2.31	2.25	2.18	2.11	2.04
22	5.79	4.38	3.78	3.44	3.22	3.05	2.93	2.84	2.76	2.70	2.60	2.50	2.39	2.33	2.27	2.21	2.14	2.08	2.00
23	5.75	4.35	3.75	3.41	3.18	3.02	2.90	2.81	2.73	2.67	2.57	2.47	2.36	2.30	2.24	2.18	2.11	2.04	1.97
24	5.72	4.32	3.72	3.38	3.15	2.99	2.87	2.78	2.70	2.64	2.54	2.44	2.33	2.27	2.21	2.15	2.08	2.01	1.94
25	5.69	4.29	3.69	3.35	3.13	2.97	2.85	2.75	2.68	2.61	2.51	2.41	2.30	2.24	2.18	2.12	2.05	1.98	1.91
26	5.66	4.27	3.67	3.33	3.10	2.94	2.82	2.73	2.65	2.59	2.49	2.39	2.28	2.22	2.16	2.09	2.03	1.95	1.88
27	5.63	4.24	3.65	3.31	3.08	2.92	2.80	2.71	2.63	2.57	2.47	2.36	2.25	2.19	2.13	2.07	2.00	1.93	1.85
28	5.61	4.22	3.63	3.29	3.06	2.90	2.78	2.69	2.61	2.55	2.45	2.34	2.23	2.17	2.11	2.05	1.98	1.91	1.83
29	5.59	4.20	3.61	3.27	3.04	2.88	2.76	2.67	2.59	2.53	2.43	2.32	2.21	2.15	2.09	2.03	1.96	1.89	1.81
30	5.57	4.18	3.59	3.25	3.03	2.87	2.75	2.65	2.57	2.51	2.41	2.31	2.20	2.14	2.07	2.01	1.94	1.87	1.79
40	5.42	4.05	3.46	3.13	2.90	2.74	2.62	2.53	2.45	2.39	2.29	2.18	2.07	2.01	1.94	1.88	1.80	1.72	1.64
60	5.29	3.93	3.34	3.01	2.79	2.63	2.51	2.41	2.33	2.27	2.17	2.06	1.94	1.88	1.82	1.74	1.67	1.53	1.48
120	5.15	3.80	3.23	2.89	2.67	2.52	2.39	2.30	2.22	2.16	2.05	1.94	1.82	1.76	1.69	1.61	1.53	1.43	1.31
∞	5.02	3.69	3.12	2.79	2.57	2.41	2.29	2.19	2.11	2.05	1.94	1.83	1.71	1.64	1.57	1.48	1.39	1.27	1.00

Degrees of freedom for the numerator (ν_1)

Degrees of freedom for the denominator (ν_2)

660

Table V Percentage Points of the F-Distribution (*continued*)

$$f_{0.01, v_1, v_2}$$

	Degrees of freedom for the numerator (v_1)																		
v_2 \ v_1	1	2	3	4	5	6	7	8	9	10	12	15	20	24	30	40	60	120	∞
1	4052	4999.5	5403	5625	5764	5859	5928	5982	6022	6056	6106	6157	6209	6235	6261	6287	6313	6339	6366
2	98.50	99.00	99.17	99.25	99.30	99.33	99.36	99.37	99.39	99.40	99.42	99.43	99.45	99.46	99.47	99.47	99.48	99.49	99.50
3	34.12	30.82	29.46	28.71	28.24	27.91	27.67	27.49	27.35	27.23	27.05	26.87	26.69	26.60	26.50	26.41	26.32	26.22	26.13
4	21.20	18.00	16.69	15.98	15.52	15.21	14.98	14.80	14.66	14.55	14.37	14.20	14.02	13.93	13.84	13.75	13.65	13.56	13.46
5	16.26	13.27	12.06	11.39	10.97	10.67	10.46	10.29	10.16	10.05	9.89	9.72	9.55	9.47	9.38	9.29	9.20	9.11	9.02
6	13.75	10.92	9.78	9.15	8.75	8.47	8.26	8.10	7.98	7.87	7.72	7.56	7.40	7.31	7.23	7.14	7.06	6.97	6.88
7	12.25	9.55	8.45	7.85	7.46	7.19	6.99	6.84	6.72	6.62	6.47	6.31	6.16	6.07	5.99	5.91	5.82	5.74	5.65
8	11.26	8.65	7.59	7.01	6.63	6.37	6.18	6.03	5.91	5.81	5.67	5.52	5.36	5.28	5.20	5.12	5.03	4.95	4.86
9	10.56	8.02	6.99	6.42	5.99	5.80	5.61	5.47	5.35	5.26	5.11	4.96	4.81	4.73	4.65	4.57	4.48	4.40	4.31
10	10.04	7.56	6.55	5.99	5.64	5.39	5.20	5.06	4.94	4.85	4.71	4.56	4.41	4.33	4.25	4.17	4.08	4.00	3.91
11	9.65	7.21	6.22	5.67	5.32	5.07	4.89	4.74	4.63	4.54	4.40	4.25	4.10	4.02	3.94	3.86	3.78	3.69	3.60
12	9.33	6.93	5.95	5.41	5.06	4.82	4.64	4.50	4.39	4.30	4.16	4.01	3.86	3.78	3.70	3.62	3.54	3.45	3.36
13	9.07	6.70	5.74	5.21	4.86	4.62	4.44	4.30	4.19	4.10	3.96	3.82	3.66	3.59	3.51	3.43	3.34	3.25	3.17
14	8.86	6.51	5.56	5.04	4.69	4.46	4.28	4.14	4.03	3.94	3.80	3.66	3.51	3.43	3.35	3.27	3.18	3.09	3.00
15	8.68	6.36	5.42	4.89	4.56	4.32	4.14	4.00	3.89	3.80	3.67	3.52	3.37	3.29	3.21	3.13	3.05	2.96	2.87
16	8.53	6.23	5.29	4.77	4.44	4.20	4.03	3.89	3.78	3.69	3.55	3.41	3.26	3.18	3.10	3.02	2.93	2.84	2.75
17	8.40	6.11	5.18	4.67	4.34	4.10	3.93	3.79	3.68	3.59	3.46	3.31	3.16	3.08	3.00	2.92	2.83	2.75	2.65
18	8.29	6.01	5.09	4.58	4.25	4.01	3.84	3.71	3.60	3.51	3.37	3.23	3.08	3.00	2.92	2.84	2.75	2.66	2.57
19	8.18	5.93	5.01	4.50	4.17	3.94	3.77	3.63	3.52	3.43	3.30	3.15	3.00	2.92	2.84	2.76	2.67	2.58	2.49
20	8.10	5.85	4.94	4.43	4.10	3.87	3.70	3.56	3.46	3.37	3.23	3.09	2.94	2.86	2.78	2.69	2.61	2.52	2.42
21	8.02	5.78	4.87	4.37	4.04	3.81	3.64	3.51	3.40	3.31	3.17	3.03	2.88	2.80	2.72	2.64	2.55	2.46	2.36
22	7.95	5.72	4.82	4.31	3.99	3.76	3.59	3.45	3.35	3.26	3.12	2.98	2.83	2.75	2.67	2.58	2.50	2.40	2.31
23	7.88	5.66	4.76	4.26	3.94	3.71	3.54	3.41	3.30	3.21	3.07	2.93	2.78	2.70	2.62	2.54	2.45	2.35	2.26
24	7.82	5.61	4.72	4.22	3.90	3.67	3.50	3.36	3.26	3.17	3.03	2.89	2.74	2.66	2.58	2.49	2.40	2.31	2.21
25	7.77	5.57	4.68	4.18	3.85	3.63	3.46	3.32	3.22	3.13	2.99	2.85	2.70	2.62	2.54	2.45	2.36	2.27	2.17
26	7.72	5.53	4.64	4.14	3.82	3.59	3.42	3.29	3.18	3.09	2.96	2.81	2.66	2.58	2.50	2.42	2.33	2.23	2.13
27	7.68	5.49	4.60	4.11	3.78	3.56	3.39	3.26	3.15	3.06	2.93	2.78	2.63	2.55	2.47	2.38	2.29	2.20	2.10
28	7.64	5.45	4.57	4.07	3.75	3.53	3.36	3.23	3.12	3.03	2.90	2.75	2.60	2.52	2.44	2.35	2.26	2.17	2.06
29	7.60	5.42	4.54	4.04	3.73	3.50	3.33	3.20	3.09	3.00	2.87	2.73	2.57	2.49	2.41	2.33	2.23	2.14	2.03
30	7.56	5.39	4.51	4.02	3.70	3.47	3.30	3.17	3.07	2.98	2.84	2.70	2.55	2.47	2.39	2.30	2.21	2.11	2.01
40	7.31	5.18	4.31	3.83	3.51	3.29	3.12	2.99	2.89	2.80	2.66	2.52	2.37	2.29	2.20	2.11	2.02	1.92	1.80
60	7.08	4.98	4.13	3.65	3.34	3.12	2.95	2.82	2.72	2.63	2.50	2.35	2.20	2.12	2.03	1.94	1.84	1.73	1.60
120	6.85	4.79	3.95	3.48	3.17	2.96	2.79	2.66	2.56	2.47	2.34	2.19	2.03	1.95	1.86	1.76	1.66	1.53	1.38
∞	6.63	4.61	3.78	3.32	3.02	2.80	2.64	2.51	2.41	2.32	2.18	2.04	1.88	1.79	1.70	1.59	1.47	1.32	1.00

Degrees of freedom for the denominator (v_2)

Chart VI   Operating Characteristic Curves

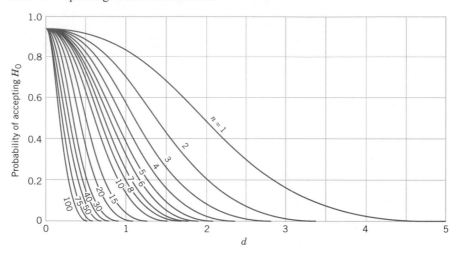

(a)   *O.C.* curves for different values of *n* for the two-sided normal test for a level of significance α = 0.05.

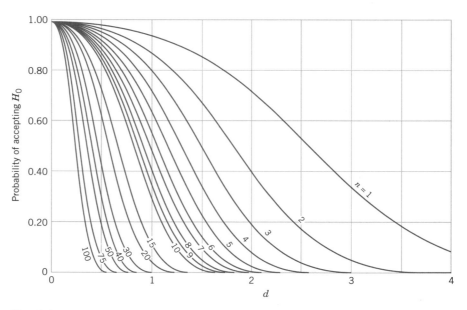

(b)   *O.C.* curves for different values of *n* for the two-sided normal test for a level of significance α = 0.01.

Source: Charts VI*a, e, f, k, m,* and *q* are reproduced with permission from "Operating Characteristics for the Common Statistical Tests of Significance," by C. L. Ferris, F. E. Grubbs, and C. L. Weaver, *Annals of Mathematical Statistics,* June 1946.
Charts VI*b, c, d, g, h, i, j, l, n, o, p,* and *r* are reproduced with permission from *Engineering Statistics*, 2nd Edition, by A. H. Bowker and G. J. Lieberman, Prentice-Hall, 1972.

Chart VI   Operating Characteristic Curves (*continued*)

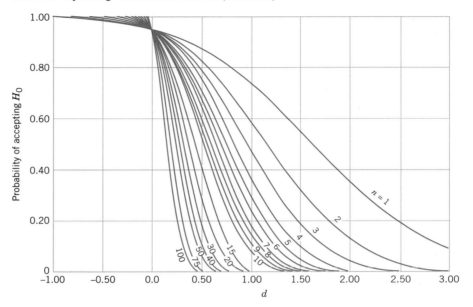

(*c*)   *O.C.* curves for different values of *n* for the one-sided normal test for a level of significance $\alpha = 0.05$.

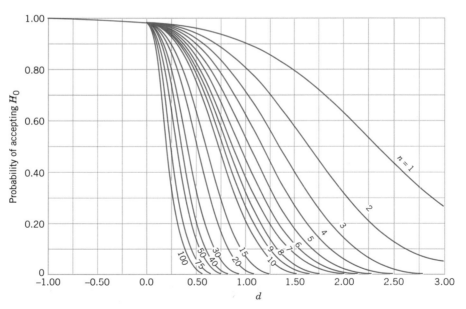

(*d*)   *O.C.* curves for different values of *n* for the one-sided normal test for a level of significance $\alpha = 0.01$.

Chart VI   Operating Characteristic Curves (*continued*)

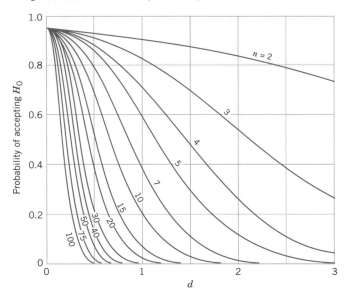

(*e*)   *O.C.* curves for different values of *n* for the two-sided *t*-test for a level of significance $\alpha = 0.05$.

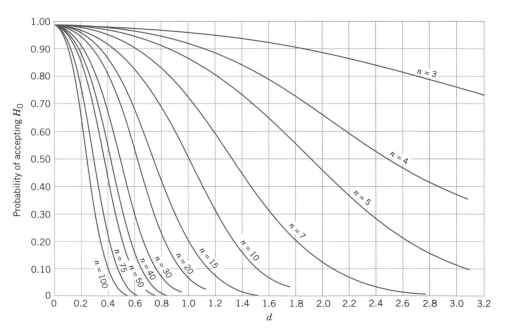

(*f*)   *O.C.* curves for different values of *n* for the two-sided *t*-test for a level of significance $\alpha = 0.01$.

Chart VI   Operating Characteristic Curves (*continued*)

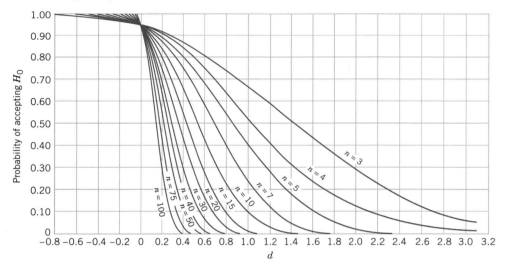

(*g*)   *O.C.* curves for different values of *n* for the one-sided *t*-test for a level of significance $\alpha = 0.05$.

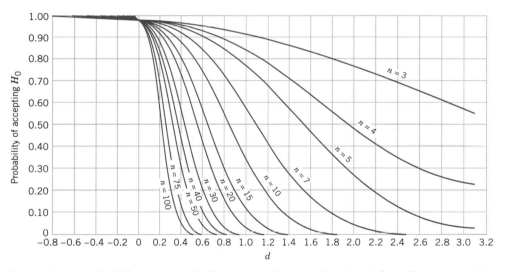

(*h*)   *O.C.* curves for different values of *n* for the one-sided *t*-test for a level of significance $\alpha = 0.01$.

Chart VI   Operating Characteristic Curves (*continued*)

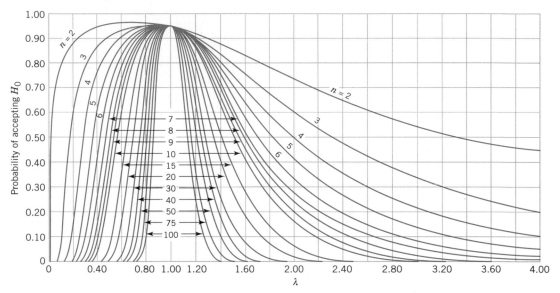

(*i*)   O.C. curves for different values of n for the two-sided chi-square test for a level of significance $\alpha = 0.05$.

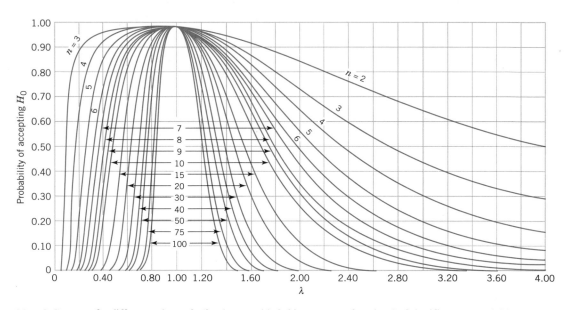

(*j*)   O.C. curves for different values of n for the two-sided chi-square test for a level of significance $\alpha = 0.01$.

Chart VI   Operating Characteristic Curves (*continued*)

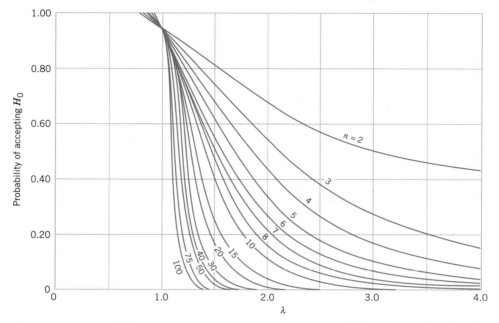

(*k*)   *O.C.* curves for different values of *n* for the one-sided (upper tail) chi-square test for a level of significance $\alpha = 0.05$.

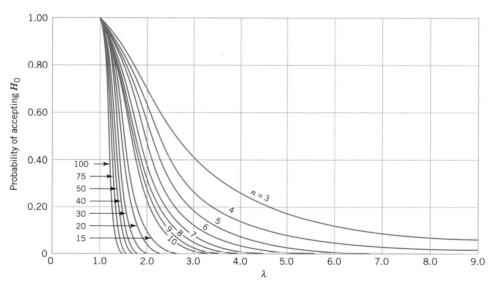

(*l*)   *O.C.* curves for different values of *n* for the one-sided (upper tail) chi-square test for a level of significance $\alpha = 0.01$.

Chart VI   Operating Characteristic Curves (*continued*)

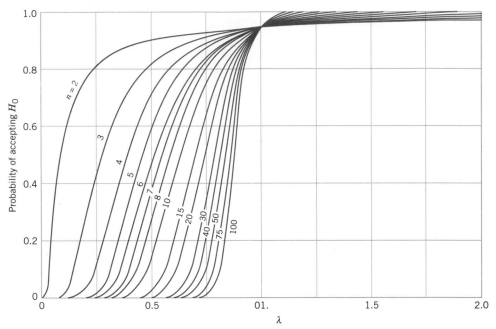

(*m*)   *O.C.* curves for different values of *n* for the one-sided (lower tail) chi-square test for a level of significance $\alpha = 0.05$.

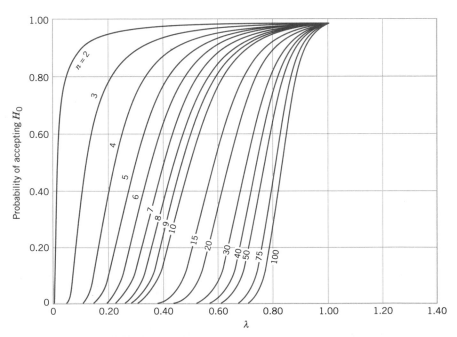

(*n*)   *O.C.* curves for different values of *n* for the one-sided (lower tail) chi-square test for a level of significance $\alpha = 0.01$.

Chart VI   Operating Characteristic Curves (*continued*)

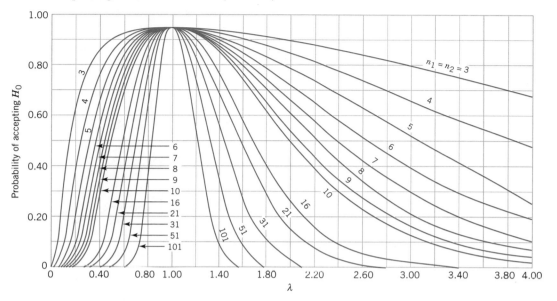

(*o*)   O.C. curves for different values of *n* for the two-sided *F*-test for a level of significance $\alpha = 0.05$.

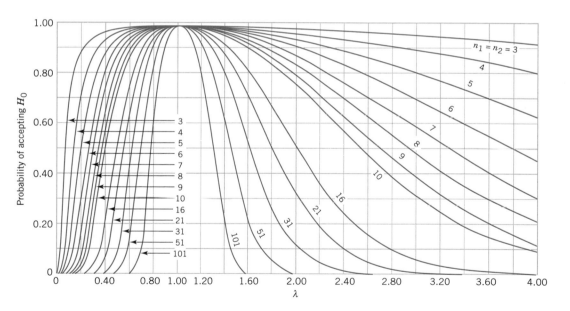

(*p*)   O.C. curves for different values of *n* for the two-sided *F*-test for a level of significance $\alpha = 0.01$.

Chart VI   Operating Characteristic Curves (*continued*)

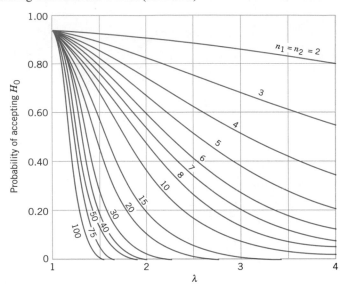

(*q*)   *O.C. curves for different values of n for the one-sided F-test for a level of significance* $\alpha = 0.05$.

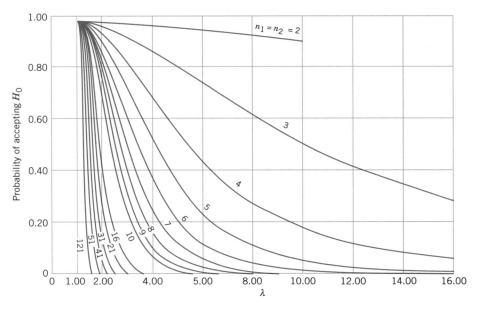

(*r*)   *O.C. curves for different values of n for the one-sided F-test for a level of significance* $\alpha = 0.01$.

Table VII   Critical Values for the Sign Test

r_α^*

α / n	0.10 / 0.05	0.05 / 0.025	0.01 / 0.005	Two-sided tests / One-sided tests	α / n	0.10 / 0.05	0.05 / 0.025	0.01 / 0.005	Two-sided tests / One-sided tests
5	0				23	7	6	4	
6	0	0			24	7	6	5	
7	0	0			25	7	7	5	
8	1	0	0		26	8	7	6	
9	1	1	0		27	8	7	6	
10	1	1	0		28	9	8	6	
11	2	1	0		29	9	8	7	
12	2	2	1		30	10	9	7	
13	3	2	1		31	10	9	7	
14	3	2	1		32	10	9	8	
15	3	3	2		33	11	10	8	
16	4	3	2		34	11	10	9	
17	4	4	2		35	12	11	9	
18	5	4	3		36	12	11	9	
19	5	4	3		37	13	12	10	
20	5	5	3		38	13	12	10	
21	6	5	4		39	13	12	11	
22	6	5	4		40	14	13	11	

Table VIII   Critical Values for the Wilcoxon Signed-Rank Test

w_α^*

α / n^*	0.10 / 0.05	0.05 / 0.025	0.02 / 0.01	0.01 / 0.005	Two-sided tests / One-sided tests
4					
5	0				
6	2	0			
7	3	2	0		
8	5	3	1	0	
9	8	5	3	1	
10	10	8	5	3	
11	13	10	7	5	
12	17	13	9	7	
13	21	17	12	9	
14	25	21	15	12	
15	30	25	19	15	
16	35	29	23	19	
17	41	34	27	23	
18	47	40	32	27	
19	53	46	37	32	
20	60	52	43	37	
21	67	58	49	42	
22	75	65	55	48	
23	83	73	62	54	
24	91	81	69	61	
25	100	89	76	68	

* If $n > 25$, W^+ (or W^-) is approximately normally distributed with mean $n(n+1)/4$ and variance $n(n+1)(2n+1)/24$.

Table IX   Critical Values for the Wilcoxon Rank-Sum Test

$$w_{0.05}$$

n_1 \\ n_2	4	5	6	7	8	9	10	11	12	13	14	15
4	10											
5	11	17										
6	12	18	26									
7	13	20	27	36								
8	14	21	29	38	49							
9	15	22	31	40	51	63						
10	15	23	32	42	53	65	78					
11	16	24	34	44	55	68	81	96				
12	17	26	35	46	58	71	85	99	115			
13	18	27	37	48	60	73	88	103	119	137		
14	19	28	38	50	63	76	91	106	123	141	160	
15	20	29	40	52	65	79	94	110	127	145	164	185
16	21	31	42	54	67	82	97	114	131	150	169	
17	21	32	43	56	70	84	100	117	135	154		
18	22	33	45	58	72	87	103	121	139			
19	23	34	46	60	74	90	107	124				
20	24	35	48	62	77	93	110					
21	25	37	50	64	79	95						
22	26	38	51	66	82							
23	27	39	53	68								
24	28	40	55									
25	28	42										
26	29											
27												
28												

*For n_1 and $n_2 > 8$, W_1 is approximately normally distributed with mean $\frac{1}{2}n_1(n_1 + n_2 + 1)$ and variance $n_1n_2(n_1 + n_2 + 1)/12$.

Table IX   Critical Values for the Wilcoxon Rank-Sum Test (*continued*)

$$w_{0.01}$$

n_1 \\ n_2	4	5	6	7	8	9	10	11	12	13	14	15
5		15										
6	10	16	23									
7	10	17	24	32								
8	11	17	25	34	43							
9	11	18	26	35	45	56						
10	12	19	27	37	47	58	71					
11	12	20	28	38	49	61	74	87				
12	13	21	30	40	51	63	76	90	106			
13	14	22	31	41	53	65	79	93	109	125		
14	14	22	32	43	54	67	81	96	112	129	147	
15	15	23	33	44	56	70	84	99	115	133	151	171
16	15	24	34	46	58	72	86	102	119	137	155	
17	16	25	36	47	60	74	89	105	122	140		
18	16	26	37	49	62	76	92	108	125			
19	17	27	38	50	64	78	94	111				
20	18	28	39	52	66	81	97					
21	18	29	40	53	68	83						
22	19	29	42	55	70							
23	19	30	43	57								
24	20	31	44									
25	20	32										
26	21											
27												
28												

Table X   Factors for Constructing Variables Control Charts

	Factor for Control Limits						
	$\overline{X}$ Chart			R Chart		S Chart	
n^*	A_1	A_2	d_2	D_3	D_4	c_4	n
2	3.760	1.880	1.128	0	3.267	0.7979	2
3	2.394	1.023	1.693	0	2.575	0.8862	3
4	1.880	.729	2.059	0	2.282	0.9213	4
5	1.596	.577	2.326	0	2.115	0.9400	5
6	1.410	.483	2.534	0	2.004	0.9515	6
7	1.277	.419	2.704	.076	1.924	0.9594	7
8	1.175	.373	2.847	.136	1.864	0.9650	8
9	1.094	.337	2.970	.184	1.816	0.9693	9
10	1.028	.308	3.078	.223	1.777	0.9727	10
11	.973	.285	3.173	.256	1.744	0.9754	11
12	.925	.266	3.258	.284	1.716	0.9776	12
13	.884	.249	3.336	.308	1.692	0.9794	13
14	.848	.235	3.407	.329	1.671	0.9810	14
15	.816	.223	3.472	.348	1.652	0.9823	15
16	.788	.212	3.532	.364	1.636	0.9835	16
17	.762	.203	3.588	.379	1.621	0.9845	17
18	.738	.194	3.640	.392	1.608	0.9854	18
19	.717	.187	3.689	.404	1.596	0.9862	19
20	.697	.180	3.735	.414	1.586	0.9869	20
21	.679	.173	3.778	.425	1.575	0.9876	21
22	.662	.167	3.819	.434	1.566	0.9882	22
23	.647	.162	3.858	.443	1.557	0.9887	23
24	.632	.157	3.895	.452	1.548	0.9892	24
25	.619	.153	3.931	.459	1.541	0.9896	25

$^*n > 25$: $A_1 = 3/\sqrt{n}$ where n = number of observations in sample.

Table XI   Factors for Tolerance Intervals

	Values of k for Two-Sided Intervals								
Confidence Level	0.90			0.95			0.99		
Percent Coverage	0.90	0.95	0.99	0.90	0.95	0.99	0.90	0.95	0.99
2	15.978	18.800	24.167	32.019	37.674	48.430	160.193	188.491	242.300
3	5.847	6.919	8.974	8.380	9.916	12.861	18.930	22.401	29.055
4	4.166	4.943	6.440	5.369	6.370	8.299	9.398	11.150	14.527
5	3.949	4.152	5.423	4.275	5.079	6.634	6.612	7.855	10.260
6	3.131	3.723	4.870	3.712	4.414	5.775	5.337	6.345	8.301
7	2.902	3.452	4.521	3.369	4.007	5.248	4.613	5.488	7.187
8	2.743	3.264	4.278	3.136	3.732	4.891	4.147	4.936	6.468
9	2.626	3.125	4.098	2.967	3.532	4.631	3.822	4.550	5.966
10	2.535	3.018	3.959	2.839	3.379	4.433	3.582	4.265	5.594
11	2.463	2.933	3.849	2.737	3.259	4.277	3.397	4.045	5.308
12	2.404	2.863	3.758	2.655	3.162	4.150	3.250	3.870	5.079
13	2.355	2.805	3.682	2.587	3.081	4.044	3.130	3.727	4.893
14	2.314	2.756	3.618	2.529	3.012	3.955	3.029	3.608	4.737
15	2.278	2.713	3.562	2.480	2.954	3.878	2.945	3.507	4.605
16	2.246	2.676	3.514	2.437	2.903	3.812	2.872	3.421	4.492
17	2.219	2.643	3.471	2.400	2.858	3.754	2.808	3.345	4.393
18	2.194	2.614	3.433	2.366	2.819	3.702	2.753	3.279	4.307
19	2.172	2.588	3.399	2.337	2.784	3.656	2.703	3.221	4.230
20	2.152	2.564	3.368	2.310	2.752	3.615	2.659	3.168	4.161
21	2.135	2.543	3.340	2.286	2.723	3.577	2.620	3.121	4.100
22	2.118	2.524	3.315	2.264	2.697	3.543	2.584	3.078	4.044
23	2.103	2.506	3.292	2.244	2.673	3.512	2.551	3.040	3.993
24	2.089	2.489	3.270	2.225	2.651	3.483	2.522	3.004	3.947
25	2.077	2.474	3.251	2.208	2.631	3.457	2.494	2.972	3.904
30	2.025	2.413	3.170	2.140	2.529	3.350	2.385	2.841	3.733
40	1.959	2.334	3.066	2.052	2.445	3.213	2.247	2.677	3.518
50	1.916	2.284	3.001	1.996	2.379	3.126	2.162	2.576	3.385
60	1.887	2.248	2.955	1.958	2.333	3.066	2.103	2.506	3.293
70	1.865	2.222	2.920	1.929	2.299	3.021	2.060	2.454	3.225
80	1.848	2.202	2.894	1.907	2.272	2.986	2.026	2.414	3.173
90	1.834	2.185	2.872	1.889	2.251	2.958	1.999	2.382	3.130
100	1.822	2.172	2.854	1.874	2.233	2.934	1.977	2.355	3.096

Table XI   Factors for Tolerance Intervals (*continued*)

Confidence Level	0.90			0.95			0.99		
Percent Coverage	0.90	0.95	0.99	0.90	0.95	0.99	0.90	0.95	0.99
2	10.253	13.090	18.500	20.581	26.260	37.094	103.029	131.426	185.617
3	4.258	5.311	7.340	6.155	7.656	10.553	13.995	17.370	23.896
4	3.188	3.957	5.438	4.162	5.144	7.042	7.380	9.083	12.387
5	2.742	3.400	4.666	3.407	4.203	5.741	5.362	6.578	8.939
6	2.494	3.092	4.243	3.006	3.708	5.062	4.411	5.406	7.335
7	2.333	2.894	3.972	2.755	3.399	4.642	3.859	4.728	6.412
8	2.219	2.754	3.783	2.582	3.187	4.354	3.497	4.285	5.812
9	2.133	2.650	3.641	2.454	3.031	4.143	3.240	3.972	5.389
10	2.066	2.568	3.532	2.355	2.911	3.981	3.048	3.738	5.074
11	2.011	2.503	3.443	2.275	2.815	3.852	2.898	3.556	4.829
12	1.966	2.448	3.371	2.210	2.736	3.747	2.777	3.410	4.633
13	1.928	2.402	3.309	2.155	2.671	3.659	2.677	3.290	4.472
14	1.895	2.363	3.257	2.109	2.614	3.585	2.593	3.189	4.337
15	1.867	2.329	3.212	2.068	2.566	3.520	2.521	3.102	4.222
16	1.842	2.299	3.172	2.033	2.524	3.464	2.459	3.028	4.123
17	1.819	2.272	3.137	2.002	2.486	3.414	2.405	2.963	4.037
18	1.800	2.249	3.105	1.974	2.453	3.370	2.357	2.905	3.960
19	1.782	2.227	3.077	1.949	2.423	3.331	2.314	2.854	3.892
20	1.765	2.028	3.052	1.926	2.396	3.295	2.276	2.808	3.832
21	1.750	2.190	3.028	1.905	2.371	3.263	2.241	2.766	3.777
22	1.737	2.174	3.007	1.886	2.349	3.233	2.209	2.729	3.727
23	1.724	2.159	2.987	1.869	2.328	3.206	2.180	2.694	3.681
24	1.712	2.145	2.969	1.853	2.309	3.181	2.154	2.662	3.640
25	1.702	2.132	2.952	1.838	2.292	3.158	2.129	2.633	3.601
30	1.657	2.080	2.884	1.777	2.220	3.064	2.030	2.515	3.447
40	1.598	2.010	2.793	1.697	2.125	2.941	1.902	2.364	3.249
50	1.559	1.965	2.735	1.646	2.065	2.862	1.821	2.269	3.125
60	1.532	1.933	2.694	1.609	2.022	2.807	1.764	2.202	3.038
70	1.511	1.909	2.662	1.581	1.990	2.765	1.722	2.153	2.974
80	1.495	1.890	2.638	1.559	1.964	2.733	1.688	2.114	2.924
90	1.481	1.874	2.618	1.542	1.944	2.706	1.661	2.082	2.883
100	1.470	1.861	2.601	1.527	1.927	2.684	1.639	2.056	2.850

Values of *k* for One-Sided Intervals

Appendix B
Bibliography

INTRODUCTORY WORKS AND GRAPHICAL METHODS

Chambers, J., Cleveland, W., Kleiner, B., and P. Tukey (1983), *Graphical Methods for Data Analysis*, Wadsworth & Brooks/Cole, Pacific Grove, CA. A very well-written presentation of graphical methods in statistics.

Freedman, D., Pisani, R., Purves R., and A. Adbikari (1991), *Statistics*, 2nd ed., Norton, New York. An excellent introduction to statistical thinking, requiring minimal mathematical background.

Hoaglin, D., Mosteller, F., and J. Tukey (1983), *Understanding Robust and Exploratory Data Analysis*, John Wiley & Sons, New York. Good discussion and illustration of techniques such as stem-and-leaf displays and box plots.

Tanur, J., et al. (eds.) (1989), *Statistics: A Guide to the Unknown*, 3rd edition, Wadsworth & Brooks/Cole, Pacific Grove, CA. Contains a collection of short nonmathematical articles describing different applications of statistics.

Tukey, J. (1977), *Exploratory Data Analysis*, Addison-Wesley, Reading, MA. Introduces many new descriptive and analytical methods. Not extremely easy to read.

PROBABILITY

Hoel, P. G., Port, S. C., and C. J. Stone (1971), *Introduction to Probability Theory*, Houghton Mifflin, Boston. A well-written and comprehensive treatment of probability theory and the standard discrete and continuous distributions.

Olkin, I., Derman, C., and L. Gleser (1994), *Probability Models and Applications*, 2nd ed., Macmillan, New York. A comprehensive treatment of probability at a higher mathematical level than this book.

Mosteller, F., Rourke, R., and G. Thomas (1970), *Probability with Statistical Applications*, 2nd ed., Addison-Wesley, Reading, MA. A precalculus introduction to probability with many excellent examples.

Ross, S. (1998), *A First Course in Probability*, 5th ed., Macmillan, New York. More mathematically sophisticated than this book, but has many excellent examples and exercises.

MATHEMATICAL STATISTICS

Efron, B., and R. Tibshirani (1993), *An Introduction to the Bootstrap*, Chapman and Hall, New York. An important reference on this useful but computer-intensive technique.

Hoel, P. G. (1984), *Introduction to Mathematical Statistics*, 5th ed., John Wiley & Sons, New York. An outstanding introductory book, well written, and generally easy to understand.

Hogg, R., and A. Craig (1995), *Introduction to Mathematical Statistics*, 5th ed., Prentice-Hall, Englewood Cliffs, NJ. Another classic work on the mathematical principles of statistics; higher level than the Hoel book, but contains excellent discussions of estimation and hypothesis testing.

Larsen, R., and M. Marx (1986), *Introduction to Mathematical Statistics*, 2nd ed., Prentice-Hall, Englewood Cliffs, NJ. Written at a relatively low mathematical level, very readable.

Larson, H. J. (1982), *Introduction to Probability Theory and Statistical Inference*, 3rd ed., John Wiley & Sons, New York. An extremely well-written book that gives broad coverage to many aspects of probability and mathematical statistics.

ENGINEERING STATISTICS

Devore, J. L. (2000), *Probability and Statistics for Engineering and the Sciences*, 5th ed., Duxburg & Brooks/Cole, Pacific Grove, CA. Covers many of the same topics as this text, but at a slightly higher mathematical level. Many of the examples and exercises involve applications to biological and life sciences.

Hines, W. W., and D. C. Montgomery (1990), *Probability and Statistics in Engineering and Management Science*, 3rd ed., John Wiley & Sons, New York. Covers many of the same topics as this book. More emphasis on probability and a higher mathematical level.

Ross, S. (1987), *Introduction to Probability and Statistics for Engineers and Scientists*, John Wiley & Sons, New York. More tightly written and mathematically oriented than this book, but contains some good examples.

Walpole, R. E., Myers, R. H., and S. L. Myers (2002), *Probability and Statistics for Engineers and Scientists*, 7th ed., Prentice-Hall, Inc., Upper Saddle River, New Jersey. A very well-written book at about the same level as this one.

REGRESSION ANALYSIS

Daniel, C., and F. Wood (1980), *Fitting Equations to Data*, 2nd ed., John Wiley & Sons, New York. An excellent reference containing many insights on data analysis.

Draper, N., and H. Smith (1998), *Applied Regression Analysis*, 3rd ed., John Wiley & Sons, New York. A comprehensive book on regression written for statistically oriented readers.

Montgomery, D. C., Peck, E. A., and G. G. Vining (2001), *Introduction to Linear Regression Analysis*, 3rd ed., John Wiley & Sons, New York. A comprehensive book on regression written for engineers and physical scientists.

Myers, R. H. (1990), *Classical and Modern Regression with Applications*, 2nd ed., PWS-Kent, Boston. Contains many examples with annotated SAS output. Very well written.

Neter, J., Wasserman, W., Nachtsheim, C., and M. Kutner (1996), *Applied Linear Statistical Models*, 4th ed., Richard D. Irwin, Homewood, Ill. The first part of the book is an introduction to simple and multiple linear regression. The orientation is to business and economics.

Younger, M. S. (1985), *A Handbook for Linear Regression*, 2nd ed., Duxburg, Boston. A good presentation of regression methods. The discussion of SAS, BMD, and SPSS computer packages is excellent.

DESIGN OF EXPERIMENTS

Box, G. E. P., Hunter, W. G., and J. S. Hunter (1978), *Statistics for Experimenters*, John Wiley & Sons, New York. An excellent introduction to the subject for those readers desiring a statistically oriented treatment. Contains many useful suggestions for data analysis.

Mason, R. L., Gunst, R. F., and J. F. Hess (1989), *Statistical Design and Analysis of Experiments*, John Wiley & Sons, New York. A comprehensive book covering basic statistics,

hypothesis testing and confidence intervals, elementary aspects of experimental design, and regression analysis.

Montgomery, D. C. (2001), *Design and Analysis of Experiments*, 5th ed., John Wiley & Sons, New York. Written at the same level as the Box, Hunter, and Hunter book, but focused on engineering applications.

NONPARAMETRIC STATISTICS

Conover, W. J. (1998), *Practical Nonparametric Statistics*, 3rd ed., John Wiley & Sons, New York. An excellent exposition of the methods of nonparametric statistics, many good examples and exercises.

Hollander, M., and D. Wolfe (1999), *Nonparametric Statistical Methods*, 2nd ed., John Wiley & Sons, New York. A good reference book, with a very useful set of tables.

STATISTICAL QUALITY CONTROL AND RELATED METHODS

Duncan, A. J. (1986), *Quality Control and Industrial Statistics*, 5th ed., Richard D. Irwin, Homewood, Illinois. A classic book on the subject.

Grant, E. L., and R. S. Leavenworth (1988), *Statistical Quality Control*, 6th ed., McGraw-Hill, New York. One of the first books on the subject; contains many good examples.

John, P. W. M. (1990), *Statistical Methods in Engineering and Quality Improvement*, John Wiley & Sons, New York. Not a methods book, but a well-written presentation of statistical methodology for quality improvement.

Montgomery, D. C. (2001), *Introduction to Statistical Quality Control*, 4th ed., John Wiley & Sons, New York. A modern comprehensive treatment of the subject written at the same level as this book.

Nelson, W. (1982), *Applied Life Data Analysis*, John Wiley & Sons, New York. Contains many examples of using statistical methods for the study of failure data; a good reference for the statistical aspects of reliability engineering and the special probability distributions used in that field.

Ryan, T. P. (2000), *Statistical Methods for Quality Improvement*, 2nd ed., John Wiley & Sons, New York. Gives broad coverage of the field, with some emphasis on newer techniques.

Wadsworth, H. M., Stephens, K. S., and A. B. Godfrey (2001), *Modern Methods for Quality Control and Improvement*, 2nd ed., John Wiley & Sons, New York. A comprehensive treatment of statistical methods for quality improvement at a somewhat higher level than this book.

Western Electric Company (1956), *Statistical Quality Control Handbook*, Western Electric Company, Inc., Indianapolis, Indiana. An oldie but a goodie.

Appendix C

Answers to

Selected

Exercises

CHAPTER 2

Section 2-1

2-1. Let a, b denote a part above, below the specification.
$S = \{aaa, aab, aba, abb, baa, bab, bba, bbb\}$

2-7. S is the sample space of 100 possible two-digit integers.

2-9. $S = \{0, 1, 2, \dots\}$ in ppb.

2-17. c = connect, b = busy, $S = \{c, bc, bbc, bbbc, bbbbc, \dots\}$

2-21. (a) $S = \{0, 1, 2, 3, \dots\}$
(b) S (c) $\{12, 13, 14, 15\}$
(d) $\{0, 1, 2, \dots, 11\}$ (e) S
(f) $\{0, 1, 2, \dots, 7\}$ (g) $\varnothing$
(h) $\varnothing$ (i) $\{8, 9, 10, \dots\}$

2-23. Let d denote a distorted bit and let o denote a bit that is not distorted.

(a) $S = \begin{Bmatrix} dddd, dodd, oddd, oodd, \\ dddo, dodo, oddo, oodo, \\ ddod, dood, odod, oood, \\ ddoo, dooo, odoo, oooo \end{Bmatrix}$

(b) No
(c) $\{dddd, dodd, dddo, dodo, ddod, dood, ddoo, dooo\}$

(d) $\{oddd, oodd, oddo, oodo, odod, oood, odoo, oooo\}$
(e) $\{dddd\}$
(f) $\{dddd, dodd, dddo, oddd, ddod, oodd, ddoo\}$

2-25. $2^{12} = 4096$

2-27. (a) $A' \cap B = 10$, $B' = 10$, $A \cup B = 92$

2-29. (a) $A' = \{x \mid x \geq 72.5\}$
(b) $B' = \{x \mid x \leq 52.5\}$
(c) $A \cap B \doteq \{x \mid 52.5 < x < 72.5\}$
(d) $A \cup B = \{x \mid x > 0\}$

2-31. Let g denote a good board, m a board with minor defects, and j a board with major defects.
(a) $S = \{gg, gm, gj, mg, mm, mj, jg, jm, jj\}$
(b) $S = \{gg, gm, gj, mg, mm, mj, jg, jm\}$

Section 2-2

2-35. (a) 0.4 (b) 0.8 (c) 0.6
(d) 1 (e) 0.2

2-37. (a) $S = \{1, 2, 3, 4, 5, 6, 7, 8\}$
(b) 2/8 (c) 6/8

2-39. (a) 0.7 (b) 0.8

2-41. (a) 0.25 (b) 0.75

2-43. 5.7×10^{-8}

2-45. (a) 0.86 (b) 0.79 (c) 0.14
(d) 0.70 (e) 0.95 (f) 0.84

2-47. (a) 0.30 (b) 0.77 (c) 0.70
(d) 0.22 (e) 0.85 (f) 0.92

Section 2-3

2-49. (a) 0.7 (b) 0.4 (c) 0.1
(d) 0.2 (e) 0.6 from part (b)
(f) 0.8

2-51. No

2-53. (a) 350/370 (b) 362/370
(c) 358/370 (d) 345/370

2-55. (a) 13/130 (b) No

Section 2-4

2-57. (a) 86/100 (b) $P(B) = 79/100$ (c) 70/79
(d) 70/86

2-59. (a) 345/357 (b) 5/13

2-61. (a) 0.15 (b) 0.153
(c) 0.72 (d) 0.733
(e) 0.11 (f) 0.76

2-63. (a) 15/40 (b) 14/39
(c) 0.135 (d) 0.615

2-65. (a) $4/499 = 0.0080$
(b) $(5/500)(4/499) = 0.000080$
(c) $(495/500)(494/499) = 0.98$

2-67. (a) 0.813 (b) 0.632
(c) 0.764

Section 2-5

2-71. 0.22
2-73. 0.023
2-75. 0.028
2-77. (a) 0.0225 (b) 0.125
2-79. (a) 0.20 (b) 0.20

Section 2-6

2-81. No
2-83. No
2-85. (a) No (b) 0.733
2-87. (a) 0.59 (b) 0.328 (c) 0.41
2-89. (a) 0.00003 (b) 0.00024
(c) 0.00107
2-91. 0.9702
2-93. (a) No (b) Yes

Section 2-7

2-95. 0.003
2-97. (a) 0.615 (b) 0.618
(c) 0.052
2-99. (a) 0.9847 (b) 0.1184

Supplemental

2-101. The sample space $S = \{A, A'D_1,$
$A'D_2, A'D_3, A'D_4, A'D_5\}$
2-103. (a) 0.19 (b) 0.15 (c) 0.99
(d) 0.80 (e) 0.158
2-105. (a) No (b) No (c) 40/240
(d) 200/240 (e) 234/400
(f) 1
2-107. (a) 0.282 (b) 0.718
2-109. 0.996
2-111. (a) 0.0037 (b) 0.811
2-113. (a) 0.0778 (b) 0.00108
(c) 0.947
2-115. (a) 0.9764 (b) 0.3159
2-117. (a) 0.207 (b) 0.625 (c) 0.6
2-119. (a) 0.453 (b) 0.262
(c) 0.881 (d) 0.547
(e) 0.783 (f) 0.687
2-121. 1.58×10^{-7}

CHAPTER 3

Section 3-1

3-1. $\{0, 1, 2, \ldots, 1000\}$
3-3. $\{0, 1, 2, \ldots, 99999\}$

Section 3-2

3-13. $f_X(0) = 1/3, f_X(1.5) = 1/3,$
$f_X(2) = 1/6, f_X(3) = 1/6$
3-15. (a) 1 (b) 7/8 (c) $\frac{3}{4}$
(d) $\frac{1}{2}$
3-17. (a) 9/25 (b) 4/25
(c) 12/25 (d) 1
3-19. $P(X = 10 \text{ million}) = 0.3,$
$P(X = 5 \text{ million}) = 0.6,$
$P(X = 1 \text{ million}) = 0.1$
3-21. $P(X = 0) = 8 \times 10^{-6},$
$P(X = 1) = 0.0012,$
$P(X = 2) = 0.0576,$
$P(X = 3) = 0.9412$
3-23. $P(X = 15 \text{ million}) = 0.6,$
$P(X = 5 \text{ million}) = 0.3,$
$P(X = -0.5 \text{ million}) = 0.1$
3-25. $P(X = 0) = 0.00001,$
$P(X = 1) = 0.00167,$
$P(X = 2) = 0.07663,$
$P(X = 3) = 0.92169$

Section 3-3

3-27. (a) 7/8 (b) 1 (c) $\frac{3}{4}$
(d) 3/8
3-29. $F(x) = 0$ for $x < 1$ million; 0.1
for 1 million $\leq x < 5$ million;
0.7 for 5 million $\leq x < 10$ million; 1 for 10 million $\leq x$
3-31. $F(x) = 0$ for $x < 0$; 0.008 for
$0 \leq x < 1$; 0.104 for $1 \leq x < 2$;
0.488 for $2 \leq x < 3$; 1 for $3 \leq x$
3-33. (a) 1 (b) 0.5 (c) 0.5
(d) 0.5
3-35. (a) 1 (b) 0.75 (c) 0.25
(d) 0.25 (e) 0 (f) 0

Section 3-4

3-37. $E(X) = 2, V(X) = 2$
3-39. $E(X) = 0, V(X) = 1.5$
3-41. $E(X) = 6.1$ million,
$V(X) = 7.89$ million^2
3-43. $E(X) = 2.4, V(X) = 0.48$
3-45. $x = 24$

Section 3-5

3-47. $E(X) = 2, V(X) = 0.667$
3-49. $E(X) = 0.17, V(X) = 0.0002$
3-51. $E(X) = 590.45, V(X) = 0.0825$

Section 3-6

3-57. (a) 0.2461 (b) 0.0547
(c) 0.0107 (d) 0.3223

3-59. (a) 2.4×10^{-8} (b) 0.99989
(c) 9.91×10^{-18}
(d) 1.138×10^{-4}
3-61. $F(x) = 0$ for $x < 0$; 0.4219 for
$0 \leq x < 1$; 0.8438 for $1 \leq x < 2$;
0.9844 for $2 \leq x < 3$; 1 for $3 \leq x$
3-63. (a) 0.3681 (b) 0.6319
(c) 0.9198 (d) $E(X) = 1$,
$V(X) = 0.999$
3-65. (a) $n = 50, p = 0.1$
(b) 0.1117 (c) 4.51×10^{-48}
3-67. (a) 0.9961 (b) 0.9886
3-69. (a) 9.677×10^{-10} (b) 0.2137

Section 3-7

3-71. (a) 0.5 (b) 0.0625
(c) 0.0039 (d) 0.75 (e) 0.25
3-73. (a) 0.0064 (b) 0.9984
(c) 0.008
3-75. (a) 0.0167 (b) 0.9224 (c) 50
3-77. (a) 3.91×10^{-19} (b) 200
(c) 2.56×10^{18}
3-79. (a) 5 (b) 5
3-81. (a) 20 (b) 0.0436
(c) 0.0459 (d) 0.0411 (e) 19
3-83. (a) 3000 (b) 1731.18

Section 3-8

3-87. (a) 0.4623 (b) 0.0002
(c) 0.9866 (d) $E(X) = 0.8$,
$V(X) = 0.539$
3-89. $F(X) = 0$ for $x < 0$; 1/6 for $0 \leq$
$x < 1$; 2/3 for $1 \leq x < 2$; 29/30
for $2 \leq x < 3$; 1 for $3 \leq x$
3-91. (a) 0.1201 (b) 0.8523
3-93. (a) 0.7069 (b) 0.0607
(c) 0.2811

Section 3-9

3-97. (a) 0.0183 (b) 0.2381
(c) 0.1954 (d) 0.0298
3-99. $E(X) = 2.996, \quad V(X) = 2.996$
3-101. (a) 0.0045 (b) 0.3679
(c) 0.1353 (d) 0.2642
3-103. (a) 4.54×10^{-5} (b) 0.6321
3-105. (a) 0.6065 (b) 0.0067
(c) 0.00504

Supplemental

3-107. 0.3714
3-109. (a) 0.0117 (b) 1.3333
3-111. (a) 0.1755 (b) 0.0858
(c) 0.2873
3-113. 0.9810

3-115.

x	2	3	4	5	6
$f(x)$	0.0025	0.01	0.03	0.065	0.13

x	7	8	9	10
$f(x)$	0.18	0.2225	0.2	0.16

3-117. 299

3-119. (a) 4.1×10^{-5} (b) 10
(c) 0.9995

3-121. (a) 0.6 (b) 0.8 (c) 0.7
(d) 3.9 (e) 3.09

3-123. (a) 0.2408 (b) 0.4913

3-125. (a) 0.3233 (b) 0.7619

3-127. 0.0739

3-129. (a) 0.3679 (b) 50.51
(c) 0.9234

CHAPTER 4

Section 4-2

4-1. (a) 0.3679 (b) 0.2858 (c) 0
(d) 0.9817 (e) 0.0498

4-3. (a) 0.4375 (b) 0.7969
(c) 0.5625 (d) 0.7031
(e) 0.5

4-5. (a) 0.5 (b) 0.4375 (c) 0.125
(d) 0 (e) 1 (f) 0.9655

4-7. (a) 0.5 (b) 49.8

4-9. (a) 0.10 (b) 2.5

Section 4-3

4-11. (a) 0.56 (b) 0.7 (c) 0 (d) 0

4-13. $1 - e^{-x}$ for $x > 0$

4-15. $1 - e^{-(x-4)}$ for $x > 4$

4-17. (a) $1.25x - 93.25$ for
$74.6 < x < 75.4$ (b) 0.5

4-21. $F(x) = 0$ for $x < 0$; $0.25x^2$ for
$0 \le x < 2$; 1 for $x \le 2$

Section 4-4

4-23. $E(X) = 2.6667, V(X) = 0.8889$

4-25. $E(X) = 4.083, V(X) = 0.3264$

4-27. (a) $E(X) = 109.39$ μm,
$V(X) = 33.19$ μm^2
(b) \$54.70

4-29. (a) $E(X) = 5.1$ mm, $V(X) =$
0.01 mm^2 (b) 0.3679

Section 4-5

4-31. (a) $E(X) = 3.5, V(X) = 1.33$,
$\sigma_X = 1.155$ (b) 0.25

4-33. (a) $E(X) = 50, V(X) = 0.0208$,
$\sigma_X = 0.144$ (b) $F(x) = 2x$
-99.5 for $49.75 < x < 50.25$
(c) 0.7

4-35. (a) $E(X) = 1.85$ min, $V(X) =$
0.0408 min^2 (b) 0.7143
(c) $F(x) = (x - 1.5)/0.7$ for
$1.5 < x < 2.2$

4-37. (b) 0.25 (c) 0.2140
(d) $E(X) = 0.2100$ μm,
$V(X) = 0.00000833$ μm^2

Section 4-6

4-39. (a) 0.90658 (b) 0.99865
(c) 0.07353 (d) 0.98422
(e) 0.95116

4-41. (a) 1.28 (b) 0 (c) 1.28
(d) -1.28 (e) 1.33

4-43. (a) 0.93319 (b) 0.69146
(c) 0.9545 (d) 0.00132
(e) 0.15866

4-45. (a) 0.93319 (b) 0.89435
(c) 0.38292 (d) 0.80128
(e) 0.54674

4-47. (a) 0.99379 (b) 0.13591
(c) 5835

4-49. (a) 0.0082 (b) 0.72109
(c) 0.564

4-51. (a) 0.00135 (b) 0.15866
(c) 71.6 min

4-53. (a) 0.02275 (b) 0.47725
(c) 0.336

4-55. (a) 0.15866 (b) 90.0
(c) 99.73%

4-57. (a) 0.15245 (b) 125.6

4-59. (a) 0.06681 (b) 0.86638
(c) 0.000214

Section 4-7

4-61. (a) 0.075 (b) 0.85

4-63. (a) 0.129 (b) 0.488

4-65. 0.013

4-67. 0.966

4-71. (b) 330 (c) 0.0089

Section 4-9

4-73. (a) 0.3679 (b) 0.1353
(c) 0.0498 (d) 29.96

4-75. (a) 0.333 min (b) 0.333 min
(c) 0.9986

4-77. (a) 0.1353 (b) 0.4866
(c) 0.2031 (d) 34.54

4-79. (a) 0.0498 (b) 0.8775

4-81. (a) 0.0025 (b) 0.6321

4-83. (a) 0.1353 (b) 0.2707 (c) 5

4-85. (a) 0.2212 (b) 0.2865
(c) 0.2212

4-87. 0.8488

4-93. (a) 5 (b) 0.1353
(d) 11.51

Section 4-10

4-97. (a) 0.1755 (b) 0.26
4-99. (a) 50,000 (b) 0.67
4-101. (a) 500,000 (b) 22.
(c) 0.0803

4-103. (a) 0.1429 (b) 0.18

4-105. (a) 120 (b) 1.32934
(c) 11.6317

Section 4-11

4-109. (a) 12,000 (b) 3.61×10^{10}

4-111. (a) 0.5273 (b) 8862.3
(c) 0.00166

4-113. (a) 0.275 (b) 0.685

4-115. (a) 443.11 (b) 53650.5
(c) 0.2212

Section 4-12

4-117. (a) 0.9332 (b) 20952.2
(c) $E(X) = 13359.7$,
$V(X) = 1.45 \times 10^{12}$

4-119. (a) 0.983 (b) 0.45

4-121. $\theta = 3.38, \omega^2 = 2.25$

Supplemental

4-125. $0.25x^2 - x + 1$ for $2 < x < 4$

4-127. (a) 0.3935 (b) 0.3834
(c) 23.03

4-129. (a) 0.423 (b) 50

4-133. (a) $\theta = 3.43, \omega^2 = 0.96$
(b) 0.946

4-135. (a) 0.6915 (b) 0.683
(c) 1.86

4-137. (a) 0.0062 (b) 0.0124
(c) 5.33

4-139. 0.0008 to 0.0032

4-141. $\mu = 11,398$

4-143. (a) 0.5633 (b) 737.5

4-145. (a) 0.984 (b) 0.834

CHAPTER 5

Section 5-1

5-1. $f(x, y) \ge 0$, $\sum f(x, y) = 1$

5-3. $E(X) = 1.8125$,
$E(Y) = 2.875$

5-5. $c = 1/36$

5-7. $E(X) = 2.167, V(X) = 0.639$,
$E(Y) = 2.167, V(Y) = 0.639$

5-9. $f(x, y) \geq 0$, $\sum f(x, y) = 1$

5-11. $E(X) = 1/8$, $E(Y) = 1/4$

5-13. $X \geq 0$, $Y \geq 0$ and $X + Y \leq 4$

5-15. (b) $f_X(0) = 0.2338$, $f_X(1) = 0.4188$, $f_X(2) = 0.2679$, $f_X(3) = 0.0725$, $f_X(4) = 0.0070$ (c) $E(X) = 1.2$
(d) $f_{Y|3}(0) = 0.857$, $f_{Y|3}(1) = 0.143$
(e) $E(Y | X = 3) = 0.143$
(f) 0.123   (g) No

Section 5-2

5-17. (a) 0.5   (b) 0.35   (c) 0.5
(d) 0.8   (e) 1.5

5-19. $P(X = 1 | Y = 1, Z = 2) = 0.4$, $P(X = 2 | Y = 1, Z = 2) = 0.6$

5-21. (a) 0   (b) 0.072   (c) 0.736
(d) 1

5-23. (a) $x \geq 0, y \geq 0, z \geq 0$, $x + y + z = 4$
(b) No

5-25. (a) 0.1758   (b) 0.2198
(c) $E(X) = 1.067$, $V(X) = 0.6146$

5-27. (a) Yes   (b) 0.1944
(c) 0.0001

5-29. (a) 0.7347   (b) 0
(c) $P(X = 0 | Y = 2) = 0.0204$, $P(X = 1 | Y = 2) = 0.2449$, $P(X = 2 | Y = 2) = 0.7347$
(d) $E(X | Y = 2) = 1.7143$

5-31. (a) binomial $p = 0.03$, $n = 3$ $E(X) = 0.03$, $V(X) = 0.0297$
(b) $P(X = 0 | Y = 2) = 0.98958$, $P(X = 1 | Y = 2) = 0.01042$
(c) $E(X | Y = 2) = 0.01042$, $V(X | Y = 2) = 0.01031$

Section 5-3

5-35. (a) 0.4444   (b) 0.6944
(c) 0.5833   (d) 0.3733   (e) 2
(f) 0

5-37. 1/24

5-39. (a) $(2x + 1)/12$ for $0 < x < 3$
(b) $(y + 1)/6$ for $1 < y < 3$
(c) 2.111
(d) 0.5833
(e) $(2 + x)/6$ for $0 < x < 2$

5-43. 10

5-45. (a) $10(e^{-2x} - e^{-5x})/3$ for $0 < x$
(b) $3.157e^{-3y}$ for $0 < y < 1$
(c) 0.2809   (d) $2e^{-2x+4}$ for $2 < x$

5-49. 2/15

5-51. (a) $(x + 1)/7.5$ for $0 < x < 1$, $2/7.5$ for $1 < x < 4$
(b) 0.5 for $0 < y < 2$
(c) 1   (d) 0.25

5-53. (a) 0.0439, 0.0019
(b) 0.065

Section 5-4

5-55. (a) 0.25   (b) 0.0625   (c) 1
(d) 1   (e) 2/3

5-57. (a) $2x$ for $0 < x < 1$
(b) 0.25

5-61. 6

5-63. (a) $3(x - 1)^2$ for $0 < x < 1$
(b) $6(1 - x - y)$ for $0 < x$, $0 < y$ and $x + y < 1$
(c) 1 for $x = 0$
(d) $4(1 - 2x)$ for $x < 0.5$

5-65. (a) 0.032   (b) 0.0267

Section 5-5

5-67. $\sigma_{XY} = 0.703$, $\rho_{XY} = 0.885$

5-69. $c = 1/36$, $\sigma_{XY} = -1/36$, $\rho_{XY} = -0.0435$

5-71. $\sigma_{XY} = -2.267$, $\rho_{XY} = -0.51$

5-73. $c = 9.5$, $\sigma_{XY} = 1.85181$, $\rho = 0.9206$

5-75. X and Y are independent and $\sigma_{XY} = \rho_{XY} = 0$

Section 5-6

5-81. 0.827

Section 5-7

5-87. (a) 30   (b) 97   (c) 0.5
(d) 0.846

5-89. (a) $E(T) = 4$, $\sigma_T = 0.1414$
(b) 0.017

5-91. (a) $E(D) = 1/8$, $\sigma_D = 0.140$
(b) 0.187   (c) 0.187

5-93. (a) 0.05   (b) 0.023
(c) 12.129
(d) 0.388   (e) 136

5-95. (a) 0.3446   (b) 6790

Supplemental

5-97. (a) 3/8   (b) 3/4   (c) 3/4
(d) 3/8
(e) $E(X) = 7/8$, $V(X) = 39/64$, $E(Y) = 7/8$, $V(Y) = 39/64$

5-99. (a) 0.0631   (b) 0.1216
(c) $E(X) = 2$, $V(X) = 1.8$
(d) $f_{X|19}(0) = 0.667$, $f_{X|19}(1) = 0.333$
(e) 1/3

5-103. (b) 1/3   (c) No

5-105. (a) 0.0093   (b) 0.5787
(c) 0.75   (d) 0.2199
(e) 2.25   (f) 1.3333

5-107. (a) 1/2   (b) 1/4   (c) $1/\pi$ for $x^2 + y^2 \leq 1$
(d) $2\sqrt{1 - x^2}/\pi$ for $-1 < x < 1$

5-109. 3/4

5-111. (a) 0.085   (b) Bin(10, 0.3)
(c) 3

5-113. (a) 0.499   (b) 0.5

5-115. (a) 0.057   (b) 0.057

5-119. (a) $E(T) = 1.5$, $V(T) = 0.078$
(b) 0.1216
(c) $E(P) = 4$, $V(P) = 0.568$

5-121. $\mu = 5$, $\sigma = \sqrt{3}$

CHAPTER 6

Section 6-1

6-1. $\bar{x} = 74.0044$, $s = 0.00473$

6-3. $\bar{x} = 7068.1$, $s = 226.5$

6-5. $\bar{x} = 43.975$, $s = 12.294$

6-7. $\mu = 5.44$

6-11. (a) $\bar{x} = 7.184$
(b) $s = 0.02066$

6-13. (a) $\bar{x} = 65.86$, $s = 12.16$
(c) $\bar{x} = 66.86$, $s = 10.74$

Section 6-3

6-19.

Variable	N	Median	Q1	Q3
cycles	70	1436.5	1097.8	1735.0

6-21.

Variable	N	Median	Q1	Q3
yield	90	89.250	86.100	93.125

6-25. $\bar{x} = 260.30$, $s = 13.41$, and $\tilde{x} = 260.85$

6-27. (b) $\bar{x} = 89.45$, $s = 2.8$, and $\tilde{x} = 90$ (c) 22/40

Section 6-5

6-43. (a) $\bar{x} = 4.0$ (b) $s^2 = 0.867$, $s = 0.931$

6-45. (a) $\bar{x} = 952.44$, $s^2 = 9.55$, $s = 3.09$
(b) $\tilde{x} = 953$, largest value can increase by any amount

6-47. (a) $\bar{x} = 48.125$, $\tilde{x} = 49$
(b) $s^2 = 7.246$, $s = 2.692$

Supplemental

6-73. (a) $\bar{x} = 65.083$
(b) $s^2 = 1.8697$, $s = 1.367$
6-75. (a) Sample 1: range $= 4$;
Sample 2: range $= 4$
(b) Sample 1: $s = 1.604$;
Sample 2: $s = 1.852$
6-79. (b) $\bar{x} = 9.325$, $s = 4.48586$

CHAPTER 7

Section 7-2

7-1. Estimator 1
7-3. Estimator 2
7-5. 2.5
7-7. Estimator 3 is most efficient; estimator 2 is the best "unbiased" estimator.
7-11. (a) 75.615 (b) 75.2
(c) $\sigma^2 = 2.738$, $\sigma = 1.655$
(d) 0.325 (e) 0.0385

7-15. (b) $se = \sqrt{\dfrac{\sigma_1^2}{n_1} + \dfrac{\sigma_2^2}{n_2}}$

7-17. (b)

$$se(\hat{\mu}) = \sigma_1 \sqrt{\frac{\alpha^2 n_2 + (1 - \alpha)^2 a n_1}{n_1 n_2}}$$

(c) $\alpha = \dfrac{a n_1}{n_2 + a n_1}$ (d) 0.10

Section 7-3

7-31. (a) 423.33, 82.4464

Section 7-5

7-33. 0.8186
7-35. 0.43055
7-37. 0.1915
7-39. 12
7-41. 0.2313
7-43. (a) 0.5885
(b) 0.1759
7-45. 0.983

Supplemental

7-49. $\bar{X}_1 - \bar{X}_2 \sim N(-5, 0.2233)$
7-51. 0.8664
7-53. 1 (approximately)
7-55. 0 (approximately)

CHAPTER 8

Section 8-2

8-1. 97.93%, 99.36%, and 96.78%
8-3. (a) 96.76% (b) 98.72%
(c) 93.56%
8-5. 3
8-7. (a) Longer (b) No (c) Yes
8-9. (87.85, 93.11)
8-11. (a) (74.0353, 74.0367)
(b) (74.0356, ∞)
8-13. (a) (3232.11, 3267.89)
(b) (3226.4, 3273.6)
8-15. 267
8-17. 4

Section 8-3

8-19. $t_{0.025,15} = 2.131$, $t_{0.05,10} = 1.812$, $t_{0.10,20} = 1.325$, $t_{0.005,25} = 2.787$, $t_{0.001,30} = 3.385$
8-21. (a) $t_{0.05,14} = 1.761$
(b) $t_{0.01,19} = 2.539$
(c) $t_{0.001,24} = 3.467$
8-23. (1.108, ∞)
8-25. (a) Yes (b) (1.094, 1.106)
8-27. (a) Yes (b) (8.216, 8.244)
8-29. (4.023, ∞)
8-31. (1.093, 1.106)

Section 8-4

8-33. $\chi^2_{0.05,10} = 18.31$,
$\chi^2_{0.025,15} = 27.49$,
$\chi^2_{0.01,12} = 26.22$,
$\chi^2_{0.005,25} = 46.93$,
$\chi^2_{0.95,20} = 10.85$,
$\chi^2_{0.99,18} = 7.01$,
$\chi^2_{0.995,16} = 5.14$
8-35. $0.00003075 < \sigma^2$
8-37. $7,975,727.09 < \sigma^2$
8-39. $0.31 < \sigma < 0.46$
8-41. $3.8 \leq \sigma$

Section 8-5

8-43. 622
8-45. 666
8-47. 5759

Section 8-6

8-49. $52131.1 \leq X_{n+1} \leq 68148.3$
8-51. $263.7 \leq X_{n+1} \leq 370.7$
8-53. $2193.5 \leq X_{n+1} \leq 2326.5$
8-55. $3.91 \leq X_{n+1} \leq 4.19$
8-57. $2.58 \leq X_{n+1} \leq 3.22$
8-59. $228.1 \leq X_{n+1} \leq 235.2$

Section 8-7

8-61. (0.408, 2.092)
8-63. (15.14, 18.82)
8-65. (8.16, 8.30)
8-67. (3.91, ∞)
8-69. (1.06, 1.14)

Supplemental

8-71. (a) $0.1 \leq P\text{-value} \leq 0.25$
(b) $0.05 \leq P\text{-value} \leq 0.1$
(c) $P\text{-value} = 0.00621$
8-75. (a) 40 (b) 23
8-77. (4.65, 45.59)
8-79. (a) Yes
(b) $0.618 \leq \mu \leq 0.630$
(c) $0.588 \leq X_{n+1} \leq 0.660$
(d) (0.582, 0.622)
8-81. (a) Yes
(b) $2.270 \leq \mu \leq 4.260$
(c) $-1.297 \leq X_{n+1} \leq 7.827$
(d) (−3.113, 10.363)
8-83. (a) (0.0021, 0.0113)
(b) No
8-85. (a) $0.210 \leq p \leq 0.274$
(b) $0.204 \leq p \leq 0.280$

CHAPTER 9

Section 9-1

9-1. (a) Yes (b) No (c) No
(d) No (e) No
9-3. (a) 0 (b) 0.02275
9-5. 11.5875
9-7. (a) 0.09296 (b) 0.04648
(c) 0.00005
9-9. (a) Reject H_0 (b) 0.00889
9-11. (a) 182.9 (b) 0.00776
9-13. (a) 0.0164 (b) 0.21186
9-17. (a) 0.08535 (b) 0.9951
9-19. (a) 0.29372 (b) 0.25721

Section 9-2

9-21. (a) $z_0 = 0.36$, do not reject H_0
(b) $P\text{-value} = 0.71884$
(c) 5 (d) 0.68054
(e) (87.85, 93.11)
9-23. (a) $z_0 = -1.69$, do not reject H_0
(b) 0.091028 (c) 0 (d) 2
9-25. (a) $z_0 = -14.43$, reject H_0
(b) 0 (c) (3205.31, 3294.69)
9-27. (a) 1.56, (b) 0.97062
(c) 5 (d) (99.888, ∞)
9-29. (a) $z_0 = 1.77$, reject H_0 (b) 1
(c) 2 (d) (4.003, ∞)

Section 9-3

9-31. (a) $t_0 = -3.48$, reject H_0,
P-value $= 0.002$
(b) 1 (c) 20
(d) (98.065, 98.463)
(e) Yes

9-33. (a) $t_0 = -1.46$, do not reject
H_0, P-value $= 0.156$
(b) Yes (c) 0.80 (d) 100
(e) (129.406, 130.100)

9-35. $t_0 = 4.47$, reject H_0,
P-value < 0.0005

9-37. (a) $t_0 = -5.35$, do not reject H_0
(b) P-value > 0.4
(c) 0.75 (d) 40

9-39. (a) $t_0 = 2.806$, reject H_0
(b) P-value $= 0.004$ (c) 1
(d) 4

9-41. (a) $t_0 = 3.018$, reject H_0
(b) P-value $= 0.0038$ (c) 0.8
(d) 75

Section 9-4

9-43. (a) $\chi_0^2 = 8.96$, do not reject H_0
(b) $0.5 <$ P-value < 0.9
(c) 50

9-45. (a) $\chi_0^2 = 4984.83$, reject H_0
(b) P-value < 0.005

9-47. (a) $\chi_0^2 = 109.52$, reject H_0
(b) $0.31 < \sigma < 0.46$

9-49. 30

Section 9-5

9-51. 0.639, 118

9-53. (a) $z_0 = -0.53$, do not reject H_0
(b) P-value $= 0.29806$

9-55. (a) $z_0 = 0.452$, do not reject H_0
(b) P-value $= 0.67364$

9-57. (a) $\alpha = 0.0853$ (b) $\beta \cong 0$

Section 9-7

9-59. (a) $\chi_0^2 = 7.2$, do not reject H_0
(b) $0.05 <$ P-value < 0.10

9-61. (a) $\chi_0^2 = 1.72$, do not reject H_0
(b) $0.5 <$ P-value < 0.9

9-63. (a) $\chi_0^2 = 1.053$, do not reject H_0
(b) $0.1 <$ P-value < 0.5

Section 9-8

9-65. (a) $\chi_0^2 = 11.65$, do not reject H_0
(b) $0.05 <$ P-value < 0.10

9-67. (a) $\chi_0^2 = 25.55$, reject H_0
(b) P-value < 0.005

9-69. (a) $\chi_0^2 = 10.71$, do not reject H_0
(b) $0.05 <$ P-value < 0.10

Supplemental

9-71. (a) $p(1 - p)/50$
(b) $p(1 - p)/80$
(c) $p(1 - p)/100$

9-73. (a) $\beta = 0.564$
(b) $\beta = 0.161$
(c) $\beta = 0.116$

9-75. (a) 0.61026 (b) 0.995
(c) 0.9988

9-79. (a) $\chi_0^2 = 5.546$, reject H_0
(b) $0.01 <$ P-value < 0.025

9-81. (a) $\chi_0^2 = 1.75$, do not reject H_0

9-83. (a) $\chi_0^2 = 17.929$, reject H_0
(b) P-value $= 0.0123$

9-85. (a) $\chi_0^2 = 63.36$, reject H_0

9-87. (a) $z_0 = -7.32$, reject H_0
(b) P-value $\cong 0$
(c) $\chi_0^2 = 12.0$, reject H_0

9-89. (b) $t_0 = 1.608$, do not reject H_0
(c) $0.1 <$ P-value < 0.2

CHAPTER 10

Section 10-2

10-1. (a) Yes, cannot reject H_0
(b) P-value $= 0.3222$
(c) 0.9967
(d) $(-0.0098, 0.00298)$
(e) 9

10-3. 1, Yes

10-5. (a) (0.0987, 0.2813)
(b) (0.0812, 0.299)
(c) $(-\infty, 0.2813)$

10-7. (a) $(-3.684, -2.116)$
(b) $z_0 - 7.25$ reject H_0
(c) P-value $\cong 0$

10-9. 11

10-11. (a) $(-5.83, -0.57)$ (b) Yes

10-13. $z_0 = -2.385$, reject H_0

10-15. Yes

Section 10-3

10-17. (a) $t_0 = 0.230$, do not reject H_0
(b) P-value > 0.80
(c) $(-0.394, 0.494)$

10-19. (a) $t_0 = -3.11$, reject H_0
(b) $(-5.688, -0.3122)$

10-21. (a) $t_0 = -2.83$, reject H_0
(b) $0.010 <$ P-value < 0.020
(c) (0.111, 0.749)

10-23. (17.175, 44.825)

10-25. $t_0 = -5.498$, P-value < 0.0010

10-27. (a) $t_0 = 3.03$, reject H_0
(b) $0.005 <$ P-value < 0.010
(c) $t_0 = 3.03$, reject H_0

10-29. $(-14.52, 22.12)$

10-31. (b) $t_0 = 2.558$, reject H_0
(c) P-value $\cong 0.020$
(d) 0.05 (e) $n = 51$
(f) (1.86, 18.94)

Section 10-4

10-33. (0.1694, 0.3778)

10-35. $t_0 = 0.357$, cannot reject H_0

10-37. $(-727.46, 2464.21)$

10-39. $t_0 = 5.465$, reject H_0

10-41. $t_0 = 8.387$, reject H_0

10-43. $t_0 = 3.45$, reject H_0

Section 10-5

10-45. (a) 1.59 (b) 2.28 (c) 2.64
(d) 0.529 (e) 0.525
(f) 0.311

10-47. $f_0 = 0.657$, cannot reject H_0

10-49. No

10-51. (a) (0.08775, 3.594)
(b) (0.0585, 5.3)
(c) $(0.137, \infty)$

10-53. $f_0 = 0.297$, cannot reject H_0

10-55. $f_0 = 0.2575$, cannot reject H_0

10-57. (0.3369, 2.640)

10-59. $f_0 = 0.640$, cannot reject H_0

Section 10-6

10-61. $z_0 = 1.49$, cannot reject H_0

10-63. (a) 0.81859 (b) 383

10-65. (a) $z_0 = 3.42$, reject H_0,
P-value $= 0.00062$

10-67. (0.0434, 0.1616)

Supplemental

10-69. (1.40, 8.36)

10-71. (a) $t_0 = 2.554$, reject H_0
(b) $t_0 = 2.554$, cannot reject H_0
(c) $t_0 = -1.986$, cannot reject H_0
(d) $t_0 = -1.986$, cannot
reject H_0

10-73. (a) $z_0 = 6.55$, reject H_0
(b) $z_0 = 6.55$, reject H_0

10-75. (a) $(-0.0335, 0.0329)$
(b) $(-0.0282, 0.0276)$
(c) $(-0.0238, 0.0232)$,
$(-0.0203, 0.0200)$

10-79. 60

10-81. 26

10-83. (a) No (b) Yes
(d) (18.114, 294.35)

10-85. (h) $t_0 = -6.06$, reject H_0

10-87. (b) $t_0 = -0.512$, cannot
reject H_0 (c) 16

10-89. (b) $t_0 = -2.74$, reject H_0
(c) 0.8   (d) 26

CHAPTER 11

Section 11-2

11-1. (a) $\hat{\beta}_0 = 48.013$, $\hat{\beta}_1 = -2.330$
(b) 37.99   (c) 39.39   (d) 6.71
11-3. (a) $\hat{\beta}_0 = 0.4631476$,
$\hat{\beta}_1 = 0.0074902$
(b) $\hat{\beta}_1 = 0.00749$
11-5. (a) $\hat{\beta}_0 = 13.3202$,
$\hat{\beta}_1 = 3.32437$,
$\hat{\sigma}^2 = 8.76775$
(b) 38.253   (c) -2.0273
11-7. (a) $\hat{\beta}_0 = 33.5348$,
$\hat{\beta}_1 = -0.0353971$,
$\hat{\sigma}^2 = 13.392$
(b) 28.226   (c) 1.50048
11-9. (b) $\hat{\beta}_0 = -9.8131$,
$\hat{\beta}_1 = 0.171484$,
$\hat{\sigma}^2 = 1.9818$
(c) 4.76301
11-11. (b) $\hat{\beta}_0 = 0.470467$,
$\hat{\beta}_1 = 20.5673$,
$\hat{\sigma}^2 = 13.81$
(c) 21.038   (d) 1.6629
11-17. $\hat{\beta}_0 = 0$, $\hat{\beta}_1 = 21.031461$

Section 11-5

11-19. (a) $f_0 = 73.95$, P-value $=$
0.000001, reject H_0
(b) $se(\hat{\beta}_1) = 0.0004839$,
$se(\hat{\beta}_0) = 0.04091$
11-21. (a) $t_0 = 8.518$, reject H_0
(b) $f_0 = 72.5563$, reject H_0
(c) $se(\hat{\beta}_1) = 0.3902$,
$se(\hat{\beta}_0) = 2.5717$
(d) $t_0 = 2.179$, reject H_0
11-23. (a) $f_0 = 4.53158$, do not reject
H_0, P-value $= 0.04734$
(b) $se(\hat{\beta}_1) = 0.0166281$,
$se(\hat{\beta}_0) = 2.61396$
(c) $t_0 = 0.87803$, P-value $=$
0.804251, do not reject H_0
(d) $t_0 = 12.8291$, P-value $\cong 0$,
reject H_0
11-25. (a) $f_0 = 53.50$, reject H_0
P-value $= 0.000009$
(b) $se(\hat{\beta}_1) = 0.0256613$,
$se(\hat{\beta}_0) = 2.13526$
(c) $t_0 = -5.709$, reject H_0,
P-value $= 0.00078$
11-27. (a) $f_0 = 155.2$, reject H_0
P-value < 0.00001

(b) $se(\hat{\beta}_1) = 45.3468$,
$se(\hat{\beta}_0) = 2.96681$
(c) $t_0 = -2.3466$, reject H_0.
P-value $= 0.0306$
(d) $t_0 = 57.8957$, reject H_0
P-value < 0.00001
(e) $t_0 = 2.7651$, reject H_0
P-value $= 0.0064$

Section 11-6 and Section 11-7

11-31. (a) $(-2.9173, -1.7423)$
(b) $(46.7145, 49.3115)$
(c) $(41.3293, 43.0477)$
(d) $(38.4289, 46.1281)$
11-33. (a) $(-0.00961, -0.00444)$
(b) $(16.2448, 27.3318)$
(c) $(7.91435, 10.37165)$
(d) $(4.07215, 14.21385)$
11-35. (a) $(9.10130, 9.31543)$
(b) $(-11.6219, -1.04911)$
(c) $(498.72024, 501.52776)$
(d) $(495.57344, 504.67456)$
11-37. (a) $(0.03689, 0.010183)$
(b) $(-47.0877, 14.0691)$
(c) $(44.0897, 49,1185)$
(d) $(37.8298, 55.3784)$
11-39. (a) $(201.552, 226.590)$
(b) $(-4.67015, -2.346960)$
(c) $(111.8339, 145.7941)$
11-41. (a) $(-43.1964, -30.7272)$
(b) $(2530.09, 2720.68)$
(c) $(1823.7833, 1948.5247)$
(d) $(1668.9013, 2103.4067)$

Section 11-8

11-43. (d) $R^2 \cong 76.73\%$
11-45. (a) $R^2 = 20.1121\%$
(c) Yes
11-47. (a) $R^2 = 71.27\%$
11-49. (a) $R^2 = 85.22\%$

Section 11-10

11-55. (a) $\hat{\beta}_0 = -0.0280411$,
$\hat{\beta}_1 = 0.990987$
(b) $f_0 = 79.838$, reject H_0
(c) 0.903   (d) $t_0 = 8.9345$,
reject H_0
(e) $z_0 = 3.879$, reject H_0
(f) $(0.7677, 0.9615)$
11-57. (a) $r = -0.738027$
(b) $t_0 = -5.577$, reject H_0,
P-value $= 0.00000738$
(c) $(-0.871, -0.504)$
(d) $z_0 = -0.394$, do not reject
H_0, P-value $= 0.6936$

11-59. (a) $t_0 = 5.47$, reject H_0,
P-value $\cong 0$
(b) $(0.3358, 0.8007)$ (c) Yes
11-61. (a) $r = 0.933203$
(b) $t_0 = 10.06$, reject H_0
(c) $\hat{\beta}_0 = 0.72538$,
$\hat{\beta}_1 = 0.498081$,
$f_0 = 101.16$, reject H_0
(d) $t_0 = 0.468345$, do not
reject H_0

Supplemental

11-65. (a) $\hat{\beta}_0 = 93.34$, $\hat{\beta}_1 = 15.64$
(b) $f_0 = 12.872$, reject H_0
(c) $(7.961, 23.322)$
(d) $(74.758, 111.923)$
(e) $(125.97, 138.91)$
11-67. (b) $\hat{\beta}_0 = -0.8819$,
$\hat{\beta}_1 = 0.00385$
(c) $f_0 = 122.03$, reject H_0
(d) No.   (e) $\hat{\beta}_0^* = 0.5967$,
$\hat{\beta}_1^* = 0.00097$
11-69. $\hat{y} = 0.7916x$
11-71. (b) $\hat{\beta}_0 = -193$, $\hat{\beta}_1 = 15.296$
(c) $(-4.912, 35.504)$
11-75. (b) $\hat{\beta}_0 = 66$, $\hat{\beta}_1 = 0.930$
(c) $f_0 = 19.79$, reject H_0
$R^2 = 71.2\%$
(d) $t_0 = -0.3354$, can not
reject H_0

CHAPTER 12

Section 12-1

12-1. (b) $\hat{\boldsymbol{\beta}} = \begin{bmatrix} 171.055 \\ 3.713 \\ -1.126 \end{bmatrix}$
(c) 189.49
12-3. (b) 2
12-5. (a) $\hat{y} = 33.4491 - 0.05435x_1 +$
$1.07822x_2$
(b) 8.03   (c) 19.30
12-7. (a) $\hat{y} = 383.80 - 3.6381x_1 -$
$0.1119x_2$
(b) $\hat{\sigma}^2 = 153.0, se(\hat{\beta}_0) = 36.22$,
$se(\hat{\beta}_1) = 0.5665$,
$se(\hat{\beta}_2) = 0.04338$
(c) 180.95
(d) $\hat{y} = 484.0 - 7.656$
$x_1 - 0.222$
$x_2 - 0.0041x_{12}$
(e) $\hat{\sigma}^2 = 147.0, se(\hat{\beta}_0) = 101.3$,
$se(\hat{\beta}_1) = 3.846$,
$se(\hat{\beta}_2) = 0.113$,
$se(\hat{\beta}_{12}) = 0.0039$
(f) 173.1

12-9. (a) $\hat{y} = 47.174 - 9.7352x_1 + 0.4283x_2 + 18.2375x_3$
(b) 12 (c) $se(\hat{\beta}_0) = 49.5815$, $se(\hat{\beta}_1) = 3.6916$, $se(\hat{\beta}_2) = 0.2239$, $se(\hat{\beta}_3) = 1.312$ (d) 91.38

12-11. $\hat{y} = -8.0119 + 0.494x_1 + 0.0018x_2 + 0.0023x_3 + 0.0383x_4 - 0.2068x_5 - 0.0128x_6 + 0.030x_7 + 0.0407x_8 - 0.2083x_9$
$\hat{\sigma}^2 = 2.32$, $se(\hat{\beta}_0) = 16.18$, $se(\hat{\beta}_1) = 0.0481$, $se(\hat{\beta}_2) = 0.0064$, $se(\hat{\beta}_3) = 0.0209$, $se(\hat{\beta}_4) = 0.0515$, $se(\hat{\beta}_5) = 0.2611$, $se(\hat{\beta}_6) = 0.0266$, $se(\hat{\beta}_7) = 0.038$, $se(\hat{\beta}_8) = 0.1483$, $se(\hat{\beta}_9) = 0.1110$

Section 12-2

12-13. (a) 184.25, reject H_0
(b) reject H_0 both significant

12-15. (a) $f_0 = 30.308$, reject H_0
P-value < 0.000001
(b) Reject H_0, all coefficients are significant

12-17. (a) $f_0 = 53.3162$, reject H_0
(b) Only β_1 is significant

12-19. (a) $f_0 = 10.08$, P-value $= 0.005$
(b) Only β_1 is significant

12-21. (a) $f_0 = 7.714$, reject H_0
(b) $f_0 = 1.11$, do not reject H_0
(c) 147.0

12-23. (a) $f_0 = 8.283$, reject H_0
(b) Regression coefficients for x_1 and x_3 are significant

12-25. (a) $f_0 = 101.79$, reject H_0
(b) Only regression coefficient for "PTS" is significant
(c) $\hat{y} = -5.531 + 0.497x_{PTS} - 0.004x_{PPG}$, $f_0 = 510.12$
reject H_0 only regressor "PTS" is significant

Section 12-3 and Section 12-4

12-27. (a) $(-0.00730, -0.00205)$
(b) 0.462 (c) $(7.35, 9.25)$

12-29. (a) 95% CI on coefficients
$-1.9646 \le \beta_2 \le 17.0026$,
$-1.7953 \le \beta_3 \le 6.7613$,
$-1.7941 \le \beta_4 \le 0.8319$

(b) $(272.44, 308.44)$
(c) $(257.25, 323.64)$

12-31. (a) $-0.595 \le \beta_2 \le 0.535$,
$0.229 \le \beta_3 \le 0.812$,
$-0.216 \le \beta_4 \le 0.013$,
$-7.298 \le \beta_5 \le 2.977$
(b) $(7.982, 10.009)$
(c) $(6.8481, 11.143)$

12-33. (a) $-0.00042 \le \beta_{Temp} \le 0.00012$, $0.00203 \le \beta_{soaktime} \le 0.00288$, $-0.02306 \le \beta_{soakpct} \le 0.05976$, $0.00501 \le \beta_{DFtime} \le 0.01056$, $-0.01969 \le \beta_{Diffpct} \le 0.01342$
(b) $(0.0206, 0.0234)$

12-35. (a) $(0.3882, 0.5998)$
(b) $y = -5.767703 + 0.496501x_{Pts}$
(c) $(0.4648, 0.5282)$

Section 12-5

12-37. (a) 0.82897 (d) No
12-39. (a) 0.985 (b) 0.99
12-41. (b) 0.9937
12-43. (a) $R^2 = 0.955$ (c) 32
12-45. (a) 0.12
(b) 17 and 18

Section 12-6

12-47. (a) $\hat{y} = -1.633 + 1.232x - 1.495x^2$
(b) $f_0 = 1858613$, reject H_0
(c) $t_0 = -601.64$, reject H_0

12-49. (a) 802.943
(b) $\hat{y} = -26204.14 + 189.09x - 0.331x^2$

12-51. (a) $\hat{y} = -1.769 + 0.421x_1 + 0.222x_2 - 0.128x_3 - 0.02x_1x_2 + 0.009x_1x_3 + 0.003x_2x_3 - 0.019x_1^2 - 0.007x_2^2 + 0.001x_3^2$
(b) $f_0 = 19.628$, reject H_0
(d) $f_0 = 1.612$, do not reject H_0

12-55. (a) Min. MS_E: $x_1, x_3, x_4, x_5, x_7, x_8, x_{10}$, $MS_E = 6.579$, $C_p = 6.1$, Min. C_p: x_5, x_8, x_{10}, $C_p = 5.02$, $MS_E = 7.97$
(b) $\hat{y} = 34.434 - 0.048x_1$, $MS_E = 8.81$, $C_p = 5.55$
(c) Same as part (b)
(d) $\hat{y} = 0.341 + 2.862x_5 + 0.246x_8 - 0.010x_{10}$, $MS_E = 7.97$, $C_p = 5.02$

12-57. (a) $y = 4.656 + 0.511x_3 - 0.124x_4$
(b) Same as part (a)
(c) Same as part (a)
(d) All models are the same

12-59. (a) $\hat{y} = -0.304 + 0.083x_1 - 0.031x_3 + 0.004x_2^2$, $C_p = 4.04$, $MS_E = 0.004$
(b) $\hat{y} = -0.256 + 0.078x_1 + 0.022x_2 - 0.042x_3 + 0.0008x_3^2$, $C_p = 4.66$, $MS_E = 0.004$

12-61. (a) Min. C_p: x_1, x_9, $C_p = -1.67$
(b) Min. MS_E: x_1, x_7, x_9, $MS_E = 1.67$, $C_p = -0.77$
(c) Max. adjusted R^2: x_1, x_7, x_9, Adj. $R^2 = 0.98448$

Supplemental

12-65. (a) $f_0 = 1321.4$, reject H_0 P-value < 0.00001
(b) Only regressor x_4 is significant H_0

12-67. (a) $\hat{y} = -1.060 + 5.509x + 1.53x_2 - 3.989x_3 - 1.102x_4$
(b) $f_0 = 116.21$, reject H_0, all regressors are significant

12-69. (a) $\hat{y} = -3982.1 + 1.0990x_1 + 0.1831x_3 + 3.741x_4 + 0.8375x_5 - 16.346x_6$, $MS_E(p) = 695.90$, $C_p = 5.67$
(b) Same as model a
(c) Same as models a + b
(d) All of the models come out the same

12-71. $VIF(\hat{\beta}_3^*) = 52.4$, $VIF(\hat{\beta}_4) = 9.3$ $VIF(\hat{\beta}_5) = 29.1$

12-73. (a) $f_0 = 18.28$, reject H_0
(b) $f_0 = 2$, do not reject H_0

CHAPTER 13

Section 13-2

13-1. (a) Reject H_0
(b) Model is satisfactory

13-3. (a) Reject H_0
(b) P-value $\cong 0$

13-5. (a) Reject H_0 (c) $(140.71, 149.29)$, $(7.36, 24.14)$

13-7. (a) Do not reject H_0
(b) P-value $= 0.214$

13-9. (a) Reject H_0
(b) P-value $= 0.002$

(d) $(69.17, 81.81)$

(e) $(8.42, 26.33)$

13-19. 5

Section 13-3

13-21. (a) Reject H_0 (b) 0.01412

(c) 0.0148

13-23. (a) Do not reject H_0 (b) 0

(c) 24

Section 13-4

13-25. (a) Reject H_0

13-27. (a) Do not reject H_0

13-29. (a) Do not reject H_0

Supplemental

13-31. (a) Reject H_0

(c) $(132.97, 147.83)$

13-35. (a) Reject H_0

(b) P-value $= 0.007$

13-37. (a) Reject H_0

13-39. (a) 0.2 (b) 50

CHAPTER 14

Section 14-4

14-1. (a) Reject H_0 for both main effects and the interaction

14-3. (a) Reject H_0 for main effects

14-7. $(-3.40, 7.64)$

14-9. (a) Reject H_0 for both main effects and the interaction

Section 14-5

14-11. (a) All these main effects are significant and the hardwood concentration-freeness interaction is significant at $\alpha - 0.05$. The P-value for the hardwood-cooking time interaction is 0.075, it is possibly an important effect as well.

Section 14-7

14-13. (a) Reject H_0 for factors B, C, and AC

(b) $\hat{y} = 413.125 + 9.125x_1 + 45.12x_2 + 35.87x_3 - 59.62x_1x_3$

14-15. $\hat{y} = 175.25 + 8.5x_1 + 5.44x_3 + 4.19x_4 + 4.56x_1x_4$

14-17. (b) Reject H_0 for factor A

14-19. (a) Factors A, B, C, and AB

14-21. (b) Factors A, B, and AB

(c) $\hat{y} = 400 + 40.124x_1 - 32.75x_2 + 26.625x_1x_2$

Section 14-8

14-23. Block 1: (1) ab ac bc

Block 2: a b c abc

There are no significant factors

14-25. Block 1: (1) ab acd bcd

Block 2: c ad bd abc

Block 3: d ac bc abd

Block 4: a b cd $abcd$

Factor A is significant

14-27. Block 1: (1) ab de acd bcd ace bce $abde$

Block 2: a b cd ce ade bde $abce$ $abcd$

Block 3: d e bc bd abd abe $acde$ $bcde$

Block 4: c ad ae bd be abc cde $abcde$

14-29. (a) Factors A, C, AB, and AC are significant

Section 14-9

14-31. (a) Factors A, B, and D are significant

(c) Factors A, B, D, AB, and AD are significant

14-33. (b) Design Generators: $D = BE$, $E = AC$; Defining Relation: $I = ACE = BDE - ABCDE$; Aliases: $A = BD = CD = ABCDE$, $B = AD = CDE = ABCE$; $C = AE = BDE = ABCD$, $D = AB = BCE = ACDE$, $E = AC = BCD = ABDE$ (c) $A = -1.525$, $B = -5.175$, $C = 2.275$, $D = -0.675$, $E = 2.275$

14-35. 2^{4-1} replicated twice

14-37. Factors A, B, and D are significant

14-39. Design Generators: $D = AB$, $E = AC$, $F = BC$; Defining Relations: $I = ABD = ACE = BCF = BCDE = ACDF = ABEF = DEF$; Aliases: $A = BD = CE$, $B = AD = CF$, $C = AE = BF$, $D = AB = EF$, $E = AC = DF$, $F = BC = DE$, $AB = EF$, $AF = BE = CD$

Supplemental

14-41. The main effect of pH and the interaction of pH and Catalyst Concentration are significant

14-43. The salts, application levels, and the interaction between salts and application levels are significant

14-45. There are no significant factors

14-47. (a) The factors V, P, G, and PG are significant. Effects $P = -10.75$, $V = 15.75$, $G = -25.00$, $PG = 19.25$

(b) $\hat{y} = 102.75 + 7.88x_1 - 5.37x_2 - 12.50x_4 + 9.62x_2x_4$

14-49. None of the factors or interactions is significant with the fractional factorial design.

14-51. Design Generators: $D = \pm AB$, $E = \pm AC$; Defining Relations: $I = ABD = ACE = BCDE$; Aliases: $A = BD = CE$, $B = AD = CDE$, $C = AE = BDE$, $D = AB = BCE$, $E = AC = BCD$, $BC = DE$, $BE = CD$

14-53. (a) $E = ABCD$ (b) Factors A, B, C, E, and interaction BE are significant (c) Factor A is significant in affecting variability

CHAPTER 15

Section 15-2

15-1. Do not reject H_0, P-value $= 0.109$

15-3. Reject H_0, P-value $= 0.0002$

15-5. (a) Do not reject H_0

(b) $z_0 = 0.577$, P-value $= 0.281$

15-7. $z_0 = -1.34$, do not reject H_0, P-value $= 0.1802$

15-9. Do not reject H_0

15-11. $z_0 = 2.83$, reject H_0

15-13. Reject H_0

15-15. $z_0 = 2.84$, reject H_0

15-17. (a) 0.025 (b) 0.115

(c) 0.011 (d) 0.1587

15-19. P-value $= 0.0075$

Section 15-3

15-21. $w = 80.5 > 52$, do not reject H_0

15-23. $w = 5 < 65$, reject H_0

15-25. $w = 27 < 27$, reject H_0

15-27. $w = 1 < 25$, reject H_0

Section 15-4

15-29. $w = 38 > 23$, do not reject H_0

15-31. $z_0 = -0.58$, do not reject H_0, P-value $= 0.5619$

15-33. $w = 73 < 78$, reject H_0

15-35. $z_0 = -0.242$, reject H_0, P-value $= 0.0155$

Section 15-5

15-37. Reject H_0

15-39. Do not reject H_0

15-41. P-value $= 0.018$

Supplemental

15-43. Do not reject H_0, P-value $\cong 1$

15-45. Do not reject H_0

15-47. Do not reject H_0

15-49. Reject H_0

15-51. Reject H_0

15-53. Reject H_0

15-55. Reject H_0, P-value $\cong 0$

15-57. Reject H_0, P-value $= 0.009$

CHAPTER 16

Section 16-5

16-1. (a) $\bar{x}$ chart: $UCL = 37.5789$, $CL = 34.32$, $LCL = 31.0611$ R chart: $UCL = 11.9461$, $CL = 5.65$, $LCL = 0$
(b) 1 point outside limits on $\bar{x}$ chart. Revised limits: $\bar{x}$ chart: $UCL = 37.4038$, $CL = 34.0947$, $LCL = 30.7857$, R chart: $UCL = 12.1297$, $CL = 5.7368$, $LCL = 0$

16-3. (a) $\bar{x}$ chart: $UCL = 17.40$, $CL = 15.09$, $LCL = 12.79$ R chart: $UCL = 5.792$, $CL = 2.25$, $LCL = 0$
(b) $\bar{x}$ chart: $UCL = 17.96$, $CL = 15.78$, $LCL = 16.62$ R chart: $UCL = 5.453$, $CL = 2.118$, $LCL = 0$
(c) $\bar{x}$ chart: $UCL = 17.42$, $CL = 15.09$, $LCL = 12.77$ s chart: $UCL = 3.051$, $CL = 1.1188$, $LCL = 0$ revised limits: $\bar{x}$ chart: $UCL = 17.95$, $CL = 15.78$, $LCL = 13.62$, s chart:

$UCL = 2.848$, $CL = 1.109$, $LCL = 0$

16-5. (a) $\bar{x}$ chart: $UCL = 242.78$, $CL = 223$, $LCL = 203.22$ R chart: $UCL = 72.51$, $CL = 34.286$, $LCL = 0$
(b) $\hat{\mu} = 223$, $\hat{\sigma} = 14.74$

16-7. (a) $\bar{x}$ chart: $UCL = 0.06347$, $CL = 0.06294$, $LCL = 0.0624$ R chart: $UCL = 0.001954$, $CL = 0.000924$, $LCL = 0$
(b) $\bar{x}$ chart: $UCL = 0.06346$, $CL = 0.06295$, $LCL = 0.06241$ s chart: $UCL = 0.000766$, $CL = 0.000367$, $LCL = 0$
(c) Remove 5 from the list of out of control points

Section 16-6

16-9. (a) I chart: $UCL = 60.8887$, $CL = 53.05$, $LCL = 45.2113$ MR chart: $UCL = 9.63382$, $CL = 2.94737$, $LCL = 0$
(b) $\hat{\mu} = 53.05$, $\hat{\sigma} = 2.613$

16-11. (a) I chart: $UCL = 10.5358$, $CL = 10.0272$, $LCL = 9.51856$ MR chart: $UCL = 0.625123$, $CL = 0.19125$, $LCL = 0$
(b) $\hat{\mu} = 10.027$, $\hat{\sigma} = 0.1696$

Section 16-7

16-13. (a) $PCR = PCR_k = 1.5$
(b) 0

16-15. (a) 0.00075 (b) $PCR = 1.13$, $PCR_k = 1.104$

16-17. (a) $PCR = PCR_k = 1.18$
(b) 0.00046

16-19. (a) 0.0009 (b) $PCR = 1.13$, $PCR_k = 1.06$

16-21. $PCR = 0.50$, $PCR_k = 0.357$

16-23. $PCR = 0.49$, $PCR_k = 0.474$

Section 16-8

16-25. (a) U chart: $UCL = 3.811$, $CL = 1.942$, $LCL = 0.0722$
(b) Revised limits: U chart: $UCL = 3.463$, $CL = 1.709$, $LCL = 0$

16-27. (c) chart: $UCL = 19.06$, $CL = 9.708$, $LCL = 0.3609$

Section 16-9

16-29. (a) 0.2177 (b) 4.6

16-31. (a) 0.1020 (b) 9.8

16-33. (a) 0.3936 (b) 2.54

16-35. (a) 0.1515 (b) 6.6

16-37. (a) 0.16603 (b) 6.02

Section 16-10

16-39. (a) $\hat{\sigma} = 0.174$

16-41. (a) ARL $= 38.0$
(b) ARL $= 10.4$

Supplemental

16-43. (a) $\bar{x}$ chart: $UCL = 64.0181$, $CL = 64.0$, $LCL = 63.982$ R chart: $UCL = 0.0453972$, $CL = 0.01764$, $LCL = 0$
(b) $\hat{\mu} = 64$, $\hat{\sigma} = 0.0104$
(c) $PCR = 0.641$
(d) $PCR_k = 0.641$
(e) $\sigma^2 = 0.0000111$
(f) 0.1705, ARL $= 5.87$

16-45. (a) p chart: $UCL = 0.20387$, $CL = 0.11$, $LCL = 0.01613$
(b) p chart: $UCL = 0.1717$, $CL = 0.106$, $LCL = 0.04092$

16-47. (a) c chart: $UCL = 7.51442$, $CL = 2.64$, $LCL = 0$
(b) c chart: $UCL = 6.50924$, $CL = 2.1304$, $LCL = 0$

16-49. (b) 6.30 (c) 2

16-51. (a) $\bar{x}$ chart: $UCL = 140.417$, $CL = 139.709$, $LCL = 139.001$ R chart: $UCL = 2.596$, $CL = 1.227$, $LCL = 0$
(b) Revised control limits: $\bar{x}$ chart: $UCL = 140.518$, $CL = 139.808$, $LCL = 139.098$ R chart: $UCL = 2.6023$, $CL = 1.237$, $LCL = 0$
(c) $PCR = 1.26$, $PCR_k = 1.08$
(d) $\sigma^2 = 0.0081$
(e) 0.1803, ARL $= 5.55$

16-53. (a) 0.96995 (b) 1

16-57. 0.000135

Glossary

Acceptance region. In hypothesis testing, a region in the sample space of the test statistic such that if the test statistic falls within it, the null hypothesis is accepted (better terminology is that the null hypothesis cannot be rejected, since rejection is always a strong conclusion and acceptance is generally a weak conclusion).

Addition rule. A formula used to determine the probability of the union of two (or more) events from the probabilities of the events and their intersection(s).

Additivity property of χ^2. If two independent random variables X_1 and X_2 are distributed as chi-square with v_1 and v_2 degrees of freedom respectively, $Y = X_1 + X_2$ is a chi-square random variable with $u = v_1 + v_2$ degrees of freedom. This generalizes to any number of independent chi-square random variables.

Adjusted R^2. A variation of the R^2 statistic that compensates for the number of parameters in a regression model. Essentially, the adjustment is a penalty for increasing the number of parameters in the model.

Alias. In a fractional factorial experiment when certain factor effects cannot be estimated uniquely, they are said to be aliased.

All possible (subsets) regressions. A method of variable selection in regression that examines all possible subsets of the candidate regressor variables. Efficient computer algorithms have been developed for implementing all possible regressions.

Alternative hypothesis. In statistical hypothesis testing, this is a hypothesis other than the one that is being tested. The alternative hypothesis contains feasible conditions, whereas the null hypothesis specifies conditions that are under test.

Analysis of variance. A method of decomposing the total variability in a set of observations, as measured by the sum of the squares of these observations from their average, into component sums of squares that are associated with specific defined sources of variation.

Analytic study. A study in which a sample from a population is used to make inference to a future population. Stability needs to be assumed. See enumerative study.

Arithmetic mean. The arithmetic mean of a set of numbers $x_1, x_2, \ldots, x_n$ is their sum divided by the number of observations, or $(1/n)\sum_{i=1}^{n} x_i$. The arithmetic mean is usually denoted by $\bar{x}$, and is often called the average.

Assignable cause. The portion of the variability in a set of observations that can be traced to specific causes, such as operators, materials, or equipment. Also called a special cause.

Attribute. A qualitative characteristic of an item or unit, usually arising in quality control. For example, classifying production units as defective or nondefective results in attributes data.

Attribute control chart. Any control chart for a discrete random variable. See variables control charts.

Average. See Arithmetic Mean.

Average run length, or ARL. The average number of samples taken in a process monitoring or inspection scheme until the scheme signals that the process is operating at a level different from the level in which it began.

Axioms of probability. A set of rules that probabilities defined on a sample space must follow. See probability.

Backward elimination. A method of variable selection in regression that begins with all of the candidate regressor variables in the model and eliminates the insignificant regressors one at a time until only significant regressors remain.

Bayes' theorem. An equation for a conditional probability such as $P(A \mid B)$ in terms of the reverse conditional probability $P(B \mid A)$.

Bernoulli trials. Sequences of independent trials with only two outcomes, generally called "success" and "failure," in which the probability of success remains constant.

Bias. An effect that systematically distorts a statistical result or estimate, preventing it from representing the true quantity of interest.

Biased estimator. *See* Unbiased estimator.

Bimodal distribution. A distribution with two modes.

Binomial random variable. A discrete random variable that equals the number of successes in a fixed number of Bernoulli trials.

Bivariate normal distribution. The joint distribution of two normal random variables.

Block. In experimental design, a group of experimental units or material that is relatively homogeneous. The purpose of dividing experimental units into blocks is to produce an experimental design wherein variability within blocks is smaller than variability between blocks. This allows the factors of interest to be compared in a environment that has less variability than in an unblocked experiment.

Box plot (or box and whisker plot). A graphical display of data in which the box contains the middle 50% of the data (the interquartile range) with the median dividing it, and the whiskers extend to the smallest and largest values (or some defined lower and upper limits).

C chart. An attribute control chart that plots the total number of defects per unit in a subgroup. Similar to a defects-per-unit or U chart.

Categorical data. Data consisting of counts or observations that can be classified into categories. The categories may be descriptive.

Causal variable. When $y = f(x)$ and y is considered to be caused by x, x is sometimes called a causal variable.

Cause-and-effect diagram. A chart used to organize the various potential causes of a problem. Also called a fishbone diagram.

Center line. A horizontal line on a control chart at the value that estimates the mean of the statistic plotted on the chart.

Center line. *See* Control chart.

Central composite design (CCD). A second-order response surface design in k variables consisting of a two-level factorial, $2k$ axial runs, and one or more center points. The two-level factorial portion of a CCD can be a fractional factorial design when k is large. The CCD is the most widely used design for fitting a second-order model.

Central limit theorem. The simplest form of the central limit theorem states that the sum of n independently distributed random variables will tend to be normally distributed as n becomes large. It is a necessary and sufficient condition that none of the variances of the individual random variables are large in comparison to their sum. There are more general forms of the central theorem that allow infinite variances and correlated random variables, and there is a multivariate version of the theorem

Central tendency. The tendency of data to cluster around some value. Central tendency is usually expressed by a measure of location such as the mean, median, or mode.

Chance cause of variation. The portion of the variability in a set of observations that is due to only random forces and which cannot be traced to specific sources, such as operators, materials, or equipment. Also called a common cause.

Chebyshev's inequality. A result that provides bounds for certain probabilities for arbitrary random variables.

Chi-square (or chi-squared) random variable. A continuous random variable that results from the sum of squares of independent standard normal random variables. It is a special case of a gamma random variable.

Chi-squared test. Any test of significance based on the chi-square distribution. The most common chi-square tests are (1) testing hypotheses about the variance or standard deviation of a normal distribution and (2) testing goodness of fit of a theoretical distribution to sample data.

Coefficient of determination. *See* R^2.

Completely randomized design. A type of experimental design in which the treatments or design factors are assigned to the experimental units in a random manner. In designed experiments, a completely randomized design results from running all of the treatment combinations in random order.

Components of variance. The individual components of the total variance that are attributable to specific sources. This usually refers to the individual variance components arising from a random or mixed model analysis of variance.

Conditional mean. The mean of the conditional probability distribution of a random variable.

Conditional probability. The probability of an event given that the random experiment produces an outcome in another event.

Conditional probability density function. The probability density function of the conditional probability distribution of a continuous random variable.

Conditional probability distribution. The distribution of a random variable given that the random experiment produces an outcome in an event. The given event might specify values for one or more other random variables.

Conditional probability mass function. The probability mass function of the conditional probability distribution of a discrete random variable.

Conditional variance. The variance of the conditional probability distribution of a random variable.

Confidence coefficient. The probability $1 - \alpha$ associated with a confidence interval expressing the probability that the stated interval will contain the true parameter value.

Confidence interval. If it is possible to write a probability statement of the form

$$P(L \leq \theta \leq U) = 1 - \alpha$$

where L and U are functions of only the sample data and θ is a parameter, then the interval between L and U is called a confidence interval (or a $100(1 - \alpha)\%$ confidence interval). The interpretation is that a statement that the parameter θ lies in this interval will be true $100(1 - \alpha)\%$ of the times that such a statement is made.

Confidence level. Another term for the confidence coefficient.

Confounding. When a factorial experiment is run in blocks and the blocks are too small to contain a complete replicate of the experiment, one can run a fraction of the replicate in each block, but this results in losing information on some effects. These effects are linked with or confounded with the blocks. In general, when two factors are varied such that their individual effects cannot be determined separately, their effects are said to be confounded.

Consistent estimator. An estimator that converges in probability to the true value of the estimated parameter as the sample size increases.

Contingency table. A tabular arrangement expressing the assignment of members of a data set according to two or more categories or classification criteria.

Continuity correction. A correction factor used to improve the approximation to binomial probabilities from a normal distribution.

Continuous distribution. A probability distribution for a continuous random variable.

Continuous random variable. A random variable with an interval (either finite or infinite) of real numbers for its range.

Continuous uniform random variable. A continuous random variable with range of a finite interval and a constant probability density function.

Contour plot. A two-dimensional graphic used for a bivariate probability density function that displays curves for which the probability density function is constant.

Control chart. A graphical display used to monitor a process. It usually consists of a horizontal center line corresponding to the in-control value of the parameter that is being monitored and lower and upper control limits. The control limits are determined by statistical criteria and are not arbitrary nor are they related to specification limits. If sample points fall within the control limits, the process is said to be in-control, or free from assignable causes. Points beyond the control limits indicate an out-of-control process; that is, assignable causes are likely present. This signals the need to find and remove the assignable causes.

Control limits. *See* Control chart.

Convolution. A method to derive the probability density function of the sum of two independent random variables from an integral (or sum) of probability density (or mass) functions.

Cook's distance. In regression, Cook's distance is a measure of the influence of each individual observation on the estimates of the regression model parameters. It expresses the distance that the vector of model parameter estimates with the ith observation removed lies from the vector of model parameter estimates based on all observations. Large values of Cook's distance indicate that the observation is influential.

Correction factor. A term used for the quantity $(1/n)(\sum_{i=1}^{n} x_i)^2$ that is subtracted from $\sum_{i=1}^{n} x_i^2$ to give the corrected sum of squares defined as $(1/n)\sum_{i=1}^{n}(x_i - \bar{x})^2$. The correction factor can also be written as $n\bar{x}^2$.

Correlation. In the most general usage, a measure of the interdependence among data. The concept may include more than two variables. The term is most commonly used in a narrow sense to express the relationship between quantitative variables or ranks.

Correlation coefficient. A dimensionless measure of the interdependence between two variables, usually lying in the interval from -1 to $+1$, with zero indicating the absence of correlation (but not necessarily the independence of the two variables). The most common form of the correlation coefficient used in practice is

$$r = \sum_{i=1}^{n}[(y_i - \bar{y})(x_i - \bar{x})]/\sqrt{\sum_{i=1}^{n}(y_i - \bar{y})^2 \sum_{i=1}^{n}(x_i - \bar{x})^2}$$

which is also called the product moment correlation coefficient. It is a measure of the linear association between the two variables y and x.

Correlation matrix. A square matrix that contains the correlations among a set of random variables, say X_1, $X_2, \ldots, X_k$. The main diagonal elements of the matrix are unity and the off diagonal elements r_{ij} are the correlations between X_i and X_j.

Counting techniques. Formulas used to determine the number of elements in sample spaces and events.

Covariance. A measure of association between two random variables obtained as the expected value of the product of the two random variables around their means; that is, $\text{Cov}(X, Y) = E[(X - \mu_X)(Y - \mu_Y)]$.

Covariance matrix. A square matrix that contains the variances and covariances among a set of random variables, say $X_1, X_2, \ldots, X_k$. The main diagonal elements of the matrix are the variances of the random variables and the off diagonal elements are the covariances between X_i and X_j. Also called the variance-covariance matrix. When the random variables are standardized to have unit variances, the covariance matrix becomes the correlation matrix.

Critical region. In hypothesis testing, this is the portion of the sample space of a test statistic that will lead to rejection of the null hypothesis.

Critical value(s). The value of a statistic corresponding to a stated significance level as determined from the sampling distribution. For example, if $P(Z \geq z_{0.05}) = P(Z \geq 1.96) = 0.05$, then $z_{0.05} = 1.96$ is the critical value of z at the 0.05 level of significance.

Crossed factors. Another name for factors that are arranged in a factorial experiment.

Cumulative distribution function. For a random variable X, the function of X defined as $P(X \leq x)$ that is used to specify the probability distribution.

Cumulative normal distribution function. The cumulative distribution of the standard normal distribution, often denoted as $\Phi(x)$ and tabulated in Appendix Table II.

Cumulative sum control chart (CUSUM). A control chart in which the point plotted at time t is the sum of the measured deviations from target for all statistics up to time t.

Curvilinear regression. An expression sometimes used for nonlinear regression models or polynomial regression models.

Decision interval. A parameter set in a Tabular CUSUM algorithm that is determined from a trade-off between false alarms and the detection of assignable causes.

Defect. Used in statistical quality control, a defect is a particular type of nonconformance to specifications or requirements. Sometimes defects are classified into types, such as appearance defects and functional defects.

Defects-per-unit control chart. See U chart.

Degrees of freedom. The number of independent comparisons that can be made among the elements of a sample. The term is analogous to the number of degrees of freedom for an object in a dynamic system, which is the number of independent coordinates required to determine the motion of the object.

Deming. W. Edwards Deming (1900–1993) was a leader in the use of statistical quality control.

Deming's 14 points. A management philosophy promoted by W. Edwards Deming that emphasizes the importance of change and quality.

Density function. Another name for a probability density function.

Dependent variable. The response variable in regression or a designed experiment.

Discrete distribution. A probability distribution for a discrete random variable.

Discrete random variable. A random variable with a finite (or countably infinite) range.

Discrete uniform random variable. A discrete random variable with a finite range and constant probability mass function.

Dispersion. The amount of variability exhibited by data.

Distribution free methods(s). Any method of inference (hypothesis testing or confidence interval construction) that does not depend on the form of the underlying distribution of the observations. Sometimes called nonparametric method(s).

Distribution function. Another name for a cumulative distribution function.

Efficiency. A concept in parameter estimation that uses the variances of different estimators; essentially, an estimator is more efficient than another estimator if it has smaller variance. When estimators are biased, the concept requires modification.

Enumerative study. A study in which a sample from a population is used to make inference to the population. *See* Analytic study.

Erlang random variable. A continuous random variable that is the sum of a fixed number of independent, exponential random variables.

β-error (or β-risk). In hypothesis testing, an error incurred by failing to reject a null hypothesis when it is actually false (also called a type II error).

α-error (or α-risk). In hypothesis testing, an error incurred by rejecting a null hypothesis when it is actually true (also called a type I error).

Error mean square. The error sum of squares divided by its number of degrees of freedom.

Error of estimation. The difference between an estimated value and the true value.

Error sum of squares. In analysis of variance, this is the portion of total variability that is due to the random component in the data. It is usually based on replication of observations at certain treatment combinations in the experiment. It is sometimes called the residual sum of squares, although this is really a better term to use only when the sum of squares is based on the remnants of a model fitting process and not on replication.

Error variance. The variance of an error term or component in a model.

Estimate (or point estimate). The numerical value of a point estimator.

Estimator (or point estimator). A procedure for producing an estimate of a parameter of interest. An estimator is usually a function of only sample data values, and when these data values are available, it results in an estimate of the parameter of interest.

Event. A subset of a sample space.

Exhaustive. A property of a collection of events that indicates that their union equals the sample space.

Expected value. The expected value of a random variable X is its long-term average or mean value. In the continuous case, the expected value of X is $E(X) = \int_{-\infty}^{\infty} x f(x) \, dx$ where $f(x)$ is the density function of the random variable X.

Exponential random variable. A continuous random variable that is the time between counts in a Poisson process.

Factorial experiment. A type of experimental design in which every level of one factor is tested in combination with every level of another factor. In general, in a factorial experiment, all possible combinations of factor levels are tested.

F-distribution. The distribution of the random variable defined as the ratio of two independent chi-square random variables each divided by their number of degrees of freedom.

Finite population correction factor. A term in the formula for the variance of a hypergeometric random variable.

First-order model. A model that contains only first-order terms. For example, the first-order response surface model in two variables is $y = \beta_0 + \beta_1 x_1 + \beta_2 x_2 + \epsilon$. A first-order model is also called a main effects model.

Fixed factor (or fixed effect). In analysis of variance, a factor or effect is considered fixed if all the levels of interest for that factor are included in the experiment. Conclusions are then valid about this set of levels only, although when the factor is quantitative, it is customary to fit a model to the data for interpolating between these levels.

Forward selection. A method of variable selection in regression, where variables are inserted one at a time into the model until no other variables that contribute significantly to the model can be found.

Fraction defective control chart. See P chart.

Fraction defective. In statistical quality control, that portion of a number of units or the output of a process that is defective.

Fractional factorial. A type of factorial experiment in which not all possible treatment combinations are run. This is usually done to reduce the size of an experiment with several factors.

Frequency distribution. An arrangement of the frequencies of observations in a sample or population according to the values that the observations take on.

F-test. Any test of significance involving the F-distribution. The most common F-tests are (1) testing hypotheses about the variances or standard deviations of two independent normal distributions, (2) testing hypotheses about treatment means or variance components in the analysis of variance, and (3) testing significance of regression or tests on subsets of parameters in a regression model.

Gamma function. A function used in the probability density function of a gamma random variable that can be considered to extend factorials.

Gamma random variable. A random variable that generalizes an Erlang random variable to noninteger values of the parameter r.

Gaussian distribution. Another name for the normal distribution, based on the strong connection of Karl F. Gauss to the normal distribution; often used in physics and electrical engineering applications.

Generating function. A function that is used to determine properties of the probability distribution of a random variable. *See* Moment generating function.

Geometric mean. The geometric mean of a set of n positive data values is the nth root of the product of the data values; that is $\bar{g} = (\prod_{i=1}^{n} x_i)^{1/n}$.

Geometric random variable. A discrete random variable that is the number of Bernoulli trials until a success occurs.

Goodness of fit. In general, the agreement of a set of observed values and a set of theoretical values that depend on some hypothesis. The term is often used in fitting a theoretical distribution to a set of observations.

Harmonic mean. The harmonic mean of a set of data values is the reciprocal of the arithmetic mean of the reciprocals of the data values; that is, $\bar{h} = \left(\frac{1}{n} \sum_{i=1}^{n} \frac{1}{x_i} \right)^{-1}$.

Hat matrix. In multiple regression, the matrix $\mathbf{H} = \mathbf{X}(\mathbf{X}'\mathbf{X})^{-1}\mathbf{X}'$. This a projection matrix that maps the vector of observed response values into a vector of fitted values by $\hat{\mathbf{y}} = \mathbf{X}(\mathbf{X}'\mathbf{X})^{-1}\mathbf{X}'\mathbf{y} = \mathbf{H}\mathbf{y}$.

Histogram. A univariate data display that uses rectangles proportional in area to class frequencies to visually exhibit features of data such as location, variability, and shape.

Hypergeometric random variable. A discrete random variable that is the number of success obtained from a sample drawn without replacement from a finite populations.

Hypothesis (as in statistical hypothesis). A statement about the parameters of a probability distribution or a model, or a statement about the form of a probability distribution.

Hypothesis testing. Any procedure used to test a statistical hypothesis.

Independence. A property of a probability model and two (or more) events that allows the probability of the intersection to be calculated as the product of the probabilities.

Independent random variables. Random variables for which $P(X \in A, Y \in B) = P(X \in A)P(Y \in B)$ for any sets A and B in the range of X and Y, respectively. There are several equivalent descriptions of independent random variables.

Independent variable. The predictor or regressor variables in a regression model.

Indicator variable(s). Variables that are assigned numerical values to identify the levels of a qualitative or categorical response. For example, a response with two categorical levels (yes and no) could be represented with an indicator variable taking on the values 0 and 1.

Individuals control chart. A Shewhart control chart in which each plotted point is an individual measurement, rather than a summary statistic. See control chart, Shewhart control chart.

Interaction. In factorial experiments, two factors are said to interact if the effect of one variable is different at different levels of the other variables. In general, when variables operate independently of each other, they do not exhibit interaction.

Intercept. The constant term in a regression model.

Interquartile range. The difference between the third and first quartiles if a sample of data. The interquartile range is less sensitive to extreme data values than the usual sample range.

Interval estimation. The estimation of a parameter by a range of values between lower and upper limits, in contrast to point estimation, where the parameter is estimated by a single numerical value. A confidence interval is a typical interval estimation procedure.

Jacobian. A matrix of partial derivatives that is used to determine the distribution of transformed random variables.

Joint probability density function. A function used to calculate probabilities for two or more continuous random variables.

Joint probability distribution. The probability distribution for two or more random variables in a random experiment. *See* Joint probability mass function and Joint probability density function.

Joint probability mass function. A function used to calculate probabilities for two or more discrete random variables.

Kurtosis. A measure of the degree to which a unimodal distribution is peaked.

Lack of memory property. A property of a Poisson process. The probability of a count in an interval depends only on the length of the interval (and not on the starting point of the interval). A similar property holds for a series of Bernoulli trials. The probability of a success in a specified number of trials depends only on the number of trials (and not on the starting trial).

Least significance difference test (or Fisher's LSD test). An application of the *t*-test to compare pairs of means following rejection of the null hypothesis in an analysis of variance. The error rate is difficult to calculate exactly because the comparisons are not all independent.

Least squares (method of). A method of parameter estimation in which the parameters of a system are estimated by minimizing the sum of the squares of the differences between the observed values and the fitted or predicted values from the system.

Least squares estimator. Any estimator obtained by the method of least squares.

Level of significance. If Z is the test statistic for a hypothesis, and the distribution of Z when the hypothesis is true are known, then we can find the probabilities $P(Z \leq z_L)$ and $P(Z \geq z_U)$. Rejection of the hypothesis is usually expressed in terms of the observed value of Z falling outside the interval from z_L to z_U. The probabilities $P(Z \leq z_L)$ and $P(Z \geq z_U)$ are usually chosen to have small values, such as 0.01, 0.025, 0.05, or 0.10, and are called levels of significance. The actual levels chosen are somewhat arbitrary and are often expressed in percentages, such as a 5% level of significance.

Likelihood function. Suppose that the random variables $X_1, X_2, \ldots, X_n$ have a joint distribution given by $f(x_1, x_2, \ldots, x_n; \theta_1, \theta_2, \ldots, \theta_p)$ where the θs are unknown parameters. This joint distribution, considered as a function of the θs for fixed x's, is called the likelihood function.

Likelihood principle. This principle states that the information about a model given by a set of data is completely contained in the likelihood.

Likelihood ratio. Let $x_1, x_2, \ldots, x_n$ be a random sample from the population $f(x; \theta)$. The likelihood function for this sample is $L = \prod_{i=1}^{n} f(x_i; \theta)$. We wish to test the hypothesis $H_0: \theta \in \omega$, where ω is a subset of the possible values Ω for θ. Let the maximum value of L with respect to θ over the entire set of values that the parameter can take on be denoted by $L(\hat{\Omega})$, and let the maximum value of L with θ restricted to the set of values given by ω be $L(\hat{\omega})$. The null hypothesis is tested by using the likelihood ratio $\lambda = L(\hat{\omega})/L(\hat{\Omega})$, or a simple function of it. Large values of the likelihood ratio are consistent with the null hypothesis.

Likelihood ratio test. A test of a null hypothesis versus an alternative hypothesis using a test statistic derived from a likelihood ratio.

Linear combination. A random variable that is defined as a linear function of several random variables.

Linear model. A model in which the observations are expressed as a linear function of the unknown parameters. For example, $y = \beta_0 + \beta_1 x + \epsilon$ and $y = \beta_0 + \beta_1 x + \beta_2 x^2 + \epsilon$ are linear models.

Location parameter. A parameter that defines a central value in a sample or a probability distribution. The mean and the median are location parameters.

Lognormal random variable. A continuous random variable with probability distribution equal to that of $\exp(W)$ for a normal random variable W.

Main effect. An estimate of the effect of a factor (or variable) that independently expresses the change in response due to a change in that factor, regardless of other factors that may be present in the system.

Marginal probability density function. The probability density function of a continuous random variable obtained from the joint probability distribution of two or more random variables.

Marginal probability distribution. The probability distribution of a random variable obtained from the joint probability distribution of two or more random variables.

Marginal probability mass function. The probability mass function of a discrete random variable obtained from the joint probability distribution of two or more random variables.

Maximum likelihood estimation. A method of parameter estimation that maximizes the likelihood function of a sample.

Mean. The mean usually refers either to the expected value of a random variable or to the arithmetic average of a set of data.

Mean square. In general, a mean square is determined by dividing a sum of squares by the number of degrees of freedom associated with the sum of squares.

Mean square(d) error. The expected squared deviation of an estimator from the true value of the parameter it estimates. The mean square error can be decomposed into the variance of the estimator plus the square of the bias; that is, $MSE(\hat{\Theta}) = E(\hat{\Theta} - \theta)^2 = V(\hat{\Theta}) + [E(\hat{\Theta}) - \theta]^2$.

Median. The median of a set of data is that value that divides the data into two equal halves. When the number of observations is even, say $2n$, it is customary to define the median as the average of the nth and $(n + 1)$st rank-ordered values. The median can also be defined for a random variable. For example, in the case of a continuous random variable X, the median M can be defined as $\int_{-\infty}^{M} f(x)\,dx = \int_{M}^{\infty} f(x)\,dx = 1/2$.

Method of steepest ascent. A technique that allows an experimenter to move efficiently towards a set of optimal operating conditions by following the gradient direction. The method of steepest ascent is usually employed in conjunction with fitting a first-order response surface and deciding that the current region of operation is inappropriate.

Mixed model. In an analysis of variance context, a mixed model contains both random and fixed factors.

Mode. The mode of a sample is that observed value that occurs most frequently. In a probability distribution $f(x)$ with continuous first derivative, the mode is a value of x for which $df(x)/dx = 0$ and $d^2f(x)/dx^2 < 0$. There may be more than one mode of either a sample or a distribution.

Moment (or population moment). The expected value of a function of a random variable such as $E(X - c)^r$ for constants c and r. When $c = 0$, it is said that the moment is about the origin. *See* Moment generating function.

Moment estimator. A method of estimating parameters by equating sample moments to population moments. Since the population moments will be functions of the unknown parameters, this results in equations that may be solved for estimates of the parameters.

Moment generating function. A function that is used to determine properties (such as moments) of the probability distribution of a random variable. It is the expected value of $\exp(tX)$. See generating function and moment.

Moving range. The absolute value of the difference between successive observations in time-ordered data. Used to estimate chance variation in an individuals control chart.

Multicollinearity. A condition occurring in multiple regression where some of the predictor or regressor variables are nearly linearly dependent. This condition can lead to instability in the estimates of the regression model parameters.

Multinomial distribution. The joint probability distribution of the random variables that count the number of results in each of k classes in a random experiment with a series of independent trials with constant probability of each class on each trial. It generalizes a binomial distribution.

Multiplication rule. For probability, A formula used to determine the probability of the intersection of two (or more) events. For counting techniques, a formula used to determine the numbers of ways to complete an operation from the number of ways to complete successive steps.

Mutually exclusive events. A collection of events whose intersections are empty.

Natural tolerance limits. A set of symmetric limits that are three times the process standard deviation from the process mean.

Negative binomial random variable. A discrete random variable that is the number of trials until a specified number of successes occur in Bernoulli trials.

Nonlinear regression model. A regression model that is nonlinear in the parameters. It is sometimes applied to regression models that are nonlinear in the regressors or predictors, but this is an incorrect usage.

Nonparametric statistical method(s). *See* Distribution free method(s).

Normal approximation. A method to approximate probabilities for binomial and Poisson random variables.

Normal equations. The set of simultaneous linear equations arrived at in parameter estimation using the method of least squares.

Normal probability plot. A specially constructed plot for a variable x (usually on the abscissa) in which

y (usually on the ordinate) is scaled so that the graph of the normal cumulative distribution is a straight line.

Normal random variable. A continuous random variable that is the most important one in statistics because it results from the central limit theorem. *See* Central limit theorem.

NP chart. An attribute control chart that plots the total of defective units in a subgroup. Similar to a fraction-defective chart or *P* chart.

Nuisance factor. A factor that probably influences the response variable, but which is of no interest in the current study. When the levels of the nuisance factor can be controlled, blocking is the design technique that is customarily used to remove its effect.

Null hypothesis. This term generally relates to a particular hypothesis that is under test, as distinct from the alternative hypothesis (which defines other conditions that are feasible but not being tested). The null hypothesis determines the probability of type I error for the test procedure.

One-way model. In an analysis of variance context, this involves a single variable or factor with a different levels.

Operating characteristic curves (OC curves). A plot of the probability of type II error versus some measure of the extent to which the null hypothesis is false. Typically, one OC curve is used to represent each sample size of interest.

Orthogonal. There are several related meanings, including the mathematical sense of perpendicular, two variables being said to be orthogonal if they are statistically independent, or in experimental design where a design is orthogonal if it admits statistically independent estimates of effects.

Orthogonal design. *See* Orthogonal.

Outcome. An element of a sample space.

Outlier(s). One or more observations in a sample that are so far from the main body of data that they give rise to the question that they may be from another population.

Overcontrol. Unnecessary adjustments made to processes that increase the deviations from target.

Overfitting. Adding more parameters to a model than is necessary.

P chart. An attribute control chart that plots the proportion of defective units in a subgroup. Also called a fraction-defective control chart. Similar to an NP chart.

Parameter estimation. The process of estimating the parameters of a population or probability distribution. Parameter estimation, along with hypothesis testing, is one of the two major techniques of statistical inference.

Parameter. An unknown quantity that may vary over a set of values. Parameters occur in probability distributions and in statistical models, such as regression models.

Pareto diagram. A bar chart used to rank the causes of a problem.

PCR. A process capability ratio with numerator equal to the difference between the product specification limits and denominator equal to six times the process standard deviation. Said to measure the potential capability of the process because the process mean is not considered. See process capability, process capability ratio, process capability study and PCR_k. Sometimes denoted as C_p in other references.

PCR_k. A process capability ratio with numerator equal to the difference between the product target and the nearest specification limit and denominator equal to three times the process standard deviation. Said to measure the actual capability of the process because the process mean is considered. See process capability, process capability ratio, process capability study, and PCR. Sometimes denoted as C_{pk} in other references.

Percentage point. A particular value of a random variable determined from a probability (expressed as a percentage). For example, the upper 5 percentage point of the standard normal random variable is $z_{0.05} = 1.645$.

Percentile. The set of values that divide the sample into 100 equal parts.

Poisson process. A random experiment with counts that occur in an interval and satisfy the following assumptions. The interval can be partitioned into subintervals such that the probability of more than one count in a subinterval is zero, the probability of a count in a subinterval is proportional to the length of the subinterval, and the count in each subinterval is independent of other subintervals.

Poisson random variable. A discrete random variable that is the number of counts that occur in a Poisson process.

Pooling. When several sets of data can be thought of as having been generated from the same model, it is possible to combine them, usually for purposes of estimating one or more parameters. Combining the samples for this purpose is usually called *pooling*.

Population standard deviation. See standard deviation.

Population variance. See variance.

Population. Any finite or infinite collection of individual units or objects.

Power. The power of a statistical test is the probability that the test rejects the null hypothesis when the null hypothesis is indeed false. Thus the power is equal to one minus the probability of type II error.

Prediction. The process of determining the value of one or more statistical quantities at some future point in time. In a regression model, predicting the response y for some specified set of regressors or predictor variables also leads to a predicted value, although there may be no temporal element to the problem.

Prediction interval. The interval between a set of upper and lower limits associated with a predicted value designed to show on a probability basis the range of error associated with the prediction.

Predictor variable(s). The independent or regressor variables in a regression model.

Probability density function. A function used to calculate probabilities and to specify the probability distribution of a continuous random variable.

Probability distribution. For a sample space, a description of the set of possible outcomes along with a method to determine probabilities. For a random variable, a probability distribution is a description of the range along with a method to determine probabilities.

Probability mass function. A function that provides probabilities for the values in the range of a discrete random variable.

Probability. A numerical measure between 0 and 1 assigned to events in a sample space. Higher numbers indicate the event is more likely to occur. See axioms of probability.

Process capability ratio. A ratio that relates the width of the product specification limits to measures of process performance. Used to quantify the capability of the process to produce product within specifications. See process capability, process capability study, *PCR* and PCR_k.

Process capability study. A study that collects data to estimate process capability. See process capability, process capability ratio, *PCR* and PCR_k.

Process capability. The capability of a process to produce product within specification limits. See process capability ratio, process capability study, *PCR*, and PCR_k.

P-Value. The exact significance level of a statistical test; that is, the probability of obtaining a value of the test statistic that is at least as extreme as that observed when the null hypothesis is true.

Qualitative (data). Data derived from nonnumeric attributes, such as sex, ethnic origin or nationality, or other classification variable.

Quality control. Systems and procedures used by an organization to assure that the outputs from processes satisfy customers.

Quantiles. The set of $n - 1$ values of a variable that partition it into a number n of equal proportions. For example, $n - 1 = 3$ values partition data into four quantiles with the central value usually called the median and the lower and upper values usually called the lower and upper quartiles, respectively.

Quantitative (data). Data in the form of numerical measurements or counts.

Quartile(s). The three values of a variable that partition it into four equal parts. The central value is usually called the median and the lower and upper values are usually called the lower and upper quartiles, respectively. *Also see* Quantiles.

R^2. A quantity used in regression models to measure the proportion of total variability in the response accounted for by the model. Computationally, $R^2 = SS_{Regression}/SS_{Total}$, and large values of R^2 (near unity) are considered good. However, it is possible to have large values of R^2 and find that the model is unsatisfactory. R^2 is also called the coefficient of determination (or the coefficient of multiple determination in multiple regression).

Random. Nondeterministic, occurring purely by chance, or independent of the occurrence of other events.

Random effects model. In an analysis of variance context, this refers to a model that involves only random factors.

Random error. An error (usually a term in a statistical model) that behaves as if it were drawn at random from a particular probability distribution.

Random experiment. An experiment that can result in different outcomes, even though it is repeated in the same manner each time.

Random factor. In analysis of variance, a factor whose levels are chosen at random from some population of factor levels.

Random order. A sequence or order for a set of objects that is carried out in such a way that every possible ordering is equally likely. In experimental design the runs of the experiment are typically arranged and carried out in random order.

Random sample. A sample is said to be random if it is selected in such a way so that every possible sample has the same probability of being selected.

Random variable. A function that assigns a real number to each outcome in the sample space of a random experiment.

Randomization. A set of objects is said to be randomized when they are arranged in random order.

Randomized block design. A type of experimental design in which treatment (or factor levels) are assigned to blocks in a random manner.

Range. The largest minus the smallest of a set of data values. The range is a simple measure of variability and is widely used in quality control.

Range (control) chart. A control chart used to monitor the variability (dispersion) in a process. *See* Control chart.

Rank. In the context of data, the rank of a single observation is its ordinal number when all data values are ordered according to some criterion, such as their magnitude.

Rational subgroup. A sample of data selected in a manner to include chance sources of variation and to exclude assignable sources of variation, to the extent possible.

Reference distribution. The distribution of a test statistic when the null hypothesis is true. Sometimes a reference distribution is called the null distribution of the test statistic.

Reference value. A parameter set in a Tabular CUSUM algorithm that is determined from the magnitude of the process shift that should be detected.

Regression. The statistical methods used to investigate the relationship between a dependent or response variable y and one or more independent variables x. The independent variables are usually called regressor variables or predictor variables.

Regression coefficient(s). The parameter(s) in a regression model.

Regression diagnostics. Techniques for examining a fitted regression model to investigate the adequacy of the fit

and to determine if any of the underlying assumptions have been violated.

Regression line (or curve). A graphical display of a regression model, usually with the response y on the ordinate and the regressor x on the abcissa.

Regression sum of squares. The portion of the total sum of squares attributable to the model that has been fit to the data.

Regressor variable. The independent or predictor variable in a regression model.

Rejection region. In hypothesis testing, this is the region in the sample space of the test statistic that leads to rejection of the null hypothesis when the test statistic falls in this region.

Relative frequency. The relative frequency of an event is the proportion of times the event occurred in a series of trial of a random experiment.

Reliability. The probability that a specified mission will be completed. It usually refers to the probability that a lifetime of a continuous random variable exceeds a specified time limit.

Replicates. One of the independent repetitions of one or more treatment combinations in an experiment.

Replication. The independent execution of an experiment more than once.

Reproductive property of the normal distribution. A linear combination of independent, normal random variables is a normal random variable.

Residual. Generally this is the difference between the observed and the predicted value of some variable. For example, in regression a residual is the difference between the observed value of the response and the corresponding predicted value obtained from the regression model.

Residual analysis. Any technique that uses the residuals, usually to investigate the adequacy of the model that was used to generate the residuals.

Residual sum of squares. *See* Error sum of squares.

Response (variable). The dependent variable in a regression model or the observed output variable in a designed experiment.

Response surface. When a response y depends on a function of k quantitative variables $x_1, x_2, \dots, x_k$, the values of the response may be viewed as a surface in $k + 1$ dimensions. This surface is called a response surface. Response surface methodology is a subset of experimental design concerned with approximating this surface with a model and using the resulting model to optimize the system or process.

Response surface designs. Experimental designs that have been developed to work well in fitting response surfaces. These are usually designs for fitting a first- or second-order model. The central composite design is a widely used second-order response surface design.

Ridge regression. A method for fitting a regression model that is intended to overcome the problems associated with using standard (or ordinary) least squares when there is a problem with multicollinearity in the data.

Rotatable design. In a rotatable design, the variance of the predicted response is the same at all points that are the same distance from the center of the design.

Run rules. A set of rules applied to the points plotted on a Shewhart control chart that are used to make the chart more sensitized to assignable causes. See control chart, Shewhart control chart.

Sample. Any subset of the elements of a population.

Sample mean. The arithmetic average or mean of the observations in a sample. If the observations are $x_1, x_2, \ldots, x_n$ then the sample mean is $(1/n) \sum_{i=1}^{n} x_i$. The sample mean is usually denoted by $\bar{x}$.

Sample moment. The quantity $(1/n) \sum_{i=1}^{n} x_i^k$ is called the kth sample moment.

Sample range. See range.

Sample size. The number of observations in a sample.

Sample space. The set of all possible outcomes of a random experiment.

Sample standard deviation. The positive square root of the sample variance. The sample standard deviation is the most widely used measure of variability of sample data.

Sample variance. A measure of variability of sample data, defined as $s^2 = [1/(n-1)] \sum_{i=1}^{n} (x_i - \bar{x})^2$, where $\bar{x}$ is the sample mean.

Sampling distribution. The probability distribution of a statistic. For example, the sampling distribution of the sample mean $\bar{X}$ is the normal distribution.

Scatter diagram. A diagram displaying observations on two variables, x and y. Each observation is represented by a point showing its x-y coordinates. The scatter diagram can be very effective in revealing the joint variability of x and y or the nature of the relationship between them.

Screening experiment. An experiment designed and conducted for the purpose of screening out or isolating a promising set of factors for future experimentation. Many screening experiments are fractional factorials, such as two-level fractional factorial designs.

Second-order model. A model that contains second-order terms. For example, the second-order response surface model in two variables is $y = \beta_0 + \beta_1 x_1 + \beta_2 x_2 + \beta_{12} x_1 x_2 + \beta_{11} x_1^2 + \beta_{22} x_2^2 + \epsilon$. The second order terms in this model are $\beta_{12} x_1 x_2$, $\beta_{11} x_1^2$, and $\beta_{22} x_2^2$.

Shewhart control chart. A specific type of control chart developed by Walter A. Shewhart. Typically, each plotted point is a summary statistic calculated from the data in a rational subgroup. See control chart.

Sign test. A statistical test based on the signs of certain functions of the observations and not their magnitudes.

Signed-rank test. A statistical test based on the differences within a set of paired observations. Each difference has a sign and a rank, and the test uses the sum of the differences with regard to sign.

Significance. In hypothesis testing, an effect is said to be significant if the value of the test statistic lies in the critical region.

Significance level. *See* Level of significance.

Skewness. A term for asymmetry usually employed with respect to a histogram of data or a probability distribution.

Standard deviation. The positive square root of the variance. The standard deviation is the most widely used measure of variability.

Standard error. The standard deviation of the estimator of a parameter. The standard error is also the standard deviation of the sampling distribution of the estimator of a parameter.

Standard normal random variable. A normal random variable with mean zero and variance one that has its cumulative distribution function tabulated in Appendix Table II.

Standardize. The transformation of a normal random variable that subtracts its mean and divides by its standard deviation to generate a standard normal random variable.

Standardized residual. In regression, the standardized residual is computed by dividing the ordinary residual by the square root of the residual mean square. This produces scaled residuals that have, approximately, a unit variance.

Statistic. A summary value calculated from a sample of observations. Usually, a statistic is an estimator of some population parameter.

Statistical Process Control. A set of problem-solving tools based on data that are used to improve a process.

Statistical quality control. Statistical and engineering methods used to measure, monitor, control, and improve quality.

Statistics. The science of collecting, analyzing, interpreting, and drawing conclusions from data.

Stem and leaf display. A method of displaying data in which the stem corresponds to a range of data values and the leaf represents the next digit. It is an alternative to the histogram but displays the individual observations rather than sorting them into bins.

Stepwise regression. A method of selecting variables for inclusion in a regression model. It operates by introducing the candidate variables one at a time (as in forward selection) and then attempting to remove variables following each forward step.

Studentized range. The range of a sample divided by the sample standard deviation.

Studentized residual. In regression, the studentized residual is calculated by dividing the ordinary residual by its exact standard deviation, producing a set of scaled residuals that have, exactly, unit standard deviation.

Sufficient statistic. An estimator is said to be a sufficient statistic for an unknown parameter if the distribution of the sample given the statistic does not depend on the unknown parameter. This means that the distribution of the estimator contains all of the useful information about the unknown parameter.

Tabular CUSUM. A numerical algorithm used to detect assignable causes on a cumulative sum control chart. See V mask.

Tampering. Another name for overcontrol.

t-distribution. The distribution of the random variable defined as the ratio of two independent random variables. The numerator is a standard normal random variable and the denominator is the square root of a chi-square random variable divided by its number of degrees of freedom.

Test statistic. A function of a sample of observations that provides the basis for testing a statistical hypothesis.

Time series. A set of ordered observations taken at difference points in time.

Tolerance interval. An interval that contains a specified proportion of a population with a stated level of confidence.

Tolerance limits. A set of limits between which some stated proportion of the values of a population must fall with specified level of confidence.

Total probability rule. Given a collection of mutually exclusive events whose union is the sample space, the probability of an event can be written as the sum of the probabilities of the intersections of the event with the members of this collection.

Treatment. In experimental design, a treatment is a specific level of a factor of interest. Thus if the factor is temperature, the treatments are the specific temperature levels used in the experiment.

Treatment sum of squares. In analysis of variance, this is the sum of squares that accounts for the variability in the response variable due to the different treatments that have been applied.

t-test. Any test of significance based on the t distribution. The most common t-tests are (1) testing hypotheses about the mean of a normal distribution with unknown variance, (2) testing hypotheses about the means of two normal distributions and (3) testing hypotheses about individual regression coefficients.

Type I error. In hypothesis testing, an error incurred by rejecting a null hypothesis when it is actually true (also called an α-error).

Type II error. In hypothesis testing, an error incurred by failing to reject a null hypothesis when it is actually false (also called a β-error).

U chart. An attribute control chart that plots the average number of defects per unit in a subgroup. Also called a defects-per-unit control chart. Similar to a C chart.

Unbiased estimator. An estimator that has its expected value equal to the parameter that is being estimated is said to be unbiased.

Uniform random variable. Refers to either a discrete or continuous uniform random variable.

Uniqueness property of moment generating function. Refers to the fact that random variables with the same moment generating function have the same distribution.

Universe. Another name for *population*.

V mask. A geometrical figure used to detect assignable causes on a cumulative sum control chart. With appropriate values for parameters, identical conclusions can be made from a V mask and a tabular CUSUM.

Variable selection. The problem of selecting a subset of variables for a model from a candidate list that contains all or most of the useful information about the response in the data.

Variables control chart. Any control chart for a continuous random variable. See attributes control charts.

Variance. A measure of variability defined as the expected value of the square of the random variable around its mean.

Variance component. In analysis of variance models involving random effects, one of the objectives is to determine how much variability can be associated with each of the potential sources of variability defined by the experimenters. It is customary to define a variance associated with each of these sources. These variances in some sense sum to the total variance of the response, and are usually called variance components.

Variance inflation factors. Quantities used in multiple regression to assess the extent of multicollinearity (or near linear dependence) in the regressors. The variance inflation factor for the ith regressor VIF_i can be defined as $VIF_i = [1/(1 - R_i^2)]$, where R_i^2 is the coefficient of determination obtained when x_i is regressed on the other regressor variables. Thus when x_i is nearly linearly dependent on a subset of the other regressors R_i^2 will be close to unity and the value of the corresponding vari-

ance inflation factor will be large. Values of the variance inflation factors that exceed 10 are usually taken as a signal that multicollinearity is present.

Warning limits. Horizontal lines added to a control chart (in addition to the control limits) that are used to make the chart more sensitive to assignable causes.

Weibull random variable. A continuous random variable that is often used to model the time until failure of a physical system. The parameters of the distribution are flexible enough that the probability density function can assume many different shapes.

Western Electric rules. A specific set of run rules that were developed at Western Electric Corporation. See run rules.

Wilcoxon signed rank test. A distribution-free test of the equality of the location parameters of two otherwise identical distributions. It is an alternative to the two-sample t-test for nonnormal populations.

With replacement. A method to select samples in which items are replaced between successive selections.

Without replacement. A method to select samples in which items are *not* replaced between successive selections.

Index

Limited Use License Agreement

This is the John Wiley and Sons, Inc. (Wiley) limited use License Agreement, which governs your use of any Wiley proprietary software products (Licensed Program) and User Manual (s) delivered with it.

Your use of the Licensed Program indicates your acceptance of the terms and conditions of this Agreement. If you do not accept or agree with them, you must return the Licensed Program unused within 30 days of receipt or, if purchased, within 30 days, as evidenced by a copy of your receipt, in which case, the purchase price will be fully refunded.

License: Wiley hereby grants you, and you accept, a non-exclusive and non-transferable license, to use the Licensed Program and User Manual (s) on the following terms and conditions only:

a. The Licensed Program and User Manual(s) are for your personal use only.
b. You may use the Licensed Program on a single computer, or on its temporary replacement, or on a subsequent computer only.
c. The Licensed Program may be copied to a single computer hard drive for playing.
d. A backup copy or copies may be made only as provided by the User Manual(s), except as expressly permitted by this Agreement.
e. You may not use the Licensed Program on more than one computer system, make or distribute unauthorized copies of the Licensed Program or User Manual(s), create by decompilation or otherwise the source code of the Licensed Program or use, copy, modify, or transfer the Licensed Program, in whole or in part, or User Manual(s), except as expressly permitted by this Agreement.
    If you transfer possession of any copy or modification of the Licensed Program to any third party, your license is automatically terminated. Such termination shall be in addition to and not in lieu of any equitable, civil, or other remedies available to Wiley.

Term: This License Agreement is effective until terminated. You may terminate it at any time by destroying the Licensed Program and User Manual together with all copies made (with or without authorization).
      This Agreement will also terminate upon the conditions discussed elsewhere in this Agreement, or if you fail to comply with any term or condition of this Agreement. Upon such termination, you agree to destroy the Licensed Program, User Manual (s), and any copies made (with or without authorization) of either.

Wiley's Rights: You acknowledge that all rights (including without limitation, copyrights, patents and trade secrets) in the Licensed Program (including without limitation, the structure, sequence, organization, flow, logic, source code, object code and all means and forms of operation of the Licensed Program) are the sole and exclusive property of Wiley. By accepting this Agreement, you do not become the owner of the Licensed Program, but you do have the right to use it in accordance with the provisions of this Agreement. You agree to protect the Licensed Program from unauthorized use, reproduction, or distribution. You further acknowledge that the Licensed Program contains valuable trade secrets and confidential information belonging to Wiley. You may not disclose any component of the Licensed Program, whether or not in machine readable form, except as expressly provided in this Agreement.

WARRANTY: TO THE ORIGINAL LICENSEE ONLY, WILEY WARRANTS THAT THE MEDIA ON WHICH THE LICENSED PROGRAM IS FURNISHED ARE FREE FROM DEFECTS IN THE MATERIAL AND WORKMANSHIP UNDER NORMAL USE FOR A PERIOD OF NINETY (90) DAYS FROM THE DATE OF PURCHASE OR RECEIPT AS EVIDENCED BY A COPY OF YOUR RECEIPT. IF DURING THE 90 DAY PERIOD, A DEFECT IN ANY MEDIA OCCURS, YOU MAY RETURN IT. WILEY WILL REPLACE THE DEFECTIVE MEDIA WITHOUT CHARGE TO YOU. YOUR SOLE AND EXCLUSIVE REMEDY IN THE EVENT OF A DEFECT IS EXPRESSLY LIMITED TO REPLACEMENT OF THE DEFECTIVE MEDIA AT NO ADDITIONAL CHARGE. THIS WARRANTY DOES NOT APPLY TO DAMAGE OR DEFECTS DUE TO IMPROPER USE OR NEGLIGENCE.
      THIS LIMITED WARRANTY IS IN LIEU OF ALL OTHER WARRANTIES, EXPRESSED OR IMPLIED, INCLUDING, WITHOUT LIMITATION, ANY WARRANTIES OF MERCHANTABILITY OR FITNESS FOR A PARTICULAR PURPOSE.
      EXCEPT AS SPECIFIED ABOVE, THE LICENSED PROGRAM AND USER MANUAL(S) ARE FURNISHED BY WILEY ON AN "AS IS" BASIS AND WITHOUT WARRANTY AS TO THE PERFORMANCE OR RESULTS YOU MAY OBTAIN BY USING THE LICENSED PROGRAM AND USER MANUAL(S). THE ENTIRE RISK AS TO THE RESULTS OR PERFORMANCE, AND THE COST OF ALL NECESSARY SERVICING, REPAIR, OR CORRECTION OF THE LICENSED PROGRAM AND USER MANUAL(S) IS ASSUMED BY YOU.
      IN NO EVENT WILL WILEY OR THE AUTHOR, BE LIABLE TO YOU FOR ANY DAMAGES, INCLUDING LOST PROFITS, LOST SAVINGS, OR OTHER INCIDENTAL OR CONSEQUENTIAL DAMAGES ARISING OUT OF THE USE OR INABILITY TO USE THE LICENSED PROGRAM OR USER MANUAL(S), EVEN IF WILEY OR AN AUTHORIZED WILEY DEALER HAS BEEN ADVISED OF THE POSSIBILITY OF SUCH DAMAGES.

General: This Limited Warranty gives you specific legal rights. You may have others by operation of law which varies from state to state. If any of the provisions of this Agreement are invalid under any applicable statute or rule of law, they are to that extent deemed omitted.
      This Agreement represents the entire agreement between us and supersedes any proposals or prior Agreements, oral or written, and any other communication between us relating to the subject matter of this Agreement.
      This Agreement will be governed and construed as if wholly entered into and performed within the State of New York. You acknowledge that you have read this Agreement, and agree to be bound by its terms and conditions.

Summary of Two-Sample Hypothesis-Testing Procedures

Case	Null Hypothesis	Test Statistic	Alternative Hypothesis	Criteria for Rejection	OC Curve Parameter	OC Curve Appendix Chart IV
1.	$H_0: \mu_1 - \mu_2 = \Delta_0$ σ_1^2 and σ_2^2 known	$z_0 = \dfrac{\bar{x}_1 - \bar{x}_2 - \Delta_0}{\sqrt{\dfrac{\sigma_1^2}{n_1} + \dfrac{\sigma_2^2}{n_2}}}$	$H_1: \mu_1 - \mu_2 \neq \Delta_0$ $H_1: \mu_1 - \mu_2 > \Delta_0$ $H_1: \mu_1 - \mu_2 < \Delta_0$	$\lvert z_0 \rvert > z_{\alpha/2}$ $z_0 > z_\alpha$ $z_0 < -z_\alpha$	$d = \dfrac{\lvert \mu_1 - \mu_2 - \Delta_0 \rvert}{\sqrt{\sigma_1^2 + \sigma_2^2}}$ $d = \dfrac{\mu_1 - \mu_2 - \Delta_0}{\sqrt{\sigma_1^2 + \sigma_2^2}}$ $d = \dfrac{\mu_2 - \mu_1 - \Delta_0}{\sqrt{\sigma_1^2 + \sigma_2^2}}$	a, b c, d c, d
2.	$H_0: \mu_1 - \mu_2 = \Delta_0$ $\sigma_1^2 = \sigma_2^2$ unknown	$t_0 = \dfrac{\bar{x}_1 - \bar{x}_2 - \Delta_0}{s_p\sqrt{\dfrac{1}{n_1} + \dfrac{1}{n_2}}}$	$H_1: \mu_1 - \mu_2 \neq \Delta_0$ $H_1: \mu_1 - \mu_2 > \Delta_0$ $H_1: \mu_1 - \mu_2 < \Delta_0$	$\lvert t_0 \rvert > t_{\alpha/2, n_1+n_2-2}$ $t_0 > t_{\alpha, n_1+n_2-2}$ $t_0 < -t_{\alpha, n_1+n_2-2}$	$d = \lvert \Delta - \Delta_0 \rvert / 2\sigma$ $d = (\Delta - \Delta_0)/2\sigma$ $d = (\Delta_0 - \Delta)/2\sigma$ where $\Delta = \mu_1 - \mu_2$	e, f g, h g, h
3.	$H_0: \mu_1 - \mu_2 = \Delta_0$ $\sigma_1^2 \neq \sigma_2^2$ unknown	$t_0 = \dfrac{\bar{x}_1 - \bar{x}_2 - \Delta_0}{\sqrt{\dfrac{s_1^2}{n_1} + \dfrac{s_2^2}{n_2}}}$ $v = \dfrac{\left(\dfrac{s_1^2}{n_1} + \dfrac{s_2^2}{n_2}\right)^2}{\dfrac{(s_1^2/n_1)^2}{n_1 - 1} + \dfrac{(s_2^2/n_2)^2}{n_2 - 1}}$	$H_1: \mu_1 - \mu_2 \neq \Delta_0$ $H_1: \mu_1 - \mu_2 > \Delta_0$ $H_1: \mu_1 - \mu_2 < \Delta_0$	$\lvert t_0 \rvert > t_{\alpha/2, v}$ $t_0 > t_{\alpha, v}$ $t_0 < -t_{\alpha, v}$	— — —	— — —
4.	Paired data $H_0: \mu_D = 0$	$t_0 = \dfrac{\bar{d}}{s_d/\sqrt{n}}$	$H_1: \mu_d \neq 0$ $H_1: \mu_d > 0$ $H_1: \mu_d < 0$	$\lvert t_0 \rvert > t_{\alpha/2, n-1}$ $t_0 > t_{\alpha, n-1}$ $t_0 < -t_{\alpha, n-1}$	— — —	— — —
5.	$H_0: \sigma_1^2 = \sigma_2^2$	$f_0 = s_1^2/s_2^2$	$H_1: \sigma_1^2 \neq \sigma_2^2$ $H_1: \sigma_1^2 > \sigma_2^2$	$f_0 > f_{\alpha/2, n_1-1, n_2-1}$ or $f_0 < f_{1-\alpha/2, n_1-1, n_2-1}$ $f_0 > f_{\alpha, n_1-1, n_2-1}$	$\lambda = \sigma_1/\sigma_2$ $\lambda = \sigma_1/\sigma_2$	o, p q, r
6.	$H_0: p_1 = p_2$	$z_0 = \dfrac{\hat{p}_1 - \hat{p}_2}{\sqrt{\hat{p}(1 - \hat{p})\left[\dfrac{1}{n_1} + \dfrac{1}{n_2}\right]}}$	$H_1: p_1 \neq p_2$ $H_1: p_1 > p_2$ $H_1: p_1 < p_2$	$\lvert z_0 \rvert > z_{\alpha/2}$ $z_0 > z_\alpha$ $z_0 < -z_\alpha$	— — —	— — —